AF251514

Ideas and Methods in Quantum and Statistical Physics

Ideas and Methods in Quantum and Statistical Physics

In Memory of Raphael Høegh-Krohn (1938-1988)
Volume 2

Edited by

Sergio Albeverio
Ruhr-Universität Bochum, Germany

Jens Erik Fenstad
University of Oslo, Norway

Helge Holden
University of Trondheim, Norway

Tom Lindstrøm
University of Oslo, Norway

CAMBRIDGE
UNIVERSITY PRESS

Published by the Press Syndicate of the University of Cambridge
The Pitt Building, Trumpington Street, Cambridge CB2 1RP
40 West 20th Street, New York, NY 10011-4211, USA
10, Stamford Road, Oakleigh, Victoria 3166, Australia

First published 1992

Printed in MEXICO

Library of Congress cataloguing in publication data applied for

British Library cataloguing in publication data applied for

ISBN   0-521-41930-1   hardback

La filosofia è scritta in questo grandissimo libro che continuamente ci sta aperto innanzi a gli occhi (io dico l'universo), ma non si può intendere se prima non s'impara a intender la lingua, e conoscer i caràtteri, ne' quali è scritto. Egli è scritto in lingua matematica, e i caràtteri son triangoli, cerchi, ed altre figure geometriche, senza i quali mezi è impossibile a intenderne umanamente parola; senza questi è un aggirarsi vanamente per un oscuro laberinto.

GALILEO GALILEI

Døyr fe;
døyr frendar;
døyr sjølv det same.
Men ordet om deg
aldri døyr
vinn du eit gjetord gjevt.

HÅVAMÅL

CONTENTS

vii

STATISTICAL PHYSICS

Volume I

Ideas and Methods in
Mathematical Analysis, Stochastics, and Applications

The contribution by F. Guerra will be published elsewhere.

Raphael Jan Høegh–Krohn
1938 – 1988

PREFACE

Raphael Høegh-Krohn, the mathematician and physicist, died suddenly on January 24, 1988, a few weeks before his fiftieth birthday. At the time of his death, Høegh-Krohn and his many collaborators were involved in a large number of projects; some near completion, but others either in rapid development or still on the starting blocks. In September of 1988, a symposium on "Ideas and Methods in Mathematics and Physics" was held at the University of Oslo with a twofold purpose; partly to honor the memory of Raphael Høegh-Krohn and to celebrate his scientific achievements, and partly to stimulate his friends and collaborators to continue the work he had left unfinished. The present volumes grew out of that symposium; many of the papers in it are based on talks given in Oslo at the time, while others were sent to us by scientists who could not attend the conference for practical reasons, but who still wanted to contribute in some way. A few were commissioned by the editors to give a more complete picture of the breadth and depth of Høegh-Krohn's contribution to science.

The most prominent characteristic of Raphael Høegh-Krohn as a scientist was probably the unique mixture of knowledge and intuition which made it possible for him to grasp the significance and true meaning of new developments almost immediately. Although his technical skills were formidable, mere technical improvements never interested him much; he was always looking for the underlying ideas and the correct physical interpretation of the results. Many of us have heard him at conferences where his comments after a lecture could in just a few seconds explain the true relevance of the speaker's result to central problems in physics and mathematics, and outline the most promising directions for future research. At the Oslo Symposium there were moments when everybody seemed to be waiting for his voice from the back of the room; waiting for him to trace the hidden connections, to tie up the loose ends, to put the emphasis where it belonged, or simply for him once more to suffuse the subject with his enthusiasm. We shall have to do without his help in the future, but we hope that these volumes dedicated to his memory will make the task a little easier – it is written by many of his closest collaborators and some of his most esteemed colleagues in a spirit close to his own. As with his own papers, the range of subjects may seem daunting, but there is a unifying theme; the quest for the right language for the formulation of physical laws.

To Raphael Høegh-Krohn physics and mathematics were intimately connected; he knew the physical significance of every little part of his formulas – the sign of a constant or the size of an exponent always had an immediate physical meaning to him – and, on the other hand, every physical phenomenon seemed to him to be waiting for a mathematical description. Particularly close to his heart were solvable models; models which are simple enough to admit an analytic solution, yet sufficiently general to describe the typical behaviour of a system. He never tired of emphasizing the wealth of invaluable information which can first be extracted from solvable models and later extended to more general situations. In solvable models he saw the interplay between the arts of physics and mathematics at its finest; the deepest insight of the physicist is needed to create a simplified model retaining all important features of the original phenomenon, and the greatest skill of the mathemati-

xiii

cian is required in solving the ensuing equations explicitly. But this love for the specific and the explicit never stopped him from generalizing and unifying; he was a Grand Master of formal calculations, and on entering his office one would usually find oneself in the midst of dazzling formulas whose physical content was bursting with significance, but whose mathematical status was far from secure. Much of his work went into making rigorous sense out of these formal calculations, and in this he was helped as much by his ebullient optimism as by his enormous technical skills; he liked to tease more timid souls by telling them that "if you can do it for analytic step functions, you can do it for anything". We hope the present volume achieves the kind of balance Raphael Høegh-Krohn was always looking for; the balance between the general and the specific, between the heuristic and the rigorous, between physical content and mathematical form.

As we dedicate this book to the memory of Raphael Høegh-Krohn, we would like to thank everybody who has helped in its production; the contributors, the participants in the Oslo Symposium, the staff at the mathematics departments in Bochum, Oslo and Trondheim, and the staff at Cambridge University Press, in particular Dr. A. Harvey. But first and foremost our thoughts go to those left behind, those who have had to come to terms with a loss even greater than ours – to all the members of his family.

Bochum, Oslo and Trondheim
May 1991

Sergio Albeverio   Jens Erik Fenstad   Helge Holden   Tom Lindstrøm

xiv

ON THE LOW DENSITY LIMIT OF BOSON MODELS:
LANGEVIN EQUATION

L. Accardi   Lu Yun Gang [*]

Centro Matematico V.Volterra

Dipartimento di Matematica

Universita' di Roma II

ABSTRACT

We obtain a quantum Langevin equation in the Boson Fock case as the low density limit of Heisenberg evolutions. We show that the low density limit can be described by a quantum stochastic differential equation driven by a quantum jump (number or Poisson) processes.

---

[*] On leave of absence from Beijing Normal University

1

§1. Introduction

In a series of papers ([1], [2], [4] and [4a]) we have studied the low density limit of nonrelativistic quantum systems coupled to a free Boson or Fermi gas in some quasi-free state. We have proved, for all these systems, the convergence (in the quantum probabilistic sense of [1] or [5]) of the basic dynamical variable, i.e. the wave operator at time t/z^2 (z^2 being the fugacity) to a unitary Markovian cocycle satisfying a quantum stochastic differential equation of Poisson type (i.e. driven only by the quantum jump martingales).

In the present paper we prove the basic result which allows to realize this program for the Heisenberg evolution.

On the basis of very plausible physical considerations, Frigerio and Maassen ([13]), Frigerio and Alicki ([12]), have suggested that, in the low density limit, one should expect a unitary Markovian cocycle satisfying a quantum stochastic differential equation driven by purely jump (i.e. number) processes. Once one has the unitary Markovian cocycle, one associates to it a quantum Markovian (or dynamical) semigroup by standard application of the quantum Feynman-Kac formula of [0], and in this way one recovers the results on the deduction of the master equation obtained by several authors (Davies, Pulé, ... in the weak coupling limit, Dümcke, Palmer, ... in the low density limit) in which, however, any detailed information on the structure of the heat bath (or reservoir) in the limit is lost.

From now on we shall concentrate on the statement and the solution of the mathematical aspects of the problem.

From [1], [2] and [3], one knows that the most important step is to deal with the problem in Fock case, which gives the basic idea and solution of the most difficul terms. Now let H_0 denote the system Hilbert space and H_1 the one particle resevoir Hilbert space . Let

$$W(H_1) \ := \ \{W(f) : \ f \in H_1\} \tag{1.1}$$

be the Weyl C^*-algebra on H_1 ; let H be a self-adjoint bounded below operator on H_1 and z, β positive real numbers interpreted respectively as density of the resevoir particles and inverse temperature. Define

$$Q_z \ := \ \left(1 + z^2 e^{-\beta H}\right) \cdot \left(1 - z^2 e^{-\beta H}\right)^{-1} \ = \coth(\beta H + \mu) \tag{1.2}$$

(with $z^2 = e^\mu$) and suppose that, for each z in an interval $[0, Z]$, Q_z is a self-adjoint operator on a domain D, independent on z. In the Fermion case (1.2) is replaced by

$$Q_z \ := \ \left(1 - z^2 e^{-\beta H}\right) \cdot \left(1 + z^2 e^{-\beta H}\right)^{-1} \tag{1.3}$$

which is a bounded operator for each z. Denote φ_{Q_z} the mean zero gauge invariant quasi-free state on $W(H_1)$ with covariance operator Q_z, characterized by:

$$\varphi_{Q_z}(W(f)) \ = \ \exp\left(-\frac{1}{2} < f, Q_z f >\right) \ , \ \forall f \in H_1 \tag{1.4}$$

and let $\{\mathcal{H}_{Q_z}, \pi_{Q_z}, \Phi_{Q_z}\}$ be the GNS - triple of $\{W(H_1), \varphi_{Q_z}\}$, so that

$$< \Phi_{Q_z}, \pi_{Q_z}(W(f))\Phi_{Q_z} > = \ \varphi_{Q_z}(W(f)) \tag{1.5}$$

We shall write W_{Q_z} for $\pi_{Q_z} \circ W$. The Fock representation corresponds to the case $Q_z = 1$, i.e. $\beta = \infty$. In this case the GNS representation will be simply denoted $\{\mathcal{H}, \pi, \Phi\}$. Let S_t be a unitary group on $B(H_1)$ (the one particle free evolution of the resevoir) and suppose that

$$S_t \cdot Q_z = Q_z \cdot S_t \quad , \quad \forall t \geq 0 \tag{1.6}$$

where the equality is meant on D. This implies that the second quantization of S_t, denoted $W(S_t)$, leaves φ_{Q_z} invariant hence it is implemented, in the GNS representation, by a unitary 1-parameter group $V_t^{(z)}$ whose generator $H_R^{(z)}$ is called the free Hamiltonian of the resevoir. As in [1] and [5], we assume that there exists a non zero subspace K of H_1 (in all the examples it is a dense subspace) such that

$$\int_{\mathbf{R}} | < f, S_t g > | dt \ < \infty \quad , \quad \forall f, g \in K \tag{1.7}$$

Moreover , we suppose that $Q_z K \subseteq K$ and $S_t K \subseteq K$ for all $t \geq 0$. For example, for the free Bose gas,

$$S_t = e^{-it\Delta}$$

where Δ is the Laplacian on $L^2(\mathbf{R}^d)$ all these conditions are satisfied if we choose $K \subset L^1 \cap L^\infty(\mathbf{R}^d)$ and $d \geq 3$. Let be given a self-adjoint operator H_S on the system space H_o, called the system Hamiltonian. The total free Hamiltonian is defined to be

$$H^{(0)} := H_S \otimes 1 + 1 \otimes H_R^{(z)} \tag{1.8}$$

We define the interaction Hamiltonian V as in [1] and [11] i.e., we fix two functions $g_1, g_o \in K$ and define

$$V \ := \ i \left(D \otimes A^+(g_0) \cdot A(g_1) \ - \ D^+ \otimes A^+(g_1) \cdot A(g_0) \right)$$

$$= i \sum_{\varepsilon \in \{0,1\}} D_\varepsilon \otimes A^+(g_\varepsilon) \cdot A(g_{1-\varepsilon}) \tag{1.9}$$

with the notations

$$D_0 \ = \ D \quad , \quad D_1 \ = \ -D^+ \tag{1.10}$$

and where D is a bounded operator on H_o satisfying

$$\exp(-itH_S) \cdot D \cdot \exp(itH_S) \ = \ D \tag{1.11}$$

Notice that in the present paper we do not use, as in [1], ... , [4], the additional assumption that g_0 and g_1 have disjoint energy spectra, i.e.

$$< g_0, S_t g_1 > = 0 \ , \qquad \forall t \in \mathbf{R}$$

The only problem that this generalization creates is the proof of unitarity of the limiting evolution. We have indications that unitarity takes place, but there are technical difficulties and, at the moment, the proof has not yet been completed.

With these notations, the total Hamiltonian is

$$H_{total} := H_S \otimes 1 + 1 \otimes H_R^{(z)} + V \tag{1.12}$$

and the wave operator at time t is defined by

$$U_t := \exp\left(-itH^{(0)}\right) \cdot \exp\left(itH_{total}\right) \tag{1.13}$$

Therefore we have the identity

$$\frac{d}{dt}U_t = \frac{1}{i}V(t)U_t \quad ; \quad U_0 = 1 \tag{1.14}$$

$$\frac{d}{dt}U_t^* = iU_t^*V(t) \quad ; \quad U_0^* = 1 \tag{1.14a}$$

where ,

$$V(t) := \exp\left(-itH_S \otimes 1\right) V \exp\left(itH_S \otimes 1\right) = i \sum_{\varepsilon \in \{0,1\}} D_\varepsilon \otimes A^+(S_t g_\varepsilon)A(S_t g_{1-\varepsilon}) \tag{1.15}$$

and the solutions of (1.14) and (1.14a) are given by the iterated series

$$U_t = \sum_{n=0}^{\infty} \int_0^t dt_1 \cdots \int_0^{t_{n-1}} dt_n (-i)^n V(t_1) \cdots V(t_n) \tag{1.16}$$

$$U_t^* = \sum_{n=0}^{\infty} \int_0^t dt_1 \cdots \int_0^{t_{n-1}} dt_n i^n V(t_n) \cdots V(t_1)$$

$$= \sum_{n=0}^{\infty} \int_0^t dt_1 \int_{t_1}^t dt_2 \cdots \int_{t_{n-1}}^t dt_n i^n V(t_1) \cdots V(t_n) \tag{1.16a}$$

which are weakly convergent on the domain of the vectors of the form

$$u \otimes \Phi_{Q_z}\left(z \int_{S/z^2}^{T/z^2} S_u f du\right)$$

where $u \in H_o$, $f \in K$, $S, T \in \mathbf{R}$ and

$$\Phi_{Q_z}\left(z \int_{S/z^2}^{T/z^2} S_u f du\right) := W_{Q_z}\left(z \int_{S/z^2}^{T/z^2} S_u f du\right)\Phi_{Q_z} \tag{1.17}$$

is a collective coherent vector in the sense of [1] and [5]. Moreover the identity (1.14) holds rigorously in the weak topology on this domain.

From Lemma (3.2) of [1], we know that the assumption (1.7) implies that the sesquilinear form $(\cdot|\cdot) : K \times K \longrightarrow \mathcal{C}$ defined by

$$(f|g) := \int_{\mathbf{R}} < f, S_t g > dt \quad , \quad f, g \in K \tag{1.18}$$

defines a pre-scalar product on K. We denote $\{K, (\cdot|\cdot)\}$, or simply K, the completion of the quotient of K by the zero $(\cdot|\cdot)$ -norm elements .

In [10], [11] and [16] it was proved that, for some "good " initial states φ and under some additional conditions , the limit

$$\lim_{z \to 0} \varphi \otimes \varphi_{Q_z} \left(U_{t/z^2}(X \otimes 1)U^*_{t/z^2} \right) \tag{1.19}$$

exists in a small time interval and is equal to

$$\varphi(T_t(X)) \tag{1.20}$$

where , $\{T_t\}_{t \geq 0}$ is a quantum Markovian semigroup . The analogy with the new techniques, developed in [1], suggests to consider the limit, as $z \to 0$ of expressions of the form

$$< u,(\ \cdot\ )v > \otimes \varphi_{Q_z} \left(1 \otimes W(f_z)\ U_{t/z^2}(X \otimes 1)U^*_{t/z^2}\ 1 \otimes W(f'_z) \right) \tag{1.21}$$

where

$$f_z = z \int_{S/z^2}^{T/z^2} S_u f du \quad , \quad f \in K \tag{1.22}$$

In the control of the limit, as $z \to 0$ of the expressions (1.21), the basic difference between the low density and the weak coupling case is that in the low density case the wave operator U_{t/z^2} depends on z (the density) only in the rescaling of the time ($t \mapsto t/z^2$), while in the weak coupling case the wave operator $U^{(z)}_{t/z^2}$ depends on z (the coupling constant) in its very structure. On the other hand, the resevoir state is independent on z in the weak coupling limit, but dependent on it in the low density limit.

$\underline{DEFINITION}(1.1)$ Let $\mathcal{K}$ be a Hilbert space, T an inteval in $\mathbf{R}$. Let $Q \geq 1$ be a self-adjoint operator on $\mathcal{K}$ and let

$$\{\mathcal{H}_Q, \pi_Q, \Phi_Q\} \tag{1.23}$$

denote the GNS representation of the CCR over $L^2(T, dt; \mathcal{K})$ with respect to the quasi-free state φ_Q on $W(L^2(T, dt; \mathcal{K}))$ characterized by

$$\varphi_Q(W(\xi)) = e^{-\frac{1}{2}<\xi, 1 \otimes Q\xi>} \quad ; \quad \xi \in L^2(T, dt; \mathcal{K}) \tag{1.24}$$

The stochastic process

$$\left\{ \Gamma(L^2(T,dt;\mathcal{K})) \ , \ A(\chi_{(s,t]} \otimes f) \ ,, \ A^+(\chi_{(s,t]} \otimes f) \ ; \ (s,t] \subseteq T \ , f \in \mathcal{K} \right\} \tag{1.25}$$

where $A(\cdot)$, $A^+(\cdot)$ denote respectively the annihilation and creation fields in the representation (1.24) , is called the **Q-Quantum Brownian Motion** on $L^2(T,dt;\mathcal{K})$.

The main result in this paper is to prove that the limit

$$\lim_{z \to 0} < u \otimes W(z \int_{S/z^2}^{T/z^2} S_u f du)\Phi, U_{t/z^2}(X \otimes 1)U^*_{t/z^2} \cdot v \otimes W(z \int_{S'/z^2}^{T'/z^2} S_u f' du)\Phi > \tag{1.26}$$

exists and is equal to

$$< u \otimes W(\chi_{[S,T]} \otimes f)\Psi, U(t)(X \otimes 1)U^*(t) \cdot v \otimes W(\chi_{[S',T']} \otimes f')\Psi > \tag{1.27}$$

where $\{\mathcal{H}, W, \Psi\}$ is the Fock Brownian motion on $L^2(\mathbf{R},dt;K)$ and $U(t)$ satisfies the quantum stochastic differential equation (4.63).

First recall from [1] that, for each $f, f' \in K, S, S', T, T' \in \mathbf{R}$, one has

$$\lim_{z \to 0} < z \int_{S/z^2}^{T/z^2} S_u f du, z \int_{S'/z^2}^{T'/z^2} S_u f' du >=< \chi_{[S,T]}, \chi_{[S',T']} >_{L^2(\mathbf{R})} \cdot \int_{\mathbf{R}} < f, S_t f' > dt$$

$$=< \chi_{[S,T]} \otimes f, \chi_{[S',T']} \otimes f' >_{L^2(\mathbf{R},dt;K)} \tag{1.28}$$

Moreover for each $n \in \mathbf{N}, \{f_k\}_{k=1}^n \subset K, \{S_k, T_k\}_{k=1}^n \subset \mathbf{R}, \{x_k\}_{k=1}^n \subset \mathbf{R}$,

$$\lim_{z \to 0} < \Phi_{Q_z}, W(x_1 z \int_{S_1/z^2}^{T_1/z^2} S_u f_1 du) \cdots W(x_n z \int_{S_n/z^2}^{T_n/z^2} S_u f_n du)\Phi_{Q_z} >$$

$$=< \Psi, W(x_1\chi_{[S_1,T_1]} \otimes f_1) \cdots W(x_n\chi_{[S_n,T_n]} \otimes f_n)\Psi > \tag{1.29}$$

and the convergence is uniform for $\{x_k\}_{k=1}^n, \{S_k, T_k\}_{k=1}^n$ in a bounded set of $\mathbf{R}$, where, Ψ is the vacuum of $\Gamma(L^2(\mathbf{R},dt;K))$.

The strategy of the paper is not the same as ones of the weak coupling limit Langevin case, i.e. we do not use the iterated series of $U_t(X \otimes 1)U^*_t$, where X is any bounded operator on the system space H_0, but consider separately the iterated series of U_t and U^*_t then deal with the interacting raction between the two series. The introduce of the new method dues to some technical difficulties on investigating iterated series of $U_t(X \otimes 1)U^*_t$, for example, in this case we are even not able to distinguish the type I terms from the type II terms.

ACKNOWLEDGEMENTS L. Accardi acknowledges supprot from Grant AFOSR 870249 and ONR N00014-86-K-0538 through the Center for Mathematical System Theory, University of Florida.

§2. The uniform estimate

As in the weak coupling limit the gist of our analysis will be a detailed study of expressions of the form

$$\int_0^{t/z^2} dt_1 \int_0^{t_1} dt_2 \cdots \int_0^{t_{n-1}} dt_n < u \otimes W(z \int_{S/z^2}^{T/z^2} S_u f du)\Phi,$$

$$(-i)^n V(t_1) \cdots V(t_n) \cdot v \otimes W(z \int_{S'/z^2}^{T'/z^2} S_u f' du)\Phi > \tag{2.1}$$

and

$$\int_0^{t/z^2} dt_1 \int_{t_1}^{t/z^2} dt_2 \cdots \int_{t_{n-1}}^{t/z^2} dt_n < u \otimes W(z \int_{S/z^2}^{T/z^2} S_u f du)\Phi,$$

$$i^n V(t_1) \cdots V(t_n) \cdot v \otimes W(z \int_{S'/z^2}^{T'/z^2} S_u f' du)\Phi > \tag{2.1a}$$

with $n \in \mathbf{N}$, $S, T, S', T' \in \mathbf{R}$ and $f, f' \in K$, which arise from the expansions of $U(t/z^2)$ and $U^*(t/z^2)$ using the iterated series. Using (1.16) these expressions can be written as

$$\sum_{\varepsilon \in \{0,1\}^n} \int_0^{t/z^2} dt_1 \int_0^{t_1} dt_2 \cdots \int_0^{t_{n-1}} dt_n < W(z \int_{S/z^2}^{T/z^2} S_u f du)\Phi,$$

$$A^+(S_{t_1} g_{\varepsilon(1)})A(S_{t_1} g_{1-\varepsilon(1)}) \cdots A^+(S_{t_n} g_{\varepsilon(n)})A(S_{t_n} g_{1-\varepsilon(n)})$$

$$\cdot W(z \int_{S'/z^2}^{T'/z^2} S_u f' du)\Phi > \cdot < u, D_{\varepsilon(1)} \cdots D_{\varepsilon(n)} v > \tag{2.2}$$

and

$$\sum_{\varepsilon \in \{0,1\}^n} \int_0^{t/z^2} dt_1 \int_{t_1}^{t/z^2} dt_2 \cdots \int_{t_{n-1}}^{t/z^2} dt_n < W(z \int_{S/z^2}^{T/z^2} S_u f du)\Phi,$$

$$(-1)^n A^+(S_{t_1} g_{\varepsilon(1)})A(S_{t_1} g_{1-\varepsilon(1)}) \cdots A^+(S_{t_n} g_{\varepsilon(n)})A(S_{t_n} g_{1-\varepsilon(n)})$$

$$\cdot W(z \int_{S'/z^2}^{T'/z^2} S_u f' du)\Phi > \cdot < u, D_{\varepsilon(1)} \cdots D_{\varepsilon(n)} v > \tag{2.2a}$$

respectively and this leads to study the quantities

$$\Delta_n^\varepsilon(1, z, t) := \int_0^{t/z^2} dt_1 \int_0^{t_1} dt_2 \cdots \int_0^{t_{n-1}} dt_n \tag{2.3}$$

$$A^+(S_{t_1} g_{\varepsilon(1)})A(S_{t_1} g_{1-\varepsilon(1)}) \cdots A^+(S_{t_n} g_{\varepsilon(n)})A(S_{t_n} g_{1-\varepsilon(n)})$$

and

$$\Delta_n^\varepsilon(2,z,t) := \int_0^{t/z^2} dt_1 \int_{t_1}^{t/z^2} dt_2 \cdots \int_{t_{n-1}}^{t/z^2} dt_n \tag{2.3a}$$

$$(-1)^n A^+(S_{t_1} g_{\varepsilon(1)}) A(S_{t_1} g_{1-\varepsilon(1)}) \cdots A^+(S_{t_n} g_{\varepsilon(n)}) A(S_{t_n} g_{1-\varepsilon(n)})$$

As usual the case of the uniform estimate is a Pulé type inequality. The inequality we need here is Lemma (9.1) of [9a] which will be stated in the following

LEMMA (2.1) Let $f : \mathbf{R} \longrightarrow \mathbf{R}_+$ be a positive integrable symmetric (i.e. $f(t) = f(-t)$) function. Let $n, m \in \mathbf{N}$, $1 \leq m \leq n-1$ and let $2 \leq q_1 < \cdots < q_m \leq n$; $1 \leq p_1 \leq \cdots \leq p_m \leq n-1$; $p_h < q_h$, $h = 1, \cdots, m$. For each p_h, q_h $h = 1, \cdots, m$ as above, let $\mathcal{S}_m$ denote the group of permutations of m elements and let

$$\mathcal{S}_m^1 := \{\sigma \in \mathcal{S}_m \; : \; p_h < q_{\sigma(h)}, \; h = 1, \cdots, m\} \tag{2.4}$$

Then, for any positive function $F \in L^1_{\text{loc}}(\mathbf{R}^{n-m})$, one has

$$\lambda^{-2m} \int_0^t dt_1 \cdots \int_0^{t_{n-1}} dt_n \sum_{\sigma \in \mathcal{S}_m^1} \prod_{h=1}^m f(\frac{t_{q_{\sigma(h)}} - t_{p_h}}{\lambda^2}) F(t_1, \cdots, \widehat{t_{q_1}}, \cdots, \widehat{t_{q_m}}, \cdots, t_n)$$

$$\leq \left[\int_{-\infty}^0 f(t)dt\right]^m$$

$$\int_{0 \leq \tau_n \leq \cdots \widehat{\tau_{q_m}} \cdots \widehat{\tau_{q_1}} \cdots \leq \tau_1} d\tau_1 \cdots \widehat{d\tau_{q_1}} \cdots \widehat{d\tau_{q_m}} \cdots d\tau_n F(\tau_1, \cdots, \widehat{\tau_{q_1}}, \cdots, \widehat{\tau_{q_m}}, \cdots, \tau_n) \tag{2.5}$$

COROLLARY (2.2) Let $f : \mathbf{R} \longrightarrow \mathbf{R}_+$ be a positive integrable symmetric function. Then, for each $n \in \mathbf{N}$, $1 \leq m \leq n-1$, $2 \leq q_1 < \cdots < q_m \leq n$, $1 \leq p_1 \leq \cdots \leq p_m \leq n-1$, if $\{p_h, q_h\}_{h=1}^m = \{1, \cdots, n\}$ and $p_h < q_h$, $h = 1, \cdots, m$ and $F(\lambda; \cdot) \in L^2(\mathbf{R}^{n-m})$ positive integrable function, one has

$$\lambda^{-2m} \int_0^t dt_1 \int_{t_1}^t dt_2 \cdots \int_{t_{n-1}}^t dt_n \sum_{\sigma \in \mathcal{S}_m^1} \prod_{h=1}^m f(\frac{t_{q_{\sigma(h)}} - t_{p_h}}{\lambda^2}) F(\lambda; t_1, \cdots, \widehat{t_{q_1}}, \cdots, \widehat{t_{q_m}}, \cdots, t_n)$$

$$\leq \left[\int_{-\infty}^0 f(t)dt\right]^m$$

$$\int_{0 \leq \tau_n \leq \cdots \widehat{\tau_{q_m}} \cdots \widehat{\tau_{q_1}} \cdots \leq \tau_1} d\tau_1 \cdots \widehat{d\tau_{q_1}} \cdots \widehat{d\tau_{q_m}} \cdots d\tau_n F(\tau_1, \cdots, \widehat{\tau_{q_1}}, \cdots, \widehat{\tau_{q_m}}, \cdots, \tau_n) \tag{2.6}$$

PROOF Using the identity

$$\int_0^t dt_1 \int_{t_1}^t dt_2 \cdots \int_{t_{n-1}}^t dt_n G(t_1, \cdots, t_n) = \int_0^t dt_1 \cdots \int_0^{t_{n-1}} dt_n G(t_n, \cdots, t_1) \tag{2.7}$$

for any $G \in L^1_{\mathrm{loc}}(\mathbf{R}^n)$, the left hand side of (2.6) becomes

$$\lambda^{-2m} \int_0^t dt_1 \int_0^{t_1} dt_2 \cdots \int_0^{t_{n-1}} dt_n \sum_{\sigma' \in \mathcal{S}^1_m} \prod_{h=1}^m f\left(\frac{t_{q'_{\sigma'(h')}} - t_{p'_{h'}}}{\lambda^2}\right) F_0(\lambda;\ t_1, \cdots, \widehat{t_{q'_1}}, \cdots, \widehat{t_{q'_m}}, \cdots, t_n)$$

(2.8)

where,

$$F_0(\lambda;\ \cdots, t_j, \cdots) := F(\lambda;\ \cdots, t_{n-j+1}, \cdots)$$

(2.9)

and $p'_{h'}, q'_{h'}$ and σ' have the same properties as those of p_h, q_h, σ. Therefore one obtains (2.6) from Lemma (2.1).

<u>LEMMA (2.3)</u> Let $f : \mathbf{R} \longrightarrow \mathbf{R}_+$ be a positive integrable symmetric function, then for each $n \in \mathbf{N}$, $1 \le m \le n-1$, $2 \le q_1, \cdots, q_m \le n$, $\{q_h\}_{h=1}^m| = m$, $\{x_h\}_{h=1}^m \subset \mathbf{R}$

$$\lambda^{-2m} \int_0^t dt_1 \int_{t_1}^t dt_2 \cdots \int_{t_{n-1}}^t dt_n \prod_{h=1}^m f\left(\frac{t_{q_h} - x_h}{\lambda^2}\right)$$

$$\le \left(\int_{-\infty}^\infty f(t)dt\right)^m \cdot \frac{t^{n-m}}{(n-m)!}$$

(2.10)

<u>PROOF</u> In order to prove (2.10) we need only to change variables

$$\frac{t_{q_h} - x_h}{\lambda^2} = s_{q_h}, \qquad h = 1, \cdots, m$$

(2.11)

<u>LEMMA (2.4)</u> For each $n \in \mathbf{N}$ and $\varepsilon \in \{0,1\}^n$, the normlly ordered form of product

$$A^+(S_{t_1} g_{\varepsilon(1)}) A(S_{t_1} g_{1-\varepsilon(1)}) \cdots A^+(S_{t_n} g_{\varepsilon(n)}) A(S_{t_n} g_{1-\varepsilon(n)})$$

(2.12)

is equal to

$$\sum_{d=0}^n \sum_{\substack{2 \le q_1 < \cdots < q_d \le n}} \sum_{\substack{1 \le p_1 < \cdots < p_d \le n \\ |\{p_h\}_{h=1}^d|=d, p_h < q_h, h=1,\cdots,d}} \sum_{\sigma \in \mathcal{S}^1_d} \prod_{h=1}^d < S_{t_{p_{\sigma(h)}}} g_{1-\varepsilon(p_h)}, S_{t_{q_h}} g_{\varepsilon(q_h)} >$$

$$\prod_{\alpha \in \{1,\cdots,n\} \setminus \{q_h\}_{h=1}^d} A^+(S_{t_\alpha} g_{\varepsilon(\alpha)}) \cdot \prod_{\alpha \in \{1,\cdots,n\} \setminus \{p_h\}_{h=1}^d} A(S_{t_\alpha} g_{1-\varepsilon(\alpha)})$$

(2.13)

and also equal to

$$\sum_{d=0}^n \sum_{\substack{2 \le p_1 < \cdots < p_d \le n}} \sum_{\substack{1 \le q_1 < \cdots < q_d \le n \\ |\{q_h\}_{h=1}^d|=d, p_h < q_h, h=1,\cdots,d}} \sum_{\sigma \in \mathcal{S}^1_d} \prod_{h=1}^d < S_{t_{p_h}} g_{1-\varepsilon(p_h)}, S_{t_{q_{\sigma(h)}}} g_{\varepsilon(q_h)} >$$

$$\prod_{\alpha\in\{1,\cdots,n\}\backslash\{q_h\}_{h=1}^d} A^+(S_{t_\alpha}g_{\varepsilon(\alpha)})\cdot \prod_{\alpha\in\{1,\cdots,n\}\backslash\{p_h\}_{h=1}^d} A(S_{t_\alpha}g_{1-\varepsilon(\alpha)}) \qquad (2.14)$$

with the convention that if $d = 0$, then the sums over $q_1 < \cdots < q_d$ and $p_1,\cdots,p_d$ and the product over h, in (2.14) and (2.14a) are equal to 1.

$\underline{PROOF}$ The proof of the product (2.12) being equal to (2.13) is the same as Lemma (3.1) of [1].

Notice that the difference between (2.13) and (2.14) is only that in the expression (2.13), the set $\{q_h\}_{h=1}^d$ is ordered and the set $\{p_h\}_{h=1}^d$ is not ordered but in the expression (2.14), $\{p_h\}_{h=1}^d$ is ordered and the set $\{q_h\}_{h=1}^d$ is not ordered. If we notice that for each $m \in \mathbf{N}$ and $\{f_h, g_h\}_{h=1}^m$, the pure scalar product term in the normlly ordered form of product

$$A(f_1)\cdots A(f_m)A^+(g_1)\cdots A^+(g_m)$$

is equal to

$$\sum_{\sigma\in\mathcal{S}_m}\prod_{h=1}^m < f_h, g_{\sigma(h)} >$$

and also equal to

$$\sum_{\sigma\in\mathcal{S}_m}\prod_{h=1}^m < f_{\sigma(h)}, g_h >$$

then (2.14) is obtained form (2.13) immediately.

The following Theorem provides the uniform estimate which allows to take the low density limit term by term.

$\underline{THEOREM}$ (2.5) For each $f, f', g_0, g_1 \in K$, $S, S', T, T' \in \mathbf{R}$, $t, s \geq 0$, $n, n' \in \mathbf{N}$, $\varepsilon \in \{0,1\}^n$ and $\varepsilon' \in \{0,1\}^{n'}$,

$$\left|< W(z\int_{S/z^2}^{T/z^2} S_u f du)\Phi, \sum_{\varepsilon\in\{0,1\}^n,\ \varepsilon'\in\{0,1\}^{n'}} \Delta_n^\varepsilon(1,z,t)\Delta_{n'}^{\varepsilon'}(2,z,s)W(z\int_{S'/z^2}^{T'/z^2} S_u f' du)\Phi >\right|$$

$$\leq 16^{n+n'}\cdot \max_{0\leq m\leq n-1, 0\leq m'\leq n', 0\leq p\leq (n-m)\wedge(n'-m')} \frac{t^{n-m}s^{n'-m'-p}}{(n-m)!\cdot(n'-m'-p)!}$$

$$\cdot\left(\|g_0, g_1\|_-\right)^{2(m+m'+p)}\left[\max_{F=f,f';G=g_0,g_1}\int_{-\infty}^\infty |< F, S_t G > |dt\right]^{2(n-m+n'-m'-p)} \qquad (2.15)$$

where, here and in the following $\|g_0, g_1\|_-$ is defined by

$$\max_{\varepsilon,\sigma\in\{0,1\}}\int_{-\infty}^\infty |< g_\varepsilon, S_t g_\sigma > | \qquad (2.15a)$$

PROOF Applying Lemma (2.4) to the products of creation and annihilation operators in $\Delta_n^\varepsilon(1,z,t)$ and $\Delta_{n'}^{\varepsilon'}(2,z,s)$, we find that

$$\left| < W(z\int_{S/z^2}^{T/z^2} S_u f du)\Phi, \sum_{\varepsilon\in\{0,1\}^n,\ \varepsilon'\in\{0,1\}^{n'}} \Delta_n^\varepsilon(1,z,t)\Delta_{n'}^{\varepsilon'}(2,z,s)W(z\int_{S'/z^2}^{T'/z^2} S_u f' du)\Phi > \right|$$

$$\leq \sum_{\varepsilon\in\{0,1\}^n,\ \varepsilon'\in\{0,1\}^{n'}} \sum_{m=0}^{n}\sum_{2\leq q_1<\cdots<q_m\leq n}\sum_{\substack{1\leq p_1<\cdots<p_m\leq n\\ |\{p_h\}_{h=1}^m|=m, p_h<q_h, h=1,\cdots,m}}\sum_{\sigma\in\mathcal{S}_m^1}$$

$$\sum_{m'=0}^{n}\sum_{2\leq q_1'<\cdots<q_{m'}'\leq n'}\sum_{\substack{1\leq p_1',\cdots,p_{m'}'\leq n'\\ |\{p_h\}_{h=1}^{m'}|=m', p_h'<q_h', h=1,\cdots,m'}}\sum_{\sigma'\in\mathcal{S}_{m'}^1}$$

$$\int_0^{t/z^2} dt_1\int_0^{t_1} dt_2\cdots\int_0^{t_{n-1}} dt_n\cdot\int_0^{s/z^2} ds_1\int_{s_1}^{s/z^2} ds_2\cdots\int_{s_{n'-1}}^{s/z^2} ds_{n'}$$

$$\left| < W(z\int_{S/z^2}^{T/z^2} S_u f du)\Phi, \prod_{h=1}^{m} < S_{t_{p_h}}g_{1-\varepsilon(p_h)}, S_{t_{q_{\sigma(h)}}}g_{\varepsilon(q_h)} > \right.$$

$$\prod_{\alpha\in\{1,\cdots,n\}\setminus\{q_h\}_{h=1}^m} A^+(S_{t_\alpha}g_{\varepsilon(\alpha)})\prod_{\alpha\in\{1,\cdots,n\}\setminus\{p_h\}_{h=1}^m} A(S_{t_\alpha}g_{1-\varepsilon(\alpha)})$$

$$\prod_{h=1}^{m'} < S_{s_{p_h'}}g_{1-\varepsilon'(p_h')}, S_{s_{q_{\sigma'(h)}'}}g_{\varepsilon'(q_h')} > \prod_{\alpha\in\{1,\cdots,n'\}\setminus\{q_h'\}_{h=1}^{m'}} A^+(S_{s_\alpha}g_{\varepsilon'(\alpha)})$$

$$\left.\prod_{\alpha\in\{1,\cdots,n'\}\setminus\{p_h'\}_{h=1}^{m'}} A(S_{s_\alpha}g_{1-\varepsilon'(\alpha)})W(z\int_{S'/z^2}^{T'/z^2} S_u f' du)\Phi > \right| \qquad (2.16)$$

Notice that the product

$$\prod_{\alpha\in\{1,\cdots,n\}\setminus\{p_h\}_{h=1}^m} A(S_{t_\alpha}g_{1-\varepsilon(\alpha)})\cdot\prod_{\alpha\in\{1,\cdots,n'\}\setminus\{q_h'\}_{h=1}^{m'}} A^+(S_{s_\alpha}g_{\varepsilon'(\alpha)}) \qquad (2.17)$$

is not normally ordered form, so firstly we should change it to normlly ordered form. Obviously it is equal to

$$\sum_{p=0}^{(n-m)\wedge(n'-m')}\sum_{\substack{y_1<\cdots<y_p,\ \{y_h\}_{h=1}^p\subset\{1,\cdots,n'\}\setminus\{q_h'\}_{h=1}^{m'}}}\sum_{\{x_h\}_{h=1}^p\subset\{1,\cdots,n\}\setminus\{p_h\}_{h=1}^m,\ |\{x_h\}_{h=1}^p|=p}$$

$$\prod_{h=1}^{p} < S_{t_{x_h}} g_{1-\varepsilon(x_h)}, S_{s_{y_h}} g_{\varepsilon'(y_h)} >$$

$$\prod_{\alpha \in \{1,\cdots,n'\}\setminus(\{q'_h\}_{h=1}^{m'} \cup \{y_h\}_{h=1}^{p})} A^{+}(S_{s_\alpha} g_{\varepsilon'(\alpha)}) \qquad \prod_{\alpha \in \{1,\cdots,n\}\setminus(\{p_h\}_{h=1}^{m} \cup \{x_h\}_{h=1}^{p})} A(S_{t_\alpha} g_{1-\varepsilon(\alpha)})$$

$$(2.17a)$$

Now we apply all creators on the coherent vectors

$$W(z \int_{S/z^2}^{T/z^2} S_u f\, du)\Phi \tag{2.18}$$

and all annihilators on the coherent vectors

$$W(z \int_{S'/z^2}^{T'/z^2} S_u f'\, du)\Phi \tag{2.18a}$$

and since the module of the coherent vector is 1, one knows that the right hand side of (2.16) is dominated by

$$\sum_{\varepsilon \in \{0,1\}^n,\ \varepsilon' \in \{0,1\}^{n'}} \sum_{m=0}^{n} \sum_{2 \leq q_1 < \cdots < q_m \leq n} \sum_{\substack{1 \leq p_1 < \cdots < p_m \leq n \\ |\{p_h\}_{h=1}^m|=m, p_h < q_h, h=1,\cdots,m}} \sum_{\sigma \in \mathcal{S}_m^1}$$

$$\sum_{m'=0}^{n} \sum_{2 \leq q'_1 < \cdots < q'_{m'} \leq n'} \sum_{\substack{1 \leq p'_1,\cdots,p'_{m'} \leq n' \\ |\{p_h\}_{h=1}^{m'}|=m', p'_h < q'_h, h=1,\cdots,m'}} \sum_{\sigma' \in \mathcal{S}_{m'}^1}$$

$$z^{2(n+n'-m-m'-p)} \cdot \int_0^{t/z^2} dt_1 \int_0^{t_1} dt_2 \cdots \int_0^{t_{n-1}} dt_n \cdot \int_0^{s/z^2} ds_1 \int_{s_1}^{s/z^2} ds_2 \cdots \int_{s_{n'-1}}^{s/z^2} ds_{n'}$$

$$\sum_{y_1 < \cdots < y_p,\ \{y_h\}_{h=1}^{p} \subset \{1,\cdots,n'\}\setminus\{q'_h\}_{h=1}^{m'}} \sum_{\{x_h\}_{h=1}^{p} \subset \{1,\cdots,n\}\setminus\{p_h\}_{h=1}^{m},\ |\{x_h\}_{h=1}^{p}|=p}$$

$$\prod_{h=1}^{m} |< S_{t_{p_h}} g_{1-\varepsilon(p_h)}, S_{t_{q_{\sigma(h)}}} g_{\varepsilon(q_h)} >|$$

$$\prod_{h=1}^{m'} |< S_{s_{p'_h}} g_{1-\varepsilon'(p'_h)}, S_{s_{q'_{\sigma(h)}}} g_{\varepsilon'(q'_h)} >|$$

$$\prod_{h=1}^{p} |< S_{t_{x_h}} g_{1-\varepsilon(x_h)}, S_{s_{y_h}} g_{\varepsilon'(y_h)} >|$$

$$\left[\max_{F=f,f';G=g_0,g_1} \int_{-\infty}^{\infty} |< F, S_t G >|\, dt \right]^{2(n-m+n'-m'-p)} \tag{2.19}$$

With the change of variables

$$z^2 t_h = t'_h, \quad h = 1, \cdots, m; \quad z^2 s_h = s'_h, \quad h = 1, \cdots, m \tag{2.19a}$$

(2.19) becomes

$$\sum_{\varepsilon \in \{0,1\}^n, \ \varepsilon' \in \{0,1\}^{n'}} \sum_{m=0}^{n} \sum_{\substack{2 \le q_1 < \cdots < q_m \le n}} \sum_{\substack{1 \le p_1 < \cdots < p_m \le n \\ |\{p_h\}_{h=1}^m|=m, p_h < q_h, h=1,\cdots,m}} \sum_{\sigma \in \mathcal{S}_m^1}$$

$$\sum_{m'=0}^{n} \sum_{\substack{2 \le q'_1 < \cdots < q'_{m'} \le n'}} \sum_{\substack{1 \le p'_1, \cdots, p'_{m'} \le n' \\ |\{p_h\}_{h=1}^{m'}|=m', p'_h < q'_h, h=1,\cdots,m'}} \sum_{\sigma' \in \mathcal{S}_{m'}^1}$$

$$z^{-2(m+m'+p)} \int_0^t dt'_1 \int_0^{t'_1} dt'_2 \cdots \int_0^{t'_{n-1}} dt'_n$$

$$\sum_{m'=0}^{n} \sum_{\substack{2 \le q'_1 < \cdots < q'_{m'} \le n'}} \sum_{\substack{1 \le p'_1, \cdots, p'_{m'} \le n' \\ |\{p_h\}_{h=1}^{m'}|=m', p'_h < q'_h, h=1,\cdots,m'}} \int_0^s ds'_1 \int_{s'_1}^s ds'_2 \cdots \int_{s'_{n'-1}}^s ds'_{n'}$$

$$\sum_{\substack{y_1 < \cdots < y_p, \ \{y_h\}_{h=1}^p \subset \{1,\cdots,n'\} \setminus \{q'_h\}_{h=1}^{m'}}} \sum_{\substack{\{x_h\}_{h=1}^p \subset \{1,\cdots,n\} \setminus \{p_h\}_{h=1}^m, \ |\{x_h\}_{h=1}^p|=p}}$$

$$\prod_{h=1}^{m} |< g_{1-\varepsilon(p_h)}, S_{(t'_{q_{\sigma(h)}} - t'_{p_h})/z^2} g_{\varepsilon(q_h)} >|$$

$$\prod_{h=1}^{m'} |< g_{1-\varepsilon'(p'_h)}, S_{(s'_{q'_{\sigma(h)}} - s'_{p'_h})/z^2} g_{\varepsilon'(q'_h)} >|$$

$$\prod_{h=1}^{p} |< g_{1-\varepsilon(x_h)}, S_{(s'_{y_h} - t'_{x_h})/z^2} g_{\varepsilon'(y_h)} >|$$

$$\left[\max_{F=f,f';G=g_0,g_1} \int_{-\infty}^{\infty} |< F, S_t G >| dt \right]^{2(n-m+n'-m'-p)} \tag{2.20}$$

Denote

$$f(\alpha) := \max_{\varepsilon, \sigma \in \{0,1\}} |< g_\varepsilon, S_\alpha g_\sigma >| \tag{2.21}$$

and

$$F(z; u_1, \cdots, u_p) := \max_{\varepsilon, \sigma \in \{0,1\}^p} \prod_{h=1}^{p} |< g_{\varepsilon(h)}, S_{u_h/z^2} g_{\sigma(h)} >| \tag{2.22}$$

(2.20) is majorized by

$$\sum_{\varepsilon\in\{0,1\}^n,\ \varepsilon'\in\{0,1\}^{n'}}\ \sum_{m=0}^{n}\ \sum_{\substack{2\le q_1<\cdots<q_m\le n}}\ \sum_{\substack{1\le p_1<\cdots<p_m\le n\\ |\{p_h\}_{h=1}^m|=m,\,p_h<q_h,h=1,\cdots,m}}\ \sum_{\sigma\in\mathcal{S}_m^1}$$

$$\sum_{m'=0}^{n}\ \sum_{\substack{2\le q_1'<\cdots<q_{m'}'\le n'}}\ \sum_{\substack{1\le p_1',\cdots,p_{m'}'\le n'\\ |\{p_h\}_{h=1}^{m'}|=m',\,p_h'<q_h',h=1,\cdots,m'}}\ \sum_{\sigma'\in\mathcal{S}_{m'}^1}$$

$$z^{-2(m+m'+p)}\int_0^t dt_1'\int_0^{t_1'}dt_2'\cdots\int_0^{t_{n-1}'}dt_n'$$

$$\sum_{m'=0}^{n}\ \sum_{\substack{2\le q_1'<\cdots<q_{m'}'\le n'}}\ \sum_{\substack{1\le p_1',\cdots,p_{m'}'\le n'\\ |\{p_h\}_{h=1}^{m'}|=m',\,p_h'<q_h',h=1,\cdots,m'}}\ \int_0^s ds_1'\int_{s_1'}^s ds_2'\cdots\int_{s_{n'-1}'}^s ds_{n'}'$$

$$\sum_{\substack{y_1<\cdots<y_p,\ \{y_h\}_{h=1}^p\subset\{1,\cdots,n'\}\backslash\{q_h'\}_{h=1}^{m'}}}\ \sum_{\substack{\{x_h\}_{h=1}^p\subset\{1,\cdots,n\}\backslash\{p_h\}_{h=1}^m,\ |\{x_h\}_{h=1}^p|=p}}$$

$$\prod_{h=1}^m f\Big(\frac{t_{q_{\sigma(h)}}'-t_{p_h}'}{z^2}\Big)\cdot\prod_{h=1}^{m'} f\Big(\frac{s_{q_{\sigma(h)}'}'-s_{p_h'}'}{z^2}\Big)$$

$$\prod_{h=1}^p |<g_{1-\varepsilon(x_h)},S_{(s_{y_h}'-t_{x_h}')/z^2}g_{\varepsilon'(y_h)}>|$$

$$\Big[\max_{F=f,f';G=g_0,g_1}\int_{-\infty}^{\infty}|<F,S_t G>|dt\Big]^{2(n-m+n'-m'-p)}\tag{2.23}$$

Using Lemma (2.1) one obtains that

$$\sum_{\sigma\in\mathcal{S}_m^1} z^{-2(m+p)}\int_0^t dt_1'\int_0^{t_1'}dt_2'\cdots\int_0^{t_{n-1}'}dt_n'$$

$$\prod_{h=1}^m f\Big(\frac{t_{q_{\sigma(h)}}'-t_{p_h}'}{z^2}\Big)\cdot\prod_{h=1}^p |<g_{1-\varepsilon(x_h)},S_{(s_{y_h}'-t_{x_h}')/z^2}g_{\varepsilon'(y_h)}>|$$

$$\le\Big[\int_{-\infty}^0 f(t)dt\Big]^m$$

$$z^{-2p}\int_{0\le t_n\le\cdots\widehat{t_{q_m}}\cdots\widehat{t_{q_1}}\cdots\le t_1} dt_1\cdots\widehat{dt_{q_1}}\cdots\widehat{dt_{q_m}}\cdots dt_n\prod_{h=1}^p |<g_{1-\varepsilon(x_h)},S_{(s_{y_h}'-t_{x_h})/z^2}g_{\varepsilon'(y_h)}>|$$

$$\tag{2.24}$$

and using Lemma (2.3) to the right hand side of (2.24) one obtains its majorization as

$$\left[\int_{-\infty}^{0} f(t)dt\right]^{m} \cdot \left[\int_{-\infty}^{\infty} f(t)dt\right]^{p} \cdot \frac{t^{n-m-p}}{(n-m-p)!} \tag{2.25}$$

Using Corollary (2.2) one finds

$$\sum_{\sigma' \in \mathcal{S}_{m'}^{1}} z^{2m'} \int_{0}^{s} ds_1' \int_{s_1'}^{s} ds_2' \cdots \int_{s_{n'-1}'}^{s} ds_{n'}' \prod_{h=1}^{m'} f\left(\frac{s_{q'_{\sigma(h)}}' - s_{p'_h}'}{z^2}\right) \tag{2.26}$$

$$\leq \left[\int_{-\infty}^{0} f(t)dt\right]^{m'} \frac{s^{n'-m'}}{(n'-m')!}$$

Summing up we get that (2.23) (so (2.20)) is majorized by

$$\sum_{\varepsilon \in \{0,1\}^n, \ \varepsilon' \in \{0,1\}^{n'}} \sum_{m=0}^{n} \sum_{2 \leq q_1 < \cdots < q_m \leq n} \sum_{\substack{1 \leq p_1 < \cdots < p_m \leq n \\ |\{p_h\}_{h=1}^{m}|=m, p_h < q_h, h=1,\cdots,m}}$$

$$\sum_{m'=0}^{n'} \sum_{2 \leq q_1' < \cdots < q_{m'}' \leq n'} \sum_{\substack{1 \leq p_1', \cdots, p_{m'}' \leq n' \\ |\{p_h\}_{h=1}^{m'}|=m', p_h' < q_h', h=1,\cdots,m'}}$$

$$\left[\int_{-\infty}^{0} f(t)dt\right]^{m} \cdot \left[\int_{-\infty}^{\infty} f(t)dt\right]^{p} \cdot \frac{t^{n-m-p}}{(n-m-p)!}$$

$$\leq \left[\int_{-\infty}^{0} f(t)dt\right]^{m'} \frac{s^{n'-m'}}{(n'-m')!}$$

$$\left[\max_{F=f,f';G=g_0,g_1} \int_{-\infty}^{\infty} |<F,S_tG>|dt\right]^{2(n-m+n'-m'-p)} \tag{2.27}$$

Since

$$\sum_{m=0}^{n} \sum_{2 \leq q_1 < \cdots < q_m \leq n} \sum_{\substack{1 \leq p_1 < \cdots < p_m \leq n \\ |\{p_h\}_{h=1}^{m}|=m, p_h < q_h, h=1,\cdots,m}} 1$$

$$\leq \sum_{m=0}^{n} \binom{n}{m}^2 \leq 2^n \cdot \sum_{m=0}^{n} \binom{n}{m} = 4^n \tag{2.28}$$

$$\sum_{m'=0}^{n'} \sum_{2 \leq q_1' < \cdots < q_{m'}' \leq n'} \sum_{\substack{1 \leq p_1', \cdots, p_{m'}' \leq n' \\ |\{p_h\}_{h=1}^{m'}|=m', p_h' < q_h', h=1,\cdots,m'}} 1$$

L. Accardi & L. Gang

$$\leq \sum_{m'=0}^{n'} \binom{n'}{m'}^2 \leq 2^{n'} \cdot \sum_{m'=0}^{n'} \binom{n'}{m'} = 4^{n'} \tag{2.28a}$$

and

$$\sum_{\varepsilon \in \{0,1\}^n, \ \varepsilon' \in \{0,1\}^{n'}} 1 = 2^{n+n'} \tag{2.29}$$

one finishes the proof.

§3. The collective terms and negliable terms

In the section we shall first investigate the collective terms and the negligible terms for each $n, n' \in \mathbf{N}$. Then, using the previous results and the uniform estimate, we shall obtain the low density limit.

First of all, we consider the decomposition of $\Delta_n^\varepsilon(j, z, t)$ as $I_n^\varepsilon(j, z, t) + II_n^\varepsilon(j, z, t)$, $j = 1, 2$ with

$$I_n^\varepsilon(1, z, t) = \sum_{m=0}^{n} \sum_{2 \le q_1 < \cdots < q_m \le n} \int_0^{t/z^2} dt_1 \int_0^{t_1} dt_2 \cdots \int_0^{t_{n-1}} dt_n$$

$$\prod_{h=1}^{m} < S_{t_{q_h-1}} g_{1-\varepsilon(q_h-1)}, S_{t_{q_h}} g_{\varepsilon(q_h)} > \prod_{\alpha \in \{1,\cdots,n\} \setminus \{q_h\}_{h=1}^m} A^+(S_{t_\alpha} g_{\varepsilon(\alpha)})$$

$$\prod_{\alpha \in \{1,\cdots,n\} \setminus \{q_h-1\}_{h=1}^m} A(S_{t_\alpha} g_{1-\varepsilon(\alpha)}) \tag{3.1}$$

and

$$II_n^\varepsilon(1, z, t) = \sum_{m=0}^{n} \sum_{2 \le q_1 < \cdots < q_m \le n} \sum_{(q_1, p_1, \cdots, q_m, p_m)} \int_0^{t/z^2} dt_1 \int_0^{t_1} dt_2 \cdots \int_0^{t_{n-1}} dt_n$$

$$\prod_{h=1}^{m} < S_{t_{p_h}} g_{1-\varepsilon(p_h)}, S_{t_{q_h}} g_{\varepsilon(q_h)} > \prod_{\alpha \in \{1,\cdots,n\} \setminus \{q_h\}_{h=1}^m} A^+(S_{t_\alpha} g_{\varepsilon(\alpha)})$$

$$\prod_{\alpha \in \{1,\cdots,n\} \setminus \{p_h\}_{h=1}^m} A(S_{t_\alpha} g_{1-\varepsilon(\alpha)}) \tag{3.2}$$

where, $\sum_{(q_1, p_1, \cdots, q_m, p_m)}$ means the sum for all $1 \le p_1, \cdots, p_m \le n$ such that $|\{p_h\}_{h=1}^m| = m$, $p_h < q_h$, $h = 1, \cdots, m$ and $p_h < q_h - 1$ for some $h = 1, \cdots, m$. With the same notations

$$I_n^\varepsilon(2, z, t) = \sum_{m=0}^{n} \sum_{2 \le q_1 < \cdots < q_m \le n} \int_0^{t/z^2} dt_1 \int_{t_1}^{t/z^2} dt_2 \cdots \int_{t_{n-1}}^{t/z^2} dt_n$$

$$(-1)^n \prod_{h=1}^{m} < S_{t_{q_h}-1} g_{1-\varepsilon(q_h-1)}, S_{t_{q_h}} m g_{\varepsilon(q_h)} > \prod_{\alpha \in \{1,\cdots,n\} \setminus \{q_h\}_{h=1}^m} A^+(S_{t_\alpha} g_{\varepsilon(\alpha)})$$

$$\prod_{\alpha \in \{1,\cdots,n\} \setminus \{q_h-1\}_{h=1}^m} A(S_{t_\alpha} g_{1-\varepsilon(\alpha)}) \tag{3.1a}$$

and

$$II_n^\varepsilon(1, z, t) = \sum_{m=0}^{n} \sum_{2 \le q_1 < \cdots < q_m \le n} \sum_{(q_1, p_1, \cdots, q_m, p_m)} \int_0^{t/z^2} dt_1 \int_{t_1}^{t/z^2} dt_2 \cdots \int_{t_{n-1}}^{t/z^2} dt_n$$

$$(-1)^n \prod_{h=1}^{m} < S_{t_{p_h}} g_{1-\varepsilon(p_h)}, S_{t_{q_h}} g_{\varepsilon(q_h)} > \prod_{\alpha \in \{1,\cdots,n\} \setminus \{q_h\}_{h=1}^m} A^+(S_{t_\alpha} g_{\varepsilon(\alpha)})$$

$$\prod_{\alpha \in \{1,\cdots,n\} \setminus \{p_h\}_{h=1}^m} A(S_{t_\alpha} g_{1-\varepsilon(\alpha)}) \tag{3.2a}$$

With these notations one has that

$$\Delta_n^\varepsilon(1,z,t)\Delta_{n'}^{\varepsilon'}(2,z,t) = (I_n^\varepsilon(1,z,t) + II_n^\varepsilon(1,z,t)) \cdot (I_{n'}^{\varepsilon'}(2,z,t) + II_{n'}^{\varepsilon'}(2,z,t))$$

$$= I_n^\varepsilon(1,z,t)I_{n'}^{\varepsilon'}(2,z,t) + I_n^\varepsilon(1,z,t)II_{n'}^{\varepsilon'}(2,z,t) + II_n^\varepsilon(1,z,t)I_{n'}^{\varepsilon'}(2,z,t) + II_n^\varepsilon(1,z,t)II_{n'}^{\varepsilon'}(2,z,t) \tag{3.3}$$

In this section we shall prove that, as $z \to 0$, only the first term of the right hand side of (3.3) gives nonzero contribution.

First of all, we show that the sum of third and fourth term in the right hand side of (3.3) gives zero contribution. The result is state in the following Theorem

<u>*THEOREM*</u> (3.1)

$$\lim_{z \to 0} < W(z \int_{S/z^2}^{T/z^2} S_u f \, du)\Phi, II_n^\varepsilon(1,z,t)\Delta_{n'}^{\varepsilon'}(2,z,t)W(z \int_{S'/z^2}^{T'/z^2} S_u f' \, du)\Phi >= 0 \tag{3.4}$$

<u>*PROOF*</u> By the definitions,

$$| < W(z \int_{S/z^2}^{T/z^2} S_u f \, du)\Phi, II_n^\varepsilon(1,z,t)\Delta_{n'}^{\varepsilon'}(2,z,t)W(z \int_{S'/z^2}^{T'/z^2} S_u f' \, du)\Phi > |$$

$$\leq \sum_{m=0}^{n} \sum_{2 \leq q_1 < \cdots < q_m \leq n} \sum_{(q_1,p_1,\cdots q_m,p_m)} z^{-2(m+m'+p)} \int_0^t dt_1' \int_0^{t_1'} dt_2' \cdots \int_0^{t_{n-1}'} dt_n'$$

$$\sum_{m'=0}^{n} \sum_{2 \leq q_1' < \cdots < q_{m'}' \leq n'} \sum_{\substack{1 \leq p_1', \cdots, p_{m'}' \leq n' \\ |\{p_h'\}_{h=1}^{m'}|=m', p_h' < q_h', h=1,\cdots,m'}}$$

$$\sum_{\substack{y_1 < \cdots < y_p, \ \{y_h\}_{h=1}^p \subset \{1,\cdots,n'\} \setminus \{q_h'\}_{h=1}^{m'}}} \sum_{\{x_h\}_{h=1}^m \subset \{1,\cdots,n\} \setminus \{p_h\}_{h=1}^p, |\{x_h\}+h=1^p|=p}$$

$$\int_0^t ds_1' \int_{s_1'}^t ds_2' \cdots \int_{s_{n'-1}'}^t ds_{n'}'$$

$$\prod_{h=1}^{m} | < g_{1-\varepsilon(p_h)}, S_{(t_{q_h}' - t_{p_h}')/z^2} g_{\varepsilon(q_h)} > |$$

$$\prod_{h=1}^{m'} | < g_{1-\varepsilon'(p'_h)}, S_{(s'_{q'_h}-s'_{p'_h})/z^2} g_{\varepsilon'}(q'_h) > |$$

$$\prod_{h=1}^{p} | < g_{1-\varepsilon(x_h)}, S_{(s'_{y_h}-t'_{x_h})/z^2} g_{\varepsilon'}(y_h) > |$$

$$\left[\max_{F=f,f';G=g_0,g_1} \int_{-\infty}^{\infty} | < F, S_t G > | dt \right]^{2(n-m+n'-m'-p)} \tag{3.5}$$

Use the notations (2.21), (2.22), Corollary (2.2) and Lemma (2.3) to the second integral (ds'-integral), one knows that the right hand side of (3.5) is majorized by

$$C(n,n') \sum_{m=0}^{n} \sum_{2 \leq q_1 < \cdots < q_m \leq n} \sum_{(q_1,p_1,\cdots q_m,p_m)} z^{-2m} \int_0^t dt'_1 \int_0^{t'_1} dt'_2 \cdots \int_0^{t'_{n-1}} dt'_n$$

$$\prod_{h=1}^{m} | < g_{1-\varepsilon(p_h)}, S_{(t'_{q_h}-t'_{p_h})/z^2} g_{\varepsilon}(q_h) > | \tag{3.6}$$

where, $C(n,n')$ is a constant independent on z. By Lemma (3.3) of [1], one finishes the proof.

Theorem (3.1) show that the limit

$$\lim_{z \to 0} < W(z \int_{S/z^2}^{T/z^2} S_u f du)\Phi, \Delta_n^\varepsilon(1,z,t)\Delta_{n'}^{\varepsilon'}(2,z,t)W(z \int_{S'/z^2}^{T'/z^2} S_u f' du)\Phi > \tag{3.7}$$

is equal to the limit (if it exsits)

$$\lim_{z \to 0} < W(z \int_{S/z^2}^{T/z^2} S_u f du)\Phi, I_n^\varepsilon(1,z,t)\Delta_{n'}^{\varepsilon'}(2,z,t)W(z \int_{S'/z^2}^{T'/z^2} S_u f' du)\Phi > \tag{3.8}$$

Notice that in $I_n^\varepsilon(1,z,t)\Delta_{n'}^{\varepsilon'}(2,z,t)$ the indices $\{x_h\}_{h=1}^{p}$ aer unordered. The following theorem shows that we need only to consider the ordered terms which correspond to $x_1 > \cdots > x_p$, since the other term has the limit zero as $z \to 0$.

THEOREM (3.2)

$$\lim_{z \to 0} < W(z \int_{S/z^2}^{T/z^2} S_u f du)\Phi, I_n^\varepsilon(1,z,t)\Delta_{n'}^{\varepsilon'}(2,z,t)W(z \int_{S'/z^2}^{T'/z^2} S_u f' du)\Phi >$$

$$= \lim_{z \to 0} \sum_{m=0}^{n} \sum_{2 \leq q_1 < \cdots < q_m \leq n} z^{-2(m+m'+p)} \int_0^t dt'_1 \int_0^{t'_1} dt'_2 \cdots \int_0^{t'_{n-1}} dt'_n$$

$$\sum_{m'=0}\sum_{2\le q'_1<\cdots<q'_{m'}\le n'}\sum_{\substack{1\le p'_1,\cdots,p'_{m'}\le n'\\ |\{p'_h\}^{m'}_{h=1}|=m',\,p'_h<q'_h,\,h=1,\cdots,m'}}\int_0^t ds'_1\int_{s'_1}^t ds'_2\cdots\int_{s'_{n'-1}}^t ds'_{n'}$$

$$\sum_{p=0}^{(n-m)\wedge(n'-m')}\ \sum_{y_1<\cdots<y_p,\ \{y_h\}^p_{h=1}\subset\{1,\cdots,n'\}\backslash\{q'_h\}^{m'}_{h=1}}\ \sum_{\{x_h\}^m_{h=1}\subset\{1,\cdots,n\}\backslash\{p_h\}^p_{h=1},\ x_1>\cdots>x_p}$$

$$\prod_{h=1}^m <g_{1-\varepsilon(q_h-1)},S_{(t'_{q_h}-t'_{q_h-1})/z^2}g_{\varepsilon(q_h)}>$$

$$\prod_{h=1}^{m'} <g_{1-\varepsilon'(p'_h)},S_{(s'_{q'_h}-s'_{p'_h})/z^2}g_{\varepsilon'(q'_h)}>$$

$$\prod_{h=1}^{p} <g_{1-\varepsilon(x_h)},S_{(s'_{y_h}-t'_{x_h})/z^2}g_{\varepsilon'(y_h)}>$$

$$<W(z\int_{S/z^2}^{T/z^2}S_u f\,du)\Phi,W(z\int_{S'/z^2}^{T'/z^2}S_u f'\,du)\Phi> \qquad (3.9)$$

<u>*PROOF*</u> By the definitions and (2.17a),

$$<W(z\int_{S/z^2}^{T/z^2}S_u f\,du)\Phi,I_n^\varepsilon(1,z,t)\Delta_{n'}^{\varepsilon'}(2,z,t)W(z\int_{S'/z^2}^{T'/z^2}S_u f'\,du)\Phi>$$

$$=\sum_{m=0}^n\sum_{2\le q_1<\cdots<q_m\le n}z^{-2(m+m'+p)}\int_0^t dt'_1\int_0^{t'_1}dt'_2\cdots\int_0^{t'_{n-1}}dt'_n$$

$$\sum_{m'=0}^n\sum_{2\le q'_1<\cdots<q'_{m'}\le n'}\sum_{\substack{1\le p'_1,\cdots,p'_{m'}\le n'\\ |\{p'_h\}^{m'}_{h=1}|=m',\,p'_h<q'_h,\,h=1,\cdots,m'}}\int_0^t ds'_1\int_{s'_1}^t ds'_2\cdots\int_{s'_{n'-1}}^t ds'_{n'}$$

$$\sum_{p=0}^{(n-m)\wedge(n'-m')}\ \sum_{y_1<\cdots<y_p,\ \{y_h\}^p_{h=1}\subset\{1,\cdots,n'\}\backslash\{q'_h\}^{m'}_{h=1}}\ \sum_{\{x_h\}^p_{h=1}\subset\{1,\cdots,n\}\backslash\{p_h\}^m_{h=1},\ |\{x_h\}^p_{h=1}|=p}$$

$$\prod_{h=1}^m <g_{1-\varepsilon(q_h-1)},S_{(t'_{q_h}-t'_{q_h-1})/z^2}g_{\varepsilon(q_h)}>$$

$$\prod_{h=1}^{m'} <g_{1-\varepsilon'(p'_h)},S_{(s'_{q'_h}-s'_{p'_h})/z^2}g_{\varepsilon'(q'_h)}>$$

$$\prod_{h=1}^{p} \; < g_{1-\varepsilon(x_h)}, S_{(s'_{y_h}-t'_{x_h})/z^2} g_{\varepsilon(y_h)} >$$

$$< W(z\int_{S/z^2}^{T/z^2} S_u f\, du)\Phi, W(z\int_{S'/z^2}^{T'/z^2} S_u f'\, du)\Phi > \tag{3.10}$$

Applying Lemma (2.1) to the second integral (ds'-integral) of (3.10), one finds that

$$z^{-2(m'+p)} \int_0^t ds'_1 \int_{s'_1}^t ds'_2 \cdots \int_{s'_{n'-1}}^t ds'_{n'} \prod_{h=1}^{m'} | < g_{1-\varepsilon'(p'_h)}, S_{(s'_{q'_h}-s'_{p'_h})/z^2} g_{\varepsilon'(q'_h)} > |$$

$$\prod_{h=1}^{p} | < g_{1-\varepsilon(x_h)}, S_{(s'_{y_h}-t'_{x_h})/z^2} g_{\varepsilon'(y_h)} > |$$

$$\leq C_n \cdot z^{-2p} \int_{0\leq s'_1\leq\cdots\leq \widehat{s'_{q_1}}\leq\cdots\leq \widehat{s'_{q_{m'}}}\leq\cdots\leq s'_{n'}\leq t} ds'_1 \cdots \widehat{ds'_{q_1}} \cdots \widehat{ds'_{q_{m'}}} \cdots ds'_{n'}$$

$$\prod_{h=1}^{p} | < g_{1-\varepsilon(x_h)}, S_{(s'_{y_h}-t'_{x_h})/z^2} g_{\varepsilon'(y_h)} > | \tag{3.11}$$

where, C_n is a constant independent on z. Notice that

$$y_1 < \cdots < y_p, \quad \{y_h\}_{h=1}^{p} \subset \{1,\cdots,n'\} \setminus \{q'_h\}_{h=1}^{m'} \tag{3.12}$$

so the right hand side of (3.11) can be written

$$z^{-2p} \int_{0\leq s'_1\leq\cdots\leq s'_{n'-m'}\leq t} ds'_1 \cdots ds'_{n'-m'} \prod_{h=1}^{p} | < g_{1-\varepsilon(x_h)}, S_{(s'_{y_h}-t'_{x_h})/z^2} g_{\varepsilon'(y_h)} > | \tag{3.13}$$

of course the $\{y_h\}_{h=1}^{p}$ in (3.13) is not same ones as in (3.8), but it is still true that $y_1 < \cdots < y_p$. Moreover the integral

$$z^{-2m} \int_0^t dt'_1 \int_0^{t'_1} dt'_2 \cdots \int_0^{t'_{n-1}} dt'_n \prod_{h=1}^{m} | < g_{1-\varepsilon(q_h-1)}, S_{(t'_{q_h}-t'_{q_h-1})/z^2} g_{\varepsilon(q_h)} > | \tag{3.14}$$

is bounded, so with the change of variables

$$s_{y_h} = (s'_{y_h}-t'_{x_h})/z^2, \quad h=1,\cdots,p \tag{3.15}$$

(3.13) becomes

$$\int_0^t ds'_1 \int_{s'_1}^t ds'_2 \cdots \int_{(s'_{y_1-1}-t'_{x_1})/z^2}^{(t-t'_{x_1})/z^2} ds_{y_1} | < g_{1-\varepsilon(x_1)}, S_{s_{y_1}} g_{\varepsilon'(y_1)} > | \int_{z^2 s_{y_1}+t'_{x_1}}^t ds'_{y_1+1}$$

$$\cdots \int_{s'_{y_p-2}}^{t} ds'_{y_p-1} \cdots \int_{(s'_{y_p-1}-t'_{x_p})/z^2}^{(t-t'_{x_p})/z^2} ds_{y_p}| < g_{1-\varepsilon(x_p)}, S_{s_{y_p}} g_{\varepsilon'(y_p)} > | \cdots \int_{s'_{n'-m'-1}}^{t} ds'_{n'-m'}$$

$$(3.16)$$

Denote

$$h := \min\{l \in \{1,\cdots,p\} : x_l < x_{l+1}\} \tag{3.17}$$

then (3.16) is majorized by

$$\int_0^t ds'_1 \int_{s'_1}^t ds'_2 \cdots \int_{(s'_{y_h-1}-t'_{x_h})/z^2}^{(t-t'_{x_h})/z^2} ds_{y_h}| < g_{1-\varepsilon(x_h)}, S_{s_{y_h}} g_{\varepsilon'(y_h)} > | \int_{z^2 s_{y_h}+t'_{x_h}}^t ds'_{y_h+1}$$

$$\cdots \int_{s'_{y_h+1-2}}^t ds'_{y_h+1-1} \cdots \int_{(s'_{y_h+1-1}-t'_{x_h+1})/z^2}^{(t-t'_{x_h+1})/z^2} ds_{y_h+1}| < g_{1-\varepsilon(x_h+1)}, S_{s_{y_h+1}} g_{\varepsilon'(y_h+1)} > | \cdots C_{n',p}$$

$$(3.18)$$

Now we consider (3.18) according the two cases: $y_{h+1} > y_h + 1$ or $y_{h+1} = y_h + 1$.

If $y_{h+1} > y_h + 1$ then $y_{h+1} - 1 \geq y_h + 1$ and $s'_{y_{h+1}-1} \geq s'_{y_h+1} \geq z^2 s_{y_h} + t'_{x_h}$, so

$$s'_{y_{h+1}-1} - t'_{x_h+1} \geq t'_{x_h} - t'_{x_h+1} + z^2 s_{y_h} \tag{3.19}$$

Thus (3.14) is majorized by

$$C_{n',p} \int_0^t ds'_1 \int_{s'_1}^t ds'_2 \cdots \int_{(s'_{y_h-1}-t'_{x_h})/z^2}^{(t-t'_{x_h})/z^2} ds_{y_h}| < g_{1-\varepsilon(x_h)}, S_{s_{y_h}} g_{\varepsilon'(y_h)} > | \int_{z^2 s_{y_h}+t'_{x_h}}^t ds'_{y_h+1}$$

$$\cdots \int_{s'_{y_h+1-2}}^t ds'_{y_h+1-1} \cdots \int_{(t'_{x_h}-t'_{x_h+1})/z^2+s_{y_h}}^{(t-t'_{x_h+1})/z^2} ds_{y_h+1}| < g_{1-\varepsilon(x_h+1)}, S_{s_{y_h+1}} g_{\varepsilon'(y_h+1)} > |$$

$$(3.20)$$

If $y_{h+1} = y_h + 1$ then

$$s'_{y_{h+1}-1} = s'_{y_h} = z^2 s_{y_h} + t'_{x_h} \tag{3.21}$$

so in any case (3.18) is equal to

$$C_{n',p} \int_0^t ds'_1 \int_{s'_1}^t ds'_2 \cdots \int_{(s'_{y_h-1}-t'_{x_h})/z^2}^{(t-t'_{x_h})/z^2} ds_{y_h}| < g_{1-\varepsilon(x_h)}, S_{s_{y_h}} g_{\varepsilon'(y_h)} > | \int_{z^2 s_{y_h}+t'_{x_h}}^t ds'_{y_h+1}$$

$$\int_{s_{y_h}+(t'_{x_h}-t'_{x_h+1})/z^2}^{(t-t'_{x_h+1})/z^2} ds_{y_h+1}| < g_{1-\varepsilon(x_h+1)}, S_{s_{y_h+1}} g_{\varepsilon'(y_h+1)} > | \tag{3.16}$$

Since $t'_{x_h} - t'_{x_h+1} > 0$ a.e., (3.20) and (3.22) tend to zero as $z \to 0$. This ends the proof.

The following Theorem shows that as $z \to 0$ only the term $I_n^\varepsilon(1,z,t)I_{n'}^{\varepsilon'}(2,z,t)$ gives the non-zero contribution, i.e. the term $I_n^\varepsilon(1,z,t)II_{n'}^{\varepsilon'}(2,z,t)$ gives zero contribution. Moreover by Theorem (3.2) we need only show that the terms of $I_n^\varepsilon(1,z,t)II_{n'}^{\varepsilon'}(2,z,t)$, which corresponds to the case $x_1 > \cdots > x_p$ and $y_1 < \cdots < y_p$ gives zero contribution.

THEOREM (3.3)

$$\lim_{z\to 0} < W(z\int_{S/z^2}^{T/z^2} S_u f\, du)\Phi,\, I_n^\varepsilon(1,z,t)\Delta_{n'}^{\varepsilon'}(2,z,t) W(z\int_{S'/z^2}^{T'/z^2} S_u f'\, du)\Phi >$$

$$= \lim_{z\to 0} \sum_{m=0}^{n} \sum_{2\le q_1<\cdots<q_m\le n} z^{-2(m+m'+p)} \int_0^t dt'_1 \int_0^{t'_1} dt'_2 \cdots \int_0^{t'_{n-1}} dt'_n$$

$$\sum_{m'=0}^{n} \sum_{2\le q'_1<\cdots<q'_{m'}\le n'} \int_0^t ds'_1 \int_{s'_1}^t ds'_2 \cdots \int_{s'_{n'-1}}^t ds'_{n'}(-1)^{n'}$$

$$\sum_{p=0}^{(n-m)\wedge(n'-m')} \sum_{\substack{y_1<\cdots<y_p,\ \{y_h\}_{h=1}^p\subset\{1,\cdots,n'\}\setminus\{q'_h\}_{h=1}^{m'}}} \sum_{\substack{\{x_h\}_{h=1}^p\subset\{1,\cdots,n\}\setminus\{q_h-1\}_{h=1}^m,\ x_1>\cdots>x_p}}$$

$$\prod_{h=1}^{m} < g_{1-\varepsilon(q_h-1)},\, S_{(t'_{q_h}-t'_{q_h-1})/z^2} g_{\varepsilon(q_h)} >$$

$$\prod_{h=1}^{m'} < g_{1-\varepsilon'(q'_h-1)},\, S_{(s'_{q'_h}-s'_{q'_h-1})/z^2} g_{\varepsilon'(q'_h)} >$$

$$\prod_{h=1}^{p} < g_{1-\varepsilon(x_h)},\, S_{(s'_{y_h}-t'_{x_h})/z^2} g_{\varepsilon'(y_h)} >$$

$$< W(z\int_{S/z^2}^{T/z^2} S_u f\, du)\Phi,\, W(z\int_{S'/z^2}^{T'/z^2} S_u f'\, du)\Phi > \tag{3.23}$$

PROOF Notice that the type II terms appear only in the second intrgral (ds'-integral) and moreover by Theorem (3.2) we need only consider the situation of $x_1 > \cdots > x_p$, $y_1 < \cdots < y_p$. Therefore it is enough to show that for each $m \in \mathbf{N}, p \in \mathbf{N}\cup\{0\}$, $f_1,\cdots,f_m,\ F_1,\cdots,F_p \in L^1(\mathbf{R})$ and positive, if there exists $h = 1,\cdots,m$ such that $p_h < q_h - 1$ then

$$z^{-2(m+p)} \int_0^t ds_1 \int_{s_1}^t ds_2 \cdots \int_{s_{n-1}}^t ds_n \prod_{h=1}^{m} f_h\Big(\frac{s_{q_h}-s_{p_h}}{z^2}\Big) \cdot \prod_{h=1}^{p} F_h\Big(\frac{s_{y_h}-t_{x_h}}{z^2}\Big) \longrightarrow 0 \tag{3.24}$$

First of all we notice that since both $\{y_h\}_{h=1}^p$ and $\{q_h\}_{h=1}^m$ label creators and in each time there is only one creator, one has

$$\{y_h\}_{h=1}^p \cap \{q_h\}_{h=1}^m = \emptyset \tag{3.25}$$

Denote the left hand side of (3.24) by $\Delta_n(z,t,\{q_h,p_h\}_{h=1}^m)$, we use induction in m to prove (3.24).

If $m = 1$ then it is natural that $q_1 - 1 > p_1$, so that

$$s_{q_1 - 1} > s_{p_1}, \qquad a.e. \tag{3.25a}$$

With the change the variable

$$s'_{q_1} = (s_{q_1} - s_{p_1})/z^2 \tag{3.26}$$

and because the formula (3.25), the change of variable (3.26) does not influnce the product

$$\prod_{h=1}^{p} F_h\left(\frac{s_{y_h} - t_{x_h}}{z^2}\right)$$

therefore $\Delta_n(z, t, \{q_1, p_1\})$ is equal to

$$z^{-2p} \int_0^t ds_1 \int_{s_1}^t ds_2 \cdots \int_{s_{q_1-2}}^t ds_{q_1-1} \int_{(s_{q_1-1}-s_{p_1})/z^2}^{(t-s_{p_1})/z^2} ds'_{q_1} f_1(s'_{q_1}) \int_{z^2 s'_{q_h}+s_{p_1}}^t ds_{q_1+1} \cdots$$

$$\cdots \int_{s_{n-1}}^t ds_n \prod_{h=1}^{p} F_h\left(\frac{s_{y_h} - t_{x_h}}{z^2}\right) \tag{3.27}$$

Moreover

$$s'_{q_1} \in [(s_{q_h-1} - s_{p_1})/z^2, (t - s_{p_1})/z^2] \implies z^2 s'_{q_h} + s_{p_1} \in [s_{q_h-1}, t] \tag{3.28}$$

so that (3.27) is majorized by

$$z^{-2p} \int_0^t ds_1 \int_{s_1}^t ds_2 \cdots \int_{s_{q_1-2}}^t ds_{q_1-1} \int_{z^2 s'_{q_h}+s_{p_1}}^t ds_{q_1+1} \cdots \int_{s_{n-1}}^t ds_n \prod_{h=1}^{p} F_h\left(\frac{s_{y_h} - t_{x_h}}{z^2}\right)$$

$$\int_{(s_{q_h-1}-s_{p_1})/z^2}^{(t-s_{p_1})/z^2} ds'_{q_1} f_1(s'_{q_1}) \tag{3.29}$$

By Lemma (2.3) one knows that the first part of (3.29), i.e.

$$z^{-2p} \int_0^t ds_1 \int_{s_1}^t ds_2 \cdots \int_{s_{q_1-2}}^t ds_{q_1-1} \int_{z^2 s'_{q_h}+s_{p_1}}^t ds_{q_1+1} \cdots \int_{s_{n-1}}^t ds_n \prod_{h=1}^{p} F_h\left(\frac{s_{y_h} - t_{x_h}}{z^2}\right)$$

$$\tag{3.30}$$

is bounded and it is clear that the second part of (3.29), i.e.

$$\int_{(s_{q_1-1}-s_{p_1})/z^2}^{(t-s_{p_1})/z^2} ds'_{q_1} f_1(s'_{q_1}) \tag{3.31}$$

has a bound

$$\int_0^\infty ds'_{q_1} f_1(s'_{q_1}) \tag{3.32}$$

Thus, by the dominated convergence and (3.25a) we obtain (3.24) for $m = 1$.

Assuming that (3.24) is true for m, let consider the situation of $m + 1$. Repeating above arguments on p_{m+1}, q_{m+1} and notice that

-the change of variable (the analogy of (3.26))

$$s'_{q_{m+1}} = (s_{q_{m+1}} - s_{p_{m+1}})/z^2 \tag{3.33}$$

influence neither $\{y_h\}_{h=1}^p$ (by (3.25)) nor $\{p_h, q_h\}_{h=1}^m$ because $q_{m+1} > q_h, p_h$ for each $h = 1, \cdots, m$

-by Lemma (2.1) and Lemma (2.3) one knows that

$$z^{-2(m+p)} \int_0^t ds_1 \int_{s_1}^t ds_2 \cdots \int_{s_{q_{m+1}-2}}^t ds_{q_{m+1}-1} \int_{z^2 s'_{q_{m+1}}+s_{p_{m+1}}}^t ds_{q_{m+1}+1} \cdots \int_{s_{n-1}}^t ds_n$$

$$\prod_{h=1}^m f_h(\frac{s_{q_h} - s_{p_h}}{z^2}) \cdot \prod_{h=1}^p F_h(\frac{s_{y_h} - t_{x_h}}{z^2}) \tag{3.34}$$

is bounded.

Therefore if $p_{m+1} < q_{m+1} - 1$ then applying the dominated convergence we end the proof; if $p_{m+1} = q_{m+1} - 1$ then the boundness of

$$\int_{(s_{q_{m+1}-1}-s_{p_{m+1}})/z^2}^{(t-s_{p_{m+1}})/z^2} ds'_{q_{m+1}} f_{m+1}(s'_{q_{m+1}}) \tag{3.35}$$

deduce the problem to the situation of $\Delta_n(z, t, \{q_h, p_h\}_{h=1}^m)$, , thus by the induction we end the proof.

Now we compute the collective terms. Form the above results one knows that we need only consider the terms with the following form:

$$< W(z \int_{S/z^2}^{T/z^2} S_u f du)\Phi, I_n^\varepsilon(1, z, t) I_{n'}^{\varepsilon'}(2, z, t) W(z \int_{S'/z^2}^{T'/z^2} S_u f' du)\Phi > \tag{3.36}$$

For computing the limit of the collective term we would like to rewrite the $I_n^\varepsilon(j, z, t)$, $j = 1, 2$ in other form. Notice that in (3.1) and (3.1a) we use the indices $\{q_h\}_{h=1}^m$ to label such creators which are used to produce scalar products in the procedure of normalizing ordered form, now we use the indices $\{q_h\}_{h=1}^m$ to label such creators which are <u>not</u> used to produce scalar products.

LEMMA(3.4) For each $n \in \mathbf{N}$, $\varepsilon \in \{0, 1\}^n$

$$I_n^\varepsilon(1, z, t) = \sum_{m=1}^n \sum_{1=q_1<q_2<\cdots<q_m \leq n} \int_0^{t/z^2} dt_1 \int_0^{t_1} dt_2 \cdots \int_0^{t_{n-1}} dt_n$$

$$\prod_{\alpha\in\{1,\cdots,n\}\setminus\{q_h\}_{h=1}^{m}} < S_{t_{\alpha-1}}g_{1-\varepsilon(\alpha-1)}, S_{t_\alpha}g_{\varepsilon(\alpha)} >$$

$$\prod_{h=1}^{m} A^{+}(S_{t_{q_h}}g_{\varepsilon(q_h)}) \cdot \prod_{h=1}^{m} A(S_{t_{q_{h+1}-1}}g_{1-\varepsilon(q_{h+1}-1)}) \tag{3.37}$$

and

$$I_n^{\varepsilon}(2,z,t) = \sum_{m=1}^{n}\ \sum_{1=q_1<q_2<\cdots<q_m\leq n}\ \int_0^{t/z^2} dt_1 \int_{t_1}^{t/z^2} dt_2 \cdots \int_{t_{n-1}}^{t/z^2} dt_n$$

$$(-1)^n \prod_{\alpha\in\{1,\cdots,n\}\setminus\{q_h\}_{h=1}^{m}} < S_{t_{\alpha-1}}g_{1-\varepsilon(\alpha-1)}, S_{t_\alpha}g_{\varepsilon(\alpha)} >$$

$$\prod_{h=1}^{m} A^{+}(S_{t_{q_h}}g_{\varepsilon(q_h)}) \cdot \prod_{h=1}^{m} A(S_{t_{q_{h+1}-1}}g_{1-\varepsilon(q_{h+1}-1)}) \tag{3.37a}$$

where, $q_{m+1} := n+1$.

PROOF The proof is the same as ones of Lemma (3.3) in [4a].

Use Lemma (3.4), one has the follwing

$$< W(z\int_{S/z^2}^{T/z^2} S_u f\,du)\Phi, I_n^{\varepsilon}(1,z,t)I_{n'}^{\varepsilon'}(2,z,t)W(z\int_{S'/z^2}^{T'/z^2} S_u f'\,du)\Phi >=$$

$$= \sum_{m=1}^{n}\ \sum_{1=q_1<q_2<\cdots<q_m\leq n}\ \sum_{m'=1}^{n'}\ \sum_{1=q_1'<q_2'<\cdots<q_{m'}'\leq n'} (-1)^{n'}$$

$$\int_0^{t/z^2} dt_1 \int_0^{t_1} dt_2 \cdots \int_0^{t_{n-1}} dt_n$$

$$\prod_{\alpha\in\{1,\cdots,n\}\setminus\{q_h\}_{h=1}^{m}} < S_{t_{\alpha-1}}g_{1-\varepsilon(\alpha-1)}, S_{t_\alpha}g_{\varepsilon(\alpha)} >$$

$$< W(z\int_{S/z^2}^{T/z^2} S_u f\,du)\Phi, \prod_{h=1}^{m} A^{+}(S_{t_{q_h}}g_{\varepsilon(q_h)}) \cdot \prod_{h=1}^{m} A(S_{t_{q_{h+1}-1}}g_{1-\varepsilon(q_{h+1}-1)})$$

$$\int_0^{t/z^2} ds_1 \int_{s_1}^{t/z^2} ds_2 \cdots \int_{s_{n'-1}}^{t/z^2} ds_{n'}$$

$$\prod_{\alpha\in\{1,\cdots,n'\}\setminus\{q_h'\}_{h=1}^{m'}} < S_{s_{\alpha-1}}g_{1-\varepsilon'(\alpha-1)}, S_{s_\alpha}g_{\varepsilon'(\alpha)} >$$

$$\prod_{h=1}^{m'} A^+(S_{s_{q'_h}} g_{\varepsilon'(q'_h)}) \cdot \prod_{h=1}^{m'} A(S_{s_{q'_{h+1}-1}} g_{1-\varepsilon'(q'_{h+1}-1)}) W(z \int_{S'/z^2}^{T'/z^2} S_u f' du) \Phi > \qquad (3.38)$$

Making the product, in the right hand side of (3.38),

$$\prod_{h=1}^{m} A(S_{t_{q_{h+1}-1}} g_{1-\varepsilon(q_{h+1}-1)}) \cdot \prod_{h=1}^{m'} A^+(S_{s_{q'_h}} g_{\varepsilon'(q'_h)}) \qquad (3.39)$$

in the normally ordered form and by Theorem (3.2), one knows that the limit of (3.38) as $z \to 0$ is equal to

$$\lim_{z \to 0} \sum_{m=1}^{n} \sum_{1=q_1<q_2<\cdots<q_m \le n} \sum_{m'=1}^{n'} \sum_{1=q'_1<q'_2<\cdots<q'_{m'} \le n'} (-1)^{n'}$$

$$\sum_{p=0}^{m \wedge m'} \sum_{y_1<\cdots<y_p, \ \{y_h\}_{h=1}^{p} \subset \{1,\cdots,m'\} \ \{x_h\}_{h=1}^{p} \subset \{1,\cdots,m\}, \ x_1>\cdots>x_p}$$

$$\int_0^{t/z^2} dt_1 \int_0^{t_1} dt_2 \cdots \int_0^{t_{n-1}} dt_n$$

$$\prod_{\alpha \in \{1,\cdots,n\} \backslash \{q_h\}_{h=1}^{m}} < S_{t_{\alpha-1}} g_{1-\varepsilon(\alpha-1)}, S_{t_\alpha} g_{\varepsilon(\alpha)} >$$

$$< W(z \int_{S/z^2}^{T/z^2} S_u f du) \Phi, \prod_{h=1}^{m} A^+(S_{t_{q_h}} g_{\varepsilon(q_h)}) \cdot \prod_{h \in \{1,\cdots,m'\} \backslash \{y_l\}_{l=1}^{p}} A^+(S_{s_{q'_h}} g_{\varepsilon'(q'_h)})$$

$$\int_0^{t/z^2} ds_1 \int_{s_1}^{t/z^2} ds_2 \cdots \int_{s_{n'-1}}^{t/z^2} ds_{n'}$$

$$\prod_{\alpha \in \{1,\cdots,n'\} \backslash \{q'_h\}_{h=1}^{m'}} < S_{s_{\alpha-1}} g_{1-\varepsilon'(\alpha-1)}, S_{s_\alpha} g_{\varepsilon'(\alpha)} >$$

$$\prod_{l=1}^{p} < S_{t_{q_{x_l}+1}-1} g_{1-\varepsilon(q_{x_l}+1-1)}, S_{s_{q'_{y_l}}} g_{\varepsilon'(q'_{y_l})} >$$

$$\cdot \prod_{h \in \{1,\cdots,m\} \backslash \{x_l\}_{l=1}^{p}} A(S_{t_{q_{h+1}-1}} g_{1-\varepsilon(q_{h+1}-1)})$$

$$\prod_{h=1}^{m'} A(S_{s_{q'_{h+1}-1}} g_{1-\varepsilon'(q'_{h+1}-1)}) W(z \int_{S'/z^2}^{T'/z^2} S_u f' du) \Phi > \qquad (3.39)$$

Notice that in (3.39), for notational convience we introduced the $\{q_{x_l}+1-1\}_{l=1}^{p}$ and $\{q'_{y_l}\}_{l=1}^{p}$ to play the rolus which previous had been played by $\{x_l\}_{l=1}^{p}$ and $\{y_l\}_{l=1}^{p}$ respectively.

In (3.39), applying the creation and annihilation operators on the coherent vectors, we can rewrite (3.39) to

$$\lim_{z\to 0}\sum_{m=1}^{n}\sum_{1=q_1<q_2<\cdots<q_m\le n}\sum_{m'=1}^{n'}\sum_{1=q'_1<q'_2<\cdots<q'_{m'}\le n'}(-1)^{n'}$$

$$\sum_{p=0}^{m\wedge m'}\ \sum_{\substack{y_1<\cdots<y_p,\ \{y_h\}_{h=1}^{p}\subset\{1,\cdots,m'\}}}\ \sum_{\{x_h\}_{h=1}^{p}\subset\{1,\cdots,m\},\ x_1>\cdots>x_p}$$

$$z^{2(m+m'-p)}\int_0^{t/z^2}dt_1\int_0^{t_1}dt_2\cdots\int_0^{t_{n-1}}dt_n$$

$$\prod_{\alpha\in\{1,\cdots,n\}\setminus\{q_h\}_{h=1}^{m}}<S_{t_{\alpha-1}}g_{1-\varepsilon(\alpha-1)},S_{t_\alpha}g_{\varepsilon(\alpha)}>$$

$$\int_0^{t/z^2}ds_1\int_{s_1}^{t/z^2}ds_2\cdots\int_{s_{n'-1}}^{t/z^2}ds_{n'}$$

$$\prod_{\alpha\in\{1,\cdots,n'\}\setminus\{q'_h\}_{h=1}^{m'}}<S_{s_{\alpha-1}}g_{1-\varepsilon'(\alpha-1)},S_{s_\alpha}g_{\varepsilon'(\alpha)}>$$

$$\prod_{l=1}^{p}<S_{t_{q_{x_l}+1}-1}g_{1-\varepsilon(q_{x_l}+1-1)},S_{s_{q'_{y_l}}}g_{\varepsilon'(q'_{y_l})}>$$

$$\prod_{h=1}^{m}\int_{S/z^2}^{T/z^2}<S_uf,S_{t_{q_h}}g_{\varepsilon(q_h)}>du\cdot\prod_{h\in\{1,\cdots,m'\}\setminus\{y_l\}_{l=1}^{p}}\int_{S/z^2}^{T/z^2}<S_uf,S_{s_{q'_h}}g_{\varepsilon'(q'_h)}>du$$

$$\cdot\prod_{h\in\{1,\cdots,m\}\setminus\{x_l\}_{l=1}^{p}}\int_{S'/z^2}^{T'/z^2}<S_{t_{q_{h+1}-1}}g_{1-\varepsilon(q_{h+1}-1)},S_uf'>du$$

$$\prod_{h=1}^{m'}\int_{S/z^2}^{T/z^2}<S_{s_{q'_{h+1}-1}}g_{1-\varepsilon'(q'_{h+1}-1)},S_uf>du$$

$$<W\!\left(z\int_{S/z^2}^{T/z^2}S_uf\,du\right)\Phi,W\!\left(z\int_{S'/z^2}^{T'/z^2}S_uf'\,du\right)\Phi> \tag{3.40}$$

For each $n\in\mathbf{N}$, $m\le n$, $1=q_1<q_2<\cdots<q_m\le n$, $q_{m+1}:=n+1$

$$\{1,\cdots,n\}\setminus\{q_h\}_{h=1}^{m}=\bigcup_{r=1}^{m}\{q_r+1,\cdots,q_{r+1}-1\} \tag{3.41}$$

and similarly for each $n' \in \mathbf{N}$, $m' \le n'$, $1 = q_1' < q_2' < \cdots < q_{m'}' \le n'$, $q_{m'+1}' := n' + 1$

$$\{1, \cdots, n'\} \setminus \{q_h'\}_{h=1}^{m'} = \bigcup_{r'=1}^{m'} \{q_{r'}' + 1, \cdots, q_{r'+1}' - 1\} \tag{3.41a}$$

With the change of variables:

$$t_1 z^2 \hookrightarrow t_1, \cdots, t_n z^2 \hookrightarrow t_n \tag{3.42}$$

and

$$s_1 z^2 \hookrightarrow s_1, \cdots, s_{n'} z^2 \hookrightarrow s_{n'} \tag{3.42a}$$

(3.40) becomes

$$\lim_{z \to 0} \sum_{m=1}^{n} \sum_{1=q_1<q_2<\cdots<q_m \le n} \sum_{m'=1}^{n'} \sum_{1=q_1'<q_2'<\cdots<q_{m'}' \le n'} (-1)^{n'}$$

$$\sum_{p=0}^{m \wedge m'} \sum_{y_1 < \cdots < y_p, \ \{y_h\}_{h=1}^{p} \subset \{1,\cdots,m'\}} \sum_{\{x_h\}_{h=1}^{p} \subset \{1,\cdots,m\}, \ x_1 > \cdots > x_p}$$

$$z^{-2(n-m+n'-m'+p)} \int_0^t dt_1 \int_0^{t_1} dt_2 \cdots \int_0^{t_{n-1}} dt_n$$

$$\prod_{\alpha \in \{1,\cdots,n\} \setminus \{q_h\}_{h=1}^{m}} < g_{1-\varepsilon(\alpha-1)}, S_{(t_\alpha - t_{\alpha-1})/z^2} g_{\varepsilon(\alpha)} >$$

$$\int_0^t ds_1 \int_{s_1}^t ds_2 \cdots \int_{s_{n'-1}}^t ds_{n'}$$

$$\prod_{\alpha \in \{1,\cdots,n'\} \setminus \{q_h'\}_{h=1}^{m'}} < g_{1-\varepsilon'(\alpha-1)}, S_{(s_\alpha - s_{\alpha-1})/z^2} g_{\varepsilon'(\alpha)} >$$

$$\prod_{l=1}^{p} < g_{1-\varepsilon(q_{x_l}+1-1)}, S_{(s_{q_{y_l}'} - t_{q_{x_l}+1-1})/z^2} g_{\varepsilon'(q_{y_l}')} >$$

$$\prod_{h=1}^{m} \int_{S/z^2}^{T/z^2} < S_u f, S_{t_{q_h}/z^2} g_{\varepsilon(q_h)} > du \cdot \prod_{h \in \{1,\cdots,m'\} \setminus \{y_l\}_{l=1}^{p}} \int_{S/z^2}^{T/z^2} < S_u f, S_{s_{q_h'}/z^2} g_{\varepsilon'(q_h')} > du$$

$$\prod_{h \in \{1,\cdots,m\} \setminus \{x_l\}_{l=1}^{p}} \int_{S'/z^2}^{T'/z^2} < S_{t_{q_h+1}-1/z^2} g_{1-\varepsilon(q_h+1-1)}, S_u f' > du$$

$$\prod_{h=1}^{m'} \int_{S'/z^2}^{T'/z^2} < S_{s_{q_{h+1}'}-1/z^2} g_{1-\varepsilon'(q_{h+1}'-1)}, S_u f' > du$$

$$< W(z \int_{S/z^2}^{T/z^2} S_u f du)\Phi, W(z \int_{S'/z^2}^{T'/z^2} S_u f' du)\Phi > \tag{3.43}$$

Now change the variables as following:

$$t_{q_h} = t'_{q_h}, \quad h = 1, \cdots, m \tag{3.44}$$

$$t_{q_h+1} - t_{q_h} = t'_{q_h+1}, \quad t_{q_h+2} - t_{q_h+1} - t_{q_h} = t'_{q_h+2}, \cdots,$$

$$t_{q_{h+1}-1} - t_{q_{h+1}-2} - \cdots - t_{q_h+1} - t_{q_h} = t'_{q_{h+1}-1}, \quad h = 1, \cdots, m \tag{3.45}$$

and

$$s_{q'_h} = s'_{q'_h}, \quad h = 1, \cdots, m' \tag{3.44a}$$

$$s_{q'_h+1} - s_{q'_h} = s'_{q'_h+1}, \quad s_{q'_h+2} - s_{q'_h+1} - s_{q'_h} = s'_{q'_h+2}, \cdots,$$

$$s_{q'_{h+1}-1} - s_{q'_{h+1}-2} - \cdots - s_{q'_h+1} - s_{q'_h} = s'_{q'_{h+1}-1}, \quad h = 1, \cdots, m' \tag{3.45a}$$

then (3.43) becomes

$$\sum_{m=1}^{n} \sum_{1=q_1<q_2<\cdots<q_m\leq n} \sum_{m'=1}^{n'} \sum_{1=q'_1<q'_2<\cdots<q'_{m'}\leq n'} (-1)^{n'}$$

$$\sum_{p=0}^{m\wedge m'} \sum_{y_1<\cdots<y_p, \ \{y_h\}_{h=1}^p\subset\{1,\cdots,m'\}} \sum_{\{x_h\}_{h=1}^p\subset\{1,\cdots,m\}, \ x_1>\cdots>x_p}$$

$$\lim_{z\to 0} z^{-2(n-m)} \int_0^t dt'_{q_1} \int_{-t'_{q_1}}^0 dt'_{q_1+1} \int_{-t'_{q_1+1}-t'_{q_1}}^0 dt'_{q_1+2} \cdots$$

$$\int_{-t'_{q_2-2}-\cdots-t'_{q_1+1}-t'_{q_1}}^0 dt'_{q_2-1} \int_0^{t'_{q_2-1}+\cdots+t'_{q_1+1}+t'_{q_1}} dt'_{q_2} \cdots$$

$$\int_0^{t'_{q_m-1}+t'_{q_m-1}+\cdots+t'_{q_{m-1}+1}} dt'_{q_m} \int_{-t'_{q_m}}^0 dt'_{q_m+1} \int_{-t'_{q_m+1}-t'_{q_m}}^0 dt'_{q_m+2} \cdots$$

$$\int_{-t'_{n-1}-\cdots-t'_{q_m+1}-t'_{q_m}}^0 dt'_n$$

$$\prod_{r=1}^{m} \prod_{h=q_r+1}^{q_{r+1}-1} < g_{1-\varepsilon(h-1)}, S_{t'_h/z^2} g_{\varepsilon(h)} > \cdot \prod_{h=1}^{m} \int_{S/z^2}^{T/z^2} < S_u f, S_{t'_{q_h}/z^2} g_{\varepsilon(q_h)} > du$$

$$\prod_{h\in\{1,\cdots,m\}\backslash\{x_l\}_{l=1}^p} \int_{S'/z^2}^{T'/z^2} < S_{(t'_{q_{h+1}-1}+t'_{q_{h+1}-2}+\cdots+t'_{q_h})/z^2} g_{1-\varepsilon(q_{h+1}-1)}, S_v f' > dv$$

$$z^{-2(n'-m'+p)} \int_0^t ds'_{q'_1} \int_0^{t-s'_{q'_1}} ds'_{q'_1+1} \int_0^{t-s'_{q'_1+1}-s'_{q'_1}} ds'_{q'_1+2} \cdots$$

$$\int_0^{t-s'_{q'_2-2}-\cdots-s'_{q'_1+1}-s'_{q'_1}} ds'_{q'_2-1} \int_{s'_{q'_2-1}+\cdots+s'_{q'_1+1}+s'_{q'_1}}^{t-s'_{q'_2-1}-\cdots-s'_{q'_1}} ds'_{q'_2} \cdots$$

$$\int_{s'_{q'_{m'-1}}+s'_{q'_{m'}-1}+\cdots+s'_{q'_{m'-1}+1}}^{t} ds'_{q'_{m'}} \int_0^{t-s'_{q'_{m'}}} ds'_{q'_{m'}+1}$$

$$\int_0^{t-s'_{q'_{m'}+1}-s'_{q'_{m'}}} ds'_{q'_{m'}+2} \cdots \int_0^{t-s'_{n'-1}-\cdots-s'_{q'_{m'}+1}-s'_{q'_{m'}}} ds'_{n'}$$

$$\prod_{r=1}^{m'} \prod_{h=q'_r+1}^{q'_{r+1}-1} < g_{1-\varepsilon'(h-1)}, S_{s'_h}/z^2 g_{\varepsilon'(h)} >$$

$$\prod_{h \in \{1,\cdots,m'\} \setminus \{y_l\}_{l=1}^p} \int_{S/z^2}^{T/z^2} < S_u f, S_{s'_{q'_h}}/z^2 g_{\varepsilon'(q'_h)} > du$$

$$\prod_{h=1}^{m'} \int_{S'/z^2}^{T'/z^2} < S_{(s'_{q'_{h+1}-1}+s'_{q'_{h+1}-2}+\cdots+s'_{q'_h})}/z^2 g_{1-\varepsilon'(q'_{h+1}-1)}, S_v f' > dv$$

$$\prod_{l=1}^{p} < g_{1-\varepsilon(q_{x_l}+1-1)}, S_{(s'_{q'_{y_l}}-t'_{q_{x_l}+1-1}-\cdots-t'_{q_{x_l}})}/z^2 g_{\varepsilon'(q'_{y_l})} >$$

$$< W(z \int_{S/z^2}^{T/z^2} S_u f du)\Phi, W(z \int_{S'/z^2}^{T'/z^2} S_u f' du)\Phi > \tag{3.46}$$

With the change the variables

$$t'_\alpha/z^2 = t_\alpha, \quad \alpha \in \{1,\cdots,n\} \setminus \{q_h\}_{h=1}^m; \quad t'_{q_h} = t_{q_h}, \quad h = 1,\cdots,m \tag{3.47}$$

and

$$s'_\alpha/z^2 = s_\alpha, \quad \alpha \in \{1,\cdots,n'\} \setminus \{q'_h\}_{h=1}^{m'}; \quad s'_{q'_h} = s_{q'_h}, \quad h = 1,\cdots,m' \tag{3.47a}$$

(3.46) becomes

$$\sum_{m=1}^{n} \sum_{1=q_1<q_2<\cdots<q_m \leq n} \sum_{m'=1}^{n'} \sum_{1=q'_1<q'_2<\cdots<q'_{m'} \leq n'} (-1)^{n'}$$

$$\sum_{p=0}^{m \wedge m'} \sum_{y_1<\cdots<y_p, \ \{y_h\}_{h=1}^p \subset \{1,\cdots,m'\}} \sum_{\{x_h\}_{h=1}^p \subset \{1,\cdots,m\}, \ x_1>\cdots>x_p}$$

$$\lim_{z \to 0} \int_0^t dt_{q_1} \int_{-t_{q_1}/z^2}^0 dt_{q_1+1} \int_{-t_{q_1+1}-t_{q_1}/z^2}^0 dt_{q_1+2} \cdots$$

$$\int_{-t_{q_2-2}-\cdots-t_{q_1+1}-t_{q_1}/z^2}^0 dt_{q_2-1} \int_0^{z^2(t_{q_2-1}+\cdots+t_{q_1+1})+t_{q_1}} dt_{q_2} \cdots$$

$$\int_0^{t_{q_{m-1}}+z^2(t_{q_{m-1}}+\cdots+t_{q_{m-1}+1})} dt_{q_m}$$

$$\int_{-t_{q_m}/z^2}^0 dt_{q_m+1} \int_{-t_{q_m+1}-t_{q_m}/z^2}^0 dt_{q_m+2} \cdots \int_{-t_{n-1}-\cdots-t_{q_m+1}-t_{q_m}/z^2}^0 dt_n$$

$$\prod_{r=1}^m \prod_{h=q_r+1}^{q_{r+1}-1} \; < g_{1-\varepsilon(h-1)}, S_{t_h} g_{\varepsilon(h)} >$$

$$\prod_{h=1}^m \int_{S/z^2}^{T/z^2} \; < S_u f, S_{t_{q_h}} g_{\varepsilon(q_h)} > du$$

$$\prod_{h \in \{1,\cdots,m\} \setminus \{x_l\}_{l=1}^p} \int_{S'/z^2}^{T'/z^2} \; < S_{t_{q_{h+1}-1}+t_{q_{h+1}-2}+\cdots+t_{q_h+1}+t_{q_h}/z^2} g_{1-\varepsilon(q_{h+1}-1)}, S_v f' > dv$$

$$z^{-2p} \int_0^t ds_{q_1'} \int_0^{(t-s_{q_1'})/z^2} ds_{q_1'+1} \int_0^{(t-s_{q_1'})/z^2-s_{q_1'+1}} ds_{q_1'+2} \cdots$$

$$\int_0^{(t-s_{q_1'})/z^2-s_{q_2'-2}-\cdots-s_{q_1'+1}} ds_{q_2'-1} \int_{z^2(s_{q_2'-1}+\cdots+s_{q_1'+1})+s_{q_1'}}^t ds_{q_2'} \cdots$$

$$\int_{s_{q_{m'}'-1}+z^2(s_{q_{m'}'-1}+\cdots+s_{q_{m'}'-1+1})}^t ds_{q_{m'}'}$$

$$\int_0^{(t-s_{q_{m'}'})/z^2} ds_{q_{m'}'+1} \int_0^{(t-s_{q_{m'}'})/z^2-s_{q_{m'}'+1}} ds_{q_{m'}'+2} \cdots \int_0^{(t-s_{q_{m'}'})/z^2-s_{n'-1}-\cdots-s_{q_{m'}'+1}} ds_{n'}$$

$$\prod_{r=1}^{m'} \prod_{h=q_r'+1}^{q_{r+1}'-1} \; < g_{1-\varepsilon'(h-1)}, S_{s_h} g_{\varepsilon'(h)} >$$

$$\prod_{h \in \{1,\cdots,m'\} \setminus \{y_l\}_{l=1}^p} \int_{S/z^2}^{T/z^2} \; < S_u f, S_{s_{q_h'}/z^2} g_{\varepsilon'(q_h')} > du$$

$$\prod_{h=1}^{m'} \int_{S'/z^2}^{T'/z^2} \; < S_{s_{q_{h+1}'-1}+s_{q_{h+1}'-2}+\cdots+s_{q_h'+1}+s_{q_h'}/z^2} g_{1-\varepsilon'(q_{h+1}'-1)}, S_v f' > dv$$

$$\prod_{l=1}^{p} < g_{1-\varepsilon(q_{x_l}+1}-1)}, S_{(s_{q'_{y_l}}-t_{q_{x_l}})/z^2 - t_{q_{x_l}+1}-1 - \cdots - t_{q_{x_l}+1}} g_{\varepsilon'}(q'_{y_l}) >$$

$$< W(z \int_{S/z^2}^{T/z^2} S_u f du)\Phi, W(z \int_{S'/z^2}^{T'/z^2} S_u f' du)\Phi > \qquad (3.48)$$

Now let investigate the second $(ds-)$ integral, i.e.

$$z^{-2p} \int_0^t ds_{q'_1} \int_0^{(t-s_{q'_1})/z^2} ds_{q'_1+1} \int_0^{(t-s_{q'_1})/z^2 - s_{q'_1}+1} ds_{q'_1+2} \cdots$$

$$\int_0^{(t-s_{q'_1})/z^2 - s_{q'_2}-2 - \cdots - s_{q'_1}+1} ds_{q'_2-1} \int_{z^2(s_{q'_2-1}+\cdots+s_{q'_1+1})+s_{q'_1}}^{t} ds_{q'_2} \cdots$$

$$\int_{s_{q'_{m'}-1} + z^2(s_{q'_{m'}-1} + \cdots + s_{q'_{m'}-1+1})}^{t} ds_{q'_{m'}}$$

$$\int_0^{(t-s_{q'_{m'}})/z^2} ds_{q'_{m'}+1} \int_0^{(t-s_{q'_{m'}})/z^2 - s_{q'_{m'}}+1} ds_{q'_{m'}+2} \cdots \int_0^{(t-s_{q'_{m'}})/z^2 - s_{n'-1} - \cdots - s_{q'_{m'}}+1} ds_{n'}$$

$$\prod_{r=1}^{m'} \prod_{h=q'_r+1}^{q'_{r+1}-1} < g_{1-\varepsilon'(h-1)}, S_{s_h} g_{\varepsilon'(h)} >$$

$$\prod_{l=1}^{p} < g_{1-\varepsilon(q_{x_l}+1}-1)}, S_{(s_{q'_{y_l}}-t_{q_{x_l}})/z^2 - t_{q_{x_l}+1}-1 - \cdots - t_{q_{x_l}+1}} g_{\varepsilon'}(q'_{y_l}) >$$

$$\prod_{h=1}^{m'} \int_{S'/z^2}^{T'/z^2} < S_{s_{q'_{h+1}-1} + s_{q'_{h+1}-2} + \cdots + s_{q'_h+1} + s_{q'_h}/z^2} g_{1-\varepsilon'(q'_{h+1}-1)}, S_v f' > dv \qquad (3.49)$$

It is clear that there exists a $\alpha \le p$ and $1 \le r_1 < \cdots < r_\alpha = p$ such that

$$y_1 + 1 = y_2, \cdots, y_{r_1-1} + 1 = y_{r_1}, \; y_{r_1} + 1 < y_{r_1+1}, \cdots, y_{r_1+1} + 1 = y_{r_1+2}, \cdots,$$

$$y_{r_2-1} + 1 = y_{r_2}, \; y_{r_2} + 1 < y_{r_2+1}, \cdots,$$

$$y_{r_{\alpha-1}+1} + 1 = y_{r_{\alpha-1}+2}, \cdots, y_{r_\alpha-1} + 1 = y_{r_\alpha} := y_p \qquad (3.50)$$

So, one can rewrite (3.49) to

$$\int_0^t ds_{q'_1} \int_0^{(t-s_{q'_1})/z^2} ds_{q'_1+1} \int_0^{(t-s_{q'_1})/z^2 - s_{q'_1}+1} ds_{q'_1+2} \cdots$$

$$\int_0^{(t-s_{q'_{y_1}-1})/z^2 - s_{q'_{y_1}-1+1} - \cdots - s_{q'_{y_1}-2}} ds_{q'_{y_1}-1}$$

$$\int_{(s_{q'_{y_1}-1}-t_{qx_1})/z^2-t_{qx_1+1}-1-\cdots-t_{qx_1}+1+s_{q'_{y_1}+1}+\cdots+s_{q'_{y_1}-1}}^{(t-t_{qx_1})/z^2-t_{qx_1+1}-1-\cdots-t_{qx_1}+1} ds_{q'_{y_1}}$$

$$\int_0^{(t-t_{qx_1})/z^2-s_{q'_{y_1}}-t_{qx_1+1}-1-\cdots-t_{qx_1}+1} ds_{q'_{y_1}+1}$$

$$\int_0^{(t-t_{qx_1})/z^2-s_{q'_{y_1}}-s_{q'_{y_1}+1}-t_{qx_1+1}-1-\cdots-t_{qx_1}+1} ds_{q'_{y_1}+2}\cdots$$

$$\int_0^{(t-t_{qx_1})/z^2-s_{q'_{y_1}}-s_{q'_{y_1}+1}-\cdots-s_{q'_{y_1+1}-2}-t_{qx_1+1}-1-\cdots-t_{qx_1}+1} ds_{q'_{y_1+1}-1}$$

$$\int_{(t_{qx_1}-t_{qx_2})/z^2-L(t,s)}^{(t-t_{qx_2})/z^2-L(t,s)} ds_{q'_{y_2}} \int_0^{(t-t_{qx_2})/z^2-L(t,s)} ds_{q'_{y_2}+1}\cdots$$

$$\int_{(t_{qx_{r_1}-1}-t_{qx_{r_1}})/z^2-L(t,s)}^{(t-t_{qx_{r_1}})/z^2-L(t,s)} ds_{q'_{yr_1}} \int_0^{(t-t_{qx_{r_1}})/z^2-L(t,s)} ds_{q'_{yr_1}+1}\cdots$$

$$\int_0^{(t-t_{qx_{r_1}})/z^2-L(t,s)} ds_{q'_{yr_1}+1}-1 \int_{t_{qx_{r_1}}+z^2L(t,s)}^t ds_{q'_{yr_1}+1}\cdots$$

$$\int_{(s_{q'_{yr_{\alpha-1}-1}}-t_{qxr_{\alpha_1}})/z^2-L(t,s)}^{(t-t_{qxr_{\alpha_1}})/z^2-L(t,s)} ds_{q'_{yr_{\alpha-1}}} \int_0^{(t-t_{qxr_{\alpha_1}})/z^2-L(t,s)} ds_{q'_{yr_{\alpha-1}}+1}\cdots$$

$$\int_0^{(t-t_{qxr_{\alpha_1}})/z^2-L(t,s)} ds_{q'_{yr_{\alpha-1}+1}}-1 \int_{(s_{q'_{yr_{\alpha-1}+1}-1}-t_{qxr_{\alpha_1}})/z^2-L(t,s)}^{(t-t_{qxr_{\alpha_1}})/z^2-L(t,s)} ds_{q'_{yr_{\alpha-1}+1}}\cdots$$

$$\int_0^{(t-t_{qxr_{\alpha-1}})/z^2-L(t,s)} ds_{q'_{yr_{\alpha-1}+1}} \cdots \int_0^{(t-t_{qxr_{\alpha-1}})/z^2-L(t,s)} ds_{q'_{yr_{\alpha-1}+1}+1}-1$$

$$\int_{(t_{qxr_{\alpha-1}+1}-t_{qxr_{\alpha-1}+1+1})/z^2-L(t,s)}^{(t-t_{qxr_{\alpha-1}+1+1})/z^2-L(t,s)} ds_{q'_{yr_{\alpha-1}+1+1}}$$

$$\int_0^{(t-t_{qxr_{\alpha-1}+1+1})/z^2-L(t,s)} ds_{q'_{yr_{\alpha-1}+1+1}+1}\cdots$$

$$\int_{(t_{qxr_{\alpha-1}+1}-t_{qxr_\alpha})/z^2-L(t,s)}^{(t-t_{qxr_\alpha})/z^2-L(t,s)} ds_{q'_{yr_\alpha}} \int_0^{(t-t_{qxr_\alpha})/z^2-L(t,s)} ds_{q'_{yr_\alpha}+1}\cdots$$

$$\int_0^{(t-t_{qxr_\alpha})/z^2-L(t,s)} ds_{q'_{yr_\alpha}+1}-1 \int_{t_{qxr_\alpha}+z^2L(t,s)}^t ds_{q'_{yr_\alpha}+1} \cdots \int_0^{(t-s_{q'_{m'}})/z^2-L(t,s)} ds_{n'}$$

$$\prod_{h=1}^{p} \int_{S'/z^2}^{T'/z^2} < S_{t_{q_{x_h}}/z^2 + L(t,s)} g_{1-\varepsilon'(q'_{y_h}+1}-1), S_v f' > dv$$

$$\prod_{h\in\{1,\cdots,m'\}\setminus\{y_h\}_{h=1}^{p}} \int_{S'/z^2}^{T'/z^2} < S_{s_{q'_{h+1}-1}+s_{q'_{h+1}-2}+\cdots+s_{q'_h+1}+s_{q'_h}/z^2} g_{1-\varepsilon'(q'_{h+1}-1)}, S_v f' > dv$$

$$\prod_{r=1}^{m'} \prod_{h=q'_r+1}^{q'_{r+1}-1} < g_{1-\varepsilon'(h-1)}, S_{s_h} g_{\varepsilon'(h)} > \prod_{l=1}^{p} < g_{1-\varepsilon(q_{x_l}+1}-1), S_{s_{q'_{y_l}}} g_{\varepsilon'(q'_{y_l})} > \tag{3.51}$$

where, $L(t,s)$ are linear functions of $t_{q_1}, \cdots, t_{q_m}$, $s_{q'_1}, \cdots, s_{q'_{m'}}$, and they are not necessery the same for different integrals. Taking the limit for $z \to 0$, (3.51) gives the term

$$\int_0^t ds_{q'_1} \int_{s_{q'_1}}^t ds_{q'_2} \cdots \int_{s_{q'_{y_1-2}}}^t ds_{q'_{y_1-1}} \chi_{[0,t_{q_{x_1}})}(s_{q'_{y_1-1}}) \int_{t_{q_{x_{r_1}}}}^t ds_{q'_{y_{r_1}+1}} \cdots$$

$$\int_{s_{q'_{y_{r_\alpha-1}-2}}}^t ds_{q'_{y_{r_\alpha-1}-1}} \chi_{[0,t_{q_{x_{r_\alpha-1}}})}(s_{q'_{y_{r_\alpha-1}-1}})$$

$$\int_{t_{q_{x_{r_\alpha}}}}^t ds_{q'_{y_{r_\alpha}+1}} \int_{s_{q'_{y_{r_\alpha}}+1}}^t ds_{q'_{y_{r_\alpha}+2}} \cdots \int_0^{s_{q'_{m'}-1}} ds_{q'_{m'}}$$

$$\prod_{h=1}^{p} \chi_{[S',T']}(t_{q_{x_h}})(g_{1-\varepsilon'(q'_{y_h}+1}-1)|f') \cdot \prod_{h\in\{1,\cdots,m'\}\setminus\{y_h\}_{h=1}^{p}} \chi_{[S',T']}(s_{q'_h})(g_{1-\varepsilon'(q'_{h+1}-1)}|f')$$

$$\prod_{r=1}^{m'} \prod_{h=q'_r+1}^{q'_{r+1}-1} (g_{1-\varepsilon'(h-1)}|g_{\varepsilon'(h)}) + \prod_{l=1}^{p} (g_{1-\varepsilon(q_{x_l}+1}-1)|g_{\varepsilon'(q'_{y_l})}) \tag{3.52}$$

therefore, we obtain the following

THEOREM (3.5) The limit of the quantity (3.36) exists and is equal to

$$\sum_{m=1}^{n} \sum_{1=q_1<q_2<\cdots<q_m\leq n} \sum_{m'=1}^{n'} \sum_{1=q'_1<q'_2<\cdots<q'_{m'}\leq n'} (-1)^{n'}$$

$$\sum_{p=0}^{m\wedge m'} \sum_{y_1<\cdots<y_p, \{y_h\}_{h=1}^{p}\subset\{1,\cdots,m'\}} \sum_{\{x_h\}_{h=1}^{p}\subset\{1,\cdots,m\}, x_1>\cdots>x_p}$$

$$\int_0^t dt_{q_1} \int_0^{t_{q_1}} dt_{q_2} \cdots \int_0^{t_{q_{m-1}}} dt_{q_m}$$

$$\prod_{\alpha\in\{1,\cdots,n\}\setminus\{q_h\}_{h=1}^m}(g_{1-\varepsilon(\alpha-1)}|g_{\varepsilon(\alpha)})-$$

$$\prod_{\alpha\in\{1,\cdots,n'\}\setminus\{q'_h\}_{h=1}^{m'}}\cdot\,(g_{1-\varepsilon'(\alpha-1)}|g_{\varepsilon'(\alpha)})+$$

$$\int_0^t ds_{q'_1}\int_{s_{q'_1}}^t ds_{q'_2}\cdots\int_{s_{q'_{y_1-2}}}^t ds_{q'_{y_1-1}}\chi_{[0,t_{qx_1})}(s_{q'_{y_1-1}})\int_{t_{qx_{r_1}}}^t ds_{q'_{y_{r_1}+1}}\cdots$$

$$\int_{s_{q'_{y_{r_{\alpha-1}}-2}}}^t ds_{q'_{y_{r_{\alpha-1}}-1}}\chi_{[0,t_{qx_{r_{\alpha-1}}})}(s_{q'_{y_{r_{\alpha-1}}-1}})$$

$$\int_{t_{qx_{r_\alpha}}}^t ds_{q'_{y_{r_\alpha}+1}}\int_{s_{q'_{y_{r_\alpha}+1}}}^t ds_{q'_{y_{r_\alpha}+2}}\cdots\int_0^{s_{q'_{m'-1}}} ds_{q'_{m'}}$$

$$\prod_{l=1}^p(g_{1-\varepsilon(q_{x_l}+1-1)}|g_{\varepsilon'(q'_{y_l})})$$

$$\prod_{h=1}^m\chi_{[S,T]}(t_{q_h})(f|g_{\varepsilon(q_h)})\cdot\prod_{h\in\{1,\cdots,m'\}\setminus\{y_l\}_{l=1}^p}\chi_{[S,T]}(s_{q'_h})(f|g_{\varepsilon'(q'_h)})$$

$$\cdot\prod_{h\in\{1,\cdots,m\}\setminus\{x_l\}_{l=1}^p}\chi_{[S',T']}(t_{q_h})(g_{1-\varepsilon(q_h+1-1)}|f')$$

$$\prod_{h=1}^p\chi_{[S',T']}(t_{q_{x_h}})(g_{1-\varepsilon'(q'_{y_h}+1-1)}|f')\cdot\prod_{h\in\{1,\cdots,m'\}\setminus\{y_h\}_{h=1}^p}\chi_{[S',T']}(s_{q'_h})(g_{1-\varepsilon'(q'_h+1-1)}|f')$$

$$< W(\chi_{[S,T]}\otimes f)\Psi, W(\chi_{[S',T']}\otimes f')\Psi > \tag{3.53}$$

where, $q'_0 := 0$ and $s_0 := 0$.

Combining together the uniform estimate and Theorem (3.5), one gets the following Theorem about low density limit

<u>*THEOREM*</u> (3.6) For each $D, X \in B(H_0)$, $f, f', g_0, g_1 \in K$, if

$$(\|g_0, g_1\|)^2 < \frac{1}{16\|D\|} \tag{3.54}$$

then the limit

$$\lim_{z\to 0} < W(z\int_{S/z^2}^{T/z^2} S_u f\, du)\Phi, U(t/z^2)(X\otimes 1)U^*(t/z^2)W(z\int_{S'/z^2}^{T'/z^2} S_u f'\, du)\Phi > \tag{3.55}$$

exists and is equal to

$$\sum_{n,n'=0}^{\infty} \sum_{\varepsilon\in\{0,1\}^n,\ \varepsilon'\in\{0,1\}^{n'}} < u, D_{\varepsilon(1)}\cdots D_{\varepsilon(n)} X D_{\varepsilon'(1)}\cdots D_{\varepsilon'(n')} v > (-1)^{n'}$$

$$\sum_{m=1}^{n} \sum_{1=q_1<q_2<\cdots<q_m\leq n} \sum_{m'=1}^{n'} \sum_{1=q_1'<q_2'<\cdots<q_{m'}'\leq n'}$$

$$\sum_{p=0}^{m\wedge m'} \sum_{y_1<\cdots<y_p,\ \{y_h\}_{h=1}^{p}\subset\{1,\cdots,m'\}} \sum_{\{x_h\}_{h=1}^{p}\subset\{1,\cdots,m\},\ x_1>\cdots>x_p}$$

$$\int_0^t dt_{q_1} \int_0^{t_{q_1}} dt_{q_2} \cdots \int_0^{t_{q_{m-1}}} dt_{q_m}$$

$$\prod_{\alpha\in\{1,\cdots,n\}\backslash\{q_h\}_{h=1}^{m}} (g_{1-\varepsilon(\alpha-1)}|g_{\varepsilon(\alpha)})_-$$

$$\prod_{\alpha\in\{1,\cdots,n'\}\backslash\{q_h'\}_{h=1}^{m'}} (g_{1-\varepsilon'(\alpha-1)}|g_{\varepsilon'(\alpha)})_+$$

$$\int_0^t ds_{q_1'} \int_{s_{q_1'}}^t ds_{q_2'} \cdots \int_{s_{q_{y_1-2}'}}^t ds_{q_{y_1-1}'} \chi_{[0,t_{qx_1})}(s_{q_{y_1-1}'}) \int_{t_{qx_{r_1}}}^t ds_{q_{y_{r_1}+1}'} \cdots$$

$$\int_{s_{q_{y_{r_{\alpha-1}}-2}'}}^t ds_{q_{y_{r_{\alpha-1}}-1}'} \chi_{[0,t_{qx_{r_{\alpha-1}}})}(s_{q_{y_{r_{\alpha-1}}-1}'})$$

$$\int_{t_{qx_{r_\alpha}}}^t ds_{q_{y_{r_\alpha}+1}'} \int_{s_{q_{y_{r_\alpha}}'}+1}^t ds_{q_{y_{r_\alpha}+2}'} \cdots \int_{s_{q_{m'-1}'}}^t ds_{q_{m'}'}$$

$$\prod_{l=1}^{p}(g_{1-\varepsilon(q_{x_l}+1-1)}|g_{\varepsilon'(q_{y_l}')})$$

$$\prod_{h=1}^{m} \chi_{[S,T]}(t_{q_h})(f|g_{\varepsilon(q_h)}) \cdot \prod_{h\in\{1,\cdots,m'\}\backslash\{y_l\}_{l=1}^{p}} \chi_{[S,T]}(s_{q_h'})(f|g_{\varepsilon'(q_h')})$$

$$\cdot \prod_{h\in\{1,\cdots,m\}\backslash\{x_l\}_{l=1}^{p}} \chi_{[S',T']}(t_{q_h})(g_{1-\varepsilon(q_{h+1}-1)}|f')$$

$$\prod_{h=1}^{p} \chi_{[S',T']}(t_{q_{x_h}})(g_{1-\varepsilon'(q_{y_h}'+1-1)}|f') \cdot \prod_{h\in\{1,\cdots,m'\}\backslash\{y_h\}_{h=1}^{p}} \chi_{[S',T']}(s_{q_h'})(g_{1-\varepsilon'(q_{h+1}'-1)}|f')$$

$$< W(\chi_{[S,T]}\otimes f)\Psi, W(\chi_{[S',T']}\otimes f')\Psi > \tag{3.56}$$

§4. The quantum Langevin equation

In the section we investigate the properties of the low density limit (3.56) and show that it is the solution of a quantum stochastic differential equation ruled by purely quantum jump processes whose form is explicitly deduced.

First of all, we show that in (3.56) the time derivative can be exchanged with the series $\sum_{n=0,n'=0}^{\infty}$. Notice that in (3.56), the second integral is not the good form for obtaining the term like $\frac{t^d}{d!}$, we shall change it as following:

$$x_{r_1} \geq x_1 \implies q_{x_{r_1}} \leq q_{x_1} \implies t_{q_{x_{r_1}}} \geq t_{q_{x_1}} \tag{4.1}$$

therefore for $0 \leq F \in L^1_{\text{loc}}([0,t]^{m'-p})$ bounded, if we enlarge $t_{q_{x_{r_1}}}$ to $t_{q_{x_1}}$ in the second $(ds-)$ integral of (3.56), i.e.

$$\int_0^t ds_{q_1'} \int_{s_{q_1'}}^t ds_{q_2'} \cdots \int_{s_{q_{y_1-2}'}}^t ds_{q_{y_1-1}'} \chi_{[0,t_{q_{x_1}})}(s_{q_{y_1-1}'}) \int_{t_{q_{x_{r_1}}}}^t ds_{q_{y_{r_1}+1}'} \cdots$$

$$\int_{s_{q_{y_{r_\alpha-1}-2}'}}^t ds_{q_{y_{r_\alpha-1}-1}'} \chi_{[0,t_{q_{x_{r_\alpha-1}}})}(s_{q_{y_{r_\alpha-1}-1}'})$$

$$\int_{t_{q_{x_{r_\alpha}}}}^t ds_{q_{y_{r_\alpha}+1}'} \int_{s_{q_{y_{r_\alpha}}'+1}}^t ds_{q_{y_{r_\alpha}+2}'} \cdots \int_{s_{q_{m'-1}'}}^t ds_{q_{m'}'} F(s_{q_1'}, \cdots, \widehat{s_{q_{y_1}'}}, \cdots, \widehat{s_{q_{y_p}'}}, \cdots, s_{q_{m'}'}) \tag{4.2}$$

we obtain the following majorization of (4.2)

$$\int_0^t ds_{q_1'} \int_{s_{q_1'}}^t ds_{q_2'} \cdots \int_{s_{q_{y_1-2}'}}^t ds_{q_{y_1-1}'} \chi_{[0,t_{q_{x_1}})}(s_{q_{y_1-1}'}) \int_{t_{q_{x_1}}}^t ds_{q_{y_{r_1}+1}'} \cdots$$

$$\int_{s_{q_{y_{r_\alpha-1}-2}'}}^t ds_{q_{y_{r_\alpha-1}-1}'} \chi_{[0,t_{q_{x_{r_\alpha-1}}})}(s_{q_{y_{r_\alpha-1}-1}'})$$

$$\int_{t_{q_{x_{r_\alpha}}}}^t ds_{q_{y_{r_\alpha}+1}'} \int_{s_{q_{y_{r_\alpha}}'+1}}^t ds_{q_{y_{r_\alpha}+2}'} \cdots \int_{s_{q_{m'-1}'}}^t ds_{q_{m'}'} F(s_{q_1'}, \cdots, \widehat{s_{q_{y_1}'}}, \cdots, \widehat{s_{q_{y_p}'}}, \cdots, s_{q_{m'}'})$$

$$\tag{4.2a}$$

Moreover, in the right hand side of (4.2a) we have the characteristic function $\chi_{[0,t_{q_{x_1}})}(s_{q_{y_1-1}'})$, so for $t_{q_{x_1}} < s_{q_{y_1-1}'}$ (4.2a) is equal to zero. It means that we need only to consider the integral in (4.2a) for the case $t_{q_{x_1}} > s_{q_{y_1-1}'}$. In this case (4.2a) is majorized by

$$\int_0^t ds_{q_1'} \int_{s_{q_1'}}^t ds_{q_2'} \cdots \int_{s_{q_{y_1-2}'}}^t ds_{q_{y_1-1}'} \int_{s_{q_{y_1-1}'}}^t ds_{q_{y_{r_1}+1}'} \cdots$$

$$\int_{s_{q'_{y_{r_{\alpha-1}}-2}}}^{t} ds_{q'_{y_{r_{\alpha-1}}-1}} \chi_{[0,t_{q_{x_{r_{\alpha-1}}})}}(s_{q'_{y_{r_{\alpha-1}}-1}})$$

$$\int_{t_{q_{x_{r_\alpha}}}}^{t} ds_{q'_{y_{r_\alpha}+1}} \int_{s_{q'_{y_{r_\alpha}+1}}}^{t} ds_{q'_{y_{r_\alpha}+2}} \cdots \int_{s_{q'_{m'-1}}}^{t} ds_{q'_{m'}} F(s_{q'_1}, \cdots, \widehat{s_{q'_{y_1}}}, \cdots, \widehat{s_{q'_{y_p}}}, \cdots, s_{q'_{m'}}) \quad (4.3)$$

Using the same change for $t_{q_{x_{r_\beta}}}$, $\beta = 2, \cdots, \alpha$, we see (4.2) is majorized by

$$\int_{0 \le s_{q'_1} \le \cdots \le \widehat{s_{q'_{y_1}}} \le \cdots \le \widehat{s_{q'_{y_p}}} \le \cdots \le s_{q'_{m'}}} ds_{q'_1} \cdots \widehat{ds_{q'_{y_1}}} \cdots \widehat{ds_{q'_{y_p}}} \cdots ds_{q'_{m'}} F(s_{q'_1}, \cdots, \widehat{s_{q'_{y_1}}}, \cdots, \widehat{s_{q'_{y_p}}}, \cdots, s_{q'_{m'}})$$

$$(4.4)$$

Thus one gets the following

$\underline{THEOREM(4.1)}$ The module of (3.56) less than or equal to

$$\sum_{n,n'=0}^{\infty} \sum_{m=1}^{n} \sum_{m'=1}^{n'} \sum_{p=0}^{m \wedge m'} \|u\| \cdot \|v\| \|D\|^{n+n'} \frac{1}{m! \cdot (m'-p)!} \|g_0, g_1\|^{2(n-m+n'-m'-p)} 8^{n+n'} C_{n,n'}$$

$$(4.5)$$

$\underline{REMARK}$ One could obtain (4.5) directly from the uniform estimate, but the proof of (4.5) also can be used to prove that the operation $\frac{d}{dt}$ can be exchanged with the series $\sum_{n=0,n'=0}^{\infty}$ in (3.56). The result of this exchange is expressed by the following Theorem:

$\underline{THEOREM(4.2)}$

$$\frac{d}{dt} \sum_{n,n'=0}^{\infty} \sum_{\varepsilon \in \{0,1\}^n, \ \varepsilon' \in \{0,1\}^{n'}} < u, D_{\varepsilon(1)} \cdots D_{\varepsilon(n)} X D_{\varepsilon'(1)} \cdots D_{\varepsilon'(n')} v > (-1)^{n'}$$

$$\sum_{m=1}^{n} \sum_{1=q_1<q_2<\cdots<q_m \le n} \sum_{m'=1}^{n'} \sum_{1=q'_1<q'_2<\cdots<q'_{m'} \le n'}$$

$$\sum_{p=0}^{m \wedge m'} \sum_{y_1<\cdots<y_p, \ \{y_h\}_{h=1}^p \subset \{1,\cdots,m'\}} \sum_{\{x_h\}_{h=1}^p \subset \{1,\cdots,m\}, \ x_1>\cdots>x_p}$$

$$\int_0^t dt_{q_1} \int_0^{t_{q_1}} dt_{q_2} \cdots \int_0^{t_{q_{m-1}}} dt_{q_m}$$

$$\prod_{\alpha \in \{1,\cdots,n\} \backslash \{q_h\}_{h=1}^m} (g_{1-\varepsilon(\alpha-1)} | g_{\varepsilon(\alpha)})_- \cdot \prod_{\alpha \in \{1,\cdots,n'\} \backslash \{q'_h\}_{h=1}^{m'}} (g_{1-\varepsilon'(\alpha-1)} | g_{\varepsilon'(\alpha)})_+$$

$$\int_0^t ds_{q_1'} \int_{s_{q_1'}}^t ds_{q_2'} \cdots \int_{s_{q_{y_1-2}'}}^t ds_{q_{y_1-1}'} \chi_{[0,t_{q_{x_1}}]}(s_{q_{y_1-1}'}) \int_{t_{q_{x_{r_1}}}}^t ds_{q_{y_{r_1}+1}'} \cdots$$

$$\int_{s_{q_{y_{r_{\alpha-1}}-2}'}}^t ds_{q_{y_{r_{\alpha-1}}-1}'} \chi_{[0,t_{q_{x_{r_{\alpha-1}}}}]}(s_{q_{y_{r_{\alpha-1}}-1}'})$$

$$\int_{t_{q_{x_{r_\alpha}}}}^t ds_{q_{y_{r_\alpha}+1}'} \int_{s_{q_{y_{r_\alpha}}+1}'}^t ds_{q_{y_{r_\alpha}+2}'} \cdots \int_{s_{q_{m'-1}'}}^t ds_{q_{m'}'}$$

$$\prod_{l=1}^p (g_{1-\varepsilon(q_{x_l+1}-1)}|g_{\varepsilon'(q_{y_l}')})$$

$$\prod_{h=1}^m \chi_{[S,T]}(t_{q_h})(f|g_{\varepsilon(q_h)}) \cdot \prod_{h\in\{1,\cdots,m'\}\setminus\{y_l\}_{l=1}^p} \chi_{[S,T]}(s_{q_h'})(f|g_{\varepsilon'(q_h')})$$

$$\cdot \prod_{h\in\{1,\cdots,m\}\setminus\{x_l\}_{l=1}^p} \chi_{[S',T']}(t_{q_h})(g_{1-\varepsilon(q_{h+1}-1)}|f')$$

$$\prod_{h=1}^p \chi_{[S',T']}(t_{q_{x_h}})(g_{1-\varepsilon'(q_{y_h}'+1)}|f') \cdot \prod_{h\in\{1,\cdots m'\}\setminus\{y_h\}_{h=1}^p} \chi_{[S',T']}(s_{q_h'})(g_{1-\varepsilon'(q_{h+1}'-1)}|f')$$

$$< W(\chi_{[S,T]}\otimes f)\Psi, W(\chi_{[S',T']}\otimes f')\Psi >$$

$$= \sum_{n,n'=0}^\infty \sum_{\varepsilon\in\{0,1\}^n,\ \varepsilon'\in\{0,1\}^{n'}} < u, D_{\varepsilon(1)}\cdots D_{\varepsilon(n)} X D_{\varepsilon'(1)}\cdots D_{\varepsilon'(n')}v > (-1)^{n'}$$

$$\sum_{m=1}^n \sum_{1=q_1<q_2<\cdots<q_m\leq n} \sum_{m'=1}^{n'} \sum_{1=q_1'<q_2'<\cdots<q_{m'}'\leq n'}$$

$$\sum_{p=0}^{m\wedge m'} \sum_{y_1<\cdots<y_p,\ \{y_h\}_{h=1}^p\subset\{1,\cdots,m'\}} \sum_{\{x_h\}_{h=1}^p\subset\{1,\cdots,m\},\ x_1>\cdots>x_p}$$

$$\frac{d}{dt}\int_0^t dt_{q_1} \int_0^{t_{q_1}} dt_{q_2} \cdots \int_0^{t_{q_{m-1}}} dt_{q_m}$$

$$\prod_{\alpha\in\{1,\cdots,n\}\setminus\{q_h\}_{h=1}^m} (g_{1-\varepsilon(\alpha-1)}|g_{\varepsilon(\alpha)})-$$

$$\prod_{\alpha\in\{1,\cdots,n'\}\setminus\{q_h'\}_{h=1}^{m'}} (g_{1-\varepsilon'(\alpha-1)}|g_{\varepsilon'(\alpha)})+$$

$$\int_0^t ds_{q_1'} \int_{s_{q_1'}}^t ds_{q_2'} \cdots \int_{s_{q_{y_1-2}'}}^t ds_{q_{y_1-1}'} \chi_{[0,t_{q_{x_1}}]}(s_{q_{y_1-1}'}) \int_{t_{q_{x_{r_1}}}}^t ds_{q_{y_{r_1}+1}'} \cdots$$

$$\int_{s_{q_{y_{r_{\alpha-1}}-2}'}}^t ds_{q_{y_{r_{\alpha-1}}-1}'} \chi_{[0,t_{q_{x_{r_{\alpha-1}}}}]}(s_{q_{y_{r_{\alpha-1}}-1}'})$$

$$\int_{t_{q_{x_{r_\alpha}}}}^t ds_{q_{y_{r_\alpha}+1}'} \int_{s_{q_{y_{r_\alpha}}'}+1}^t ds_{q_{y_{r_\alpha}+2}'} \cdots \int_{s_{q_{m'-1}'}}^t ds_{q_{m'}'}$$

$$\prod_{l=1}^p (g_{1-\varepsilon(q_{x_l}+1-1)} | g_{\varepsilon'(q_{y_l}')})$$

$$\prod_{h=1}^m \chi_{[S,T]}(t_{q_h})(f|g_{\varepsilon(q_h)}) \cdot \prod_{h\in\{1,\cdots,m'\}\setminus\{y_l\}_{l=1}^p} \chi_{[S,T]}(s_{q_h'})(f|g_{\varepsilon'(q_h')})$$

$$\cdot \prod_{h\in\{1,\cdots,m\}\setminus\{x_l\}_{l=1}^p} \chi_{[S',T']}(t_{q_h})(g_{1-\varepsilon(q_h+1-1)}|f')$$

$$\prod_{h=1}^p \chi_{[S',T']}(t_{q_{x_h}})(g_{1-\varepsilon'(q_{y_h}'+1-1)}|f') \cdot \prod_{h\in\{1,\cdots m'\}\setminus\{y_h\}_{h=1}^p} \chi_{[S',T']}(s_{q_h'})(g_{1-\varepsilon'(q_h'+1-1)}|f')$$

$$< W(\chi_{[S,T]} \otimes f)\Psi, W(\chi_{[S',T']} \otimes f')\Psi >$$

$$+ \sum_{n,n'=0}^\infty \sum_{\varepsilon\in\{0,1\}^n,\ \varepsilon'\in\{0,1\}^{n'}} < u, D_{\varepsilon(1)} \cdots D_{\varepsilon(n)} X D_{\varepsilon'(1)} \cdots D_{\varepsilon'(n')}$$

$$\sum_{m=1}^n \sum_{1=q_1<q_2<\cdots<q_m\leq n} \sum_{m'=1}^{n'} \sum_{1=q_1'<q_2'<\cdots<q_{m'}'\leq n'}$$

$$\sum_{p=0}^{m\wedge m'} \sum_{y_1<\cdots<y_p,\ \{y_h\}_{h=1}^p\subset\{1,\cdots,m'\}} \sum_{\{x_h\}_{h=1}^p\subset\{1,\cdots,m\},\ x_1>\cdots>x_p}$$

$$\int_0^t dt_{q_1} \int_0^{t_{q_1}} dt_{q_2} \cdots \int_0^{t_{q_{m-1}}} dt_{q_m}$$

$$\prod_{\alpha\in\{1,\cdots,n\}\setminus\{q_h\}_{h=1}^m} (g_{1-\varepsilon(\alpha-1)}|g_{\varepsilon(\alpha)})-$$

$$\prod_{\alpha\in\{1,\cdots,n'\}\setminus\{q_h'\}_{h=1}^{m'}} (g_{1-\varepsilon'(\alpha-1)}|g_{\varepsilon'(\alpha)})+$$

$$\frac{d}{dt}\int_0^t ds_{q_1'}\int_{s_{q_1'}}^t ds_{q_2'}\cdots\int_{s_{q_{y_1-2}'}}^t ds_{q_{y_1-1}'}\chi_{[0,t_{q_{x_1}})}(s_{q_{y_1-1}'})\int_{t_{q_{x_{r_1}}}}^t ds_{q_{y_{r_1}+1}'}\cdots$$

$$\int_{s_{q_{y_{r_\alpha-1}-2}'}}^t ds_{q_{y_{r_\alpha-1}-1}'}\chi_{[0,t_{q_{x_{r_\alpha-1}}})}(s_{q_{y_{r_\alpha-1}-1}'})$$

$$\int_{t_{q_{x_{r_\alpha}}}}^t ds_{q_{y_{r_\alpha}+1}'}\int_{s_{q_{y_{r_\alpha}}'}+1}^t ds_{q_{y_{r_\alpha}+2}'}\cdots\int_{s_{q_{m'-1}'}}^t ds_{q_{m'}'}$$

$$\prod_{l=1}^p (g_{1-\varepsilon(q_{x_l}+1-1)}|g_{\varepsilon'(q_{y_l}')})$$

$$\prod_{h=1}^m \chi_{[S,T]}(t_{q_h})(f|g_{\varepsilon(q_h)})\cdot\prod_{h\in\{1,\cdots,m'\}\setminus\{y_l\}_{l=1}^p}\chi_{[S,T]}(s_{q_h'})(f|g_{\varepsilon'(q_h')})$$

$$\cdot\prod_{h\in\{1,\cdots,m\}\setminus\{x_l\}_{l=1}^p}\chi_{[S',T']}(t_{q_h})(g_{1-\varepsilon(q_{h+1}-1)}|f')$$

$$\prod_{h=1}^p \chi_{[S',T']}(t_{q_{x_h}})(g_{1-\varepsilon'(q_{y_h+1}'-1)}|f')\cdot\prod_{h\in\{1,\cdots m'\}\setminus\{y_h\}_{h=1}^p}\chi_{[S',T']}(s_{q_h'})(g_{1-\varepsilon'(q_{h+1}'-1)}|f')$$

$$<W(\chi_{[S,T]}\otimes f)\Psi, W(\chi_{[S',T']}\otimes f')\Psi> \tag{4.6}$$

where, the operation $\frac{d}{dt}$ is meant almost everywhere.

In the following, we denote (3.47) by $<u,G(t)v>$ and the right hand side of (4.6) by $<u,G(1,t)v> + <u,G(2,t)v>$. Clearly,

$$<u,G(1,t)v> = \sum_{n=1,n'=0}^\infty \sum_{\varepsilon\in\{0,1\}^n,\ \varepsilon'\in\{0,1\}^{n'}} <u,D_{\varepsilon(1)}\cdots D_{\varepsilon(n)}X D_{\varepsilon'(1)}\cdots D_{\varepsilon'(n')}v>$$

$$\sum_{m=1}^n \sum_{1=q_1<q_2<\cdots<q_m\leq n} \sum_{m'=1}^{n'} \sum_{1=q_1'<q_2'<\cdots<q_{m'}'\leq n'}$$

$$\sum_{p=0}^{m\wedge m'} \sum_{y_1<\cdots<y_p,\ \{y_h\}_{h=1}^p\subset\{1,\cdots,m'\}} \sum_{\{x_h\}_{h=1}^p\subset\{1,\cdots,m\},\ x_1>\cdots>x_p}$$

$$\chi_{[S,T]}(t)(f|g_{\varepsilon(q_h)})\int_0^t dt_{q_2}\int_0^{t_{q_2}} dt_{q_3}\cdots\int_0^{t_{q_{m-1}}} dt_{q_m}$$

$$\prod_{\alpha\in\{1,\cdots,n\}\setminus\{q_h\}_{h=1}^m}(g_{1-\varepsilon(\alpha-1)}|g_{\varepsilon(\alpha)})-$$

$$\prod_{\alpha \in \{1,\cdots,n'\} \setminus \{q'_h\}_{h=1}^{m'}} (g_{1-\varepsilon'(\alpha-1)}|g_{\varepsilon'(\alpha)})+$$

$$\int_0^t ds_{q'_1} \int_{s_{q'_1}}^t ds_{q'_2} \cdots \int_{s_{q'_{y_1-2}}}^t ds_{q'_{y_1-1}} \chi_{[0,t_{q_{x_1}})}(s_{q'_{y_1-1}}) \int_{t_{q_{x_{r_1}}}}^t ds_{q'_{y_{r_1}+1}} \cdots$$

$$\int_{s_{q'_{y_{r_{\alpha-1}}-2}}}^t ds_{q'_{y_{r_{\alpha-1}}-1}} \chi_{[0,t_{q_{x_{r_{\alpha-1}}}})}(s_{q'_{y_{r_{\alpha-1}}-1}})$$

$$\int_{t_{q_{x_{r_\alpha}}}}^t ds_{q'_{y_{r_\alpha}+1}} \int_{s_{q'_{y_{r_\alpha}}+1}}^t ds_{q'_{y_{r_\alpha}+2}} \cdots \int_{s_{q'_{m'-1}}}^t ds_{q'_{m'}}$$

$$\prod_{l=1}^p (g_{1-\varepsilon(q_{x_l}+1-1)}|g_{\varepsilon'(q'_{y_l})})$$

$$\prod_{h=2}^m \chi_{[S,T]}(t_{q_h})(f|g_{\varepsilon(q_h)}) \cdot \prod_{h \in \{1,\cdots,m'\} \setminus \{y_l\}_{l=1}^p} \chi_{[S,T]}(s_{q'_h})(f|g_{\varepsilon'(q'_h)})$$

$$\cdot \prod_{h \in \{1,\cdots,m\} \setminus \{x_l\}_{l=1}^p} \chi_{[S',T']}(t_{q_h})(g_{1-\varepsilon(q_{h+1}-1)}|f')$$

$$\prod_{h=1}^p \chi_{[S',T']}(t_{q_{x_h}})(g_{1-\varepsilon'(q'_{y_h}+1-1)}|f') \cdot \prod_{h \in \{1,\cdots m'\} \setminus \{y_h\}_{h=1}^p} \chi_{[S',T']}(s_{q'_h})(g_{1-\varepsilon'(q'_{h+1}-1)}|f')$$

$$< W(\chi_{[S,T]} \otimes f)\Psi, W(\chi_{[S',T']} \otimes f')\Psi > \qquad (4.7)$$

Slitting the right hand side of (4.7) to two terms according to $x_p > 1$ or $x_p = 1$ and denoting them by $< u, G(1,a,t)v >$, $< u, G(1,b,t)v >$ respectively, we obtain that

$$< u, G(1,a,t)v >= \sum_{n=1,n'=0}^\infty \sum_{\varepsilon \in \{0,1\}^n, \ \varepsilon' \in \{0,1\}^{n'}} < u, D_{\varepsilon(1)} \cdots D_{\varepsilon(n)} X D_{\varepsilon'(1)} \cdots D_{\varepsilon'(n')} v >$$

$$\sum_{m=1}^n \sum_{1=q_1 < q_2 < \cdots < q_m \le n} \sum_{m'=1}^{n'} \sum_{1=q'_1 < q'_2 < \cdots < q'_{m'} \le n'}$$

$$\sum_{p=0}^{m \wedge m'} \sum_{y_1 < \cdots < y_p, \ \{y_h\}_{h=1}^p \subset \{1,\cdots,m'\}} \sum_{\{x_h\}_{h=1}^p \subset \{1,\cdots,m\}, \ x_1 > \cdots > x_p > 1}$$

$$\chi_{[S,T]}(t)(f|g_{\varepsilon(q_h)}) \int_0^t dt_{q_2} \int_0^{t_{q_2}} dt_{q_3} \cdots \int_0^{t_{q_{m-1}}} dt_{q_m}$$

$$\prod_{\alpha\in\{1,\cdots,n\}\setminus\{q_h\}_{h=1}^m}(g_{1-\varepsilon(\alpha-1)}|g_{\varepsilon(\alpha)})-$$

$$\prod_{\alpha\in\{1,\cdots,n'\}\setminus\{q'_h\}_{h=1}^{m'}}(g_{1-\varepsilon'(\alpha-1)}|g_{\varepsilon'(\alpha)})+$$

$$\int_0^t ds_{q'_1}\int_{s_{q'_1}}^t ds_{q'_2}\cdots\int_{s_{q'_{y_1-2}}}^t ds_{q'_{y_1-1}}\chi_{[0,t_{qx_1})}(s_{q'_{y_1-1}})\int_{t_{qx_{r_1}}}^t ds_{q'_{y_{r_1}+1}}\cdots$$

$$\int_{s_{q'_{y_{r_{\alpha-1}}-2}}}^t ds_{q'_{y_{r_{\alpha-1}}-1}}\chi_{[0,t_{qx_{r_{\alpha-1}}})}(s_{q'_{y_{r_{\alpha-1}}-1}})$$

$$\int_{t_{qx_{r_\alpha}}}^t ds_{q'_{y_{r_\alpha}+1}}\int_{s_{q'_{y_{r_\alpha}}+1}}^t ds_{q'_{y_{r_\alpha}+2}}\cdots\int_{s_{q'_{m'-1}}}^t ds_{q'_{m'}}$$

$$\prod_{l=1}^p(g_{1-\varepsilon(q_{x_l}+1-1)}|g_{\varepsilon'(q'_{y_l})})$$

$$\prod_{h=2}^m\chi_{[S,T]}(t_{q_h})(f|g_{\varepsilon(q_h)})\cdot\prod_{h\in\{1,\cdots,m'\}\setminus\{y_l\}_{l=1}^p}\chi_{[S,T]}(s_{q'_h})(f|g_{\varepsilon'(q'_h)})$$

$$\cdot\prod_{h\in\{1,\cdots,m\}\setminus\{x_l\}_{l=1}^p}\chi_{[S',T']}(t_{q_h})(g_{1-\varepsilon(q_{h+1}-1)}|f')$$

$$\prod_{h=1}^p\chi_{[S',T']}(t_{q_{x_h}})(g_{1-\varepsilon'(q'_{y_h}+1)-1}|f')\cdot\prod_{h\in\{1,\cdots m'\}\setminus\{y_h\}_{h=1}^p}\chi_{[S',T']}(s_{q'_h})(g_{1-\varepsilon'(q'_{h+1}-1)}|f')$$

$$< W(\chi_{[S,T]}\otimes f)\Psi, W(\chi_{[S',T']}\otimes f')\Psi > \tag{4.8a}$$

and

$$< u, G(1,b,t)v >= \sum_{n=1,n'=0}^{\infty}\sum_{\varepsilon\in\{0,1\}^n,\ \varepsilon'\in\{0,1\}^{n'}} < u, D_{\varepsilon(1)}\cdots D_{\varepsilon(n)}X D_{\varepsilon'(1)}\cdots D_{\varepsilon'(n')}v >$$

$$\sum_{m=1}^n\sum_{1=q_1<q_2<\cdots<q_m\leq n}\sum_{m'=1}^{n'}\sum_{1=q'_1<q'_2<\cdots<q'_{m'}\leq n'}$$

$$\sum_{p=1}^{m\wedge m'}\sum_{y_1<\cdots<y_p,\ \{y_h\}_{h=1}^p\subset\{1,\cdots,m'\}}\sum_{\{x_h\}_{h=1}^p\subset\{1,\cdots,m\},\ x_1>\cdots>x_p=1}$$

$$\chi_{[S,T]}(t)(f|g_{\varepsilon(q_h)})\int_0^t dt_{q_2}\int_0^{t_{q_2}} dt_{q_3}\cdots\int_0^{t_{q_{m-1}}} dt_{q_m}$$

$$\prod_{\alpha\in\{1,\cdots,n\}\setminus\{q_h\}_{h=1}^{m}}(g_{1-\varepsilon(\alpha-1)}|g_{\varepsilon(\alpha)})-$$

$$\prod_{\alpha\in\{1,\cdots,n'\}\setminus\{q'_h\}_{h=1}^{m'}}(g_{1-\varepsilon'(\alpha-1)}|g_{\varepsilon'(\alpha)})+$$

$$\int_0^t ds_{q'_1}\int_{s_{q'_1}}^t ds_{q'_2}\cdots\int_{s_{q'_{y_1-2}}}^t ds_{q'_{y_1-1}}\chi_{[0,t_{q_{x_1}})}(s_{q'_{y_1-1}})\int_{t_{q_{x_{r_1}}}}^t ds_{q'_{y_{r_1}+1}}\cdots$$

$$\int_{s_{q'_{y_{r_{\alpha-1}}-2}}}^t ds_{q'_{y_{r_{\alpha-1}}-1}}\chi_{[0,t_{q_{x_{r_{\alpha-1}}}})}(s_{q'_{y_{r_{\alpha-1}}-1}})$$

$$\int_t^t ds_{q'_{y_{r_\alpha}+1}}\int_{s_{q'_{y_{r_\alpha}}+1}}^t ds_{q'_{y_{r_\alpha}+2}}\cdots\int_{s_{q'_{m'-1}}}^t ds_{q'_{m'}}$$

$$\prod_{l=1}^{p}(g_{1-\varepsilon(q_{x_l}+1-1)}|g_{\varepsilon'(q'_{y_l})})$$

$$\prod_{h=2}^{m}\chi_{[S,T]}(t_{q_h})(f|g_{\varepsilon(q_h)})\cdot\prod_{h\in\{1,\cdots,m'\}\setminus\{y_l\}_{l=1}^{p}}\chi_{[S,T]}(s_{q'_h})(f|g_{\varepsilon'(q'_h)})$$

$$\cdot\prod_{h\in\{1,\cdots,m\}\setminus\{x_l\}_{l=1}^{p}}\chi_{[S',T']}(t_{q_h})(g_{1-\varepsilon(q_{h+1}-1)}|f')$$

$$\prod_{h=1}^{p}\chi_{[S',T']}(t_{q_{x_h}})(g_{1-\varepsilon'(q'_{y_h}+1-1)}|f')\cdot\prod_{h\in\{1,\cdots m'\}\setminus\{y_h\}_{h=1}^{p}}\chi_{[S',T']}(s_{q'_h})(g_{1-\varepsilon'(q'_{h+1}-1)}|f')$$

$$<W(\chi_{[S,T]}\otimes f)\Psi,W(\chi_{[S',T']}\otimes f')\Psi> \tag{4.8b}$$

where the appearing of the integral $\int_t^t ds_{q'_{y_{r_\alpha}+1}}$ dues to $r_\alpha = p$, $x_p = 1$ and $q_1 = 1$. So (4.8b) is equal to zero if $y_p < m'$, this shows that if we consider the case of $x_p = 1$, then we can put the condition of $y_p = m'$.

First of all, let consider the term $<u, G(1, a, t)v>$. Changing the order of the sums $\sum_{n=1}^{\infty}\sum_{m=1}^{n}$, we obtain that

$$<u, G(1, a, t)v>=\sum_{m=1}^{\infty}\sum_{n=m}^{\infty}\sum_{n'=0}^{\infty}\sum_{\varepsilon\in\{0,1\}^n}\sum_{\varepsilon'\in\{0,1\}^{n'}}$$

$$\sum_{\beta=1}^{n-m+1}\sum_{\beta+1=q_2<\cdots<q_m\leq n}\sum_{m'=1}^{n'}\sum_{1=q'_1<q'_2<\cdots<q'_{m'}\leq n'}(-1)^{n'}$$

$$\sum_{p=0}^{m\wedge m'}\quad \sum_{\substack{y_1<\cdots<y_p,\ \{y_h\}_{h=1}^p\subset\{1,\cdots,m'\}}}\quad \sum_{\substack{\{x_h\}_{h=1}^p\subset\{1,\cdots,m\},\ x_1>\cdots>x_p>1}}$$

$$< u, D_{\varepsilon(1)}\cdots D_{\varepsilon(\beta)}D_{\varepsilon(\beta+1)}\cdots D_{\varepsilon(n)}X D_{\varepsilon'(1)}\cdots D_{\varepsilon'(n')}v >$$

$$\chi_{[S,T]}(t)(f|g_{\varepsilon(1)})\int_0^t dt_{q_2}\int_0^{t_{q_2}}dt_{q_3}\cdots\int_0^{t_{q_{m-1}}}dt_{q_m}$$

$$\prod_{\alpha\in\{2,\cdots,q_2-1\}}(g_{1-\varepsilon(\alpha-1)}|g_{\varepsilon(\alpha)})-\cdot\prod_{\alpha\in\{q_2,\cdots,n\}\setminus\{q_h\}_{h=2}^m}(g_{1-\varepsilon(\alpha-1)}|g_{\varepsilon(\alpha)})-$$

$$\prod_{\alpha\in\{1,\cdots,n'\}\setminus\{q'_h\}_{h=1}^{m'}}(g_{1-\varepsilon'(\alpha-1)}|g_{\varepsilon'(\alpha)})+$$

$$\int_0^t ds_{q'_1}\int_{s_{q'_1}}^t ds_{q'_2}\cdots\int_{s_{q'_{y_1-2}}}^t ds_{q'_{y_1-1}}\chi_{[0,t_{qx_1})}(s_{q'_{y_1-1}})\int_{t_{qx_{r_1}}}^t ds_{q'_{y_{r_1}+1}}\cdots$$

$$\int_{s_{q'_{y_{r_{\alpha-1}}-2}}}^t ds_{q'_{y_{r_{\alpha-1}}-1}}\chi_{[0,t_{qx_{r_{\alpha-1}}})}(s_{q'_{y_{r_{\alpha-1}}-1}})$$

$$\int_{t_{qx_{r_\alpha}}}^t ds_{q'_{y_{r_\alpha}+1}}\int_{s_{q'_{y_{r_\alpha}}+1}}^t ds_{q'_{y_{r_\alpha}+2}}\cdots\int_{s_{q'_{m'-1}}}^t ds_{q'_{m'}}$$

$$\prod_{l=1}^p(g_{1-\varepsilon(q_{x_l}+1-1)}|g_{\varepsilon'(q'_{y_l})})$$

$$\prod_{h=2}^m \chi_{[S,T]}(t_{q_h})(f|g_{\varepsilon(q_h)})\cdot\prod_{h\in\{1,\cdots,m'\}\setminus\{y_l\}_{l=1}^p}\chi_{[S,T]}(s_{q'_h})(f|g_{\varepsilon'(q'_h)})$$

$$\cdot\chi_{[S',T']}(t)(g_{1-\varepsilon(q_2-1)}|f')\prod_{h\in\{2,\cdots,m\}\setminus\{x_l\}_{l=1}^p}\chi_{[S',T']}(t_{q_h})(g_{1-\varepsilon(q_{h+1}-1)}|f')$$

$$\prod_{h=1}^p\chi_{[S',T']}(t_{q_{x_h}})(g_{1-\varepsilon'(q'_{y_h+1}-1)}|f')\cdot\prod_{h\in\{1,\cdots m'\}\setminus\{y_h\}_{h=1}^p}\chi_{[S',T']}(s_{q'_h})(g_{1-\varepsilon'(q'_{h+1}-1)}|f')$$

$$< W(\chi_{[S,T]}\otimes f)\Psi, W(\chi_{[S',T']}\otimes f')\Psi > \tag{4.9}$$

Notice that

$$\sum_{n=m}^{\infty}\sum_{\beta=1}^{n-m+1}=\sum_{\beta=1}^{\infty}\sum_{n=m+\beta-1}^{\infty} \tag{4.10}$$

Introduce the new indice as the following:

$$\bar m := m-1,\quad \bar n := n-\beta$$

$$\bar{q}_1 := q_2 - \beta \, , \cdots, \, \bar{q}_{\bar{m}} := q_m - \beta$$

$$\bar{x}_h := x_h - 1, \quad h = 1 \cdots, p \tag{4.11}$$

then,

$$\sum_{m=1}^{\infty} = \sum_{\bar{m}=0}^{\infty} \, , \qquad \sum_{n=m+\beta-1}^{\infty} = \sum_{\bar{n}=\bar{m}}^{\infty} \tag{4.12}$$

$$\sum_{\beta+1=q_2<q_3<\cdots<q_m} = \sum_{1=\bar{q}_1<\bar{q}_2<\cdots<\bar{q}_{\bar{m}}\leq\bar{n}} \tag{4.13}$$

$$\sum_{p=0}^{m\wedge m'} = \sum_{p=0}^{\bar{m}\wedge m'} \tag{4.14}$$

$$\{x_h\}_{h=1}^p \subset \{2,\cdots,m\}, \; x_1 > \cdots > x_p > 1 \iff \{\bar{x}_h\}_{h=1}^p \subset \{1,\cdots,\bar{m}\}, \; \bar{x}_1 > \cdots > \bar{x}_p \tag{4.15}$$

$$\bar{q}_{\bar{x}_h} = \bar{q}_{x_h-1} = q_{x_h} - \beta \tag{4.16}$$

Moreover, if we denote

$$\mathcal{M}_{0,1;\beta+1,\cdots,n} := \{\epsilon; \; \epsilon : \{\beta+1,\cdots,n\} \longrightarrow \{0,1\}\} \tag{4.17}$$

then

$$\{0,1\}^n = \{0,1\}^\beta \times \mathcal{M}_{0,1;\beta+1,\cdots,n} \tag{4.18}$$

with the identification given by:

$$\big(\varepsilon(1),\cdots,\varepsilon(n)\big) = \big(\sigma(1),\cdots,\sigma(\beta),\epsilon(\beta+1),\cdots,\epsilon(n)\big) \tag{4.19}$$

It is clear that

$$\mathcal{M}_{0,1;\beta+1,\cdots,n} = \{0,1\}^{\bar{n}} \tag{4.20}$$

if we put, for each $\epsilon \in \mathcal{M}_{0,1;\beta+1,\cdots,n}$,

$$\bar{\epsilon}(1) = \epsilon(\beta+1),\cdots,\bar{\epsilon}(\bar{n}) = \epsilon(n) \tag{4.21}$$

Thus,

$$\sum_{\varepsilon\in\{0,1\}^n} = \sum_{\sigma\in\{0,1\}^\beta} \sum_{\bar{\varepsilon}\in\{0,1\}^n} \tag{4.22}$$

$$< u, D_{\varepsilon(1)} \cdots D_{\varepsilon(\beta)} D_{\varepsilon(\beta+1)} \cdots D_{\varepsilon(n)} X D_{\varepsilon'(1)} \cdots D_{\varepsilon'(n')} v >$$

$$= < u, D_{\sigma(1)} \cdots D_{\sigma(\beta)} D_{\bar{\varepsilon}(1)} \cdots D_{\bar{\varepsilon}(\bar{n})} X D_{\varepsilon'(1)} \cdots D_{\varepsilon'(n')} v > \tag{4.23}$$

$$\sum_{\{x_h\}_{h=1}^p\subset\{1,\cdots,m\}, \; x_1>\cdots>x_p>1} = \sum_{\{\bar{x}_h\}_{h=1}^p\subset\{1,\cdots,\bar{m}\}, \; \bar{x}_1>\cdots>\bar{x}_p} \tag{4.24}$$

$$\chi_{[S,T]}(t)(f|g_{\varepsilon(1)}) = \chi_{[S,T]}(t)(f|g_{\sigma(1)}) \tag{4.25}$$

$$\int_0^t dt_{q_2} \int_0^{t_{q_2}} dt_{q_3} \cdots \int_0^{t_{q_{m-1}}} dt_{q_m} = \int_0^t dt_{\bar{q}_1} \int_0^{t_{\bar{q}_1}} dt_{\bar{q}_2} \cdots \int_0^{t_{\bar{q}_{\bar{m}-1}}} dt_{\bar{q}_{\bar{m}}} \tag{4.26}$$

$$\prod_{\alpha \in \{2,\cdots,q_2-1\}} (g_{1-\varepsilon(\alpha-1)}|g_{\varepsilon(\alpha)})_- = \prod_{h=1}^{\beta-1} (g_{1-\sigma(h)}|g_{\sigma(h+1)})_- \tag{4.27}$$

$$\prod_{\alpha \in \{q_2,\cdots,n\}\setminus\{q_h\}_{h=2}^m} (g_{1-\varepsilon(\alpha-1)}|g_{\varepsilon(\alpha)})_- = \prod_{\alpha \in \{1,\cdots,\bar{n}\}\setminus\{\bar{q}_h\}_{h=1}^{\bar{m}}} (g_{1-\bar{\varepsilon}(\alpha-1)}|g_{\bar{\varepsilon}(\alpha)})_- \tag{4.28}$$

$$\prod_{h=2}^m \chi_{[S,T]}(t_{q_h})(f|g_{\varepsilon(q_h)}) = \prod_{h=1}^{\bar{m}} \chi_{[S,T]}(t_{\bar{q}_h})(f|g_{\bar{\varepsilon}(\bar{q}_h)}) \tag{4.29}$$

$$\prod_{l=1}^p (g_{1-\varepsilon(q_{x_l}+1-1)}|g_{\varepsilon'(q'_{y_l})}) = \prod_{l=1}^p (g_{1-\bar{\varepsilon}(\bar{q}_{\bar{x}_l}+1-1)}|g_{\varepsilon'(q'_{y_l})}) \tag{4.30}$$

$$\chi_{[S',T']}(t)(g_{1-\varepsilon(q_2-1)}|f') = \chi_{[S',T']}(t)(g_{1-\sigma(\beta)}|f') \tag{4.31}$$

$$\prod_{h\in\{2,\cdots,m\}\setminus\{x_l\}_{l=1}^p} \chi_{[S',T']}(t_{q_h})(g_{1-\varepsilon(q_{h+1}-1)}|f')$$

$$= \prod_{h\in\{1,\cdots,\bar{m}\}\setminus\{\bar{x}_l\}_{l=1}^p} \chi_{[S',T']}(t_{\bar{q}_h})(g_{1-\bar{\varepsilon}(\bar{q}_{h+1}-1)}|f') \tag{4.32}$$

These show that the right hand side of (4.9) is equal to

$$\sum_{\beta=1}^\infty \sum_{\sigma\in\{0,1\}^\beta} \sum_{\bar{m}=0}^\infty \sum_{\bar{n}=\bar{m}}^\infty$$

$$\sum_{n'=0}^\infty \sum_{\bar{\varepsilon}\in\{0,1\}^{\bar{n}}} \sum_{\varepsilon'\in\{0,1\}^{n'}} \sum_{1=\bar{q}_1<\bar{q}_2<\cdots<\bar{q}_{\bar{m}}\leq\bar{n}} \sum_{m'=1}^{n'} \sum_{1=q'_1<q'_2<\cdots<q'_{m'}\leq n'}$$

$$\sum_{p=0}^{\bar{m}\wedge m'} \sum_{y_1<\cdots<y_p,\ \{y_h\}_{h=1}^p\subset\{1,\cdots,m'\}} \sum_{\{\bar{x}_h\}_{h=1}^p\subset\{1,\cdots,\bar{m}\},\ \bar{x}_1>\cdots>\bar{x}_p}$$

$$<u, D_{\sigma(1)}\cdots D_{\sigma(\beta)} D_{\bar{\varepsilon}(1)}\cdots D_{\bar{\varepsilon}(\bar{n})} X D_{\varepsilon'(1)}\cdots D_{\varepsilon'(n')}v>(-1)^{n'}$$

$$\chi_{[S,T]}(t)(f|g_{\sigma(1)}) \int_0^t dt_{\bar{q}_1} \int_0^{t_{\bar{q}_1}} dt_{\bar{q}_2} \cdots \int_0^{t_{\bar{q}_{\bar{m}-1}}} dt_{\bar{q}_{\bar{m}}}$$

$$\prod_{h=1}^{\beta-1} (g_{1-\sigma(h)}|g_{\sigma(h+1)})_- \cdot \prod_{\alpha\in\{1,\cdots,\bar{n}\}\setminus\{\bar{q}_h\}_{h=1}^m} (g_{1-\bar{\varepsilon}(\alpha-1)}|g_{\bar{\varepsilon}(\alpha)})_-$$

$$\prod_{\alpha\in\{1,\cdots,n'\}\setminus\{q'_h\}_{h=1}^{m'}} (g_{1-\varepsilon'(\alpha-1)}|g_{\varepsilon'(\alpha)})_+$$

$$\int_0^t ds_{q_1'} \int_{s_{q_1'}}^t ds_{q_2'} \cdots \int_{s_{q_{y_1-2}'}}^t ds_{q_{y_1-1}'} \chi_{[0,t_{q_{x_1}})}(s_{q_{y_1-1}'}) \int_{t_{q_{x_{r_1}}}}^t ds_{q_{y_{r_1}+1}'} \cdots$$

$$\int_{s_{q_{y_{r_\alpha-1}-2}'}}^t ds_{q_{y_{r_\alpha-1}-1}'} \chi_{[0,t_{q_{x_{r_\alpha-1}}})}(s_{q_{y_{r_\alpha-1}-1}'})$$

$$\int_{t_{q_{x_{r_\alpha}}}}^t ds_{q_{y_{r_\alpha}+1}'} \int_{s_{q_{y_{r_\alpha}}+1}'}^t ds_{q_{y_{r_\alpha}+2}'} \cdots \int_{s_{q_{m'-1}'}}^t ds_{q_{m'}'}$$

$$\prod_{l=1}^p (g_{1-\bar\varepsilon(\bar q_{\bar x_l+1}-1)} | g_{\varepsilon'(q_{y_l}')})$$

$$\prod_{h=1}^{\bar m} \chi_{[S,T]}(t_{\bar q_h})(f|g_{\bar\varepsilon(\bar q_h)}) \cdot \prod_{h\in\{1,\cdots,m'\}\backslash\{y_l\}_{l=1}^p} \chi_{[S,T]}(s_{q_h'})(f|g_{\varepsilon'(q_h')})$$

$$\cdot \chi_{[S',T']}(t)(g_{1-\sigma(\beta)}|f') \prod_{h\in\{1,\cdots,\bar m\}\backslash\{\bar x_l\}_{l=1}^p} \chi_{[S',T']}(t_{\bar q_h})(g_{1-\bar\varepsilon(\bar q_h+1)-1)}|f')$$

$$\prod_{h=1}^p \chi_{[S',T']}(t_{q_{x_h}})(g_{1-\varepsilon'(q_{y_h}'+1)-1)}|f') \cdot \prod_{h\in\{1,\cdots m'\}\backslash\{y_h\}_{h=1}^p} \chi_{[S',T']}(s_{q_h'})(g_{1-\varepsilon'(q_{h+1}'-1)}|f')$$

$$< W(\chi_{[S,T]} \otimes f)\Psi, W(\chi_{[S',T']} \otimes f')\Psi >$$

$$= \sum_{\beta=1}^\infty \sum_{\sigma\in\{0,1\}^\beta} \prod_{h=1}^{\beta-1} (g_{1-\sigma(h)}|g_{\sigma(h+1)})_-$$

$$\chi_{[S,T]}(t)(f|g_{\sigma(1)}) \cdot \chi_{[S',T']}(t)(g_{1-\sigma(\beta)}|f') \cdot < u, D_{\sigma(1)} \cdots D_{\sigma(\beta)} G(t)v > \tag{4.33}$$

where

$$\prod_{h=1}^0 (g_{1-\sigma(h)}|g_{\sigma(h+1)})_- := 1 \tag{4.34}$$

Now, let distinguish the case $\sigma(1) = \sigma(\beta)$ from the case $\sigma(1) \neq \sigma(\beta)$.

For each $\beta \geq 3$, put $\epsilon(h) = \sigma(h-2)$ for $h = 3, \cdots, \beta$, then $\epsilon \in \{0,1\}^{\beta-2}$ and moreover

$$\sum_{\beta=1}^\infty \sum_{\sigma\in\{0,1\}^\beta} \prod_{h=1}^{\beta-1} (g_{1-\sigma(h)}|g_{\sigma(h+1)})_-$$

$$(f|g_{\sigma(1)}) \cdot (g_{1-\sigma(\beta)}|f') \cdot D_{\sigma(1)} \cdots D_{\sigma(\beta)}$$

$$= \sum_{\varepsilon\in\{0,1\}} D_\varepsilon(f|g_\varepsilon) \cdot (g_{1-\varepsilon}|f') + \sum_{\varepsilon\in\{0,1\}} D_\varepsilon D_\varepsilon (g_{1-\varepsilon}|g_{1-\varepsilon})_- (f|g_\varepsilon) \cdot (g_{1-\varepsilon}|f')$$

$$+ \sum_{\varepsilon \in \{0,1\}} D_\varepsilon D_{1-\varepsilon}(g_{1-\varepsilon}|g_{1-\varepsilon})_- (f|g_\varepsilon) \cdot (g_\varepsilon|f')$$

$$+ \sum_{\beta=3}^{\infty} \sum_{\varepsilon \in \{0,1\}} \sum_{\epsilon \in \{0,1\}^{\beta-2}} (g_{1-\varepsilon}|g_{\epsilon(1)})_- \cdot \prod_{h=1}^{\beta-3}(g_{1-\epsilon(h)}|g_{\epsilon(h+1)})_-$$

$$\left[D_\varepsilon D_{\epsilon(1)} \cdots D_{\epsilon(\beta-2)} D_\varepsilon (g_{1-\epsilon(\beta-2)}|g_\varepsilon)_- (f|g_\varepsilon) \cdot (g_{1-\varepsilon}|f') \right.$$

$$\left. + D_\varepsilon D_{\epsilon(1)} \cdots D_{\epsilon(\beta-2)} D_{1-\varepsilon} (g_{1-\epsilon(\beta-2)}|g_{1-\varepsilon})_- (f|g_\varepsilon) \cdot (g_\varepsilon|f') \right] \tag{4.35}$$

Defining

$$D_1(\varepsilon) := D_\varepsilon + D_\varepsilon D_\varepsilon (g_{1-\varepsilon}|g_\varepsilon)_-$$

$$+ \sum_{\beta=3}^{\infty} \sum_{\epsilon \in \{0,1\}^{\beta-2}} (g_{1-\varepsilon}|g_{\epsilon(1)})_- \cdot \prod_{h=1}^{\beta-3}(g_{1-\epsilon(h)}|g_{\epsilon(h+1)})_-$$

$$D_\varepsilon D_{\epsilon(1)} \cdots D_{\epsilon(\beta-2)} D_\varepsilon (g_{1-\epsilon(\beta-2)}|g_\varepsilon)_- \tag{4.36}$$

and

$$D_2(\varepsilon) := D_\varepsilon D_{1-\varepsilon}(g_{1-\varepsilon}|g_{1-\varepsilon})_- + \sum_{\beta=3}^{\infty} \sum_{\epsilon \in \{0,1\}^{\beta-2}} (g_{1-\varepsilon}|g_{\epsilon(1)})_- \cdot \prod_{h=1}^{\beta-3}(g_{1-\epsilon(h)}|g_{\epsilon(h+1)})_-$$

$$D_\varepsilon D_{\epsilon(1)} \cdots D_{\epsilon(\beta-2)} D_{1-\varepsilon} (g_{1-\epsilon(\beta-2)}|g_{1-\varepsilon})_- \tag{4.37}$$

Summing up (4.33), (4.35), (4.36) and (4.37) we obtain the following

$THEOREM(4.3)$

$$< u, G(1,a,t)v > = \sum_{\varepsilon\{0,1\}} \chi_{[S,T]}(t)\chi_{[S',T']}(t) \Big(< u, D_1(\varepsilon)G(t)v > (f|g_\varepsilon) \cdot (g_{1-\varepsilon}|f') +$$

$$+ < u, D_2(\varepsilon)G(t)v > (f|g_\varepsilon) \cdot (g_\varepsilon|f') \Big) \tag{4.37}$$

Now let us deal with $< u, G(1,b,t)v >$. Since this term corresponds to $x_p = 1$ and $y_p = m'$, the second (ds-) integral in (4.8b) is in fact

$$\int_0^t ds_{q_1'} \int_{s_{q_1'}}^t ds_{q_2'} \cdots \int_{s_{q_{y_1-2}'}}^t ds_{q_{y_1-1}'} \chi_{[0,t_{qx_1})}(s_{q_{y_1-1}'}) \int_{t_{qxr_1}}^t ds_{q_{y_{r_1}+1}'} \cdots$$

$$\int_{s_{q_{y_{r_\alpha-1}-2}'}}^t ds_{q_{y_{r_\alpha-1}-1}'} \chi_{[0,t_{qxr_{\alpha-1}})}(s_{q_{y_{r_\alpha-1}-1}'}) \tag{4.38}$$

Notice that in this case $n' \geq 1$, introducing the new indices $\bar{n}$, $\bar{m}$, $\{\bar{q}_h\}_{h=1}^{\bar{m}}$, $\bar{\varepsilon}$ the same ones as above $((4.11), (4.21))$ and $\bar{p} := p - 1$, $\bar{x}_h := x_h - 1$ for $h = 1, \cdots, \bar{p} = p - 1$, one has

$$\sum_{p=1}^{m \wedge m'} = \sum_{\bar{p}=0}^{\bar{m} \wedge (m'-1)} \tag{4.40}$$

and

$$\sum_{\{\bar{x}_h\}_{h=1}^{\bar{p}} \subset \{1, \cdots, m\},\ x_1 > \cdots > x_p = 1} = \sum_{\{x_h\}_{h=1}^{p} \subset \{1, \cdots, \bar{m}\},\ \bar{x}_1 > \cdots > \bar{x}_{\bar{p}}} \tag{4.40a}$$

Moreover, $x_p = 1$ results in

$$\prod_{h \in \{1, \cdots, m\} \backslash \{x_l\}_{l=1}^{p}} \chi_{[S', T']}(t_{q_h})(g_{1 - \varepsilon(q_{h+1}-1)}|f')$$

$$= \prod_{h \in \{2, \cdots, m\} \backslash \{x_l\}_{l=1}^{\bar{p}}} \chi_{[S', T']}(t_{q_h})(g_{1 - \varepsilon(q_{h+1}-1)}|f')$$

$$= \prod_{h \in \{1, \cdots, \bar{m}\} \backslash \{\bar{x}_l\}_{l=1}^{\bar{p}}} \chi_{[S', T']}(t_{\bar{q}_h})(g_{1 - \bar{\varepsilon}(\bar{q}_{h+1}-1)}|f') \tag{4.41}$$

where, the situation is not the same as before, we can not get the term

$$\chi_{[S', T']}(t)(g_{1 - \varepsilon(q_2 - 1)}|f') \tag{4.42}$$

But, on the other hand side,

$$\prod_{l=1}^{p}(g_{1 - \varepsilon(q_{x_l+1}-1)}|g_{\varepsilon'}(q'_{y_l}))$$

$$= (g_{1 - \varepsilon(q_{x_p+1}-1)}|g_{\varepsilon'}(q'_{y_p})) \prod_{l=1}^{\bar{p}}(g_{1 - \bar{\varepsilon}(\bar{q}_{x_l+1}-1)}|g_{\varepsilon'}(q'_{y_l}))$$

$$= (g_{1 - \sigma(\beta)}|g_{\varepsilon'}(q'_{y_p})) \cdot \prod_{l=1}^{\bar{p}}(g_{1 - \bar{\varepsilon}(\bar{q}_{x_l+1}-1)}|g_{\varepsilon'}(q'_{y_l})) \tag{4.43}$$

Thus, one has

$$< u, G(1, b, t)v> = \sum_{\beta=1}^{\infty} \sum_{\sigma \in \{0,1\}^{\beta}} \sum_{\bar{m}=0}^{\infty} \sum_{\bar{n}=\bar{m}}^{\infty}$$

$$\sum_{n'=1}^{\infty} \sum_{\bar{\varepsilon} \in \{0,1\}^{\bar{n}}} \sum_{\varepsilon' \in \{0,1\}^{n'}} \sum_{1 = \bar{q}_1 < \bar{q}_2 < \cdots < \bar{q}_{\bar{m}} \leq \bar{n}} \sum_{m'=2}^{n'} \sum_{\beta=m'}^{n'} \sum_{1 = q'_1 < q'_2 < \cdots < q'_{m'} = \beta' \leq n'}$$

$$\sum_{\bar p=0}^{\bar m \wedge (m'-1)} \quad \sum_{y_1<\cdots<y_p,\ \{y_h\}_{h=1}^{p}\subset\{1,\cdots,m'\}} \quad \sum_{\{\bar x_h\}_{h=1}^{\bar p}\subset\{1,\cdots,\bar m\},\ \bar x_1>\cdots>\bar x_{\bar p}}$$

$$<u, D_{\sigma(1)}\cdots D_{\sigma(\beta)} D_{\bar\varepsilon(1)}\cdots D_{\bar\varepsilon(\bar n)} X D_{\varepsilon'(1)}\cdots D_{\varepsilon'(n')} v > (-1)^{n'}$$

$$\chi_{[S,T]}(t)(f|g_{\sigma(1)}) \int_0^t dt_{\bar q_1} \int_0^{t_{\bar q_1}} dt_{\bar q_2} \cdots \int_0^{t_{\bar q_{\bar m -1}}} dt_{\bar q_{\bar m}}$$

$$\prod_{h=1}^{\beta-1}(g_{1-\sigma(h)}|g_{\sigma(h)})-\ \cdot \prod_{\alpha\in\{1,\cdots,\bar n\}\setminus\{\bar q_h\}_{h=1}^{\bar m}}(g_{1-\bar\varepsilon(\alpha-1)}|g_{\bar\varepsilon(\alpha)})-$$

$$\prod_{\alpha\in\{1,\cdots,n'\}\setminus\{q'_h\}_{h=1}^{m'}}(g_{1-\varepsilon'(\alpha-1)}|g_{\varepsilon'(\alpha)})+$$

$$\int_0^t ds_{q'_1}\int_{s_{q'_1}}^t ds_{q'_2}\cdots \int_{s_{q'_{y_1-2}}}^t ds_{q'_{y_1-1}}\chi_{[0,t_{qx_1})}(s_{q'_{y_1-1}})\int_{t_{qx_{r_1}}}^t ds_{q'_{y_{r_1}+1}}\cdots$$

$$\int_{s_{q'_{y_{r_{\alpha-1}}-2}}}^t ds_{q'_{y_{r_{\alpha-1}}-1}}\chi_{[0,t_{qx_{r_{\alpha-1}}})}(s_{q'_{y_{r_{\alpha-1}}-1}})$$

$$(g_{1-\sigma(\beta)}|g_{\varepsilon'(q'_{y_p})})\cdot \prod_{l=1}^{\bar p}(g_{1-\bar\varepsilon(\bar q_{\bar x_l+1}-1)}|g_{\varepsilon'(q'_{y_l})})$$

$$\prod_{h=1}^{\bar m}\chi_{[S,T]}(t_{\bar q_h})(f|g_{\bar\varepsilon(\bar q_h)})\cdot \prod_{h\in\{1,\cdots,m'-1\}\setminus\{y_l\}_{l=1}^{\bar p}}\chi_{[S,T]}(s_{q'_h})(f|g_{\varepsilon'(q'_h)})$$

$$\prod_{h\in\{1,\cdots,\bar m\}\setminus\{\bar x_l\}_{l=1}^{\bar p}}\chi_{[S',T']}(t_{\bar q_h})(g_{1-\bar\varepsilon(\bar q_{h+1}-1)}|f')$$

$$\prod_{h=1}^{p}\chi_{[S',T']}(t_{qx_h})(g_{1-\varepsilon'(q'_{y_h}+1)-1)}|f')\cdot \prod_{h\in\{1,\cdots m'\}\setminus\{y_h\}_{h=1}^{p}}\chi_{[S',T']}(s_{q'_h})(g_{1-\varepsilon'(q'_{h+1}-1)}|f')$$

$$<W(\chi_{[S,T]}\otimes f)\Psi, W(\chi_{[S',T']}\otimes f')\Psi>$$

$$+\sum_{\beta=1}^{\infty}\sum_{\sigma\in\{0,1\}^{\beta}}\sum_{\bar m=1}^{\infty}\sum_{\bar n=\bar m}^{\infty}$$

$$\sum_{n'=1}^{\infty}\sum_{\bar\varepsilon\in\{0,1\}^{\bar n}}\sum_{\varepsilon'\in\{0,1\}^{n'}}\sum_{1=\bar q_1<\bar q_2<\cdots<\bar q_{\bar m}\leq\bar n}$$

$$<u, D_{\sigma(1)}\cdots D_{\sigma(\beta)} D_{\bar\varepsilon(1)}\cdots D_{\bar\varepsilon(\bar n)} X D_{\varepsilon'(1)}\cdots D_{\varepsilon'(n')} v >$$

$$\chi_{[S,T]}(t)(f|g_{\sigma(1)})\int_0^t dt_{\bar{q}_1}\int_0^{t_{\bar{q}_1}} dt_{\bar{q}_2}\cdots\int_0^{t_{\bar{q}_{\bar m -1}}} dt_{\bar{q}_{\bar m}}$$

$$\prod_{h=1}^{\beta-1}(g_{1-\sigma(h)}|g_{\sigma(h+1)})-\cdot\prod_{\alpha\in\{1,\cdots,\bar n\}\backslash\{\bar q_h\}_{h=1}^m}(g_{1-\bar\varepsilon(\alpha-1)}|g_{\bar\varepsilon(\alpha)}')-$$

$$\prod_{\alpha\in\{2,\cdots,n'\}}(g_{1-\varepsilon'(\alpha-1)}|g_{\varepsilon'(\alpha)})+$$

$$(g_{1-\sigma(\beta)}|g_{\varepsilon'(1)})\cdot\prod_{h=1}^{\bar m}\chi_{[S,T]}(t_{\bar q_h})(f|g_{\bar\varepsilon(\bar q_h)})\cdot\prod_{h\in\{1,\cdots,\bar m\}}\chi_{[S',T']}(t_{\bar q_h})(g_{1-\bar\varepsilon(\bar q_{h+1}-1)}|f')$$

$$\chi_{[S',T']}(t)(g_{1-\varepsilon'(n'-1)}|f') < W(\chi_{[S,T]}\otimes f)\Psi, W(\chi_{[S',T']}\otimes f')\Psi > \qquad (4.44)$$

where, the second term of (4.44) corresponds to the term of $m'=1$. Now let consider the indices with prime. First of all

$$\sum_{n'=1}^{\infty}\sum_{m'=2}^{n'} = \sum_{m'=2}^{\infty}\sum_{n'=m'}^{\infty} \qquad (4.45)$$

Defining

$$\beta' := q'_{m'}$$

then

$$m' \leq \beta' \leq n' \qquad (4.46)$$

Defining

$$\bar m' := m' - 1 \qquad (4.47)$$

one knows that

$$< u, G(1,b,t)v >= \sum_{\beta=1}^{\infty}\sum_{\sigma\in\{0,1\}^\beta}\sum_{\bar m=0}^{\infty}\sum_{\bar n=\bar m}^{\infty}$$

$$\sum_{\bar\varepsilon\in\{0,1\}^{\bar n}}\sum_{1=\bar q_1<\bar q_2<\cdots<\bar q_{\bar m}\leq\bar n}\sum_{\bar m'=1}^{\infty}\sum_{n'=\bar m'+1}^{\infty}\sum_{\varepsilon'\in\{0,1\}^{n'}}$$

$$\sum_{\beta'=\bar m'+1}^{n'}\sum_{1=q'_1<q'_2<\cdots<q'_{\bar m'}\leq\beta'-1}$$

$$\sum_{\bar p=0}^{\bar m\wedge\bar m'}\sum_{y_1<\cdots<y_{\bar p},\ \{y_h\}_{h=1}^{\bar p}\subset\{1,\cdots,\bar m'\}}\sum_{\{\bar x_h\}_{h=1}^{\bar p}\subset\{1,\cdots,\bar m\},\ \bar x_1>\cdots>\bar x_{\bar p}}$$

$$< u, D_{\sigma(1)}\cdots D_{\sigma(\beta)}D_{\bar\varepsilon(1)}\cdots D_{\bar\varepsilon(\bar n)}X D_{\varepsilon'(1)}\cdots D_{\varepsilon'(\beta'-1)}D_{\varepsilon'(\beta')}\cdots D_{\varepsilon'(n')}v > (-1)^{n'}$$

$$\chi_{[S,T]}(t)(f|g_{\sigma(1)})\int_0^t dt_{\bar{q}_1}\int_0^{t_{\bar{q}_1}} dt_{\bar{q}_2}\cdots\int_0^{t_{\bar{q}_{\bar{m}-1}}} dt_{\bar{q}_{\bar{m}}}$$

$$\prod_{h=1}^{\beta-1}(g_{1-\sigma(h)}|g_{\sigma(h+1)})-\cdot\prod_{\alpha\in\{1,\cdots,\bar{n}\}\setminus\{\bar{q}_h\}_{h=1}^{\bar{m}}}(g_{1-\bar{\varepsilon}(\alpha-1)}|g_{\bar{\varepsilon}(\alpha)})-$$

$$\prod_{\alpha\in\{1,\cdots,n'\}\setminus\{q'_h\}_{h=1}^{m'}}(g_{1-\varepsilon'(\alpha-1)}|g_{\varepsilon'(\alpha)})+$$

$$\int_0^t ds_{q'_1}\int_{s_{q'_1}}^t ds_{q'_2}\cdots\int_{s_{q'_{y_1-2}}}^t ds_{q'_{y_1-1}}\chi_{[0,t_{\bar{q}_{\bar{x}_1}}]}(s_{q'_{y_1-1}})\int_{t_{\bar{q}_{\bar{x}r_1}}}^t ds_{q'_{y_{r_1}+1}}\cdots$$

$$\int_{s_{q'_{y_{r_{\alpha-1}}-2}}}^t ds_{q'_{y_{r_{\alpha-1}}-1}}\chi_{[0,t_{\bar{q}_{\bar{x}r_{\alpha-1}}}]}(s_{q'_{y_{r_{\alpha-1}}-1}})$$

$$(g_{1-\sigma(\beta)}|g_{\varepsilon'(\beta')})\cdot\prod_{l=1}^{\bar{p}}(g_{1-\bar{\varepsilon}(\bar{q}_{\bar{x}_l}+1-1)}|g_{\varepsilon'(q'_{y_l})})$$

$$\prod_{h=1}^{\bar{m}}\chi_{[S,T]}(t_{\bar{q}_h})(f|g_{\bar{\varepsilon}(\bar{q}_h)})\cdot\prod_{h\in\{1,\cdots,\bar{m}'\}\setminus\{y_l\}_{l=1}^{\bar{p}}}\chi_{[S,T]}(s_{q'_h})(f|g_{\varepsilon'(q'_h)})$$

$$\cdot\chi_{[S',T']}(t)(g_{1-\varepsilon'(n')}|f')\prod_{h\in\{1,\cdots,\bar{m}\}\setminus\{\bar{x}_l\}_{l=1}^{\bar{p}}}\chi_{[S',T']}(t_{\bar{q}_h})(g_{1-\bar{\varepsilon}(\bar{q}_{h+1}-1)}|f')$$

$$\prod_{h=1}^{\bar{p}}\chi_{[S',T']}(t_{\bar{q}_{\bar{x}_h}})(g_{1-\varepsilon'(q'_{y_h}+1-1)}|f')\cdot\prod_{h\in\{1,\cdots,\bar{m}'\}\setminus\{y_h\}_{h=1}^{\bar{p}}}\chi_{[S',T']}(s_{q'_h})(g_{1-\varepsilon'(q'_{h+1}-1)}|f')$$

$$<W(\chi_{[S,T]}\otimes f)\Psi, W(\chi_{[S',T']}\otimes f')\Psi>$$

$$+\sum_{\beta=1}^{\infty}\sum_{\sigma\in\{0,1\}^{\beta}}\sum_{\bar{m}=0}^{\infty}\sum_{\bar{n}=\bar{m}}^{\infty}$$

$$\sum_{n'=1}^{\infty}\sum_{\bar{\varepsilon}\in\{0,1\}^{\bar{n}}}\sum_{\varepsilon'\in\{0,1\}^{n'}}\sum_{1=\bar{q}_1<\bar{q}_2<\cdots<\bar{q}_{\bar{m}}\leq\bar{n}}$$

$$<u, D_{\sigma(1)}\cdots D_{\sigma(\beta)}D_{\bar{\varepsilon}(1)}\cdots D_{\bar{\varepsilon}(\bar{n})}XD_{\varepsilon'(1)}\cdots D_{\varepsilon'(n')}v>(-1)^{n'}$$

$$\chi_{[S,T]}(t)(f|g_{\sigma(1)})\int_0^t dt_{\bar{q}_1}\int_0^{t_{\bar{q}_1}} dt_{\bar{q}_2}\cdots\int_0^{t_{\bar{q}_{\bar{m}-1}}} dt_{\bar{q}_{\bar{m}}}$$

$$\prod_{h=1}^{\beta-1}(g_{1-\sigma(h)}|g_{\sigma(h+1)})-\cdot\prod_{\alpha\in\{1,\cdots,\bar{n}\}\setminus\{\bar{q}_h\}_{h=1}^{\bar{m}}}(g_{1-\bar{\varepsilon}(\alpha-1)}|g_{\bar{\varepsilon}(\alpha)})-$$

$$\prod_{\alpha \in \{2,\cdots,n'\}} (g_{1-\varepsilon'(\alpha-1)}|g_{\varepsilon'(\alpha)})+$$

$$(g_{1-\sigma(\beta)}|g_{\varepsilon'(1)}) \cdot \prod_{h=1}^{\bar{m}} \chi_{[S,T]}(t_{\bar{q}_h})(f|g_{\bar{\varepsilon}(\bar{q}_h)}) \cdot \prod_{h \in \{1,\cdots,\bar{m}\}} \chi_{[S',T']}(t_{\bar{q}_h})(g_{1-\bar{\varepsilon}(\bar{q}_{h+1}-1)}|f')$$

$$\chi_{[S',T']}(t)(g_{1-\varepsilon'(n')}|f') < W(\chi_{[S,T]} \otimes f)\Psi, W(\chi_{[S',T']} \otimes f')\Psi > \tag{4.48}$$

Moreover, we have

$$\sum_{\bar{m}'=1}^{\infty} \sum_{n'=\bar{m}'+1}^{\infty} \sum_{\beta'=\bar{m}'+1}^{n'} = \sum_{\bar{m}'=1}^{\infty} \sum_{\beta'=\bar{m}'+1}^{\infty} \sum_{n'=\beta'}^{\infty} = \sum_{\beta'-1=1}^{\infty} \sum_{\bar{m}'=1}^{\beta'-1} \sum_{n'=\beta'}^{\infty} \tag{4.49}$$

$$\sum_{\varepsilon' \in \{0,1\}^{n'}} = \sum_{\varepsilon' \in \{0,1\}^{\beta'-1}} \sum_{\sigma' \in \{0,1\}^{n'-\beta'+1}} \tag{4.50}$$

where,

$$\sigma'(1) := \varepsilon'(\beta'), \cdots, \sigma'(n'-\beta'+1) := \varepsilon'(n') \tag{4.51}$$

$$\prod_{\alpha \in \{1,\cdots,n'\} \backslash \{q'_h\}_{h=1}^{m'}} (g_{1-\varepsilon'(\alpha-1)}|g_{\varepsilon'(\alpha)})+ \ =$$

$$= \prod_{\alpha \in \{1,\cdots,\beta'-1\} \backslash \{q'_h\}_{h=1}^{m'}} (g_{1-\varepsilon'(\alpha-1)}|g_{\varepsilon'(\alpha)})+ \cdot \prod_{h=1}^{n'-\beta'} (g_{1-\sigma'(h)}|g_{\sigma'(h+1)})+ \tag{4.52}$$

So, the first term of the right hand side of (4.48) is equal to

$$\sum_{\beta=1}^{\infty} \sum_{\sigma \in \{0,1\}^{\beta}} \sum_{\bar{m}=1}^{\infty} \sum_{\bar{n}=\bar{m}}^{\infty} \sum_{\bar{\varepsilon} \in \{0,1\}^{\bar{n}}} \sum_{1=\bar{q}_1<\bar{q}_2<\cdots<\bar{q}_{\bar{m}}\leq\bar{n}} \sum_{\beta'-1=1}^{\infty} \sum_{\bar{m}'=0}^{\beta'-1} \sum_{n'-\beta'+1=1}^{\infty}$$

$$\sum_{\varepsilon' \in \{0,1\}^{\beta'-1}} \sum_{\sigma' \in \{0,1\}^{n'-\beta'+1}} \sum_{1=q'_1<q'_2<\cdots<q'_{\bar{m}'}\leq\beta'-1}$$

$$\sum_{\bar{p}=0}^{\bar{m}\wedge\bar{m}'} \sum_{y_1<\cdots<y_{\bar{p}}, \ \{y_h\}_{h=1}^{\bar{p}}\subset\{1,\cdots,\bar{m}'\}} \sum_{\{\bar{x}_h\}_{h=1}^{\bar{p}}\subset\{1,\cdots,\bar{m}\}, \ \bar{x}_1>\cdots>\bar{x}_{\bar{p}}}$$

$$< u, D_{\sigma(1)} \cdots D_{\sigma(\beta)} D_{\bar{\varepsilon}(1)} \cdots D_{\bar{\varepsilon}(\bar{n})} X D_{\varepsilon'(1)} \cdots D_{\varepsilon'(\beta'-1)} D_{\sigma'(1)} \cdots D_{\sigma'(n'-\beta'+1)}v > (-1)^{n'}$$

$$\chi_{[S,T]}(t)(f|g_{\sigma(1)}) \int_0^t dt_{\bar{q}_1} \int_0^{t_{\bar{q}_1}} dt_{\bar{q}_2} \cdots \int_0^{t_{\bar{q}_{\bar{m}-1}}} dt_{\bar{q}_{\bar{m}}}$$

$$\prod_{h=1}^{\beta-1}(g_{1-\sigma(h)}|g_{\sigma(h+1)})_- \cdot \prod_{\alpha\in\{1,\cdots,\bar{n}\}\backslash\{\bar{q}_h\}_{h=1}^{\bar{m}}}(g_{1-\bar{\varepsilon}(\alpha-1)}|g_{\bar{\varepsilon}(\alpha)})_-$$

$$\prod_{\alpha\in\{1,\cdots,\beta'-1\}\backslash\{q_h'\}_{h=1}^{\bar{m}'}}(g_{1-\varepsilon'(\alpha-1)}|g_{\varepsilon'(\alpha)})_+ \cdot \prod_{h=1}^{n'-\beta'}(g_{1-\sigma'(h)}|g_{\sigma'(h+1)})_+$$

$$\int_0^t ds_{q_1'}\int_{s_{q_1'}}^t ds_{q_2'}\cdots\int_{s_{q_{y_1-2}'}}^t ds_{q_{y_1-1}'}\chi_{[0,t_{\bar{q}\bar{x}_1})}(s_{q_{y_1-1}'})\int_{t_{\bar{q}\bar{x}_{r_1}}}^t ds_{q_{y_{r_1}+1}'}\cdots$$

$$\int_{s_{q_{y_{r_\alpha-1}-2}'}}^t ds_{q_{y_{r_\alpha-1}-1}'}\chi_{[0,t_{\bar{q}\bar{x}_{r_\alpha-1}})}(s_{q_{y_{r_\alpha-1}-1}'})$$

$$(g_{1-\sigma(\beta)}|g_{\sigma'(1)})\cdot\prod_{l=1}^{\bar{p}}(g_{1-\bar{\varepsilon}(\bar{q}\bar{x}_l+1-1)}|g_{\varepsilon'(q_{y_l}')})$$

$$\prod_{h=1}^{\bar{m}}\chi_{[S,T]}(t_{\bar{q}_h})(f|g_{\bar{\varepsilon}(\bar{q}_h)})\cdot\prod_{h\in\{1,\cdots,\bar{m}'\}\backslash\{y_l\}_{l=1}^{\bar{p}}}\chi_{[S,T]}(s_{q_h'})(f|g_{\varepsilon'(q_h')})$$

$$\cdot\chi_{[S',T']}(t)(g_{1-\sigma(n'-\beta'+1)}|f')\prod_{h\in\{1,\cdots,\bar{m}\}\backslash\{\bar{x}_l\}_{l=1}^{\bar{p}}}\chi_{[S',T']}(t_{\bar{q}_h})(g_{1-\bar{\varepsilon}(\bar{q}_{h+1}-1)}|f')$$

$$\prod_{h=1}^{\bar{p}}\chi_{[S',T']}(t_{\bar{q}\bar{x}_h})(g_{1-\varepsilon'(q_{y_h}'+1-1)}|f')\cdot\prod_{h\in\{1,\cdots,\bar{m}'\}\backslash\{y_h\}_{h=1}^{\bar{p}}}\chi_{[S',T']}(s_{q_h'})(g_{1-\varepsilon'(q_{h+1}'-1)}|f')$$

$$< W(\chi_{[S,T]}\otimes f)\Psi, W(\chi_{[S',T']}\otimes f')\Psi > \tag{4.52}$$

and the second term of the right hand side of (4.48) is equal to

$$\sum_{\beta=1}^\infty\sum_{\sigma\in\{0,1\}^\beta}\sum_{\bar{m}=0}^\infty\sum_{\bar{n}=\bar{m}}^\infty$$

$$\sum_{n'=1}^\infty\sum_{\bar{\varepsilon}\in\{0,1\}^{\bar{n}}}\sum_{\sigma'\in\{0,1\}^{n'}}\sum_{1=\bar{q}_1<\bar{q}_2<\cdots<\bar{q}_{\bar{m}}\leq\bar{n}}$$

$$< u, D_{\sigma(1)}\cdots D_{\sigma(\beta)}D_{\bar{\varepsilon}(1)}\cdots D_{\bar{\varepsilon}(\bar{n})}XD_{\sigma'(1)}\cdots D_{\sigma'(n')}v >(-1)^{n'}$$

$$\chi_{[S,T]}(t)(f|g_{\sigma(1)})\int_0^t dt_{\bar{q}_1}\int_0^{t_{\bar{q}_1}}dt_{\bar{q}_2}\cdots\int_0^{t_{\bar{q}_{\bar{m}-1}}}dt_{\bar{q}_{\bar{m}}}$$

$$\prod_{h=1}^{\beta-1}(g_{1-\sigma(h)}|g_{\sigma(h+1)})_- \cdot \prod_{\alpha\in\{1,\cdots,\bar{n}\}\backslash\{\bar{q}_h\}_{h=1}^{\bar{m}}}(g_{1-\bar{\varepsilon}(\alpha-1)}|g_{\bar{\varepsilon}(\alpha)})_-$$

$$\prod_{\alpha\in\{2,\cdots,n'\}}(g_{1-\sigma'(\alpha-1)}|g_{\sigma'(\alpha)})_+$$

$$(g_{1-\sigma(\beta)}|g_{\varepsilon'(1)}) \cdot \prod_{h=1}^{\bar{m}}\chi_{[S,T]}(t_{\bar{q}_h})(f|g_{\bar{\varepsilon}(\bar{q}_h)}) \cdot \prod_{h\in\{1,\cdots,\bar{m}\}}\chi_{[S',T']}(t_{\bar{q}_h})(g_{1-\bar{\varepsilon}(\bar{q}_{h+1}-1)}|f')$$

$$\chi_{[S',T']}(t)(g_{1-\sigma'(n')}|f') < W(\chi_{[S,T]}\otimes f)\Psi, W(\chi_{[S',T']}\otimes f')\Psi > \tag{4.53}$$

These show that, if we introduce the new indices

$$\bar{\beta}' := \beta' - 1, \quad \bar{n}' := n' - \beta' \tag{4.54}$$

then

$$< u, G(1,b,t)v >= \sum_{\beta=1}^{\infty}\sum_{\sigma\in\{0,1\}^\beta}\sum_{\bar{m}=0}^{\infty}\sum_{\bar{n}=\bar{m}}^{\infty}$$

$$\sum_{\bar{\varepsilon}\in\{0,1\}^{\bar{n}}}\sum_{1=\bar{q}_1<\bar{q}_2<\cdots<\bar{q}_{\bar{m}}\leq\bar{n}}\sum_{\bar{\beta}'=0}^{\infty}\sum_{\bar{m}'=0}^{\bar{\beta}'}\sum_{\bar{n}'=0}^{\infty}$$

$$\sum_{\varepsilon'\in\{0,1\}^{\bar{\beta}'}}\sum_{\sigma'\in\{0,1\}^{\bar{n}'+1}}\sum_{1=q'_1<q'_2<\cdots<q'_{\bar{m}'}\leq\bar{\beta}'}$$

$$\sum_{\bar{p}=0}^{\bar{m}\wedge\bar{m}'}\sum_{y_1<\cdots<y_{\bar{p}},\ \{y_h\}_{h=1}^{\bar{p}}\subset\{1,\cdots,\bar{m}'\}}\sum_{\{\bar{x}_h\}_{h=1}^{\bar{p}}\subset\{1,\cdots,\bar{m}\},\ \bar{x}_1>\cdots>\bar{x}_{\bar{p}}}$$

$$< u, D_{\sigma(1)}\cdots D_{\sigma(\beta)}D_{\bar{\varepsilon}(1)}\cdots D_{\bar{\varepsilon}(\bar{n})}X D_{\varepsilon'(1)}\cdots D_{\varepsilon'(\beta'-1)}D_{\sigma'(1)}\cdots D_{\sigma'(\bar{n}'+1)}v > (-1)^{n'}$$

$$\chi_{[S,T]}(t)(f|g_{\sigma(1)})\int_0^t dt_{\bar{q}_1}\int_0^{t_{\bar{q}_1}} dt_{\bar{q}_2}\cdots\int_0^{t_{\bar{q}_{\bar{m}-1}}} dt_{\bar{q}_{\bar{m}}}$$

$$\prod_{h=1}^{\beta-1}(g_{1-\sigma(h)}|g_{\sigma(h+1)})_- \cdot \prod_{\alpha\in\{1,\cdots,\bar{n}\}\backslash\{\bar{q}_h\}_{h=1}^{\bar{m}}}(g_{1-\bar{\varepsilon}(\alpha-1)}|g_{\bar{\varepsilon}(\alpha)})_-$$

$$\prod_{\alpha\in\{1,\cdots,\bar{\beta}'\}\backslash\{q'_h\}_{h=1}^{\bar{m}'}}(g_{1-\varepsilon'(\alpha-1)}|g_{\varepsilon'(\alpha)})_+ \cdot \prod_{h=1}^{\bar{n}'}(g_{1-\sigma'(h)}|g_{\sigma'(h+1)})_+$$

$$\int_0^t ds_{q'_1}\int_{s_{q'_1}}^t ds_{q'_2}\cdots\int_{s_{q'_{y_1-2}}}^t ds_{q'_{y_1-1}}\chi_{[0,t_{\bar{q}_{\bar{x}_1}})}(s_{q'_{y_1-1}})\int_{t_{\bar{q}_{\bar{x}_{r_1}}}}^t ds_{q'_{y_{r_1}+1}}\cdots$$

$$\int_{s_{q'_{y_{r_{\alpha-1}}-2}}}^{t} ds_{q'_{y_{r_{\alpha-1}}-1}} \chi_{[0,t_{\bar{q}x_{r_{\alpha-1}}})}(s_{q'_{y_{r_{\alpha-1}}-1}}) \cdots \int_{s_{q'_{\bar{m}'-1}}}^{t} ds_{q'_{\bar{m}'}}$$

$$(g_{1-\sigma(\beta)}|g_{\sigma'(1)}) \cdot \prod_{l=1}^{\bar{p}}(g_{1-\bar{\varepsilon}(\bar{q}_{x_l}+1)-1}|g_{\varepsilon'(q'_{y_l})})$$

$$\prod_{h=1}^{\bar{m}} \chi_{[S,T]}(t_{\bar{q}_h})(f|g_{\bar{\varepsilon}(\bar{q}_h)}) \cdot \prod_{h\in\{1,\cdots,\bar{m}'\}\setminus\{y_l\}_{l=1}^{\bar{p}}} \chi_{[S,T]}(s_{q'_h})(f|g_{\varepsilon'(q'_h)})$$

$$\cdot\chi_{[S',T']}(t)(g_{1-\sigma(\bar{n}'+1)}|f') \prod_{h\in\{1,\cdots,\bar{m}\}\setminus\{\bar{x}_l\}_{l=1}^{\bar{p}}} \chi_{[S',T']}(t_{\bar{q}_h})(g_{1-\bar{\varepsilon}(\bar{q}_{h+1}-1)}|f')$$

$$\prod_{h=1}^{\bar{m}'} \chi_{[S',T']}(s_{q'_h})(g_{1-\varepsilon'(q'_{h+1}-1)}|f') < W(\chi_{[S,T]}\otimes f)\Psi, W(\chi_{[S',T']}\otimes f')\Psi >$$

$$= \sum_{\beta=1}^{\infty}\sum_{\sigma\in\{0,1\}^{\beta}}\prod_{h=1}^{\beta-1}(g_{1-\sigma(h)}|g_{\sigma(h+1)}) - \sum_{\beta''=1}^{\infty}\sum_{\sigma'\in\{0,1\}^{\beta''}}\prod_{h=1}^{\beta''-1}(g_{1-\sigma'(h)}|g_{\sigma'(h+1)})+$$

$$\chi_{[S,T]}(t)(f|g_{\sigma(1)}) \cdot \chi_{[S',T']}(t)(g_{1-\sigma'(\beta'')}|f') \cdot (g_{1-\sigma(\beta)}|g_{\sigma'(\beta'')})$$

$$< u, D_{\sigma(1)}\cdots D_{\sigma(\beta)}G(t)(D_{\sigma'(1)}\cdots D_{\sigma'(\beta'')})^{+}v > \tag{4.55}$$

Therefore, we obtain the following

$THEOREM(4.4)$

$$< u, G(1,b,t)v >= \chi_{[S,T]}(t)\chi_{[S',T']}(t)\sum_{\varepsilon,\varepsilon'\in\{0,1\}}(f|g_\varepsilon)\cdot(g_{\varepsilon'}|f')$$

$$\Big(< u, D_1(\varepsilon)G(t)D_1^{+}(\varepsilon')v > (g_{1-\varepsilon}|g_{1-\varepsilon'}) + < u, D_1(\varepsilon)G(t)D_2^{+}(\varepsilon')v > (g_{1-\varepsilon}|g_{\varepsilon'})+$$

$$+ < u, D_2(\varepsilon)G(t)D_1^{+}(\varepsilon')v > (g_\varepsilon|g_{1-\varepsilon'}) + < u, D_2(\varepsilon)G(t)D_2^{+}(\varepsilon')v > (g_\varepsilon|g_{\varepsilon'})\Big) \tag{4.56}$$

Now let investigate $< u, G(2,t)v >$.
Notice that

$$\frac{d}{dt}\int_0^t ds_{q'_1}\int_{s_{q'_1}}^t ds_{q'_2}\cdots\int_{s_{q'_{y_1-2}}}^t ds_{q'_{y_1-1}}\chi_{[0,t_{q x_1})}(s_{q'_{y_1-1}})\int_{t_{q x_{r_1}}}^t ds_{q'_{y_{r_1}+1}}\cdots$$

$$\int_{s_{q'_{y_{r_{\alpha-1}}-2}}}^t ds_{q'_{y_{r_{\alpha-1}}-1}}\chi_{[0,t_{q x_{r_{\alpha-1}}})}(s_{q'_{y_{r_{\alpha-1}}-1}})$$

$$\int_{t_{q_x r_\alpha}}^{t} ds_{q'_{y_{r_\alpha}+1}} \int_{s_{q'_{y_{r_\alpha}}+1}}^{t} ds_{q'_{y_{r_\alpha}+2}} \cdots \int_{s_{q'_{m'-1}}}^{t} ds_{q'_{m'}} \qquad (4.57)$$

is equal to the operation $\frac{d}{dt}$ is applied on the last integral. Moreover, if $y_p = m'$, the last integral is

$$\int_{s_{q'_{y_{r_\alpha-1}}-2}}^{t} ds_{q'_{y_{r_\alpha-1}}-1} \chi_{[0,t_{q_x r_\alpha-1})}(s_{q'_{y_{r_\alpha-1}}-1}) \qquad (4.58)$$

and

$$\frac{d}{dt} \int_{s_{q'_{y_{r_\alpha-1}}-2}}^{t} ds_{q'_{y_{r_\alpha-1}}-1} \chi_{[0,t_{q_x r_\alpha-1})}(s_{q'_{y_{r_\alpha-1}}-1}) = \chi_{[0,t_{q_x r_\alpha-1})}(t) = 0 \quad a.e. \qquad (4.58a)$$

So we can consider only the case $y_p < m'$. In this case, (4.57) is equal to

$$\int_{0}^{t} ds_{q'_1} \int_{s_{q'_1}}^{t} ds_{q'_2} \cdots \int_{s_{q'_{y_1-2}}}^{t} ds_{q'_{y_1-1}} \chi_{[0,t_{q_x 1})}(s_{q'_{y_1-1}}) \int_{t_{q_x r_1}}^{t} ds_{q'_{y_{r_1}+1}} \cdots$$

$$\int_{s_{q'_{y_{r_\alpha-1}}-2}}^{t} ds_{q'_{y_{r_\alpha-1}}-1} \chi_{[0,t_{q_x r_\alpha-1})}(s_{q'_{y_{r_\alpha-1}}-1})$$

$$\int_{t_{q_x r_\alpha}}^{t} ds_{q'_{y_{r_\alpha}+1}} \int_{s_{q'_{y_{r_\alpha}}+1}}^{t} ds_{q'_{y_{r_\alpha}+2}} \cdots \int_{s_{q'_{m'-2}}}^{t} ds_{q'_{m'-1}} \frac{d}{dt} \int_{s_{q'_{m'-1}}}^{t} ds_{q'_{m'}} \qquad (4.59)$$

$$= \int_{0}^{t} ds_{q'_1} \int_{s_{q'_1}}^{t} ds_{q'_2} \cdots \int_{s_{q'_{y_1-2}}}^{t} ds_{q'_{y_1-1}} \chi_{[0,t_{q_x 1})}(s_{q'_{y_1-1}}) \int_{t_{q_x r_1}}^{t} ds_{q'_{y_{r_1}+1}} \cdots$$

$$\int_{s_{q'_{y_{r_\alpha-1}}-2}}^{t} ds_{q'_{y_{r_\alpha-1}}-1} \chi_{[0,t_{q_x r_\alpha-1})}(s_{q'_{y_{r_\alpha-1}}-1})$$

$$\int_{t_{q_x r_\alpha}}^{t} ds_{q'_{y_{r_\alpha}+1}} \int_{s_{q'_{y_{r_\alpha}}+1}}^{t} ds_{q'_{y_{r_\alpha}+2}} \cdots \int_{s_{q'_{m'-2}}}^{t} ds_{q'_{m'-1}}$$

So this case is the same as to consider the term $< u, G(1,a,t)v >$. Therefore, we have the following

$THEOREM(4.5)$

$$< u, G(2,t)v > = \chi_{[S,T]}(t)\chi_{[S',T']}(t) \sum_{\varepsilon \in \{0,1\}} \Big(< u, G(t)D_1^+(\varepsilon)v > (f|g_{1-\varepsilon}) \cdot (g_\varepsilon|f') +$$

$$+ < u, G(t)D_2^+(\varepsilon)v > (f|g_\varepsilon) \cdot (g_\varepsilon|f') \Big) \qquad (4.60)$$

Summing up, one obtains

THEOREM(4.6)

$$< u, G(t)v > = < u, Xv > < W(\chi_{[S,T]} \otimes f)\Psi, W(\chi_{[S',T']} \otimes f')\Psi > +$$

$$\int_0^t \chi_{[S,T]}(s)\chi_{[S',T']}(s) \sum_{\varepsilon \in \{0,1\}}$$

$$\Big(< u, D_1(\varepsilon)G(s)v > (f|g_\varepsilon) \cdot (g_{1-\varepsilon}|f') + < u, D_2(\varepsilon)G(s)v > (f|g_\varepsilon) \cdot (g_\varepsilon|f')$$

$$+ < u, G(s)D_1^+(\varepsilon)v > (f|g_{1-\varepsilon}) \cdot (g_\varepsilon|f') + < u, G(s)D_2^+(\varepsilon)v > (f|g_\varepsilon) \cdot (g_\varepsilon|f') +$$

$$+ \sum_{\varepsilon' \in \{0,1\}} (f|g_\varepsilon) \cdot (g_{\varepsilon'}|f') \cdot \Big(< u, D_1(\varepsilon)G(s)D_1^+(\varepsilon')v > (g_{1-\varepsilon}|g_{1-\varepsilon'}) +$$

$$+ < u, D_1(\varepsilon)G(s)D_2^+(\varepsilon')v > (g_{1-\varepsilon}|g_{\varepsilon'}) +$$

$$+ < u, D_2(\varepsilon)G(s)D_1^+(\varepsilon')v > (g_\varepsilon|g_{1-\varepsilon'}) + < u, D_2(\varepsilon)G(s)D_2^+(\varepsilon')v > (g_\varepsilon|g_{\varepsilon'}) \Big) \Big) \qquad (4.61)$$

Moreover, we have

THEOREM(4.7) Under the condition (3.54), the low density limit (1.26) exists and is equal to

$$< u \otimes W(\chi_{[S,T]} \otimes f)\Psi, U(t)(X \otimes 1)U^*(t)v \otimes W(\chi_{[S',T']} \otimes f')\Psi > \qquad (4.62)$$

where, $U(t)$ is the solution of quantum stochastic differential equation

$$U(t) = 1 + \int_0^t \sum_{\varepsilon \in \{0,1\}} \Big(D_1(\varepsilon) \otimes dN_s(g_\varepsilon, g_{1-\varepsilon}) + D_2(\varepsilon) \otimes dN_s(g_\varepsilon, g_\varepsilon) \Big) U(s) \qquad (4.63)$$

on $H_0 \otimes \Gamma(L^2(\mathbf{R}_+) \otimes K)$, where,

$$N_s(f,g) := N(\chi_{[0,s]} \otimes |f > < g|) \qquad (4.64)$$

is the quantum number processes and $D_1(\varepsilon)$, $D_2(\varepsilon)$ are defined in (4.36) and (4.37) respectively.

PROOF Denoting (4.62) by $< u, F(t)v >$, then by quantum stochastic differential equation (4.63), it is easy to check that $< u, F(t)v >$ satisfies integral equation (4.61). From the uniqueness of (4.63) we know that (4.61) has a unique solution, therefore we finish the proof.

REFERENCES

[0] . L.Accardi: On the quantum Feymann-Kac formula. Rendiconti del seminario Mate. e Fis., Milano 48 (1978), 135-180.

[1] . L.Accardi , Lu Yun-Gang : The Low Density Limit: the Boson Fock case, to appear .

[2] . L.Accardi , Lu Yun-Gang : The Low Density Limit: the finite temperature case, to appear.

[3] . L.Accardi , Lu Yun-Gang : The Low Density Limit: the quasi-free case without Rotating wave approximation, to appear .

[4] . L.Accardi , Lu Yun-Gang : The Low Density Limit of Boson Model, to appear in : Quantum Probability and Applications V,

[4a] . L.Accardi , Lu Yun-Gang : The Low Density Limit for Fermion Fock Case, to appear.

[5] . L.Accardi , A.Frigerio , Lu Yun-Gang : On the weak coupling limit , Quantum Probability and Applications IV, LNM 1396, pp. 20-58.

[6] . L.Accardi , A.Frigerio , Lu Yun-Gang : On the weak coupling limit (II) , submittied to RIMS.

[7] . L.Accardi , A.Frigerio , Lu Yun-Gang : The weak coupling limit for Fermions, submittied to J. Math. Phys.

[8] . L.Accardi, Lu Yun-Gang : The weak coupling limit without rotating wave approximation, to appear.

[9] . L.Accardi, Lu Yun-Gang : The weak coupling limit for non-linear interaction, to appear .

[9a] . L.Accardi, Lu Yun-Gang, R. Alicki, A. Frigerio: An invitation to the weak coupling limit and low density limit, to appear .

[10] . R.Dümcke : The low density limit for n-level systems , Quantum Probability and Applications II, LNM 1136, pp. 151-161.

[11] . R.Dümcke : The low density limit for an N-level systems intracting with a Free Bose or Fermi Gas, to appear in Comm. Math. Phys.

[12]. A.Frogerio and R. Alicki: Quantum Poisson noise and linear Boltzmann equation. to appear.

[13] . A.Frogerio , H.Maassen : Quantum Poisson processes and dilations of dunamical semigroups . to appear .

[14] . H.Grad : Principles of the knietic theory of Gases . Handbuch der Physik , vol . 12 , Springer (1958).

[15] . R.L.Hudson , K.R.Pathasarathy : Quantum Ito's formula and stochastic evolutions. Comm. Math. Phys. (93) , 301-323 (1984) .

[16] . P.F.Palmer : Thesis , Oxford University .

Schrödinger operators with potentials supported by null sets

by

S. Albeverio[1,2,3,6], J.E. Fenstad[4],

R. Høegh-Krohn[4,†], W. Karwowski[2,3,5],

T. Lindstrøm[4]

Abstract

We study perturbations of the Laplacian in $I\!R^n$ by potentials supported by sets of Lebesgue measure zero. We exhibit self adjoint realizations of such operators. The results have applications in the construction of the singular Schrödinger operators which arise in models of solid state physics and nuclear physics, as well as in such problems as the study of quantum mechanical scattering from polymers and in the study of certain quantum fields.

1. Institut für Mathematik, Ruhr-Universität Bochum, D–4630 Bochum 1, FRG
2. BiBoS Research Centre, Universität Bielefeld, D–4800 Bielefeld, FRG
3. SFB 237, Bochum–Essen–Düsseldorf
4. Matematisk Institutt, Universitetet i Oslo, N–0316 Oslo 3, Norway
5. Institute of Theoretical Physics, University of Wroclaw, Wroclaw, Poland
6. CERFIM, Locarno
† Deceased on January 24, 1988

1. Introduction

Schrödinger operators for particles in $I\!R^n$ involving potentials supported in measure zero subsets of $I\!R^n$ are of great importance in physics (quantum mechanics, nuclear physics, optics and electromagnetism, solid state physics, impurity scattering and disordered media, see e.g. [AGHKH], [AFGHKHK], [Pa], [ExS1-3], [ExSS], [DeO] and references therein).

In recent years the mathematical study of such operators have been undertaken quite systematically. The case where $n = 1, 2, 3$ and the support is a discrete, finite or infinite, subset of $I\!R^n$ is studied extensively in [AGHKH] (and references therein). The case $n = 3$ with support on a 2-dimensional surface is studied in [AGS], see also [BrT], [DSh], [Koc],

[Kur]. The case where the support is the typical trajectory of a Brownian motion in $I\!\!R^n$, $n \leq 5$ is studied in [AFHKKL], [AFHKL], [L1,2]. Latter case is of importance in view of two problems. One is the quantum mechanical study of a particle in the stochastic force field created by a "polymer source", modelized by the path of a Brownian motion, a model which arises in work by Edwards. The other is the relation which exists between such Hamiltonians created by polymers and "polymer measures" for the probabilistic distribution of classical statistical mechanical polymers, the relation being obtained via a (stochastic) Feynman-Kac formula, see [AFHKL].

In [AFHKL] the Hamiltonians associated with perturbations supported by paths of Brownian motion arise actually as special cases of perturbations of generators of symmetric Markov semigroups by potentials with support on measure zero subsets. The problem is formulated and solved in that reference using methods of nonstandard analysis.

In the present paper we take up this problem anew, avoiding any use of nonstandard analytic tools.

In fact the method used in the present paper is rather different from the ones used in the nonstandard analysis approach. Let us mention that the problem studied in the present paper has been approached recently using other techniques by J. Brasche, W. Karwowski and A. Teta [Br1-3], [BrK], [BrT], [Te] and Cheremtsatsev [Ch]. Some of the results of these authors are in fact complementary to ours. Let us discuss shortly the contents of this paper.

We study the Schrödinger operators formally given by $-\Delta - \lambda\mu$, with μ a positive Radon measure on $I\!\!R^n$ and $\lambda \in L^1(I\!\!R^n, \mu)$. We give a meaning to such operators as self-adjoint lower bounded operators in $L^2(I\!\!R^n, dx)$, different from $-\Delta$ in the cases where the support of μ is a set of Lebesgue measure zero (of course the case of perturbations $-\Delta - \lambda\mu + V$, for some multiplicative operators V can also be treated).

We first construct regularized versions H_s of $-\Delta - \lambda\mu$, by replacing $-\lambda\mu$ by a regularized potential acting as an integral kernel $V_s(x,y)$ in $L^2(I\!\!R^n, dx)$, s being so chosen that $(g, V_s f)_{L^2(dx)} \longrightarrow (g, f)_{L^2(\lambda\mu)}$ (with $L^2(dx) \equiv L^2(I\!\!R^n; dx)$, $L^2(\lambda\mu) \equiv L^2(I\!\!R^n; \lambda d\mu)$) as $s \longrightarrow 0$. H_s is self-adjoint, lower bounded and its resolvent has the form, for E sufficiently large negative,

$$(H_s - E)^{-1} = G_E + \check{G}_E^s(\lambda^{-1} - \bar{G}_E^s)^{-1}\hat{G}_E^s \tag{1.1}$$

with $G_E \equiv (-\Delta - E)^{-1}$, G_E^s the regularized version of G_E, and $\check{}$ resp. $\hat{}$ resp. $\bar{}$ indicating the action of G_E^s as kernel from $L^2(I\!\!R^n, \mu)$ into $L^2(I\!\!R^n, dx)$ resp. from $L^2(I\!\!R^n, dx)$ into $L^2(I\!\!R^n, \mu)$ resp. from $L^2(I\!\!R^n, \mu)$ into itself. Assumptions on μ are found in order that the strong limit of (1.1) as $s \longrightarrow 0$ exists and defines the resolvent of a self-adjoint operator H.

The results are essentially different according to whether the dimension n of the space is larger than 3 or less or equal 3.

a) In the case $n \leq 3$ we show (Prop. 2) that $\hat{G}_E^s$ resp. $\check{G}_E^s$ converge to Hilbert-Schmidt operators $\hat{G}_E$ resp. $\check{G}_E$ from $L^2(I\!\!R^n, dx))$ into $L^2(M; \mu)$ and from $L^2(M; \mu)$ into $L^2(I\!\!R^n; dx)$ respectively, where M is some closed subset of $I\!\!R^n$.

b) In order to obtain results in the case $n \geq 4$ we have first to adapt to our situation some potential theoretic results of Pustylnik [Pu]. The concept of dimensionality m

of a subset M of $I\!\!R^n$ with respect to a Radon probability measure μ supported by M is introduced (Prop. 4).

For $m > n - 4$ and for measures μ having a certain property A it is then proven (Prop. 5) that the kernel G_E defines compact operators $\hat{G}_E$ resp. $\check{G}_E$ from $L^2(I\!\!R^n, dx)$ into $L^2(M; \mu)$ and from $L^2(M; \mu)$ into $L^2(I\!\!R^n; dx)$ respectively. $\hat{G}_E$ and $\check{G}_E$ are the operator limits of $\hat{G}_E^s$ and $\check{G}_E^s$ respectively (Prop. 8). In this case it is remarked that the expression

$$\int_M \int_M \left(\int_{I\!\!R^n} G_E(x - z) G_E(z - y) dz \right) \mu(dz) \mu(dy) \tag{1.2}$$

is finite.

Moreover taking $V_s = -\lambda_s \psi_s^* \psi_s$, with ψ_s the operator in $L^2(I\!\!R^n, dx)$ defined by

$$(\psi_s f)(x) \equiv (\lambda_s(x))^{\frac{1}{2}} e^{-\sqrt{-\Delta} s} f(x) , \quad s > 0$$

with $\lambda_s(x)^{-1} = \alpha(x) + \int_M G_{E'}^s(x - y) \mu(dy)$ for some bounded μ-measurable α s.t. $\alpha(x) > \delta > 0$ for some δ (independent of x) and some E', we get (Th. 10) that H_s converges in strong resolvent sense as $s \longrightarrow 0$ to a positive self-adjoint operator $H \neq -\Delta$ with a resolvent which can be described quite explicitly. In a precise sense H is obtained from $-\Delta$ by a perturbation supported by M.

A concept of non admissibility of a number for M is also introduced and it is shown that if $n - 4$ is not admissible for M then (1.2) is infinite.

Moreover if $n \geq 4$ and $m > n - 2$ and moreover μ satisfies a sharpened version of property A then it is shown that

$$\int_M \int_M G_E(x - y) \mu(dx) \mu(dy) \tag{1.3}$$

is finite.

If on the other hand $n - 2$ is not admissible for M then (1.3) is infinite.

In case (1.3) is finite a finer distinction can be done, as discussed in Sect. 3.

If $\int G_E(x - y) \mu(dy)$ is bounded and goes uniformly to zero for $E \longrightarrow -\infty$, then one can get convergence to $H \neq -\Delta$ without having to restrict λ to be small. If we only assume

$$g = \inf_x \sup \int_M G_E(x - y) \mu(dy) < \infty ,$$

(so that μ is potentially bounded in the sense of [FeP]) then the convergence result holds but we have to restrict λ to be small.

If (1.3) is still finite but $\int_M G_E(x - y) \mu(dy)$ is not bounded, the convergence holds but λ cannot be chosen to be bounded away from zero.

Prop. 12 discusses the case where the perturbation supported by M can be realized as a form perturbation of $-\Delta$ given by $\lambda \mu$. This is certainly realized when λ is a positive μ-measurable function and $m > n - 2$.

We close the paper by giving some examples.

In the case of an interaction V supported by a point, Theor. 10 is only applicable when $n \leq 3$. For $n = 1$ one has the situation of Prop. 12. For $n = 2, 3$ the construction of H by Theor. 10 requires the choice of a sequence $\lambda_s \longrightarrow 0$ ("infinitesimal coupling constant", in the terminology of [AFHKL]). In the case of an interaction V supported by a piecewise smooth manifold M in $I\!R^n$ and μ absolutely continuous with respect to the volume measure of M, we have that the geometrical dimension of M is equal to the dimension m of M with respect to μ. If $m \geq n - 3$ then Theor. 10 applies and one has a perturbation supported by M.

For $m = n - 1$ there is no restriction on smallness of the coupling constant.

A third example is given by the choice of M being the intersection of a Brownian path $b(s)$, $s \in [0, t]$ with the closure of an open bounded set and μ the normalized Lebesgue measure restricted to M. In this case the dimension m of M with respect to μ is equal to the Hausdorff dimension of M, which is known to be equal 2. For $n = 2, 3$ Prop. 12 applies (and λ need not be small), for $n = 4, 5$ Theor. 10 still applies, with an infinitesimal choice of λ.

In all these cases the constructed Hamiltonian H gives a meaning to the heuristic expression

$$-\Delta - \lambda \int_0^t \delta(x - b(s)) ds \ .$$

As discussed in [AFHKL], [AFHKKL], [AHK], [ABlHK] these Hamiltonians occur both in the study of polymer measures and in Symanzik's polymers representation of certain quantum fields.

2. Construction of the perturbed Schrödinger operators

Let μ be a positive Radon measure on $I\!R^n$. We are concerned with the following problem: What are the conditions on μ so that the formal operator

$$H = -\Delta - \mu$$

can be given sense as a self-adjoint operator in the complex Hilbert space $L^2(I\!R^n; dx)$? It will be convenient for us to assume μ finite, say a probability measure, and put

$$H = -\Delta - \lambda\mu \tag{2.1}$$

where $\lambda \in L^1(I\!R^n, \mu)$ and $\lambda \geq 0$, $\mu-$ a.e. The expression (2.1) can be understood in the sense of a quadratic form on $C_0^2(I\!R^n)$ by

$$\mathcal{E}(f, f) = (f, Hf) = \int \bar{f}(x)(-\Delta f)(x) dx - \int |f(x)|^2 \lambda(x)\mu(dx) \ ,$$

$\bar{f}$ meaning complex conjugate. If λ and μ are such that $\mathcal{E}$ is a closable quadratic form which is bounded below, and its closure defines a self-adjoint lower semibounded operator H. Namely, if for some $C > 0$ one has

$$\bar{\mathcal{E}}(f, f) + C\|f\|^2 \geq 0 \text{ for all } f \in Q(\bar{\mathcal{E}})$$

where $\bar{\mathcal{E}}$ is the closure of $\mathcal{E}$ and $Q(\bar{\mathcal{E}})$ the domain of $\bar{\mathcal{E}}$, then H is defined by

$$\left((H + C)^{\frac{1}{2}}f, (H + C)^{\frac{1}{2}}f\right) = \bar{\mathcal{E}}(f, f) + C\|f\|^2 \ .$$

One has then

$$H + C \geq 0 \text{ and } D\left((H + C)^{\frac{1}{2}}\right) = Q(\bar{\mathcal{E}})$$

(where $D(\cdot)$ denotes the domain of the operator.). Thus our problem can be formulated as follows:

What are the conditions on λ and μ so that $\mathcal{E}$ is closable and bounded below?

In what follows we are solely interested in the cases when μ is supported by a subset of $\mathbb{R}^n$ of Lebesgue measure zero. If M is such a set then the resulting Hamiltonian is referred to as the <u>Hamiltonian with a potential concentrated on the set M</u>. Note that indeed the perturbation of the free Hamiltonian $H_0 = -\Delta$ takes place only on M since $\bar{\mathcal{E}}(f, f) = \bar{\mathcal{E}}_0(f, f) \equiv \left(H_0^{\frac{1}{2}}f, H_0^{\frac{1}{2}}f\right)$ for all $f \in C_0^2(\mathbb{R}^n \setminus M)$. We shall see that the solution to our problem requires rather restrictive conditions. However one may ask a more general question:

When is it possible to construct an operator H which is the free Hamiltonian H_0 perturbed by a potential type perturbation concentrated on M?

The answer gives much less restrictive conditions. The intuitive similarity with the previous case is that if one insisted on describing the resulting Hamiltonian by a form $\mathcal{E}$ then λ would be an infinitesimal quantity.

We begin with some preparations. Assume μ to be a Radon probability measure. Unless stated differently we shall also choose λ to be a function $\lambda_s(x)$ of s (and x), which is μ-measurable in x, bounded in both variables and $\lambda_s(x) > 0$ for μ-a.e.x.

For any $s > 0$ and $f \in L^2(\mathbb{R}^n; dx)$ put

$$(\psi_s f)(x) = (\lambda_s(x))^{\frac{1}{2}} e^{-\sqrt{-\Delta}s} f(x) = (2\pi)^{-\frac{n}{2}}(\lambda_s(x))^{\frac{1}{2}} \int e^{-ipx} e^{-|p|s} \tilde{f}(p) dp \qquad (2.2)$$

where $\tilde{f}$ is the Fourier transform of f.

The family of operators $e^{-\sqrt{-\Delta}s}$ form the so called <u>Poisson semigroup</u>.

The result of integration in (2.2) is a bounded continuous function which together with our assumptions on λ and μ implies

$$\psi_s f \in L^2(\mathbb{R}^n; \mu) \ .$$

Moreover ψ_s viewed as an operator from $L^2(\mathbb{R}^n, dx)$ to $L^2(\mathbb{R}^n, \mu)$ is bounded with norm less than

$$C_s = \left(\int e^{-2|p|^s} dp \right)^{\frac{1}{2}} ess\sup(\lambda_s^{\frac{1}{2}})$$

Indeed

$$\|\psi_s f\|_{L^2(\mu)} = (2\pi)^{-n} \int \left| \int e^{-ipx} e^{-|p|^s} \tilde{f}(p) dp \right|^2 \lambda(x)\mu(dx)$$
$$\leq (2\pi)^{-n} \left(\int e^{-2|p|^s} dp \right) ess\sup(\lambda_s) \int |f(p)|^2 dp = C_s^2 \|f\|^2 \tag{2.3}$$

Put ψ_s^* for the adjoint of ψ_s. Then $V_s = \psi_s^* \psi_s$ is an $L^2(\mathbb{R}^n; dx)$ operator. In fact (2.3) implies:

Lemma 1. V_s is a bounded operator in $L^2(\mathbb{R}^n, dx)$ and $\|V_s\| = \|\psi_s\|^2 = \|\psi_s^*\|^2 \leq C_s^2$.
$\blacksquare$

Thus

$$H_s = -\Delta - V_s \tag{2.4}$$

is a self-adjoint lower bounded operator in $L^2(\mathbb{R}^n; dx) \equiv L^2(dx)$. Formally $H_s \longrightarrow -\Delta - \lambda\mu$ as $s \longrightarrow 0$ so that if the limit exists in any reasonable sense then it is consistent with our intuitive expectation.

Let E_0^s be the infimum of the spectrum of H_s, and $E < \min(0, E_0^s)$.

Then we have

$$H_s - E = (-\Delta - E)^{\frac{1}{2}} \left[1 - (-\Delta - E)^{-\frac{1}{2}} V_s (-\Delta - E)^{-\frac{1}{2}} \right] (-\Delta - E)^{\frac{1}{2}} . \tag{2.5}$$

Set $G_E \equiv (-\Delta - E)^{-1}$, then the corresponding kernel is

$$G_E(x - y) = (2\pi)^{-\frac{n}{2}} \int \frac{e^{-ip(x-y)}}{p^2 - E} dp .$$

This is a bounded operator from $L^2(dx)$ to $L^2(dx)$ with the norm

$$\|G_E\| = |E|^{-1} . \tag{2.6}$$

We shall also need

$$A_s \equiv G_E^{\frac{1}{2}} V_s G_E^{\frac{1}{2}} \tag{2.7}$$

which is bounded with

$$\|A_s\| \leq |E|^{-1} C_s^2 . \tag{2.8}$$

Define also

$$B_s \equiv \psi_s G_E \psi_s^* . \tag{2.9}$$

This is an operator from $L^2(\mu)$ to $L^2(\mu)$ and

$$\|B_s\| \leq C_s^2 |E|^{-1} .$$

If $E < E_0^s$ then (2.5) and (2.7) yield

$$(H_s - E)^{-1} = G_E^{\frac{1}{2}} [1 - A_s]^{-1} G_E^{\frac{1}{2}} . \tag{2.10}$$

If moreover $-E > C_s^2$ then by (2.8) $\|A_s\| < 1$ and $[1 - A_s]^{-1}$ can be expanded in the norm convergent series

$$[1 - A_s]^{-1} = \sum_{l=0}^{\infty} A_s^l .$$

Now observe that for $l = 1, 2, \ldots$

$$A_s^l = G_E^{\frac{1}{2}} \psi_s^* B_s^{l-1} \psi_s G_E^{\frac{1}{2}} .$$

Thus

$$(H_s - E)^{-1} = G_E + G_E \psi_s^* \sum_{l=0}^{\infty} B_s^l \psi_s G_E = G_E + G_E \psi_s^* [1 - B_s]^{-1} \psi_s G_E . \tag{2.11}$$

According to our assumption we have $\lambda_s(x) > 0$ and we can define

$$\lambda_s^{-\frac{1}{2}} \psi_s G_E = \hat{G}_E^s$$

with the kernel

$$\hat{G}_E^s(x - y) = (2\pi)^{-\frac{n}{2}} \int e^{-ip(x-y)} e^{-|p|s} \frac{dp}{p^2 - E} .$$

We use $\hat{\ }$ to indicate that $\hat{G}_E^s$ is an operator from $L^2(dx)$ to $L^2(\mu) \equiv L^2(\mathbb{R}^n; \mu)$ and $\check{\ }$ for an operator from $L^2(\mu)$ to $L^2(dx)$. Thus

$$\left(\hat{G}_E^s \right)^* = \check{G}_E^s$$

and

$$\check{G}_E^s(x - y) = (2\pi)^{-\frac{n}{2}} \int e^{-ip(x-y)} e^{-|p|s} \frac{dp}{p^2 - E} .$$

The operator $\lambda_s^{-\frac{1}{2}} B_s \lambda_s^{-\frac{1}{2}}$ from $L^2(\mu)$ to $L^2(\mu)$ we denote by $\bar{G}_E^s$. Its kernel is

$$\bar{G}_E^s(x - y) = (2\pi)^{-\frac{n}{2}} \int e^{-ip(x-y)} e^{-2|p|s} \frac{dp}{p^2 - E}$$

With this notation we can rewrite (2.11) in a different manner. Namely

$$G_E \psi_s^* \sum_{l=0}^{\infty} B_s^l \psi_s G_E = \check{G}_E^s \sum_{l=0}^{\infty} \lambda \left(\bar{G}_E^s \lambda \right)^l \hat{G}_E^s = \check{G}_E^s \left(\lambda_s^{-1} - \bar{G}_E^s \right)^{-1} \hat{G}_E^s$$

so that

$$(H_s - E)^{-1} = G_E + \check{G}_E^s \left(\lambda_s^{-1} - \bar{G}_E^s \right)^{-1} \hat{G}_E^s \ . \tag{2.12}$$

Formula (2.12) has been obtained as a limit of power series convergent for $-E > (C_s)^2$. Since $C_s \longrightarrow \infty$ as $s \longrightarrow 0$ it might be impossible to find E so that the series converges for all s. However as we shall see this is not a drawback. Namely, given s, we obtain formula (2.12) for $-E > (C_s)^2$ and extend it by analytic continuation to the whole resolvent set of H_s i.e. possibly to the points where the series is not convergent.

We intend to show that under some assumptions on the measure μ there is E_1 so that formula (2.12) remains valid for all $E < E_1$ independently of s. Moreover the strong limit of (2.12) as $s \longrightarrow 0$ exists and defines the resolvent of a self-adjoint operator.

We shall be able to obtain this result provided the unregularized kernel $G_E(x - y)$ also defines bounded operators $\hat{G}_E$ and $\check{G}_E$ from $L^2(dx)$ to $L^2(\mu)$ and from $L^2(\mu)$ to $L^2(dx)$ respectively.

Let us first quote some properties of the kernel $G_E(x - y)$. This kernel describes the resolvent of the Hamiltonian $-\Delta$. Thus it can be expressed by $T^t = e^{t\Delta}$, $t \geq 0$, the semigroup generated by Δ, according to the formula

$$G_E = \int_0^{\infty} T^t e^{Et} dt \ .$$

Since T^t is given by the Gaussian kernel

$$T^t(x - y) = (4\pi t)^{-\frac{n}{2}} \exp\left(-\frac{|x - y|^2}{4t} \right)$$

we get

$$G_E(x - y) = \int_0^{\infty} (4\pi t)^{-\frac{n}{2}} \exp\left(-\frac{|x - y|^2}{4t} \right) \exp(Et) dt \tag{2.13}$$

and

$$\int G_E(x - z) G_E(z - y) dz = \int_0^{\infty} \frac{1}{4\pi} (4\pi t)^{-\frac{n}{2}+1} \exp\left(-\frac{|x - y|^2}{4t} \right) \exp(Et) dt \ .$$

Put $|x - y| = r$ then, writing now $G_E^{(n)}$ for G_E to stress the dependence on the dimension:

$$\int G_E^{(n)}(x - z) G_E^{(n)}(z - y) dz = \frac{1}{4\pi} G_E^{(n-2)}(r) \ .$$

Formula (2.13) yields (see [AFHKL])

1) If $n = 1$ then $G_E^{(1)}(r)$ is continuous, bounded and converges uniformly to zero as $E \longrightarrow -\infty$.

2) If $n \geq 2$ then $G_E^{(n)}$ is continuous away from zero and the asymptotic expression for $r \longrightarrow \infty$ is

$$G_E^{(n)}(r) = \frac{1}{2}(2\pi)^{-\frac{(n-1)}{2}}(-E)^{\frac{(n-3)}{4}}r^{-\frac{(n-1)}{2}}e^{-\sqrt{-E}r}\{1 + o(r)\} .$$

3) If $n = 2$ then the asymptotic behaviour as $r \longrightarrow 0$ is

$$G_E^{(2)}(r) = -\frac{1}{2\pi}\ln(\sqrt{-E}r) + o(r)$$

4) If $n > 2$ then the asymptotic behaviour as $r \longrightarrow 0$ is

$$G_E^{(n)}(r) = \frac{\Gamma[(n-2)/2]}{4\pi^{\frac{n}{2}}}r^{-n+2}\{1 + o(r)\} .$$

Thus $\int G_E^{(n)}(x - z)G_E^{(n)}(z - y)dz$ is bounded and continuous for $n = 1, 2, 3$. It has a logarithmic divergence at zero when $n = 4$ and behaves near zero as $|x - y|^{-n+4}$ when $n > 4$. Another important fact following from (2.13) is

$$G_E(x - y) > 0 . \tag{2.14}$$

Thus the behaviour of $G_E(x - y)$ for large values of $|x - y|$ is "good". On the other hand the singularities at $x = y$ will have to be compensated by appropriate assumptions on μ. We begin with the case $n \leq 3$.

Proposition 2.   If $n \leq 3$ then for any probability Radon measure in $\mathbb{R}^n$ the kernel $G_E(x - y)$ defines Hilbert-Schmidt operators $\hat{G}_E$ and $\check{G}_E$ from $L^2(\mathbb{R}^n; dx)$ to $L^2(M; \mu)$ and from $L^2(M; \mu)$ to $L^2(\mathbb{R}^n; dx)$ respectively. Moreover $\hat{G}_E^s$ and $\check{G}_E^s$ converge to $\hat{G}_E$ and $\check{G}_E$ respectively in corresponding norms, as $s \downarrow 0$.

Proof.   Since $n \leq 3$ the expression $\int_{\mathbb{R}^n} G_E(x - z)G_E(z - y)dz$ is bounded. Hence

$$\int_M \left(\int_{\mathbb{R}^n} (G_E(x - y))^2\, dy\right)\mu(dx) < \infty \tag{2.15}$$

which implies that both $\hat{G}_E$ and $\check{G}_E$ are Hilbert-Schmidt.
We also have

$$\int_M \int_M \left(\int_{\mathbb{R}^n} G_E(x - z)G_E(z - y)dz\right)^2 \mu(dx)\mu(dy) < \infty \tag{2.16}$$

If we put

$$M(p) = (2\pi)^{-\frac{n}{2}} \int_M e^{ipx} \mu(dx)$$

then (2.16) is equivalent to

$$\int_{\mathbb{R}^n} \int_{\mathbb{R}^n} \frac{M^2(p-q)}{(p^2 - E)^2(q^2 - E)^2} dpdq < \infty \; .$$

Similarly

$$\gamma(s) \equiv \int_M \int_M \left(\int_{\mathbb{R}^n} (G_E - G_E^s)(x-z)(G_E - G_E^s)(z-y)dz \right)^2 \mu(dx)\mu(dy)$$

$$= \int_{\mathbb{R}^n} \int_{\mathbb{R}^n} \frac{(1 - e^{-|p|s})^2}{(p^2 - E)^2} \frac{(1 - e^{-|q|s})^2}{(q^2 - E)^2} M^2(p-q)dpdq \; .$$

Since M is bounded we have

$$\gamma(s) \leq \sup_u M^2(u) \left[\int_{\mathbb{R}^n} \frac{(1 - e^{-|p|s})^2}{(p^2 - E)^2} dp \right]^2 \longrightarrow 0 \text{ as } s \longrightarrow 0 \; .$$

This and the fact that for any $f \in L^2(\mathbb{R}^n; dx)$ we have the bound

$$\left\| \left(\hat{G}_E - \hat{G}_E^s \right) f \right\|_{L^2(\mu)}^2 \leq (\gamma(s))^{\frac{1}{2}} \|f\|_{L^2(dx)}^2$$

proves the desired norm convergence of $\hat{G}_E^s$ to $\hat{G}_E$. Since $\check{G}_E^s = \left(\hat{G}_E^s \right)^*$ and $\check{G}_E = \left(\hat{G}_E \right)^*$ the corresponding convergence $\check{G}_E^s \longrightarrow \check{G}_E$ follows.

$\blacksquare$

If $n \geq 4$ the situation is more delicate. Here we shall profit from results of E.I. Pustyl'nik [Pu] (see also [He], [Mak]) which we present here in a simplified version adequate to our needs.

Let M be a closed subset of $\mathbb{R}^n$ and μ a Radon probability measure supported by M. Set $M_{x,\rho} = M \cap K_{(x,\rho)}$ where $K_{(x,\rho)}$ is the ball of radius ρ centered at x. A number $N \geq 0$ is called <u>admissible</u> for M if there is a constant $C(N)$ such that

$$\mu(M_{x,\rho}) \leq C(N)\rho^N \; , \tag{2.17}$$

for μ-a.e. $x \in M$, and all $\rho > 0$.

The least upper bound of all admissible numbers for M is called <u>dimensionality of M with respect to measure μ</u>. We denote it by m. Note that m may be admissible or not.

According to [Pu] we have

Lemma 3. If M is bounded with diameter d, N is an admissible number for M, and $0 < \lambda < N$, then

$$\int_M |x - y|^{-\lambda} \mu(dy) \leq C d^{N-\lambda} \tag{2.18}$$

for μ-a.a. $x \in M$, where the constant C depends only on the constant $C(N)$ occurring in (2.17) and on λ.

If N is not admissible for M then there are $L \subset M$ and $C_1 > 0$ such that $\mu(L) > 0$ and $\mu(M_{x,\rho}) \geq C_1 \rho^N$ for all $x \in L$.

$\blacksquare$

We also remark that as a consequence of [Kr] (3.1. Theorem 1) the norms of these operators can be majorised by constants depending on μ only through an admissible N and $C(N)$ of (2.18).

Following proposition is a specification of more general results described in Theorem 2 of [Pu].

Proposition 4. Let $\Omega \subset \mathbb{R}^n$ be open and bounded, $M \subset \Omega$ compact. If μ is a probability Radon measure supported by M such that $m > n-4$, then the kernel $|x-y|^{-n+2}$ defines compact operators from $L^2(\Omega; dx)$ to $L^2(M; \mu)$ and from $L^2(M; \mu)$ to $L^2(\Omega; dx)$.

$\blacksquare$

In the situation we are discussing neither Ω nor M are assumed to be bounded. Also the kernel $G_E(x - y)$ is not exactly $|x - y|^{-n+2}$. Hence we have to generalize Proposition 4 to cover our case.

We have $M = \text{supp } \mu$. Let $\{M^i\}$ denote a family of closed n-dimensional unit cubes such that

$$M \subset \bigcup_i M^i \ , \ \mu(M^i) > 0 \ \text{and} \ \text{int } M^i \cap \text{int } M^j = \emptyset \ , \ i \neq j \ .$$

Suppose $m > n - 4$. Then any q satisfying $n - 4 < q < m$ is an admissible number for M and there is $C(q)$ such that

$$\mu(M_{x,\rho}) \leq C(q)\rho^q$$

for a.e. $x \in M$.

We say that the measure μ has property A if there are $q \in (n - 4, m)$ and a family $\{M^i\}$ described above such that

$$\mu_i\left(M^i_{x,\rho}\right) \leq \mu(M^i)C(q)\rho^q \tag{2.19}$$

for a.a. $x \in M^i$, where μ_i is the Radon measure in $\mathbb{R}^n$ supported by $M \cap M^i$ and such that $\mu_i(B) = \mu(B \cap M^i)$ for any Borel B.

Proposition 5.   If the probability Radon measure μ supported by $M \subset \mathbb{R}^n$, $n \geq 4$, is such that m, the dimensionality of M with respect to μ, satisfies

$$m > n - 4$$

and moreover μ has property A, then the kernel $G_E(x - y)$ defines compact operators $\hat{G}_E$ and $\check{G}_E$ from $L^2(\mathbb{R}^n; dx)$ to $L^2(M; \mu)$ and from $L^2(M; \mu)$ to $L^2(\mathbb{R}^n; dx)$ respectively.

Proof.   The proof is given in the Appendix.

$\blacksquare$

Lemma 6.   Under the assumptions of Proposition 5 the expression

$$\int_M \int_M \left(\int_{\mathbb{R}^n} G_E(x - z) G_E(z - y) dz \right) \mu(dx)\mu(dy) \tag{2.20}$$

is finite. If $n \geq 4$, $m \leq n - 4$ and $n - 4$ is not admissible for M, then (2.20) is infinite.

Proof.   Note first that

$$\int_{\mathbb{R}^n} G_E(x - z) G_E(z - y) dz = |x - y|^{-n+4} \varphi(x - y)$$

where φ is a bounded function. Thus Lemma 3 is valid if we substitute the kernel $|x - y|^{-\lambda}$ with $\lambda = n - 4$ by $\int_{\mathbb{R}^n} G_E(x - z) G_E(z - y) dz$. This is a standard generalization of the theory of potential operators [Kr].

We have

$$\int_M \left(\int_{\mathbb{R}^n} G_E(x - z) G_E(z - y) dz \right) \mu(dy)$$
$$\leq \sum_i \int_{M^i} \left(\int_{\mathbb{R}^n} G_E(x - z) G_E(z - y) dz \right) \mu(dy) \ .$$

To estimate this expression we shall show that there is a constant $C_2 > 0$ such that

$$\int_{M^i} \left(\int_{\mathbb{R}^n} G_E(x - z) G_E(z - y) dz \right) \mu(dy) \leq C_2 \mu(M^i)$$

for all $i \in \mathbb{N}$ and $x \in \mathbb{R}^n$. Since

$$\int_{\mathbb{R}^n} G_E(x - z) G_E(z - y) dz = \varphi(x - y)|x - y|^{4-n}$$

where φ is a bounded function we have

$$\int_{M^i} \left(\int_{\mathbb{R}^n} G_E(x - z) G_E(z - y) dz \right) \mu(dy)$$
$$\leq F \int_{M^i} |x - y|^{4-n} \mu(dy)$$

with $F = \sup_u \varphi(u)$.

Let $x \in M^i \cap M$. Then by property A and the fact that diameter of M^i is $\sqrt{n}$ we get

$$\int_{M^i} |x-y|^{4-n} \mu(dy)$$

$$\leq q C(q) \mu(M^i) \int_0^{\sqrt{n}} \rho^{n-4+q-1} d\rho$$

$$= q C(q)(q-n+4)^{-1} n^{\frac{1}{2}(q-n+4)} \mu(M^i) \, .$$

Let now $x \notin M^i \cap M$ and put ρ_0 for distance between x and $M^i \cap M$.
If $\rho_0 > \sqrt{n}$ then $|x-y|^{4-n} \leq n^{\frac{1}{2}(4-n)}$ for all $y \in M^i \cap M$. Thus

$$\int_{M^i} |x-y|^{4-n} \mu(dy) \leq n^{\frac{1}{2}(4-n)} \mu(M^i) \, .$$

If $\rho_0 \leq \sqrt{n}$ then

$$\mu_i(K_{(x,\rho)}) \leq C(q)(2\rho)^q \mu(M^i)$$

for all $\rho > 0$.

Indeed, if $\rho < \rho_0$ then $K_{(x,\rho)} \cap M^i \cap M = \emptyset$ and $\mu_i(K_{(x,\rho)}) = 0$. If $\rho \geq \rho_0$ then $K_{(x,\rho)} \subset K_{(x_0,2\rho)}$ for any $x_0 \in M \cap M^i$. Hence

$$\mu_i(K_{(x,\rho)}) \leq \mu_i(K_{(x_0,2\rho)}) \leq C(q)(2\rho)^q \mu(M^i) \, .$$

Note however that if $\rho > 2\sqrt{n}$ then $\mu_i(K_{(x,\rho)}) = \mu(M^i)$ independently of ρ. Thus if $\rho_0 \leq \sqrt{n}$ we have

$$\int_{M^i} |x-y|^{4-n} \mu(dy) \leq 2^q q C(q) \mu(M^i) \int_0^{2\sqrt{n}} \rho^{q-n+3} d\rho$$

$$= 2^q (q-n+4)^{-1} (4n)^{\frac{1}{2}(q-n+4)} C(q) \mu(M^i) \, .$$

As a consequence of the above estimations we conclude that there is a positive constant C_2 such that

$$F \int_{M^i} |x-y|^{4-n} \mu(dy) \leq C_2 \mu(M^i) \, .$$

Since the number of cubes overlapping a cube M^i is finite and does not depend on i the sum $\sum_i (\mu(M^i)$ is finite which shows that

$$\int_M \left(\int_{\mathbb{R}^n} G_E(x-z) G_E(z-y) dz \right) \mu(dy)$$

is bounded and integrated over x with the probability measure μ yields a finite result.

If $n-4$ is not admissible for M there are $L \subset M$, with $\mu(L) > 0$ and a constant B as well as a $\rho > 0$ such that

$$\mu(M_{x,\rho}) \geq B\rho^{n-4}$$

for all $x \in L$. Thus

$$\int_M \left(\int_{\mathbb{R}^n} G_E(x-z) G_E(z-y) dz \right) \mu(dy) \geq a \int_{M_{x,\rho}} |x-y|^{-n+4} \mu(dy)$$
$$\geq bB \int_0^\rho |y|^{-n+4} |y|^{n-4-1} dy = \infty$$

for all $x \in L$. Since $\mu(L) > 0$ the integral with respect to μ is infinite.

■

Similarly one can prove

Lemma 7.   If M has dimensionality (with respect to μ) $m > n-2$ and μ has property A for q satisfying $n-2 < q < m$ then

$$\int_M \int_M G_E(x-y) \mu(dx) \mu(dy) < \infty \ .$$

If $m \leq n-2$ and $n-2$ is not admissible for M then

$$\int_M \int_M G_E(x-y) \mu(dx) \mu(dy)$$

is divergent.

■

Remark 1.   The condition in Lemma 6 that (2.20) is finite (which follows under the assumption $m > n-4$ in Prop. 5) should be compared with the corresponding condition discussed in [AFHKL] and [Br2]. The condition in Lemma 7 that

$$\int_M \int_M G_E(x-y) \mu(dx) \mu(dy) < \infty$$

also arises in the non standard analysis approach, see [AFHKL]. One should also compare with [Br2].

Now we are prepared to discuss convergence of $\hat{G}_E^s$ and $\check{G}_E^s$ as $s \longrightarrow 0$.

Proposition 8.   If $n \geq 4$, $m > n-4$ and the measure μ has property A then the operators $\hat{G}_E^s$ and $\check{G}_E^s$ converge to $\hat{G}_E$ and $\check{G}_E$ respectively, in corresponding operator norms.

Proof. We shall give this proof following arguments used in the proof of Theorem 5.12 of [Kr].

Suppose the statement is false. Then there are sequences $s_i \longrightarrow 0$ and $f_i \in L^2(\mathbb{R}^n; dx)$ such that $\|f_i\|_{L^2(dx)} = 1$ and

$$\left\| \left(\hat{G}_E - \hat{G}_E^{s_i} \right) f_i \right\|_{L^2(\mu)} \geq \delta \tag{2.21}$$

for some $\delta > 0$.

We have

$$\left\| \left(\hat{G}_E - \hat{G}_E^{s_i} \right) f \right\|_{L^1(\mu)}$$
$$= \int_M \left[\int_{\mathbb{R}^n} (G_E - G_E^{s_i})(x - y) f(y) dy \right] \mu(dx)$$
$$\leq \left[\int_M \int_M \left(\int_{\mathbb{R}^n} (G_E - G_E^{s_i})(x - z)(G_E - G_E^{s_i})(z - y) dz \right) \mu(dx)\mu(dy) \right]^{\frac{1}{2}} \|f\|_{L^2(dx)}$$

for all $f \in L^2(\mathbb{R}^n; dx)$.

By Lemma 6 the expression in the bracket is finite. Taking Fourier transforms we observe that it is proportional to

$$\int_{\mathbb{R}^n} \frac{(1 - e^{-|p|s})^2}{(p^2 - E)^2} |M(p)|^2 dp , \tag{2.22}$$

where

$$M(p) = (2\pi)^{-\frac{n}{2}} \int_M e^{ipx} \mu(dx) .$$

By an $\frac{1}{2}\varepsilon$ argument (2.22) goes to zero as $s \longrightarrow 0$. This shows that $\hat{G}_E^s$ converges to $\hat{G}_E$ in norm as operators from $L^2(\mathbb{R}^n; dx)$ to $L^1(M; \mu)$.

As a consequence

$$\left\| \left(\hat{G}_E - \hat{G}_E^{s_i} \right) f_i \right\|_{L^1(M;\mu)} \longrightarrow 0$$

and hence $\left(\hat{G}_E - \hat{G}_E^{s_i} \right) f_i \longrightarrow 0$ in measure. Since the operation $e^{-\sqrt{-\Delta}s}$ is a contraction we have $G_E(x) > G_E^s(x)$ for all $x \neq 0$ and $s > 0$.

Thus

$$\left| \left[\left(\hat{G}_E - \hat{G}_E^{s_i} \right) f_i \right](x) \right|$$
$$= \left| \int_{\mathbb{R}^n} (G_E - G_E^{s_i})(x - y) f_i(y) dy \right|$$
$$\leq \int_{\mathbb{R}^n} G_E(x - y) |f_i(y)| \, dy$$
$$= \left(\hat{G}_E |f_i| \right)(x) .$$

Since $\hat{G}_E$ is compact the set $\hat{G}_E|f_i|$ is compact and hence by Lemma 1.1 of [Kr] has equi-absolutely continuous $L^2(\mu)$ norms. The above inequality implies that the set $\left(\hat{G}_E - \hat{G}_E^{s_i}\right) f_i$ has equi-absolutely continuous $L^2(\mu)$ norms. This together with convergence to zero in measure implies convergence to zero in $L^2(\mu)$, which contradicts (2.21).

Thus $\hat{G}_E^s$ converges to $\hat{G}_E$ in norm, since $\check{G}_E = \hat{G}_E^*$ and $\check{G}_E^s = \hat{G}_E^{s*}$ we have

$$\left\|\hat{G}_E - \hat{G}_E^s\right\|_{L^2(dx)\longrightarrow L^2(\mu)} = \left\|\check{G}_E - \check{G}_E^s\right\|_{L^2(\mu)\longrightarrow L^2(dx)}$$

so that $\check{G}_E^s$ converges to $\check{G}_E$ in norm.

$\blacksquare$

Let now α be a bounded μ-measurable function such that $\alpha(x) > \delta > 0$. Let

$$\lambda_s^{-1}(x) = \alpha(x) + \int_M G_{E'}^s(x-y)\mu(dy) \ . \tag{2.23}$$

for some $E' < 0$ (independent of s, x). Then

$$(\mathcal{G}_s f)(x) \equiv \left(\alpha(x) + \int_M G_{E'}^s(x-y)\mu(dy)\right) f(x) - \int_M G_E^s(x-y)f(y)\mu(dy) \tag{2.24}$$

defines a bounded linear operator from $L^2(M;\mu)$ to $L^2(M;\mu)$. If $E < E'$ then $\mathcal{G}_s$ is a positive operator.

Indeed then

$$\begin{aligned}(f, \mathcal{G}_s f)_{L^2(\mu)} = &\int_M \left(\alpha(x) + \int_M (G_{E'}^s - G_E^s)(x-y)\mu(dy)\right) |f(x)|^2 \mu(dx) \\ &+ \frac{1}{2}\int_M \int_M G_E^s(x-y)(f(x) - f(y))^2 \mu(dx)\mu(dy) \\ &\geq \delta\|f\|_{L^2(\mu)}^2 \ .\end{aligned} \tag{2.25}$$

Put

$$\Gamma_s(f) \equiv (f, \mathcal{G}_s f)_{L^2(\mu)} \ , \ f \in L^2(M;\mu) \ .$$

For any $s > 0$ this is a bounded, positive, closed quadratic form. For any $f \in L^2(M;\mu)$ we have

$$\Gamma_s(f) \leq \Gamma_{s'}(f) \ \text{if} \ s' < s \ . \tag{2.26}$$

If f is the restriction to M of some $h \in C_0^1(\mathbb{R}^n)$ then $\Gamma(f) \equiv \Gamma_0(f)$ is also well defined as a positive quadratic form. Indeed $(G_{E'} - G_E)(x-y) = (-E+E')\int_{\mathbb{R}^n} G_{E'}(x-z)G_E(z-y)dz$ so that $\int_M(G_{E'} - G_E)(x-y)\mu(dy)$ is bounded and

$$\int_M \int_M (G_{E'} - G_E)(x-y)\mu(dy)|f(x)|^2 \mu(dx) < \infty \ .$$

The function $\frac{(f(x)-f(y))^2}{|x-y|^2}$ is bounded. Hence

$$\int_M \int_M G_E(x-y)\,(f(x)-f(y))^2\,\mu(dx)\mu(dy)$$

$$= \int_M \int_M (x-y)^2 G_E(x-y)\frac{(f(x)-f(y))^2}{|x-y|^2}\mu(dx)\mu(dy) < \infty .$$

In fact we know much more about the form Γ, namely we have

Proposition 9. The quadratic form Γ defined on $C_0^1(I\!\!R^n)$ by

$$\Gamma(f) = \int_M (\alpha(x)+(-E+E'))\int_M \left(\left(\int_{I\!\!R^n} G_{E'}(x-z)G_E(z-y)dz\right)\mu(dy)\right)|f(x)|^2\mu(dx)$$

$$+\frac{1}{2}\int_M \int_M G_E(x-y)\,(f(x)-f(y))^2\,\mu(dx)\mu(dy)$$

is closable in $L^2(M;\mu)$. If $\bar\Gamma$ denotes the closure of Γ and $Q(\bar\Gamma)$ the domain of $\bar\Gamma$ then for every $f \in Q(\bar\Gamma)$ and every $s > 0$ we have $\bar\Gamma(f) \geq \Gamma_s(f)$ and moreover $\bar\Gamma(f) = \lim_{s \longrightarrow \infty} \Gamma_s(f)$.

Proof. The proof is given in the Appendix.

∎

The facts established in Proposition 9 together with (2.26) permit us to verify all assumptions of a theorem by Kato [Ka] and Simon [Si] (Th. 3.1), see also [ABrR]. According to that theorem the positive self adjoint operators $\mathcal{G}_s$ defined by Γ_s converge in the strong resolvent sense as $s \longrightarrow 0$ to the positive self adjoint operator $\mathcal{G}$ defined by $\bar\Gamma$.

Since all forms in question are bounded away from zero by $\delta > 0$ the resolvent sets of all $\mathcal{G}_s$ and $\mathcal{G}$ contain zero. Thus $\mathcal{G}_s^{-1}$ converges to $\mathcal{G}^{-1}$ strongly in $L^2(M;\mu)$.

With this we can prove

Theorem 10. Let μ be a Radon probability measure supported by a closed subset M of $I\!\!R^n$, $n \in I\!\!N$. If $n \geq 4$ we assume in addition that μ is such that dimensionality of M with respect to μ is strictly greater than $n-4$ and moreover μ has property A, then the family of positive self adjoint operators $H_s = -\Delta+V_s$, $V_s = -\psi_s^*\psi_s$, with ψ_s given by (2.2) and λ_s by (2.23) converges in strong resolvent sense as $s \longrightarrow 0$, to a positive self-adjoint operator H with resolvent given by the formula

$$(H-E)^{-1} = (-\Delta-E)^{-1} + \check{G}_E\mathcal{G}_E^{-1}\hat{G}_E \tag{2.27}$$

where $\hat{G}_E$, $\check{G}_E$ are compact operators from $L^2(I\!\!R^n;dx)$ to $L^2(M;\mu)$ and from $L^2(M;\mu)$ to $L^2(I\!\!R^n;dx)$ respectively given by the same kernel

$$G_E(x-y) = (2\pi)^{-\frac{n}{2}}\int_{I\!\!R^n}\frac{e^{-ip(x-y)}}{p^2-E}dp .$$

$\mathcal{G}_E$ is the positive self adjoint operator in $L^2(M; \mu)$ defined by the quadratic form $\bar{\Gamma}$ which is the closure of the form Γ defined on $C_0^1(\mathbb{R}^n)$ by

$$\Gamma(f) = \int_M \left(\alpha(x) + (-E + E') \int_M \left(\int_{\mathbb{R}^n} G_{E'}(x - z) G_E(z - y) dz \right) \mu(dy) \right) |f(x)|^2 \mu(dx)$$
$$+ \frac{1}{2} \int_M \int_M G_E(x - y) (f(x) - f(y))^2 \, \mu(dx)\mu(dy) \ .$$

$$(2.28)$$

The proof of this theorem is based on the following well known result (see e.g.[Ka])

Theorem 11.   (Kato, Trotter). Let A_i be a sequence of self adjoint bounded below operators with the common lower bound E_0, and R_E^i their resolvents. If there is a point $E < E_0$ such that for every $f \in H$ the sequence $R_E^i f$ converges strongly to $T_E f$ and T_E has a dense range, then there is a self adjoint operator A, the strong resolvent limit of A_i, and T_E is its resolvent.

∎

Proof of Theorem 10.   We have

$$H_s = -\Delta + V_s$$

where V_s is a positive operator. Hence 0 is a common lower bound for all H_s. Let $E < E'$, then for any $f \in L^2(\mathbb{R}^n, dx)$

$$\left\| \left((-\Delta - E)^{-1} + \check{G}_E \mathcal{G}_E^{-1} \hat{G}_E - (H_s - E)^{-1} \right) f \right\|_{L^2(dx)}$$
$$= \left\| \left(\check{G}_E \mathcal{G}_E^{-1} \hat{G}_E - \check{G}_E^s (\mathcal{G}_E^s)^{-1} \hat{G}_E^s \right) f \right\|_{L^2(dx)} \leq \left\| \left(\check{G}_E \mathcal{G}_E^{-1} \hat{G}_E - \check{G}_E^s \mathcal{G}_E^{-1} \hat{G}_E \right) f \right\|_{L^2(dx)}$$
$$+ \left\| \left(\check{G}_E^s \mathcal{G}_E^{-1} \hat{G}_E - \check{G}_E^s (\mathcal{G}_E^s)^{-1} \hat{G}_E \right) f \right\|_{L^2(dx)} + \left\| \left(\check{G}_E^s (\mathcal{G}_E^s)^{-1} \hat{G}_E - \check{G}_E^s (\mathcal{G}_E^s)^{-1} \hat{G}_E^s \right) f \right\|_{L^2(dx)}$$

We have

$$\left\| \left(\check{G}_E \mathcal{G}_E^{-1} \hat{G}_E - \check{G}_E^s \mathcal{G}_E^{-1} \hat{G}_E \right) f \right\|_{L^2(dx)} \leq \left\| \check{G}_E - \check{G}_E^s \right\|_{L^2(\mu) \longrightarrow L^2(dx)} \left\| \mathcal{G}_E^{-1} \hat{G} f \right\|_{L^2(\mu)} \longrightarrow 0$$

as $s \longrightarrow 0$ by Proposition 8, for $n \geq 4$, and Proposition 2 for $n < 4$.

$$\left\| \left(\check{G}_E^s \mathcal{G}_E^{-1} \hat{G}_E - \check{G}_E^s (\mathcal{G}_E^s)^{-1} \hat{G}_E \right) f \right\|_{L^2(dx)}$$
$$\leq \left\| G_E^s \right\|_{L^2(\mu) \longrightarrow L^2(dx)} \left\| \left(\mathcal{G}_E^{-1} - (\mathcal{G}_E^s)^{-1} \right) \hat{G}_E f \right\|_{L^2(\mu)} \longrightarrow 0 \text{ as } s \longrightarrow 0$$

because the first term is bounded as a function of s and the second goes to zero by the strong convergence of $(\mathcal{G}_E^s)^{-1}$ to $\mathcal{G}_E^{-1}$.

$$\left\| \left(\check{G}_E^s (\mathcal{G}_E^s)^{-1} \hat{G}_E - \check{G}_E^s (\mathcal{G}_E^s)^{-1} \hat{G}_E^s \right) f \right\|_{L^2(dx)}$$
$$\leq \delta^{-2} \left\| \check{G}_E^s \right\|_{L^2(\mu) \longrightarrow L^2(dx)} \left\| \hat{G}_E - \hat{G}_E^s \right\|_{L^2(dx) \longrightarrow L^2(\mu)} \left\| f \right\|_{L^2(dx)} \longrightarrow 0 \text{ as } s \longrightarrow 0 \ .$$

This follows from Proposition 8 and the fact that by (2.25)

$$\left\|\left(\mathcal{G}_E^s\right)^{-1}\right\|_{L^2(\mu)\longrightarrow L^2(\mu)} \leq \delta^{-2} \text{ for all } s > 0 \ .$$

It is also obvious that $(-\Delta - E)^{-1} + \check{G}_E \mathcal{G}_E^{-1}\hat{G}_E$ has dense range. Indeed for any $0 \neq f \in L^2(\mathbb{R}^n; dx)$ we have

$$\left(f,(-\Delta - E)^{-1}f\right) + \left(f,\check{G}_E \mathcal{G}_E^{-1}\hat{G}_E f\right) \geq \left(f,(-\Delta - E)^{-1}f\right) > 0 \ .$$

Thus all conditions of Theorem 11 (Kato-Trotter Theorem) are satisfied and Theorem 10 is proven.

∎

In [AFHKL] (see also [AFHK], [ABlHK], [L1]) nonstandard methods have been used to handle the perturbation problem. The conditions in [AFHKL] (Theor. 6.2.6 and Theor. 6.2.10, 6.2.11, Coroll. 6.2.12) (as well as those in [Br1–3]) involve essentially the condition in Lemma 6 that (2.20) is finite. The results obtained by nonstandard analytic methods give the existence of a non trivial part, $\check{G}_E \left(\frac{1}{\lambda} - \bar{G}_E\right)^{-1}\hat{G}_E$, with λ infinitesimal, in the resolvent expression for $E < E'$ (λ is a realization of the standard sequence λ_s for $s \downarrow 0$, with λ_s so that (2.23) holds and $\alpha(x) \geq \delta > 0$).

It is to be remarked that the nonstandard analysis methods yield a stronger existence result for $H \neq -\Delta$ (only the condition that (2.20) is finite is used to obtain Coroll. 6.2.12 a)); on the other hand it should be remarked that there is no claim on how the resolvent of H looks like in standard terms, whereas by the standard methods we discussed we automatically obtain also a (standard) expression for the resolvent of H. It would of course be very interesting to see whether the result obtained by standard analysis can be refined to cover also those obtained by nonstandard analysis. We also remark that the condition in Coroll. 6.2.12 b in [AFHKL] corresponds to Condition (3.1) below. In this case the standard and non standard methods yield the same results. The precise relation of our results to those of [Br3] should also be further investigated, the conditions of [Br3] coincide in fact with those of the nonstandard analytic approach.

3. Discussion of the results and examples

Let us reflect on the result we have obtained.

Formula (2.27) gives the resolvent of a self-adjoint operator H. Since $\check{G}_E \mathcal{G}_E^{-1}\hat{G} \neq 0$ we have $H \neq -\Delta$. Moreover, if f is in the domain of $-\Delta$ and $\ \text{supp } f \cap M = \emptyset$ then

$$(H - E)^{-1}(-\Delta - E)f = f + \check{G}_E \mathcal{G}_E^{-1}f = f$$

because $\mathcal{G}_E^{-1}$ acts in $L^2(M;\mu)$ and $f_{|M} = 0$. Thus the action of H and $-\Delta$ is the same away from M and H is obtained from $-\Delta$ by a perturbation supported by M.

As we already mentioned, if we wished to maintain analogy with the intuitive expression (2.1) we might have to think of λ as an infinitesimal quantity. Now we can explain this better.

Suppose the dimensionality m of M with respect to μ is greater than $n - 4$, but $n - 2$ is not admissible (in particular thus condition A is not satisfied). Then, by Lemma 7,

$$\int_M \int_M G_E(x - y)\mu(dx)\mu(dy)$$

is infinite and the form

$$\Gamma(f) = \int_M \lambda^{-1}(x)|f(x)|^2 \mu(dx) + \int \int G_E(x - y)f(x)f(y)\mu(dx)\mu(dy)$$

is not densely defined.

However if we assume (2.23) then $\lambda_s(x)$ goes to zero as $s \longrightarrow 0$ in the prescribed way on a set whose μ-measure is strictly positive. As we already remarked the convergence of λ_s to zero does not prevent the resulting Hamiltonian to differ from $-\Delta$. This case corresponds to the non standard "infinitesimal coupling", see [AFHKL].

Let us now discuss three cases with less singular measures. The construction of H is of course given by the general Theorem 10, but we shall analyze which choices of λ_s have to be made and how the quantities Γ and $\mathcal{G}$ appearing in the resolvent look like in these cases. Suppose that

$$\int_M \int_M G_E(x - y)\mu(dx)\mu(dy) < \infty . \tag{3.1}$$

Then we have following possibilities:

1). The measure μ is such that

$$\int G_E(x - y)\mu(dy)$$

is bounded and goes uniformly to zero as $E \longrightarrow -\infty$. Such measures have been studied in [BlM], where they were said to belong to the generalized Kato class of measures. In this case, for any λ_s which is bounded and bounded away from zero, there are $\delta > 0$ and E_1 such that

$$\lambda_s^{-1} - \int_M G_E(x - y)\mu(dy) > \delta \tag{3.2}$$

for all $E < E_1$. We can take λ_s independent of s i.e. substitute (2.24) by

$$(\mathcal{G}_s f)(x) = \left(\alpha(x) + \int_M G_{E_1}(x - y)\mu(dy)\right) f(x) - \int_M G_E^s(x - y)f(y)\mu(dy) \tag{3.3}$$

and Proposition 9 is still valid. The proof requires only a minor adjustment. We then also have the validity of (2.27), which determines the Hamiltonian H, in this case. Here we are allowed to take λ arbitrary large.

2). Let us assume μ is such that

$$g \equiv \inf_E \sup_x \int_M G_E(x - y)\mu(dy) < \infty \tag{3.4}$$

We observe that this is, in general, a weaker assumption than the one made above at the beginning of 1). The measures fulfilling this assumption are sometimes called potentially bounded [FeP].

In this case one can find E_1, δ and λ such that (3.2) holds for $E < E_1$. Then (3.3) can be used but we must require $\lambda(x) < \delta^{-1}$. In particular only small coupling constants are allowed.

3). Let us now assume (3.1) holds but $\int_M G_E(x - y)\mu(dy)$ is not bounded. If $\alpha(x)$ is a bounded μ measurable function such that $\alpha(x) > \delta$ for some $\delta > 0$ then we can choose $E_1 < 0$ and define

$$\lambda^{-1}(x) = \alpha(x) + \int_M G_{E_1}(x - y)\mu(dy)$$

Again (3.3) is applicable with $E < E_1$ but we cannot have λ bounded away from zero, in particular we cannot have that λ be a constant.

We may also ask when the perturbation supported by M can be realized as a form perturbation i.e. ask when the quadratic form defined on $C_0^2(\mathbb{R}^n) \subset L^2(\mathbb{R}^n; dx)$ by

$$h(f) \equiv (f, -\Delta f) - \int_M f^2(x)\lambda(x)\mu(dx) + (f, -Ef) \tag{3.5}$$

is closable. This happens of course under stronger assumptions, in fact the answer is given by

Proposition 12. Let $\lambda(x)$ be a positive μ-measurable function and $m > n - 2$, then there is $E < 0$ and $C > 0$ such that if $\lambda(x) \leq C$ then h is a positive closable form. The closure of h defines then a positive self adjoint operator H.

Remark. In this case it is natural to write $H = -\Delta - \lambda\mu$ and look upon H as the realization of the perturbation of $-\Delta$ by $-\lambda\mu$.

Proof. Let $E < 0$ and $G_E^{\frac{1}{2}}(x - y)$ be the kernel of $(-\Delta - E)^{-\frac{1}{2}}$. If $n = 1$ then $G_E^{\frac{1}{2}}$ is logarithmic divergent at zero. For $n \geq 2$ the singularity of $G_E^{\frac{1}{2}}$ at zero is $|x - y|^{-n+1}$. By the quoted theorem of Pustylnik if $m > n - 2$ then $G_E^{\frac{1}{2}}$ is a compact operator from $L^2(\mathbb{R}^n; dx)$ to $L^2(M; \mu)$.

Thus if $f \in C_0^2(\mathbb{R}^n)$ then

$$\|f\|_{L^2(\mu)}^2 = \left\|G_E^{\frac{1}{2}}(-\Delta - E)^{\frac{1}{2}}f\right\|_{L^2(\mu)}^2 \leq \left\|G_E^{\frac{1}{2}}\right\|_{L^2(dx) \longrightarrow L^2(\mu)}^2 \left\|(-\Delta - E)^{\frac{1}{2}}f\right\|_{L^2(dx)}^2$$

$$= \left\|G_E^{\frac{1}{2}}\right\|_{L^2(dx) \longrightarrow L^2(\mu)}^2 ((f, -\Delta f) + (f, -Ef)) \ .$$

If $\lambda(x) < \left\|G_E^{\frac{1}{2}}\right\|^2_{L^2(dx)\longrightarrow L^2(\mu)}$ then $(f, (-\Delta - E)f) \geq \int_M f^2(x)\lambda(x)\mu(dx)$ i.e. $h(f) \geq 0$, $h(f) = 0$ only for $f = 0$.

Moreover if $f_i \in C_0^2(\mathbb{R}^n)$ is such that $\|f_i\|_{L^2(dx)} \longrightarrow 0$ and $h(f_i - f_j) \longrightarrow 0$ then $(f_i, (-\Delta - E)f_i) \longrightarrow 0$, which implies, $-\Delta - E$ being closable, that $h(f_i) \longrightarrow 0$, which means closability of h.

$\blacksquare$

Examples.

A. Point interactions. Suppose μ is supported by a point set in $\mathbb{R}^n$, say $\mu(\{0\}) = 1$. Since $\dim\{0\} = 0$, the assumptions of Theorem 10 are only satisfied for $n = 1, 2, 3$. It is only for $n = 1$ that the form given by (3.5) is closable. In this case $n = 1$ one has $\left\|G_E^{\frac{1}{2}}\right\|_{L^2(dx)\longrightarrow L^2(\mu)} = \frac{1}{\sqrt{-E}}$ and we take $\lambda(x) < -E$. Thus for any bounded λ we can find $E < 0$ so that the form (3.5) is positive and closable. The closure of h defines a positive self-adjoint operator, which can be written as $-\Delta_\delta - E$, where $-\Delta_\delta$ is the free Hamiltonian perturbed by a point interaction at 0. In this case $\int G_E(x - y)\mu(dy) = G_E(x)$ falls into the above case 1), in particular "large coupling constants" are admitted (in this case one has a particular case of an extensive theory of perturbations by Radon measures, see e.g. [ABrR], [BHH], [BlZ], [Br2], [Ko], [Sm], [So1–2], [StV], [Stu], [Voi]).

If $n = 2, 3$ then 3.1 is not satisfied but Theorem 10 is applicable and (2.27) defines the resolvent of the perturbed operator with infinitesimal coupling constant. If $n > 3$ then $0 = m \leq n - 4$ and our procedure does not work, but it is known that in this case the Laplacian defined on $C_0^2(\mathbb{R}^n \backslash \{0\})$ is essentially self-adjoint so there is no other self-adjoint extension but the free Hamiltonian $-\Delta$. For further results in this case see [AGHKH] (the above procedure was used first in [BeF], [AFHK], [GrHKM], another method is the one of Dirichlet forms [AHKS], which also allows singular perturbations of point interactions [AM1–3], [ABlM]; for point interactions in L^p see [CaC]).

B. Interactions supported by manifolds. Let M be a piecewise smooth manifold in $\mathbb{R}^n$ of codimension d. Suppose the probability measure μ is supported by M and is absolutely continuous with respect to the volume measure on M, then the dimensionality of M is $n - d$. If M is unbounded we assume that the Radon-Nikodym derivative goes to zero at infinity. Then m – the dimension of M with respect to μ-equals to the Hausdorff dimension of M. Moreover condition A of Theorem 10 is satisfied. Thus if $m \geq n - 3$ there is a perturbation supported by M.

If $m = n - 1(d = 1)$ we can have large coupling constant. In the other cases $d = 2, 3$ λ has to be infinitesimal. For further discussions in this case see [AGS], [BrT], [DSh], [Koc], [Kur].

C. Interactions supported by Brownian paths.   Let $n > 1$ and M be a Brownian path (i.e. a typical path of a Brownian motion $b(t)$, $t \in [0,1]$ in $I\!\!R^n$, started at the origin) intersected with the closure of an open bounded set Λ. Then we can define a probability measure supported by M by

$$\mu_\omega(B) \equiv L\{t \in [0,1]; b(\omega,t) \in B\} \, , \ B \subset \Lambda \cap \{b(\omega,s) \, , \ s \in [0,t]\} \, ,$$

where $L(\cdot)$ means Lebesgue measure. Then m equals the Hausdorf dimension of M which is 2. Hence for $n = 2,3$ the form (3.5) is closable (by Proposition 12) and we can choose an arbitrary coupling constant (in particular the coupling constant can be taken to be finite).

If $n = 4,5$ then we have $m > n-4$ and the perturbation with infinitesimal coupling constant is defined by Theorem 10. These results correspond to Lemma 6.4.1. resp. Lemma 6.4.3 (see also Theorem 6.4.5) proven by nonstandard analysis in [AHKFL]. See also [Ch] for further discussions of this case. The same method works also in the case of potentials supported by other fractal subsets of $I\!\!R^n$, with Hausdorff dimension greater than $n - 4$. This is of interest in quantum mechanical problems involving fractals sets, see e.g. [Li2], [FuS], [Me].

Appendix.

Proof of Proposition 5.   Let $\{M^j\}$ be the family of cubes described in connection with the formulation of property A for the measure μ, and assume that (2.19) holds. Denote by Ω_j the sphere of radius $2\sqrt{n}$ and center identical with that of M^j. For any $f \in L^2(I\!\!R^n; dx)$ we have the estimate

$$\left\| \hat{G}_E f \right\|_{L^2\mu}^2 = \int_M \left(\int_{I\!\!R^n} G_E(x-y)f(y)dy \right)^2 \mu(dx)$$

$$\leq \sum_{j=1}^\infty \int_{M^j} \left(\int_{I\!\!R^n} G_E(x-y)f(y)dy \right)^2 \mu(dx)$$

$$\leq 2 \sum_{j=1}^\infty \left[\int_{M^j} \left(\int_{I\!\!R^n \setminus \Omega_j} G_E(x-y)f(y)dy \right)^2 \mu(dx) \right.$$

$$\left. + \int_{M^j} \left(\int_{\Omega_j} G_E(x-y)f(y)dy \right)^2 \mu(dx) \right] .$$

With our choice of M^j and Ω_j the function

$$\int_{I\!\!R^n \setminus \Omega_j} (G_E(x-y))^2 \, dy$$

is bounded for $x \in M^j$ with bound B independent of j.

Thus

$$\int_{M^j} \left(\int_{\mathbb{R}^n \setminus \Omega_j} G_E(x-y)f(y)dy \right)^2 \mu(dx) \leq \left(B \int_{\mathbb{R}^n \setminus \Omega_j} f^2(y)dy \right) \mu(M^j)$$

$$\leq B\mu(M^j)\|f\|^2_{L^2(dx)}$$

By Lemma 3 and Proposition 4 the kernel $G_E(x-y)$ defines a continuous operator from $L^2(\Omega_j; dx)$ to $L^2(M^j; \mu)$ i.e.

$$\|\hat{G}f\|^2_{L^2(\mu)} \leq D_j \|f\|^2_{L^2(\Omega_j; dx)}$$

and moreover the constant D_j depends only on the constant $C_j(q)$ for which

$$\mu(M^j{}_{x,\rho}) \leq C_j(q)\rho^q$$

holds for almost all $x \in M^j$.

By property A we have $C_j(q) = \mu(M^j)C(q)$.

Hence if we substitute the measure μ on M^j by the measure $(\mu(M^j))^{-1}\mu$ then we get

$$\left(\mu(M^j)\right)^{-1}\mu(M^j{}_{x,\rho}) \leq C(q)\rho^q \; ,$$

which according to Proposition 4 and the remark following it implies

$$\|\hat{G}f\|^2_{L^2((\mu(M^j))^{-1}\mu)} \leq D\|f\|^2_{L^2(\Omega_j; dx)} \; ,$$

where D is independent of j.

In terms of the corresponding integrals this inequality reads

$$\int_{M^j} \left(\int_{\Omega_j} G_E(x-y)f(y)dy \right)^2 (\mu(M^j))^{-1}\mu(dx) \leq D \int_{\Omega_j} |f(x)|^2 dx$$

which implies

$$\int_{M^j} \left(\int_{\Omega_j} G_E(x-y)f(y)dy \right)^2 \mu(dx) \leq \mu(M^j)D\|f\|^2_{L^2(\mathbb{R}^n; dx)} \; .$$

Let $\{f_i\} \subset L^2(\mathbb{R}^n; dx)$ be a sequence which converges weakly to zero, then the numerical sequence $\|f_i\|^2_{L^2(\mathbb{R}^n; dx)}$ is bounded.

Thus for any $r = 1, 2, 3, \ldots$ we have

$$\left\| \hat{G}_E f_i \right\|_{L^2 \mu}^2 \leq 2 \sum_{j=1}^{r} \left[\int_{M^j} \left(\int_{\mathbb{R}^n \backslash \Omega_j} G_E(x-y) f_i(y) dy \right)^2 \mu(dx) \right.$$

$$\left. + \int_{M^j} \left(\int_{\Omega_j} G_E(x-y) f_i(y) dy \right)^2 \mu(dx) \right]$$

$$+ K \sum_{j=r+1}^{\infty} \mu(M^j) \, , \text{ for some constant } K \, .$$

As we already remarked in the proof of Lemma 6 the latter sum is convergent.
Thus, given $\varepsilon > 0$ there is an $r \in \mathbb{N}$ such that

$$K \sum_{j=r+1}^{\infty} \mu(M^j) < \frac{1}{2} \varepsilon \, .$$

Since the distance between M^j and $\mathbb{R}^n \backslash \Omega_j$ is finite and $G_E(x-y)$ decreases exponentially as $|x - y| \to \infty$, the function

$$\int_{M^j} G_E(z-y) G_E(z-x) \mu(dz)$$

is bounded in $(\mathbb{R}^n \backslash \Omega_j) \times (\mathbb{R}^n \backslash \Omega_j)$ and decreases rapidly as $|x|$ and (or) $|y|$ goes to infinity. Hence it belongs to $L^2((\mathbb{R}^n \backslash \Omega_j) \times (\mathbb{R}^n \backslash \Omega_j); dx \, dy)$.
Let $h_i = f_{i|(\mathbb{R}^n \backslash \Omega_j)}$ i.e. h_i is the restriction of f_i to $\mathbb{R}^n \backslash \Omega_j$. Then $h_i \times h_i$ converges weakly to zero in

$$L^2 \left((\mathbb{R}^n \backslash \Omega_j) \times (\mathbb{R}^n \backslash \Omega_j) ; dx \, dy \right) \, .$$

Thus

$$\sum_{j=1}^{r} \int_{M^j} \left(\int_{\mathbb{R}^n \backslash \Omega_j} G_E(x-y) f_i(y) dy \right)^2 \mu(dx) \text{ goes to zero as } i \to \infty \, .$$

By Proposition 4, G_E defines a compact operator from $L^2(\Omega_j; dx)$ to $L^2(M^j; \mu)$. Hence the sequence $g_i = f_{i|\Omega_j}$ is weakly convergent to zero in $L^2(\Omega_j; dx)$ and is mapped into the sequence $\int_{\Omega_j} G_E(x-y) f_i(y) dy$ which is strongly convergent to zero in $L^2(M^j; \mu)$, hence

$$\int_{M^j} \left(\int_{\Omega_j} G_E(x-y) f_i(y) dy \right)^2 \mu(dx) \to 0 \, , \text{ as } i \to \infty \, .$$

As a conclusion we can find i_0 such that for any $i > i_0$ we have

$$2 \sum_{j=1}^{r} \left[\int_{M^j} \left(\int_{\mathbb{R}^n \backslash \Omega_j} G_E(x-y) f_i(y) dy \right)^2 \mu(dx) + \int_{M^j} \left(\int_{\Omega_j} G_E(x-y) f_i(y) dy \right)^2 \mu(dx) \right]$$

$$< \frac{1}{2} \varepsilon \, .$$

Hence $\|\hat{G}_E f_i\|_{L^2(\mu)}$ goes to zero and $\hat{G}_E$ is compact.

∎

Proof of Proposition 9.   We first show that for any $f \in C_0^1(\mathbb{R}^n)$ one has

$$\Gamma(f) = \lim_{s \to 0} \Gamma_s(f) \ . \tag{A.1}$$

We have

$$\int_M \int_M [(G_{E'} - G_E)(x - y) - (G_{E'}^s - G_E^s)(x - y)] \, \mu(dy) |f(x)|^2 \mu(dx)$$

$$\leq \sup_x |f(x)|^2 (2\pi)^{-\frac{n}{2}} (-E + E') \int_{\mathbb{R}^n} \frac{\left(1 - e^{-|p|s}\right) |M(p)|^2}{(p^2 - E')(p^2 - E)} dp$$

$$\leq C \int_{|p| > \omega} \frac{|M(p)|^2}{(p^2 - E')(p^2 - E)} dp + C \int_{|p| \leq \omega} \frac{1 - e^{-|p|s}}{(p^2 - E')(p^2 - E)} dp \ ,$$

where $M(p) = (2\pi)^{-\frac{n}{2}} \int e^{ipx} \mu(dx)$ is bounded and continuous. Given $\varepsilon > 0$, there is $\omega > 0$ so that the first term is less than $\frac{1}{2}\varepsilon$. Then there is $s_0 > 0$ so that the second term is less than $\frac{1}{2}\varepsilon$ for all $s < s_0$.

Now

$$\int_M \int_M (G_E(x - y) - G_E^s(x - y)) (f(x) - f(y))^2 \mu(dx)\mu(dy)$$

$$= \left(\int \int_{|x-y|<\eta} + \int \int_{|x-y|\geq\eta} \right) (G_E(x - y) - G_E^s(x - y)) (f(x) - f(y))^2 \, \mu(dx)\mu(dy)) \ .$$

We first observe that

$$\int \int_{|x-y|<\eta} (G_E(x - y) - G_E^s(x - y)) (f(x) - f(y))^2 \mu(dx)\mu(dy)$$

$$\leq 2 \sup_{|x-y|<\eta} \frac{(f(x) - f(y))^2}{|x - y|^2} \int \int_{|x-y|<\eta} G_E(x - y)|x - y|^2 \mu(dx)\mu(dy) \ .$$

To estimate this integral we proceed as follows. Let $a \in \mathbb{R}^n$ have integer components i.e. $a = (a_1, ..., a_n)$ where $a_i \in \mathbb{Z}$ (the set of all integers). For any such a we define the family of half open boxes

$$s_a(\eta) = \prod_{i=1}^{n} (\eta a_i, \ \eta(a_i + 1)] \ .$$

Let $I(\eta)$ denote family of a's such that $s_a(\eta) \cap M \neq \emptyset$. To simplify the notation we shall drop the letter η when possible.

We have $M \subset \bigcup_{a \in I} s_a$ and $\mu(M) = \mu\left(\bigcup_{a \in I} s_a\right) = \sum_{a \in I} \mu(s_a)$. Define

$$S_a = \prod_{i=1}^{n} (\eta(a_i - 1), \eta(a_i + 2)] \ .$$

We then have $M \subset \bigcup_{a \in I} S_a$ and

$$\mu(M) \leq \mu\left(\bigcup_{a \in I} S_a\right) \leq \sum_{a \in I} \mu(S_a) \leq 3^n \mu(M) \ . \tag{A.2}$$

Now

$$\{(x,y) \in M^l; \ |x - y| < \eta\} \subset \bigcup_{a \in I} \{(x,y); \ x,y \in S_a\}$$

Thus

$$\int\int_{|x-y|<\eta} G_E(x - y)|x - y|^2 \mu(dx)\mu(dy) \leq D \int\int_{|x-y|<\eta} |x - y|^{-n+4} \mu(dx)\mu(dy)$$
$$\leq D \sum_{a \in I} \int_{S_a} |x - y|^{-n+4} \mu(dx)\mu(dy) \ .$$

To estimate the latter sum we shall use Lemma 3. For any $a \in I$ the diameter of S_a equals $3\sqrt{n}\eta$. The dimensionality of S_a with respect to μ is not less than m, the dimensionality of M with respect to μ.

Then, choosing $m - n + 4 > \xi > 0$, we get by Lemma 3

$$\int_{S_a} |x - y|^{-n+4} \mu(dy) \leq C(3\sqrt{n}\eta)^\xi \ .$$

This together with (A.2) implies

$$\sum_{a \in I} \int_{S_a} \int_{S_a} |x - y|^{-n+4} \mu(dx)\mu(dy) \leq 3^{\xi+n} n^{\frac{1}{2}\xi} C \eta^\xi$$

which shows that

$$\int\int_{|x-y|<\eta} \left(G_E(x - y) - G_E^s(x - y)\right)\left(f(x) - f(y)\right)^2 \mu(dx)\mu(dy) \tag{A.3}$$

goes to zero as $\eta \to 0$ and the convergence is independent of s. Thus given $\varepsilon > 0$ there is $\eta > 0$ such that (A.3) is smaller than $\frac{1}{3}\varepsilon$.

Let $\gamma > \eta$ then

$$\int\!\!\int_{|x-y|>\gamma} (G_E(x-y) - G_E^s(x-y))\,(f(x) - f(y))^2\,\mu(dx)\mu(dy)$$

$$\leq 8 \sup_x |f(x)|^2 \int\!\!\int_{|x-y|>\gamma} G_E(x-y)\mu(dx)\mu(dy) \leq 8 \sup_x |f(x)|^2 \sup_{|x|>\gamma} G_E(x) \ . \tag{A.4}$$

Since $G_E(x) \to 0$ as $|x| \to \infty$ there is γ so that (A.4) is less than $\frac{1}{3}\varepsilon$. With the above choice of η and γ consider

$$4 \sup_x |f(x)|^2 \int\!\!\int_{\eta \leq |x-y| \leq \gamma} (G_E(x-y) - G_E^s(x-y))\,\mu(dx)\mu(dy) \ . \tag{A.5}$$

As we know, $G_E(\cdot) \in L^1(\mathbb{R}^n; dx)$. On the other hand G_E^s can be expressed as $(G_E * P)(x,s)$, the convolution of G_E and the Poisson kernel $P(y,s)$. According to the proof of Theorem 1.25 chapter I of [St] one has $(G_E * P)(x,s) \to G_E(x)$ as $s \to 0$ for all x in the Lebesgue set of G_E. But since all points of continuity belong to the Lebesgue set this convergence is pointwise for all $x \in \mathbb{R}^n \backslash \{0\}$. Hence by monotonicity of G_E^s this convergence is uniform for $\eta \leq |x| \leq \gamma$, which implies that (A.5) is smaller than $\frac{1}{3}\varepsilon$ for all $s < s_0$. This completes the proof of (A.1).

Now we shall prove that the quadratic form defined on $C_0^1(\mathbb{R}^n)$ by

$$\Gamma(f) = \int_M \left(\alpha(x) + \int_M \int_M (G_{E'} - G_E)(x-y)\mu(dy) \right) |f(x)|^2 \mu(dx)$$

$$+ \frac{1}{2} \int_M \int_M G_E(x-y)(f(x) - f(y))^2 \mu(dx)\mu(dy)$$

is closable.

First we recall that the quadratic forms $\Gamma_s(f)$ are defined by positive, bounded operators and hence they are closed forms on $L^2(M;\mu)$. Let now $f_i \in C_0^1(\mathbb{R}^n)$ be such that $\|f_i\|_{L^2(\mu)} \to 0$ and $\Gamma(f_i - f_j) \to 0$ as $i,j \to \infty$.

Suppose Γ is not closable. Then there is $a > 0$ such that $\Gamma(f_i) > a^2$ for all $i \in \mathbb{N}$ (more precisely $\Gamma(f_i) > a^2$ follows only for some subsequence of f_i, but there is no loss of generality saying that it holds for all $i \in \mathbb{N}$)

Let N be such that for all $i,j > N$ one has

$$\Gamma^{\frac{1}{2}}(f_i - f_j) < \frac{1}{3}a \ .$$

Then

$$\frac{1}{3}a > \Gamma^{\frac{1}{2}}(f_i - f_j) \geq \Gamma_s^{\frac{1}{2}}(f_i - f_j) \geq |\Gamma_s^{\frac{1}{2}}(f_i) - \Gamma_s^{\frac{1}{2}}(f_j)|$$

Since Γ_s is closed we have $\Gamma_s(f_j) \to 0$ as $j \to \infty$, hence $\Gamma_s^{\frac{1}{2}}(f_i) < \frac{1}{3}a$ if $i > \tilde{N}$, for some $\tilde{N}$. By (A.1) there is s_0 such that

$$\Gamma^{\frac{1}{2}}(f_i) - \Gamma_s^{\frac{1}{2}}(f_i) < \frac{1}{3}a$$

for $0 < s < s_0$. Then

$$\Gamma^{\frac{1}{2}}(f_i) < \frac{1}{3}a + \Gamma^{\frac{1}{s}}(f_i) < \frac{2}{3}a$$

in contradiction to our assumption. Thus Γ is closable.

Denote by $\bar{\Gamma}$ the closure of Γ and by $Q(\bar{\Gamma})$ the domain of $\bar{\Gamma}$. We want to show that for any $f \in Q(\bar{\Gamma})$ we have

$$\bar{\Gamma}(f) = \lim_{s \to 0} \Gamma_s(f) \ .$$

First we show that $\Gamma(f) \geq \Gamma_s(f)$.

Suppose ad absurdum that for some $f \in Q(\bar{\Gamma})$ there are $q, s_0 > 0$ such that

$$\bar{\Gamma}_{s_0}(f) \geq \bar{\Gamma}(f) + q \ .$$

Then we have also $\bar{\Gamma}_s(f) \geq \bar{\Gamma}(f) + q$ for all $s < s_0$. Let $f_i \in C_0^1(\mathbb{R}^n)$, $\|f_i - f\|_{L^2(M;\mu)} \to 0$ and $\bar{\Gamma}(f - f_i) \to 0$.

Given $\varepsilon > 0$, there is N such that for all $i > N$ we have

$$\varepsilon > \bar{\Gamma}^{\frac{1}{2}}(f - f_i) \geq \left| \bar{\Gamma}^{\frac{1}{2}}(f) - \bar{\Gamma}^{\frac{1}{2}}(f_i) \right|$$

i.e.

$$\bar{\Gamma}^{\frac{1}{2}}(f_i) - \varepsilon \leq \bar{\Gamma}^{\frac{1}{2}}(f) \leq \bar{\Gamma}^{\frac{1}{2}}(f_i) + \varepsilon \ .$$

Then

$$\bar{\Gamma}^{\frac{1}{2}}(f_i) - \varepsilon \leq \Gamma^{\frac{1}{s}}_s(f) - q$$

and

$$\bar{\Gamma}^{\frac{1}{2}}(f_i) - \Gamma^{\frac{1}{s}}_s(f_i) - \varepsilon \leq \Gamma^{\frac{1}{s}}_s(f) - \Gamma^{\frac{1}{s}}_s(f_i) - q \ .$$

Now fix $s < s_0$, then there is N' such that $i > N'$ implies

$$\Gamma^{\frac{1}{s}}_s(f) - \Gamma^{\frac{1}{s}}_s(f_i) < \varepsilon \ .$$

Hence for $i > \max (N, N')$ we have

$$\bar{\Gamma}^{\frac{1}{2}}(f_i) - \Gamma^{\frac{1}{s}}_s(f_i) \leq 2\varepsilon - q \ .$$

If the choice of $\varepsilon > 0$ was such that $2\varepsilon < q$ then we have a contradiction, because as we know

$$\bar{\Gamma}^{\frac{1}{2}}(f_i) - \Gamma^{\frac{1}{s}}_s(f_i) \geq 0$$

for $f_i \in C_0^1(\mathbb{R}^n)$.

Let now $f \in Q(\bar{\Gamma})$ and f_i as above. We have

$$0 \leq \bar{\Gamma}^{\frac{1}{2}}(f) - \Gamma^{\frac{1}{s}}_s(f) \leq \left| \bar{\Gamma}^{\frac{1}{2}}(f) - \bar{\Gamma}^{\frac{1}{2}}(f_i) \right| + \left| \bar{\Gamma}^{\frac{1}{2}}(f_i) - \Gamma^{\frac{1}{s}}_s(f_i) \right| + \left| \Gamma^{\frac{1}{s}}_s(f_i) - \Gamma^{\frac{1}{s}}_s(f) \right| \ ,$$

since

$$\bar{\Gamma}^{\frac{1}{2}}(f - f_i) \geq \left| \bar{\Gamma}^{\frac{1}{2}}(f) - \bar{\Gamma}^{\frac{1}{2}}(f_i) \right|$$

and

$$\bar{\Gamma}^{\frac{1}{2}}(f - f_i) \geq \Gamma_s^{\frac{1}{2}}(f - f_i) \geq \left| \Gamma_s^{\frac{1}{2}}(f) - \Gamma_s^{\frac{1}{2}}(f_i) \right| \ .$$

Given $\varepsilon > 0$ let i be such that $\bar{\Gamma}^{\frac{1}{2}}(f - f_i) < \frac{1}{3}\varepsilon$.
Then

$$\left| \bar{\Gamma}^{\frac{1}{2}}(f) - \bar{\Gamma}^{\frac{1}{2}}(f_i) \right| < \frac{1}{3}\varepsilon \ ,$$

$$\left| \Gamma_s^{\frac{1}{2}}(f_i) - \Gamma_s^{\frac{1}{2}}(f) \right| < \frac{1}{3}\varepsilon \text{ for all } s > 0 \ .$$

Now, there is s_0 such that

$$\bar{\Gamma}^{\frac{1}{2}}(f_i) - \Gamma_s^{\frac{1}{2}}(f_i) < \frac{1}{3}\varepsilon \text{ for all } a < s < s_0 \ .$$

This implies

$$\bar{\Gamma}^{\frac{1}{2}}(f) - \Gamma_s^{\frac{1}{2}}(f) < \varepsilon \ .$$

Thus the proof of Proposition 9 is complete.

Acknowledgements.

Very stimulating discussions with Zhi Ming Ma, Johannes Brasche, Rodolfo Figari, Shigeo Kusuoka, Sandro Teta are gratefully acknowledged.

The financial support of the Centre de Physique Théorique, CNRS Marseille, Université d'Aix-Marseille II, Luminy, and Université de Provence, where the work was initiated, the SFB 237 and BiBoS-projects at the Ruhr-Universität Bochum and Bielefeld University, as well as of the Norwegian Science Foundation (NAVF) is also gratefully acknowledged. We thank Martin Jarrath and Frank Nitzschner for their competent help with the setting of the manuscript.

Note.

This work was initiated many years ago (ca. '81) and greatly stimulated by the enthusiasm and insight of Raphael Høegh-Krohn. When Raphael departed January 24, 1988, the paper was still not written up in its present form (although the essential content was reported upon on several occasions).

The other authors completed the work as well as they could. They would like to mark here their great indebtedness and gratitude to Raphael, for all he has given them, as a scientist and as a friend.

References

[A] S. Albeverio, *Non standard analysis; polymer models, quantum fields*, pp. 233–254 in Proc. Schladming '84, 23. Internat. Universitätswochen für Kernphysik, Acta Phys. Suppl. XXVI (1984): "Stochastic Methods and Computer Techniques in Quantum Dynamics", Edts. H. Mitter, L. Pittner

[AB1HK] S. Albeverio, Ph. Blanchard, R. Høegh-Krohn, *Newtonian diffusions and planets, with a remark on non-standard Dirichlet forms and polymers*, pp. 1-24 in "Stochastic Analysis and Applications", Proc. Swansea 1983, Edts. A. Truman, D. Williams, Lect. Notes Math. 1095, Springer, Berlin (1984)

[AB1M] S. Albeverio, Ph. Blanchard, Z.M. Ma, *Feynman-Kac semigroups in terms of signed smooth measures*, BiBoS Preprint 424 (1990), to appear in Proc. Oberwolfach Meeting, Ed. Hornung, Kotelenez, Lect. Notes Maths., Springer, Berlin (1991)

[ABrR] S. Albeverio, J. Brasche, M. Röckner, *Dirichlet forms and generalized Schrödinger operators*, in Proc. Sønderborg Conf. "Schrödinger operators", Edts. H. Holden, A. Jensen, Lect. Notes Phys., Springer, Berlin (1989)

[AFGHKHK] S. Albeverio, R. Figari, F. Gesztesy, R. Høegh-Krohn, H. Holden, W. Kirsch, *Point interaction Hamiltonians for crystals with random defects*, pp. 87–99 in [ExS2]

[AFHK] S. Albeverio, J.E. Fenstad, R. Høegh-Krohn, *Singular perturbations and non-standard analysis*, Trans. Am. Math. Soc. 252, 275–295 (1979)

[AFHKKL] S. Albeverio, J.E. Fenstad, R. Høegh-Krohn, W. Karwowski, T. Lindstrøm, *Perturbations of the Laplacian supported by null sets, with applications to polymer measures and quantum fields*, Phys. Letts. A 104, 396–400 (1984)

[AFHKL] S. Albeverio, J.E. Fenstad, R. Høegh-Krohn, T. Lindstrøm, *Nonstandard methods in stochastic analysis and mathematical physics*, Acedemic Press, New York (1986)

[AGHKH] S. Albeverio, F. Gesztesy, R. Høegh-Krohn, H. Holden, *Solvable models in quantum mechanics*. Springer, Berlin (1988)

[AGS] J.P. Antoine, F. Gesztesy, J. Shabani, *Exactly solvable models of sphere interactions in quantum mechanics*, J. Phys. A20, 3687–3712 (1987)

[AHKS] S. Albeverio, R. Høegh-Krohn, L. Streit, *Energy forms, Hamiltonians and distorted Brownian paths*, Journal of Mathematical Physics 18, 907–917 (1977)

[AM1] S. Albeverio, Z.M. Ma, *Perturbation of Dirichlet forms – lower semiboundedness, closability and form cores*, SFB 237–Preprint, to appear in J. Funct. Anal.

[AM2] S. Albeverio, Z.M. Ma, *Additive functionals, nowhere Radon and Kato class smooth measure associated with Dirichlet forms*, SFB 237–Preprint

[AM3] S. Albeverio, Z.M. Ma, *Nowhere Radon smooth measures, perturbations of Dirichlet forms and singular quadratic forms*, pp. 3–45 in Proc. Bad Honnef Conf. 1988, Edts. N. Christopeit, K. Helmes, M. Kohlmann, Lect. Notes Control and Inform. Sciences 126, Springer, Berlin 1989

[AR] S. Albeverio, M. Röckner, *New developments in theory and applications of Dirichlet forms*, pp. 27–76 in "Stochastic Processes, Physics and Geometry", Proc. 2nd Int. Conf. Ascona – Locarno – Como 1988, Edts. S. Albeverio, G. Casati, U. Cattaneo, D. Merlini, R. Moresi, World Scientific (1990)

[BHH] A. Boukricha, W. Hansen, H. Hueber: *Continuous solutions of the generalized Schrödinger equation and perturbation of harmonic spaces*, Exp. Math. 5 (1987), 97–135

[BeF] F.A. Berezin, L.D. Faddeev, *A remark on Schrödinger's equation with a singular potential*, Sov. Math. Dokl. $\underline{2}$ (1961), 372–375

[BlM] Ph. Blanchard, Zhiming Ma, *Semigroup of Schrödinger operators with potentials given by Radon measures.* BiBoS Preprint Nr. 262, 1987

[Br1] J. Brasche, *Dirichlet forms and generalized Schrödinger operators*, pp. 43–58 in [ExS3]

[Br2] J. Brasche, *Perturbations of self-adjoint operators supported by null sets*, Ph. Thesis 1988, Universität Bielefeld, and paper in preparation

[Br3] J. Brasche, *An inverse problem in spectral analysis and the Efimov effect*, to appear in "Stochastic Processes, Physics and Geometry", Proc. Ascona–Locarno 1988, Edts. S. Albeverio, G. Casati, U. Cattaneo, D. Merlini, R. Moresi. World Scientific, Singapore (1990)

[BrK] J. Brasche, W. Karwowski, *On boundary theory for Schrödinger operators and stochastic processes*, Oper. Th. Adv. and Appl. $\underline{46}$ (1990), 199–208

[BrT] J.F. Brasche, A. Teta, *Spectral analysis and scattering theory for Schrödinger operators with an interaction supported by a regular curve*, in this volume

[CaC] W. Caspers, Ph. Clément, *Point interactions in L^p*, Delft Preprint (1991)

[Ch] S.E. Cheremshantsev, *Hamiltonians with zero-range interactions supported on a Brownian path*, LOMI-Preprint, Leningrad (1989)

[DSh] L. Dabrowski, J. Shabani, *Finitely many sphere interactions in quantum mechanics – non separated boundary conditions*, J. Math. Phys. $\underline{29}$, 2241–2244 (1988)

[DeO] Y.N. Demkov, V.N. Ostrovski, *Zero-Range Potentials and Their Applications in Atomic Physics*, Plenum Press, New York (1988)

[ExS1] P. Exner, P. Šeba, *A new type of quantum interference transistors*, Phys. Letts. $\underline{A}$, $\underline{129}$, 477–480 (1988)

[ExS2] P. Exner, P. Šeba, Edts., *"Applications of Self-adjoint Extensions in Quantum Physics"*, Proc. Dubna 1987, Lect. Notes Phys. 324, Springer, Berlin (1989)

[ExS3] P. Exner, P. Šeba, Edts., *Schrödinger Operators, Standard and Non-Standard*, Proc. Dubna, World Scientific, Singapore (1989)

[ExSS] P. Exner, P. Šeba, P. Šlŏvicèk, *Quantum interference on graphs controlled by an external electric field*, J. Phys. A 21, 4009–4019 (1988)

[FeP] D. Feyel, A. de la Pradelle, *Étude de l'équation $\frac{1}{2}\Delta u - u\mu = 0$ où μ est une mesure positive*, C.R. Acad. Sci. Paris Ser I Math. $\underline{303}$ (1986), 5–6

[FiT] R. Figari, A. Teta, *A boundary value problem of mixed type on perforated domains*, SFB 237–Preprint, Bochum (1990)

[Fu] M. Fukushima, *Dirichlet forms and Markov processes*, North Holland, Amsterdam (1980)

[FuS] M. Fukushima, T. Shima, *On a spectral analysis for the Sierpinski gasket*, Osaka Prepr.

[GrHKM] A. Grossmann, R. Høegh-Krohn, M. Mebkhout, *A class of explicitly soluble, local, many-center Hamiltonians for one-particle quantum mechanics in two and three dimensions*, I, J. Math. Phys. $\underline{21}$, 2376–2385 (1980)

[He] L.I. Hedberg, *Spectral synthesis in Sobolev spaces and uniqueness of solutions of the Dirichlet problem*, Acta Math. $\underline{147}$, 237 – 264 (1982)

[Ka] T. Kato, *Perturbation theory for linear operators.* Springer, Berlin 1966

[Ko] V.D. Koshmanenko, *Singular perturbations defined by forms*, pp. 55–66 in [ExS2]

[Koc] A.N. Kochubej, *Elliptic operators with boundary conditions on a subset of measure zero*, Funct. Anal. Appl. 16 (1982), 137–139

[Kr] M.A. Krasnoselskii, P.P. Zabreiko, E.I. Pustylnik, P.E. Sobolewskii, *Integral operators in spaces of summable functions*. Noordhoff, Leiden 1976

[Ku] S. Kusuoka, *Dirichlet forms on fractals and products of random matrices*, RIMS Kyoto Prepr.

[Kur] Y.V. Kurylev, *Boundary condition on a curve for a three-dimensional Laplace operator*, J. Sov. Math. 22 (1983), No. 7

[L1] T. Lindstrøm, *Nonstandard analysis and perturbations of the Laplacian along Brownian paths*, pp. 180–200 in "Stochastic processes – mathematics and physics", Edts. S. Albeverio, Ph. Blanchard, L. Streit. Lect. Notes Maths. 1158, Springer, Berlin (1986)

[L2] T. Lindstrøm, *Nonstandard energy forms and diffusions on manifolds and fractals*, pp. 363–380 in "Stochastic Processes in Classical and Quantum Systems", Proc. Ascona 1985, Edts. S. Albeverio, G. Casati, D. Merlini. Lect. Notes Phys. 262, Springer, Berlin (1986)

[MaK] V.G. Maz'ya, V.P. Khavin, *Non-linear Potential Theory*, Uspiekhi Matiematicheskich Nauk 27, 67–138, (1972)

[Me] V. Metz, *Potentialtheorie auf dem Sierpinski gasket*, Math. Ann. 289 (1991), 207–237

[Pa] B.S. Pavlov, *The theory of extensions and explicitly soluble models*, Russ. Math. Surv. 42, 127–168 (1987)

[Pu] Pustylnik, E.I., *Operators of potential type for the case of sets with abstract measures*. Dokt. Akad. Nauk SSSR 156 (1964) 519–520

[Si] B. Simon, *A canonical decomposition for quadratic forms with applications to monotone convergence theorems*. J. Funct. Anal. 28 (1978), 377–385

[Sm] L. Smits, Ph.D. Thesis, University of Antwerp (1990)

[St] E. Stein, *Topics in harmonic analysis related to the Littlewood–Paley theory*, Princeton Univ. Press, Princeton (1970)

[Sto1] P. Stollmann, *Admissible and regular potentials for Schrödinger forms*, J. Oper. Th. 18 (1987), 139–151

[Sto2] P. Stollmann, *Smooth perturbations of regular Dirichlet forms*, Oldenburg Preprint (1990)

[StV] P. Stollmann, J. Voigt, *A regular potential which is nowhere in L_1*, Letts. Math. Phys. 9 (1985), 227–230

[Stu] T. Sturm, *Störung von Hunt-Prozessen durch signierte additive Funktionale*, Diss., Erlangen (1988)

[Te] A. Teta, *Quadratic forms for singular perturbations of the Laplacian*, Publ. RIMS Kyoto Univ. 26 (1990), 803–819

[Voi] J. Voigt, *Absorption semigroups, their generators and Schrödinger semigroups*, J. Funct. Anal. 67 (1986), 167–205

LANDAU HAMILTONIANS ON SYMMETRIC SPACES

J. E. AVRON AND A. PNUELI

Department of Physics, Technion, IIT, Haifa, 32000, ISRAEL

1 INTRODUCTION AND OVERVIEW

A classical particle on a manifold moves on geodesics: Straight lines on the plane, great circles on the sphere and semi-circles on the hyperbolic-plane, represented as the upper-half plane with the usual metric[1].

The corresponding Schrödinger operators are (minus) the Laplacians on the manifolds and their spectra are $[0, \infty)$ for the plane, $\{0, 2, 6, ..., n(n+1), ..., \infty\}$ for the sphere, and $[\frac{1}{4}, \infty)$ for the hyperbolic plane [see e.g. McKean (1970) or Terras (1985)].

Magnetic fields are 2-forms, and a constant magnetic field is a multiple, B, of the area form, with $B > 0$, the magnetic field strength. Constant magnetic fields are therefore natural for orientable two dimensional manifolds.

In contrast with magnetic fields, constant electric fields are not natural in general: Electric fields are vector fields and on a curved manifold zero is the only constant vector field; Constant electric field are defined only for flat spaces. For this reason the dynamics in constant magnetic fields is the essentially unique generalisation of the free dynamics which preserves the underlying symmetry of the manifold.

In classical mechanics a magnetic field leads to a Lorentz force and the orbit is no longer a geodesic. For example, the orbits in the plane are circles (rather than straight lines). As we shall explain in section 3, for the sphere, the orbits can be described geometrically as the circle of tangency of the sphere, (embedded in three dimensional Euclidean space) with a cone tangent to it, see Fig. 1. The larger the energy of the particle, the thinner the cone. Similarly, for the hyperbolic plane the orbits are the lines of tangency of the hyperboloid, embedded in three dimensional Minkowski space, with cones. If the energy is in $[0, B^2)$, the cone is time-like and the orbit is closed, see Fig. 2; If the energy is in (B^2, ∞), the cone is space like, the orbit is open and runs to infinity, see Fig. 3. More on this, in section 3.

96

We call (minus) the "Magnetic Laplacians" with constant magnetic fields "*Landau Hamiltonians*"[2]. Like the ordinary Laplacians, they represent pure kinetic energy. For the plane and the hyperbolic plane B can be any real number. For the unit sphere $2B$ must be an integer [Dirac (1931), Wu and Yang (1976), Coleman (1982)].

The spectral analysis of the Landau Hamiltonian for the plane goes back to Landau in 1930. The spectrum is

$$\text{Spect } (H_P) = \{B(2n+1) \mid n = 0, 1, 2, \ldots \infty\}, \tag{1.1}$$

each point is infinitely degenerate[3]. This is, of course, quite unlike the spectrum of the Laplacian.

The Landau Hamiltonian and the Laplacian on the sphere turn out to have related spectral properties. The spectral analysis of the former also goes back to the 30's [Tamm, (1931), Wu and Yang (1976)] and

$$\text{Spect } (H_S) = \{(B+n)(B+n+1) - B^2 \mid n = 0, 1, \ldots \infty\}, \tag{1.2}$$

with degeneracy $2(n+B)+1$.

Landau Hamiltonians for the Hyperbolic plane were analyzed by [Comtet and Houston (1985), Comtet (1987) , Grosche (1988)] and turn out to have rich spectral properties:

$$\text{Spect}(H_H) = \left\{B^2 - (n-B)(n-B+1) \mid n = 0, 1, \ldots, [B - \tfrac{1}{2}]\right\} \cup [\tfrac{1}{4} + B^2, \infty), \tag{1.3}$$

where $[x]$ is the integer part of x. See Fig. 4. There are no isolated spectral points for $0 \leq B \leq \frac{1}{2}$. For $B > \frac{1}{2}$, there are $[B + \frac{1}{2} - 0]$ isolated spectral points in the interval $(\frac{1}{2}, B^2 + \frac{1}{4})$ each infinitely degenerate. In addition, there is continuous spectrum on the semi-infinite line $[\frac{1}{4} + B^2, \infty)$. An elementary derivation of these results is given in section 4.

With the vector spaces associated to the isolated spectral intervals of Schrödinger operators on two dimensional (noncompact) manifolds, one can associate an index defined as follows: Cut the manifold from the origin to infinity[4] and impose $\exp 2\pi i\theta$ boundary conditions across the cut. This gives a circle of operators, $H(\theta)$. Since $H(0) = H(1)$, the spectrum is periodic in θ and in particular, dim $P(0) = $ dim $P(1)$ with $P(\theta)$ the spectral projection on the interval, and we allow dim $P = \infty$. Define the index by[5] :

$$\text{Index} \equiv \#(\text{Arriving}) - \#(\text{Departing}),$$

where $\#(Arriving)$ is the number of states that joint the spectral interval as θ increases by a period, and similarly for $\#(Departing)$[6].

Finite dimensional projections have index zero. In particular, this is the case for all the spectral points of Landau Hamiltonian on the sphere.

Spectral intervals with infinite dimensional projection can have a nonzero index and the Landau Hamiltonians on the Plane and Hyperbolic plane are examples. The $n-th$ spectral point has:

$$\text{Index}_P(P_n) = \text{Index}_H(P_n) = 1. \tag{1.6}$$

The calculation is given in section 5. The results for the plane are known from other considerations that we discuss in the next paragraph. For the Hyperbolic plane the results are new.

The germs of the definition of the index given above go back to Laughlin [Laughlin (1987)] who was interested in the Hall conductance, and in particular wanted to explain why, under appropriate conditions, the Hall conductance assumes integer values. The credit for identifying the Hall conductance with an Index, and saying so in so many words, goes to J. Bellissard [Bellisard (1988)]. He realized this in the context of Connes noncommutative differential geometry [Connes (1986)]. The index defined above is distinct from, although closely related to, the index of certain Fredholm operators that arise in Connes theory [Avron, Seiler and Simon, (1990)]. Connes went part of the way in computing the indices for the plane and the hyperbolic plane.

2. NOTATIONS AND CONVENTIONS

2.1 Units:
We absorb the charge of the particle and the velocity of light in the definition of the strength of the magnetic field. We choose units of length so that the curvatures of the sphere and the hyperbolic plane are ± 1, and the remaining units so that $\hbar$ and the mass of the particle are unity. Since there is a factor of 2 between (minus) the Laplacian and the (free) Schrödinger operator in these units, we absorb this factor into the definition of the energy scale.

2.2 Spaces:
P, S and H will denote the Euclidean plane, the 2-sphere, and the Hyperbolic plane respectively. E^n denotes Euclidean n-space, and M^3 Minkowski space.

Z is the integers and Z_+ the non-negative integers.

2.3 Metrics:

g denotes the metric 2-form:

$$\mathbf{g} \equiv \pm(dx^0)^2 + (dx^1)^2 + (dx^2)^2 = \pm(dr)^2 + r^2((d\tau)^2 + k^2(\tau)\,(d\phi)^2), \qquad (2.1)$$

where the upper signs are for E^3 and the lower signs for M^3; x^j denote cartesian coordinates and (r, τ, ϕ) spherical coordinates (for the forward time-like cone in M^3), $0 \le r \le \infty$, $0 \le \phi < 2\pi$ and

$$k(\tau) \equiv \begin{cases} \tau, & 0 \le \tau < \infty, & \text{for P;} \\ \sin(\tau), & 0 \le \tau \le \pi, & \text{for S;} \\ \sinh(\tau), & 0 \le \tau < \infty, & \text{for H.} \end{cases} \qquad (2.2)$$

2.4 Signs:
Equation with $\pm$ or $\mp$ signs, the upper sign usually refers to S (or E) and the lower sign to H (or M), except in cases where it is clear from the context that the meaning is otherwise.

Since B and $-B$ are related (by time reversal) we restrict ourselves to $B > 0$.

2.5 Geometry:
We use lower case boldface to denote 2-forms and uppercase boldface to denote 1-forms. $\mathbf{b} = \mathbf{dA}$ stands for the basic relation between the magnetic-field 2-form **b** and the vector-potential 1-form **A**. Vectors are denoted by arrows.

$\langle \mathbf{b} | \vec{v} \otimes \bullet \rangle = \mathbf{W}$ is the contraction of a 2-form **b** with a vector $\vec{v}$ to yield a 1-form **W**. We use lower indices for the components of forms and upper indices for the components of vectors. In index notation, the previous equation reads $b_{ij}\, v^j = W_i$.

We write $\mathbf{g}(\vec{v}) = \mathbf{V}$ and the converse relation $\mathbf{g}^{-1}(\mathbf{V}) = \vec{v}$, for the correspondence between vectors and forms. Scalar products are denoted by dots, so $\vec{x} \cdot \vec{x} = \langle \mathbf{g} | \vec{x} \otimes \vec{x} \rangle$. We set $0 \le r \equiv \sqrt{|\vec{x} \cdot \vec{x}|}$.

3 THE CLASSICAL ORBITS: A GEOMETRIC DESCRIPTION

In this section we describe the geometry of the classical orbits. Since all three cases, P, S and H are described by (almost) the same set of equations, except that the various symbols mean slightly different things, we start with detailing the notation.

The 2-sphere S and the Hyperbolic-plane H are:

$$\vec{x} \cdot \vec{x} = \pm 1, \qquad (3.1)$$

where $\vec{x} \in E^3$ for S and $\vec{x} \in M^3$ for H. The magnetic 2-form

$$\mathbf{b} \equiv \begin{cases} B\, dx^1 \wedge dx^2, & \text{on P;} \\ \frac{B}{r^3}\,(x^0 dx^1 \wedge dx^2 + x^1 dx^2 \wedge dx^0 + x^2 dx^0 \wedge dx^1), & \text{on S, H,} \end{cases} \qquad (3.2)$$

satisfies $\mathbf{db} = 0$ (away from $r = 0$). Its restriction to the S and H has $r = 1$.

The Lorentz force is

$$\vec{F} \equiv \mathbf{g}^{-1}(\langle \mathbf{b}|\vec{v} \otimes \bullet \rangle), \tag{3.3a}$$

and in components

$$F^j = g^{jk} b_{km} v^m, \tag{3.3b}$$

where $\vec{v} \equiv \dot{\vec{x}}$ is the velocity. (Just a fancy way to write $\vec{v} \times \vec{B}$). Newton's equation in the plane is

$$\dot{\vec{v}} = \vec{F}. \tag{3.4a}$$

On S and H the motion is constrained so that $\vec{x} \cdot \vec{v} = 0$ and $\vec{v} \cdot \vec{v} + \vec{x} \cdot \dot{\vec{v}} = 0$. Since

$$\vec{x} \cdot \vec{F} = \langle \mathbf{b}|\vec{v} \otimes \vec{x} \rangle = 0, \tag{3.5}$$

the Lorentz force is incompatible with the constraint and the equation of motion (3.4a) is modified to:

$$\dot{\vec{v}} = \mp(\vec{v} \cdot \vec{v})\vec{x} + \vec{F}. \tag{3.4b}$$

The first term on the right hand side is the centripetal force applied by the surface. In index notation, Eq. (3.4b) reads

$$\dot{v}^i = \mp(\vec{v} \cdot \vec{v})\, x^i + g^{il} b_{lj} v^j. \tag{3.4b'}$$

Equations (3.4) have two constants of motion that will play a central role throughout. The first is

$$\vec{c} \equiv \vec{x} + \left(\frac{1}{B}\right)^2 \vec{F}. \tag{3.6}$$

$\dot{\vec{c}} = 0$ is easily verified for the plane by differentiation and substituting the equation of motion. For S and H one needs to use:

$$\langle \mathbf{b}|\vec{x} \otimes \bullet \rangle = \langle \dot{\mathbf{b}}|\vec{v} \otimes \bullet \rangle = 0, \tag{3.7}$$

and the geometric fact that a tangent vector to the manifold can be flipped with the help of the area form:

$$B^2 \vec{v} = -\mathbf{g}^{-1}\left(\langle \mathbf{b}|\vec{F} \otimes \bullet \rangle\right). \tag{3.8}$$

Both equations (3.7) and (3.8) rely on the special form of $\mathbf{b}$.

The second constant of motion is (twice) the kinetic energy:

$$E \equiv \vec{v} \cdot \vec{v} = B^2(\vec{c} - \vec{x})^2 \tag{3.9a}$$

$$= B^2(\vec{c} \cdot \vec{c} - \vec{x} \cdot \vec{x}) = B^2(\vec{c} \cdot \vec{c} \mp 1). \tag{3.9b}$$

and Eq. (3.9b) holds for S and H only. The constancy of E follows from the fact that the Lorentz force, and the constraint force for S and H, are both perpendicular to the velocity. (3.9a) comes from (3.6) and (3.8) and (3.9b) from (3.5) and (3.6). The kinetic energy E is non-negative. This is manifest for P and S. For H it can be seen by noting that $\vec{x}$ is time-like, and therefore $\vec{v}$ is space-like.

The two constants give a geometric description of the motion: In the Plane the orbit is a circle with center $\vec{c}$ whose radius increases with the energy. The group of rigid motions of E^2 takes an orbit to an orbit with the same energy, of course. This is the sense in which the classical dynamics in a constant magnetic field retains the symmetry of the manifold.

For S, the positivity of E implies that $\vec{c}$ is outside the unit sphere, and Eq. (3.5) and (3.6) that the orbit is the circle of tangency of a cone, whose apex is at $\vec{c}$, with the unit sphere. This is illustrated in Fig 1. The rotations in E^3 take orbits to orbits with the same energy; the classical dynamics preserves the symmetry of the manifold.

On H, the orbits are the lines of tangency of cones whose apex is at $\vec{c}$ with the hyperboloid. The positivity of E implies that $-1 \leq \vec{c} \cdot \vec{c}$. There are two cases: 1) $\vec{c}$ is time like: $-1 \leq \vec{c} \cdot \vec{c} \leq 0$, which is the case if $0 \leq E < B^2$. The orbits in this case are all closed. 2) $\vec{c}$ is space like, which is the case if the energy is high, $E > B^2$, and the orbits are all open. The two kinds of orbits are illustrated in Figs. 2 and 3. Lorentz transformations in M^3 take orbits to orbits with the same energy, as is the case for the free dynamics.

4. SPECTRA OF LANDAU HAMILTONIANS

Landau Hamiltonians are the self-adjoint operators obtained from the kinetic energy E by replacing the velocity by the operator valued 1-form: $\mathbf{V} \equiv -\mathrm{id} - \mathbf{A}(\vec{x})$, where $\mathbf{b} = \mathrm{d}\mathbf{A}$. It turns out that much can be said about their spectra using elementary algebraic methods, and solving (at most three) first order ordinary differential equations.

The spectra of Landau Hamiltonians in the plane have been derived in $n + 1$ ways by 2^n authors. We reproduce one such derivation here, because it is a prototype of the methods we shall use below for the other cases.

The Landau Hamiltonian is the (self-adjoint, differential) operator corresponding to the energy (3.9a),

$$H_P \equiv \mathbf{V} \cdot \mathbf{V}. \tag{4.1}$$

The relevant commutations are:

$$[V_1, V_2] = iB, \quad [\mathbf{V}, \vec{c}] = 0, \quad [c^1, c^2] = -i/B. \tag{4.2}$$

The two components of $\mathbf{V}$ satisfy the canonical commutation up to a multiple B. $H = V_+ V_- + B$ with $V_\pm \equiv V_1 \mp iV_2$ being creation and annihilation operators. The spectrum is therefore a sequence of points separated by $2B$ whose *vacuum state* solves the first order ordinary differential equation:

$$V_- \, \psi_- = 0. \tag{4.3}$$

In the gauge $\mathbf{A} = \frac{B}{2}(-x^2\,dx_1 + x^1\,dx_2)$, $V_- = i(-\frac{d}{dz} + \frac{B}{2}\bar{z})$, with $z \equiv x^1 + ix^2$, the vacuum is $\psi_-(z) = \exp(-B\,z\bar{z}/2) \in L^2(d^2x)$. The sequence obtained by successive applications of V_+ on ψ_- does not terminate because V_+ has no (square integrable) vacuum. The infinite degeneracy comes from the fact that H_P commutes with the generator of (magnetic) translations[7], $B\vec{c}$.

To determine the spectra of $H_{S,r}$, it is convenient to extend the operators to the embedding space, i.e. as operators on $L^2(E^3) = L^2(S) \otimes L^2(r)$ and $L^2(M^3) = L^2(H) \otimes L^2(r)$. This is done as follows. Let $\vec{J} \equiv \mp B\vec{c}$ on the surface $r = 1$, and extend $\vec{J}$ to all r by *scale invariance*:

$$\begin{aligned}
J^0 &= \mp B\,x^0/r + (x^1 V_2 - x^2 V_1), \\
J^1 &= \mp B\,x^1/r \pm (\pm x^2 V_0 - x^0 V_2), \\
J^2 &= \mp B\,x^2/r \mp (x^0 V_1 \mp x^1 V_0).
\end{aligned} \tag{4.4}$$

The Landau Hamiltonian, from Eq. (3.8b), is

$$H_{S,H} = \vec{J} \cdot \vec{J} \mp B^2. \tag{4.5}$$

Because of scale invariance the spectra of the operators restricted to the manifold and the operators in the embedding space are the same.

Besides scale invariance, the $\vec{J}$'s have additional interesting properties: They satisfy $su(2)$ algebra for S, and $su(1,1)$ algebra for H, and operate on vectors like the generators of rotation for S and like the generators of Lorentz transformation for H. The relevant commutation relations are:

$$\begin{aligned}
[V_0, V_1] &= i\,b_{01}, \quad [V_1, V_2] = i\,b_{12}, \quad [V_2, V_0] = i\,b_{20} \\
[J^0, J^1] &= i\,J^2, \quad [J^1, J^2] = \pm i\,J^0, \quad [J^2, J^0] = i\,J^1 \\
[J^1, x^1] &= 0, \quad [J^1, x^2] = \pm i\,x^0, \quad [J^1, x^0] = i\,x^2
\end{aligned} \tag{4.6}$$

The representations of these groups are labeled by (j, ℓ), with ℓ the eigenvalues of J^0 and $\pm j(j+1)$ the eigenvalues of $\vec{J} \cdot \vec{J}$ [V. Bargmann (1947), N. Ja. Vilenkin (1968), Wybourne (1974)]. $J_\pm \equiv J^1 \pm iJ^2$ are ladder operators on ℓ.

The point spectra can be determined by solving an eigenvalue problem associated with the first order differential operator J^0 and two first order ordinary differential equations for *the vacuum states* $J_\pm\,\psi_\pm = 0$.

We choose a gauge (in spherical coordinates)

$$\mathbf{A} = \mathrm{B}\,\mathrm{K}(\tau)\,\mathbf{d}\phi, \quad \mathrm{K}'(\tau) = k(\tau), \quad \mathrm{K}(0) = 0, \tag{4.7}$$

$k(\tau)$ given in 2.2. This gauge field is smooth on H and on S it is smooth away from the south pole. (One can not do better on S if $B \neq 0$). One finds

$$J^0 = -i\partial_\phi \mp B. \tag{4.8}$$

From this it follows that

$$\mathrm{Spect}\,(\mathrm{J}^0) = \big\{\mathrm{m} \mp \mathrm{B} \mid \mathrm{m} \in \mathrm{Z}\big\}. \tag{4.9}$$

4.1 $SU(2)$:

The unitary representations of $SU(2)$ are such that $m \in \{-j, -j+1, \ldots, j-1, j\}$, with $2j$ a (non-negative) integer. It follows that $2B$ must be integer, which gives an easy way of seeing Dirac quantization. Furthermore,

$$\mathrm{Spect}_S(\vec{J}\cdot\vec{J}) \subseteq \left\{ j(j+1) \mid j = \begin{cases} 0, 1, \ldots, \infty, & \text{if } B = 0 \ (\mathrm{mod}1) \\ \frac{1}{2}, \frac{3}{2}, \ldots, \infty, & \text{if } B = \frac{1}{2} \ (\mathrm{mod}1) \end{cases} \right\}.$$

The positivity of H_S restricts the spectrum[8] to that given in Eq. (1.2).

4.2 $SU(1,1)$:

Fix $m \in Z$. The vacua equations have the solutions:

$$\psi_{m,\pm}(\tau,\phi) = \Big(\exp im\phi \Big) \Big(\sinh(\tau/2) \Big)^{\pm m} \Big(\cosh(\tau/2) \Big)^{\pm(m+2B)}. \tag{4.10}$$

Smoothness at $\tau = 0$ imposes $\pm m > 0$. $\int d\tau\ k(\tau) \mid \psi_{m,\pm}(\tau,\phi)\mid^2 < \infty$ forces $\pm 2(m+B) < -1$. These inequalities rule out the $\psi_{m,+}$ for all m (recall that $B \geq 0$), and give the eigenfunctions $\psi_{m,-}$ with m a *negative integer* and $M > \frac{1}{2} - B$. From $\vec{J}\cdot\vec{J} = -J^0(J^0 - 1) + 2J_+J_-$ the vacuum states are square integrable eigenfunctions of $\vec{J}\cdot\vec{J}$ and J^0. On each of these one builds a semi-infinite tower of states by applying J_+. As a consequence:

$$\mathrm{Spect}_H(\vec{J}\cdot\vec{J}) \supseteq \left\{ -(\mathrm{n}-\mathrm{B})(\mathrm{n}-\mathrm{B}+1) \mid \mathrm{n} \in \mathrm{Z}_+, \ \mathrm{n} < \mathrm{B} - \frac{1}{2} \right\}, \tag{4.11}$$

and each point in the set is infinitely degenerate. This gives the isolated points in the spectrum.

We can not get complete handle on the continuous part of the spectrum in these elementary ways. Since anyway it is for the isolated points that we define and compute an index for, we leave this at that.

That the spectrum has no continuous spectrum at low energies, (in the interval $[\frac{1}{4}, \frac{1}{4} + B^2]$), is in agreement with the fact that the orbits at low energies (i.e. $B^2[0, 1]$) are all closed. Conversely, that there is continuous spectrum at high energies is in agreement with the fact that at high energies classical orbits go to infinity.

5. THE INDEX

5.1 A circle of Landau Hamiltonians:

Landau Hamiltonians are defined directly on P, S and H, by the quadratic form:

$$\langle \psi | H | \psi \rangle = \langle \mathbf{V}\psi | \cdot | \mathbf{V}\psi \rangle. \tag{5.1}$$

This gives, in index notation, the formal differential operator:

$$H = \frac{1}{\sqrt{g}}(-i\partial_i - A_i)\sqrt{g}\,g^{ij}(-i\partial_j - A_j), \tag{5.2}$$

with

$$\mathbf{g} = (d\tau)^2 + k^2(\tau)\,(d\phi)^2, \tag{5.3}$$

$k(\tau)$ given in Eq. (2.2). We define a circle of Landau Hamiltonians, $H(\theta)$, by the boundary condition

$$\Psi(\tau, \phi = 0) = (\exp 2\pi i\theta)\,\Psi(\tau, \phi = 2\pi). \tag{5.4}$$

5.2 A circle of eigenvalue problems:

Associated with the circle of Landau Hamiltonians, $H(\theta)$, $0 < \theta < 1$, we have a circle of eigenvalue problems that we now proceed to analyze by separation of variables. Take the vector potential as in Eq. (4.7). Let,

$$\Psi(\tau, \phi) = \left(\exp im\phi\right)\psi_m(\tau), \ m \in \{-\infty, \ldots, -1 + \theta, \theta, \theta + 1, \ldots \infty\}. \tag{5.5}$$

We get the family of eigenvalue problems associated with ordinary differential operators on $L^2(k(\tau)d\tau)$ with Dirichlet boundary conditions at $\tau = 0$ (and $\tau = \pi$ for S) if $m \neq 0$:

$$H_m = -k^{-1}(\tau)\frac{d}{d\tau}\,k(\tau)\frac{d}{d\tau} + \left(\frac{m - BK(\tau)}{k(\tau)}\right)^2. \tag{5.6}$$

Let:

$$x \equiv \begin{cases} \tau^2/2; \\ -\cos(\tau); \\ \cosh(\tau), \end{cases} \ p(x) \equiv k^2(\tau); \quad q_m(x) \equiv \begin{cases} m - Bx; \\ m - Bx - B; \\ m - Bx + B, \end{cases} \tag{5.7}$$

for P, S and H respectively. We find

$$H_m = -\frac{d}{dx}p(x)\frac{d}{dx} + \frac{q_m^2(x)}{p(x)}, \qquad (5.8)$$

as an operator on (an appropriate domain in) $L^2(dx)$. This gives an hypergeometric eigenvalue problem, and it can be solved by a machine which we outline in the next section:

5.3 A Hypergeometric Machine [Nikiforov and Uvarov (1988)]:
One starts with the observation that the ordinary differential equation (in the complex):

$$p(x)y''(x) + \ell(x)y'(x) + \lambda y(x) = 0, \qquad (5.9)$$

with $p(x)$ a quadratic polynomial, $\ell(x)$ linear and λ a number, is solved by:

$$\lambda_n = -n\ell' - \frac{n(n-1)}{2}p''$$
$$y_n(x) = \frac{1}{\rho^2(x)}\frac{d^n}{dx^n}(p^n(x)\rho^2(x)), \qquad (5.10)$$

where $\rho(x)$ is a solution of:

$$(p(x)\rho^2(x))' = \ell(x)\rho^2(x), \qquad (5.11)$$

and $n \in Z_+$. $y_n(x)$ are polynomials[9]. (5.10) is known as a generalized Rodrigues formula.

The L^2 solutions of the differential equation $H_m\psi = E\psi$ reduces to the study of solutions of 5.9 by the following procedure: Choose μ so that

$$q_m^2(x) - \mu p(x) \equiv P^2(x), \qquad (5.12)$$

with $P(x)$ *a linear function* of x. There are, in general, two complex values, μ_1 and μ_2 that will do that. Set $\psi(x) = \rho(x)y(x)$. $y(x)$ solves 5.9 with

$$\lambda = P'(x) + E - \mu, \quad \ell(x) = p'(x) + 2P(x), \quad \rho(x) = \exp \pm \int^x dx' \left(\frac{P(x')}{p(x')}\right). \quad (5.13a)$$

In particular, the (generalised) eigenvalues, E_n, of (5.8) are

$$E_n = -(2n+1)P'(x) - \frac{n(n+1)}{2}p''(x) + \mu. \qquad (5.13b)$$

Not all the solutions satisfy the boundary conditions. This has to be checked separately by examining ρ (recall that $y_n(x)$ are polynomials). The solutions that do, solve the eigenvalue problem[10].

5.4 Stable and Moving States

The solutions to the circle of eigenvalue problems are of two kinds. Most of the states are such that their eigenvalues are independent of θ. We call them the stable states; *Ipso facto* they do not contribute to the index. On the other hand, the few eigenstates whose eigenvalues do depend on θ, are those we shall have to follow in order to determine the index. We call these the moving states. The hypergeometric machine separates the stable from the moving states according to whether $\mu = 0$ or not. The stable states all have $\mu = 0$ and we describe them first, even though, as we have said, for our purpose these states are redundant. The interesting, moving states are analyzed in the following subsection.

5.5 The Stable States

Eq. (5.12) has the immediate solution

$$\mu_1 = 0, \quad P_{1\pm}(x) = \pm q_m(x), \tag{5.14}$$

Plugging (5.14) in Eq. (5.13), one finds:

$$E_{nm}^{1,\pm} = \pm B\,(2n+1), \quad \rho(x) = \exp \mp (Bx/2)\, x^{\pm m/2}, \quad \text{for } P; \tag{5.15a}$$

$$E_{nm}^{1,\pm} = n(n+1) \pm B\,(2n+1), \quad \rho(x) = (1+x)^{\pm m/2}\,(1-x)^{\pm(B-m/2)}, \quad \text{for } S; \tag{5.15b}$$

$$E_{nm}^{1,\pm} = -\,n(n+1) \pm B\,(2n+1), \quad \rho(x) = (x-1)^{\pm m/2}\,(1+x)^{\mp(B+m/2)}, \quad \text{for } H. \tag{5.15c}$$

The $\pm$ signs correspond to those in 5.14.

5.5.1 The Plane

From 5.15a and the behavior of ρ at $x = 0$ and $x = \infty$, the eigenvalues are:

$$E_{nm}^{1,+} = B\,(2n+1), \quad n \in Z_+, \quad m \in \{\theta, \theta+1, \ldots, \infty\}. \tag{5.16a}$$

There are *infinitely many* such eigenstates whose eigenvalue is independent of θ. They do not go anywhere as θ varies and are irrelevant for the index. The stability of infinitely many states is a consequence of the stability of the essential spectrum [Weyl (1909), Kato (1967)].

5.5.2 The Sphere

For the regularity of ρ at $x = \pm 1$ we need $\pm m \geq 0$, $\pm(2B - m) \geq 0$. The lower sign in 5.15b is rejected for all m (recall $B \geq 0$) and the upper sign is constrained by $0 \leq m \leq 2B$. Hence the eigenvalues are:

$$E_{nm}^{1,+} = n(n+1) + B\,(2n+1),$$
$$n \in Z_+, \quad m \in \{\theta, 1+\theta, \ldots, \}, \, m \leq 2B. \tag{5.16b}$$

With each spectral point $(\theta \neq 0)$ there are $[2B]$ eigenstates whose energy is θ independent and are irrelevant for the index.

5.5.3 The Hyperbolic Plane

The regularity of ρ at $x = 1$ and square integrability at infinity require $\pm m \geq 0$ and $0 \leq n < -\frac{1}{2} \pm B$ respectively. The lower sign is always rejected (since $B \geq 0$), and the upper sign survives for $n < B - \frac{1}{2}$ and m non-negative. The eigenvalues are:

$$E_{nm}^{1+} = -n(n+1) + B\,(2n+1),$$

$$m \in \{\theta, 1+\theta, \ldots, \infty\}, \ n \in Z_+, \ n < B - \frac{1}{2} \cdot \tag{5.16c}$$

At each spectral point, there are *infinitely many* eigenstates whose eigenvalues are independent of θ that do not go anywhere and are irrelevant for the index. The stability of these states is a consequence of the stability of the essential spectrum, as was the case for the plane.

5.6 The Moving States

The states that are relevant for the index, are those states whose energies depend on θ. They arise from the solutions of Eq. (5.12) with non-zero μ's, which we denote μ_2. There is a finite number of such states for each spectral point of the $(\theta = 0)$ Landau Hamiltonian.

5.6.1 The Plane:

From 5.12 and 5.13:

$$\mu_2 = -2mB; \ P_2(x) = \pm(Bx + m); \ \rho_{2\pm}(x) = \exp\pm(Bx/2)\ x^{\pm m/2}. \tag{5.17a}$$

The upper sign is rejected because of bad behavior at infinity, and the lower sign gives admissible eigenfunctions for $m \leq 0$. Since $P'(x) = -B$, $\ell'(x) = -2B$, Eq. (5.10) and (5.13) give

$$E_{nm}^{2-} = B(1 + 2n - 2m), \ n \in Z_+, \ m \in \{-\infty, \ldots, -1 + \theta\}. \tag{5.18a}$$

The lowest spectral point $E = B$ has no moving states, the next one, $E = 3B$, has one moving state *etc.* The energy of each moving state is a decreasing linear function of θ with slope $2B$. As a consequence, as θ increases from zero to 1, n states from the the (n+1)-st spectral point move down to the spectral point just below. For each point in the Landau spectrum, there is one more state arriving (from above) than departing (below) and the index is unity. A flow diagram is shown in Fig. 5.

5.6.2 The Sphere:

Set $j \equiv m - B$, m given in 5.5. From 5.12 and 5.13 we get:

$$\mu_2 = j^2 - B^2; \quad P_2(x) = \pm(jx - B); \quad \rho_{2\pm}(x) = (1-x)^{\pm\frac{B-j}{2}}(1+x)^{\mp\frac{B+j}{2}}. \quad (5.17b)$$

Regularity of the solutions at $x = \pm 1$ gives: $\pm(B-j) \geq 0$, $\mp(B+j) \geq 0$. The upper sign holds if $j \leq -B$ and the lower sign if $j \geq B$

$$E^{\pm 2}_{n,m} = (n + |j|)(n + 1 + |j|) - B^2, \, for \, |j| \geq |B|. \quad (5.18b)$$

Figure 6 show the flow of the states with θ. The index is zero, as it should be.

5.6.3 The Hyperbolic Plane:

From 5.12 and 5.13 we get:

$$\mu_2 = -(m+B)^2 + B^2; \quad P_2(x) = \pm\big((m+B)x - B\big);$$
$$\rho_{2\pm}(x) = (x-1)^{\pm\frac{m}{2}}(1+x)^{\pm(B+\frac{m}{2})} \quad (5.17c)$$

Regularity of the solutions at $x = 1$ requires $\pm m \geq 0$, and square integrability requires $n + \frac{1}{2} < \mp(B+m)$. The two inequalities rule out the upper sign for non-negative B. The lower sign holds provided $n - B + \frac{1}{2} < m \leq 0$. We therefore find for the moving states, from 5.13:

$$E^{2,-}_{n,m} = -(n - m - B)(n - m - B + 1) + B^2,$$
$$m \in \{\ldots, -2 + \theta - 1 + \theta\}; \, n \in Z_+, n < m + B - \frac{1}{2}. \quad (5.18c)$$

The energy is constant on the lines $n - m = const$. On such a line there are $[n - m]$ (degenerate) moving states, all moving together with θ. The lowest spectral point has no moving state, and the next spectral point has one moving states, etc. until the highest spectral point just below the continuum has $[B - \frac{1}{2}]$ moving states. As θ increases, the states all move down to the spectral point just below. For each spectral point there is always one more state arriving from above than departing below. The index is therefore 1. (The spectral point just below the continuum too has index 1, and the states that arrive all have their birth at the bottom of the continuum). A flow diagram is shown in Fig. 7.

ACKNOWLEDGEMENT

We thank Dr. Andreas Knauff for bringing Comtet work to our attention. The reasearch is supported by BSF, the Israeli Academy of Sceience, and the Fund for the promotion of research at the Technion.

FOOTNOTES

1. That is: $y^{-2}\big((dx)^2 + (dy)^2\big)$.

2. Schrödinger operators with magnetic fields in R^n have been extensively studied and a good part of the mathematical results is reviewed in [Cycon, Froese, Kirsch and Simon, (1986)]. Schrödinger operators with magnetic fields on general Riemanian manifolds have not attracted a lot of attention, for obvious reasons.

3. Landau computed the density of states associated with the infinite degeneracy. A bare hand computation of this density of states is given in [Avron and Seiler (1983)].

4. For the sphere we take a cut from the north to the south pole. This is an *ad-hoc* definition, because there appears to be no natural notion of this index for compact manifolds.

5. In order to follow the eigenvalues one needs to assume some mild continuity properties.

6. *Index* $= +\infty$ means that there are infinitely many arriving and finitely many departing states (similarly for Index $= -\infty$). Infinite indices however, do not occur for the specific operators we study here.

7. The Magnetic translation group is due to [Zak (1964)] who applied it in a solid state physics context. In the context of separation of center of mass in atomic physics see [Avron, Herbst and Simon, (1978)] and [Johnson, Hirschfelder and Yang (1983)].

8. In order to show that one gets *all* the spectrum this way one needs to solve the vacuum equations in $L^2(S)$, for all j's in Eq. (1.2). We shall not do that here.

9. $y_0(x) = 1$, $y_1(x) = \ell(x)$, etc.

10. It is not clear from this outline that one gets a complete spectral resolution in this way. For a discussion of this point see [Nikiforov and Uvarov (1988)].

REFERENCES

J. E. Avron, I. W. Herbst and B. Simon, Ann. Phys., **114**, 431, (1978)

J. E. Avron and R. Seiler, Phys Rev. **27B**, 7763, (1983)

J.E. Avron, R. Seiler and B. Simon, Phys. Rev. Lett., **65**, 2185, (1990).

J. Bellisard, "Ordinary Quantum Hall effect and Noncommutative Cohomology", Bad–Schandau Conference on Localization, W. Weller and P. Ziesche Eds., Teubner, Leipzig, (1988).

V. Bargmann, Ann. Math., **48**, 568, (1947)

H. L. Cycon, R. G. Froese, W. Kirsch and B. Simon, *"Schrödinger Operators"*, Springer, (1986)

S. Coleman, Proceedings of the 1982 International School of Subnuclear Physics, Erice, Zichichi Ed., Plenum, (1983)

A. Comtet, Ann. Phys., **173**, 185, (1987).

A. Comtet and P. J. Houston, J. Math. Phys. **26**, 185, (1985).

A. Connes, Pub. Math. IHES, **62**, 257, (1986)

P.A.M. Dirac, Proc. Roy. Soc. **133A**, 60, (1931)

C. Grosche, Ann. Phys. **187**, 110, (1999)

B. R. Johnson, J. Hirschfelder and Kuo-Ho Yang, Rev. Mod. Phys., **55**, 109, (1983)

T. Kato, *"Perturbation Theory for Linear Operators"*, Springer, 1966.

L.D. Landau in L.D. Landau and E. M. Lifshitz, *"Quantum Mechanics"*, Pergamon, (1977).

R. Laughlin, in *"The Quantum Hall Effect"*, R.E. Prange and S.M. Girvin, Eds., Springer, (1987).

H.P. McKean, J. Diff. Geometry, **4**, 359, (1970).

A. F. Nikiforov and V. B. Uvarov, *"Special Functions of Mathematical Physics"*, Birkhäuser, Basel, (1988)

Ig. Tamm, Z. Phys., **71**, 141, (1931)

A. Terras, *"Harmonic Analysis on Symmetric Spaces"*, Springer, (1985)

N.J. Vilenkin *"Special Functions and the Theory of Group Representations"*, Trans. A.M.S., Providence, (1968)

H. Weyl, Rend. Circ. Mat. Palermo, **27**, 373, (1909)

T.T. Wu and C.N. Yang, Nucl. Phys., **B107**, 365, (1976)

B. G. Wybourne, *"Classical Groups for Physicists"*, Wiley, (1974)

J. Zak, Phys. Rev. **134**, A1602, (1964)

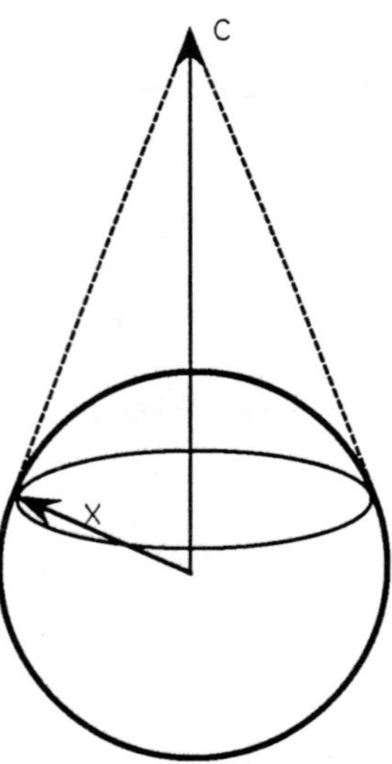

fig. 1

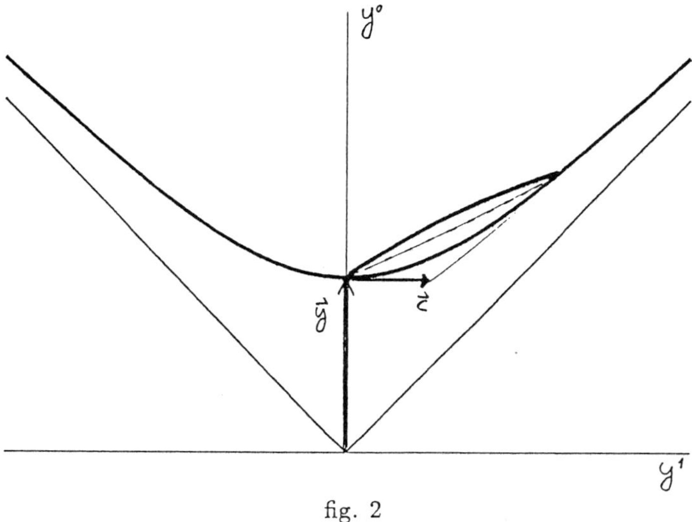

fig. 2

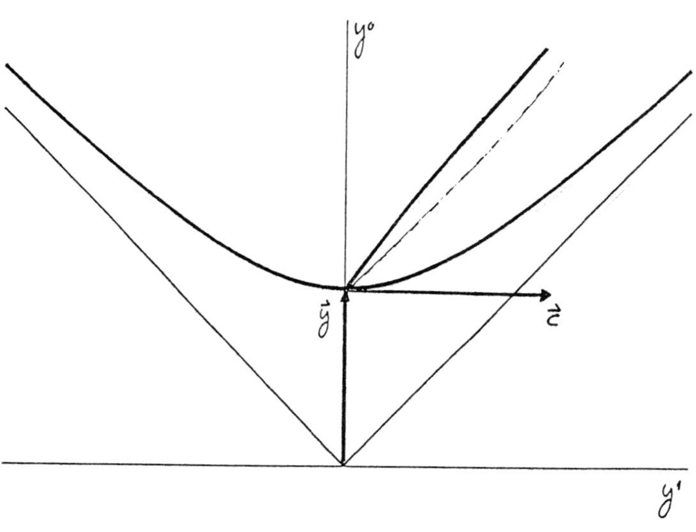

fig. 3

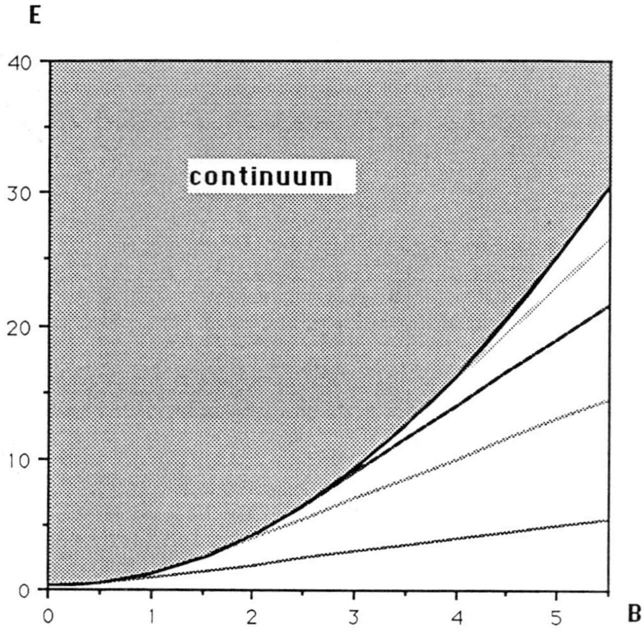

fig. 4

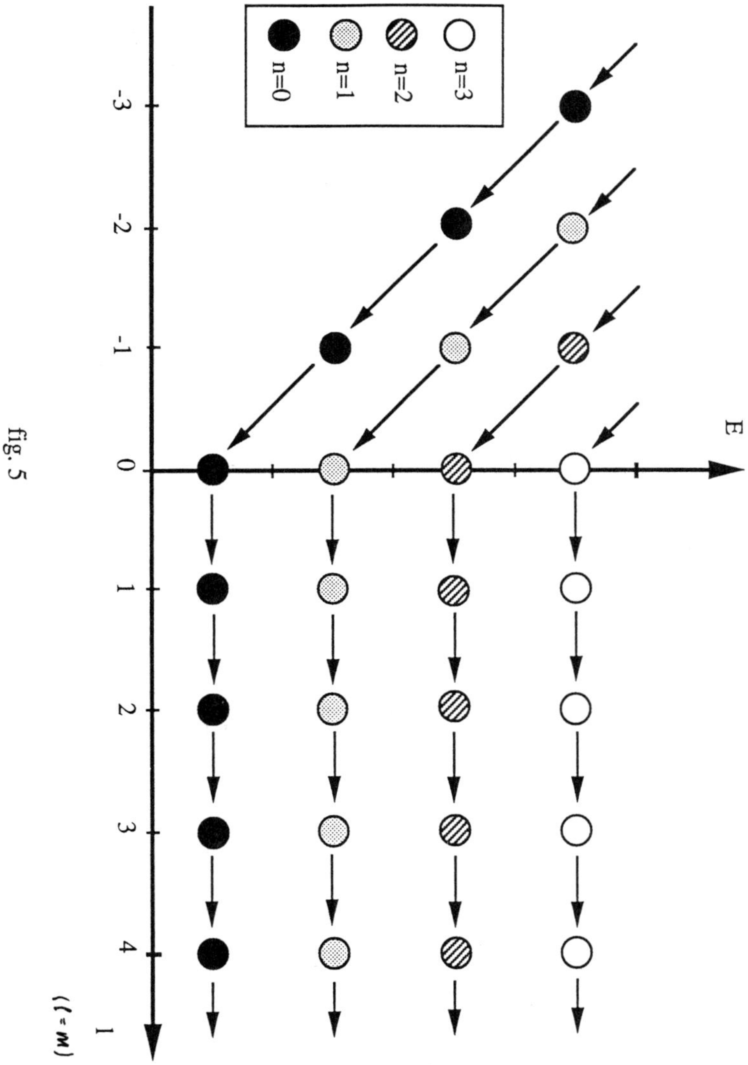
n=0
n=1
n=2
n=3
E
-3
-2
-1
0
1
2
3
4
(m = l)
fig. 5

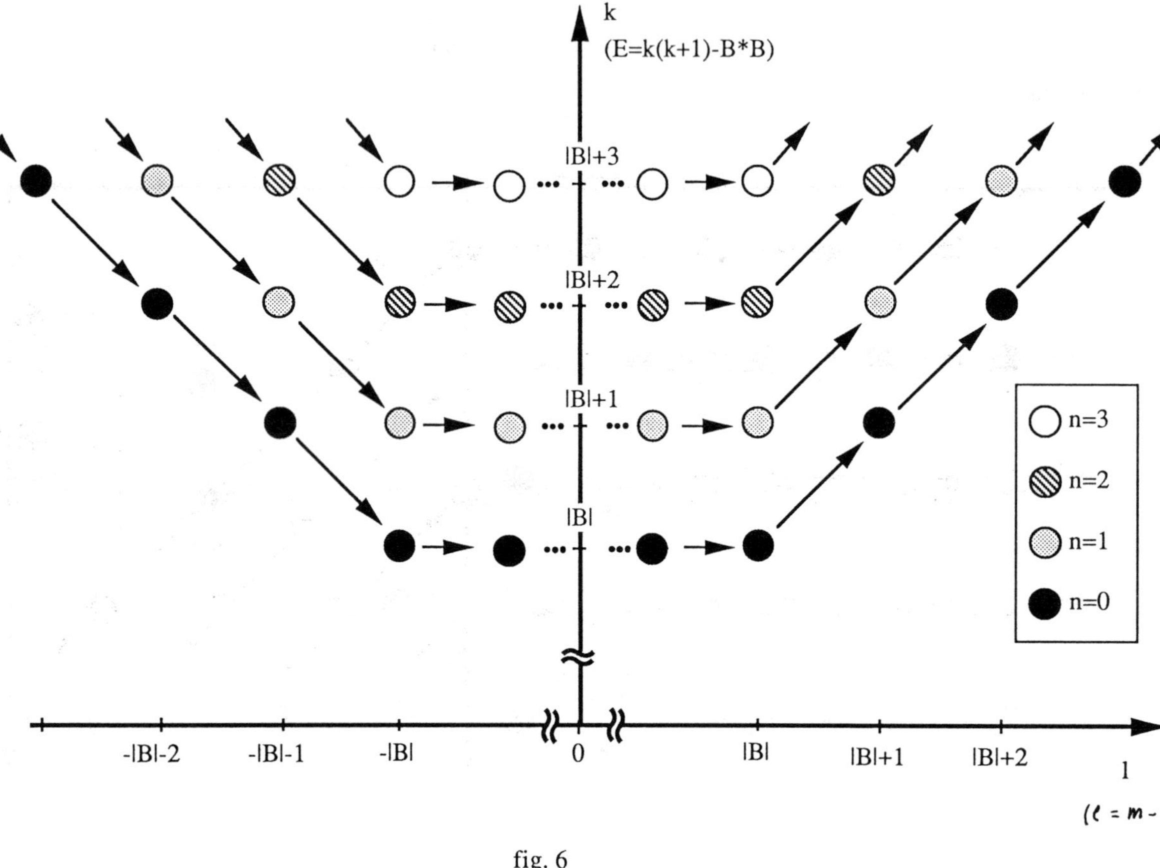

fig. 6

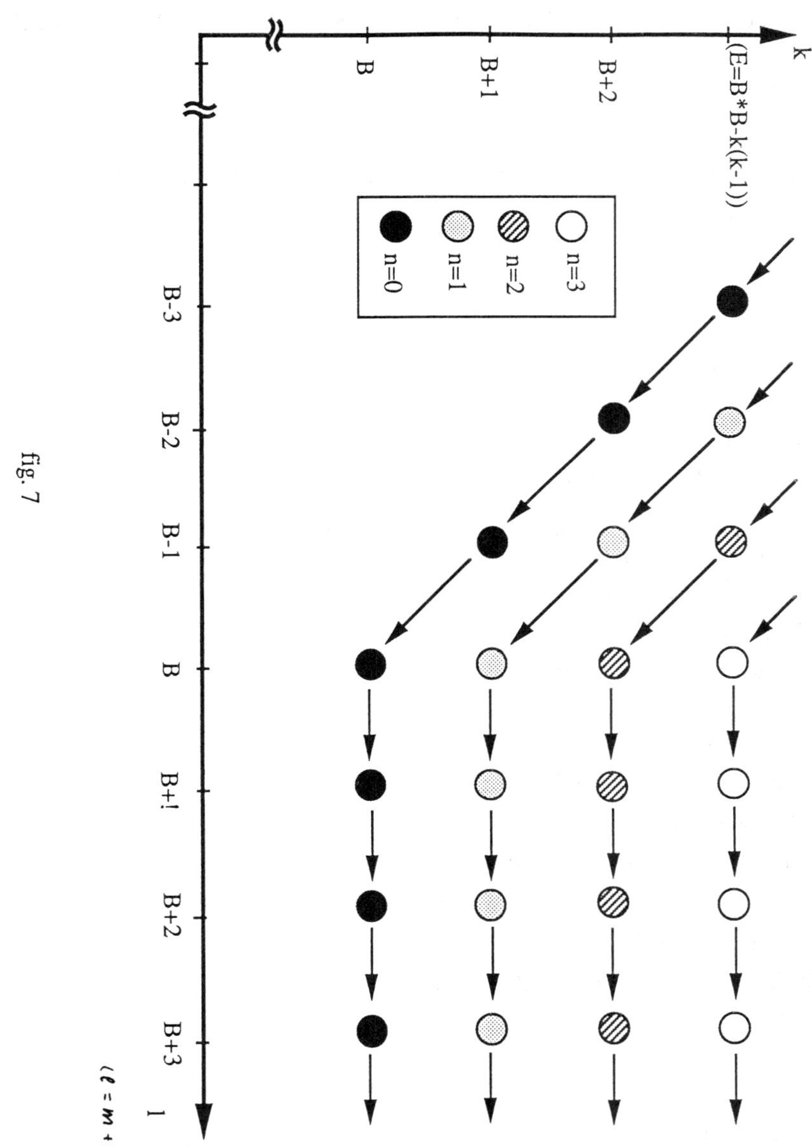

fig. 7

Renormalization Group Analysis and Quasicrystals

J. BELLISSARD

Wissenschaftskolleg zu Berlin & Centre de Physique Théorique de Marseille,CNRS.
On leave from the "Université de Provence", Marseille.

Table of Content

1.INTRODUCTION

Several quantal systems involving scale invariant properties, have been studied during the last few years by means of a Renormalization Group (RG) method. The most useful type of models is probably the hamiltonian describing the motion of a particle, phonon or

118

electron, in a quasicrystal. The first quantity to be calculated is the energy spectrum, from which we usually get others like the density of states (DOS), thermodynamical information, like the heat capacity or the magnetic susceptibility, or even various transport coefficients, like the conductivity. Using the spatial macroscopic symmetries, translations and scale invariance, it is possible to get equations satisfied by the model which happen to be sufficient to compute the spectrum in many cases. In particular the scale invariance will produce fractal spectra and scaling laws for the physical quantities.

The main difficulty is that unlike the 1D case for which the calculation can usually be performed by means of the transfer matrix method, the higher dimensional cases are far from being under control yet. In this short paper we want to give an account of a new strategy using operator algebras which should permit to extend the analysis to higher dimension. Eventhough the method is not yet completely developed, it has already given a certain number of convincing results, and we believe it should be the most efficient way of studying these problems. In this paper we compare it with the transfer matrix formulation for 1D chain and we show that both point of view are equivalent. We will only give an insight of what happens for higher dimensional quasicrystals, for this part of the work is still under progress.

2.JACOBI MATRIX OF A JULIA SET

2.1 The Julia Set of a Polynomial

The simplest model was designed in 1982 [Bellissard(82)], to get a new class of hamiltonians with Cantor spectra. It is the Jacobi matrix associated to a Julia set. Let $P(z)$ $= z^N+p_1z^{N-1}+...+p_{N-1}z+p_N$ be a polynomial with real coefficients. We then consider the dynamical system on the complex plane defined by $z_{n+1}=P(z_n)$. Clearly the point at infinity is fixed by P , and it is attractive, for there is $R>0$ big enough, such that whenever $|z| \geq R$, then $|P(z)| \geq R^N/2$. Let ζ be a fixpoint, namely a solution of $P(\zeta) = \zeta$, and let $D(\zeta)$ be the "domain of attraction of ζ", namely the open set of points z_0 such that $z_n \to \zeta$ as $n \to \infty$. The Julia set $J(P)$ of P is the complement of the union of the attraction domain of all fixpoints. Since the point at infinity is always attractive, $J(P)$ is always compact. A famous theorem by Julia and Fatou [Julia(18), Fatou(19), Douady(82)] asserts that $J(P)$ is completely disconnected whenever all critical points of P are attracted by the point at infinity.

In this section we will restrict ourselves to the case for which the Julia set is contained in the real line and completely disconnected. Let $J = [a,b]$ be its convex hull. Then b is automatically an unstable fixpoint and a belongs to the Julia set. Now each preimage of a and b is also in the Julia set. Therefore, the equations $P(\xi) = a$ or $P(\xi) = b$ have exactly N solutions in the interval $[a,b]$ .It follows that P has exactly $N-1$ critical points in $[a,b]$ and that at each of them, P takes on values outside the interval $[a,b]$. In particular, given $x \in [a,b]$, the equation $P(\xi) = x$ has exactly N real solutions $a \leq w_N(x) < w_{N-1}(x) < ... < w_1(x) \leq b$, such that the intervals $J_k = w_k([a,b])$ $(1 \leq k \leq N$) are disjoint from each other. One can also easily see from this argument that $P(a) = a$ or b . It follows that if a point is not in the union of the J_k 's it cannot be in the Julia set. So that using this argument for the iterated preimages, the Julia set is contained into the intersection J_∞ defined by:

$$J_\infty = \bigcap_{s>0} \bigcup_{k(1),...,k(s)\in[1,N]} w_{k(1)} \circ w_{k(2)} \circ ... \circ w_{k(s)} ([a,b]) . \tag{1}$$

It is easy to check that J_∞ is a Cantor set, namely it is compact, nowhere dense and has no isolated point. Moreover, it is P -invariant so it must also be contained in the Julia set of P, showing that it is actually equal to $J(P)$.

From the previous construction, it follows that there is a unique probability measure μ on the real line, supported by $J(P)$, such that :

$$\mu\{ w_{k(1)} \circ w_{k(2)} \circ ... \circ w_{k(s)} ([a,b])\} = N^{-s} . \tag{2}$$

That this measure is singular continuous is a result of a theorem by Brölin [Brölin(65), Barnsley(83), Barnsley(85)]. Let us now consider the Hilbert space $\mathcal{H} = L^2(J(P), d\mu)$, and let H be the operator of multiplication by x in $\mathcal{H}$. Then H is a self adjoint bounded operator with spectrum given by $J(P)$ and spectral measure μ . So its spectrum is singular continuous and a Cantor set.

2.2 The RG Equation [Bellissard(85)]

Let now D be the operator on $\mathcal{H}$ defined by :

$$DF(x) = (1/N) \sum_{0 \leq k \leq N} F \circ w_k(x) \qquad \Rightarrow \qquad D^*F(x) = F \circ P(x) \tag{3}$$

Then one can easily see that $DD^*=1$, and that H is a solution of the following RG equation:

$$D\{z\mathbf{1}-H\}^{-1}D^* = (P'(z)/N)\,\{P(z)\mathbf{1}-H\}^{-1}\,. \tag{4}$$

In addition, it is easy to check that the set of vectors of the form $H^n D^*\phi$ with $\phi\in\mathcal{H},\, n\in N$ generates a dense subspace of $\mathcal{H}$, namely we will say that D^* is H-cyclic.
Conversely, we get the following result:

Theorem 1: Let $\mathcal{H}$ be a Hilbert space with a selfadjoint bounded operator H and a partial isometry D satisfying $DD^*=1$ and being H-cyclic. If H satisfies (4), the spectrum of H is the Julia set of the polynomial P.   ◊

Proof : Let us denote by $\rho(H)$ the resolvent set of H. If $z_0\in\rho(H)$, then $z\mathbf{1}-H$ is invertible with a bounded inverse. Therefore the right hand side of (4) is bounded which implies that $P(z_0)\in\rho(H)$. Conversely, let $z_0\in C$ be such that $P(z_0)\in\rho(H)$. There is a small disc $\mathcal{D}$ centered at z_0 contained in $P^{-1}(\rho(H))$. Since H is selfadjoint, without loss of generality, we can assume that z_0 is real, otherwise we know that it belongs to $\rho(H)$. Using (4), if f is a complex function holomorphic in a neighbourhood of $Sp(H)$, a Cauchy formula applied to (4) leads to :

$$D\,f(H)\,D^* = (1/N)\sum_{1\leq k\leq N} f\mathrm{ow}_k(H)\,, \tag{5}$$

where the w_k's are the solutions of $P(w)=z$. It is a standard result that the w's are holomorphic in a cut plane, the thresholds being given by the values $P(\zeta)$ where ζ varies in the set of critical points of P. Moreover, the w's are continuous at the threshold. Therefore $w_k(H)$ is a well defined normal operator. Let Q_1,Q_2 be two polynomials. Applying (5) to the map $\zeta\to Q_1 Q_2(\zeta)\,(z-\zeta)^{-1}$ for $z\in\mathcal{D}$, we get :

$$D\,Q_1(H)\,\{z\mathbf{1}-H\}^{-1}\,Q_2(H)\,D^* = (1/N)\sum_{1\leq k\leq N}Q_1\mathrm{ow}_k(H)\,\{z\mathbf{1}-w_k(H)\}^{-1}\,Q_2\mathrm{ow}_k(H)\,. \tag{6}$$

Since the right hand side is holomorphic in $\mathcal{D}$, so does the left hand side. In particular the spectral projection χ_δ on the interval $[z_0-\delta, z_0+\delta]$, and given by

$$\chi_\delta = \text{s-lim}_{\varepsilon\downarrow 0^+} \int_{[-\delta,\delta]} dx/2i\pi \left\{ (z_0+x-i\varepsilon - H)^{-1} - (z_0+x+i\varepsilon - H)^{-1} \right\} ,$$

must satisfy $D\, Q_1(H)\, \chi_\delta\, Q_2(H)\, D^* = 0$ for any choice of Q_1, Q_2. Since D^* is H-cyclic, this implies that $\chi_\delta = 0$ showing that $z_0 \in \rho(H)$.

Thus we conclude that $\rho(H)$ is both P- and P^{-1}-invariant. Since it contains a neighbourhood of the point at infinity, and since it is not equal to the entire complex plane, it is automatically equal to the complement of the Julia set.   ◊

2.3 The Jacobi Matrix of P

What makes this example interesting is that it can be written as a Jacobi semi infinite matrix. For indeed, the monomial functions $X^n : x \in J(P) \to x^n \in C$, are linearly independent in $\mathcal{H} = L^2(J(P), d\mu)$, and generate a dense linear space by the Stone-Weierstrass theorem. By a Gram-Schmidt procedure, we get a total sequence of mutually orthogonal monic polynomials $P_0 = 1, \ldots, P_n = X^n + O(X^{n-1}), \ldots$, such that [Barnsley(83)]:

$$P_{n+1} = (X - a_n) P_n - b_n^2 P_{n-1} , \tag{7}$$

where the a_n's are real, and the b_n's are positive numbers. If we denote by p_n the unit vector in $\mathcal{H}$ parallel to P_n, the matrix of H is tridiagonal and real symmetric, with the a_n's on the main diagonal, and the b_n's as the off-diagonal elements. So that H acts on $l^2(N)$ through the Jacobi matrix H_P as follows:

$$H_P \psi(n) = b_{n+1} \psi(n+1) + a_n \psi(n) + b_n \psi(n-1) , \qquad n \geq 0 . \tag{8}$$

Moreover, the operator D becomes the dilation operator

$$D\psi(n) = \psi(Nn) , \tag{9}$$

which comes from the fact that $D^* P_n = P_n oP = P_{nN}$. In particular we get the following theorem [Barnsley(85), Bellissard(85)]:

Theorem 2: if D is defined by (9) on $l^2(N)$, there is a unique Jacobi matrix H_P solution of the RG equation (4).   ◊

Remark: the coefficients of H_P are computable by a recursion formula [Bellissard(82)]. In the case for which $P(z)=z^2-\lambda$ (with $\lambda>2$) the diagonal coefficients vanish, whereas the off-diagonal ones satisfy :

$$b_0=0, \qquad b_{2n}^2 + b_{2n+1}^2 = \lambda, \qquad\qquad b_{2n-1}b_{2n} = b_n . \qquad (10)$$

In this latter case it has been proved that for $\lambda>3$, this sequence is the uniform limit of periodic sequences of period 2^N [Bellissard(82,85), Barnsley(83,85)].

More generally, a sequence is limit periodic if it is the uniform limit of periodic sequences [Bohr(47), Simon(82)]. The main property of the Jacobi matrix H_P associated to the polynomial P is summarized in the following conjecture :

Conjecture: let P be a polynomial with completely disconnected Julia set contained in the real line. Then the diagonals of the associated Jacobi matrix H_P are uniform limits of periodic sequences of periods N^n . $\qquad\qquad\Diamond$

Remark: the conjecture has been proved in [Barnsley(85)] in the following cases :
(i) if $P(z) = z^2-\lambda$, with $\lambda>3$.
(ii) if $P(z) = a^N T_N(z/a)$ where T_N is the Tchebyshev polynomial of degree N , and $a>\sqrt{3}/2$.
(iii) if $P(z) = a^N T_N(z/a) + b$ where $N=3, a\geq 5, |b|\leq 5$, or $N=4, a\geq 2, |b|\leq 22$. $\qquad\Diamond$

3. SIERPINSKY LATTICE IN A MAGNETIC FIELD

3.1 The 2D Sierpinsky Lattice [Alexander(83,84), Rammal(84)]

The Sierpinsky lattice S in 2D is usually constructed according to the fig.1 below. Namely, let e_1, e_2 be two unit vectors making an angle of $60°$. Then S is contained in the set $Ne_1 + Ne_2$. Let then S_k be the subset of points $x \in S$ with $x = me_1 + ne_2$ and $0 < m+n \leq 2^k$. S_k is recursively constructed as $S_1 = \{me_1 + ne_2 ; 0 \leq m+n \leq 2\}$, $S_{k+1} = S_k \cup \{S_k + 2^k e_1\} \cup \{S_k + 2^k e_2\}$ for $k \geq 1$, and $S = \cup_{k \geq 1} S_k$.

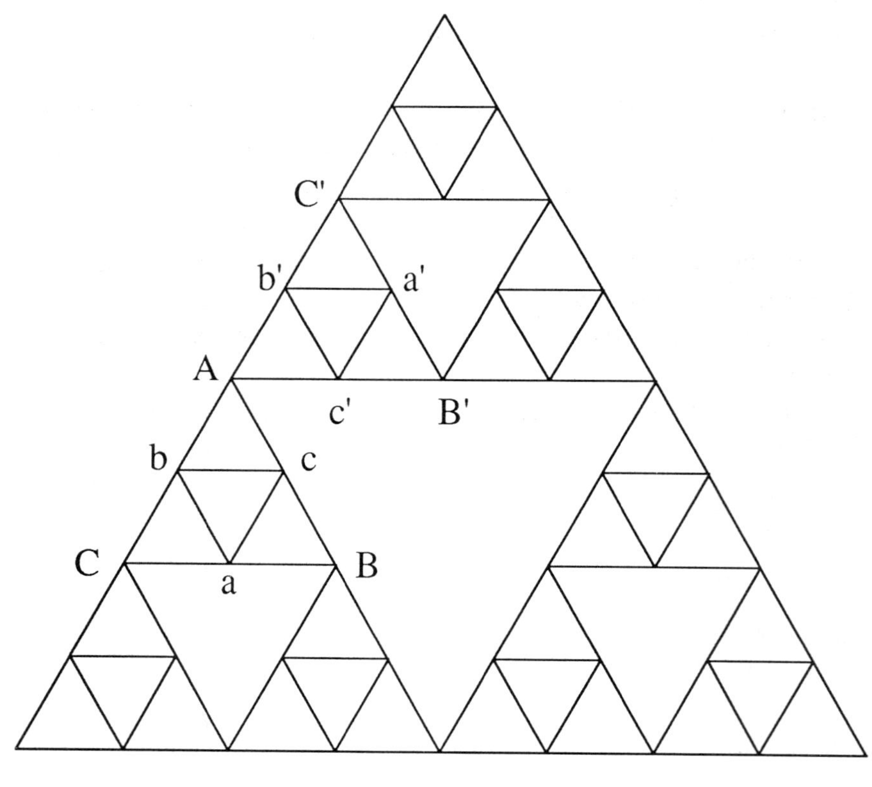

Fig.1- The subset S_3 of the Sierpinsky lattice in 2D -

From this construction it follows that $2S$ is included in S . A site in $2S$ is called "even", the others "odd". Any odd site admits the decomposition $2x+y$ where $x \in S$ and $y \in T=\{e_1, e_2, e_1+e_2\}$. The subsets $T(x) = T+2x$ are called "blocks". If $x \in S$, its nearest neigbours are all points in S within a distance 1 of x .

3.2 The Laplacean on S

The Laplace operators Δ_+ and Δ_- are defined on the Hilbert spaces $l^2(S)$ and $l^2(S \setminus \{0\})$ respectively by:

$$\Delta_+\phi\,(0) = \sqrt{2}\,\textstyle\sum_{x';|x'|=1}\,\phi(x') \,, \qquad\qquad \Delta_+\phi\,(x) = \sqrt{2}\psi(0)+\textstyle\sum_{x'\neq 0;|x'-x|=1}\,\phi(x') \,,$$

if $|x| = 1$, and

$$\Delta_+\phi\,(x) = \textstyle\sum_{x';|x'-x|=1}\,\phi(x') \,, \qquad\qquad \text{if } |x| > 1, \;\; \phi \in l^2(S) \,, \qquad\qquad (11a)$$

$$\Delta_-\psi\,(x) = \textstyle\sum_{x';|x'-x|=1}\,\psi(x') \,, \qquad\qquad x \in S \setminus \{0\} \,, \;\; \psi \in l^2(S \setminus \{0\}) \,. \qquad (11b)$$

Our goal is to compute the spectrum of Δ . In order to do so we will use the scale invariance of the Sierpinsky lattice. The main result is the following [Rammal(84), Bellissard(85)]

Theorem 3: The spectrum of $\Delta_{\pm}$ is made of two infinite sequences of eigenvalues of infinite multiplicity accumulating on the Julia set of the polynomial $P(z) = z(z-3)$. The first sequence consists of one isolated eigenvalue in each gap of $J(P)$, whereas the other consists of one edge of each gap of $J(P)$. $\lozenge$

Proof: Let us introduce the dilation operator D defined by

$$D\psi\,(x) = \psi(2x) \qquad\qquad x \in S, \qquad\qquad \psi \in l^2(S) \qquad\qquad (12)$$

It is a partial isometry such that $DD^*=1$. Then we claim that $\Delta_{\pm}$ are solutions of the following RG equation [Bellissard(85)]:

$$D\{z1-\Delta\}^{-1}D^* = (z-2)(z+1)/(z+2)\,\{P(z)1-\Delta\}^{-1} \,, \qquad P(z) = z(z-3) \qquad (13)$$

It follows immediately that if $z \in \rho(\Delta)$ then $z \neq -2$, otherwise the right hand side of (13) would be unbounded. Thus $z = -2$ belongs to the spectrum. Let χ_{-2} be the corresponding eigenprojection. From (13), it follows that $D \chi_{-2} D^* = 4\{10-\Delta\}^{-1}$. Note that by (11), Δ satisfies the a priori bound $\|\Delta\| \leq 4$ so that χ_{-2} is a non zero projection. Now if $z \in \rho(\Delta)$ and $z \neq -1,2$, then (13) shows that $P(z) \in \rho(\Delta)$.

Hence the spectrum is "essentially" P^{-1} -invariant in that there are only two possible exceptions. Thus we expect that the Julia set $J(P)$ of P plays a role. It is easy to check that the convex hull of $J(P)$ is the interval $[-1,4]$. By the Julia-Fatou theorem, it is completely disconnected for its critical point $z = 3/2$ satisfies $P(3/2) = 9/4 \notin [-1,4]$, and does not belong to $J(P)$. In particular, -2 does not belong to $J(P)$, nor does its inverse images $z = 1, 2$.

From (13), $z = 2$ requires a special treatment. An elementary calculation then shows that if χ_2 is the corresponding eigenprojection, then $D\chi_2 D^* = 0$. This means that any eigenvector ψ of Δ with eigenvalue 2 is localized on the odd sites of S . From the eigenvalue equation it follows that $\psi(a) = \psi(b) = \psi(c)$ on each block (see fig.1). Applying the eigenvalue equation at each even site, we can check that on each triangle $T_2(x) = 4x + S_2$ the eigenvector must vanish, showing that $\chi_2 = 0$. Thus $2 \notin Sp(\Delta)$. On the other hand one can check that each inverse image of 1 is actually an eigenvalue, giving then one isolated eigenvalue in each gap of $J(P)$. In addition, if χ_ζ is the eigenprojection corresponding to the eigenvalue ζ , then

$$D\chi_\zeta D^* = c(\zeta)\chi_{P(\zeta)} , \qquad\qquad c(\zeta) = (\zeta-2)(\zeta+1)/(\zeta+2)P'(\zeta) . \qquad (14)$$

The value $z = -1$ also requires a special treatment. By the same argument if χ_{-1} is the corresponding eigenprojection, then $D\chi_{-1}D^* = 0$. Again, any eigenvector ψ of Δ with eigenvalue -1 is localized on the odd sites of S . From the eigenvalue equation it follows that $\psi(a) + \psi(b) + \psi(c) = 0$ on each block (see fig.1). But now it is a simple matter of calculation to check that such a vector indeed exists, showing that χ_{-1} is non zero. Thus any inverse image of -1 is also an eigenvalue with eigenprojector again satisfying (14). In this way we get another countable family of eigenvalues given by the gap edges of the Julia set of P .

Both families of eigenvalues actually accumulate on $J(P)$ showing that the spectrum contains the union of $J(P)$ and $\{-2\} \cup \{P^{-n}(1); n \geq 0\}$.

Would the dilation operator be Δ-cyclic, the argument in §2 could be used to conclude that the rest of the spectrum is actually the entire Julia set. However here, the subspace $\mathcal{H}_\perp$ of vectors ϕ in the Hilbert space for which $D\Delta^n\phi = 0$ for all $n\in N$, is closed and not reduced to zero for it contains the image of the projection χ_{-1} . However, $\mathcal{H}_\perp$ is Δ-invariant, and if $\phi\in \mathcal{H}_\perp$, ϕ vanishes on the even sites. So that $\Delta\phi= \oplus_T \Delta_T\phi$ where the sum runs over the set of blocks, and Δ_T represents then restriction of the Laplacean on the block T . Clearly then, Δ_T is a 3x3 matrix with spectrum $\{2,-1\}$. We have seen that 2 is not in the spectrum of Δ whereas -1 is, showing that the restriction of Δ to $\mathcal{H}_\perp$ is exactly the projection χ_{-1} . With this last argument the proof is achieved. $\Diamond$

3.2 Effect of a Magnetic Field

The same kind of properties have been studied with the occurence of a uniform magnetic field perpendicular to the plane of the lattice [Alexander(83,84), Rammal(83), Ghez(87)]. The Laplace operator is then modified by means of a phase factor according to :

$$\Delta_-\psi (x) = \sum\nolimits_{x';|x'-x|=1} e^{2\pi ieA(x,x')/h} \psi(x') , \qquad x\in S\backslash\{0\} , \ \psi\in l^2(S\backslash\{0\}) , \qquad (15)$$

and a similar formula for Δ_+ , where h is the Planck constant, e is the electric charge of the particle involved, and $A(x,x')=\int_x{}^x dl.A$ where $A=(A_1,A_2)$ is the vector potential defined as a solution of $\partial_1 A_2- \partial_2 A_1=B$. This Laplace operator does depend upon the specific solution chosen for A (choice of a gauge), but changing the gauge results in a unitary equivalent operator having thus the same spectrum. Thus the spectrum of $\Delta_\pm$ depends only upon the magnetic field B , namely upon the flux ϕ in a unit triangle. The normalized flux is $\alpha=e\phi/h$ and is a dimensionless parameter. We then remark that changing α in $\alpha+1$ does not change the spectrun, so that α is defined modulo 1 .

In much the same way as in the previous section, both Δ_+ are solution of a RG equation. However, in order to get a closed relation we must introduce two normalized fluxes α, α' the first corresponding to elementary triangles like abc in fig.1, the second to triangles like Cab . The flux per unit cell is then $\alpha+\alpha'$. Then one gets:

$$D\{z\mathbf{1}-\Delta(\alpha,\alpha')\}^{-1}D^* = G(z,\alpha,\alpha') \{E\mathbf{1}-\Delta(\beta,\beta')\}^{-1} \qquad (16)$$

where [Ghez(87)]:

$$G(z,\alpha,\alpha') = \{z^3 - 3z - 2(XU+YV)\}/\{S^2 + C^2\}^{1/2} ,$$

$$E = \{z^4 - 7z^2 - [2(XU+YV)+ 4X]z + 4(1-U)\}/\{S^2 + C^2\}^{1/2} ,$$

$$(17)$$

$$\beta + \beta' = 4(\alpha + \alpha') \qquad\qquad \beta - \beta' = 2(\alpha - \alpha') - 3/\pi \; \text{Arctan} \; (S/C)$$

with:

$$S = [(XV + YU) + 2Y]z + 2(V + XY) , \; C = z^2 + [(XY - YV) + 2X]z + 2(U - Y^2),$$

$$(18)$$

$$X = \cos 2\pi\alpha , Y = \sin 2\pi\alpha , \quad U = \cos 2\pi(\alpha+\alpha') \qquad V = \sin 2\pi \; (\alpha+\alpha')$$

Following the intuition provided by the last section, the "dynamical spectrum" is defined as the invariant set of the map $F(z,\alpha,\alpha') = (E,\beta,\beta')$ of $\mathbf{R}x\mathbf{T}^2$. Since $\beta + \beta' = 4(\alpha + \alpha')$ in (17), only one of the two normalized fluxes is actually relevant, leading to an effective 2D map. *Is the dynamical spectrum equal to the actual spectrum of the original operator ? This is a question with no answer yet.* Nevertheless the numerical calculation of the dynamical spectrum given in fig.2 below [Ghez(87)], shows that it should be.

One should point out there that this calculation has been compared to an experiment performed in Grenoble, on a superconducting network designed according to fig. 1. Landau-Ginzburg's theory [de Gennes(81), Alexander(83)] shows that the transition between the normal metal and the super conducting phases occurs in the *(T,B)* plane (where *T* is the temperature) on a curve which is simply related to the edge of the dynamical spectrum as calculated above [Ghez(87)]. The comparison between theory and experiment is actually very accurate as shown in fig.3 below.

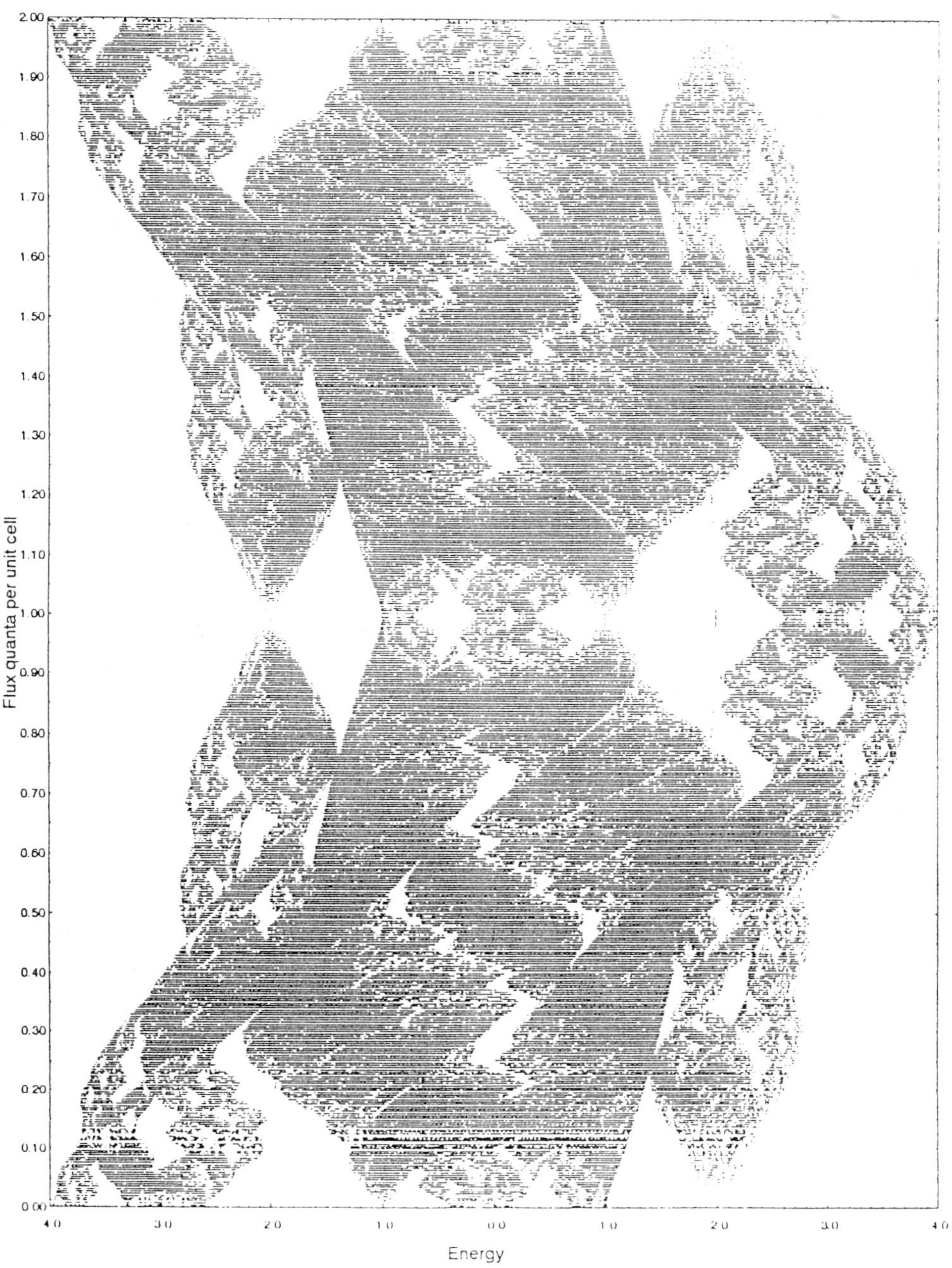

Fig.2- The dynamical spectrum of the Sierpinsky Laplacean in a magnetic field. The magnetic flux is represented on the horizontal axis, whereas the energy is represented on the vertical one. (Picture designed by C.Kreft) -

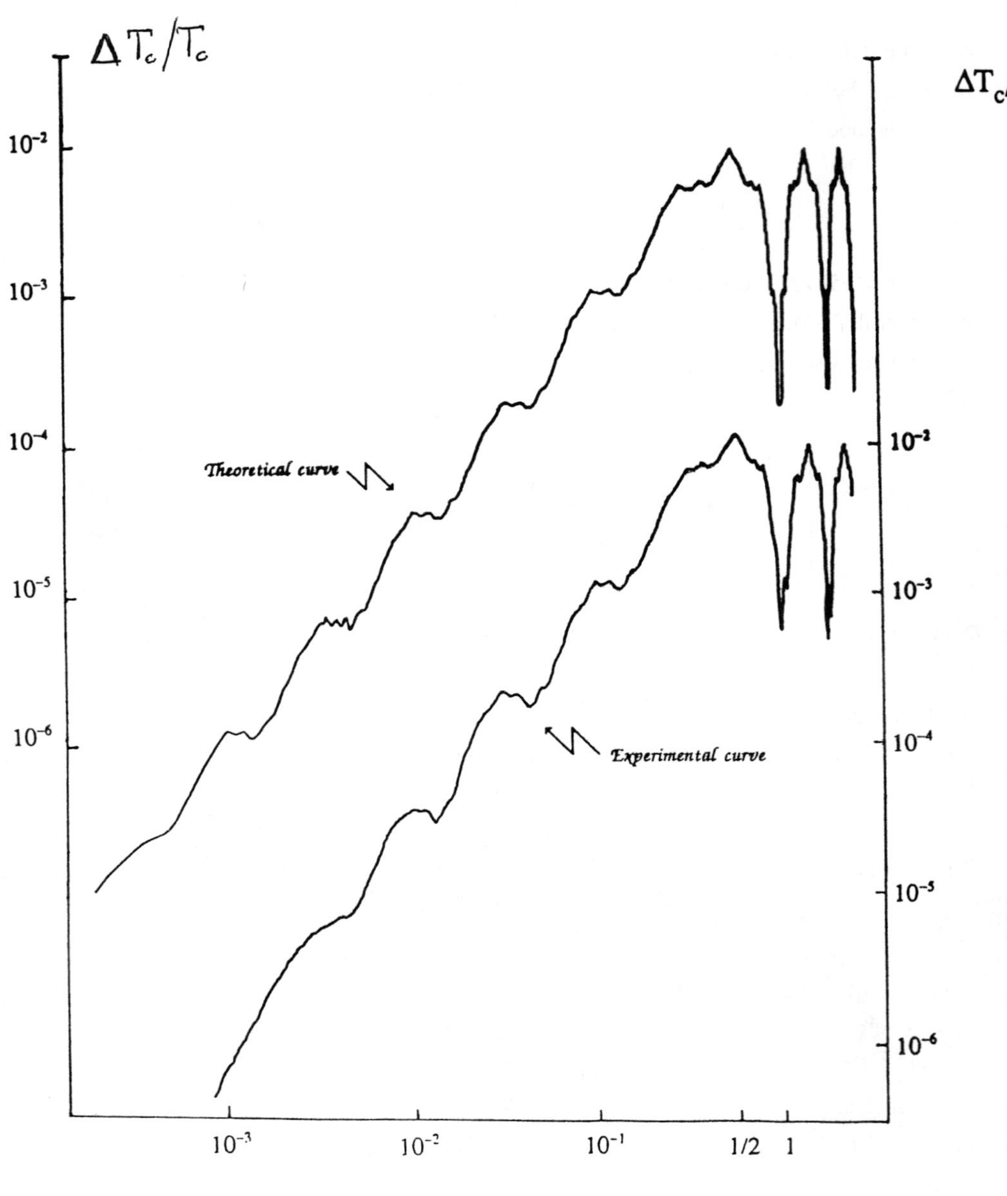

Fig.3- Comparison between the boundary value of the Sierpinsky spectrum in a magnetic field and the experimental measurement of the normal-metal-superconductor transition line for a Sierpinsky network of superconductors [Ghez(87)] -

4.ONE DIMENSIONAL QUASICRYSTALS

4.1 The Khomoto Model in 1D

In 1983, Khomoto et al [Kadanoff(83)] and Ostlund et al [Ostlund(83)] considered the following model on the 1D lattice:

$$H_{\alpha,x}\psi(n) = \psi(n+1) + \psi(n-1) + V\,\chi_{[1-\alpha,1)}(n\alpha+x)\,\psi(n)\,, \qquad (19)$$

where $\alpha=(\sqrt{5}-1)/2$ is the golden mean, and $\chi_{[1-\alpha,1)}(x)$ is the characteristic function of the interval $[1-\alpha,1)$ on the unit circle normalized to 1, namely x is defined modulo 1. To study the spectrum of this operator, they proposed to look at the transfer matrix method, namely we write the equation $H\psi=E\psi$ as:

$$\begin{bmatrix}\psi(n+1)\\[1em]\psi(n)\end{bmatrix} = \begin{bmatrix} E-V\chi_{[1-\alpha,1)}(n\alpha+x) & -1\\[1em] 1 & 0\end{bmatrix}\begin{bmatrix}\psi(n)\\[1em]\psi(n-1)\end{bmatrix} = T(n)\begin{bmatrix}\psi(n)\\[1em]\psi(n-1)\end{bmatrix}, \qquad (20)$$

where, $T(n)$ is called the transfer matrix. It is a 2x2 matrix with determinant equal to 1. We see that $T(n)$ takes on two possible values namely $T(0)=B$, and $T(1)=A$ .Let F_s be the Fibonacci numbers defined by $F_0=1,\ F_1=1,\ F_{s+1}=F_s+F_{s-1}$, for $s\geq 1$, and we set $M(s) = T(F_s)T(F_s-1)...T(2)T(1)$. Then one can show that if $x=0$ the following recursion law is valid:

$$M(s+1) = M(s-1)\,M(s)\,, \quad s\geq 1, \qquad\qquad M(0) = A\,,\ M(1) = B\,. \qquad (21)$$

Taking the trace $x_s= Tr\{M(s)\}$, and using the identity $M^2-M.Tr(M)+1=0$ valid for 2x2 matrices with determinant 1, we also get $x_{s+1} = x_s x_{s-1} - x_{s-2}$. The main remark is that $M(s)$ can be seen as the Floquet matrix for the periodic model similar to (15) where the period is F_s and the potential coincides with $V\chi_{(1-\alpha,1]}(n\alpha-x)$ for $1\leq n \leq F_s$. Floquet's theory then implies that the spectrum of the auxiliary problem is the set of energies E such that $|x_s(E)| \leq 2$. Since $x_s(E)$ is a polynomial of degree F_s in E this gives a band spectrum with at most F_s bands.

The recursion (17) can be seen as a dynamical system acting on the space Ω of pairs (X,Y) of matrices in $SL(2,R)$ by

$$\Phi(X,Y) = (YX, X) \qquad\qquad (X,Y)\in \Omega = SL(2,\mathbf{R})^{\times 2} . \qquad (22)$$

For indeed we get $(M(s+1), M(s)) = \Phi^s(A,B)$. It defines an induced mapping on the traces, acting on $\mathbf{R}^3$ by $F(w,x,y) = (xw-y, w, x)$ and giving $(x_{s+1}, x_s, x_{s-1}) = F^s(x_2, x_1, x_0)$. The "dynamical spectrum" is then defined as the set of energies for which $limsup_{s\to\infty} |x_s(E)| < \infty$, and has been guessed to coincide with the spectrum of H [Kadanoff(83), Ostlund(83,85), Casdagli(86)].

More generally, if α is now considered as a varying parameter in the Khomoto model, two cases have to be studied separately. If $\alpha=p/q$ is a rational number, then the Hamiltonian (19) is periodic and its spectrum is given by the requirement that the trace of the transfer matrix over one period be smaller than or equal to 2 . If α is irrational, let $\alpha=[a_1,a_2,\ldots,a_s,\ldots]$ be its continuous fraction. The modifications are as follows: the Fibonacci sequence is replaced by the pair of rational approximants (p_s,q_s) such that $p_s/q_s= [a_1,a_2,\ldots,a_s]$ namely $p_{s+1}=a_{s+1}p_s+p_{s-1}$ and $q_{s+1}=a_{s+1}q_s+q_{s-1}$, with $p_0=0$, $p_1=1$, $q_0=1$, $q_1=a_1$. Then we set $M(s) = T(q_s)T(q_s-1)\ldots T(2)T(1)$ to get [Ostlund(85), Bellissard(89)]:

$$M(s+1) = M(s-1)\, M(s)^{a_{s+1}} , \quad s\geq 0, \qquad M(0) = A , \; M(-1) = BA^{-1} . \qquad (23)$$

To express this recursion in term of a dynamical system similar to (22), we introduce the mappings $S(X,Y)=(Y,X)$, and $\Gamma(X,Y)=(X,YX)$ on Ω . Setting $\Phi(a) = S o \Gamma^a$ we obtain the following identity $(M(s+1), M(s)) = \Phi(a_s) o \ldots o \Phi(a_1)(M(0),M(-1))$. This system is non autonomous for it depends upon the data of the continuous fraction of α . To take this fact into account we introduce the space $\Omega'=(0,1]\times\Omega$ and the mappings $G(\alpha)=\{1/\alpha\}$, $a(\alpha)=[1/\alpha]$ where $\{x\}$ denotes the fractional part of x and $[x]$ its integer part, and we set $\Phi'(\alpha,(X,Y))=(G(\alpha),\Phi(a[\alpha])(X,Y))$. If we define $\alpha_s=[a_s,a_{s+1},\ldots,a_{s+n},\ldots]$ we can write (23) as $(\alpha_{s+1}, (M(s+1),M(s))) = \Phi'(\alpha_s,(M(s),M(s-1)))$.

As before, this mapping induces a trace map as follows: we set $x=Tr(X)$, $y=Tr(Y)$, $w=Tr(XY)$ and $s(w,x,y)=(w,y,x)$, $\gamma(w,x,y) = (wx-y,x,w)$. Then $\Phi(a)$ generates the map $\phi(a)=s o \gamma^a$ namely $\phi(a)(w,x,y,) = (wS_a(x)-yS_{a-1}(x), wS_{a-1}(x)-yS_{a-2}(x), x)$ where S_a is the Tchebyshev polynomial defined by $S_a(2cos\theta) = sin(a+1)\theta/sin\theta$. This dynamical system admits one first integral as was remarked first in [Kadanoff(83), Ostlund(83)] for the golden mean and by Ostlund &Kim in general [Ostlund(85)] namely

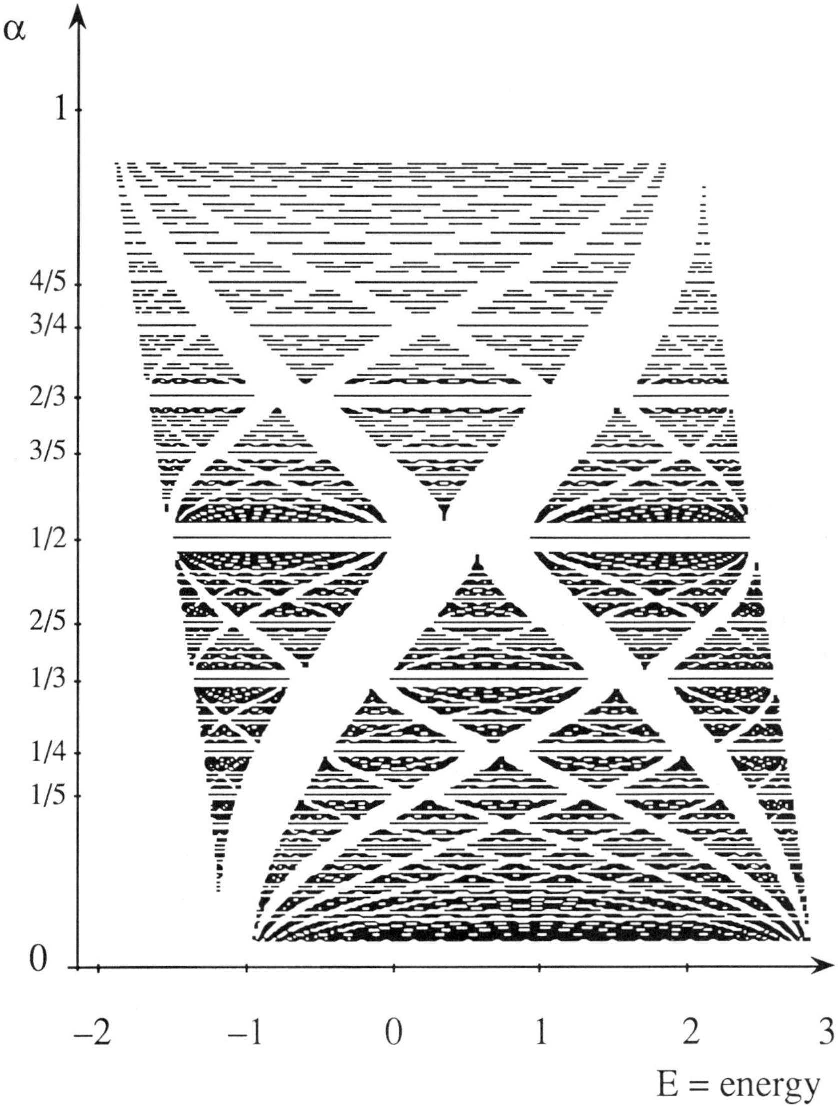

Fig.4- Spectrum of the Khomoto Hamiltonian for $V=1$ as a function of α (numerical computation by B. Iochum; see [Ostlund(85)]) -

the function I such that $I(w,x,y) = w^2 + x^2 + y^2 - xyw = Tr(X^{-1}Y^{-1}XY) + 2$ which is invariant by s,S and γ,Γ as can be checked directly. The main result proved in [Sütò(87,89), Bellissard(89)] is the following theorem:

Theorem 4: (i) For any α in $(0,1]$, and any $V>0$, the spectrum of the Khomoto Hamiltonian $H_{\alpha,x}$ does not depend on x.

(ii) For any irrational α and any $V>0$, the spectrum of $H_{\alpha,x}$ is the set of energies E for which the sequence $x_s(E) = Tr(M(s))$ is bounded.

(iii) For any irrational α, and any $V>0$, the spectrum of $H_{\alpha,x}$ is a Cantor set of zero Lebesgue measure, and the spectral measure is singular continuous.   ◊

In 1985, Ostlund & Kim computed numerically this spectrum using the previous result, which is displayed in fig.4 . For more results one can look at [Delyon(84), Kalugin(86), Kohmoto(84), Levitov(89), Ostlund(84), Sire(89, 90a, 90c), Tang(86), Wijnands(89)].

4.2 The Cut and Projection Method

As we will see the Khomoto model represents also the Hamiltonian of a quantum particle on a 1D quasicrystal. This was remarked by Luck & Petritis [Luck(86)]. In order to see this let us first describe the construction of such a lattice.

The cut and projection method [Duneau(85), Gratias(86), Levine(84), Steinhardt(87), Janots(88)] consists in seeing a quasiperiodic chain as some projection of a 2D lattice Z^2 according to fig.5. Let $E_{//}$ be a real line in R^2 with slope $\omega = \alpha/(1-\alpha)$ and let C be the semi-open unit square $(0,1]x[0,1)$. $E_\perp$ denote the line orthogonal to $E_{//}$. The strip Σ is obtained by translating C along $E_{//}$ namely $\Sigma = C + E_{//}$. We will set $\Sigma_d = \Sigma \cap Z^2$ and we will identify any substrip Σ_0 of Σ with its "acceptance zone", namely its intersection with $E_\perp$. The quasiperiodic lattice is obtained as the orthogonal projection of the points of Σ_d on $E_{//}$. We can label each site $(x,y) \in \Sigma_d$ by the integer $n = x + y$. Conversely, if $n \in Z$ we get a unique site in Σ_d namely $x = [n/(1+\omega)]$, $y = n - x$, where $[X]$ is the integer part of X . From the previous construction it follows that the distance between two consecutive sites is either $c = cos\theta$ or $s = sin\theta$ where $tan\theta = \omega$ depending whether the corresponding bond in Z^2 is horizontal or vertical. The sequence of successive bonds $I_\infty = csccsccs...$ is generated by concatenation as follows: we first assume $0 < \omega < 1$ with a continuous fraction expansion $\omega = [b_1, b_2, ..., b_s, ...]$; if $I_{-1} = s$, $I_0 = c$, the sequence I_∞ is obtained as

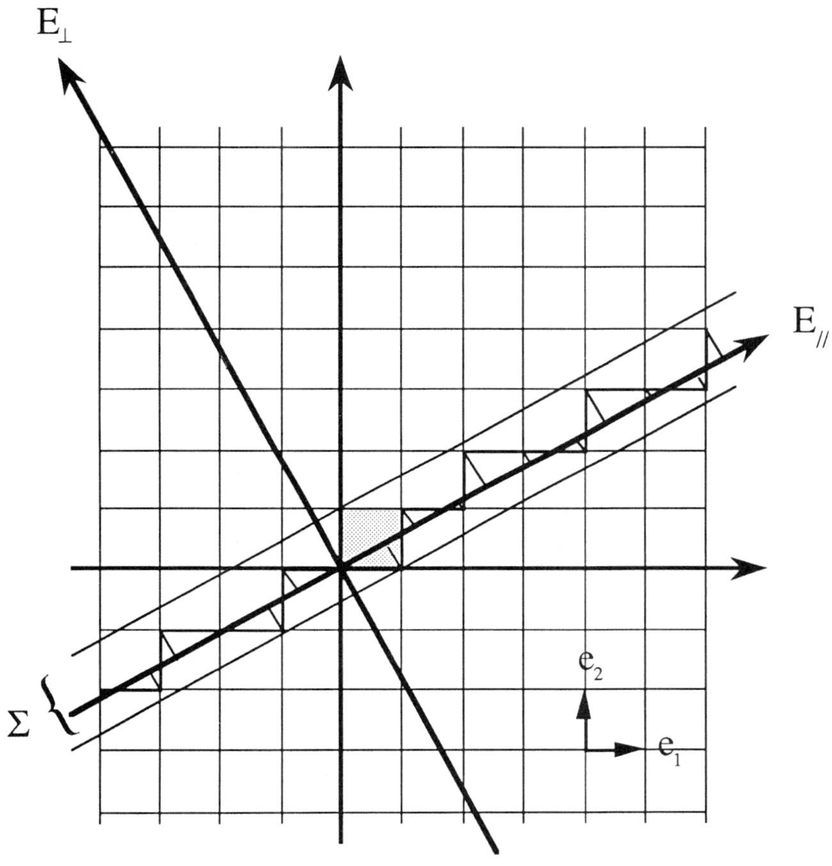

Fig.5- Construction of a 1D quasicrystal by the cut and projection

method -

the limit of the sequence defined recursively by $I_{k+1} = I_k{}^{b_{k+1}} I_{k-1}$, whenever $k \geq 0$. If $\omega > 1$ it is sufficient to remark that then $b_1 = 0$, and the same rules occur . It is then easy to see that the density of s -bonds is $\alpha = \omega/(1+\omega)$ whereas the density of c -bonds is $1-\alpha$. On the other hand, a site in the lattice is surrounded by $sc,\ cc$ or cs bonds, partitioning them into three subsets. It is remarkable that these three subsets have projection in three non intersecting intervals of the "acceptance zone". Indeed a point (x,y) belongs to the strip Σ if and only if $-\omega \leq y-x\omega < 1$. Accordingly point (x,y) in the strip Σ with "local environment" sc is defined by the property that both $(x+1,y)$ and $(x,y-1)$ also belong to Σ, namely if e_1, e_2 is the canonical basis of $\mathbf{Z}^2$ the set of these points is the substrip $\Sigma \cap \Sigma - e_1 \cap \Sigma + e_2$, given by $1 - \omega \le y - x\omega < 1$. In much the same way , the cc -points are in $\Sigma \cap \Sigma - e_1 \cap \Sigma + e_1$, namely $0 \le y - x\omega < 1 - \omega$, and the cs -points are in $\Sigma \cap \Sigma + e_1 \cap \Sigma - e_2$, namely $\omega \le y - x\omega < 0$ (see fig.6 below). If $\omega > 1$, cc -points are replaced by ss -points. This representation is actually quite convenient to compute the density of sites with a given local environment, for if it is the relative length of the corresponding interval in the acceptance zone, as one can check easily in the example given in fig.6 .

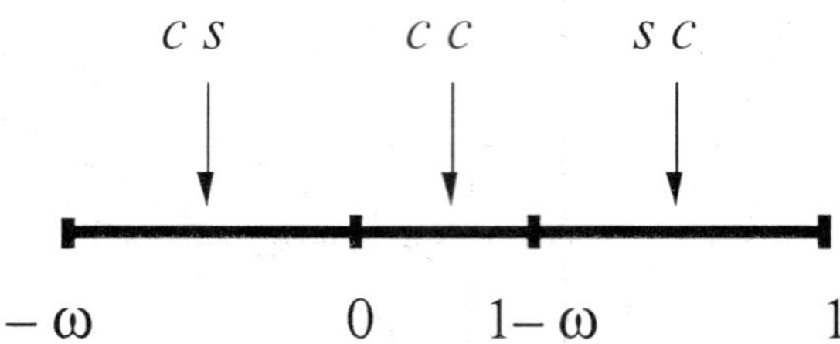

Fig.6- The acceptance zone for the local environments of a quasiperiodic chain in the variable $y - x\omega$ -

A "path" γ is a sequence $(a(1), a(2), ..., a(L))$ of vectors such that $a(k) = \pm e_\mu$, $\mu = 1,2,$ and that there is a site $\xi = (x,y)$ in the strip Σ_d for which $\xi + a(1) + ... + a(k) \in \Sigma_d$ for all k . L is the length of the path, and the substrip $\Sigma(\gamma) = \cap_{1 \le k \le L} \{ \Sigma - (a(1) + ... + a(k)) \}$ is the acceptance zone of the set of initial sites ξ 's. For convenience we will set $e_{\mu+2} = -e_\mu$ and μ is defined mod.4 .

4.3 Quantal Observable Algebra [Bellissard(90b), Sire(90c)]

The previous construction allows us to get a canonical description of the observable algebra. On a periodic lattice, the Hamiltonian H is usually constructed from the operator T_μ of translation by the periods of the lattice. If the period is not 1 , H can be constructed as a finite matrix of size given by the number of sites in the unit cell, and the entries of which being functions of the translation operators T_μ . In much the same way here, we will consider the two translations T_1, T_2 on the lattice $\mathbf{Z}^2$ namely two commuting unitary operators, which generate a C*-algebra isomorphic to $C(T^2)$ namely the space of continuous functions over the two-torus. However, because the physical lattice lives in the strip Σ , we must restrict this algebra by projecting every observable on to the strip. We will do this by introducing a family χ_Σ of commuting projections where Σ runs into the family of substrips of Σ generated by finite union and finite intersections of $\Sigma(\gamma)$'s where γ is a path. These projections satisfy $\chi_\Sigma \chi_{\Sigma''} = \chi_{\Sigma \cap \Sigma''}$ and $T_\mu \chi_\Sigma T_\mu^{-1} = \chi_{\Sigma + e(\mu)}$ where $T_{\mu+2} = T_\mu^{-1}$ according to our convention. Thus we define the "non-commutative" translation operator as:

$$S_\mu = \chi_\Sigma\, T_\mu\, \chi_\Sigma \qquad \Rightarrow \qquad S_{\mu+2} = S_\mu{}^* \qquad (24)$$

The C^*-algebra generated by S_1, S_2 will be called $\mathcal{B}_\omega$. The projection χ_Σ plays the role of the identity in $\mathcal{B}_\omega$. We emphasize the following algebraic relations :

$$S_\mu S_\mu{}^* = \chi_{\Sigma \cap \Sigma + e(\mu)}\,, \qquad S_\mu{}^* S_\mu = \chi_{\Sigma \cap \Sigma - e(\mu)}\,, \qquad S_1 S_2{}^* = S_1{}^* S_2 = 0$$

$$(25)$$

$$S_1 S_1{}^* + S_2 S_2{}^* = S_1{}^* S_1 + S_2{}^* S_2 = \chi_\Sigma = \mathbf{1}\,.$$

In particular it follows that if $T = S_1 + S_2$ then $T T^* = T^* T = 1$. We set $\chi = S_2 S_2{}^* = \chi_{\Sigma \cap \Sigma + e(2)}$ in such a way that $\chi T = S_2$, $(1-\chi)T = S_1$, showing that $\mathcal{B}_\omega$ is also the C^*-algebra generated by the unitary operator T and the projection χ . A faithfull representation π_x [Pedersen(79)] of this algebra is obtained on the Hilbert space $l^2(\Sigma_d) \approx l^2(\mathbf{Z})$ by means of (where $\psi \in l^2(\mathbf{Z})$):

$$\pi_x(T)\psi\,(n) = \psi(n-1)\,, \qquad\qquad \pi_x(\chi)\psi\,(n) = \chi_{[1-\alpha,1)}(n\alpha + x)\,\psi(n)\,. \quad (26)$$

This shows that the Khomoto Hamiltonian can be written as:

$$H_{\alpha,x} = \pi_x\{T + T^* + V\chi\} = \pi_x\{S_1 + S_1{}^* + S_2 + S_2{}^* + V\, S_2 S_2{}^*\} \qquad (27)$$

This description allows us to introduce the most general Hamiltonian with nearest neighbour interactions namely the 5-parameters family where $\chi_{cs} = S_1{}^* S_1 S_2 S_2{}^*$, $\chi_{sc} = S_2{}^* S_2 S_1 S_1{}^*$ and $\chi_{cc} = 1 - \chi_{cs} - \chi_{sc}$

$$H(t_c, t_s, V_{sc}, V_{cc}, V_{cs}) = t_c(S_1 + S_1{}^*) + t_s(S_2 + S_2{}^*) + V_{cs}\chi_{cs} + V_{sc}\chi_{sc} + V_{cc}\chi_{cc}\,, \qquad (28)$$

which takes into account the possibility that the overlap depends upon the bond and the local environment. Note that when $\omega > 1$, V_{cc} becomes V_{ss} for then no cc-site occurs, but ss-ones do. Any Hamiltonian on this lattice will be in general a convergent series of products of the operators S_1, S_1^*, S_2, S_2^* , the coefficient of which being corrections to the tight-binding approximation taking into account the possibility of long range interactions. Beside the algebra $\mathcal{B}_\omega$ let us introduce the "universal" algebra $\mathcal{B} = \cup_\omega \mathcal{B}_\omega$. More technically we consider in $L^\infty(T^2)$ the sub C^*-algebra C generated by the functions $\chi \circ \phi^n$ $(n \in \mathbf{Z})$ where we set $\chi(\alpha, x) = \chi_{[1-\alpha, 1)}(x)$ and $\phi(\alpha, x) = (\alpha, x-\alpha)$. Then $\mathcal{B}$ is the crossed product [Pedersen(79)] of C by $\mathbf{Z}$ via the action given by ϕ , namely the C^*-algebra generated by χ and T with the commutation relation $T\chi T^* = \chi \circ \phi$. It is actually quite easy to show that there is a canonical *-homomorphism η_α between $\mathcal{B}$ and $\mathcal{B}_\omega$ (with $\omega = \alpha/(1-\alpha)$) such that $\eta_\alpha(T) = T$ and $\eta_\alpha(\chi) = \chi_{[1-\alpha, 1)}$. Then the main result is the following (see [Bellissard(90a)]):

Theorem 5:   Let $H = H^*$ be a self adjoint element of $\mathcal{B}$. The spectrum $\Sigma(\alpha)$ of $\eta_\alpha(H)$ is continuous at each irrational value of α , namely the gap edges are continuous functions of α . In particular, for any element A of $\mathcal{B}$, the norm $||\eta_\alpha(A)||$ is continuous at the irrational values of α .

Near a rational value $\alpha = p/q$, the left and right limits of $\Sigma(\alpha)$ exist, but they do not necessarily coincide with $\Sigma(p/q)$. $\Diamond$

This theorem permits to justify the numerical calculation by Ostlund & Kim (see fig.4) performed for rational values of α , using the transfer matrix method.

4.4 Scaling and RG

The Ostlund & Kim spectrum shown in fig.4, exhibits an obvious fractal shape. It is therefore natural to ask whether a renormalization group analysis gives an account of it. The transfer matrix method indeed gives rise to a dynamical system and the spectrum appears as the "Julia set" on the map then defined. However, the method is typically one-dimensional and there is no chance that it extend to higher dimensional quasicrystals. The cut and projection method generalizes in any dimension, and therefore the construction of the corresponding C^*-algebra can be also performed for any quasicrystal. It is therefore a natural question to address whether there is a formulation of the RG equation in the algebraic set-up. The operator theoretic method developped in sections 2&3 suggests something similar in higher dimensions. The purpose of this section is to describe such a formalism in the 1D case.

The group $GL(2,\mathbf{Z})$ acts on the superlattice $\mathbf{Z}^2$ in a natural way, namely if A is a 2x2 matrix with integer coefficients and determinant ± 1, and if $\xi=(x,y)\in\mathbf{Z}^2$, then $A\xi=(A_{11}x+A_{12}y, A_{21}x+A_{22}y)$ where the A_{ij} 's are the coefficients of A . This action is implemented by a unitary "dilation" operator on the Hilbert space $l^2(\mathbf{Z}^2)$ by means of

$$D_A\phi\,(\xi) = \phi(A^{-1}\xi) \qquad\qquad \phi\in l^2(\mathbf{Z}^2), \qquad \xi\in\mathbf{Z}^2 , \qquad\qquad (29)$$

This dilation transforms usually the strip Σ into a new strip $A\Sigma$ because $D_A\chi_\Sigma D_A{}^* = \chi_{A\Sigma}$, for any substrip. Let us remark that if $det(A) = -1$, the orientation is reversed and we must change the convention concerning which boundary of Σ is in Σ. To take this into account, we will denote by $\Sigma_+(\omega)$ (resp. $\Sigma_-(\omega)$)the strip defined by $-\omega \le y-\omega x < 1$ (resp. $-\omega < y-\omega x \le 1$). Then the sign of the strip will change for a matrix of determinant -1 and will remain otherwise. In what follows the sign is implicit.

Among these 2x2 matrices, we specify the matrix $A(b)$ for $b\in N$ defined by :

$$A\,(b) \;=\; \begin{bmatrix} b & 1+b \\ 1 & 1 \end{bmatrix}. \qquad\qquad (30)$$

Again if $\omega=[b_1,b_2,\ldots,b_s,\ldots]$ is the continuous fraction expansion of ω , with $b_1=0$ iff $\omega\ge 1$, and $b_k\ge 1$ for $k>1$, the matrix $A(b_1)$ transform the strip $\Sigma(\omega_1)$ of slope ω_1 into a substrip of $\Sigma(\omega)$ of slope $\omega=(1+\omega_1)/\{b_1+(b_1+1)\omega_1\}$, with $\omega_1=[b_2-1,b_3,\ldots,b_{s+1},\ldots]$. Correspondingly if $\alpha=\omega/(1+\omega)$ we get $\alpha_1= \omega_1/(1+ \omega_1) = \{1/\alpha\}$ where again $\{\xi\}$ denotes the fractional part of ξ . In particular if $\alpha=[a_1,a_2,\ldots,a_s,\ldots]$, we get $a_1 = b_1+1$ and $a_k=b_k$ for $k \ge 2$. So that $A(b_1) = A(a_1-1)$ and $\alpha_1=[a_2, a_3,\ldots,a_{s+1},\ldots]$.

We then define $D(b) = \chi_{\Sigma(\omega')}D_{A(b)^{-1}}\chi_{\Sigma(\omega)}$ whenever $\omega= (1+\omega')/\{b+(b+1)\omega'\}$. This gives rise to the following identities :

$$D(b)\,D(b)^* = \chi_{\Sigma(\omega')} = \mathbf{1}_{\omega'}, \qquad\qquad D(b)^*\,D(b) = \chi_{A(b)\Sigma(\omega')} \le \mathbf{1}_\omega , \qquad (31)$$

where $\mathbf{1}_\omega$ denotes the unit in $\mathcal{B}_\omega$.

The main result in this respect is the following RG equation due to Ph. Hauguel [Hauguel(91)] :

Hauguel's formulæ : Let $H(t_c, t_s, V_{sc}, V_{cc}, V_{cs})$ be the self adjoint element of $\mathcal{B}$ defined in (28) . Then, with the previous notations, the following RG equation holds :

$$D(b)\, H(t_c,t_s,V_{sc},V_{cc},V_{cs})^{-1}\, D(b)^* = g_{b+1}\, H(t_c',t_s',V_{sc}',V_{cc}',V_{cs}')^{-1}$$

where :

$$t_c' = t_c^{\,b+1}\, t_s \, , \quad t_s' = t_c\, g_{b+1} \, ,$$

$$V_{cs}' = g_{b+1}\, V_{cc} + g_b t_c^{\,2} \, , \qquad\qquad V_{sc}' = g_{b+1}\, V_{cs} + g_b t_s^{\,2} \, ,$$

$$V_{cc}' = g_{b+1}\, V_{cs} + g_b t_c^{\,2} + g_b'\, t_s^{\,2} \, ,$$

and g_b , g_b' are defined by :

$$\begin{bmatrix} g_{b+1} & g_{b+1}' \\[1mm] g_b & g_b' \end{bmatrix} = \begin{bmatrix} V_{cc} & -t_c^{\,2} \\[1mm] 1 & 0 \end{bmatrix}^{b} \begin{bmatrix} V_{sc} & V_{cc} \\[1mm] 1 & 1 \end{bmatrix} .$$

$$\Diamond$$

This result is actually connected to the transfer matrix result in that it gives a representation of the map $\Phi(a+1)$ in a new set of variables. The connection between these two sets is given by :

$$-V_{cc}/t_c = \mathrm{Tr}\,(X) \, , \qquad\qquad -V_{sc}/t_c = (XY)_{21}/Y_{21} \, ,$$

$$\tag{32}$$

$$-V_{cs}/t_c = (YX)_{21}/Y_{21} \, , \qquad\qquad t_s/t_c = X_{21}/Y_{21} \, ,$$

where $(X,Y) \in \Omega = SL(2,\mathbf{R})^{x2}$ are the transfer matrix variables.

Actually the general structure of these equations is simpler that it appears at first sight. For indeed the matrix $A(b)$ can be written as $A(b) = B^b A(0)$ where

$$ B = \begin{bmatrix} 1 & 1 \\ 0 & 1 \end{bmatrix} . $$

(33)

Correspondingly, we get partial a isometry $D = \chi_{\Sigma(\omega')} \mathcal{D}_{B^{-1}} \chi_{\Sigma(\omega)}$ with $D(b) = D(0)D^b$ and $B\omega' = \omega$ showing that the corresponding RG map admits the same structure as in the transfer matrix case.

Thanks to these formulæ there is a complete equivalence between the two representations. The operator theoretic one being more general, it will be quite useful in view of the application to the study of higher dimensional quasicrystals.

5. HIGHER DIMENSIONAL QUASICRYSTALS

We will not give here a detailed account of the higher dimensional case. We will refer to [Sire(90b)] for the description of the octagonal lattice in 2D, and to a work in preparation [Dahmani(91)] for the Penrose lattice (see also [Sire(90c)]) .

The cut and projection method permits again to describe the octagonal lattice as follows . Let $\mathbf{Z}^4$ be the hyperlattice replacing now $\mathbf{Z}^2$. We consider the action of the 8-fold symmetry in this lattice given by the integer coefficient matrix R such that if e_μ $(\mu=1,...,4)$ are the unit vector of the canonical basis, then $Re_1 = e_3$, $Re_2 = e_4$, $Re_3 = e_2$, $Re_4 = -e_1$. This matrix splits into a direct sum of two rotations of angle $\pi/4$; correspondingly $\mathbf{R}^4$ splits into a direct sum of two real 2D R -invariant subspaces $E_{//}$ and $E_\perp$. They automatically intersect the hyperlattice only at 0 for otherwise they would contain a 2D lattice with an 8-fold symmetry. We also consider the strip Σ in $\mathbf{R}^4$ obtained by translating the unit cube along $E_{//}$. Again there will be a convention about which face of this strip belongs to it. We will not give here the details. The octagonal lattice $\mathcal{L}$ is the projection on $E_{//}$ of $\Sigma_d = \Sigma \cap \mathbf{Z}^4$. Both $\mathcal{L}$ and Σ_d are R -invariant. In fig.7 below, the acceptance zone of the octagonal lattice is shown, whereas the lattice itself is given in fig.8. The set Γ of matrices in $GL(4,\mathbf{Z})$ leaving the strip invariant admits a subgroup G , the symmetry group of $\mathcal{L}$. The other matrices are the "inflation" rules. Among them the multiplication by $\sqrt{2}+1$ leaves $\mathcal{L}$ invariant, and is represented by a matrix A in $GL(4,\mathbf{Z})$.

As before, we can introduce the non commutative translations $S_\mu = \chi_\Sigma T_\mu \chi_\Sigma$, where T_μ

are the translation operators by e_μ acting on the Hilbert space $l^2(\mathbf{Z}^4)$. The Laplace operator can be seen as

$$\Delta_L = \Sigma_{1\leq\mu\leq4} \left\{ S_\mu{}^*S_\mu - S_\mu - S_\mu{}^* + S_\mu S_\mu{}^* \right\} . \tag{34}$$

We can also define as before the dilation operator D_A associated to A , and also the rotation symmetry $U(R)$ associated to R . They both commute. More generally if H is a Hamiltonian expressed as a polynomial in the S_μ 's, we can associate its renormalized "companion" H' defined by

$$D_A H^{-1} D_A{}^* = H'^{-1} . \tag{35}$$

In general H' is no longer a polynomial, against what happens in 1D. However, the long range interactions are exponentially decaying with the distance, and if we truncate H' to get a polynomial, we do not make a large error. So this equation can be written in term of a finite number (but quite large in practice) of parameters, and the same technics can be applied to compute the spectrum . This has been done in [Sire(90b)] for the first time, and eventhough it is not yet mathematically fully justified, the numerical calculation fits well with the calculations previously performed by diagonalizing directly the Hamiltonian on a finite box [Kohmoto(86), Odagaki(86), Tsunetsugu(86)], or by comparison with various perturbative results.

One can easily extend the formalism to the case of a quasicrystal in a uniform magnetic field [Bellissard(88)]. Numerical calculation of the spectrum in this latter case, can be found in [Behrooz(86), Hatakeyama(87)].

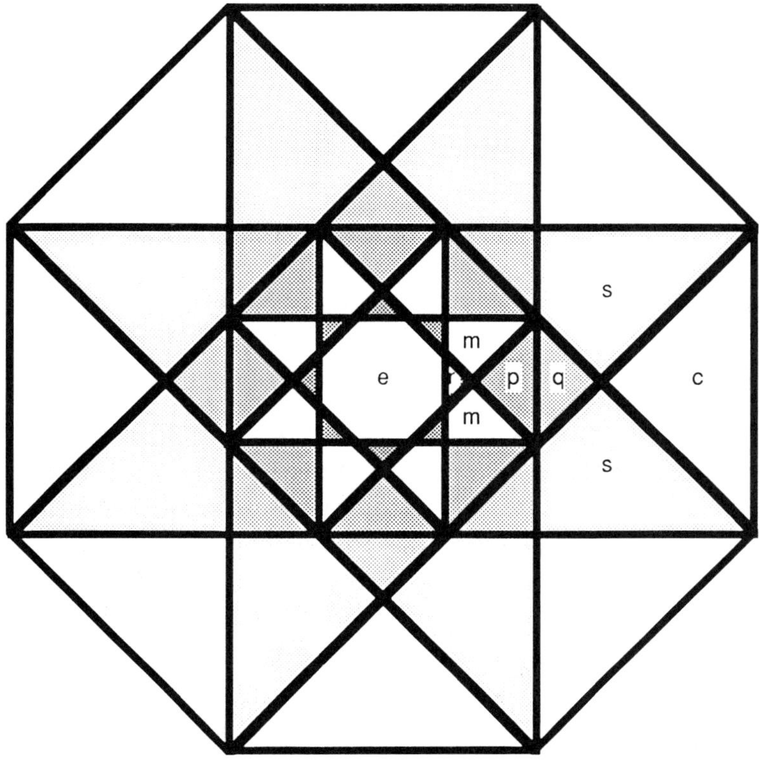

Fig.7- The acceptance zone of the 2D octagonal lattice [Sire(90b) -

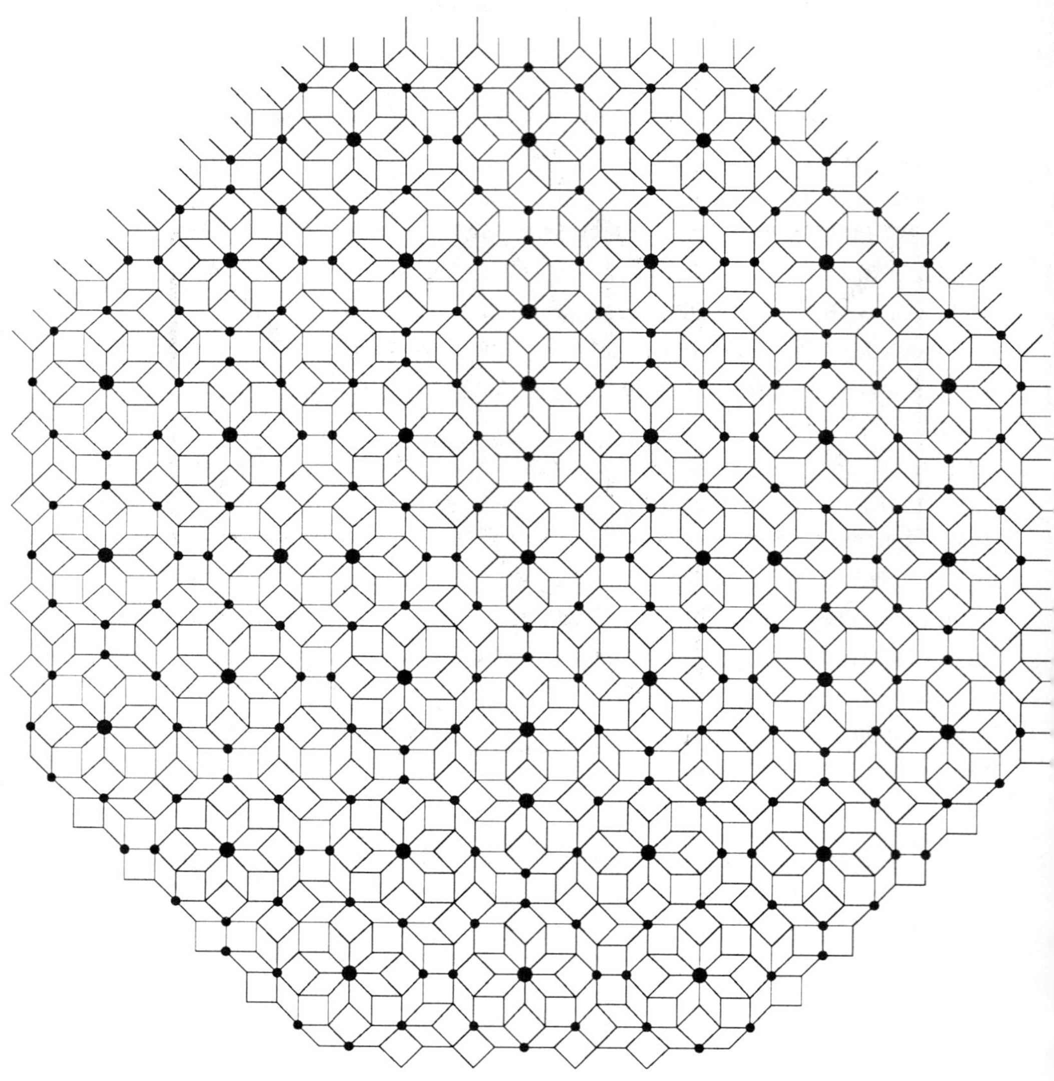

Fig.8- The 2D octagonal lattice -

6.REFERENCES

S.ALEXANDER(83), Superconductivity of networks. A percolation approach to the effect of disorder, *Phys. Rev., B27,* (1983), 1541-57.

S.ALEXANDER(84), Some properties of the spectrum of the Sierpinsky gasket in a magnetic field, *Phys. Rev., B29,* (1984), 5504-5508.

M.F.BARNSLEY(83), J.S.GERONIMO, A.N.HARRINGTON, Infinite Dimensional Jacobi Matrices associated with Julia sets, *Proc. AMS, 88 ,* (1983), 625-630.

M.F.BARNSLEY(85), J.S.GERONIMO, A.N.HARRINGTON, Almost Periodic Jacobi Matrices associated with Julia sets for Polynomials, *Commun. Math. Phys., 99 ,* (1985), 303-317.

A.BEHROOZ(86), M.BURNS, H.DECKMAN, D.LEVINE, B.WHITEHEAD, P.M.CHAIKIN, Flux Quantization on Quasicrystalline Networks, *Phys. Rev. Letters, 57,* (1986), 368-371.

J.BELLISSARD(88), C*-Algebras in Solid State Physics, *Operator Algebras and Applications*, D.E.Evans & M.Takesaki Eds., Cambridge Univ. Press, Cambridge (1988).

J.BELLISSARD(82), D.BESSIS, P.MOUSSA, Chaotic states of almost periodic Schrödinger operators, *Phys. Rev. Lett., 49,* (1982), 701-704.

J.BELLISSARD(85), Stability and instabilities in Quantum Mechanics, in *"Trends and developmentqs in the eighties "*, S.Albeverrio, Ph. Blanchard Eds., *World Sc. Pub. ,* (1985), 1-106.

J.BELLISSARD(89), B.IOCHUM, E.SCOPPOLA, D.TESTARD, Spectral Properties of One Dimensional Quasi-Crystals, *Commun. Math. Phys., 125,* (1989), 527-543.

J.BELLISSARD(90a), B.IOCHUM, D.TESTARD, Continuity properties of electronic spectrum of 1D quasicrystals, *CPT Preprint, Marseille,* October 1990, subm. to *Commun. Math. Phys.*.

J.BELLISSARD(90b), Gap Labelling Theorems for Schrödinger's Operators, to be published in *"Number Theory and Physics,* J.M.Luck, P.Moussa, M.Waldschmidt Eds, Les Houches 1989, Springer, (1991).

H.BOHR(47), *Almost Periodic Functions ,* Chelsea Pub. Co., New-York (1947).

H.BRÖLIN(65), Invariant sets under iteration of rational functions, *Arkiv för Mat. , Bd6 nr 6,* (1965), 103-144.

M.CASDAGLI(86), Symbolic Dynamics for the Renormalization Map of a Quasiperiodic Schrödinger Equation, *Commun. Math. Phys. 107,* (1986), 295-318.

L.DAHMANI(91), Ph. HAUGUEL, C.SIRE, J. BELLISSARD, Renormalization Group Calculation of Electronic Spectrum of a Quasicrystal, in preparation.

F.DELYON(86), D.PETRITIS, Absence of Localization in a class of Schrödinger Operators with Quasiperiodic Potential, *Commun. Math. Phys. 103,* (1984), 441-444.

A.DOUADY(82), Systèmes dynamiques holomorphes, *Séminaire Bourbaki, 599,* (1982), 39-63.

M.DUNEAU(85), A.KATZ, Quasiperiodic Patterns, *Phys. Rev. Letters, 54,* (1985), 2688-2691.

P.FATOU(19), Mémoire sur les Equations Fonctionelles, *Bull. Soc. Math. France, 47,* (1919), 161-271; *47,* (1920), 3-94, and 208-314.

P.G. de GENNES(81), 1)- Diamagnétisme de grains supraconducteurs près d'un seuil de percolation, *C.R.Acad. Sci., B292,* (1981), 9-12. 2)- Champ critique d'une boucle supraconductrice ramifiée, *C.R.Acad. Sci., B292,* (1981), 279-282.

J.M.GHEZ(87), W.WANG, R.RAMMAL, B.PANNETIER, J.BELLISSARD, Band Spectrum for an Electron on a Sierpinsky Gasket in a Magnetic Field, *Solid State Commun., 64,* (1987), 1291-1294.

D.GRATIAS(86), L.MICHEL, International Workshop on Aperiodic Crystals , *J. de Physique, C3, 47,* (1986).

T.HATAKEYAMA(87), H.KAMIMURA, Electronic properties of a Penrose tiling lattice in a magnetic field, *Solid State Comm., 62,* (1987), 79-83.

Ph. HAUGUEL(91), Thèse, in preparation.

Ch.JANOT(88), J.M.DUBOIS, Eds., *Quasicrystalline materials,* Grenoble 21-25 March 1988, World Scientific Pub., (1988), Singapore.

G.JULIA(18), Itération des applications fonctionelles, *J. Math. Pures et Appl.,* (1918), 47-245.

L.P.KADANOFF(83), M.KOHMOTO, C.TANG, Localization problem in One dimension: Mapping and Escape, *Phys. Rev. Lett., 50 ,* (1983), 1870-1872.

P.A.KALUGIN(86), A.Yu.KITAEV, S.LEVITOV, Electron spectrum of a one dimensional quasi-crystal, *Sov. Phys. JETP, 64,* (1986), 410-415.

M.KOHMOTO(84), Y.OONO, Cantor spectrum for an almost periodic Schrödinger operator and a dynamical map, *Phys. Lett., 102A,* (1984), 145-148.

M.KOHMOTO(86), B.SUTHERLAND, Electronic States on a Penrose Lattice, *Phys. Rev. Lett., 56,* (1986), 2740-2743.

D.LEVINE(84), P.STEINHARDT, Quasicrystals: a New Class of Ordered Structure, *Phys. Rev. Lett., 53,* (1984), 2477-2480.

L.S.LEVITOV(89), Renormalization group for a quasiperiodic Schrödinger operator, *J. of Phys., 50,* (1989), 707-716.

J.M.LUCK(86), D.PETRITIS, Phonon spectra in one dimensional quasicrystal, *J. Stat. Phys., 42,* (1986), 289-310.

T.ODAGAKI(86), D.NGUYEN, Electronic and vibrational spectra of two- dimensional quasicrystals, *Phys. Rev., B33,* (1986), 2184-2190.

S.OSTLUND(85), S.KIM, Renormalization of quasiperiodic mappings, *Physica Scripta, 9,* (1985), 193-198.

S.OSTLUND(83), R.PANDIT, D.RAND, H.J.SCHELLNHUBER, E.D.SIGGIA, One dimensional Schrödinger equation with an almost periodic potential, *Phys. Rev. Lett., 50,* (1983), 1873-1877.

S.OSTLUND, R.PANDIT(84), Renormalization group analysis of a discrete quasiperiodic Schrödinger equation, *Phys. Rev., B29,* (1984), 1394-1414.

G.PEDERSEN(79), *C*-algebras and their automorphism groups* , New-York, Academic Press, (1979).

R.RAMMAL(83), T.C.LUBENSKY, G.TOULOUSE(83), Superconducting networks in a magnetic field, *Phys. Rev., B27,* (1983), 2820-2829.

R.RAMMAL(84), Spectrum of harmonic excitations on fractals, *J. de Physique, 45,* (1984), 191-206.

R.RAMMAL(85), Landau Level spectrum of Bloch electrons in a honeycomb lattice, *J. de Physique, 46,* (1985), 1345-1354.

B.SIMON(82), Almost periodic Schrödinger operators: A Review, *Adv. in Appl. Math., 3,* (1982), 463-490.

C.SIRE(89), R.MOSSERI, Electronic spectrum of 1D quasicrystals near the periodic chain, *J. de Physique, 50,* (1989), 3447.

C.SIRE(90a), R.MOSSERI, Extended states, gap closing: some exact results for codimension one quasicrystals, *J. de Physique, 51,* (1990), 1569.

C.SIRE(90b), J.BELLISSARD, Renormalization group for the octagonal Quasi-periodic tiling, *Europhys. Lett., 11,* (1990), 439-443.

C.SIRE(90c), *Etudes des excitations élémentaires et de la stabilité de structures quasipériodiques* , Thèse Université de Paris VI, Paris, Novembre 1990.

P.J.STEINHARDT(87), S.OSTLUND, *The Physics of quasicrystals,* World Scientific Pub., Singapore (1987).

A.SÜTÓ(87), The spectrum of a quasi-periodic Schrödinger operator, *Commun. Math. Phys., 111,* (1987), 409-415.

A.SÜTÓ(89), Singular continuous spectrum on a Cantor set of zero Lebesgue measure for the Fibonacci Hamiltonian, *J. Stat. Phys., 56,* (1989), 525-531.

C.TANG(86), M.KOHMOTO, Global scaling properties of the spectrum for a quasiperiodic Schrödinger equation, *Phys. Rev., B34,* (1986), 2041-2044.

H.TSUNETSUGU(86), T.FUJIWARA, K.UEDA, T.TOKIHIRO, Eigenstates in a 2-dimensional Penrose tiling, *J. Phys. Soc. Japan, 55,* (1986), 1420-1423.

F.WIJNANDS(89), Energy spectra for one-dimensional quasiperiodic potentials: bandwidth, scaling, mapping and relation with local isomorphism, *J. Phys., A22,* (1989), 3267-3282.

THE MEASUREMENT PROBLEM IN THE STOCHASTIC FORMU-LATION OF QUANTUM MECHANICS

PHILIPPE BLANCHARD
Research Center BiBoS, Universität Bielefeld
4800 Bielefeld 1, Germany

MARCELLO CINI*
Laboratoire de Physique Théorique, Université de Paris XI
91405 Orsay Cedex, France

MAURIZIO SERVA
Dipartimento di Matematica, Università dell'Aquila
67010 Coppito, L'Aquila, Italy,
also INFN, sezione dell'Aquila

Abstract: The description of physical systems obtained in the classical limit of Quantum Mechanics is not consistent with the one given by Classical Statistical Mechanics. In Quantum Mechanics the complete description of a physical system is given by specifying the wave function, which evolves completely deterministically. Probability comes into play only when a measurement is made. Classical Statistical Mechanics, on the contrary, concerns systems whose description is not complete and where probability enters as a consequence of a lack of knowledge. In this paper we show that Nelson's Stochastic Mechanics, whose predictions are identical with those of Quantum Mechanics, provides nevertheless a unified description of the deterministic and random aspects of Quantum Mechanics which eliminates the inconsistency outlined above and gives a simple and satisfactory solution to the measurement problem.

* On leave from *Dipartimento di Fisica, Università di Roma "La Sapienza",*
piazzale A. Moro 2, Roma 00185 Italy

149

1. Introduction

In Quantum Mechanics (Q.M) the complete description of the state of a physical system with configuration space $\mathbb{R}^n$ is given by specifyng the wave function ψ, which is a square integrable function. The dynamics is given by the Schrödinger equation and so the state evolves completely deterministically.

Randomness comes into play only when a measurement is made. The manner in which it is incorporated into the formalism of Q.M. appears for many people to be not very satisfactory. Indeed physical systems are described in Q.M. as evolving according to one or the other of two apparently incompatible ways: completely deterministic Schrödinger equation and probabilistic state vector reduction.

On the other side it is almost impossible to be dissatisfied with the very brilliant successes obtained by using Schrödinger equation for predicting the outcomes of experiments.

In this paper we give an answer to this problem using Nelson's Stochastic Mechanics (S.M.). This theory gives a more detailed description of physical systems with respect to Q.M. but is completely equivalent to Q.M. for predictions of the outcomes of experiments. Nevertheless, we show that S.M. is able to describe in a unifed language the deterministic and random aspects of quantum evolution giving a simple and completely satisfactory solution to the measurement problem.

In section 2 we discuss the measurement process in Q.M. showing that the problems are essentially a consequence of a more fundamental problem connected with the classical limit of the theory.

In section 3 we give a short description of Stochastic Mechanics. We show that the theory is a kind of minimal randomization of Classical Mechanics which reconciles the individual particle random trajectories with the Schrödinger equation.

In section 4 we rediscuss the classical limit and the measurement process in the framework of S.M. showing that both have a simple and coherent description in this theory.

In section 5 we test our results against a simple model of Stern-Gerlach measuring apparatus.

Finally, in the appendix, we briefly recall how to extend S.M. in order to describe also discrete processes and we produce another example based on a simple model of spin-bosons interaction.

2. Quantum Mechanics: the problems

2.1. *The classical limit*

The classical limit of Q.M. is generally discussed by exhibiting the connection between quantum and classical equations of motion. It is shown for example, that the phase S of a wave function satisfying Schrödinger equation becomes the classical action function S_c, which satisfies the Hamilton-Jacobi equation. In this way one concentrates on the emergence of classical trajectories and remains satisfied with this result, without considering further the limiting behaviour of the statistical properties of quantum mechanical states.

Generally speaking, quantum probabilities (which refer to the output of a measurement) for a given state become, in the limit, the probabilities associated to a classical statistical ensemble. For example the quantum probabilities of an energy eigenstate become those of the classical microcanonical ensemble with the same energy.

Two main difficulties however arise if one looks carefully into this limit [1]. The first is a consequence of the superposition principle. Consider a state

$$|\psi> = \frac{1}{\sqrt{2}} \left(|\psi_r> + |\psi_l> \right). \tag{2.1}$$

and suppose that in configuration space the wave functions $\psi_r(x,t) = <x|\psi_r>$ and $\psi_l(x,t) = <x|\psi_l>$ are "classically" separated. As an example for a free particle ψ_r and ψ_l can be chosen as two gaussians centered in $x_r = p_0 t$ and $x_l = -p_0 t$ respectively. The variance of any single gaussian has the well known form $\sigma = [\frac{(\Delta p)^2 t^2}{m^2} + \frac{\hbar^2}{4(\Delta p)^2}]^{\frac{1}{2}}$ while the separation of the two peaks is $2p_0 t$. Therefore classically separated means $2p_0 t \gg 2\sigma$, which for large t reduces to $p_0 \gg \Delta p$. The probability density for the x variable from (2.1) is

$$r(x,t) = \frac{1}{2}[|\psi_r|^2 + |\psi_l|^2 + 2Re(\psi_r^* \psi_l)] \tag{2.2}$$

since ψ_r and ψ_l have a very small overlap one can write

$$r(x,t) \approx \tilde{r}(x,t) = \frac{1}{2}[|\psi_r|^2 + |\psi_l|^2] \tag{2.3}$$

The approximate equality $r \approx \tilde{r}$ becomes exact in the limit $p_0/\Delta p \to \infty$.

The probability $\tilde{r}$ (which corresponds to a mixture) coincides with a classical statistical description in which the particle is on the right of the origin with probability 1/2 and on the left with the same probability. One would be tempted to say that also the probability r, well approximated by $\tilde{r}$ describes the same situation at least when $p_0/\Delta p$ is very large. This is not possible because the standard interpretation of Q.M. denies that a particle has a definite position before interacting with a "classical instrument".

In other words, the classical probability $\tilde{r}$ is interpreted as follows: "It is meaningful to say that the particle is on the right or on the left of the origin, but since we don't know we assign probability 1/2 to both these possibilities". On the other hand we interpret r (even in the limit $r \to \tilde{r}$) by saying: "It is meaningless to say that the particle is on the left or on the right. *Only if we look with a classical instrument* (mind?) we find it on the right or on the left with equal probability".

In conclusion even if $r \to \tilde{r}$ *we do not* obtain a classical statistical interpretation. There is a fundamental gap which is not eliminated in this limit.

The second difficulty arises from the interference terms. If we consider the density matrix $\rho = |\psi><\psi|$ and the mixture density matrix $\tilde{\rho}$

$$\tilde{\rho} = \frac{1}{2}[|\psi_r><\psi_r| + |\psi_l><\psi_l|] \tag{2.4}$$

we can always find an operator A for which

$$Tr[A(\rho - \tilde{\rho})] \neq 0 \tag{2.5}$$

even when $p_0/\Delta p \gg 1$. An obvious example is the operator

$$A = \frac{1}{2}(|\psi_r><\psi_l| + |\psi_l><\psi_r|) \tag{2.6}$$

This is not astonishing. It only means that some "quantum behaviour" can always be detected, in principle, also for variables which refer to macroscopic systems.

In our example the macroscopic variable is the distance between the two gaussian peaks or, equivalently, the variable $p_0/\Delta p$. In this case the limit $Tr[(\rho - \tilde{\rho})A]$ can be different from 0 only if the operator A introduces a dependence on the variable $p_0/\Delta p$ which cancels the effect of the vanishing of $Tr(\rho - \tilde{\rho})$ in the limit $p_0/\Delta p \to \infty$.

For a system made of a very large number of particles the macroscopic properties are described by the collective variables, while the microscopic details are represented by the variables of each single particle. However we will not discuss this question in the present work and we refer to [1,2] for more details.

An orthodox physicist is therefore confirmed in his idea that the situation in which interference terms are present, even if they are almost vanishing, is qualitatively different from the situation in which they are totally absent.

2.2. *The measurement*

The measurement in Q.M. proceeds as follows [3-8]. Before the measurement the micro+macro system is described by the state vector

$$\frac{1}{\sqrt{2}}|\psi_0>[|\phi_+> + |\phi_->]$$ (2.7)

where $|\phi_+>$ and $|\phi_->$ are two eigenvectors corresponding to the variable to be measured. The apparatus is described before the measurement at $t = 0$ by a state vector $|\psi_0>$ corresponding to a "neutral" position of the pointer.

Assume that the interaction produces a one-to-one correspondence between the states $|\psi_r>$ and $|\psi_l>$ of the apparatus introduced in the previous subsection and the states $|\phi_+>$ and $|\phi_->$ of the microscopic object. Then the total wave function $\chi(x,y,t) \equiv <x,y|\chi>$, at any time t sufficiently large after the measurement, will be given by

$$\chi(x,y,t) = \frac{1}{\sqrt{2}}[\psi_r(x,t)\phi_+(y,t) + \psi_l(x,t)\phi_-(y,t)]$$ (2.8)

where y is the configurational variable corresponding to the microsystem. The probability density of the micro+macro system will be correspondingly

$$r(x,y,t) = \tilde{r}(x,y,t) + Re[\psi_r\psi_l^*\phi_+\phi_-^*]$$ (2.9)

where

$$\tilde{r}(x,y,t) = \frac{1}{2}[|\psi_r|^2|\phi_+|^2 + |\psi_l|^2|\phi_-|^2]$$ (2.10)

Since the pointer interference terms $\psi_r\psi_l^*$ and $\psi_r^*\psi_l$ become negligible when $p_0/\Delta p \to \infty$ the probability density r reduces to the probability density matrix $\tilde{r}$ of the statistical mixture describing the microscopic object in the state $|\phi_+>$ when the pointer "is" at the right of the origin or in the state $|\phi_->$ when the pointer is at the left. Unfortunately our preceding discussion about the classical limit shows that: (a) one cannot say that the pointer a "is" in a given position before it is, at its turn, measured by another apparatus (infinite regression); (b) had we considered, instead of the pointer position x, another variable A not commuting with x, we would have obtained a probability density $r(A, y, t) \equiv |<A, y|\chi>|^2$ for variables A and y in which appears the interference product $<A|\psi_r><\psi_l|A>$ which might have not been negligible. Of course this would be possible only if A is a non local operator and this would correspond to possible "quantum behaviour" of the apparatus.

If the interference terms $\psi_r\psi_l^*$ and conjugate are not negligible, the "macroscopic system" is not a good measuring apparatus, becouse the main requisite that a measuring apparatus must fulfil is that it should exhibit a one-to-one probability correlation betwwen the values of the pointer and the values of the quantity to be measured.

The difficulty however remains that the position x is no longer an "objective" property of the pointer as it is for the position of a classical object. It is therefore clear that the difficulties of the mesurement problem are simply a consequence of those encountered in the classical limit of Q.M..

3. Basic facts about Stochastic Mechanics

In order to show that S.M. [9-12] is a minimal randomization of Classical Mechanics (C.M.) it is useful to formulate the last in a convenient form.

C.M. consists essentially in two assumptions, a kinematical and a dynamical one. We can say that a particle moving in Euclidean space $\mathbb{R}^n$ follows *smooth* trajectories (the kinematical assumption) which satisfy the Newton equation (the dynamical assumption).

We can put this statement in the form of two equations. The first equation corresponds to a definition of the kinematics

$$dx(t) = b(x(t), t)dt \tag{3.1}$$

Where b is a velocity field. The second equation is simply the Newton equation

$$m\ddot{x}(t) = -\nabla V(x(t)) \tag{3.2}$$

This can be regarded as a constraint on the allowed velocity fields appearing in (3.1). In order to see this fact (and also for future comparisons) it is useful to recall that the total derivative of a smooth real valued function $f(x, t)$ is defined by

$$Df(x, t) = \lim_{\Delta t \to 0} \frac{1}{\Delta t}[f(x(t + \Delta t), t + \Delta t) - f(x(t), t)] \tag{3.3}$$

together with the condition $x(t) = x$ which means that the particle is in x at time t. A simple calculation, which takes into account formula (3.1), leads to

$$Df(x,t) = b(x,t)\nabla f(x,t) + \frac{\partial f(x,t)}{\partial t} \qquad (3.4)$$

this formula permits to calculate the acceleration field, infact, the i-th component of the acceleration is

$$a_i(x,t) \equiv DDx_i = Db_i(x,t) = b(x,t)\nabla b_i(x,t) + \frac{\partial b_i(x,t)}{\partial t}$$

Let us now assume that b is a gradient function

$$b(x,t) = \frac{1}{m}\nabla S_c(x,t) \qquad (3.5)$$

(where S_c is the classical action), a short calculation shows that the acceleration takes the simpler form

$$a(x,t) = \frac{1}{2}\nabla b^2(x,t) + \frac{\partial b(x,t)}{\partial t} \qquad (3.6)$$

and therefore the Newton equation becomes

$$m[\frac{1}{2}\nabla b^2(x,t) + \frac{\partial b(x,t)}{\partial t}] = -\nabla V(x) \qquad (3.7)$$

which can be integrated in order to obtain the Hamilton-Jacobi equation

$$\frac{1}{2m}[\nabla S(x,t)]^2 + \frac{\partial S(x,t)}{\partial t} + V(x) = 0 \qquad (3.8)$$

On the other hand, it is an obvious fact that equation (3.1) is equivalent to the continuity equation

$$\frac{\partial r(x,t)}{\partial t} = -\mathrm{div}[r(x,t)b(x,t)] \qquad (3.9)$$

which gives the evolution of the probability density r once its initial value is known. The single trajectory can be eventually obtained simply by assuming that the initial r is a Dirac delta.

The three equations (3.5), (3.8) and (3.9) give a "hydrodynamical" formulation of C.M. completely equivalent to the standard one.

S.M. can be viewed as a mimimal randomization of C.M.. The kinematical law of S.M. asserts that the paths of the particle are the *continuous but not differentiable* realizations of a diffusion process. In other words, the trajectories are given by the solution of a stochastic differential equation of the type

$$dx(t) = b^+(x(t), t)dt + \sqrt{\sigma}dw(t) \tag{3.10}$$

This equation is the random counterpart of (3.1). The trajectories are not differentiable becouse $w(t)$ is the standard Brownian motion and therefore $dw(t)$, roughly speaking, is proportional to $(dt)^{\frac{1}{2}}$. σ is the diffusion constant whose value we do not specify at the moment. This stochastic equation corresponds to the Fokker-Planck equation

$$\frac{\partial r(x, t)}{\partial t} = -\mathrm{div}[r(x, t)b^+(x, t)] + \frac{\sigma}{2}\Delta r(x, t) \tag{3.11}$$

which reduces to the continuity equation when σ is put equal 0.

Since the trajectories $x(t)$ are not differentiable it is necessary to define a kind of mean derivative in order to extend the Newton equation (2.2). This can be done in the following way: given a smooth real-valued function $f(x, t)$, we define a forward mean derivative as

$$D^+ f(x, t) = \lim_{\Delta t \to 0} \frac{1}{\Delta t} E[f(x(t + \Delta t), t + \Delta t) - f(x(t), t)|x(t) = x] \tag{3.12}$$

where $E[\cdot|x(t) = x]$ means conditional expectation with respect to $x(t) = x$. In other words the expectation is taken under the condition that the particle is in the position x at time t. Taking into account (2.4) we easily obtain

$$D^+ f(x, t) = b^+(x, t)\nabla f(x, t) + \frac{\sigma}{2}\Delta f(x, t) + \frac{\partial f(x, t)}{\partial t} \tag{3.13}$$

We also give another definition which is time-symmetric with respect to the first one. We define a backward mean derivative as

$$D^- f(x, t) = \lim_{\Delta t \to 0} \frac{1}{\Delta t} E[f(x(t + \Delta t), t + \Delta t) - f(x(t), t)|x(t + \Delta t) = x] \tag{3.14}$$

where we have changed the condition in the past with the condition in the future that the particle is in x at time $t + \Delta t$. One can easily see that

$$D^- f(x, t) = b^-(x, t)\nabla f(x, t) - \frac{\sigma}{2}\Delta f(x, t) + \frac{\partial f(x, t)}{\partial t} \tag{3.15}$$

where $b^-(x,t) \equiv b^+(x,t) - \sigma \nabla \log r(x,t)$. Both these definitions reduce to the ordinary total derivative when the trajectory is classical ($\sigma = 0$).

We can now define a mean acceleration which is the counterpart of the classical one; for the i-th component we write

$$a_i(x,t) \equiv \frac{1}{2}[D^+ D^- x_i + D^- D^+ x_i] = \frac{1}{2}[D^+ b_i^-(x,t) + D^- b_i^+(x,t)] \qquad (3.16)$$

We assume, as we have done in C.M., that b^+ is a gradient function (in this case the definition of b^- implies that b^- is also a gradient function). We can therefore introduce a scalar function $S(x,t)$ (the quantum action) such that

$$b^+(x,t) = \frac{1}{m}\nabla S(x,t) + \frac{\sigma}{2}\nabla \log r(x,t) \qquad (3.17)$$

and

$$b^-(x,t) = \frac{1}{m}\nabla S(x,t) - \frac{\sigma}{2}\nabla \log r(x,t) \qquad (3.18)$$

equation (3.16) can be integrated and becomes

$$\frac{1}{2m}[\nabla S(x,t)]^2 + \frac{\partial S(x,t)}{\partial t} - \frac{\sigma^2}{2}\frac{\Delta r^{\frac{1}{2}}(x,t)}{r^{\frac{1}{2}}(x,t)} + V(x) = 0 \qquad (3.19)$$

The equations (3.11), (3.17) and (3.19) are the counterpart of the classical equations (3.5), (3.8) and (3.9). It is also clear that assuming that σ vanishes we recover the last three from the first three.

Stochastic Mechanics is therefore simply a generalization of C.M. which is obtained with a mininimal randomization of the trajectories.

Now comes the question where Q.M. enters in all that since we have, apparently, somethting which looks much more similar to C.M. than Q.M.. In order to answer the question we construct the function ψ in the following way

$$\psi(x,t) \equiv r^{\frac{1}{2}}(x,t)\exp[\frac{iS(x,t)}{m\sigma}] \qquad (3.20)$$

we easily check that it satisfies the equation

$$i\sigma m\frac{\partial \psi(x,t)}{\partial t} = -\frac{m\sigma^2}{2}\Delta\psi(x,t) + V(x)\psi(x,t) \qquad (3.21)$$

it is therefore sufficient to put $\sigma = \frac{\hbar}{m}$ in order to obtain the Schrödinger equation which, in this context appears simply as a "trick" to linearize the equations of our perturbed version of C.M..

We can now reverse our construction and say: given a solution $\psi(x,t)$ of Schröedinger equation we can associate a diffusion whose probability density is $r(x,t) = |\psi(x,t)|^2$ and whose drift field is

$$b^+(x,t) = \frac{\hbar}{m}(Re + Im)[\nabla \log \psi(x,t)] \tag{3.22}$$

The conclusion is therefore that S.M. embeds the computational formalism of conventional Q.M. into a more detailed description of nature so that ψ no longer fully describes the state of the system. Nevertheless, since the probability densities of S.M. and Q.M. coincide, the two theories are equivalent for what concerns the preduction of outputs of measurements.

Before ending this section we make two remarks. The first concerns the behaviour of trajectories in the proximity of the the nodal set $N_\psi = \{x \in \mathbb{R}^n | \psi(x,t) = 0\}$. It has been shown [see 13,14] that a path never crosses a point of this set. This leads to a decomposition of $\mathbb{R}^n$ into invariant regions C_i separated by N_ψ. One can speak in this case of "trapping" or "confinement" in C_i since a process starting in C_i does not leaves this region with probability one. In other words N_ψ acts as an impenetrable barrier.

The second remark is that the dynamics of S.M. can be alternatively obtained from variational principles [15-18].

4. Stochastic Mechanics: solution of the problems

4.1. *The classical limit*

Consider again the case of the second section. In S.M. one associates to the wave function

$$<x|\psi> = \frac{1}{\sqrt{2}}[\psi_r(x,t) + \psi_l(x,t)] \tag{4.1}$$

a process with probability density

$$r(x,t) = \frac{1}{2}[|\psi_r|^2 + |\psi_l|^2 + 2Re(\psi_r\psi_l^*)] \tag{4.2}$$

and drift

$$b^+(x,t) = \frac{\hbar}{m}(Re + Im)\nabla \log(\psi_r + \psi_l) \tag{4.3}$$

In the limit $p_0/\Delta p \to \infty$ ψ_r and ψ_l tend to have disjoint supports and therefore the probability of crossing of the origin becomes very small, furthermore, the density and the drift at right of the origin are well approximated by

$$r_r(x,t) = |\psi_r|^2 \tag{4.4}$$

and

$$b_r^+(x,t) = \frac{\hbar}{m}(Re + Im)\nabla \log \psi_r \tag{4.5}$$

while on the left they are approximated by

$$r_l(x,t) = |\psi_l|^2 \tag{4.6}$$

and

$$b_l^+(x,t) = \hbar(Re + Im)\nabla \log \psi_l \tag{4.7}$$

As a consequence of this fact the process splits in two separate processes [see also 19,20], the first will live on the right of the origin and will have density $r_r(x,t)$ and drift $b_r^+(x,t)$ and the second will live on the left and will have density $r_l(x,t)$ and drift $b_l^+(x,t)$. The system belongs to the first process or to the second one with probability $\frac{1}{2}$.

It should be remarked that the presence of nodes around the origin would improve the mechanism because the supports of the two processes would be disjoint even before the limit. Obviusly the limit $p_0/\Delta p \to \infty$ is still necessary in order to have convergence of densities and drifts.

We can compare these two processes originated from the pure state with the other two which originated from the mixture. By definition the first process of the mixture will have density $r_r(x,t)$ and drift $b_r^+(x,t)$, furthermore, in the limit $p_0/\Delta p \to \infty$ it will be confined on the right of the origin ($r_r(x,t)$ vanishes on the left). An analogous discussion holds for the second process of the mixture. We conclude that the couple originated from the pure state and the couple associated to the mixture coincide in the limit.

The problems of Q.M. have now disappeared because the statement that the system is on the right (left) is always fully meanigful, therefore in the limit *there is not* any conceptual difference between a pure state and a mixture.

Now comes the question: what happens to the interference terms when the limit is not exact? They simply modify the behaviour of the process in the interference region. In principle one can detect this difference which corresponds to a semiclassical behavior.

In S.M., as in Q.M., one can define variables which are particularly sensitive to the details of the wave function (and therefore of the drift) in the interference region. These variables show, therefore, a typical quantum behaviour (an example of this fact will be given in section 5).

There is, however, no one-to-one correspondence with the operators of Q.M., because all variables in S.M. are defined in terms of the stochastic variable x_t. This is not a disadvantage, in our opinion, for two reasons. Firstly because an operator such as A of eq. (2.6) does not correspond to any operational definition, on the contrary, operators which are operationally well defined can be constructed from x_t (for example the output momenta of particles which scatters on a target [21,22]). Secondly, because no problems of interpretation arise in S.M., since the position always has a physical meaning.

4.2. *The measurement*

From eq.(2.7)we have, denoting, as before, by x the configurational macroscopic variable, and by y the configurational microscopic variable, that the wave function before the measurement is

$$<x,y|\chi> = \frac{1}{\sqrt{2}}<x|\psi_0>(<y|\phi_+> + <y|\phi_->) \tag{4.8}$$

The corresponding process for the two stochastic variables $x(t)$, $y(t)$ has density

$$r(x,y,t) = \frac{1}{2}[|\psi_0(x,t)|^2 |\phi_+(y,t) + \phi_-(y,t)|^2] \qquad (4.9)$$

and drift

$$b^+(x,y,t) = \frac{\hbar}{m}(Re + Im)[\nabla_y \log(\phi_+(y,t) + \phi_-(y,t)) \ , \ \nabla_x \log \psi_0(x,t)] \qquad (4.10)$$

The drift has two components, the first one for the microscopic variable, the second for the macroscopic one. Since the density for the two components is a product and the two components of the drift are uncorrelated, the stochastic variables $x(t)$ and $y(t)$ are statistically and dynamically independent.

After the measurement we have

$$<x,y|\chi> = \frac{1}{\sqrt{2}}[<x|\psi_r><y|\phi_+> + <x|\psi_l><y|\phi_->] \qquad (4.11)$$

Since ψ_r and ψ_l have disjoint supports the process splits into two separated processes [19,20]. The first one has density

$$r_r(x,y,t) = |\psi_r(x,t)|^2 |\phi_+(y,t)|^2 \qquad (4.12)$$

and drift

$$b_r^+(x,y,t) = \frac{\hbar}{m}(Re + Im)[\nabla_y \log \phi_+(y,t) \ , \ \nabla_x \log \psi_r(x,t)] \qquad (4.13)$$

The second has density

$$r_l(x,y,t) = |\psi_l(x,t)|^2 |\phi_-(y,t)|^2 \qquad (4.14)$$

and drift

$$b_l^+(x,y,t) = \frac{\hbar}{m}(Re + Im)[\nabla_y \log \phi_-(y,t) \ , \ \nabla_x \log \psi_l(x,t)] \qquad (4.15)$$

The total system belongs to one of them with probability $1/2$. The effect of a measurement on the microscopic variable alone is therefore the following [23]. Before the measurement we have the process whith density and drift

$$\frac{1}{2}|\phi_-(y,t) + \phi_+(y,t)|^2 \ , \quad \frac{\hbar}{m}(Re + Im)\nabla_y \log(\phi_+(y,t) + \phi_-(y,t)) \qquad (4.16)$$

while after the measurement we have two processes, the first with density and drift

$$|\phi_+(y,t)|^2 \quad , \quad \frac{\hbar}{m}(Re + Im)\nabla_y \log \phi_+(y,t) \qquad (4.17)$$

the second with

$$|\phi_-(y,t)|^2 \quad , \quad \frac{\hbar}{m}(Re + Im)\nabla_y \log \phi_-(y,t) \qquad (4.18)$$

The microsystem belongs to one of these two with probability 1/2. The measurement has the effect of modifying the process of the microscopic variable. This fact leads to the conclusion that the trajectories of S.M. are not observable [23] since any observation modifies the process.

There is only one case in which this modification does not occur namely the case in which the "microscopic" variable to be measured is indeed classical. This is the same as saying that ϕ_+ and ϕ_- have disjoint supports. Therefore one has that the ante-measurement process (4.16) coincides with the couple of post-measurement processes (4.17) and (4.18) In other words, if the microsystem is classical, we can look at the dichotomic variable corresponding to the states ϕ_{pm} without disturbing it. Looking has the only effect of increasing our knowledge. Classical trajectories are observable and have an objective meaning.

This kind of reasoning can also be applied to measurements of the total system (microscopic + macroscopic). Since it is globally classical, the associated process is not affected by any further measuring interaction. This means that we can look at the pointer without fear of changing its state or the state of the microsystem.

It is clear at this point that the measurement in S.M. proceeds as follows:

(a) We produce an interaction which links the microscopic variable to a macro-scopic one

(b) Since the total system is now classical we can look at it (measure, interact) without disturbing the micro+macro system. Therefore we look at it with only result the of a gain of information.

5. A model: the Stern-Gerlach apparatus

We consider a Stern-Gerlach apparatus in which a beam of atoms is split in two separated beams by a suitable interaction between their internal angular momentum and an external magnetic field.

A simple model for this system can be constructed by coupling the internal angular momentum $J = (J_1, J_2, J_3)$ with an inhomogeneous magnetic field $B(x_1, x_2, x_3) = (0, x_1, 0)$ acting in the time interval between $t = 0$ and $t = T$. The coordinates of the center of mass are x_1, x_2 and x_3 but we ignore x_2 and x_3 since they are not involved in the interaction and we simply use $x \equiv x_1$. The internal configurational variable is y. This leads to the quantum mechanical Hamiltonian

$$H = -\frac{\hbar^2}{2m}\frac{d^2}{dx^2} - \lambda J_2 x \theta(t)\theta(T - t) \tag{5.1}$$

We take the initial state to be a product of a gaussian

$$\psi_0(x) = (2/\pi\alpha)^{\frac{1}{4}} \exp\{-x^2/\alpha\} \tag{5.2}$$

times a superpostion of two eigenstates $\phi_+(y)$ and $\phi_-(y)$ corresponding to the eigenvalues $+n\hbar$ and $-n\hbar$ of J_2:

$$<x, y|\chi> = \psi_0(x)\frac{1}{\sqrt{2}}[\phi_+(y) + \phi_-(y)] \tag{5.3}$$

where $\alpha \equiv \hbar^2/(\Delta p_0)^2$ and Δp_0 is the indeterminacy of momentum. The total wave function at time t will be easily obtained once the following Schrödinger equation for the x variable only is solved

$$i\hbar\frac{d\psi}{dt} = \frac{\hbar^2}{2m}\frac{d^2\psi}{dx^2} - Lx\psi \tag{5.4}$$

with $L = \pm\lambda\hbar n$ and initial condition ψ_0. The method is standard and one obtains a solution which for t larger than T is

$$\psi(x, t) = (\frac{2}{\pi\alpha})^{\frac{1}{4}}\frac{1}{1 + i\frac{2\hbar}{m\alpha}t}\exp\{-\frac{(x + x_0 - \frac{p_0 t}{m})^2}{\alpha + i\frac{2\hbar}{m}t}\}\exp\{i\frac{p_0 x}{\hbar}\} \tag{5.5}$$

which is a gaussian whose parameters are

$$<p>_t = LT \equiv p_0 \ , \quad <x>_t = \frac{p_0}{m}(t - \frac{T}{2}) \equiv \frac{p_0}{m}t - x_0 \tag{5.6}$$

$$<\Delta p>_t = \Delta p_0 \ , \quad <\Delta x>_t = [\frac{(\Delta p_0)^2 t^2}{m^2} + \frac{\hbar^2}{4(\Delta p_0)^2}]^{\frac{1}{2}} \tag{5.7}$$

The classical condition is, therefore, $LT \gg \Delta p_0$. Of course there will be two different wave functions ψ_r and ψ_l depending on the two values $p_0 = LT = \pm \lambda \hbar n T$ which have to be substituted in the above expression.

The total quantum mechanical wave function has the form (2.8). Now we associate to this wave function a stochastic process with two components $x(t)$ and $y(t)$ whose probability density at time $t > T$ will coincide with the quantum mechanical expression

$$r(x,y,t) = \tilde{r}(x,y,t) + Re\{\psi_r^*(x,t)\psi_l(x,t)\phi_+^*(y)\phi_-(y)\} \tag{5.8}$$

with

$$\tilde{r}(x,y,t) = \frac{1}{2}(|\psi_r(x,t)|^2|\phi_+(y)|^2 + |\psi_l(x,t)|^2|\phi_-(y)|^2) \tag{5.9}$$

We will now show that there is a vanishing probability for the macrosystem to cross the origin. In fact, if we consider the region around the origin for which $|x| < \epsilon t$, we have from (5.5) that for large t and small ϵ

$$|\psi_{r,l}(x,t)| < [\frac{m^2}{2\pi(\Delta p_0)^2 t^2}]^{\frac{1}{4}} \exp\{-\frac{(p_0)^2}{4(\Delta p_0)^2}\} \tag{5.10}$$

Therefore the total probability $P(\epsilon,x) = \int_{-\epsilon t}^{+\epsilon t} r(x,y,t)dx$ that the microsystem is in y and the macrosystem in the region $|x| < \epsilon t$ satisfies

$$P(\epsilon,x) < [|\phi_-(y)|^2 + |\phi_+(y)|^2 + 2|\phi_-(y)||\phi_+(y)|]\frac{\epsilon m}{\sqrt{2\pi}\Delta p_0} \exp\{-\frac{(p_0)^2}{2(\Delta p_0)^2}\} \tag{5.11}$$

this proves our statement for large $p_0/\Delta p_0$. The process therefore splits into two processes. Since a similar inequality holds for

$$\tilde{P}(\epsilon,x) = \int_{-\epsilon t}^{+\epsilon t} \tilde{r}(x,y,t)dx \tag{5.12}$$

we can say that also the right and left process associated with the wave functions

$$\chi_r(x,y,t) = \psi_r(x,t)\phi_+(y) \tag{5.13}$$

$$\chi_l(x,y,t) = \psi_l(x,t)\phi_-(y) \tag{5.14}$$

never cross the origin and remain on the right and on the left of the origin respectively.

One can easily prove that the pair of processes originating from the pure state and the pair originating from the mixture coincide. One has, in fact, for any x_0 and sufficiently large times

$$\int_{x_0-\epsilon t}^{x_0+\epsilon t} |r(x,y,t) - \tilde{r}(x,y,t)| dx < |\phi_-(y)||\phi_+(y)| \frac{2\epsilon m}{\sqrt{2\pi}\Delta p_0} \exp\{-\frac{(p_0)^2}{2(\Delta p_0)^2}\} \quad (5.15)$$

Furthermore, since there are no processes in the region $|x| < \epsilon t$, we can restrict the calculations of the drift to the regions $|x| > \epsilon t$. The right and left processes associated to the pure state have, in the classical limit, drifts

$$b_{r,l}^+(x,y) = [(Re + Im)\{\frac{2i(x + x_0 \mp |p_0|t/m)}{m(\alpha + i2\hbar t/m)} \pm |p_0|\} \ , \ \frac{\hbar}{m}(Re + Im)\nabla_y \log \phi_\pm(y)]$$

$$(5.16)$$

They both coincide with the drifts of the two processes associated to the mixture.

Finally, we wish to show that in spite of this conclusions, there is, at least in principle, a way to detect the difference between the processes associated to the pure state and those associated to the mixture. In fact, if we evaluate the conditional probability density to find the microsystem in y after having found the macrosystem in the region $|x| < \epsilon t$, we obtain

$$r(y|\epsilon) - \tilde{r}(y|\epsilon) = \frac{2Re \int_{-\epsilon t}^{+\epsilon t} \phi_-(y)\phi_+^*(y)\psi_r^*(x)\psi_l(x)dx}{\int_{-\epsilon t}^{+\epsilon t} |\psi_r(x)|^2 + |\psi_l(x)|^2)dx} \quad (5.17)$$

where, as usual, r and $\tilde{r}$ refer to the pure state and to the mixture respectively. Now both the numerator and the denominator are of order $\exp\{-(p_0)^2/(\Delta p_0)^2\}$ and therefore the difference of the conditional probabilities has the form

$$r(y|\epsilon) - \tilde{r}(y|\epsilon) = a(\epsilon)Re[\phi_-(y)\phi_+^*(y)] + b(\epsilon)Im[\phi_-(y)\phi_+^*(y)] \quad (5.18)$$

with $a(\epsilon)$ and $b(\epsilon)$ of order 1 whith respect to $\exp\{-(p_0)^2/(\Delta p_0)^2\}$. The difference, which may be experimentally detected, does not vanish in the classical limit. Nevertheless these "non classical" events are extremely rare and one has to wait a time which grows as $\exp\{(p_0)^2/(\Delta p_0)^2\}$ to see them.

6. Discussion

The basic inconsistency between the properties attributed to physical systems by the classical limit of Quantum Mechanics and those described by Classical Statistical Mechanics is eliminated if we adopt the point of view that a representation of their state more complete than the one provided by the quantum mechanical wave function is possible. In spite of the widespread belief that this possibility does not exist, it is however a fact that Nelson's Stochastic Mechanics accomplishes this task by embedding the computational formalism of Quantum Mechanics into a more detailed description of physical systems. We have seen that the additional stochastic variable, far from being a meaningless complication of the picture, is instead necessary if a smooth and internally consistent transition from the quantum to the classical domain is required. As an example, we have shown that, according to this picture, in a Stern-Gerlach apparatus a given particle is always located in one of the two beams irrespectively of whether its position is measured or not. This localization occurs, of course, only when there is a clearcut separation in space between the two beams. As soon as they begin to overlap the usual interference phenomena start to show up.

This solution eliminates one of the most striking paradoxes of the conventional interpretation of Q.M., according to which each particle materializes instantaneously into one of the two beams, no matter how far away they may be, only when the corresponding counter is triggered at random.

Last, but not least, this picture of reality supports our cherished belief that we can look at macroscopic object without modifying its state. Classical trajectories have an objective meaning: looking, far from killing cats, only increases the knowledge of the observer.

Appendix
Quantum Mechanics and Discrete Stochastc Processes

Nelson's S.M. represents the position of a quantum mechanical system in terms of a diffusion associated to its position but it is unable to describe discrete variables like spin or photon number [24-26]. In order to extend the theory to this more general case many people have proposed to introduce also jump processes.

We briefly recall here one of these extensions, which is particularly suitable for treating the evolution of systems which provide models of macroscopic measuring instruments such as counters.

The attempt is to construct a stochastic field theory based on the assumption that for any normal mode of a given classical field with amplitude $q_k(t)$ the corresponding action variable J_k becomes a stochastic variable assuming only positive integer values n_k in units of $\hbar$.

The dynamical evolution of the field is described by means of the infinite set of discrete stochastic processes $n_k(t)$ representing the occupation number of each mode k. The state of the field at time t will be therefore specified by the probability $r(n_1, n_2, ...n_k, ..., t)$ that the stochastic variables $n_1(t)$, $n_2(t)$, ..., $n_k(t)$,..have the values $n_1, n_2, ...n_k, ...$ at time t. To obtain the time evolution of r the jump probability rates are needed. These quantities are defined, as in the case for the drift in Nelson's S.M., in terms of the quantum mechanical wave function of the state considered.

In order to show how the method works it is sufficient to treat a very simple model, namely the case of a single field oscillator in interaction with a fixed source [see 25]. The Hamiltonian is

$$H = \frac{1}{2}(p^2 + \omega^2 q^2) + \lambda\sqrt{2\omega}q \tag{A.1}$$

which reduces to

$$H = \omega a^+ a + \lambda(a + a^+) \tag{A.2}$$

when the Dirac oscillator variables

$$a = \frac{1}{\sqrt{2\omega}}(\omega q + ip) \quad a^+ = \frac{1}{\sqrt{2\omega}}(\omega q - ip) \tag{A.3}$$

are introduced. The Schrödinger equation in the occupation number representation reads ($\hbar = 1$)

$$i\frac{d}{dt}<n|\psi> = n\omega<n|\psi> + \lambda\sqrt{n+1}<n+1|\psi> + \lambda\sqrt{n}<n-1|\psi> \tag{A.4}$$

From (A.4) one obtains the time derivative of the probability

$$r(n,t) = |<n|\psi>|^2 \tag{A.5}$$

in the form

$$\frac{dr(n,t))}{dt} = -2[Im\Delta_-(n,t) + Im\Delta_+(n,t)]r(n,t) \tag{A.6}$$

with

$$\Delta_\pm(n,t) = \lambda\sqrt{n + (1 \pm 1)/2}\frac{<\psi|n \pm 1>}{<\psi|n>} \tag{A.7}$$

Eq. (A.6) does not yet have the form of a Fokker-Planck equation for the jump process $n(t)$. It is however possible to define positive jump probability rates as follows

$$p_\pm(n,t) = |\Delta_\pm(n,t)| + Im\Delta_\pm(n,t) \tag{A.8}$$

In terms of which the equation (A.6) reads

$$\frac{dr(n,t)}{dt} = p_-(n+1,t)r(n+1,t) + p_+(n-1,t)r(n-1,t) - [p_-(n,t) + p_+(n,t)]r(n,t) \tag{A.9}$$

which is now a Fokker-Planck equation for the probability density $r(n,t)$. The transition probability density $p(n,t;n_0,t_o)$ to find the system in n at time t if it was in n_0 at time t_0 satisfies the same equation. These are all the ingredients we need to discuss from the point of view of discrete stochastic processes a simple model of measuring apparatus which was proposed in reference [3].

In this model the microsystem is represented by a dichotomic variable (spin) and the macrosystem by an ensemble of N particles (bosons) which may be in one of two states, the state 0 or the state 1. Introducing creation and destruction operators a_0^+, a_0 and a_1^+, a_1 respectively, a generic state of the counter with n particles in the state 0 and $N - n$ particles in the state 1 is given by

$$|n, N-n> = \frac{1}{\sqrt{n!(N-n)!}}(a_0^+)^n(a_1^+)^{N-n}|0,0> \tag{A.10}$$

The generic spin state of the microsystem is represented by

$$|w> = c_+|u_+> + c_-|u_-> \tag{A.11}$$

where $|u_\pm>$ are the two eigenstates of the spin corresponding to the eigenvalues ± 1.

The hamiltonian H is constructed in such a way that the counter interacts only with the spin state $|u_+>$, while the spin state $|u_->$ leaves the counter state unperturbed

$$H = \frac{\lambda}{2}(1 + \sigma_z)[a_0^+ a_1 + a_1^+ a_0] \tag{A.12}$$

If the initial state of the total system is the product

$$\frac{1}{2}(|u_+> + |u_->)|N, 0> \tag{A.13}$$

after a time t the total system will be in a state

$$|\chi> = \frac{1}{2}(\sum_{n=0}^{N} f_n(t)|u_+>|n, N - n> + |u_->|N, 0>) \tag{A.14}$$

with [see 3]

$$f_n(t) = \frac{\sqrt{N!}}{\sqrt{n!(N - n)!}}(-i)^{N-n}(\cos \alpha(t))^n(\sin \alpha(t))^{N-n} \tag{A.15}$$

where $\alpha(t) = \lambda t/\sqrt{N}$ and $\tau \equiv 1/\lambda$ is the typical one-particle discharge time.

The state of the counter corresponding to $|u_+>$ is characterized by an average number $\bar{n}(t)$ of particles in the state 0 (and correspondingly $N - \bar{n}(t)$ particles in the state 1) while the counter's state corresponding to $|u_->$ has not changed from its initial state with N particles in the state 0 and no particles in the state 1.

The average number $\bar{n}(t)$ is easily obtained from the probability distribution $P(n, t)$ obtained from (A.14) and (A.15)

$$P(n, t) = \frac{N!}{(N - n)!n!}(\cos^2 \alpha(t))^n(\sin^2 \alpha(t))^{N-n} \tag{A.16}$$

This distribution is strongly peaked around the value

$$\bar{n}(t) = N\cos^2\alpha(t) \tag{A.17}$$

with a relative width $\Delta n/N$ which tends to 0 as $1/\sqrt{N}$.

The two states, the first corresponding to N particles in the state 0 and the second to $\bar{n}(t)$, becomes macroscopically different as $\bar{n}(t)/N$ become different from 1. The whole discussion made in Sections 2 can be completely repeated in this case.

We now introduce the stochastic interpretation sketched above. The bosons' state corresponding to $|u_-\rangle$ remains undischarged and therefore no jumps occur in this case. The discharged boson's state, on the contrary, has a density $P(n,t)$ given by (A.16). The jump probabilities $p_{\pm}(n,t)$ are easily calculated by using the formulas (A.7) and (A.8). One gets

$$p_+(n,t) = (N-n)\{|\text{ctg}\alpha(t)| - \text{ctg}\alpha(t)\} \tag{A.18}$$

$$p_-(n,t) = (N-n)\{|\text{tg}\alpha(t)| + \text{tg}\alpha(t)\} \tag{A.19}$$

For $t < \frac{\sqrt{N}}{\lambda}\frac{\pi}{2}$, one has

$$p_+(n,t) = 0 \quad p_-(n,t) = 2nt\text{g}\alpha(t) \tag{A.20}$$

The solution of the Fokker-Planck equation, given that the counter has initially N bosons in the state ϕ_0, coincides, of course, with the probability $P(n,t)$ of equation (A.16). In other words, one has

$$P(n,t) \equiv P(n,t;N,0) \tag{A.21}$$

We can now speak of trajectories in the discrete n space. The probability that a trajectory of the discharged counter starts at N at time zero and ends in N at time t is

$$P(N,t;N,0) = (\cos^2\alpha(t))^N \simeq \exp -\frac{t^2}{\tau^2} \tag{A.22}$$

the second approximate equality holds for $t \ll \tau\sqrt{N}$. Therefore, looking at the counter at a time t which satisfies $\tau\sqrt{N} \gg t \gg \tau$ the probability to find all the N bosons in the state 0 becomes negligible. On the contrary, the counter remains in N all the time in correspondence of $|u_-\rangle$. Trajectories which tend to go in the "wrong" beam are therefore negligible. This shows that also in this case, if we ascribe a meaning to the process $n(t)$, all interpretational problems disappear. The counter *is* either discharged or not even *before* looking at it.

Acknowledgements

One of us (M.C.) is grateful to Prof. R.Omnès for reading this manuscript and for his useful comments. He is also grateful to him and to C.N.R.S. for supporting his stay at L.P.T.H.E. Orsay. We also thank S. Nicolis for a critical reading of the manuscript.

References

[1] M. Cini and M. Serva, *Found. Phys. Lett.*, **3** (1990) 129

[2] M. Cini and M. Serva, Contribution to *Quantum Theory without reduction*, A. Hilger, IOP Pubblishing Ldt. London, 1990

[3] M. Cini, *Il Nuovo Cimento*, **73B** (1983) 27

[4] M. Cini and J. M. Levy-Leblond, *Quantum Theory without reduction*, A. Hilger, IOP Pubblishing Ldt. London, 1990

[5] N. G. van Kampen, *Physica A*, **153** (1988) 47

[6] J. S. Bell, *CERN* TH 5611/89 Contribution to *62 Years of Uncertanty*, Erice 5-14, August 1989

[7] R. Omnès, *Annals of Physics*, **2d** (1990) 354

[8] K.Hepp, *Hel. Phys. Acta*, **45** (1972) 237

[9] E. Nelson, *Dynamical Theories of Brownian Motion*, Princeton University Press, Princeton N.J. (1967)

[10] E.Nelson, *Quantum Fluctuations*, Princeton University Press, Princeton N.J. (1985)

[11] Ph. Blanchard, Ph. Combe and W. A. Zheng, *Mathematical and Phisical Aspects of Stochastic Mechanics*, Lecture Notes in Physics **281**, Springer Verlag (1987)

[12] F. Guerra, *Phys. Rep.*, **77** (1981) 263

[13] Ph. Blanchard and S. Golin, *Comm. Math. Phys.*, **109** (1987) 421

[14] S. Albeverio and R. Høegh-Krohn, *J. Math. Phys.*, **15** (1974)

[15] K. Yasue, *J. Math. Phys.*, **22** (1981) 1010

[16] K. Yasue, *J. Funct. Anal.*, **41** (1981) 327

[17] F. Guerra and L. M. Morato, *Phys. Rev. D*, **27** (1983) 1774

[18] L. M. Morato, *Phys Rev. D*, **31** (1985) 1982

[19] S. Goldstein, *J. Stat. Phys.*, **47** (1987) 645

[20] M. Jibu, T. Misawa and K. Yasue, *Phys. Lett. A*, **150** (1990) 59

[21] E. Carlen, *Ann. Inst. Henri Poincaré*, **A42** (1985) 407

[22] M. Serva, *Lett. il Nuovo Cimento*, **41** (1984) 198

[23] Ph. Blanchard, S. Golin and M. Serva, *Phys. Rev. D*, **34** (1986) 3732

[24] M. Cini and M. Serva, *J. Phys. A*, **19** (1986)1163

[25] G. F. de Angelis and G.Jona-Lasinio, *J. Phys. A*, **15** (1982) 205

[26] F. Guerra and R. Marra, *Phys. Rev. D*, **29** (1984) 647

Schrödinger operators at threshold

D. Bollé

Instituut voor Theoretische Fysica

K.U.Leuven

B-3001 Leuven,Belgium

Abstract

Recent results are reviewed on the zero-energy threshold scattering for Schrödinger operators in $d \leq 3$ dimensions. The possibility of zero-energy resonances and/or zero-energy bound states of the Hamiltonian is taken into account explicitly. No spherical symmetry of the interaction is assumed. In particular, Laurent expansions around threshold are provided for the transition operator, the scattering amplitude, the scattering operator and the trace of the resolvent difference. Low energy parameters like e.g. the scattering length are defined and related directly to the threshold behavior of the scattering amplitude. Spectral sum rules, in particular Levinson's theorem, are discussed.

1   Introduction

Knowledge of the low-energy behavior of non-relativistic quantum mechanics is important and useful in different branches of physics, as has been realized since a long time. To mention only a few examples we list e.g. partial-wave analysis of nuclear scattering, the Efimov effect in many-particle systems, propagation of cold neutrons in a crystal, spin-polarized atomic hydrogen adsorbed on a surface, the zero-energy modes in supersymmetric quantum mechanical models, etc. .... In a lot of these phenomena the occurence of zero-energy resonances or zero-energy bound states plays a significant role.

Much work on the detailed study of low-energy scattering has been done in classical references in the radial symmetric case (see e.g. [11], [69]). More recently, extensions of some of this work have appeared, centered around a generalization of Levinson's theorem to the non-central case (see [17] and references therein). Furthermore, low-energy asymptotic expansions for the resolvent, the spectral density and the on-shell scattering operator, as well as large-time results for the evolution operator have been obtained using methods of weighted Sobolev spaces [49], [50], [51], [67], [68]. Using Hilbert space methods, similar but more detailed results have been derived for $d \leq 3$ dimensions under different assumptions ([5]-[7], [9], [13], [23]-[30], [41], [42] and references therein). In particular, analytic respectively Laurent expansions have been obtained and the explicit calculation of lower order coefficients has been performed. Moreover off-shell scattering, low-energy parameters (scattering length, effective range) and trace relations, in particular Levinson's theorem, have been discussed. In this context, we finally also mention the work on the coupling constant threshold behavior i.e. the study of limit situations where eigenvalues approach zero as the coupling constant approaches a critical value [10], [41], [43], [48], [53], [56], [57], [76], [81].

In this paper we review the systematic Hilbert space analysis of low-energy scattering in $d \leq 3$ dimensions, taking into account explicitly the possibility of zero-energy resonances and/or zero-energy bound states of the Schrödinger Hamiltonian. This analysis is more involved in $d = 1, 2$ than in $d = 3$, due to the well-known additional difficulty that the free resolvent has a singularity as the energy tends to zero. In $d = 1$ this singularity is of the square root type, while in $d = 2$ the technical situation is even more complex because of the logarithmic nature of this singularity.

We will mainly discuss the methods and techniques and emphasize the common structure of the results as well as particular differences depending on the dimension. For the extensive and precise details we refer to [5], [7], [9], [23]-[30], [41], [42].

In Section 2 we discuss the occurence and properties of the zero-energy resonances and zero-energy bound states of the Schrödinger Hamiltonian $H = -\Delta \dot{+} V$ in $L^2(\mathbb{R}^d)$. No spherical symmetry of V is assumed. Section 3 describes the low-energy behavior of the transition operator $T(k)$, assuming roughly exponential fall off for V at infinity. The first few coefficients in the Taylor, respectively Laurent expansion around zero-energy threshold are obtained explicitly. Based on these results we present in Section 4 Taylor expansions for the scattering amplitude and S-operator (reflection and transmission coefficients for $d = 1$). Section 5 gives an appropriate generalization of the definition of scattering length $(d = 2, 3)$ to nonspherically symmetric interactions. This parameter is directly related to the threshold behavior of the scattering amplitude and explicit formulas are provided.

Finally in Section 6 we derive Laurent expansions for the trace of the difference between the full and free resolvents. We then discuss trace relations, in particular Levinson's theorem, and compare the contributions from zero-energy bound states and/or zero-energy resonances in different dimensions.

2   Zero-energy properties of H

Let

$$H_0 = -\Delta \quad , \quad \mathcal{D}(H_0) = H^{2,2}(\mathbb{R}^d) \quad , \quad 1 \le d \le 3 \tag{2.1}$$

denote the unperturbed Hamiltonian in $L^2(\mathbb{R}^d)$. Introduce its resolvent $R_0(k)$ by

$$R_0(k) = (H_0 - k^2)^{-1} \quad , \quad Im\, k > 0 . \tag{2.2}$$

Depending on d the kernel $R_0(k, \underline{x}, \underline{y})$ is given by

$$R_0(k, \underline{x}, \underline{y}) = \begin{cases} (i/2k)\exp(ik|x-y|) & , \ d = 1 \\ (i/4)\, H_0^{(1)}(k|\underline{x} - \underline{y}|) & , \ \underline{x} \ne \underline{y} \ , \ d = 2 \\ (4\pi|\underline{x} - \underline{y}|)^{-1}\exp(ik|\underline{x} - \underline{y}|) & , \ \underline{x} \ne \underline{y} \ , \ d = 3 , \end{cases} \tag{2.3}$$

$H_0^{(1)}(z)$ being the Hankel function of first kind and order zero [1].

Next, assume the potential V to be a real-valued measurable function, satisfying

$$\begin{aligned} &e^{2a|x|}V \in L^1(\mathbb{R}) \quad \text{for some} \quad a > 0 \ , \ d = 1 \\ &V \in L^{1+\delta}(\mathbb{R}^2) \ , \ e^{2a|\underline{x}|}V \in L^1(\mathbb{R}^2) \quad \text{for some} \quad \delta, a > 0 \ , \ d = 2 \\ &e^{2a|\underline{x}|}V \in R \quad \text{for some} \quad a > 0 \ , \ d = 3 , \end{aligned} \tag{2.4}$$

where R denotes the class of Rollnik potentials [79]. Then the Schrödinger operator H in $L^2(\mathbb{R}^d)$, $1 \le d \le 3$ is defined by the method of forms [52], [79]

$$H = -\Delta \dotplus \lambda_0 V \ , \quad \lambda_0 \in \mathbb{R}\backslash\{0\} . \tag{2.5}$$

Introducing

$$v(\underline{x}) = |V(\underline{x})|^{1/2} \ , \ u(\underline{x}) = |V(\underline{x})|^{1/2}\,\text{sign}\, V(\underline{x}) \ , \ u \cdot v = V \tag{2.6}$$

the transition operator $T(k)$ in $L^2(\mathbb{R}^d)$ is defined as

$$\begin{aligned} T(k) &= (1 + \lambda_0\, u\, R_0(k)v)^{-1} \quad , \quad Im\, k \ge 0 \ , \ k^2 \notin \sigma_p(H) \\ &\qquad\qquad\qquad\qquad\qquad k \ne 0 \quad \text{for} \quad d = 1,2 \end{aligned} \tag{2.7}$$

with $\sigma_p(\cdot)$ denoting the point spectrum.

We want to study the low-energy behavior of $T(k)$ in (2.7). From (2.3) however we see that $R_0(k)$ has a singularity as $k \to 0$ for $d = 1, 2$. In order to isolate this singularity we follow [80] and decompose

$$u\, R_0(k)v = \begin{cases} (i/2k)(v, \cdot)u + M(k) \ , \ d = 1 \\ (2\pi)^{-1}[-\ln(k) + (i\pi/2) + \ln(2) + \Psi(1)](v, \cdot)u + M(k) \ , \ d = 2 , \end{cases}$$

$$Im\, k \geq 0 \ , \quad k \neq 0 \ , \tag{2.8}$$

where $M(k) \in \mathcal{B}_2(L^2(\mathrm{I\!R}^d))$, $d = 1,2$ (the set of Hilbert-Schmidt operators in $L^2(\mathrm{I\!R}^d)$) for all $Im\, k \geq 0$ and $\Psi(z)$ represents the digamma function [1]. In particular the integral kernel of $M(0)$ reads

$$M_0(\underline{x},\underline{y}) = \begin{cases} -2^{-1}\, u(x)\, |x - y|v(y) \ , \ d = 1 \\ -(2\pi)^{-1}\, u(\underline{x})\, \ln|\underline{x} - \underline{y}|v(\underline{y}) \ , \ \underline{x} \neq \underline{y} \ , \ d = 2 \,. \end{cases} \tag{2.9}$$

Defining

$$P = (v,u)^{-1}(v,\cdot)u \ , \quad Q = 1 - P \tag{2.10}$$

we thus obtain for $T(k)$

$$T(k) = \begin{cases} [1 + (i\,\lambda_0(v,u)/2k)P + \lambda_0\, M(k)]^{-1} \ , \ d = 1 \\ \{1 + (2\pi)^{-1}\lambda_0[-\ln(k) + (i\pi/2) + \ln(2) + \Psi(1)](v,u)P + \lambda_0\, M(k)\}^{-1} \ , \ d = 2 \\ [1 + \lambda_0\, u\, R_0(k)v]^{-1} \ , \ d = 3 \end{cases}$$

$$Im\, k \geq 0 \ , \quad k^2 \notin \sigma_p(H) \ , \quad k \neq 0 \ \text{for} \ d = 1,2 \,. \tag{2.11}$$

Obviously the low-energy behavior of $T(k)$ strongly depends on the zero-energy behavior of H. Therefore we study the possible occurence of zero-energy resonances and zero-energy bound states of H by looking at the eigenvalue equation

$$\lambda_0\, u\, R_0(0)v\, \phi = -\phi \ , \ d = 3 \tag{2.12}$$
$$\lambda_0\, Q\, M_0\, Q\, \phi = -\phi \ , \ d = 1,2 \ , \ \ \phi \in L^2(\mathrm{I\!R}^d) \,. \tag{2.13}$$

For $d = 3$ the results are standard and well-known. We recall [7], [51], [56], [70]

Lemma 2.1 *Let V satisfy the condition (2.4), $d = 3$. Introduce*

$$\psi(\underline{x}) = (R_0(0)v\phi)(\underline{x}) \tag{2.14}$$

then $\psi \in L^2_{loc}(\mathrm{I\!R}^3)$ and $H\psi = 0$ in the sense of distributions. Furthermore $\psi \in L^2(\mathrm{I\!R}^3)$ is equivalent to

$$(v,\phi) = -\int d^3x\, V(\underline{x})\,\psi(\underline{x}) = 0 \,. \tag{2.15}$$

If $\psi \notin L^2(\mathrm{I\!R}^3)$ we say that H has a zero-energy resonance. If $\psi \in L^2(\mathrm{I\!R}^3)$ then H clearly has a zero-energy bound state. We now distinguish the following cases :

CASE 3.I : There exist no resonance function ψ (i.e. (-1) is not an eigenvalue of $u\, R_0(0)v$)

CASE 3.II : There exist precisely one resonance function ψ (i.e. (-1) is a simple eigenvalue of $u\, R_0(0)v$) but $\psi \notin L^2(\mathrm{I\!R}^3)$

CASE 3.III : There exist N functions ψ_j, $j = 1,2\ldots N$, which are all in $L^2(\mathrm{I\!R}^3)$

CASE 3.IV : There exist $N \geq 2$ functions ψ_j, $j = 1, 2 \ldots N$, and at least one of them is not in $L^2(\mathbb{R}^3)$

Remark 2.1. a) The condition (2.4), $d = 3$ is obviously too strong but it serves our purposes since we are interested in a strong fall off for V at infinity in the following sections. For less restrictive conditions on V see [56][70]. For a discussion of some long-range systems we refer to [9][27]. Non-local interactions have been treated in [26],[71].
b) In case 3.IV we can always choose a particular set of functions ϕ_j such that $(v, \phi_1) \neq 0$ but $(v, \phi_j) = 0$, $j = 2, \ldots N$. From now on we always adopt this convention. Thus in case 3.IV zero is an eigenvalue of H with multiplicity $(N - 1)$ whereas in case 3.III its multiplicity is N.
c) If V is spherically symmetric, then $(v, \phi) = 0$ for all functions ϕ belonging to angular momentum $l \geq 1$. Consequently, for spherically symmetric interactions case 3.II (i.e., a zero-energy resonance) occurs in s-waves only.

For $d = 1, 2$ a first step in the analysis of possible zero-energy resonances and/or zero-energy bound states is given by [23], [29], [56]

Lemma 2.2 *Let V satisfy the condition (2.4) $d = 1, 2$ and let $(v, u) \neq 0$. Assume that -1 is an eigenvalue of $\lambda_0 Q M_0 Q$ and let*

$$\mathcal{V} = \left\{\phi \in L^2(\mathbb{R}^d) | \lambda_0 Q M_0 Q \, \phi = -\phi\right\}$$
$$\mathcal{W} = \left\{\chi \in \mathcal{V} | \left(v, (z - \lambda_0 M_0)^{-1}\chi\right) = 0 \ \text{for some} \ |z| > \|\lambda_0 M_0\|\right\}$$

Then

i) $\mathcal{W}$ *is independent of z*

ii) *$\dim \mathcal{W} = \dim \mathcal{V}$ or $\dim \mathcal{W} = \dim \mathcal{V} - 1$*

iii) *$(z - \mu P - \lambda_0 M_0)^{-1}\chi = (z - \lambda_0 M_0)^{-1}\chi$ for all $\chi \in \mathcal{W}$, $\mu \in \mathbb{C}$*

iv) *$\lambda_0 M_0 \chi = -\chi$ for all $\chi \in \mathcal{W}$*

v) *If $\phi_0 \in \mathcal{V}\backslash\mathcal{W}$ then*
$$\lambda_0 M_0 \phi_0 = -\phi_0 + (v, u)^{-1} \lambda_0(v, M_0 \phi_0)u$$

vi) *If $\chi \in \mathcal{V}$ then $(v, (z - \lambda_0 M_0)^{-1}\chi) = 0$ is equivalent to $(v, M_0 \chi) = 0$.*
Consequently
$$\mathcal{W} = \{\chi \in \mathcal{V} | (v, M_0 \chi) = 0\} .$$

Next we state [23], [29].

D. Bollé

Lemma 2.3 *Let V satisfy the condition (2.4) $d = 1, 2$ and let $(v, u) \neq 0$. Assume that $\lambda_0 Q M_0 Q \phi = -\phi$ for some $\phi \in L^2(\mathbb{R}^d)$ and define the function ψ by*

$$\psi(x) \;=\; -(v, u)^{-1} \lambda_0 (v, M_0 \phi) - 2^{-1} \lambda_0 \int_{\mathbb{R}} dy \, |x - y| v(y) \, \phi(y) \;,\; d = 1 \tag{2.16}$$

$$\psi(\underline{x}) \;=\; -(v, u)^{-1} \lambda_0 (v, M_0 \phi) - (2\pi)^{-1} \lambda_0 \int_{\mathbb{R}^2} d^2y \, \ln |\underline{x} - \underline{y}| v(\underline{y}) \, \phi(\underline{y}) \;,\; d = 2 \tag{2.17}$$

Then

i) $d = 1 : \psi \in L^\infty(\mathbb{R})$, $d = 2 : \psi \in L^2_{loc}(\mathbb{R}^2)$ *and in both cases $H \psi = 0$ in the sense of distributions*

ii) $u(x) \psi(x) = -\phi(x)$ *a.e.*

iii) $d = 1 : \psi + (v, u)^{-1} \lambda_0 (v, M_0 \phi) - 2^{-1} \lambda_0 \, sign(\cdot)((\cdot)v, \phi) \in L^2(\mathbb{R})$, $\psi \notin L^2(\mathbb{R})$

$d = 2 : \psi + (v, u)^{-1} \lambda_0 (v, M_0 \phi) - (2\pi)^{-1} \lambda_0 \, |\underline{x}|^{-2} \underline{x}((\underline{\cdot})v, \phi) \in L^2(\mathbb{R}^2)$, *in particular*
$\psi \in L^2(\mathbb{R}^2) \iff (v, M_0 \phi) = ((\underline{\cdot})v, \phi) = 0$

iv) *for $d = 1$ ψ is unique and thus the nonzero eigenvalues of $\lambda_0 Q M_0 Q$ are simple.*

With the help of Lemmas 2.2-2.3 we are able to distinguish the following cases in the zero-energy behavior of H for $d = 1, 2$.

$d = 1 :$ introducing
$$c_1 = (v, u)^{-1}(v, M_0 \phi) \;,\; c_2 = 2^{-1}((\cdot)v, \phi) \tag{2.18}$$

we have

CASE 1.I : There exists no resonance function ψ (i.e. (-1) is not an eigenvalue of $\lambda_0 Q M_0 Q$)

CASE 1.II : There exists precisely one resonance function ψ (i.e. (-1) is a simple eigenvalue of $\lambda_0 Q M_0 Q$), $\psi \notin L^2(\mathbb{R})$ and
a) $c_1 = 0$, $c_2 \neq 0$
b) $c_1 \neq 0$, $c_2 = 0$
c) $c_1 \neq 0$, $c_2 \neq 0$

$d = 2 :$ introducing
$$c_1^{(j)} = (v, u)^{-1}(v, M_0 \phi_j) \;,\; c_2^{(j)} = (2\pi)^{-1}((\underline{\cdot})v, \phi_j) \tag{2.19}$$

we have

CASE 2.I : (-1) is not an eigenvalue of $\lambda_0 Q M_0 Q$

CASE 2.II : (-1) is an eigenvalue of $\lambda_0 Q M_0 Q$ of multiplicity $1 \leq M \leq 3$, and

   a) $M = 1$, $j = 0$, $c_1^{(0)} \neq 0$ and $\psi_j \notin L^2(\mathbb{R}^2)$

     or

   b) $1 \leq M \leq 2$, $j = 1$ if $M = 1$ and $j = 1, 2$ if $M = 2$, $c_1^{(j)} = 0$, $\underline{c}_2^{(j)} \neq 0$, if $M = 2$
     then $\underline{c}_2^{(1)}$, $\underline{c}_2^{(2)}$ are linearly independent, $\psi_j \notin L^2(\mathbb{R}^2)$

     or

   c) $2 \leq M \leq 3$, $j = 0, 1$ if $M = 2$ and $j = 0, 1, 2$ if $M = 3$, $c_1^{(0)} \neq 0$, $c_1^{(j)} = 0$,
     $\underline{c}_2^{(j)} \neq 0$, $j \geq 1$, if $M = 3$ then $\underline{c}_2^{(1)}$, $\underline{c}_2^{(2)}$ are linearly independent, $\psi_j \notin L^2(\mathbb{R}^2)$

CASE 2.III : (-1) is an eigenvalue of $\lambda_0 \, Q \, M_0 \, Q$ of multiplicity $N \in \mathbb{N}$ and $c_1^{(j)} = \underline{c}_2^{(j)} = 0$,
    $3 \leq j \leq 2 + N$, $\psi_j \in L^2(\mathbb{R}^2)$

CASE 2.IV : possible admixtures of cases 2.II-2.III.

   Remark 2.2. a) For $d = 1$, H has no zero-energy bound states (or equivalently c_1 and c_2 do not vanish simultaneously). In cases 1.IIa-c we have $(v, \phi) = 0$. For $d = 2$, cases 2.IIa-c describe the various possibilities of zero-energy resonances of H. If V is spherically symmetric then case 2.IIa and case 2.IIb correspond to zero-energy resonances in the s-wave and p-wave respectively [42] [73]. Case 2.III denotes the zero-energy bound state case. In particular, if V is spherically symmetric this case only happens to occur in d- and higher partial waves. In all cases 2.II-2.IV we have $(v, \phi_j) = 0$.
b) Again, the condition (2.4) $d = 1, 2$ can be relaxed [23] [29]. Also the condition $(v, u) \neq 0$ in Lemmas 2.2-2.3 can be removed as shown in [25] for $d = 1$ (The case $d = 2$ is under study.). The precise details of the next sections, especially concerning the low-energy expansion for $T(k)$, are different for $(u, v) = 0$ but the main results (S-matrix expansion, Levinson's theorem ...) take the same form [25]. In the rest of this paper we will take $(v, u) \neq 0$.

3   Low-energy expansions of $T(k)$

To discuss the low-energy expansion of $T(k)$ for the different cases presented in Section 2, we start with ([7], Lemma 3.1, [49] Theorem 4.3)

Lemma 3.1 *Assume (2.4) and let $\epsilon \in \mathbb{C} \backslash \{0\}$ be small enough. Then the norm convergent expansions*

$$(1 + \lambda_0 u R_0(0)v + \epsilon)^{-1} = \frac{P_0}{\epsilon} + \sum_{m=0}^{\infty} (-\epsilon)^m T_0^{m+1}, d = 3 \tag{3.1}$$

$$(1 + \lambda_0 Q M_0 Q + \epsilon)^{-1} = \frac{P_0}{\epsilon} + \sum_{m=0}^{\infty} (-\epsilon)^m T_0^{m+1}, d = 1, 2 \tag{3.2}$$

hold. Here P_0 denotes the projection onto the eigenspace of $\lambda_0 u R_0(0)v$ for $d = 3$ and of $\lambda_0 Q M_0 Q$ for $d = 1, 2$ to the eigenvalue -1, i.e.

$$P_0 = \sum_j (\tilde{\phi}_j, \phi_j)^{-1}(\tilde{\phi}_j, .)\phi_j, \tag{3.3}$$

(the range of j depends on the cases and is specified in Section 2) where for $d = 3$

$$\lambda_0 u R_0(0)v\phi_j = -\phi_j, \tilde{\phi}_j \in L^2(\mathrm{I\!R}^3), \tilde{\phi}_j = (signV)\phi_j, (\tilde{\phi}_j, \phi_l) = (\tilde{\phi}_j, \phi_j)\delta_{jl} \tag{3.4}$$

and for $d = 1, 2$

$$\lambda_0 Q M_0 Q\phi_j = -\phi_j, \tilde{\phi}_j \in L^2(\mathrm{I\!R}^n), \tilde{\phi}_j = (signV)\phi_j, (\tilde{\phi}_j, \phi_l) = (\tilde{\phi}_j, \phi_j)\delta_{jl}. \tag{3.5}$$

The operator T_0 denotes the corresponding reduced resolvent, viz.

$$T_0 = \begin{cases} n - \lim_{\epsilon \to 0}(1 + \lambda_0\, u\, R_0(0)v + \epsilon)^{-1}(1 - P_0) \;, & d = 3 \\ n - \lim_{\epsilon \to 0}(1 + \lambda_0\, Q\, M_0\, Q + \epsilon)^{-1}(1 - P_0) \;, & d = 1, 2\,. \end{cases} \tag{3.6}$$

In Section 2 we have already indicated that $u\, R_0(k)v$ for $d = 3$ respectively $M(k)$ for $d = 1, 2$ have expansions valid in $\mathcal{B}_2(L^2(\mathrm{I\!R}^d))$ norm. We now state more precisely (recall eqs. (2.3) and (2.8))

Lemma 3.2 *Assume (2.4). Then*

a) $d = 1$: *$M(k)$ is analytic in $\mathcal{B}_2(L^2(\mathrm{I\!R}))$ with respect to k in $Im\, k > -a$ and*

$$M(k) = \sum_{n=0}^{\infty}(ik)^n M_n\,, \tag{3.7}$$

where the M_n are Hilbert-Schmidt operators with kernels

$$M_n(x, y) = -2^{-1}u(x)\frac{|x - y|^{n+1}}{(n + 1)!}v(y)\;, \quad n = 0, 1, 2\ldots \tag{3.8}$$

b) $d = 2$:

$$M(k) = M_1(k^2) + M_2(k^2)\ln k \tag{3.9}$$

where the $M_j(k^2)$, $j = 1, 2$, are analytic in $\mathcal{B}_2(L^2(\mathrm{I\!R}^2))$ norm with respect to k^2 in $Im\, k > -a$. Moreover

$$M(k) = \sum_{m,n=0}^{\infty}(1/\ln k)^m(k^2\ln k)^n M_{mn} \tag{3.10}$$

where the coefficients M_{mn} have integral kernels

$$\begin{aligned} M_{m-1,m}(\underline{x}, \underline{y}) &= (2\pi)^{-1}4^{-m}(m!)^{-2}(-1)^{m-1}u(\underline{x})|\underline{x} - \underline{y}|^{2m}v(\underline{y})\;, \quad m \in \mathrm{I\!N} \\ M_{m,m}(\underline{x}, \underline{y}) &= (4\pi)^{-1}4^{-m}(m!)^{-2}(-1)^m\left\{-2u(\underline{x})|\underline{x} - \underline{y}|^{2m}\ln|\underline{x} - \underline{y}|v(\underline{y})\right. \\ &\quad + \left. [i\pi + \ln 4 + 2\Psi(m + 1)]u(\underline{x})|\underline{x} - \underline{y}|^{2m}v(\underline{y})\right\}\;, \underline{x} \neq \underline{y}, m \in \mathrm{I\!N} \\ M_{m,n}(\underline{x}, \underline{y}) &= 0, \quad m \leq n - 2 \text{ or } m \geq n + 1\;, \quad m, n \in \mathrm{I\!N} \end{aligned} \tag{3.11}$$

c) $d = 3$: $u\,R_0(k)v$ *is analytic in* $\mathcal{B}_2(L^2(\mathbb{R}^3))$ *norm with respect to* k *in* $\mathrm{Im}\,k > -a$ *and*

$$u\,R_0(k)v = \sum_{n=0}^{\infty} (ik)^n r_n \tag{3.12}$$

where the r_n *are Hilbert-Schmidt operators with kernel*

$$r_n(\underline{x}, \underline{y}) = (4\pi\,n!)^{-1} u(\underline{x})|\underline{x} - \underline{y}|^{n-1} v(\underline{y}) \ , \ \ n = 0\,(\underline{x} \neq \underline{y}), 1, 2 \ldots \tag{3.13}$$

The expansions (3.8), (3.10) and (3.13) converge in $\mathcal{B}_2(L^2(\mathbb{R}^d))$ *norm.*

Next, using these results we determine the low-energy behavior of $T(k)$ starting from (2.11).

Theorem 3.1 *Assume (2.4). Then* $T(k)$ *has the following norm covergent Laurent (respectively Taylor) expansion around* $k = 0$

$d = 1$:

$$T(k) = \sum_{n=q}^{\infty} (ik)^n t_n \tag{3.14}$$

where $q = 0$ *in case 1.I and* $q = -1$ *in case 1.IIa-c.*

$d = 2$:

$$T(k) = \sum_{m=p'}^{\infty} \sum_{n=q'}^{\infty} (1/\ln k)^m (k^2 \ln k)^n t_{mn} \tag{3.15}$$

where

$$p' = \begin{cases} 0 & \text{in cases 2.I, 2.IIb} \\ -1 & \text{in cases 2.III, and in case 2.IIc} \quad \text{if } c_2^{(0)} = 0 \\ -\infty & \text{in case 2.IIa, and in case 2.IIc} \quad \text{if } c_2^{(0)} \neq 0 \end{cases}$$

$$q' = \begin{cases} 0 & \text{in cases 2.I, 2.IIa} \\ -1 & \text{in cases 2.IIb,c, 2.III} \end{cases}$$

and

$$\begin{aligned} t_{m,0} &= 0, & m \leq -2 \text{ in case 2.IIa} \\ t_{m,-1} &= 0, & m \geq 1 \text{ in case 2.IIb} \\ t_{-1,-1} &= 0, & t_{m,-1} = 0, \quad m \geq 1 \text{ in case 2.IIc if } c_2^{(0)} = 0 \\ t_{m,-1} &= 0, & m \geq 1 \text{ or } m \leq -1 \text{ in case 2.IIc if } c_2^{(0)} \neq 0 \\ t_{m,0} &= 0, & m \leq -2 \text{ in case 2.IIc if } c_2^{(0)} \neq 0 \end{aligned}$$

$d = 3$:

$$T(k) = \sum_{n=q''}^{\infty} (ik)^n t_n \tag{3.16}$$

where $q'' = 0$ *in case 3.I,* $q'' = -1$ *in case 3.II and* $q'' = -2$ *in cases 3.III-IV.*

We list the first coefficients in the expansions (3.14) - (3.16) explicitly

$$
\begin{aligned}
\text{1.I} \quad &: \quad T(k) = T_0 Q + O(k) \\[4pt]
\text{1.II} \quad &: \quad T(k) = [2ik\lambda_0(|c_1|^2 + |c_2|^2)]^{-1}(\tilde{\phi},.)\phi + O(1) \\[4pt]
\text{2.I} \quad &: \quad T(k) = T_0 Q + O(1/\ln k) \\[4pt]
\text{2.IIa} \quad &: \quad T(k) = \ln k (2\pi\lambda_0 |c_1^{(0)}|^2)^{-1}(\tilde{\phi}_0,.)\phi_0 + O(1) \\[4pt]
\text{2.IIb} \quad &: \quad T(k) = -(k^2 \ln k)^{-1}(\pi\lambda_0)^{-1} \sum_{j,l=1}^{N} (\underline{c}_2^{(j)}.\underline{c}_2^{(l)})^{-1}(\tilde{\phi}_l,.)\phi_j + O(1) \\[4pt]
\text{2.III} \quad &: \quad T(k) = k^{-2}\lambda_0^{-1} \sum_{j,l=3}^{N+2} (\tilde{\phi}, M_{11}\phi)_{jl}^{-1}(\tilde{\phi}_l,.)\phi_j \\[4pt]
& \qquad\qquad - \ln k \sum_{j,l=3}^{N+2} (\tilde{\phi}, M_{11}\phi)_{jl}^{-1}(Q^\star T_0^\star M_{01}^\star \tilde{\phi}_l,.)\phi_j + O(1) \\[4pt]
\text{3.I} \quad &: \quad T(k) = T_0 Q + O(k) \\[4pt]
\text{3.II} \quad &: \quad T(k) = 4\pi[ik|(v,\phi|^2]^{-1}(\tilde{\phi},.)\phi + O(1) \\[4pt]
\text{3.III} \quad &: \quad T(k) = -k^{-2} \sum_{j,l=1}^{N} (\tilde{\phi}, r_2\phi)_{jl}^{-1}(\tilde{\phi}_l,.)\phi_j + O(k^{-1})
\end{aligned}
\tag{3.17}
$$

In this list the same symbol T_0 (recall eq. (3.6)) and ϕ (eqs. (3.3)-(3.5)) are used for the reduced resolvent respectively the zero-energy eigenfunctions in different dimensions.

Remark 3.1. a) Comparing the $d = 1, 2, 3$ results above we see that $T(k)$ is analytic in all generic cases i.e. 1.I, 2.I., 3.I. In the cases where only zero-energy resonances appear $T(k) = O(k^{-1})$ except for $d = 2$. The latter allows "weak" resonances with $T(k) = O(\ln k)$ and "strong" resonances with $T(k) = O((k^2 \ln k)^{-1})$. All cases where zero-energy bound states appear behave like $T(k) = O(k^{-2})$.
b) For all cases in $d = 1$ and $d = 3$ and case 2.I recursion relations have been derived to obtain the coefficients t_n respectively t_{mn} [23],[25],[29],[30].

4   Expansions of on-shell scattering quantities

Given the explicit threshold behavior of $T(k)$ near $k = 0$, we now discuss the low-energy expansions of the on-shell scattering amplitude and of the on-shell scattering operator. For simplicity we assume condition (2.4) and $(v, u) \neq 0$ $(d = 1, 2)$ to be valid throughout this section.

We first recall the corresponding definitions (cf. e.g. [11], [32], [34], [35], [39], [63], [72], [77]). The on-shell scattering amplitude is given by

$d = 1$:

$$
f_{\epsilon_1\epsilon_2}(k) = (2ik)^{-1}\lambda_0(v\,e^{i\epsilon_1 k\cdot}, T(k)u\,e^{i\epsilon_2 k\cdot}), \quad \epsilon_j = \pm 1, \; j = 1, 2
\tag{4.1}
$$

$d = 2$:
$$f(k,\underline{\omega},\underline{\omega}') = -\lambda_0 e^{i\pi/4}(8\pi k)^{-1/2}(v\, e^{ik\underline{\omega}\cdot}, T(k)u\, e^{ik\underline{\omega}'\cdot}), \quad \underline{\omega},\underline{\omega}' \in S^1 \tag{4.2}$$

$d = 3$:
$$f(k,\underline{\omega},\underline{\omega}') = -(4\pi)^{-1}\lambda_0(v\, e^{ik\underline{\omega}\cdot}, T(k)u\, e^{ik\underline{\omega}'\cdot}), \quad \underline{\omega},\underline{\omega}' \in S^2 \tag{4.3}$$

where $k > 0$, $k^2 \notin \mathcal{E}_+$ with

$$\mathcal{E}_+ = \{k^2 > 0 | \lambda_0 u R_0(k)v\phi = -\phi, \; 0 \neq \phi, \; \phi \in L^2(\mathbb{R}^d)\}. \tag{4.4}$$

The on-shell scattering operator is then defined by

$d = 1$:
$$S_{\epsilon_1\epsilon_2}(k) = \delta_{\epsilon_1\epsilon_2} + f_{\epsilon_1\epsilon_2}(k)$$

or equivalently

$$S(k) = \begin{pmatrix} T^l(k) & R^r(k) \\ R^l(k) & T^r(k) \end{pmatrix}, \tag{4.5}$$

where the matrix elements denote the transmission coefficients (T^l, T^r) and reflection coefficients (R^l, R^r) for left and right incidence.

$d = 2$:
$$(S(k)\phi)(\underline{\omega}) = \phi(\underline{\omega}) + (ik/2\pi)^{1/2}\int_{S^1} d\omega' f(k,\underline{\omega},\underline{\omega}')\phi(\omega'), \tag{4.6}$$

$d = 3$:
$$(S(k)\phi)(\underline{\omega}) = \phi(\underline{\omega}) + (ik/2\pi)\int_{S^2} d\omega' f(k,\underline{\omega},\underline{\omega}')\phi(\omega'), \quad \phi \in L^2(S^{d-1}). \tag{4.7}$$

Expanding $T(k)$ and $\exp(ik\underline{\omega}\cdot\underline{x})$ in eqs. (4.2) and (4.3) yields a Laurent expansion for the scattering amplitude in $d = 2,3$. (For $d = 1$ we will directly consider the S-operator later on.) We have in more detail

Theorem 4.1 *Assume (2.4). Then, for $d = 2$, $k^{1/2}f(k,\underline{\omega},\underline{\omega}')$ is analytic in $(1/\ln k, k\ln k)$ near $(0,0)$ with the Taylor expansion*

$$k^{1/2}f(k,\underline{\omega},\underline{\omega}') = -\lambda_0 e^{i\pi/4}(8\pi)^{-1/2}\sum_{m=r}^{\infty}\sum_{n=s}^{\infty}(1/\ln k)^m(k\ln k)^n f_{mn}(\underline{\omega},\underline{\omega}')$$

$$(r,s) = \begin{cases} (1,0) & \text{in cases} \quad 2.I,\; 2.IIb\text{-}c,\; 2.III,\; 2.IVa\text{-}c \\ (1,1) & \text{in case} \quad 2.IIa. \end{cases} \tag{4.8}$$

For $d = 3$, $f(k,\underline{\omega},\underline{\omega}')$ has the following Laurent expansion around $k = 0$

$$f(k,\underline{\omega},\underline{\omega}') = -\lambda_0(4\pi)^{-1}\sum_{n=s'}^{\infty}(ik)^n f_n(\underline{\omega},\underline{\omega}')$$

$$s' = \begin{cases} 0 & \text{in cases} \quad 3.I,\; 3.III \\ -1 & \text{in cases} \quad 3.II,\; 3.IV. \end{cases} \tag{4.9}$$

To be able to discuss the low-energy parameters like the scattering length etc. . . in the next section we list the first coefficients explicitly

$$3.\text{I} \quad : \quad f(k,\underline{\omega},\underline{\omega}') = -(4\pi)^{-1}(v,T_0 u) + O(k) \tag{4.10}$$

$$3.\text{II} \quad : \quad f(k,\underline{\omega},\underline{\omega}') = i/k + 4\pi \, |(v,\phi)|^{-2}(\tilde{\phi},r_2\phi)$$
$$-(\phi,v)^{-1}(\phi,v\underline{\omega}'\cdot\underline{x}) + (v,\phi)^{-1}(v\underline{\omega}\cdot\underline{x},\phi) + O(k) \tag{4.11}$$

$$3.\text{III} \quad : \quad f(k,\underline{\omega},\underline{\omega}') = -(4\pi)^{-1}(v,T_0 u) + (4\pi)^{-1}(v\underline{\omega}\cdot\underline{x},a_{-2}u\underline{\omega}'\cdot\underline{x}) + O(k)$$

$$a_{-2} = \sum_{j,l=1}^{N} (\tilde{\phi},r_2\phi)_{jl}^{-1}(\tilde{\phi}_l,\cdot)\phi_j \tag{4.12}$$

$$2.\text{I}, 2.\text{III} \quad : \quad k^{1/2}f(k,\underline{\omega},\underline{\omega}') = (\pi/2)^{1/2}e^{i\pi/4}\,\{(1/\ln k)$$
$$+ [(i\pi/2) + \ln 2 + \Psi(1) + (2\pi/\lambda_0(v,u))]\,(1/\ln k)^2$$
$$+ (2\pi/(v,u)^2)\,[(v,M_0 u) - \lambda_0(v,M_0 T_0 Q M_0 u)]\,(1/\ln k)^2$$
$$O((1/\ln k)^3)\} \tag{4.13}$$

$$2.\text{IIb} \quad : \quad k^{1/2}f(k,\underline{\omega},\underline{\omega}') = (\pi/2)^{1/2}e^{i\pi/4}\,\{(1/\ln k)$$
$$-2\pi \sum_{j,l=1}^{M} (\tilde{\phi},M_{01}\phi)_{jl}^{-1}\left(\underline{c}_2^{(j)}\cdot\underline{\omega}\right)\left(\overline{\underline{c}_2^{(l)}}\cdot\underline{\omega}'\right)(1/\ln k)$$
$$+ [(i\pi/2) + \ln 2 + \Psi(1) + (2\pi/\lambda_0(v,u))]\,(1/\ln k)^2$$
$$+ (2\pi/(v,u)^2)\,[(v,M_0 u) - \lambda_0(v,M_0 T_0 Q M_0 u)]\,(1/\ln k)^2$$
$$+ O((1/\ln k)^3)\} \tag{4.14}$$

Here T_0 is given by eq. (3.6), M_0 by (2.9), M_{01} by (3.11), r_2 by (3.13), the $\underline{c}_2$ by (2.19) and Ψ is the digamma function [1].

Employing for $d = 1$ the low-energy expansion for $T(k)$ in (4.1) and combining for $d = 2,3$ eqs. (4.8) and (4.6) respectively (4.9) and (4.7) we arrive at the following results for the S-matrix.

Theorem 4.2 *Assume (2.4). Then the on-shell scattering matrix $S(k)$ for $d = 1$ is analytic in k near $k = 0$ with the following Taylor expansion*

$$S_{\epsilon_1\epsilon_2}(k) = \sum_{n=0}^{\infty}(ik)^n s_{\epsilon_1\epsilon_2}^{(n)} \tag{4.15}$$

where in case 1.I

$$s_{\epsilon_1\epsilon_2}^{(0)} = \delta_{\epsilon_1\epsilon_2} - 1 \tag{4.16}$$

$$s_{\epsilon_1\epsilon_2}^{(1)} = -2\lambda_0^{-1}(v,u)^{-2}\left[(v,u) + \lambda_0(v,M_0 u) + \lambda_0^2(v,u)^{-1}(v,M_0 u)^2 - \lambda_0^2(v,M_0 T_0 M_0 u)\right]$$
$$+ (v,u)^{-1}(\epsilon_1 - \epsilon_2)\left[((\cdot)v,u) - \lambda_0((\cdot)v,T_0 M_0 u) + \lambda_0(v,u)^{-1}((\cdot)v,u)(v,M_0 u)\right]$$
$$-2^{-1}\lambda_0\epsilon_1\epsilon_2((\cdot)v,T_0(\cdot)u) + 2^{-1}\lambda_0\epsilon_1\epsilon_2(v,u)^{-1}((\cdot)v,u)^2 \tag{4.17}$$

and in case 1.IIa-c

$$s^{(0)}_{\epsilon_1\epsilon_2} = \delta_{\epsilon_1\epsilon_2} - 1 + \left[|c_1|^2 - \epsilon_1 c_1^* c_2 + \epsilon_2 c_1 c_2^* - \epsilon_1\epsilon_2|c_2|^2\right] / \left[|c_1|^2 + |c_2|^2\right] \tag{4.18}$$

where T_0 and M_0 are given by eq. (3.6) and (2.9) respectively and the c_1, c_2 are given by eq. (2.18).

For $d = 2$, $S(k)$ is analytic in $(1/\ln k,\ k\ln k)$ near $(0,0)$ with the following Taylor expansions

$$2.I\ :\ \ S(k) = 1 + i\pi(Y_0, \cdot)Y_0(1/\ln k) + O((1/\ln k)^2) \tag{4.19}$$

$$2.IIa\ :\ \ S(k) = 1 - 2^{-1/2}(c_1^{(0)})^{-1}(Y_0, \cdot)Y_{10}k + 2^{-1/2}(\bar{c}_1^{(0)})^{-1}(Y_{10}, \cdot)Y_0 k$$
$$+O(k^2\ln k) \tag{4.20}$$

$$2.IIb\ :\ \ S(k) = 1 + i\pi(Y_0, \cdot)Y_0(1/\ln k) - i\sum_{j,l=1}^{M}(\tilde{\phi}, M_{01}\phi)^{-1}_{jl}(Y_{1l}, \cdot)Y_{1j}(1/\ln k)$$
$$+O((1/\ln k)^2) \tag{4.21}$$

$$2.III\ :\ \ S(k) = 1 + i\pi(Y_0, \cdot)Y_0(1/\ln k) + O((1/\ln k)^2) \tag{4.22}$$

where

$$Y_0(\underline{\omega}) = (2\pi)^{-1/2}, Y_{1j}(\underline{\omega}) = (1/2\pi^{1/2})(v\underline{\omega} \cdot, \phi_j),\ j = 0, 1\ldots M$$

and M_{01} and $c_1^{(0)}$ are given by eq. (3.11) and (2.19) respectively.

For $d = 3$, $S(k)$ is analytic in k near $k = 0$ with the Taylor expansions

$$3.I\ :\ \ S(k) = 1 + (2\pi i)^{-1}k(v, T_0 u)(Y_0, \cdot)Y_0 + O(k^2) \tag{4.23}$$

$$3.II\ :\ \ S(k) = 1 - 2(Y_0, \cdot)Y_0 + O(k) \tag{4.24}$$

$$3.III\ :\ \ S(k) = 1 + (2\pi i)^{-1}k(v, T_0 u)(Y_0, \cdot)Y_0 + (8\pi i)^{-1}k\sum_{j,l=1}^{N}(\tilde{\phi}, r_2\phi)^{-1}_{jl}(Y_{1l}, \cdot)Y_{1j}$$
$$+O(k^2) \tag{4.25}$$

$$3.IV\ :\ \ S(k) = 1 - 2(Y_0, \cdot)Y_0 + O(k) \tag{4.26}$$

where

$$Y_0(\underline{\omega}) = (4\pi)^{-1/2},\ Y_{1j}(\underline{\omega}) = \pi^{-1/2}(v\underline{\omega} \cdot, \phi_j)$$

and T_0 and r_2 are given by eq. (3.6) and (3.13) respectively.

In these expansions we have listed a few coefficients to be able to discuss their most important features. For additional coefficients we refer to [5], [7], [23], [29].

For $d = 1$ the transmission and reflection coefficients (4.5) in case 1.I are given by

$$\begin{aligned}
T^l(k) &= T^r(k) = ik\, s^{(1)}_{++} + O(k^2) \\
R^l(k) &= -1 + ik\, s^{(1)}_{-+} + O(k^2) \\
R^r(k) &= -1 + ik\, s^{(1)}_{+-} + O(k^2)
\end{aligned} \tag{4.27}$$

where the $s^{(1)}_{\epsilon_1 \epsilon_2}$ can be read off from eq. (4.17). In case 1.IIa-c the transmission and reflection coefficients are

$$\begin{aligned}
T^l(k) &= T^r(k) = (|c_1|^2 - |c_2|^2)(|c_1|^2 + |c_2|^2)^{-1} + O(k) \\
R^l(k) &= 2\,c_1 c_2^*(|c_1|^2 + |c_2|^2)^{-1} + O(k) \\
R^r(k) &= -2\,c_1 c_2^*(|c_1|^2 + |c_2|^2)^{-1} + O(k)
\end{aligned}$$

(4.28)

where the c_1, c_2 are given by (2.18). The results (4.27) confirm the less detailed ones in [32], [39], [72]. For the case $(v, u) = 0$ the expressions (4.27) and (4.28) remain formally valid [25]. The condition (2.4) can be relaxed to $(1 + |\cdot|^{2m+2\epsilon})\, V \in L^1(\mathbb{R})$, $\epsilon \geq 0$ and $m = 3$ in case 1.I respectively $m = 4$ in case 1.IIa-c, if one is only interested in asymptotic expansions. A discussion of reflection and transmission coefficients for potentials with a behavior of the type $x^{-2-\epsilon}$ respectively $(-x)^{-2-\delta}$ as $x \to \infty$ respectively $x \to -\infty$ where $0 < \epsilon, \delta < 1$ has been given recently in [54] (cf. also [4], [38]).

For $d = 3$ Theorems 4.1-4.2 generalize known results for the special case where V is spherically symmetric (see e.g. [11], [17], [69] and references therein). For the latter $Y_1(\underline{\omega}) = 0$, and $Y_{1j}(\underline{\omega}) \neq 0$ only if ϕ_j is a p-wave state. General non-local interactions can also be treated [26]. Some of the results have been derived in less explicit form under weaker conditions on V, since one only aims at asymptotic expansions, with the help of weighted Sobolev spaces [51]. Also here the condition (2.4) can be relaxed to $V \in R$ and $(1 + |\cdot|^n)\, V \in L^1(\mathbb{R}^3)$, $n > 2$, e.g. $n = 4$ in case 3.I, if one only wants this type of expansions.

In case 3.I the low-energy limit is independent of the detailed shape of the potential V and determined by $(4\pi)^{-1}\,(v, T_0 u)$, the scattering length (as will be explained in the next section). In case 3.III, if the second term on the r.h.s. of (4.12) is non-vanishing, the scattering cross section never becomes isotropic. For spherically symmetric V this anisotropy is known to occur precisely at zero-energy p-wave bound states. In the cases 3.II and 3.IV where a zero-energy resonance is present $S(k)$ converges to -1, a fact which is also known in the case of spherical symmetry.

For $d = 2$ analogous results for the resolvent and the evolution group of general elliptic differential operators have been obtained in [67] [68] using weighted Sobolev space methods. The results of Theorems 4.1-4.2 considerably simplify if V is spherically symmetric, then e.g. $Y_{10}(\underline{\omega}) = 0$, and $Y_{1j}(\underline{\omega}) \neq 0$ only for the p-wave [42] [73]. Again the exponential fall-off condition (2.4) can be relaxed to a suitable finite moment condition if one is interested in asymptotic expansions [23]. Finally we emphasize that $S(k)$ converges to 1 in all cases 2.I-2.IV in sharp contrast with $d = 1$ and $d = 3$ as we have seen above.

5   Low-energy parameters

We give an appropriate generalization of the low-energy parameters (scattering length, effective range) to the non-spherically symmetric case and at the same time relate these parameters directly to the threshold behavior of the scattering amplitude. Of course the discussion is only relevant for $d = 2, 3$. We start with $d = 3$ and recall the so-called effective range expansion [69]. If V is spherically symmetric and satisfies e.g.

$$|V(r)| \leq \text{const.}\ r^{-1} e^{ar} \quad \text{for some } a > 0$$

(5.1)

then the scattering amplitude is represented by the absolutely convergent series

$$f(k,\underline{\omega},\underline{\omega}') = 4\pi \sum_{l=0}^{\infty} \sum_{m=-l}^{l} \frac{e^{2i\delta_l(k)} - 1}{2ik}\overline{Y_{l,m}(\underline{\omega})}\, Y_{l,m}(\underline{\omega}'), \quad k > 0 \tag{5.2}$$

where δ_l denotes the partial wave phase shift and Y_{lm} are the spherical harmonics. Furthermore the effective range expansion reads [69]

$$k^{2l+1}\cot\delta_l(k) = -(a_l)^{-1} + r_l\, k^2/2 + O(k^4), \quad l = 0, 1, 2\dots \tag{5.3}$$

where a_l, r_l denote the partial wave scattering lengths, respectively the effective range parameters. Under assumption (5.1) the l.h.s. of eq. (5.3) is real analytic in k^2 at $k^2 = 0$. For simplicity we restrict ourselves to the case $l = 0$. Assuming $(a_0)^{-1} \neq 0$ we obtain

$$(4\pi)^{-2}\int d\omega\, d\omega'\, f(k,\underline{\omega},\underline{\omega}') = (2ik)^{-1}(e^{2i\delta_0(k)} - 1)$$
$$= -a_0 + i\, a_0^2\, k - a_0^2\, r_0\, k^2/2 + a_0^3\, k^2 + O(k^3). \tag{5.4}$$

If $a_0^{-1} = 0$ we get

$$(4\pi)^{-2}\int d\omega\, d\omega'\, f(k,\underline{\omega},\underline{\omega}') = (2ik)^{-1}(e^{2i\delta_0(k)} - 1)$$
$$= i/k + r_0/2 + O(k). \tag{5.5}$$

For a discussion of these concepts in arbitrary dimension $d \geq 2$ with general, possibly long-range, spherically symmetric interactions we refer to [21], [22] (cf. also [14]). The connection with point interactions has been analyzed in [8].

We then introduce the following generalizations

Definition 5.1 *Assume (2.4) for $d = 3$ and $(0, k_0) \cap \mathcal{E}_+ = \emptyset$ for some $k_0 > 0$. In cases 3.I and 3.III the scattering length a and the effective range parameter r are defined by, respectively*

$$a = -\lim_{k\to 0+}(4\pi)^{-2}\int d\omega\, d\omega'\, f(k,\underline{\omega},\underline{\omega}'), \quad k \in (0, k_0) \tag{5.6}$$

$$r = 2a + \lim_{k\to 0+}2k^{-2}\left\{-a^{-2}(4\pi)^{-2}\int d\omega\, d\omega'\, f(k,\underline{\omega},\underline{\omega}') - a^{-1} + ik\right\},$$
$$k \in (0, k_0). \tag{5.7}$$

In cases 3.II and 3.IV the effective range parameter is defined to be

$$r = \lim_{k\to 0+}2(4\pi)^{-2}\int d\omega\, d\omega'[f(k,\underline{\omega},\underline{\omega}') - ik^{-1}], \quad k \in (0, k_0) \tag{5.8}$$

($a^{-1} = 0$ in these cases).

Remark 5.1. a) In case 3.I, definition (5.6) may be replaced by the usual definition $a = -\lim_{k\to 0+} f(k,\underline{\omega},\underline{\omega}')$.
b) If V is spherically symmetric then a and r clearly coincide with a_0 and r_0. It is

188   D. Bollé

obvious from the above considerations how the higher-order parameters (the so-called shape parameters) can be defined in terms of the scattering amplitude.
c) Generalizations of the definitions (5.6)-(5.8) to non-local two-body interactions have been given in [26]. Coulomb modified parameters have been described in [5], [27]. Three-body generalizations have been discussed in [44].

Definition 5.2 *Assume (2.4) for $d = 2$ and $(0, k_0^2) \cap \mathcal{E}_+ = \emptyset$ for some $k_0 > 0$. In cases 2.I, 2.IIb, 2.III and 2.IVb we define the scattering length $\tilde{a}$ by*

$$\tilde{a}^{-1} \;=\; -\Psi(1) - (i\pi/2) + \lim_{\substack{k \to 0 \\ 0 < k < k_0}} \left\{ e^{-i\pi/4} (2k/\pi)^{1/2} (\ln(k/2))^2 \times \right.$$

$$\left. \times \quad (2\pi)^{-2} \int d\omega\, d\omega'\, f(k, \underline{\omega}, \underline{\omega}') - \ln(k/2) \right\}. \tag{5.9}$$

In cases 2.IIa,c and cases 2.IVa,c the scattering length does not exist i.e. $\tilde{a}^{-1} = 0$.

Remark 5.2. a) If V is spherically symmetric then $\tilde{a}$ reduces precisely to the ordinary two-dimensional scattering length discussed extensively in [2], [21], [22], [42].
b) In exactly the same way one can define the effective range parameter for non-spherically symmetric potentials.
c) As states already before, the conditions (2.4) can be relaxed also here.

Given the threshold expansion of $f(k, \underline{\omega}, \underline{\omega}')$ in section 4 it is straightforward to show

Theorem 5.3 *Assume (2.4), $d = 2, 3$. Then*

$$3.I \;\; : \;\; a = (4\pi)^{-1}(v, T_0 u) = (4\pi)^{-1}(v, \Phi), \tag{5.10}$$

$$\Phi = u - u R_0(0) v \Phi \;, \;\; \Phi \in L^2(\mathbb{R}^3) \tag{5.11}$$

$$3.II \;\; : \;\; a^{-1} = 0$$

$$r = 8\pi |(v, \phi)|^{-2} (\tilde{\phi}, r_2 \phi) \tag{5.12}$$

$$\text{with } \phi \text{ satisfying (2.12) and } r_2 \text{ given by (3.13)}$$

$$3.III \;\; : \;\; a = (4\pi)^{-1}(v, T_0 u) = (4\pi)^{-1}(v, (1 - P_0)\Phi) \tag{5.13}$$

$$3.IV \;\; : \;\; a^{-1} = 0$$

$$r = 8\pi |(v, \phi_1)|^{-2} (\tilde{\phi}_1, r_2 \phi_1) \tag{5.14}$$

$$\text{with } \phi_1 \text{ satisfying (2.12) (see Remark 2.1b)}$$

2.I, 2.IIb, 2.III, 2.IVb :

$$\tilde{a}^{-1} = (2\pi / \lambda_0(v, u)) + (2\pi / (v, u)^2) \left[(v, M_0 u) - \lambda_0 (v, M_0 T_0 Q M_0 u) \right]. \tag{5.15}$$

Remark 5.3. a) For explicit formulas for r in cases 3.I and 3.III we refer to [5].
b) Expression (5.10) is a familiar result. If $V \leq 0$ and $\|u R_0(0) v\| < 1$ (i.e. if H has no bound states) then eq. (5.11) can be iterated to give $\Phi \leq 0$. Consequently $r > 0$ in this

case.

c) If $V \leq 0$ and $\|uR_0(0)v\| = 1$ in case 3.II then, using positivity improving arguments [78], one gets that $\phi > 0$ (or $\phi < 0$) almost everywhere (if one chooses ϕ to be real). Thus $r > 0$ in this case. This situation describes e.g. the absorption of the ground states of H into the continuous spectrum [56], [81]. A discussion of physical applications of case 3.II in connection with point interactions can be found in [8].

d) A simple square well [5] and Yamaguchi potential example [26] shows that in general r is by no means comparable with the range of $V(\underline{x})$. In fact, only if H has a zero-energy resonance or at least if the strength of the interaction V is almost critical (in the sense that a slight change in the strength of V turns a negative bound state into a zero-energy resonance), then the magnitude of r represents a measure for the range of V. This fact again explains the relevance of case 3.II (or 3.IV).

e) Eq. (5.15) implies that the scattering length $\tilde{a}$ in cases 2.I, 2.IIb, 2.III and 2.IVb never vanishes.

f) The conditions (2.4) are certainly too strong. In fact, for $d = 3$, $V \in R$, $(1 + |\cdot|)^n V \in L^1(\mathbb{R}^3)$ for some $n > 4$ and $\mathcal{E}_+ \cap (0, \delta) = \emptyset$ for some $\delta > 0$ are sufficient for the results (5.10)-(5.14) to hold. If V has e.g. a long range tail $V(\underline{x}) \sim |\underline{x}|^{-4}, |\underline{x}|^{-6}, |\underline{x}|^{-8}$ like in certain problems in atomic and molecular physics, the scattering length formula in case 3.I still remains valid but an asymptotic expansion of the scattering amplitude in arbitrary high powers of k^2 is lost. Then, a modification of the above approach, relating the low-energy parameters directly to the threshold behavior of the scattering amplitude still works but strongly depends on the particular asymptotic decrease of $V(\underline{x})$ as $|\underline{x}| \to \infty$. (cf. [12]-[15], [33], [47], [59], [62], [82], [83] and references therein).

6   Trace relations, Levinson's theorem

A recent review of trace relations in scattering theory has been given in [17]. Here we restrict ourselves to a discussion of Levinson's theorem (i.e. the zeroth order relation).

First we need the threshold behavior of the trace of the difference between the full and free resolvent. From the resolvent equation

$$R(k) - R_0(k) = -\lambda_0 R_0(k) v\, T(k) u R_0(k), \quad Im\, k > 0 \tag{6.1}$$

one obtains [17], [23], [29]

$$
\begin{aligned}
d = 1 \quad &: \quad Tr\,[R(k) - R_0(k)] = -\lambda_0(2k)^{-1} Tr\,[uR_0'(k)vT(k)] \\
d = 2 \quad &: \quad Tr\,[R(k) - R_0(k)] \\
&\qquad = \lambda_0(v, u)/4\pi k^2 + (\lambda_0^2/2k) Tr\,[uR_0'(k)vT(k)uR_0(k)v] \\
d = 3 \quad &: \quad Tr\,[R(k) - R_0(k)] \\
&\qquad = \lambda_0(v, u)/8\pi i k + (\lambda_0^2/2k) Tr\,[uR_0'(k)vT(k)uR_0(k)v]
\end{aligned}
\tag{6.2}
$$

where the prime denotes the derivative with respect to k. Using then the relevant expansions for $T(k)$, $uR_0(k)v$ and its derivative (see section 3) we have

Theorem 6.1 *Assume conditions (2.4) and additionally for $d = 2$, $V \in L^{4/3}(\mathbb{R}^2)$ [35], [75]. Then $Tr[R(k) - R_0(k)]$ has the following Laurent expansion around $k = 0$*

$d = 1$:

$$Tr\,[R(k) - R_0(k)] = \sum_{n=q-2}^{\infty} (ik)^n \Delta_n \tag{6.3}$$

where

$$\Delta_{q-2} = 4^{-1}\lambda_0(v, t_{q+1}u)$$

$$\Delta_n = 4^{-1}\lambda_0(v, t_{n+3}u) + 2^{-1}\lambda_0 \sum_{l=0}^{n+1-q} (n+2-l-q)Tr\,[M_{n+2-l-q}t_{l+q}],$$

$$n \geq q - 1 \tag{6.4}$$

with $q = 0$ in case 1.I and $q = -1$ in cases 1.IIa-c.

$d = 2$:

$$Tr\,[R(k) - R_0(k)] = \sum_{m=\mu}^{\infty} \sum_{n=\nu}^{\infty} (1/\ln k)^m (k^2 \ln k)^n \Delta_{mn} \tag{6.5}$$

where

$$(u, v) = \begin{cases} (-1, -1) & \text{in cases 2.I, 2.IIb} \\ (-2, -1) & \text{in cases 2.II and 2.IIc if } \underline{c}_2^{(0)} = 0 \\ (-\infty, -1) & \text{in cases 2.IIa and 2.IIc if } \underline{c}_2^{(0)} \neq 0 \end{cases} \tag{6.6}$$

and

$$\Delta_{m,-1} = 0 \ , \ m \leq -2 \ \text{in cases 2.I-2.III}$$

$d = 3$:

$$Tr\,[R(k) - R_0(k)] = \sum_{n=q'-1}^{\infty} (ik)^n \Delta_n \tag{6.7}$$

where

$$\Delta_n = \frac{\delta_{n,-1}}{8\pi}(v, u) - \sum_{l=0}^{n-q'+1} \sum_{j=0}^{n-l-q'+1} Tr\left[\frac{j+1}{2}r_{j+1}t_{n-l-j+1}r_l\right] \tag{6.8}$$

with $q = 0$ in case 3.I, $q = -1$ in case 3.II and $q = -2$ in cases 3.II-3.IV.

It is interesting to compare the most singular behavior of $Tr[R(k) - R_0(k)]$ given by (6.3)-(6.8) for different dimensions

$$1.I \quad \Delta_{-2} = -1/2$$
$$1.IIa\text{-}c \quad \Delta_{-2} = 0$$

$$
\begin{aligned}
\text{2.I, 2.IIa} \quad & \Delta_{-1,-1} = 0 \\
\text{2.IIb} \quad & \Delta_{-1,-1} = -M \\
\text{2.IIc} \quad & \Delta_{-1,-1} = -(M-1) \\
\text{2.III, 2.IVa} \quad & \Delta_{-1,-1} = -N \\
\text{2.IVb} \quad & \Delta_{-1,-1} = -(N+M) \\
\text{2.IVc} \quad & \Delta_{-1,-1} = -(N+M-1) \\
\text{3.I} \quad & \Delta_{-2} = 0 \ , \ \Delta_{-1} = (8\pi)^{-1}(v, T_0 u) \\
\text{3.II} \quad & \Delta_{-2} = 1/2 \ , \ \Delta_{-1} = 2\pi|(v,\phi)|^{-2}(\tilde{\phi}, r_2 \phi) \\
\text{3.III} \quad & \Delta_{-2} = N \\
\text{3.IV} \quad & \Delta_{-2} = (N-1) + 1/2
\end{aligned}
\tag{6.9}
$$

Here N is the number of zero-energy bound states and M the number of zero-energy resonances, counting multiplicity. We remark that for cases 3.I and 3.II, Δ_{-1} is proportional to the scattering length (recall eq. (5.10)) respectively the effective range parameter (eq. (5.12)). For the explicit form of other coefficients we refer to [17], [23], [29], [30].

Finally we apply the contour integration method, using the high- and low-energy properties of $Tr[R - R_0]$ to derive Levinson's theorem.

Theorem 6.2 *We assume $(1 + |\cdot|^n)\, V \in L^1(\mathbb{R}^d)$ for appropriate n, and $V \in L^{4/3}(\mathbb{R}^2)$ for $d = 2$, such that the absence of positive embedded eigenvalues and of the singular continuous spectrum of H is guaranteed [3], [40], [77], [78], then*

$$
d = 1 \ : \ \int_0^\infty dk^2 \, Im\, Tr\,[R(k+io) - R_0(k+io)] = -\pi N_- - \pi \Delta_{-2}
\tag{6.10}
$$

$$
d = 2 \ : \ \int_0^\infty dk^2 \, Im\, Tr\,[R(k+io) - R_0(k+io)] = -\pi N_- + \pi \Delta_{-1,-1}
$$

$$
-\frac{\lambda_0}{4}(v, u)
\tag{6.11}
$$

$$
d = 3 \ : \ \int_0^\infty dk^2 \left\{ Im\, Tr\,[R(k+io) - R_0(k+io)] + (8\pi k)^{-1}(v, u) \right\}
$$

$$
= -\pi N_- - \pi \Delta_{-2}
\tag{6.12}
$$

where N_- denotes the number of (strictly) negative bound states of H counting multiplicity and where the Δ_{-2}, $\Delta_{-1,-1}$ are given by (6.9).

Remark 6.1. a) For an appropriate choice of n in the different cases and for higher-order relations we refer to [16], [23], [25], [29], [30]. For an extensive list of references on the subject of sum rules we refer to [17]. New recent results for Schrödinger operators can be found in [18], [19], [23]-[26], [29], [36], [37], [42], [45], [46], [54], [55], [58], [60], [61], [64]-[66], [73].
b) The results (6.10)-(6.12) do not change for $(v, u) = 0$.
c) Theorem 6.2 can be formulated in terms of the scattering matrix $S(k)$ by noting that

[29], [31], [35]

$$2\, Im\, Tr\, [R(k+io) - R_0(k+io)] = \begin{cases} -i\frac{d}{dk^2}\ln[\det S(k)], & d = 1 \\[2mm] -i\frac{d}{dk^2}Tr[\ln S(k)], & d = 2,3 \end{cases} \tag{6.13}$$

Recalling (6.9), we see that for $d = 3$ the r.h.s. of (6.12) can be written as $-\pi(N_- + N_0 + q)$ where N_0 is the number of zero-energy bound states counting degeneracy and $q = 1/2$ when there is a zero-energy resonance (cases 3.II, 3.IV) and zero otherwise. Compared with spherically symmetric scattering [69] the additional high-energy correction term $(8\pi k)^{-1}(v,u)$ appears, as explained in [16], [31], [70], [74].

For $d = 1$, the structure of Levinson's theorem is completely different. When there is no zero-energy resonance present (case 2.I) we not only get on the r.h.s. of (6.10) the term $(-\pi N_-)$ but an additional factor $\pi/2$ (recall (6.9)) appears. In the case of a zero-energy resonance (cases 2.II) we simply obtain the term $(-\pi N_-)$. This difference is of course due to the additional Dirichlet boundary conditions at the origin when considering Schrödinger operators on the half line.

For $d = 2$, recalling the value of $\Delta_{-1,-1}$ (cf. (6.9)), we see from (6.11) that a zero-energy resonance of the s-wave type (case 2.IIa) does not contribute to Levinson's theorem, while zero-energy resonances of the p-wave type (case 2.IIb) contribute exactly like bound states. This is in sharp contrast with $d = 1$ and $d = 3$. This fact remains valid in $d \geq 4$ dimensions since for $d = 4$ the zero-energy resonance contributes like a zero-energy bound state and for $d \geq 5$ there exist no zero-energy resonances of H at all [49], [50], [56], [67], [68]. If V is spherically symmetric, the term $-\lambda_0(v,u)/4$ on the r.h.s. of (6.11) is absent [42].

Acknowledgment

It is a pleasure to thank S. Albeverio, C. Danneels, F. Gesztesy, M. Klaus, C. Nessmann, T. Osborn, W. Schweiger, L. Streit and S.F.J. Wilk for joint collaborations which led to most of the results presented here and the National Fund for Scientific Research, Belgium for financial support as an "Onderzoeksdirecteur".

References

1. M. Abramowitz and I.A. Stegun, <u>Handbook of mathematical functions</u> (Dover, New York 1972).

2. S.K. Adhikari, W.G. Gibson and T.K. Lim, J. Chem. Phys. **85**, 5580 (1986).

3. S. Agmon, Ann. Scuola Norm. Sup. Pisa **IV 2**, 151 (1975).

4. T. Aktosun, Inverse Problems **3**, L1 (1987).

5. S. Albeverio, D. Bollé, F. Gesztesy, R. Høegh-Krohn and L. Streit, Ann. Phys. (N.Y.) **148**, 308 (1983).

6. S. Albeverio, F. Gesztesy and R. Høegh-Krohn, Acta Phys. Austriaca Suppl. **23**, 577 (1981).

7. S. Albeverio, F. Gesztesy and R. Høegh-Krohn, Ann. Inst. H. Poincaré A **37**, 1 (1982).

8. S. Albeverio, F. Gesztesy, R. Høegh-Krohn and H. Holden, Solvable models in quantum mechanics (Springer, New York 1988).

9. S. Albeverio, F. Gesztesy, R. Høegh-Krohn and L. Streit, Ann. Inst. H. Poincaré A **38**, 263 (1983).

10. S. Albeverio and R. Høegh-Krohn, J. Math. Anal. Appl. **101**, 491 (1984).

11. W.O. Amrein, J.M. Jauch and K.B. Sinha, Scattering theory in quantum mechanics (Benjamin, Reading 1977).

12. V.V. Babikov, The method of phase functions in quantum mechanics, 2nd ed. (Nauka, Moscow, 1976) 132.

13. A.M. Badalyan, L.P. Kok, M.I. Polikarpov and Yu.A. Simonov, Phys. Rep. **82**, 31 (1982).

14. Gy Bencze, C. Chandler, J.L. Friar, A.G. Gibson and G.L. Payne, Phys. Rev. C **35**, 1188 (1987).

15. R.O. Berger and L. Spruch, Phys. Rev. **138**, B1106 (1965).

16. D. Bollé, Ann. Phys. (N.Y.) **121**, 131 (1979).

17. D. Bollé, in Mathematics + physics : lectures on recent results, Vol. 2, ed. L. Streit (World Scientific, Singapore 1986), 84.

18. D. Bollé, in Strongly coupled plasma physics, eds. F.J. Rogers, H.E. DeWitt (Plenum, New York 1987), 215.

19. D. Bollé, in Few-body physics, eds. L.S. Ferreira, A.C. Fonseca, L. Streit (Springer, Berlin 1987), 597.

20. D. Bollé, Phys. Rev. A **36**, 3259 (1987).

21. D. Bollé and F. Gesztesy, Phys. Rev. Lett. **52**, 1469 (1984).

22. D. Bollé and F. Gesztesy, Phys. Rev. A **30**, 1279 (1984); Phys. Rev. A **33**, 3517 (1986).

23. D. Bollé, F. Gesztesy and C. Danneels, Ann. Inst. H. Poincaré A **48**, 175 (1988).

24. D. Bollé, F. Gesztesy, C. Danneels and S.F.J. Wilk, Phys. Rev. Lett. **56**, 900 (1986).

25. D. Bollé, F. Gesztesy and M. Klaus, J. Math. Anal. Appl. **122**, 496 (1987).

26. D. Bollé, F. Gesztesy, C. Nessmann and L. Streit, Rep. Math. Phys. **23**, 373 (1986).

27. D. Bollé, F. Gesztesy and W. Schweiger, J. Math. Phys. **26**, 1661 (1985).

28. D. Bollé, F. Gesztesy and S.F.J. Wilk, Phys. Lett. A **97**, 30 (1983).

29. D. Bollé, F. Gesztesy and S.F.J. Wilk, J. Operator Theory **13**, 3 (1985).

30. D. Bollé and S.F.J. Wilk, J. Math. Phys. **24**, 1555 (1983).

31. V.S. Buslaev, Sov. Phys. Dokl. **7**, 295 (1962); in Topics in mathematical physics, ed. Mh.Sh. Birman (Consultants Bureau, New York, 1967), Vol. 1, p. 69.

32. K. Chadan and P.C. Sabatier, Inverse problems in quantum scattering theory (Springer, New York, 1977).

33. E.S. Chang, J. Phys. B **14**, 893 (1981).

34. M. Cheney, J. Math. Phys. **25**, 94 (1984).

35. M. Cheney, J. Math. Phys. **25**, 1449 (1984).

36. S.G. Chung and T.F. George, J. Math. Phys. **28**, 1103 (1987).

37. G. Cognola and S. Zerbini, Extended Levinson theorems from the heat kernel expansion, Preprint Trento, Italy 1987.

38. A. Degasperis and P.C. Sabatier, Inverse Problems, **3**, 73 (1987).

39. P. Deift and E. Trubowitz, Comm. Pure Appl. Math. **32**, 121 (1979).

40. R. Froese, I. Herbst, M. Hoffmann-Ostenhof and T. Hoffmann-Ostenhof, J. d'Anal. Math. **41**, 272 (1982).

41. F. Gesztesy, in Resonances, models and phenomena, eds. S. Albeverio, L.S. Ferreira and L. Streit, Lecture Notes in Physics 211 (Springer, New York, 1984), 78.

42. F. Gesztesy, in Models and methods in few-body physics, eds. L.S. Ferreira, A.C. Fonseca and L. Streit, Lecture Notes in Physics 273 (Springer, New York 1987), 609.

43. F. Gesztesy and H. Holden, J. Math. Anal. Appl. **123**, 181 (1987).

44. F. Gesztesy and G. Karner, SIAM J. Math. Anal. **18**, 1064 (1987).

45. W.G. Gibson, Phys. Lett. A **117**, 107 (1986).

46. W.G. Gibson, Phys. Rev. A **36**, 564 (1987).

47. O. Hinckelmann and L. Spruch, Phys. Rev. A **3**, 642 (1971).

48. H. Holden, J. Operator Theory **14**, 263 (1985).

49. A. Jensen, Duke Math. J. **47**, 57 (1980).

50. A. Jensen, J. Math. Anal. Appl. **101**, 397 (1984).

51. A. Jensen and T. Kato, Duke Math. J. **46**, 583 (1979).

52. T. Kato, <u>Perturbation theory for linear operators</u> (Springer, New York, 1980).

53. M. Klaus, Helv. Phys. Acta **55**, 49 (1982).

54. M. Klaus, Inverse Problems **4**, 505 (1988).

55. M. Klaus, J. Math. Phys. **29**, 148 (1988).

56. M. Klaus and B. Simon, Ann. Phys. (N.Y.) **130**, 251 (1980).

57. M. Klaus and B. Simon, Commun. Math. Phys. **78**, 153 (1980).

58. V.V. Kostrykin and S.P. Merkuriev, Trace formula for three-body system in an external electric field, preprint Leningrad, USSR, E-16-87.

59. A.A. Kvitsinskii, Theor. Math. Phys. **65**, 1123 (1986).

60. A.A. Kvitsinskii, Theor. Math. Phys. **68**, 801 (1987).

61. A.A. Kvitsinskii, Theor. Math. Phys. **70**, 72 (1987).

62. A.A. Kvitsinskii, I.V. Komarov and S.P. Merkuriev, Sov. J. Nucl. Phys. **38**, 58 (1983).

63. S.T. Kuroda, <u>An introduction to scattering theory</u>, Aarhus Lecture Notes Series No. 51, 1980.

64. Z.Q. Ma, Phys. Rev. D **33**, 1745 (1986).

65. Z.Q. Ma and A.Y. Dai, J. Phys. A **21**, 2085 (1988).

66. S.P. Merkuriev and A.A. Kvitsinsky, Sum rules in three-body problem at fixed angular momentum, preprint Trieste, Italy IC/88/326.

67. M. Murata, J. Func. Anal. **49**, 10 (1982).

68. M. Murata, J. Func. Anal. **56**, 300 (1984).

69. R.G. Newton, <u>Scattering theory of waves and particles</u>, 2nd ed. (Springer, New York, 1982).

70. R.G. Newton, J. Math. Phys. **18**, 1348 (1977).

71. R.G. Newton, J. Math. Phys. **18**, 1582 (1977).

72. R.G. Newton, J. Math. Phys. **21**, 493 (1980).

73. R.G. Newton, J. Math. Phys. **27**, 2720 (1986).

74. T.A. Osborn and D. Bollé, J. Math. Phys. **18**, 432 (1977).

75. T.A. Osborn, K.B. Sinha, D. Bollé and C. Danneels, J. Math. Phys. **26**, 2796 (1985).

76. J. Rauch, J. Func. Anal. **35**, 304 (1980).

77. M. Reed and B. Simon, Methods of modern mathematical physics III : Scattering theory (Academic, New York, 1979).

78. M. Reed and B. Simon, Methods of modern mathematical physics IV : Analysis of operators (Academic, New York, 1978).

79. B. Simon, Quantum mechanics for Hamiltonians defined as quadratic forms (Univ. Press Princeton, 1971).

80. B. Simon, Ann. Phys. (N.Y.) **97**, 279 (1976).

81. B. Simon, J. Func. Anal. **25**, 338 (1977).

82. H. van Haeringen and L.P. Kok, Phys. Rev. A **26**, 1218 (1982).

83. D.R. Yafaev, Commun. Math. Phys. **85**, 177 (1982).

SPECTRAL ANALYSIS AND SCATTERING THEORY FOR SCHRÖDINGER OPERATORS WITH AN INTERACTION SUPPORTED BY A REGULAR CURVE

J.F. Brasche

Institut für Mathematik,

Ruhr–Universität Bochum,

D–4630 Bochum 1, FRG

A. TETA

S.I.S.S.A.,

Strada Costiere 11,

34014 Trieste, ITALY

ABSTRACT

We study the spectrum and the scattering theory for the Schrödinger operator with a δ–like potential supported by a regular curve. Moreover an example of explicitly solvable model (i.e. when the curve is a circle) is presented.

197

1. Introduction

In this paper we study the Schrödinger operator in $L^2(\mathbb{R}^3)$ with a δ–like potential supported by a smooth curve of finite length.

From the physical point of view such hamiltonian can be employed to describe the interaction of a quantum particle with a polymer or even in the study of the scattering of classical waves from obstakles (see also the discusion in [1]).

The rigorous construction of the operator was first given in [2] using a method essentially based on application of embedding theorems; the resolvent was explicitly written in term of an operator acting in the space of the Fourier transforms.

Here we will follow another approach ([3]) based on the theory of quadratic forms, which allows us to characterize the operator and its resolvent directly in the x-space.

Starting from results of [3], here we will give a description of the (continuous and point) spectrum (section 2) and we calculate the generalized eigenfunctions and the scattering quantities (section 3). Finally we will give an example (i.e. the case of an interaction of constant strength supported by a circle) which is explicitly solvable (section 4).

It should be emphasized that the methods used and the results obtained are a direct generalization of the work done for the Schrödinger operator with point interactions, i.e. interactions supported by points ([4]). We refer to [5] – [11] and references given therein for other methods to discuss Schrödinger operators with an interaction supported by a set with Lebesgue measure zero.

We start briefly recalling the definition of the operator as given in [3]. Let C be a curve in $\mathbb{R}^3$ of class C^1, without multiple points and of finite length L, and let $y = y(s)$, $s \in I$, be the parametric representation of C in terms of the arc–length relative to a chosen initial point.

Moreover we assume that there exists a positive constant ϵ_0 such that the following condition is satisfied:

$$|y(s) - y(s')| \geq |s - s'|\,(1 - c|s - s'|^\tau) \quad \text{whenever} \quad |s - s'| \leq \epsilon_0$$

where τ, c are (henceforth fixed) positive constants satisfying $c\epsilon_0^\tau < 1$. For a given continuous bounded real valued function β defined on C and for λ greater than a fixed $\lambda_0(\beta, C) > 0$ then the following quadratic form in $L^2(I)$ is closed and coercive (see lemma 1 and proposition 2 in [3])

$$D(\Phi^\lambda_{\beta,C}) = \{\xi \in L^2(I) \mid \Phi^\lambda_{\beta,C}(\xi, \xi) < +\infty\} \tag{1.1}$$

$$\Phi^\lambda_{\beta,C}(\xi, \xi) = \frac{1}{2} \int_{I \times I} |\xi(s) - \xi(s')|^2 G_{i\sqrt{\lambda}}(y(s) - y(s')) ds\, ds'$$

$$+ \int_I |\xi(s)|^2 (a_{i\sqrt{\lambda}}(s) + \beta(s)) ds \tag{1.2}$$

where

$$G_k(x - x') = (-\Delta - k^2)^{-1}(x, x') = \frac{e^{ik|x-x'|}}{4\pi|x - x'|},$$

$$Im\, k > 0, \quad x, x' \in \mathbb{R}^3, \quad x \neq x' \tag{1.3}$$

and

$$a_{i\sqrt{\lambda}}(s) = -\int_{|s-s'|>\epsilon_0} G_{i\sqrt{\lambda}}(y(s) - y(s')) ds'$$

$$+ \int_{|s-s'|<\epsilon_0} \left[\frac{1}{4\pi|s - s'|} - G_{i\sqrt{\lambda}}(y(s) - y(s'))\right] ds'$$

$$- \frac{1}{2\pi} \log 2\epsilon_0 \tag{1.4}$$

The corresponding positive and selfadjoint operator in $L^2(I)$ defined by $\Phi^\lambda_{\beta,C}$ is denoted by $\Gamma_{\beta,C}(i\sqrt{\lambda})$. It should be remarked that the domains of $\Phi^\lambda_{\beta,C}$ and of $\Gamma_{\beta,C}(i\sqrt{\lambda})$ do not depend on the choice of $\lambda > \lambda_0(\beta, C)$.

For later use it is convenient to introduce the analytic continuation of $\Gamma_{\beta,C}(i\sqrt{\lambda})$

$$\Gamma_{\beta,C}(k) = \Gamma_{\beta,C}(i\sqrt{\lambda}) + \tilde{G}_{i\sqrt{\lambda}} - \tilde{G}_k =$$
$$= [1 + (\tilde{G}_{i\sqrt{\lambda}} - \tilde{G}_k)\Gamma_{\beta,C}(i\sqrt{\lambda})^{-1}]\Gamma_{\beta,C}(i\sqrt{\lambda}) \qquad (1.5)$$

where $k \in \mathbb{C}$, $\lambda > \lambda_0(\beta, C)$ and $\tilde{G}_{i\sqrt{\lambda}} - \tilde{G}_k$ is an integral operator of Hilbert–Schmidt type in $L^2(I)$ with kernel given by

$$(\tilde{G}_{i\sqrt{\lambda}} - \tilde{G}_k)(s, s') = \frac{e^{-\sqrt{\lambda}|y(s)-y(s')|} - e^{ik|y(s)-y(s')|}}{4\pi|y(s) - y(s')|} \qquad (1.6)$$

It is straightforward to show that the definition (1.5) is independent of the choice of $\lambda > \lambda_0(\beta, C)$; moreover $D(\Gamma_{\beta,C}(k)) = D(\Gamma_{\beta,C}(i\sqrt{\lambda}))$, $\Gamma_{\beta,C}(k)$ is closed and satisfies $\Gamma_{\beta,C}(k)^* = \Gamma_{\beta,C}(-\overline{k})$. Now the Schrödinger operator $-\Delta_{\beta,C}$ with δ−interaction supported by C of strength β is given by (see Proposition 4 in [3])

$$D(-\Delta_{\beta,C}) = \{u \in L^2(\mathbb{R}^3) \mid \exists \xi_u \in D(\Gamma_{\beta,C}),\ u - \breve{G}_k\xi_u \in H^2(\mathbb{R}^3),$$
$$(u - \breve{G}_k\xi_u)(y(s)) = (\Gamma_{\beta,C}(k)\xi_u)(s)\forall s \in I\} \qquad (1.7)$$

$$(-\Delta_{\beta,C} - k^2)u = (-\Delta - k^2)(u - \breve{G}_k\xi_u) \qquad (1.8)$$

where $Im\,k > 0$ and $\breve{G}_k$ is an integral operator acting from $L^2(I)$ to $L^2(\mathbb{R}^3)$, with integral kernel $G_k(x - y(s))$, $x \in \mathbb{R}^3$, $s \in I$ ($-\Delta_{\beta,C}$ is well–defined since for each $u \in L^2(\mathbb{R}^3)$ there exists at most one $\xi_u \in L^2(I)$ satisfying $u - \breve{G}_k\xi_u \in H^2(\mathbb{R}^3)$).

Since we know that $\Gamma_{\beta,C}(k)$ is invertible at least for $\mathrm{Re}\,k = 0$ and $Im\,k > \sqrt{\lambda_0(\beta, C)}$ then a direct application of the analytic Fredholm theorem ([12] page 201) gives us that $\Gamma_{\beta,C}$ has a bounded inverse in $L^2(I)$ for each $k \in \mathbb{C}\backslash\mathcal{E}$, where $\mathcal{E}$ is a discrete set and that the mapping $k \to \Gamma_{\beta,C}(k)^{-1}$ from $\mathbb{C}\backslash\mathcal{E}$ to $\mathcal{L}(L^2(I))$ is analytic. Moreover the points of $\mathcal{E}$ in the upper half complex plane can lie only on the imaginary axis, between 0 and $\sqrt{\lambda_0(\beta, C)}$, otherwise, reasoning as for iii) of proposition 2 of the next section, we could exhibit a complex eigenvalue of $-\Delta_{\beta,C}$.

Using the invertibility of $\Gamma_{\beta,C}(k)$ it is also possible to write the resolvent of $-\Delta_{\beta,C}$ (see proposition 5 in [3])

$$(-\Delta_{\beta,C} - k^2)^{-1} = G_k + \breve{G}_k \Gamma_{\beta,C}(k)^{-1} \hat{G}_k \,,$$
$$k^2 \in \rho(-\Delta_{\beta,C}) \,, \quad Imk > 0 \tag{1.9}$$

where $\hat{G}_k$ is an integral opearator acting from $L^2(\mathrm{I\!R}^3)$ to $L^2(I)$ with the integral kernel $G_k(x - y(s))$, $x \in \mathrm{I\!R}^3$, $s \in I$.

2. The spectrum

The description of the spectrum is based on the analysis of the resolvent. In particular the following lemma shows that the resolvent of $-\Delta_{\beta,C}$ is a small perturbation of the free resolvent.

Lemma

The bounded operator in $L^2(\mathrm{I\!R}^3)$ defined by

$$R_{\beta,C}^\lambda = (-\Delta_{\beta,C} + \lambda)^{-1} - (-\Delta + \lambda)^{-1} \tag{2.1}$$

is a trace class operator for $\lambda > \lambda_0(\beta, C)$.

Proof

The action of the operator $R_{\beta,C}^\lambda$ is explicitly given, for any $g \in L^2(\mathrm{I\!R}^3)$, by

$$(R_{\beta,C}^\lambda g)(x) = < G_{i\sqrt{\lambda}}(x - y(\cdot)), \ \Gamma_{\beta,C}(i\sqrt{\lambda})^{-1}(\hat{G}_{i\sqrt{\lambda}}g) > ,$$
$$x \in \mathrm{I\!R}^3 \tag{2.2}$$

where $< \cdot, \cdot >$ indicates the scalar product in $L^2(I)$.

The first step of the proof is to consider the approximating operator

$$[(f_\delta R_{\beta,C}^\lambda f_\delta)g](x)$$
$$= < f_\delta(x)G_{i\sqrt{\lambda}}(x - y(\cdot)), \Gamma_{\beta,C}(i\sqrt{\lambda})^{-1}[\hat{G}_{i\sqrt{\lambda}}(f_\delta g)] > \tag{2.3}$$

where f_δ is the characteristic function of $\{x \in \mathbb{R}^3 : \mathrm{dist}\,(x, C) \geq \delta\}$. Clearly one has

$$R^\lambda_{\beta,C} = s - \lim_{\delta \to 0} f_\delta R^\lambda_{\beta,C} f_\delta \tag{2.4}$$

An application of the Fubini theorem shows that $f_\delta R^\lambda_{\beta,C} f_\delta$ is an integral operator in $L^2(\mathbb{R}^3)$ with kernel explicitly given by

$$(f_\delta R^\lambda_{\beta,C} f_\delta)(x, x') \tag{2.5}$$
$$= <f_\delta(x')G_{i\sqrt{\lambda}}(x' - y(\cdot)),\ \Gamma_{\beta,C}(i\sqrt{\lambda})^{-1}(f_\delta(x)G_{i\sqrt{\lambda}}(x - y(\cdot))>$$

Using the explicit expression (2.5) it is easy to check continuity of the kernel in $\mathbb{R}^3 \times \mathbb{R}^3$ and positivity of the corresponding operator; furthermore we have the following uniform bound for the trace

$$\int_{\mathbb{R}^3} dx <f_\delta(x)G_{i\sqrt{\lambda}}(x - y(\cdot)),\ \Gamma_{\beta,C}(i\sqrt{\lambda})^{-1}(f_\delta(x)G_{i\sqrt{\lambda}}(x - y(\cdot)))>$$
$$\leq \frac{L}{8\pi\sqrt{\lambda}}\left\|\Gamma_{\beta,C}(i\sqrt{\lambda})^{-1}\right\| \tag{2.6}$$

Applying the lemma in [13], page 65, we conclude that $f_\delta R^\lambda_{\beta,C} f_\delta$ is a trace class operator.

By (2.4), (2.6) and [14], exercise 167, $R^\lambda_{\beta,c}$ is trace class.   ■

We are now in a position to characterize the continuous spectrum of $-\Delta_{\beta,C}$

Proposition 1

$$\sigma_{ess}(-\Delta_{\beta,C}) = \sigma_{ac}(-\Delta_{\beta,C}) = [0, +\infty) \tag{2.7}$$
$$\sigma_{sing}(-\Delta_{\beta,C}) = \emptyset \tag{2.8}$$

Proof

Using the above lemma $\sigma_{ess}(-\Delta_{\beta,C}) = [0, \infty)$ is a direct consequence of Weyl's theorem ([14] page 112). Using the explicit expression (1.9) for the resolvent and the continuity of the mapping $k \mapsto \Gamma_{\beta,C}(k)^{-1}$ from $\mathbb{C}\backslash\mathcal{E}$ to $\mathcal{L}(L^2(I))$ where $\mathcal{E}$ is a discrete set one can easily show that the assertions $\sigma_{sing}(-\Delta_{\beta,C}) = \emptyset$ and $\sigma_{ac}(-\Delta_{\beta,C}) = [0, \infty)$ follow from the limiting absorption principle ([14], theorem XIII.20).   ■

Now we turn to the analysis of the point spectrum. The result, summarized in the proposition below, has a strict analogy with the corresponding characterization of eigenvalues and eigenfunctions of the Schrödinger operator with point interactions (see [4] page 116).

Proposition 2

(i) $\sigma_p(-\Delta_{\beta,C}) \subset (-\infty, 0)$ $\qquad\qquad$ (2.9)

(ii) $-\Delta_{\beta,C}$ has a finite number of eigenvalues counting multiplicity and for $Imk > 0$ it results

$$k^2 \in \sigma_p(-\Delta_{\beta,C}) \ \text{ iff zero is an eigenvalue of } \Gamma_{\beta,C}(k) \quad (2.10)$$

and the multiplicity of the eigenvalue k^2 equals the multiplicity of the eigenvalue zero of $\Gamma_{\beta,C}(k)$.

(iii) If $E_0 = k_0^2 < 0$ is an eigenvalue of $-\Delta_{\beta,C}$ then the corresponding eigenfunctions Ψ_0 have the form

$$\Psi_0 = G_{k_0}\xi_0 , \quad Imk_0 > 0 \qquad\qquad (2.11)$$

where ξ_0 is an eigenfunction of $\Gamma_{\beta,C}(k_0)$ corresponding to the eigenvalue zero.

(iv) If $-\Delta_{\beta,C}$ has a ground state it is degenerate and the corresponding eigenfunctions can be chosen strictly positive.

Proof

Assertion i) means that $-\Delta_{\beta,C}$ cannot have positive eigenvalues and the proof is essentially based on a unique continuation theorem for Schrödinger operators (see e.g. the proof of theorem XIII.57 of [14]). Suppose there exists $E > 0$ and $\Psi \in D(-\Delta_{\beta,C})$ such that $-\Delta_{\beta,C}\Psi = E\Psi$. Enclose C in a large ball $B_R = \{x \in \mathbb{R}^3 \mid |x| < R\}$, $R > 0$. Then in $\mathbb{R}^3 \backslash B_R$ the eigenvalue equation reads $-\Delta\Psi = E\Psi$.

Expanding Ψ in spherical harmonics and reasoning as in [14] page 225 one obtains $\Psi = 0$ in $\mathbb{R}^3 \backslash B_R$.

Now fix $x_0 \in \mathbb{R}^3 \backslash B_R$, $y \in B_R \backslash C$ and a smooth curve γ of length L_γ joining x_0 and y and such that $\text{dist}(\gamma, C) > 0$.

Choose an integer n and R_0, $0 < R_0 < \frac{dist\,(\gamma,C)}{3}$, so that $\frac{R_0}{2}n \geq L_\gamma > \frac{R_0}{2}(n-1)$ and $x_1, \ldots, x_{n-1}, x_n = y$ on γ such that the length of γ between x_i and x_{i-1} $(i = 1, \ldots, n)$ is smaller than or equal to $\frac{R_0}{2}$. Then, in particular, $|x_i - x_{i-1}| \leq \frac{R_0}{2}$.

If we cover γ with balls $B_{R_0}(x_i)$ of radius R_0 centered in x_i, $i = 0, 1, \ldots, n$, then the proof of assertion i) follows from the following fact (proved in [14] page 243): $\Psi = 0$ in a small neighborhood of x_i and $|\Delta \Psi| \leq E|\Psi|$ in $B_{R_0}(x_i)$ imply $\Psi = 0$ in $B_{R_0}(x_i)$.

We now prove assertions ii) and iii). Let $\xi_0 \in D(\Gamma_{\beta,C})$ be a solution of

$$\Gamma_{\beta,C}(k_0)\xi_0 = 0 \tag{2.12}$$

for some $E = k_0^2 < 0$, $Imk_0 > 0$. Then one has $\Psi_0 = \check{G}_{k_0}\xi_0 \in D(-\Delta_{\beta,C})$; moreover

$$\begin{aligned}
-\Delta_{\beta,C}\check{G}_{k_0}\xi_0 &= (-\Delta - k^2)(\check{G}_{k_0}\xi_0 - \check{G}_k\xi_0) + k^2\check{G}_k\xi_0 \\
&= E_0\check{G}_{k_0}\xi_0
\end{aligned} \tag{2.13}$$

where $k^2 \in \rho(-\Delta_{\beta,C})$, $Imk > 0$.

Hence Ψ_0 is an eigenfunction of $-\Delta_{\beta,C}$ corresponding to the eigenvalue E_0. On the other hand let $E_0 = k_0^2 < 0$, $Imk_0 > 0$, be an eigenvalue of $-\Delta_{\beta,C}$ and Ψ_0 a corresponding eigenfunction. Then there exists $\xi_0 \in D(\Gamma_{\beta,C})$ such that

$$w = \Psi_0 - \check{G}_k\xi_0 \in H^2(\mathbb{R}^3), \quad k^2 \in \rho(-\Delta_{\beta,C}), \, Imk > 0 \tag{2.14}$$

$$w(y(s)) = (\Gamma_{\beta,C}(k)\xi_0)(s), \quad s \in I \tag{2.15}$$

Using (1.8) and the eigenvalue equation we obtain

$$(-\Delta - k^2)w = (k_0^2 - k^2)\Psi_0 \tag{2.16}$$

Solving (2.16) for w and then applying $-\Delta - k_0^2$ we get

$$(-\Delta - k_0^2)w = (k_0^2 - k^2)\breve{G}_k\xi_0 \, . \qquad (2.17)$$

From the above equation and (2.14) we find that

$$\Psi_0 = \breve{G}_{k_0}\xi_0 \, . \qquad (2.18)$$

Moreover using (2.15) and (2.18) we finally get

$$\Gamma_{\beta,C}(k_0)\xi_0 = 0 \qquad (2.19)$$

Assertion ii) and iii) are then proved if we observe that, by the Fredholm theorem, there exists only a finite multiplicity of solutions of $\Gamma_{\beta,C}(k)\xi = 0$ for each $k \in \mathcal{E}$ and $\mathcal{E} \cap \{k \in \mathbb{C} \mid \mathrm{Re}\, k = 0,\, 0 < Imk \leq \sqrt{\lambda_0(\beta,C)}\}$ is finite.

Concerning the properties of the ground state it is easily verified that the unit contraction operates on $\Phi_{\beta,C}^\lambda$, $\lambda > \lambda_0(\beta,C)$, (for the definition of unit contraction see [15] page 5), and so $\Gamma_{\beta,C}(i\sqrt{\lambda})^{-1}$ is positivity preserving for $\lambda > \lambda_0(\beta,C)$ (see [15], theorem 1.4.1).

Hence we have that the resolvent of $-\Delta_{\beta,C}$ is positivity improving for $\lambda > \lambda_0(\beta,C) = -inf\, \sigma(-\Delta_{\beta,C})$ and this in particular implies assertion iv) ([14], page 204). ∎

We end this section with the definition of resonances of $-\Delta_{\beta,C}$ (see the corresponding definition in the case of point interactions [4], page 119). $k_0 \in \mathbb{C}$, $Imk_0 \leq 0$ is a resonance of $-\Delta_{\beta,C}$ iff there exists a non–trivial solution of the equation

$$\Gamma_{\beta,C}(k_0)\xi = 0\,, \quad \xi \in D(\Gamma_{\beta,C}) \qquad (2.20)$$

The multiplicity of the resonance k_0 equals the multiplicity of the solution ξ of (2.20).

3. Stationary scattering theory

The aim of this section is to describe the stationary scattering theory for the pair $(-\Delta_{\beta,C}, -\Delta)$.

First we calculate the generalized eigenfunctions

$$
\Psi_{\beta,C}(k\omega, x) = \lim_{\varepsilon \to 0} \lim_{\substack{|x'| \to \infty \\ \omega = -\frac{x'}{|x'|}}} 4\pi |x'| e^{-i(k+i\varepsilon)|x'|} [-\Delta_{\beta,C} - (k+i\varepsilon)^2]^{-1}(x, x')
$$

$$
= e^{ik\omega x} + \int_I G_k(x - y(s))(\Gamma_{\beta,C}(k)^{-1} e^{ik\omega y(\cdot)})(s)\,ds \qquad (3.1)
$$

$k \notin \mathcal{E}$, $k \geq 0$, $x \in \mathbb{R}^3$, $\omega \in S^2$ and S^2 is the unit sphere in $\mathbb{R}^3$. It is easily seen that $\Psi_{\beta,C}(k\omega, x)$ satisfies

$$
-\Delta_{\beta,C}\Psi_{\beta,C}(k\omega, \cdot) = k^2 \Psi_{\beta,C}(k\omega, \cdot) \qquad (3.2)
$$

in the distributional sense.

Using the generalized eigenfunctions we can compute all relevant scattering quantities. The on–shell scattering amplitude is given by

$$
f_{\beta,C}(k, \omega, \omega') = \lim_{\substack{|x| \to \infty \\ \omega = \frac{x}{|x|}}} |x| e^{-ik|x|} [\Psi_{\beta,C}(k\omega', x) - e^{ik\omega' x}]
$$

$$
= \frac{1}{4\pi} \int_I e^{-ik\omega y(s)} (\Gamma_{\beta,C}(k)^{-1} e^{ik\omega' y(\cdot)})(s)\,ds \qquad (3.3)
$$

$k \notin \mathcal{E}$, $k \geq 0$, $\omega, \omega' \in S^2$.

The unitary on–shell scattering operator $S_{\beta,C}(k)$ in $L^2(S^2)$ has integral kernel

$$
S_{\beta,C}(k)(\omega, \omega')
$$

$$
= \delta(\omega - \omega') - \frac{k}{8\pi^2 i} \int_I e^{-ik\omega y(s)} (\Gamma_{\beta,C}(k)^{-1} e^{ik\omega' y(\cdot)})(s)\,ds \qquad (3.4)
$$

$k \notin \mathcal{E}$, $k \geq 0$, $\omega, \omega' \in S^2$.

Finally the low–energy limits of the scattering amplitude and the scattering operator are respectively given by

$$n - \lim_{k \to 0} S_{\beta,C}(k) = 1 \tag{3.5}$$

$$- \lim_{k \to 0} f_{\beta,C}(k,\omega,\omega') = -\frac{1}{4\pi} \int_I (\Gamma_{\beta,C}(0)^{-1}1)(s)ds \tag{3.6}$$

for $0 \notin \mathcal{E}$ and $1(\cdot)$ the unit function on I.

The limit (3.6) defines the scattering length $a_{\beta,C}$ associated to $-\Delta_{\beta,C}$.

We remark that the time–dependent scattering theory for $(-\Delta_{\beta,C}, -\Delta)$ can also be developed following the line of appendix E of [4]. In particular by the invariance principle ([13] page 27) and the trace class property of $R_{\beta,C}^\lambda$ we immediately get existence and asymptotic completeness of wave operators together with existence and unitarity of the associated scattering operator.

Moreover, using the eigenfunction expansion of $-\Delta_{\beta,C}$, the usual correspondence betweeen time–dependent and time–independent scattering theory can also be established.

4. An example

Here we briefly discuss the special case in which C is the circle given by $y = y(s) \equiv (r \cos s,\, r \sin s,\, 0)$, $s \in [-\pi, \pi]$, and β is a constant. This case is interesting because it is an example of solvable model, in the sense that all the relevant quantities (e.g. eigenvalues, resonances, scattering amplitude etc.) can be explicitly computed and additional informations on the properties of the operator can be obtained.

Using the expansion in Fourier series the quadratic form $\Phi_{\beta,C}^\lambda$ can be written as

$$\Phi_{\beta,C}^\lambda(\xi,\xi) = \sum_{m=-\infty}^{\infty} |\xi_m|^2 \left[r^2 \int_{-\pi}^{\pi} (1 - \cos 2ms)\frac{e^{-2\sqrt{\lambda}r \sin s}}{4\pi r \sin s}ds+ \right.$$

$$\left. +r\beta + \frac{r}{2\pi}\log\frac{1}{2\pi r} + \frac{r}{2\pi}\int_0^{\pi/2}\left(\frac{1}{s} - \frac{e^{-2\sqrt{\lambda}r \sin s}}{\sin s}\right)ds \right] \tag{4.1}$$

where $\xi_m = \frac{1}{\sqrt{2\pi}} \int_{-\pi}^{\pi} \xi(s) e^{-ims} ds$.

From (4.1) it results that in the space of the (discrete) Fourier transforms the operator $\Gamma_{\beta,C}(k)$ is diagonalized so it can be characterized by

$$D(\Gamma_{\beta,C}(k)) = \{\xi \in L^2(I) \mid \Gamma_{\beta,C}(k)_m \xi_m \in l^2\} \quad (4.2)$$

$$(\Gamma_{\beta,C}(k)\xi)(\varphi) = \frac{1}{\sqrt{2\pi}} \sum_{m=-\infty}^{\infty} \Gamma_{\beta,C}(k)_m \xi_m e^{im\varphi} \quad (4.3)$$

where

$$\Gamma_{\beta,C}(k)_m = r\beta + \frac{r}{2\pi}\left[\log\frac{1}{2\pi r} + \int_0^{\pi/2}\left(\frac{1}{s} - \frac{e^{2ikr\sin s}}{\sin s}\cos 2ms\right)ds\right] \quad (4.4)$$

Using the above expression we can also write the resolvent as

$$(g,(-\Delta_{\beta,C} - k^2)^{-1}f) = (g, G_k f) + r^2 \sum_{m=-\infty}^{\infty} \frac{(G_k g)_m^*(G_k f)_m}{\Gamma_{\beta,C}(k)_m},$$

$$k^2 \in \rho(-\Delta_{\beta,C}),\ Imk > 0 \quad (4.5)$$

where $f, g \in L^2(\mathbb{R}^3)$ and $(G_k h)_m = \frac{1}{\sqrt{2\pi}}\int_{-\pi}^{\pi}(G_k h)(y(s))e^{-ims}ds$, $\forall h \in L^2(\mathbb{R}^3)$

For small values of the radius only the term $m = 0$ is relevant in the series in (4.5); moreover

$$\Gamma_{\beta,C}(k)_0 \underset{r\to 0}{\sim} r\beta + \frac{r}{2\pi}\log\frac{1}{8r} - \frac{ik}{2}r^2 \quad (4.6)$$

This means that in the limit $r \to 0$ we have resolvent convergence to point interaction of strength $\alpha \in \mathbb{R}$ placed at the origin iff the parameter β is explicitly dependent on r according to the formula

$$\beta(r) = -\frac{1}{2\pi}\log\frac{1}{8r} + 2\pi r\alpha \quad (4.7)$$

As we will see in a moment the divergent term in (4.7) is determined by imposing that $-\Delta_{\beta,C}$ has a zero–energy resonance without bound states while the coefficient of α is simply the measure of the shrinking circle.

We remark that exactly the same situation occurs if one tries to obtain a point interaction as the limit of an interaction supported by a shrinking sphere ([16]).

Another question which can be easily investigated is the occurence of bound states.

From Proposition 2.2 and symmetry properties of $-\Delta_{\beta,C}$ we have that $E_0 < 0$ is the ground state energy iff it solves the equation

$$\beta + \frac{1}{2\pi} \log \frac{1}{8r} + \frac{1}{2\pi} \int_0^{\pi/2} \left(1 - e^{-2\sqrt{-E_0}\,r \sin s}\right) \frac{ds}{\sin s} = 0 \quad (4.8)$$

Now the integral in (4.8) is a function $F(x)$, where $x = 2\sqrt{-E_0}\,r$, satisfying $F(0) = 0$; $\lim_{x\to\infty} F(x) = +\infty$ and $F'(x) > 0$, $\forall x \geq 0$. So we conclude that $-\Delta_{\beta,C}$ has at least one bound state iff

$$\beta + \frac{1}{2\pi} \log \frac{1}{8r} < 0 \quad (4.9)$$

(while equality in (4.9) leads to the occurence of a zero–energy resonance) In particular the condition (4.9) implies for $\beta \geq 0$ (i.e. "repulsive" interaction) that we have at least one bound state for a sufficiently large r while for $\beta < 0$ (i.e. "attractive" interaction) we have no bound state for a sufficiently small r.

We conclude considering the scattering amplitude which, as usual, can be explicitly computed using the Fourier expansion

$$f_{\beta,C}(k,\omega,\omega') = \frac{r^2}{4\pi} \sum_{m=-\infty}^{\infty} \frac{(e_k)^*_m (e_{k'})_m}{\Gamma_{\beta,C}(k)_m} \quad (4.10)$$

where $(e_k)_m = \frac{1}{\sqrt{2\pi}} \int_{-\pi}^{\pi} e^{ik\omega y(s)-ims} ds$

From (4.10) we have that the scattering length (3.6) is given by

$$a_{\beta,C} = -\frac{r}{2} \frac{1}{\beta + \frac{1}{2\pi} \log \frac{1}{8r}} \quad (4.11)$$

Formula (4.11) shows that for $r \to 0$ point interaction is obtained keeping the scattering length constant.

Finally it is an easy exercise to show that the scattering amplitude has a particularly simple form in the special case of an incident plane wave propagating along the $z-$axis (i.e. $\omega = (0,0)$, $\omega' = (\vartheta', \varphi')$)

$$f_{\beta,C}(k, \vartheta') = \frac{r^2}{2\Gamma_{\beta,C}(k)_0} J_0(kr \sin \vartheta') \qquad (4.12)$$

where $J_0(\cdot)$ is the Bessel function of zero order ([17]).

ACKNOWLEDGEMENTS

The authors would like to thank Professors S. Albeverio, G.F. Dell' Antonio and R. Figari for helpful discussions and suggestions and F. Nitzschner and Dr. P. Šeba for hints to the literature. We are also grateful to Professor S. Albeverio for the kind hospitality attend to one of us (A.T.) at Institut für Mathematik, Ruhr–Universität, Bochum (FRG).

REFERENCES

[1] A. Grossman, T.T. Wu: "A class of potentials with extremely narrow resonances", Chin. J. Phys. **25** (1987), 129–139

[2] Y.V. Kurylev: "Boundary conditions on a curve for a three–dimensional Laplace operator", J. Soviet Math. **22** (1983), no 1

[3] A. Teta: "Quadratic forms for singular perturbations of the Laplacian", Preprint S.I.S.S.A. 165 FM (dec. 1988), to appear in Publ. of the R.I.M.S., Kyoto Univ.

[4] S. Albeverio, F. Gesztesy, R. Høegh–Krohn, H. Holden: "Solvable models in Quantum Mechanics", Springer–Verlag, 1988

[5] S. Albeverio, J.E. Fenstad, R. Høegh–Krohn, W. Karwowski, T. Lindström: "Perturbations of the Laplacian supported by null sets, with applications to polymer measures and quantum fields", Phys. Lett. 104 A (1984), 396–400

[6] S. Albeverio, J.E. Fenstad, R. Høegh–Krohn, T. Lindström: "Non-standard Methods in Stochastic Analysis and Mathematical Physics", Academic Press, 1986

[7] S. Albeverio, R. Høegh–Krohn, L. Streit: "Energy forms Hamiltonians and distorted Brownian paths", J. Math. Phys. **18** (1977), 907–917

[8] P. Exner, P. Šeba (Eds.): "Applications of self–adjoint Extensions in Quantum Physics". Lecture Notes in Physics 324, Springer Verlag, 1989

[9] P. Exner, P. Šeba (Eds.): "Schrödinger Operators – Standard and Non–standard", World Scientific Publishing Company, 1989

[10] J.F. Brasche: "Perturbations of self–adjoint operators supported by null sets", doctoral thesis, Bielefeld, 1988

[11] J.F. Brasche: "Generalized Schrödinger Operators, an inverse problem in spectral analysis and the Efimov effect", to apppear in Proceedings of the international conference in Ascona–Locarno on "Stochastic Processes, Physics and Geometry", eds. S. Albeverio, G. Casati, U. Cattaneo, D. Merlini, R. Moresi, World Scientific Publishing Company, 1989

[12] M. Reed, B. Simon; "Methods of Modern Mathematical Physics. I. Functional Analysis", Academic Press, 1972

[13] M. Reed, B. Simon: "Methods of Modern Mathematical Physics. III Scattering Theory", Academic Press, 1979

[14] M. Reed, B. Simon: "Methods Of Modern Mathematical Physics. IV Analysis of Operators", Academic Press, 1978

[15] M. Fukushima: "Dirichlet forms and Markov Processes", North–Holland/ Kodansha, 1980

[16] J.P. Antoine, F. Gesztesy, J. Shabani: "Exactly solvable models of sphere interaction in Quantum Mechanics", J. Phys. **A20** (1987) 3687–3712

[17] M. Abramowitz, I.A. Stegun: "Handbook of Mathematical functions", Dover Publications, 1972

Lattice Models of Solids

V.V. EVSTRATOV, B.S. PAVLOV

Institute for Physics, Leningrad State University, 198904 Leningrad,
USSR

ABSTRACT
Starting from the solvable one–body Hamiltonian, the solvable many–body
Hamiltonian is constructed by means of the extension theory. This Hamiltonian is available for spectral analysis in explicit form, and appears to have a
wave–guide part, which corresponds to bounded states of the particles. The
wave–guide spectrum can have gaps, which may be interpreted as the energetic gaps in the theory of Superconductivity. Some sort of virial theorem is
proved.

INTRODUCTION
If the quantized Hamiltonian of a solid has some invariant subspaces, "sectors" with fixed number of particles in the Fock space, then the whole
quantum–field problem can be reduced to a series of many–body quantum–
mechanical problems. One can also use an opposite approach: starting from
the one–body Hamiltonians H_1, H_2 construct the two–body Hamiltonian (in
two–body sectors) as a perturbation of the tensor sum:

$$H_{1,2} = H_1 \otimes I_2 + I_1 \otimes H_2 \longrightarrow H_1 \otimes I_2 + I_1 \otimes H_2 + V_{12} \ .$$

The simplest way to construct the interaction is given by the extension theory. We discuss here only one step in this general program. This step leads
from the one–body lattice model to the wave–guide part of the two–body
Hamiltonian, which is also represented by the one–body quasiparticle lattice
Hamiltonian. In order to construct quasiparticles in many–body sectors this
procedure can be repeated many times.

We demonstrate here some interesting properties of the constructed lattice
models. All of these models are available for the spectral analysis in explicit form. We construct a spectral decomposition of the one–body lattice
model in the quasimomentum representation. We prove also some sort of

212

virial theorem. An analogous fact has been found for the one–dimensional periodic Schrödinger equation in [4,5]. The corresponding result for the multidimensional situation was unknown even formally. We have established the existence of a series of formulae which connect the effective masses in the extrema of the bands with the global characteristics in the interaction and inner Hamiltonian.

1. ONE–BODY LATTICE HAMILTONIAN.

Let $\mathbb{Z}^\nu$ be a ν–dimensional cubic lattice. We associate the infinite number of equivalent auxillary spaces $E_l = E_0 = E$ with the nodes of the lattice, $l \in \mathbb{Z}^\nu$, and consider the lattice sum $\mathscr{E} = \oplus \sum_{l \in \mathbb{Z}^\nu} E_l$. In every E_l we consider an operator $A^{(l)}$, supposing that $A^{(l)}$ is unitary equivalent to $A^{(o)}$, $A^{(o)} = A$. We take the background one–body lattice Hamiltonian in the form $\mathscr{A} = \oplus \sum_{l \in \mathbb{Z}^\nu} A^{(l)}$. The interaction will be switched on by means of the extension theory. Namely, let $N_l = N_0 = N$ be some finite dimensional generating subspace of the operator A, and N_l are the copies of N in other spaces E_l. First of all we restrict $A^{(l)}$ onto $D_l = \{(A^{(l)} - iI)^{-1}(E_l \ominus N_l)\}$ and set $\mathscr{A}_0 = \mathscr{A}|_{D^o}$, where $D^o \equiv \oplus \sum_{l \in \mathbb{Z}^\nu} D_l$. The deficiency subspace of the operator A_0 is an orthogonal sum $\mathscr{N} = \oplus \sum_{l \in \mathbb{Z}^\nu} N_l$ and thus it is equivalent to $l^2(\mathbb{Z}^\nu, N)$. The elements from $\mathscr{N}$ we denote by ξ, $\xi = \oplus \sum_{l \in \mathbb{Z}^\nu} \xi_l$, $\xi_l \in N_l$.

Let Γ be some bounded selfadjoint operator in $\mathscr{N}$. Then the formula

$$\mathscr{A}_\Gamma u = \mathscr{A}\left(\mathscr{A} - iI\right)^{-1}\left(v + \Gamma\xi\right) - \left(\mathscr{A} - iI\right)^{-1}\xi$$

defines a selfadjoint operator $\mathscr{A}_\Gamma$ with the domain D_Γ in $\mathscr{E}$:

$$D_\Gamma = \{u : u = (\mathscr{A} - iI)^{-1}v + \mathscr{A}(\mathscr{A} - iI)^{-1}\xi + (\mathscr{A} - iI)^{-1}\Gamma\xi, v \perp \mathscr{N}\}$$

If we choose Γ in the form of the simplest (constant) Jacobian matrix with matrix elements

$$(\Gamma\xi)_l = \sum_{|l-s| \leq R} \gamma_{l-s}\xi_s, \quad \gamma_{-s} = \gamma_s^*$$

$$\gamma_s : N_l \longrightarrow N_{l-s}, \forall l$$

then we get the one–body Hamiltonian in a homogeneous lattice. The operator parameters γ_s play the role of the overlapping integrals in a one–body sector of the corresponding Hubbard model. Let us insert the following notations: $\mathscr{P}_0$ is projector on the eigenspace corresponding to eigenvalue $\lambda_0 \in \sigma(\mathscr{A})$, $\mathscr{P}_0^\perp = I \ominus \mathscr{P}_0$ and $\mathscr{P}_\mathscr{N}$ is projector on the deficiency subspace $\mathscr{N}$. Moreover, let $\mathscr{N}_0^\perp$ be the maximal subspace of $\mathscr{N}$ such that $\mathscr{P}_0\mathscr{N}_0^\perp = 0$, and $\mathscr{N}_0$ be $\mathscr{N}\ominus\mathscr{N}_0^\perp$; P and $P^\perp$ are projectors on $\mathscr{N}_0$ and $\mathscr{N}_0^\perp$: $P\mathscr{N} = \mathscr{N}_0, P^\perp\mathscr{N} = \mathscr{N}_0^\perp$.

<u>Theorem 1.</u> If the spectrum of the Hamiltonian A is discrete, then the spectrum $\sigma(\mathscr{A}_\Gamma)$ coincides with the closure of the union of families

$$\left\{ \lambda : \operatorname{Ker}\left[\Gamma - \mathscr{P}_{\mathscr{N}}\frac{I + \lambda\mathscr{A}}{\mathscr{A} - \lambda I}\mathscr{P}_{\mathscr{N}}\right] \neq 0 \right\} \bigcup$$

$$\left\{ \lambda_0 : \lambda_0 \in \sigma(\mathscr{A}), \ \operatorname{Ker} P^\perp\left[\Gamma - \frac{I + \lambda_0\mathscr{A}}{\mathscr{A} - \lambda_0 I}\mathscr{P}_0^\perp\right]P^\perp \neq 0 \right\}$$

In the case dim $N = 1$ the second family in $\sigma(\mathscr{A}_\Gamma)$ is empty.

Proof. We start from the Krein's formula for the resolvents

$$(A_\Gamma - \lambda I)^{-1} = (A - \lambda I)^{-1} + \frac{\mathscr{A} + iI}{\mathscr{A} - \lambda I}\left[\Gamma - \mathscr{P}_{\mathscr{N}}\frac{I + \lambda\mathscr{A}}{\mathscr{A} - \lambda I}\mathscr{P}_{\mathscr{N}}\right]^{-1}\mathscr{P}_{\mathscr{N}}\frac{\mathscr{A} - iI}{\mathscr{A} - \lambda I}$$

It is clear that $\sigma(\mathscr{A}_\Gamma) \subset \sigma(\mathscr{A}) \cup \left\{\lambda : \operatorname{Ker}\left[\Gamma - \mathscr{P}_{\mathscr{N}}\frac{I+\lambda\mathscr{A}}{\mathscr{A}-\lambda I}\mathscr{P}_{\mathscr{N}}\right] \neq 0\right\}$.
Let $\lambda_0, \lambda_0 \in \sigma(\mathscr{A})$ be some eigenvalue of the operator $\mathscr{A}_\Gamma$. To investigate resolvent $(\mathscr{A}_\Gamma - \lambda I)^{-1}$ one should consider the following inhomogeneous equation:

$$\left[\Gamma - \mathscr{P}_{\mathscr{N}}\frac{I + \lambda\mathscr{A}}{\mathscr{A} - \lambda I}\mathscr{P}_{\mathscr{N}}\right]u = \mathscr{P}_{\mathscr{N}}\frac{\mathscr{A} - iI}{\mathscr{A} - \lambda I}f, \quad u \in \mathscr{N}$$

Using the equality $\frac{I}{\mathscr{A}-\lambda I} = \frac{\mathscr{R}}{\lambda - \lambda_0} + \frac{I}{\mathscr{A}-\lambda I}\mathscr{P}_0^\perp$ one can obtain

$$\left[\Gamma + \mathscr{P}_{\mathscr{N}}\mathscr{A}\mathscr{P}_{\mathscr{N}} - \mathscr{P}_{\mathscr{N}}\frac{I + \mathscr{A}^2}{\mathscr{A} - \lambda I}\mathscr{P}_0^\perp\mathscr{P}_{\mathscr{N}}\right]u - \mathscr{P}_{\mathscr{N}}\mathscr{P}_0\mathscr{P}_{\mathscr{N}}\frac{(1 + \lambda_0^2)}{\lambda_0 - \lambda}u =$$
$$= \mathscr{P}_{\mathscr{N}}\mathscr{P}_0\frac{\lambda_0 - i}{\lambda_0 - \lambda}f + \mathscr{P}_{\mathscr{N}}\frac{\mathscr{A} - iI}{\mathscr{A} - \lambda I}\mathscr{P}_0^\perp f \tag{1}$$

The solution $u = u(\lambda)$ is an analytic function on the complement of $\sigma(\mathscr{A}_\Gamma) \cup \sigma(\mathscr{A})$. It can have poles in the points of spectrum. We search for the solution in the form $u(\lambda) = \sum\limits_{n=-1}^{\infty}(\lambda - \lambda_0)^n u_n$. To find the coefficients $u_n = v_n + w_n$, $v_n = Pu_n$, $w_n = P^\perp u_n$ one should solve the equation (1) in n-th order of perturbation theory. In the leading order we have $v_{-1} = 0$. The two next orders give two systems:

$$\begin{cases} P\mathscr{R}w_{-1} - \mathscr{P}_{\mathscr{N}}\mathscr{P}_0(1 + \lambda_0^2)v_0 = \mathscr{P}_{\mathscr{N}}\mathscr{P}_0(\lambda_0 - i)f \\ P^\perp\mathscr{R}w_{-1} = 0 \end{cases}$$

and

$$\begin{cases} P\mathscr{R}v_0 + P\mathscr{R}w_0 - P\mathscr{P}_{\mathscr{N}}\dfrac{I + \mathscr{A}^2}{(\mathscr{A} - \lambda_0)^2}\mathscr{P}_0^\perp\mathscr{P}_{\mathscr{N}}w_{-1} - \\ \qquad - \mathscr{P}_{\mathscr{N}}\mathscr{P}_0\mathscr{P}_{\mathscr{N}}(1 + \lambda_0^2)v_1 = f_v(\lambda_0) \\ P^\perp\mathscr{R}v_0 + P^\perp\mathscr{R}w_0 - P^\perp\mathscr{P}_{\mathscr{N}}\dfrac{I + \mathscr{A}^2}{(\mathscr{A} - \lambda_0)^2}\mathscr{P}_0^\perp\mathscr{P}_{\mathscr{N}}w_{-1} = f_w(\lambda_0) \end{cases}$$

where the following notations are used

$$f_v(\lambda) = P\mathscr{P}_N \frac{\mathscr{A} - iI}{\mathscr{A} - \lambda I}\mathscr{P}_0^\perp f$$

$$f_w(\lambda) = P^\perp \mathscr{P}_N \frac{\mathscr{A} - iI}{\mathscr{A} - \lambda I}\mathscr{P}_0^\perp f$$

$$\mathscr{R} = \left[\Gamma + \mathscr{P}_N\mathscr{A}\mathscr{P}_N - \mathscr{P}_N\frac{I + \mathscr{A}^2}{\mathscr{A} - \lambda I}\mathscr{P}_0^\perp\mathscr{P}_N\right]$$

If the operator $P^\perp\mathscr{R}P^\perp$ is invertible, then $w_{-1} = 0$ and one can find v_0 from the first equation of the first system, and w_0 from the second system. The expression for v_0 has the following form:

$$v_0 = -\frac{\lambda_0 - i}{1 + \lambda_0^2}(\mathscr{P}_N\mathscr{P}_0\mathscr{P}_N)^{-1}\mathscr{P}_N\mathscr{P}_0 f$$

The Krein's formula can be written in the form:

$$(\mathscr{A}_\Gamma - \lambda I)^{-1}f = (\mathscr{A} - \lambda I)^{-1}f + \frac{\mathscr{A} + iI}{\mathscr{A} - \lambda I}u(\lambda)$$

The r.h.s. of this equation has the following residue in the point λ_0:

$$\mathscr{P}_0 f - \mathscr{P}_0\mathscr{P}_N(\mathscr{P}_N\mathscr{P}_0\mathscr{P}_N)^{-1}\mathscr{P}_N\mathscr{P}_0 f$$

The operator $\mathscr{P} \equiv \mathscr{P}_0\mathscr{P}_N(\mathscr{P}_N\mathscr{P}_0\mathscr{P}_N)^{-1}\mathscr{P}_N\mathscr{P}_0$ is selfadjoint and $\mathscr{P}^2 = \mathscr{P}$. Moreover: $\mathscr{P}\mathscr{P}_0\mathscr{E} = \mathscr{P}_0\mathscr{E}$. Therefore $\mathscr{P} = \mathscr{P}_0$ and the residue is equal to zero. It means that $\lambda_0 \notin \sigma(\mathscr{A}_\Gamma)$.

If $\mathrm{Ker}\, P^\perp\mathscr{R}P^\perp \neq \{0\}$, then $w_{-1} \neq 0$. In this case the residue is equal to:

$$\left[(\lambda_0 - i)\mathscr{P}_0(\mathscr{P}_N\mathscr{P}_0\mathscr{P}_N)^{-1}P\mathscr{R}P^\perp - \frac{\mathscr{A} + iI}{\mathscr{A} - \lambda_0 I}\right]w_{-1}$$

and, therefore $\lambda_0 \in \sigma(\mathscr{A}_\Gamma)$. The theorem is proved.

The inverse operator $\left[\Gamma - \mathscr{P}_N\frac{I + \lambda\mathscr{A}}{\mathscr{A} - \lambda I}\mathscr{P}_N\right]^{-1}$ can be calculated using the Fourier–representation $\xi \longrightarrow \xi(p) = \sum \xi_l \exp(ipl)$. This representation turns the operator $\left[\Gamma - \mathscr{P}_N\frac{I + \lambda\mathscr{A}}{\mathscr{A} - \lambda I}\mathscr{P}_N\right]$ into the matrix $Q(\lambda, p) = \Gamma(p) - P_N\frac{I + \lambda\mathscr{A}}{\mathscr{A} - \lambda I}P_N$, $P_N \equiv \Gamma(p) - Q(\lambda)$. Thus the spectrum σ_Γ is determined by the dispersion equation:

$$\det Q(\lambda, p) = 0 , \quad p \in \Pi^\nu \tag{2}$$

Let $\lambda_r(p)$ be the r–th branch of the solution of the dispersion equation (2), and K_p^r be the $\mathrm{Ker}\, Q(\lambda_r(p), p)$. The matrix $Q(\lambda, p)$ can be represented in the form

$$Q(\lambda, p) = Q_0(\lambda, p)\prod_r \left\{(\lambda - \lambda_r(p))\, P_p^r + I - P_p^r\right\} \tag{3}$$

Here P_p^r is orthogonal projector on K_p^r. The matrix $Q_0(\lambda, p)$ is invertible.

Theorem 2.   The eigenfunction corresponding to the energy $\lambda_r(p)$ is given by the following formula:

$$\chi_r(p) = \oplus \sum_{m \in \mathbb{Z}^\nu} e^{-imp} \frac{A + iI}{A - \lambda_r(p)I} \theta^m \left[P_p^r Q'(\lambda_r(p)) P_p^r \right]^{-\frac{1}{2}} \qquad (4)$$

Here $\theta_m \in N_m$ is the element of Ker $Q(\lambda_r(p), p)$. The following conditions of orthogonality and completeness are valid (we assume $\sigma(\mathscr{A}_\Gamma) = \sigma_\Gamma$).

$$\frac{1}{(2\pi)^\nu} \langle \chi_r(p), \chi_{r'}(p) \rangle_{\mathscr{E}} = \delta(p - p') \delta_{rr'} ,$$

$$\frac{1}{(2\pi)^\nu} \sum_r \int \chi_r(p) \langle \cdot, \chi_r(p) \rangle_{\mathscr{E}} d^\nu p = I_{\mathscr{E}} \qquad (5)$$

Proof   From the spectral decomposition theorem one gets by the method of the Riesz–integral:

$$I_{\mathscr{E}} = -\frac{1}{2\pi i} \oint (\mathscr{A}_\Gamma - \lambda I)^{-1} d\lambda = -\sum_r \frac{1}{2\pi i} \oint_{(\sigma_\Gamma^r)} (\mathscr{A}_\Gamma - \lambda I)^{-1} d\lambda =$$

$$= -\sum_r \frac{1}{2\pi i} \oint_{(\sigma_\Gamma^r)} \frac{\mathscr{A} + iI}{\mathscr{A} - \lambda I} \mathscr{P}_N F^{-1} \left[\Gamma(p) - Q(\lambda) \right]^{-1} F \mathscr{P}_N \frac{\mathscr{A} - iI}{\mathscr{A} - \lambda I} d\lambda .$$

Here integration is performed over the contour encircling the band σ_Γ^r; F, F^{-1} are the Fourier transformation and the inverse respectively, $F : \left\{ \oplus \sum_n \xi_n \right\}$ $\longrightarrow \sum_n \xi_n e^{inp}$. Using the representation (3) one can calculate the residue in the first order pole:

$$\text{res} \left[\Gamma(p) - Q(\lambda) \right]^{-1} |_{\lambda = \lambda_r(p)} = - \left[P_p^r Q'(\lambda_r(p)) P_p^r \right]^{-1}$$

Therefore:

$$\left[-\frac{1}{2\pi i} \oint (\mathscr{A}_\Gamma - \lambda I)^{-1} d\lambda \right]_{ml} = \sum_r \frac{1}{(2\pi)^\nu} \oint_{\Pi^\nu} e^{-ipm} \frac{A + iI}{A - \lambda_r(p)I} \times$$

$$\times \left[P_p^r Q'(\lambda_r(p)) P_p^r \right]^{-1} P_N \frac{A - iI}{A - \lambda_r(p)I} e^{ipl} d^\nu p .$$

One can see that the last equality gives the spectral decomposition over the eigenfunctions (4). The orthogonality condition follows from the corresponding Parseval equality.

Corollary. The spectral representation of the resolvent of operator A_Γ looks as follows:

$$G_{ml}(\lambda) = \sum_r \frac{1}{(2\pi)^\nu} \oint_{\Pi^\nu} e^{-ipm} \frac{A+iI}{A-\lambda_r(p)I} \frac{\left[P_p^r Q'(\lambda_r(p))P_p^r\right]^{-1}}{\lambda_r(p)-\lambda} \times$$
$$\times P_N \frac{A-iI}{A-\lambda_r(p)I} e^{ipl} d^\nu p \ .$$

The expression for resolvent poles in the point $\lambda_0 \in \sigma(\mathscr{A}_\Gamma)\backslash\sigma_\Gamma$ gives the following formula for corresponding eigenfunction:

$$\chi(\lambda_0) = \left[\mathscr{P}_0(\mathscr{P}_N\mathscr{P}_0\mathscr{P}_N)^{-1}P\mathscr{R}P^\perp - \frac{\mathscr{A}+iI}{\mathscr{A}-\lambda I}\right]w \ ,$$
$$w \in \mathrm{Ker}\ P^\perp\mathscr{R}P^\perp \ .$$

2. THE EFFECTIVE MASS AND THE VIRIAL THEOREM.

The classical mechanical virial theorem links usually kinetic and potential energy of systems. In our lattice models the potential energy is the Hamiltonian of the internal channel, and the kinetic energy is the boundary operator Γ, which is linked with the internal Hamiltonian by the dispersion equation (2). The essential properties of the dispersion function $\lambda(p)$ in the vicinity of critical points, where $\nabla_p\lambda$ vanishes, are given by the tensor of second derivatives

$$\nabla^2\lambda \equiv \nabla\nabla\lambda = \left\{\frac{\partial^2\lambda}{\partial p_i\partial p_k}\right\}_{i,k=1}^\nu \equiv M^{-1}(p)$$

which is called the inverse effective mass. If we fix the point $\lambda_0 = \lambda_r(p_0)$ and the vector $\Theta_0 = \Theta_r(p_0)$, which is the solution of the equation (2), then we have

$$< \nabla_p\Gamma\Theta_0, \Theta_0 > -\left\langle \frac{d}{d\lambda}\left(P_N\frac{I+\lambda A}{A-\lambda I}P_N\right)\Theta_0, \Theta_0 \right\rangle \nabla_p\lambda +$$
$$+\left\langle \left(\Gamma - P_N\frac{I+\lambda A}{A-\lambda I}P_N\right)\nabla\Theta_0, \Theta_0 \right\rangle = < \nabla_p\Gamma\Theta_0, \Theta_0 > -$$
$$-\left\langle P_N\frac{I+A^2}{(A-\lambda I)^2}P_N\Theta_0, \Theta_0 \right\rangle \nabla_p\lambda = 0$$

Hence:

$$\nabla_p\lambda|_{P_0} = \left\langle P_N\frac{I+A^2}{(A-\lambda I)^2}P_N\Theta_0, \Theta_0 \right\rangle^{-1} < \nabla_p\Gamma\Theta_0, \Theta_0 > \tag{6}$$

If $\dim N = 1$, then the critical points $\lambda(p)$ coincide with the critical points of $\Gamma(p)$ and do not depend on r.

Let us consider the case $\dim \mathrm{Ker}\,\{\Gamma(p) - Q(\lambda)\} = 1$. If we normalize the vector $\Theta_r(p), |\Theta_r(p)| = 1$, then $\nabla\Theta_r(p_0) \perp \Theta_r(p_0) = \mathrm{Ker}\,\{\Gamma(p) - Q(\lambda)\}$ and $\nabla_p\Gamma\Theta_0 - \frac{dQ}{d\lambda}\Theta_0\nabla_p\lambda \in N_\perp = N \ominus \mathrm{Ker}\,\{\Gamma(p_0) - Q(\lambda_0)\}$. Therefore:

$$\nabla_p\Theta|_{p_0} = [\Gamma(p_0) - Q(\lambda_0)]_\perp^{-1}\left\{\nabla_p\Gamma\Theta_0 - \frac{dQ}{d\lambda}\Theta_0\nabla\lambda\right\} =$$

$$= [\Gamma(p_0) - Q(\lambda_0)]_\perp^{-1}\left\{\nabla_p\Gamma\Theta_0 - \frac{dQ}{d\lambda}\Theta_0\frac{<\nabla_p\Gamma\Theta_0, \Theta_0>}{\left\langle P_N\frac{I+A^2}{(A-\lambda I)^2}P_N\Theta_0, \Theta_0\right\rangle}\right\} \tag{7}$$

The calculation of the next derivatives of the dispersion matrix gives

$$\left(\nabla_2\Gamma - \nabla_2\lambda\frac{dQ}{d\lambda}\right)\Theta_r(p) + 2\left(\nabla\Gamma - \nabla\lambda\frac{dQ}{d\lambda}\right)\nabla\Theta_r(p) + (\Gamma - Q)\nabla_2\Theta_r(p) = 0$$

Using (7) one can find the inverse effective mass at the critical point $p, \nabla\lambda_r(p) = 0$:

$$\nabla_2\lambda = \frac{\langle\nabla^2\Gamma\Theta_r(p), \Theta_r(p)\rangle + 2\langle(\Gamma - Q)_\perp^{-1}\nabla\Gamma\Theta_r(p), \Theta_r(p)\rangle}{\left\langle\frac{dQ}{d\lambda}\Theta_r(p), \Theta_r(p)\right\rangle} \tag{8}$$

The extrema of the spectral bands $\min\limits_{p\in\Pi^\nu}\lambda_r(p) \equiv \lambda_r(p_{min})$ and $\max\limits_{p\in\Pi^\nu}\lambda_r(p) \equiv \lambda_r(p_{max})$ are the critical points. Despite of the calculations of $M^{-1}(p_{min})$ and $M^{-1}(p_{max})$ demand to solve the matrix transcendental equations, there appears to exist direct relations between these values and values of $Q(\lambda)$ at infinity. Let us consider some function $\Phi(\lambda, p) : N \longrightarrow N$, meromorphic in λ, which have the simple poles λ_s on $R\backslash\sigma(\mathscr{A}_\Gamma)$,

$$\Phi(\lambda) = \frac{\Phi_1^s}{\lambda - \lambda_s} + \Phi_0^s + \dots$$

The integral

$$-\frac{1}{2\pi i}\oint \Phi(\lambda, p)\left[\Gamma(p) - Q(\lambda)\right]^{-1} d\lambda \equiv I$$

can be worked out by means of a calculation of the residues at infinity (we assume that $\mathscr{A}_\Gamma$ is a bounded operator):

$$I = -\frac{1}{2\pi i}\oint\left[\Phi_0 - \frac{1}{\lambda}\Phi_1\right]\hat{\Gamma}^{-1}\left[1 - \frac{1}{\lambda}P_N(1 + A^2)P_N\right]d\lambda =$$

$$= \Phi_0\hat{\Gamma}^{-1}P_N(1 + A^2)P_N + \Phi_1\hat{\Gamma}^{-1},$$

$$\hat{\Gamma} \equiv \Gamma(p) + P_N A P_N.$$

On the other hand, the expression under the integral has poles in the points λ_s and $\lambda_r(p)$. Let us insert the following notations:

$$\{\Gamma(p) - Q(\lambda)\} = \left\{(\lambda - \lambda_r(p))P_p^r + I - P_p^r\right\}\hat{Q}_r,$$

$\hat{Q}_r(\lambda_r(p))$ is an invertible operator. Using the following equality

$$\frac{d}{d\lambda}\big|_{\lambda_r(p)}\left\{\Gamma(p) - Q(\lambda)\right\} = P_p^r \hat{Q}_r(\lambda_r(p)) + (I - P_p^r)\frac{d\hat{Q}_r(\lambda_r(p))}{d\lambda} =$$

$$= \hat{Q}_r^+(\lambda_r(p))P_p^r + \frac{d}{d\lambda}\hat{Q}_r^+(\lambda_r(p))(I - P_p^r)$$

one can see that

$$-\frac{d}{d\lambda}\hat{Q}_r(\lambda_r(p))P_p^r = \frac{d}{d\lambda}\left\{\Gamma(p) - Q(\lambda)\right\}P_p^r = P_p^r\hat{Q}_r(\lambda_r(p))P_p^r\ ,$$

and therefore:

$$\mathrm{res}\ (\Gamma - Q)^{-1}\big|_{\lambda_r(p)} = P_p^r\hat{Q}_r^{-1}(\lambda_r(p))P_p^r =$$

$$= -P_p^r\left(P_p^r\frac{d}{d\lambda}\hat{Q}_r(\lambda_r(p))P_p^r\right)^{-1}\big|_{N_{P_p^r}}P_p^r\ .$$

The equivalence of these two methods of integration gives the following formula:

$$-\frac{1}{2\pi i}\oint \Phi(\lambda, p)\left[\Gamma(p) - Q(\lambda)\right]^{-1}d\lambda =$$

$$= \Phi_0\left\{\Gamma(p) + P_N A P_N\right\}^{-1}P_N(1 + A^2)P_N + \Phi_1\left\{\Gamma(p) + P_N A P_N\right\}^{-1} =$$

$$= -\sum_s \Phi_1^s\left[\Gamma(p) - Q(\lambda_s)\right]^{-1} + \sum_r \Phi(\lambda_r(p))P_p^r\left\langle\frac{dQ}{d\lambda}\Theta_r(p), \Theta_r(p)\right\rangle^{-1}$$

This formula is especially interesting at the critical points, where $\left\langle\frac{dQ}{d\lambda}\Theta_r(p), \Theta_r(p)\right\rangle$ is connected with the effective mass tensor by the equation (8). If $\dim N = 1$ then this equation gives

$$\nabla^2\lambda = \frac{\nabla^2\Gamma}{\frac{dQ}{d\lambda}} = M^{-1}$$

and the virial formula turns into

$$\nabla^2\Gamma\left\{\frac{\Phi_0\left\langle(I + A^2)\Theta, \Theta\right\rangle + \Phi_1}{\Gamma(p_0) + <A\Theta, \Theta>} + \sum_s\frac{\Phi_1^s}{\Gamma(p_0) - \left\langle\frac{I + \lambda_s A}{A - \lambda_s I}\Theta, \Theta\right\rangle}\right\} =$$

$$= \sum_r \Phi(\lambda_r(p_0))\ M^{-1}(\lambda_r(p_0))$$

or, after the trace calculation:

$$\sum_r \Phi(\lambda_r(p_0))\ Sp\ M^{-1}(\lambda_r(p_0)) =$$

$$= \Delta\Gamma(p_0) \left\{ \frac{\Phi_0 \langle (I + A^2)\Theta, \Theta \rangle + \Phi_1}{\Gamma(p_0) + <A\Theta, \Theta>} + \sum_s \frac{\Phi_1^s}{\Gamma(p_0) - \left\langle \frac{I + \lambda_s A}{A - \lambda_s I}\Theta, \Theta \right\rangle} \right\} .$$

In the case $\Phi(\lambda) = Q(\lambda)$ we obtain the simplest variant of the formula

$$\Gamma(p_0) \frac{\langle (I + A^2)\Theta, \Theta \rangle}{[\Gamma(p_0) + <A\Theta, \Theta>]^2} \Delta\Gamma(p_0) = \sum_r Sp\ M^{-1}(\lambda_r(p_0)) .$$

3. TWO–BODY HAMILTONIAN WITH RESONANCE INTERACTION.

The background two–body Hamiltonian $\mathscr{B}$ without any interaction can be chosen acting in the Hilbert space $\mathscr{E}_1 \otimes \mathscr{E}_2 = G$ (the product of two copies of $\mathscr{E}$), as a sum of two copies of $\mathscr{A}_\Gamma$:

$$\mathscr{A}_\Gamma^1 \otimes I^2 + I^2 \otimes \mathscr{A}_\Gamma^2 \equiv \mathscr{B} .$$

Eigenfunctions of $\mathscr{B}$ are parametrized by the pair of quasimomenta $(p^1, p^2) \equiv p$ and are just a product of corresponding eigenfunctions of $\mathscr{A}_\Gamma^1, \mathscr{A}_\Gamma^2$

$$\chi_{r,s}(p) = \chi_r^1(p^1)\chi_s^2(p^2)$$
$$\mathscr{B}\chi_{r,s}(p) = [\lambda_r^1(p^1) + \lambda_s^2(p^2)]\,\chi_{r,s}(p)$$

The spectral decomposition formula for $\mathscr{B}$ can be combined from the corresponding formulae for the components, e.g.

$$I_G = I_{\mathscr{E}} \otimes I_{\mathscr{E}} =$$

$$= \frac{1}{(2\pi)^{2\nu}} \sum_{r,s} \int \chi_{r,s}(p) < \cdot, \chi_{r,s}(p) > d^{2\nu}p \tag{9}$$

$$G\,_{n_1 m_1, n_2 m_2}(\lambda) =$$

$$= \frac{1}{(2\pi)^{2\nu}} \sum_{r,s} \int \chi_{r,s}^{n_1, n_2}(p) \frac{\left\langle \cdot, \chi_{r,s}^{m_1, m_2}(p) \right\rangle}{\lambda_r(p^1) + \lambda_s(p^2) - \lambda} d^{2\nu}p$$

The space G can be represented as an orthogonal sum of two invariant subspaces of G, $G = G_{sym} \oplus G_{Asym}$, which contains symmetric and antisymmetric combinations of the Bloch waves χ, for example for $\lambda = \lambda_r(p^1) + \lambda_s(p^2)$:

$$\chi_{r,s}^{sym}(p) = \frac{1}{2!}\{\chi_r^1(p^1) \otimes \chi_s^2(p^2) + \chi_s^1(p^2) \otimes \chi_r^2(p^1)\} ,$$

$$\chi_{r,s}^{Asym}(p) = \frac{1}{2!}\{\chi_r^1(p^1) \otimes \chi_s^2(p^2) - \chi_s^1(p^2) \otimes \chi_r^2(p^1)\} .$$

In accordance with statistics we have to use in further calculations symmetric or antisymmetric components of $\mathscr{B} = \mathscr{B}^{sym} \oplus \mathscr{B}^{Asym}$. We shall write both of them in general form (9).

The interaction will be constructed by means of extension theory. Namely, let $M = M_0$ be the finite dimensional subspace in $E_0 \otimes E_0$, and M_l are the copies of M in $E_l \otimes E_l$. Moreover, let P_l be the projecting onto M_l operator. Then, the restriction $\mathscr{B}_0$ of the operator $\mathscr{B}$ onto $D(\mathscr{B}_0) = \{u : u \in G, P_l u = 0\}$ has the following deficiency subspaces $\mathscr{N}^{(0)} = V_l\{N_l^{(0)}\}, N_l^{(0)} = (B + iI)^{-1}M_l$. Attaching to every diagonal node (m, m), $m \in \mathbb{Z}^\nu$, an auxiliary space H_m, which is a copy of $H_0 = H$, we consider an orthogonal sum $\mathscr{H} = \oplus \sum H_m$. In every H_m we consider selfadjoint operator K_m, which is a copy of $K_0 = K$, assuming the spectrum of K to be discrete. Each of K_m will be restricted onto the domain $D_m = \{u : u = (K - iI)^{-1}v^\perp, v^\perp \perp N_m^{(1)}\}$. Here $N_m^{(1)}$ is a copy of a some finite–dimensional subspace $N_0^{(1)} = N^{(1)} \subset H$. Thus the definition subspace of internal Hamiltonian $\oplus \sum K_m = \mathscr{K}$ is $\mathscr{N}^{(1)} = V_m\{N_m^{(1)}\}$. The restriction of $\mathscr{B} \oplus \mathscr{K}$ is defined onto $D_0 = \{u_0, u_1\}$:

$$u_0 = (\mathscr{B} - iI)^{-1}v_0^\perp \ , \ u_1 = (\mathscr{K} - iI)^{-1}v_1^\perp \ , \ v_{0,1}^\perp \perp \mathscr{N}^{(0),(1)} \ .$$

Instead of the usual boundary values $\xi_{0l}^\pm(u)$ it is convenient now to take vectors $\rho_{0l} = (\mathscr{B} + iI)\xi_{0l}^+, \rho_{0l} \in M_l$, which play a role of densities of ξ_{0l}^+, and the values of the rest part of the deficient component, at the given point (l, l) in the subspace M_l

$$P_l(\mathscr{B} - iI)^{-1}\xi_{0l}^- \equiv u_l^- \ .$$

The boundary form in terms $\{u^-, \rho\}$ has an usual view

$$\sum_l \langle \xi_{0l}^-(u_1), \xi_{0l}^+(u_2)\rangle - \langle \xi_{0l}^+(u_1), \xi_{0l}^-(u_2)\rangle$$

$$= \sum_l \left\langle u_l^{1-}, \rho_l^2 \right\rangle - \left\langle \rho_l^1, u_l^{2-} \right\rangle \ .$$

Denoting the boundary values for the inner channel by $\xi_1^\pm$, we can write the boundary condition in the form

$$\begin{pmatrix} u_{0l} \\ \xi_{1l}^- \end{pmatrix} = \sum_{|s|<R} \tau_s \begin{pmatrix} \rho_{0l-s} \\ \xi_{1l-s}^+ \end{pmatrix} \quad , \quad \tau_s = \tau_{-s}^+ \ .$$

Here τ_s is an operator acting in $M_l \oplus N_l^1$,

$$\tau_s = \left\{ \begin{matrix} \tau_{00}^s \tau_{01}^s \\ \tau_{10}^s \tau_{11}^s \end{matrix} \right\} \ .$$

__Theorem 3.__   The operator $\mathcal{B}_\tau$, defined on the domain:

$$D_\tau = \left\{ u^0 : u^0 = (\mathcal{B} - iI)^{-1} v_\perp^0 + (\mathcal{B} - iI)^{-1} \mathcal{B}\xi_+^0 + (\mathcal{B} - iI)^{-1}\xi_-^0, \; v_\perp^0 \perp \mathcal{N}^{(0)} \right\}$$

$$\oplus \left\{ u^1 : u^1 = (\mathcal{K} - iI)^{-1} v_\perp^1 + (\mathcal{K} - iI)^{-1} \mathcal{K}\xi_+^1 + (\mathcal{K} - iI)^{-1}\xi_-^1, \; v_\perp^1 \perp \mathcal{N}^{(1)} \right\},$$

$$\begin{pmatrix} u^- \\ \xi^- \end{pmatrix} = \tau \begin{pmatrix} \rho \\ \xi^+ \end{pmatrix},$$

by the formula

$$\mathcal{B}_\tau \begin{pmatrix} u_0 \\ u_1 \end{pmatrix} = \begin{pmatrix} \mathcal{B}(\mathcal{B} - iI)^{-1}\{v + \xi_-^0\} - (\mathcal{B} - iI)^{-1}\xi_+^0 \\ \mathcal{K}(\mathcal{K} - iI)^{-1}\{v + \xi_-^1\} - (\mathcal{K} - iI)^{-1}\xi_+^1 \end{pmatrix}$$

is selfadjoint in $G \oplus \mathcal{H}$. The spectrum σ_τ of operator $\mathcal{B}_\tau$ consists of two branches: the scattering branch σ_τ^s, which coincides with the spectrum of operator $\mathcal{B}$, and the wave–quide branch σ_τ^w of the spectrum, which is defined as a solution $\lambda = \lambda(q)$ of the equation

$$0 = \det \left\{ \sum_{|s| < R} \tau_s e^{iqs} - \begin{pmatrix} \sum_l P_l \dfrac{I + \lambda \mathcal{B}}{(\mathcal{B} - \lambda I)(\mathcal{B}^2 + 1)} P_0 e^{iql} & 0 \\ 0 & P_{N_0^1} \dfrac{I + \lambda K}{K - \lambda I} P_{N_0^1} \end{pmatrix} \right\}$$

$$\equiv \det Q(q, \lambda) \; , \quad q \in \mathbb{R} \; .$$

$$(10)$$

The corresponding eigenfunction can be taken in the form

$$\Pi_q = \left\{ \begin{array}{c} \sum_l e^{iql} \dfrac{1}{\mathcal{B} - \lambda(q) I} \rho_l^w \\ e^{iql} \dfrac{K + iI}{K - I} \varphi_l^w \end{array} \right\}$$

$$(11)$$

where ρ_l^w, φ_l^w are produced by translation of the vectors ρ^w, φ^w from $M \oplus N_0^1$ along the diagonal of the lattice, $l_1 = l_2 = l$; vectors ρ^w, φ^w fullfill the equation

$$Q(q, \lambda) \begin{pmatrix} \rho^w \\ \varphi^w \end{pmatrix} = 0 \; .$$

The spectral density of waveguide–eigenfunction (11) is given by the formula

$$\left\langle \frac{d}{d\lambda} Q \begin{pmatrix} \rho^w \\ \varphi^w \end{pmatrix}, \begin{pmatrix} \rho^w \\ \varphi^w \end{pmatrix} \right\rangle_{M \oplus N_0^1} \equiv d^{-1}(p) \; .$$

The scattered waves can be taken in the form

$$\begin{pmatrix} \chi_p + \sum_l e^{i(p_1 + p_2)l} \dfrac{1}{\mathcal{B} - (\lambda + io)I} \rho^s \\ e^{i(p_1 + p_2)l} \dfrac{K + iI}{K - \lambda I} \varphi^s \end{pmatrix}$$

$$(12)$$

where (ρ_i^s, φ_i^s) is the vector from $(M_l \cdot N_l^1)$, which is produced by translation of fixed vector (ρ^s, φ^s) along the lattices diagonal. These vectors can be calculated from the equation, which we get submitting the ansatz (12) to the boundary condition

$$\left(\begin{array}{c} P_0 \chi_p + \sum_l e^{i(p_1+p_2)l} P_0 \dfrac{1}{\mathscr{B}^2 + 1} \dfrac{I + \lambda \mathscr{B}}{\mathscr{B} - \lambda - io} \rho_i^s \\ \dfrac{i + \lambda K}{K - \lambda I} \varphi_0 \end{array} \right) = $$
$$= \sum_{|l| < R} \tau_l e^{i(p_1+p_2)l} \left(\begin{array}{c} \rho_i^s \\ \varphi_i^s \end{array} \right) . \tag{13}$$

The spectral density for scattered waves is trivial, the same as it is for the plane waves χ_p.

The system $\{\Pi_q, \psi_p\}$ is a complete orthogonal system of eigenfunctions of the Hamiltonian $\mathscr{B}_\tau$.

The corresponding spectral decomposition lookes like follows for $\dim M = \dim N_0^1 = 1$:

$$I = \int \Pi_q \langle \cdot, \Pi_q \rangle \, d^\nu q + \int \psi_p \langle \cdot, \psi_p \rangle \, d^{2\nu} p$$

In the case of higher dimensions of M, N_0^1, one must perform a summation over any orthogonal basis in $M \oplus N_0^1$.

The proof is based on the Krein formula for resolvents and repeats essentially the proof of the spectral decomposition in the one–particle case. Note, that the lattice–sum in (10,13) can be essentially calculated in the unperturbed spectral representation. We do it exploiting the quasiperiodicity of Bloch waves

$$\frac{1}{\mathscr{B}^2 + I} \cdot \frac{I + \lambda \mathscr{B}}{\mathscr{B} - \lambda I} \rho_i^s = \frac{1}{(2\pi)^{2\nu}} \int \int \frac{1 + \lambda \left(\lambda_1(p_1) + \lambda_2(p_2) \right)}{\lambda_1(p_1) + \lambda_2(p_2) - \lambda} \cdot$$

$$\cdot \frac{1}{\left[\lambda_1(p_1) + \lambda_2(p_2) \right]^2 + 1} \cdot \chi(p_1, p_2) e^{-i(p_1+p_2)l} \langle \rho_0, \chi(p_1, p_2) \rangle \, dp_1^\nu dp_2^\nu \, ,$$

$$\sum_l \left\langle \frac{1 + \lambda \mathscr{B}}{\mathscr{B} - \lambda I} \cdot \frac{1}{\mathscr{B}^2 + I} \rho_l, \rho_0 \right\rangle e^{iql} =$$

$$= \frac{1}{(2\pi)^\nu} \int \int \frac{1 + \lambda(\lambda_1 + \lambda_2)}{\lambda_1 + \lambda_2 - \lambda} \cdot \frac{|\langle \rho_0, \chi(p_1, p_2) \rangle|^2}{1 + (\lambda_1 + \lambda_2)^2} \times$$

$$\times \sum_l \delta(p_1 + p_2 - q + 2\pi l) dp_1^\nu dp_2^\nu \equiv D(\lambda, q) \, .$$

Let us analyse the equations for eigenfunctions in a symplest case $\dim M = \dim N_0^1 = 1$, $M = \{f_0 e_0\}$, $N_0^1 = \{f_1, \varphi_0\}$. Then the equation for scattered waves look as follows for $q = p_1 + p_2$:

$$\begin{pmatrix} - <\chi_p, l_0 > \\ 0 \end{pmatrix} = \left\{ \begin{pmatrix} D(\lambda, q) & 0 \\ 0 & \left\langle \frac{I+\lambda K}{K-\lambda I}\varphi_0, \varphi_0 \right\rangle \end{pmatrix} - \tau(q) \right\} \begin{Bmatrix} f_0 \\ f_1 \end{Bmatrix} .$$

Here $\tau(q)$ is the matrix of the operator $\sum_{|l| < R} \tau_l e^{iql}$ with respect to the base (e_0, φ_0). Then we have for $q = p_1 + p_2$:

$$\begin{pmatrix} f_0 \\ f_1 \end{pmatrix} = - \left\{ \begin{matrix} D(\lambda, q) - \tau_{00}(q) & \tau_{01}(q) \\ \tau_{10}(Q) & \left\langle \frac{I+\lambda K}{K-\lambda I}\varphi_0, \varphi_0 \right\rangle - \tau_{11}(q) \end{matrix} \right\}^{-1} \begin{pmatrix} <\chi_p, l_0 > \\ 0 \end{pmatrix} .$$

We see, that the singularities of amplitudes $f_{0,1}$, lie just at zeros of determinant (10), which coincide with waveguide band. If the nondiagonal elements of τ are small with respect to the diagonal ones, then the waveguide spectrum can be represented in rough features as a joined band spectrum for both channels with typical quasicrossings and spectral gaps caused by them.

Obviously, the eigenvalues of the inner hamiltonian always lie in two–body spectral gaps. Those gaps are of resonant nature (see [2]). The hybridization of wave–guide functions of both channels takes the place close to these gaps.

Concerning the scattering waves (12), the following asymptotic formula

$$\left[\frac{1}{\mathscr{B} - \lambda I} \rho_l \right]_{n \; \substack{n=a\mu \\ \mu \to \infty}} \approx \sum_j \frac{e^{i<n, p^{(j)}>}}{|\mu|^{\frac{2\nu-1}{2}}} e^{-i\left\langle p_1^{(j)} + p_2^{(j)}, l \right\rangle} c(p^{(j)}) \tag{14}$$

allows to calculate the scattering amplitude and S–matrix. Here $p^{(j)}$ are the points lying on the isoenergetic surface $\lambda \equiv \lambda(p) = \lambda_1(p_1) + \lambda_2(p_2)$, in which the vectors a and $\nabla_p \lambda(p)$ are codirected.

$$c(p^{(j)}) = \frac{e^{i\frac{\pi}{4}(2\nu-3)}}{(2\pi)^{\frac{2\nu-1}{2}}} \cdot \frac{\langle \rho_0, \chi(p^{(j)}) \rangle [\chi(p^{(j)})]_0}{\sqrt{K(p^{(j)})|\nabla_p \lambda(p^{(j)})|}} ,$$

where $K(p)$ is the Gaussian curvature of the isoenergetic surface $\lambda(p) = \lambda$ at the point p.

This asymptotics is analogous to well–known Laplacian resolvent property:

$$\frac{e^{ik|x-y|}}{4\pi|x - y|} \text{ for } x \longrightarrow V^\infty e^{-i<\nu, y>} \frac{e^{ik|x|}}{4\pi|x|} .$$

The asymptotic expression (14) can be easily proved with the help of the decomposition (9) and the expression (4) by the method of stationary phase.

Let us assume that the summation in (14) is degenerated, $j = 1$, and dim $M = 1$.

In this case the asymptotic formula (14) is more simple written as

$$\left(\frac{1}{\mathscr{B} - \lambda I}\rho_l\right)_n \xrightarrow{n \to \infty} a \cdot \infty \frac{\overline{\chi_l(p_\lambda)}}{\chi_0(p_\lambda)}\left(\frac{1}{\mathscr{B} - \lambda I}\rho_0\right)_n ,$$

and we have for $\rho_l = T'\rho_0$, $\rho_0 = f_0 e_0$:

$$
\begin{aligned}
[\psi_p]_\iota &= [\chi_p]_\iota + \left(\sum_{n'} e^{i(p_1+p_2)n'}\frac{\overline{[\chi_{p_\lambda}]_{n'}}}{[\chi_{p_\lambda}]_0}\right)\left[\frac{f_0}{\mathscr{B} - \lambda I}e_0\right]_\iota \\
&= [\chi_p]_\iota + \left((2\pi)^\nu \sum_{n'} \delta\left(p^1 + p^2 - p^1_\lambda - p^2_\lambda + 2\pi n'\right)\right)\left[\frac{f_0}{\mathscr{B} - \lambda I}l_0\right]_\iota .
\end{aligned}
$$

Thus the scattering amplitude f_a is defined by the expression:

$$f_a = (2\pi)^\nu \sum_{n'} \delta(p^1 + p^2 - p^1_\lambda - p^2_\lambda + 2\pi n')\frac{f_0(p)}{\sqrt{K(p_\lambda)|\nabla_p\lambda(p_\lambda)|}}$$

One can see, that the following quasimomentum conservation law is valid:

$$p^1 + p^2 - p^1_\lambda - p^2_\lambda + 2\pi n = 0 \quad , \ n \in \mathbb{Z}^\nu .$$

CONCLUSION. MANY BODY SECTOR.

The method described above gives also a series of solvable Hamiltonians in wave–guide channels. In fact, let $\mathscr{B}_2$ be the wave–guide part of the two–body Hamiltonian, $\mathscr{B}_2 = \mathscr{B}_\tau|_{G_2}$, wher G_2 is the wave–guide channel in G. Using corresponding spectral representation in quasimomentum space

$$
\begin{aligned}
G_2 &\longrightarrow L_2\left([-\pi, \pi]^\nu, C^r\right) \\
\mathscr{B}_2 &\longrightarrow \lambda(q)
\end{aligned}
$$

we get a lattice model of the two–body bound states Hamiltonian. Then, proceeding like in section 2, we can construct four–body Hamiltonian $\mathscr{B}_2 \otimes I' + I \otimes \mathscr{B}'_2$ without interaction between pairs. The interaction also could be switched like in section 2. It gives the four–body bound states of two–pairs Hamiltonian. Going further, we can construct wave–guide bound states for many particles.

ACKNOWLEDGEMENTS.

The interest of the authors to the construction of two–body solvable models of solids with resonance interactions was stimulated by Raphael Høegh–Krohn when discussing these matters on the conference in Dubna 1987, the last which he took part in. We dedicate this work to his memory.

One of the authors is grateful to professor S. Albeverio and to the mathematical department of Bochum University, where this work was finished.

REFERENCES.

1. B.S. Pavlov: Usp. Mat. Nauk, $\underline{42}$, 99–131 (1987)

2. P.B. Kurasov, B.S. Pavlov: Teor. Mat. Fiz., $\underline{74}$, 82–93 (1988)

3. B.S. Pavlov: Vestnik Leningr. Univ., $\underline{3}$, 15, 43–49 (1987)

4. O.B. Firsov: Sov. J. Exp. and Theor., $\underline{68}$, 568–575 (1975)

5. N.E. Firsova: Teor. Mat. Fiz., $\underline{37}$, 281–288 (1978)

<u>Schroedinger operators on unusual manifolds</u>

Pavel Exner and Petr Seba

Nuclear Physics Institute, Czechoslovak Academy of Sciences,
25068 Rez near Prague, Czechoslovakia
and
Laboratory of Theoretical Physics, Joint Institute for Nuclear
Research, 141980 Dubna, USSR

There are only few potentials in the nonrelativistic quantum
mechanics which lead to exactly solvable models. One of them is
the δ-function potential which was introduced to quantum physics
in the classical work of Kronig and Penney in 1931 [1]. As soon as
this type of potentials appeared (note that this was done
originally only for a one-dimensional system) it turned out that
it represented an extremely useful object for constructing
solvable models in a wide class of one-dimensional physical
situations. Three years after the δ-function potential made its
way into solid-state physics, E.Fermi applied a similar idea also
to the description of low-energy neutron scattering [2]. In this
case, however, the delta function potential was used in a
three-dimensional situation which meant a number of mathematical
difficulties. They were removed at that time formally by
introducing the so-called Fermi pseudopotential (which was in some
sense a renormalized δ-function). From the rigorous point of view
the Fermi pseudopotential did not , of course, solve the problem
and the situation remained unclear up to the sixties when Berezin
and Fadeev showed that the pseudopotential can be mathematically
described using the self-adjoint extension theory. Approximately
at the same time it has been realized that the Fermi model suited
very well not only to neutron scattering but also other low-energy
phenomena in the atomic and molecular physics as well. For
instance, it was successfully used in modelling of the electron
detachment in slow collisions between negative ions and atoms etc.
(see the book of Demkov and Ostrovski [3]). In such a way the
three dimensional δ-potential penetrated slowly from the nuclear

227

to atomic and solid-state physics and its more thorough mathematical analysis was needed.

This stage in the development of the point-like interaction theory has been already connected with the name of Raphael Hoegh-Krohn who started to study it in late seventies together with S.Albeverio, A.Grossmann and others. The basic physical idea inherent to these rather mathematical papers was that one can describe the low-energy scattering phenomena by replacing the actual potential by an abstract but easily tractable object constructed with the help of the self-adjoint extension theory. This line of his scientific work which can be clearly distinguished on his publication list led finally to the monograph published recently by Springer [4] which exposed the present state-of-arts in this field. It is not necessary to discuss here the scientific merits of Raphael's contributions to this area of mathematical physics since it is well known and widely recognized. Our aim in the present lecture is to demonstrate that the mentioned idea of describing the quantum low-energy/long-wave phenomena by means of the von Neumann's theory is still stimulating and finds interesting applications in many other parts of physics.

1.Quantum mechanics on graphs

The question of describing a quantum motion on a graph is not new. It appeared for the first time, up to our knowledge, in connection with models attempting to describe spectroscopical data of complicated aromatic hydrocarbons and conjugated polyenes. The first step in what is now called in chemistry the free-lectron or metallic model of a molecule was done by L.Pauling, later a Nobel-prize winner, who discovered that the motion of π-electrons in aromatic molecules can be regarded as being constrained to a graph which represented the molecule skeleton [5] (let us remark that Pauling was interested primarily in describing the diamagnetic anisotropy in such substances). Approximately ten years after the Pauling's paper it became clear that this model described satisfactorily also the spectral properties of such molecules [6,7] and the interest to this type of models appropriately increased (see for instance [8-11]).

From the mathematical point of view the key problem behind all these physical calculations was to find appropriate boundary conditions which matched the wavefunctions at the branching points of the graph and which could yield a reasonable quantum dynamics of the system. These conditions were usually guessed on intuitive grounds. The mathematical attempt to justify the matching conditions has been done at that time by Ruedenberg and Sherr [12], but their analysis was not fully mathematically consistent.

The success in applying this model to particular molecules stimulated attempts to apply it also to other systems - see for instance [13] for the calculations of the diamond band structure or [14] for modelling of metals. Recently the free-electron model, which describes the π-electrons only has been replaced by more complicated (ab initio) calculations (see [15] for the relation between the free-electron model and LCAO).

Another motivation to study quantum mechanics on graphs has appeared very recently in microelectronic. The progress in microstructure fabrication opened a new frontier in the electron transport theory - the so called mesoscopic physics (see for instance [16]) - where the non-local nature of an electron during the transport is clearly manifested [17]. One particular problem which appears when dealing with such a microelectronical system is how the electrons "split" at the branching points of the supporting structure. The electron "splitting" is particularly important in structures where the interference of the conduction electrons plays a decisive role in global transport properties say if the structure consists of two leads with a loop in the middle etc.[18],[19]. To describe the splitting process means to describe electron scattering on the branching point and this is actually the way in which the problem has been approached: each branching point was associated with an ideal device - a "splitter" - which mediated the transport of the incoming electrons to the outgoing branches [20],[21]. The device has been described by a momentum independent S-matrix chosen in such a way that the incoming electron can never be reflected back. In this approach the underlying quantum Hamiltonian (which is certainly a more fundamental object then the S-matrix) has not been included. The only attempt to describe the electron transport in submicron metallic rings using a Hamiltonian has been undertaken recently by Bulka [22] but this work was, unfortunately, wrong.

From the mathematical point of view the problem of the quantum dynamics on branched graphs is in the both above cases the same ; the only difference is that one has to deal with compact graphs while describing molecules and with noncompact ones (because of the outgoing leads) in the case of the microelectronical systems. In the rest if this section we will outline a method of constructing the corresponding Hamiltonians in the most simple case of a three-port junction using the self-adjoint extension theory.

It may seem that using this method we burden the model with a complicated procedure which is actually not needed if one has only to write down appropriate boundary conditions. Even in simple cases, however, intuition may be a false guide and boundary conditions which look reasonable at a glance can lead to a non self-adjoint Hamiltonian, or on physical terms, to probability nonconservation (except of the above mentioned paper of Bulka see for instance [23] for such inconsistent boundary conditions).

Let us start with the most simple case when the configuration manifold (the graph) consists of three halflines (Fig.1).The state Hilbert space of the problem is of the form

$$\mathcal{H} = L^2(\mathbb{R}_+) \oplus L^2(\mathbb{R}_+) \oplus L^2(\mathbb{R}_+)$$

Our aim is to construct all admissible Hamiltonians on $\mathcal{H}$ which describe the free quantum dynamics for the particle under consideration everywhere except at the branching point P . We start therefore with an operator H which describes three independent halflines

$$H = H_1 \oplus H_2 \oplus H_3$$

where each H_i acts as

$$H_i f = -f''$$

with the domain

$$D(H_i) = \{ f : f, f' \in AC(\mathbb{R}_+) , f(0) = 0 \}$$

In order to "connect" the halflines together we remove
their endpoints restricting the operator H to

$$H_0 := H \mid D_0$$

with

$$D_0 = C_0^\infty(\mathbb{R}_+) \oplus C_0^\infty(\mathbb{R}_+) \oplus C_0^\infty(\mathbb{R}_+) .$$

The admissible Hamiltonians are then obtained as self-adjoint
extensions of the operator H_0. A straightforward calculation shows
that H_0 has deficiency indices equal to $(3,3)$, and consequently,
we have a nine-parameter family of its self-adjoint extensions.
All these operators are according to the von Neumann theory [24]
parametrized by 3×3 unitary matrices U. The extension
corresponding to a particular U is then given by

$$H_U (f_1, f_2, f_3) = (-f_1'', -f_2'', -f_3'')$$

$$D(H_U) = \{ f \equiv (f_1, f_2, f_3) \in \mathcal{H} ; f = \varphi + \sum_{j=1}^{3} c_j (\varphi_j^+ + \sum_{k=1}^{3} u_{jk}\varphi_k^-): c_j \in \mathbb{C} ;$$

$$\varphi = (\varphi_1, \varphi_2, \varphi_3) \in D_0 \} \tag{1}$$

where

$$\varphi_1^+ = (f_+, 0, 0) ; \varphi_2^+ = (0, f_+, 0) ; \varphi_3^+ = (0, 0, f_+)$$

with

$$f_+ = \exp(-\bar{\varepsilon}x) ; \varepsilon = \exp(i\pi/4)$$

are the vectors spanning the deficiency subspace $\mathcal{K}^+$,

$$\mathcal{K}^+ = \mathrm{Ker}(\, H_0^* - i\mathbb{1}\,)$$

and $\varphi_j^- = \overline{\varphi_j^+}$ are the corresponding vectors spanning the deficiency subspace $\mathcal{K}^-$,

$$\mathcal{K}^- = \mathrm{Ker}(\, H^* + i\mathbb{1}\,)$$

Finally, u_{ij} are elements of U.

These relations represents a complete solution to our problem – we find that in principle one can associate with the graph on Fig.1 a nine parameter family of Hamiltonians. The result is, however, not very suitable from the viewpoint of practical applications since the Hamiltonians are defined in a rather abstract form. What is needed in fact is to rewrite the definition of H_U in terms of boundary conditions at the point P.

From the viewpoint of applications the most interesting situation corresponds to Hamiltonians H_U which are invariant under permutation of the "wires". In what follows we restrict ourselves therefore primarily to this type of Hamiltonians. Mathematically speaking this means that we are going to consider only the operators H_U commuting with the operators P_{jk} representing the transposition of a pair of wires

$$P_{12}\, (f_1, f_2, f_3) = (f_2, f_1, f_3)$$

etc. We require

$$P_{jk} H_U \subset H_U P_{jk} \qquad\qquad (2)$$

for all $j,k = 1,2,3$. Since the differential expression corresponding to H_U is obviously permutation-invariant the condition (2) can be rewritten as

$$P_{jk} D(H_U) \subset D(H_U) \text{ for } j,k = 1,2,3.$$

Using the form of $D(H_U)$ we find finally that the Hamiltonian H_U is permutation invariant iff the corresponding matrix U fulfils

$$u_{ij} = u_{ji}$$
$$u_{ii} = u_{jj}$$

for all i,j, i.e., iff U has the form

$$U = \begin{pmatrix} u & v & v \\ v & u & v \\ v & v & u \end{pmatrix} \qquad (3)$$

The unitarity conditions for U lead to the following additional constraints on u,v

$$|u|^2 + 2|v|^2 = 1$$

$$2\mathrm{Re}(\overline{u}\,v) + |v|^2 = 1$$

The subclass under consideration is therefore parametrized by two real parameters. Characterization of the corresponding Hamiltonians H_U by boundary conditions is expressed by the following assertion (for the boundary conditions referring to permutation-noninvariant cases see [25]) :

Theorem 1:
(i) Suppose that u ≠ i-2v and u ≠ i+v. Then f ∈ D(H$_U$) if and only if f ∈ D(H$_0^*$) and

$$f_1(0) = Af_1'(0) + Bf_2'(0) + Bf_3'(0)$$

$$f_2(0) = Bf_1'(0) + Af_2'(0) + Bf_3'(0)$$

$$f_3(0) = Bf_1'(0) + Bf_2'(0) + Af_3'(0)$$

where A,B are two independent real constants related to the matrix elements u,v by

$$A = \frac{-\varepsilon(1+u+iu+iv+iu^2 +iuv-2iv^2)}{(u+2v-i)(1+iu-iv)}$$

$$B = \frac{i\sqrt{2}\,v}{(u+2v-i)(1+iu-iv)}$$

(ii) If u = i-2v, then f ∈ D(H$_U$) iff f ∈ D(H$_0^*$) and

$$f_1'(0) + f_2'(0) + f_2'(0) = 0$$

$$f_2(0) - f_3(0) = C\,(f_3'(0) - f_2'(0))$$

$$f_1(0) - f_3(0) = C\,(f_3'(0) - f_1'(0))$$

with C being a real constant given by

$$C = \frac{3\varepsilon v - \sqrt{2}}{v}$$

(iii) If $u = i+v$, then $f \in D(H_U)$ iff $f \in D(H_0^*)$ and

$$f_1'(0) = f_2'(0) = f_3'(0) \equiv f'(0)$$

$$f_1(0) + f_2(0) + f_3(0) = D f'(0)$$

where D is given by

$$D = - \frac{3\varepsilon v + \sqrt{2}}{v}$$

Theorem 1 describes <u>all</u> admissible boundary conditions which
yield a self-adjoint Hamiltonian invariant under permutations.
This implies, in particular, that the operators used by Weger et
al in [23] to describe the electron density in metals are not
self-adjoint. These operators were characterized in the three-port
case by the boundary conditions

$$f_1(0) = f_2(0) = f_3(0) \equiv f(0)$$

$$v_1 f_1'(0) + v_2 f_2'(0) + v_3 f_3'(0) = \lambda\, f(0)$$

where v_1, v_2, v_3 were real constants describing the bond strengths
between different atomic chains. From the Theorem 1 it follows,
however, that this type of boundary conditions yields self-adjoint
operator iff $v_1 = v_2 = v_3 = 1$ (note that this type of boundary
conditions has been introduced for the first time in [26]).

In order to compare the above obtained Hamiltonians with the "splitters" used in microelectronics for the description of ultrathin branched wires we pass now to the scattering theory for such systems. We shall work in the time-independent framework choosing the wavefunctions as

$$f_j(x) = a_{j,in}\, e^{-ikx} + a_{j,out}\, e^{ikx}$$

where $a_{j,out}$, $a_{j,in}$ refer to the outgoing and incoming wave in the j-th branch, respectively. Requiring the vector $f=(f_1,f_2,f_3)$ to belong locally to $D(H_U)$ we find a relation between the incomingand outgoing states in the form

$$a_{out} = S a_{in}$$

where a_{in} , a_{out} are column vectors of $a_{j,in}$, $a_{j,out}$ respectively. The matrix S is just what is called "splitter" in the microelectronical transport theory. Using the boundary conditions from Theorem 1 we find that S is given by

$$S(k) = \begin{pmatrix} a(k) & b(k) & b(k) \\ b(k) & a(k) & b(k) \\ b(k) & b(k) & a(k) \end{pmatrix} \tag{4a}$$

with

$$a(k) = \frac{-1+ikB+k^2(2B^2-A^2-AB)}{1-ik(2A+B)+k^2(2B^2-A^2-AB)}$$

$$b(k) = \frac{-2ikB}{1-ik(2A+B)+k^2(2B^2-A^2-AB)}$$

in the case (i),

$$S(k) = \frac{1}{3(1+ikC)} \begin{pmatrix} 3ikC-1 & 2 & 2 \\ 2 & 3ikC-1 & 2 \\ 2 & 2 & 3ikC-1 \end{pmatrix} \tag{4b}$$

in the case (ii) and

$$S(k) = \frac{1}{3-ikD} \begin{pmatrix} 1-ikD & -2 & -2 \\ -2 & 1-ikD & -2 \\ -2 & -2 & 1-ikD \end{pmatrix} \qquad (4c)$$

in the case (iii).

It can be immediately seenthat none of these splitters is reflectionless. It implies that the reflectionless splitters used in the theoretical analysis cannot refer to a permutation-invariant situation. This is an important observation because in most cases the wires attached together at a branching point are regarded as equivalent being prepared by the same technique (they are usually of the same thickness, cross section etc.). In such a way the quantum dynamics leads in the permutation-invariant case to an inescapable recoil of the incoming electrons at the branching point. Moreover, the S-matrix is in the general case k-dependent which implies the reflection/transmission of the electron to be dependent on the shape of the incoming wave (for the scattering theory on graphs see also [27]).

The next natural question is, of course, to what extent do these "splitters" actually describe the scattering process in real wires which have a nonzero thickness. Unfortunately the mesoscopic transport measurements are not up to now sophisticated enough to make it possible to distinguish the scattering effects coming from the splitters from those which are of another origin (elastic and inelastic impurity scattering etc.) Therefore to check the formulae (4a-c) we have to use data obtained from the electromagnetic and acoustic waveguide measurements. It is obvious that in the stationary situation the scattering on a junction must lead to analogous results in the quantum and the electromagnetic/acoustic case. The reason is that in the stationary case the corresponding equations (although they describe a completely different physical situations) have the same form. For instance in the most simple two-dimensional case the Hamiltonian of a quantum particle moving within a branched wire is given by

$$H = - \frac{h^2}{2m^*} \Delta$$

$$D(H) = \{ \, f \in H^{2,2}(\Omega) \, : \, f|_{\delta\Omega} = 0 \, \}$$

where Ω denotes the inner region of the wire and $\delta\Omega$ is the corresponding wire surface (cf.Fig.2). Furthermore, m^* is the effective electron mass ; its magnitude depends on the particular lattice structure of the material used. In this way the electron scattering on the junction is governed by the equation

$$H \, \psi = k^2 \psi$$

which coincides precisely with the equation for the propagation of stationary transverse electric modes through a bifurcated waveguide with perfectly conducting walls [28]. In this analogy the constant $-h^2/2m^*$ in the front of the Laplacian is replaced by the effective dielectric constant of the medium inside the waveguide, and the square of the momentum, k^2, by the squared frequency. One expects, of course, that in the long-wave limit (or equivalently for d small enough) the transmission/reflection coefficients for the waveguide of a finite width turn into the coefficients described by the splitters (4) with the constants A,B,C,D being determined by the angles between the different branches of the waveguide. It is not easy to prove this heuristic statement in a rigorous way, and we have not been able to handle this task up to now.There are, however, indications that in the symmetric case ($\alpha=\beta$) the transmission/reflection coefficients are for d $\longrightarrow$ 0 described by the splitters (ii) or (iii).In order to illustrate this fact we compare the results which we obtained from (4) with those calculated numerically for a finite width waveguide by Mehran [29] (Fig.3). The similarity between these two plots is obvious. One can see for instance that to the case with $\alpha = \beta = 2\pi/3$ either the splitter (ii) with C=0 or (iii) with D=0 corresponds. In the non-symmetric case there are similar indications that the long-wave limit is described by the splitters of the type (i) where the two constants A,B are related to the angles α,β which specify the given junction.

This results (although only heuristic up to now) are in a close correspondence with the results obtained by R.Hoegh-Krohn and his co-workers for the low-energy scattering [30] ; as we have already mentioned they demonstrated that in the low-energy (long-wave) region the scattering data did not depend on the detailed shape of the potential and were described fully by the appropriate point interaction Hamiltonian. The situation described here is in some sense similar since because the long-wave junction scattering depends also only of the global characteristics of the graph. This fact can have a great practical value, in particular, in the situations when we have to compute the wave propagation in more complicated waveguide systems (for instance, sound propagation in an air-conditioning system).

Let us now turn to compact graphs. We are going to discuss here the most simple model of a naphthalene molecule which is represented by the following graph (Fig.4). We choose the state Hilbert space in this case as

$$\mathcal{H} = L^2(0,l) \oplus L^2(0.5l) \oplus L^2(0,5l)$$

where l is the distance between the neighboring carbon nuclei ($l=1.39$ A in the aromatic hydrocarbons). The corresponding Hamiltonian represents a self-adjoint extension of the operator H_0 which is now defined as

$$H_0 = H_{0,1} \oplus H_{0,2} \oplus H_{0,3}$$

where

$$H_{0,i} = -\frac{\hbar^2}{2m_e}\frac{d^2}{dx^2}$$

and

$$D(H_{0,1}) = C_0^\infty(0,l)$$

$$D(H_{0,j}) = C_0^\infty(0,5l), \quad j=2,3$$

Taking into account that all bonds between nuclei are equivalent and that the dynamics of the π-electrons is local (i.e. that the corresponding Hamiltonian cannot be represented by an operator with boundary conditions mixing together two branching points of the graph) we get that the Hamiltonian must be described by the boundary conditions from Theorem 1 which are now imposed at each branching point. Moreover, since the angles between bonds at the branching points are equal approximately to $120°$ we choose the boundary conditions of the type (ii) with C=0 (the boundary conditions (iii) are ruled out because of the wavefunction continuity) In this way we obtain finally the Hamiltonian corresponding to the naphthalene graph. Solving the corresponding eigenvalue problem we can calculate the transition energies. We have only to take into account that each carbon atom in the molecule supplies one π-electron which, in correspondence with the Pauli principle, occupies the lowest possible energy level (Fig.5).

Table 1 (borrowed from [31]) shows the results computed for the transition between the highest filled level and the lowest empty energy level in comparison with the experimental data.

molecule	graph	calculated transition [cm^{-1}]	observed transition [cm^{-1}]
Naphthalene		31 400	28 200
Anthracene		20 500	20 700
Naphtracene		14 420	15 500
Pentacene		10 690	11 800
Phenantrene		28 800	27 500
Triphenyle		29 500	32 800

This results demonstrate clearly that the described model of quantum mechanical motion on graphs is relevant for the observed π electron transitions. Moreover, it seems that the graph dynamics of the electron is in some sense decisive for the whole system. In other words the quantum mechanics on graphs with a given topological and geometrical properties contains already all the essential information about the involved system and represents in this a suitable abstraction.

2. Point contacts

As a next model, we would like to demonstrate an application of the self-adjoint extension theory to the systems which are called in the solid-state physics usually point contacts.

Let us first outline the physical background of the model. For a metallic contact, the common wishdom suggests a linear relation between the applied voltage and the current (the Ohm's law). This remains true, however, only if the size of the contact is large enough. Once its diameter becomes comparable with the mean free path of the electrons in the metal, scattering effects appear which add a nonlinear contribution to the current.

Effects of this type were measured first at Kharkhov in the pioneering experiment by Igor Yanson [32], which gave rise to a new research branch called now point-contact spectroscopy. The measured nonlinearity represents usually a few promile to a few percent of the total current and is visible in the differential resistance dU/dI. The second derivative d^2U/dI^2 exhibits typically a more complicated shape with peaks corresponding to the electron-phonon interaction in the metal involved.

There are two basic types of the point-contact experiments. In the first of them, dubbed a pressure-type contact, a sharply tipped wire is adjusted by a screw against a flat metallic surface. The second type consists of two thin metallic films separated by an insulating (oxide) layer which is perforated at one point. In both two types the contact region is typically a few micrometers in diameter.

The "microscopic" theory of these contacts is a very complicated matter and we are not going to discuss it here. We shall restrict our attention again to the long-wave-limit situation only, when the de Broglie wavelength of the electrons is much longer than the contact diameter. In this case it is reasonable to expect that the contact scattering would not depend on the detailed shape of the contact region, but rather on the global geometry of the experiment. For brevity, we restrict ourselves only to the thin-film case (the pressure-type contacts are discussed in [33]). Our aim here is to show once more that the self-adjoint extension theory represents a powerful method to solve problems of this type.

We consider here the simplest possible quantum-mechanical model [34] in which a free electron moves on a manifold consisting of two planes connected at one point which can be regarded as the limiting case of a thin-film contact when the thickness of the films and the insulating layer approaches zero (Fig.6). The state Hilbert space $\mathcal{H}$ of our problem is the sum of the spaces corresponding to the upper and lower plane, respectively : $\mathcal{H} = L^2(\mathbb{R}^2) \oplus L^2(\mathbb{R}^2)$. To construct the quantum Hamiltonian we use the above described strategy starting with the operator

$$H = H_{0,1} \oplus H_{0,2}$$

where $H_{0,j} = -\Delta$ with $D(H_{0,j}) = C_0^\infty(\mathbb{R}^2 \setminus \{0\})$, $j = 1,2$, assuming the connection point to be placed at the origin of coordinates on each of the planes. The deficiency indices of the operators $H_{0,j}$ are known to be $(1,1)$ [24], hence the deficiency indices of H_0 are $(2,2)$ and it has a four parameter family of self-adjoint extensions. To construct these extensions we proceed in the standard way. The deficiency subspaces $\mathcal{K}^\pm$ are spanned by the vectors

$$\varphi_k^\pm \ , \quad k = 1,2,$$

where

$$\varphi_k^+ = (f_0, 0) \ , \qquad \varphi_k^- = (0, f_0)$$

with

$$f_0(x) = H_0^{(1)}(\varepsilon x) \ , \qquad \varepsilon = e^{i\pi/4}$$

and φ_k^- are complex conjugated to φ_k^+. Introducing the polar coordinates in each of the two planes and decomposing the Hilbert space

$$L^2(\mathbb{R}^2) = \bigoplus_{m=-\infty}^{\infty} L^2(\mathbb{R}^+, r\,dr) \otimes \{Y_m\}_{lin}$$

where

$$Y_m(\varphi) = (2\pi)^{-1/2} e^{im\varphi}$$

we can decompose the operators $H_{0,j}$ as

$$H_{0,j} = \bigoplus_{m=-\infty}^{\infty} h_{m,j} \otimes \mathbb{1}$$

where

$$h_{m,j} = -\frac{d^2}{dr^2} - \frac{1}{r}\frac{d}{dr} + \frac{m^2}{r^2} \quad ; \quad D(h_{m,j}) = C_0^{\infty}(\mathbb{R}^+ \setminus \{0\}).$$

All the operators $h_{m,j}$ are e.s.a. for $m \neq 0$ (see [24], Sec. X.1). We get therefore

Theorem 2:

All self-adjoint extensions of the operator H_0 are of the form

$$H_U = K_U \oplus \bar{h} ,$$

where

$$h = \left(\bigoplus_{m \neq 0} h_{m,1} \otimes \mathbb{1} \right) \oplus \left(\bigoplus_{l \neq 0} \mathbb{1} \otimes h_{l,2} \right)$$

and K_U is a self-adjoint extension of the operator

$$K_0 = (h_{0,1} \otimes \mathbb{1}) \oplus (\mathbb{1} \otimes h_{0,2})$$

with the domain

$$D(K_0) = \{ \varphi = (f_1, f_2): f_j(x) = f_j(|x|), f_j \in C_0^{\infty}(\mathbb{R}^+ \setminus \{0\}) \}$$

Hence it is only necessary to find the operators K_U. The most simple way to do it is to relate them to suitable boundary conditions. The deficiency functions are, however, singular at the origin, and therefore the boundary conditions must be written in terms of regularized boundary values [35]

$$L_0(f) = \lim_{r \to 0} \frac{f(r)}{\ln r} \quad ; \quad L_1(f) = \lim_{r \to 0} [f(r) - L_0(f) \ln r] \ .$$

__Theorem 3:__

Every self-adjoint extension K_U is uniquely specified by the following boundary conditions: $f = (f_1, f_2) \in D(K_U)$ iff

$$f_i \in \{ f \in L^2(\mathbb{R}^+, rdr); \ f, f' \in AC(\mathbb{R}^+) \ \text{and} \ f'' + \frac{1}{r} f' \in L^2(\mathbb{R}^+, rdr) \}$$

for $i = 1,2$ and

(i)
$$L_0(f_1) = aL_0(f_2) + bL_1(f_2)$$
$$L_1(f_1) = cL_0(f_2) + dL_1(f_2)$$

where the coefficients are given by

$$a = u_{12}^{-1}[\chi(u_{11}-1) + \bar{\chi}(\det(U) - u_{22})]$$
$$b = \frac{2i}{\pi} u_{12}^{-1}[1 - \text{tr}(U) + \det(U)]$$
$$c = \frac{\pi i}{2} u_{12}^{-1}[\chi^2 + \chi\bar{\chi} \ \text{tr}(U) + (\bar{\chi})^2 \det(U)]$$
$$d = u_{12}^{-1}[\chi(1-u_{22}) + \bar{\chi} (u_{11} - \det(U))]$$

and u_{ij} are matrix elements of a nondiagonal unitary 2x2 matrix describing the self adjoint extension of H_0 ; or in the case with diagonal U,

(ii)
$$L_0(f_1) = AL_1(f_1)$$
$$L_0(f_2) = BL_1(f_2)$$

with $A, B \in \mathbb{R}$. Here $\chi = L_1(f_0) = \frac{1}{2} + \frac{2i}{\pi} (\gamma - \ln 2)$ and $\gamma = 0.577..$ is the Euler's constant.

Proof: See [34]

__Remark:__ The extensions defined by the boundary conditions (ii) are physically not interesting because they yield Hamiltonians which

are a direct sum, and hence describe systems in which the two
planes are completely separated. In what follows we restrict
ourselves only to the extensions given by (i).

We suppose further that the two planes are physically equivalent,
i.e., we restrict our attention to the Hamiltonians H_U which
commute with the modified parity operator P exchanging the planes

$$P : (f_1, f_2) \longrightarrow (f_2, f_1) \quad ; \quad (f_1, f_2) \in \mathcal{H} \quad .$$

It can be shown [36] that the Hamiltonians which fulfil

$$PH_U \subset H_U P$$

form a two parameter subfamily corresponding to symmetric matrices
U. This subfamily can be described by the boundary conditions (i)
from the Theorem 3 with the coefficients a,b,c,d given by

$$a = -d = \frac{\cos \beta + \cos \xi - \sin \xi}{\sin \xi}$$

$$b = 2^{1/2} \frac{\cos \beta + \cos \xi}{\sin \beta}$$

$$c = 2^{1/2} \frac{\sin \xi - \cos \beta}{\sin \beta}$$

where $\beta, \xi \in (0, 2\pi]$ are two real parameters.

Let us now investigate scattering of the particle on the
connection point. Our aim is to find the transmission probability
from the upper to the lower plane. Using the time-independent
approach we start with the function $f = (f_1, f_2)$, where

$$f_1(r) = H_0^{(2)}(kr) + A(k) H_0^{(1)}(kr)$$

$$f_2(r) = B(k) H_0^{(1)}(kr)$$

and demand it to belong locally to $D(H_U)$. A simple calculation
yields the coefficients $A(k), B(k)$ which are expressed as

$$A(k) = \frac{c - 2a\left(\gamma + \ln\frac{k}{2}\right) - b\left[\frac{\pi^2}{4} + \left(\gamma + \ln\frac{k}{2}\right)^2\right]}{c - 2a\left(\frac{\pi}{2i} + \gamma + \ln\frac{k}{2}\right) - b\left(\frac{\pi}{2i} + \gamma + \ln\frac{k}{2}\right)^2}$$

$$B(k) = \frac{-i\pi}{c - 2a\left(\frac{\pi}{2} + \gamma + \ln\frac{k}{2}\right) - b\left(\frac{\pi}{2i} + \gamma + \ln\frac{k}{2}\right)^2}$$

A(k) and B(k) are the reflection and transmission coefficients, respectively. It can be easily seen that they fulfil

$$|A(k)|^2 + |B(k)|^2 = 1.$$

Thus we have obtained a nontrivial particle transmission from the upper to the lower plane.

Before comparing our results to the experimental data we have to relate the quantum transition coefficients to the measured resistance of the point contact. This can be done using a formula well known from the electrical transport theory [37]. If the metals involved have the same Fermi energies then the current through the contact is given by

$$I = -\frac{2e}{h} \int_0^\infty T(E)\left[f_T(E) - f_T(E - eU)\right]\,dE \qquad (5)$$

where e is the electron charge, U is the applied voltage and

$$f_T(E) = \left[1 + \exp\left(\frac{E - E_F}{kT}\right)\right]^{-1}$$

is the metal electron-gas density at the temperature T and Fermi energy E_F. Furthermore, T(E) is the quantum transition probability which is given simply by

$$T(E) = |B(k^2)|^2$$

where B(k) is the transmission coefficient.

The formula (5) becomes particularly simple in the zero-temperature limit, when $f_T(E)$ becomes a step function

$$I = - \frac{2e}{h} \int_{E_F}^{E_F+eU} T(E) \, dE \quad .$$

Differenciating it by U we get finally the sought formula for the differential resistance

$$\frac{dU}{dI} = \frac{h}{2e^2} \, T(E_F +eU)^{-1}$$

In the models described in this section there are two adjustable parameters with the help of which we can fit the basic nonlinearity of the current-voltage characteristic measured in the experiment. In order to give an example we plot on Fig.7 the resistance corresponding via the formula (5) to the constructed quantum model . The same plots are transformed on Fig.8 to the logarithmic scale. One has to compare these curves with the voltage dependence of the resistance measured in the point-contact experiment [38] which is plotted on the Figures 9 and 10. We see that the shape of the resistance curves obtained in our model for a suitable choice of the parameters fits well with the experimental data.

On the other hand, the models under consideration cannot describe a more complicated structure observed in the second derivatives of the current-voltage characteristic which is connected with the particular electron-phonon interaction in the metal used unless an appropriate interaction is added. One can, in particular, add an internal Hilbert space $\mathcal{H}_{in}$ to to the state Hilbert space $\mathcal{H}$ and model in such a way, for instance, the polaron states localized in the contact region. Constructing self-adjoint extensions in the larger Hilbert space one obtains a more complicated structure of the transition coeficients and hence also a more complicated structure of the resistance plot. Adding the internal space, it is possible to mathematically combine the global manifold geometry, which is responsible for the dU/dI plot with the detailed electron-phonon interaction being responsible for the complicated structure of the d^2U/dI^2 plot.

References

1. R.L.Kronig, W.G.Penney: Proc.Roy.Soc.(London) A130 (1931) 499
2. E.Fermi: Ric.Scientifica 7 (1936) 13
3. Yu.N.Demkov, V.N.Ostrovskij : The zero-range potential methods in the atomic physics, Leningrad University Press, Leningrad 1975 (in Russian)
4. S.Albeverio, F.Gesztesy, H.Holden, R.Hoegh-Krohn: Solvable models in quantum mechanics, Springer Verlag, Berlin 1988
5. L.Pauling: J.Chem Phys. 4 (1936) 673
6. N.S.Bayliss: J.Chem Phys. 16 (1948) 287
7. H.Kuhn: Helv.Chimica Acta 31 (1948) 1441
8. S.Basu: J.Chem.Phys. 22 (1954) 2170
9. C.A.Coulson: Proc.Phys.Soc. A 66 (1953) 652
10. J.S.Griffith: Trans.Faraday Soc. 49 (1953) 345
11. A.A.Frost: J.Chem.Phys.23 (1955) 985
12. K.Ruedenberg, Ch.Scherr: J.Chem.Phys. 21 (1953) 1565
13. J.A.Hoerni: J.Chem.Phys. 34 (1961) 508
14. C.A.Coulson: Proc.Roy.Soc. A68 (1955) 74
15. K.Ruedenberg: J.Chem.Phys. 22 (1954) 1878
16. M.Buettiker: IBM J.Res.Develop. 32 (1988) 317
17. C.P.Umbach et al: Appl.Phys.Lett. 50 (1987) 1289
18. M.Buettiker, Y.Imry, Ya.Azbel: Phys.Rev.A 30 (1984) 1982
19. B.Doucot, R.Rammal: J.Physique 47 (1987) 973
20. B.Shapiro: Phys.Rev.Lett. 50 (1983) 747
21. Y.Gefen, M.Ya.Azbel: Surf.Science 142 (1984) 203
22. B.R.Bulka: Phys.Stat.Sol. b 141 (1987) 239
23. M.Weger, S.Alexander, G.D.Riccia: J.Math.Phys. 14 (1973) 259
24. M.Reed, B.Simon: Methods of Modern Mathematical Physics II, Academic Press, New York 1975
25. P.Exner, P.Seba: Free quantum motion on a branching graph, Preprint JINR Dubna E2-87-213, Reports on Math.Phys. in press
26. A.A.Frost: J.Chem.Phys. 23 (1955) 985
27. N.I.Gerasimenko, B.S.Pavlov: Teor.Mat.Fiz. 74 (1988) 345

28. R.Mitra, S.W.Lee: Analytical Techniques in the Theory of Guided Waves; Macmillan Comp.,New York 1971

29. R.Mehran: IEEE Trans.Microwave Theory and Technique 26 (1978) 400

30. S.Albeverio, F.Gesztesy R.Hoegh-Krohn: On the universal low energy limit in nonrelativistic scattering theory. Acta Phys.Austriaca Suppl. 23 (1981) 577

31. C.W.Sherr: J.Chem Phys. 21 (9) (1953) 1582

32. I.K.Yanson: Zh.Eksp.Teor.Fiz. 66 (1974) 1035

33. P.Exner, P.Seba: J.Math.Phys. 28 (1987) 386

34. P.Exner, P.Seba: Lett.Math.Phys.12 (1986) 193

35. W.Bulla, F.Gesztesy: J.Math.Phys. 26 (1985) 2520

36. P.Exner, P.Seba: A simple model of thin-film contacts in two and three dimensions. Czech.J.Phys. B38 (1988) in press

37. K.C.Kao, W.Huang: Electrical Transport in Solids. Pergamon Press, New York, 1981

38. A.G.M. Jansen, A.P. van Gelder, P. Wyder: J.Phys.C 13 (1980) 6073

Figure captions

Fig.1 : Connection of three halflines

Fig.2 : The three port-junction of wires of a finite thickness

Fig.3a: The transmission/reflection coefficients calculated by Mehran

Fig.3b: The transmission/reflection coefficients for a symmetric three port-junction compared with those obtained from the splitter (iii)

Fig.4 : A naphtalene molecule

Fig.5 : The occupied energy levels for naphthalene

Fig.6 : A thin-film point contact

Fig.7 : A resistance plot

Fig.8 : A resistance plot - logarithmic scale

Fig.9 : A measured resistance nonlinearity

Fig.10: A measured resistance nonlinearity -logarithmic scale

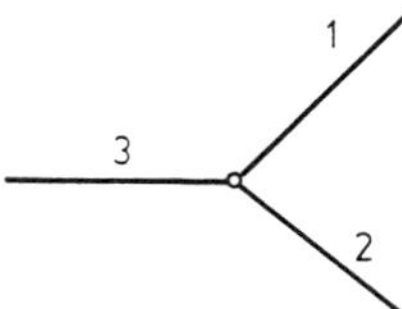

Fig.1

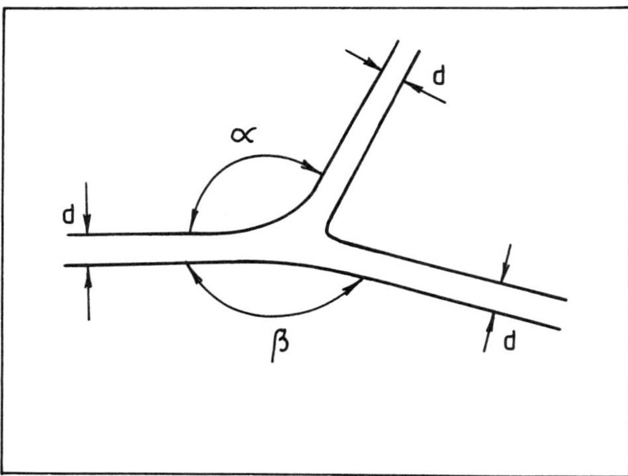

Fig. 2

|S_{12}|
2/3
180° 160° 140°
120°
120°
140°
160°
180°
1/3 |S_{11}|
0 4 8 12 16 20 24 28 32 36 40 44
Frequency [GHz]

Fig 3a

1
2/3 |S_{12}|
|S_{11}|
1/3
0
0 k .5

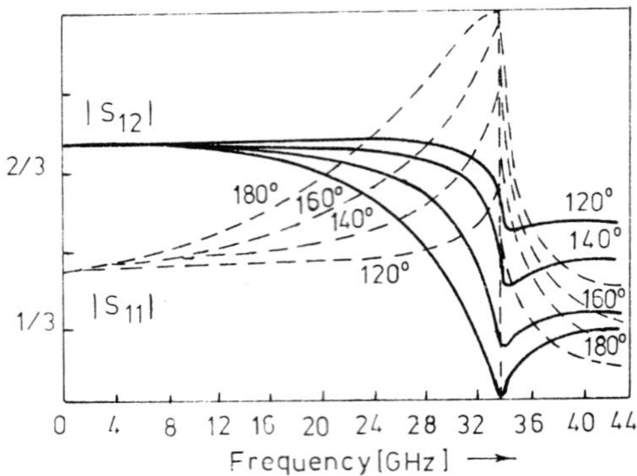

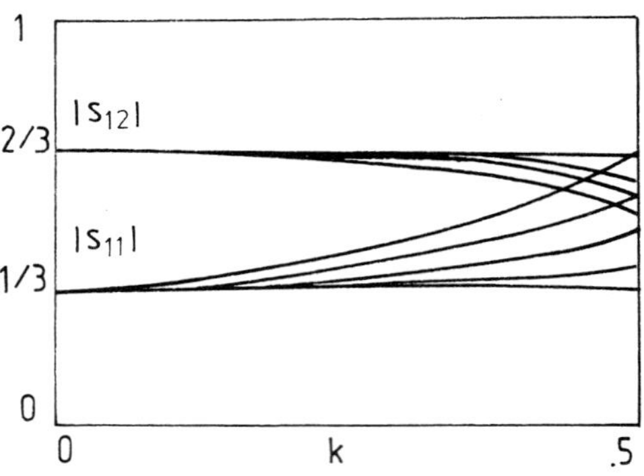

Fig. 3b

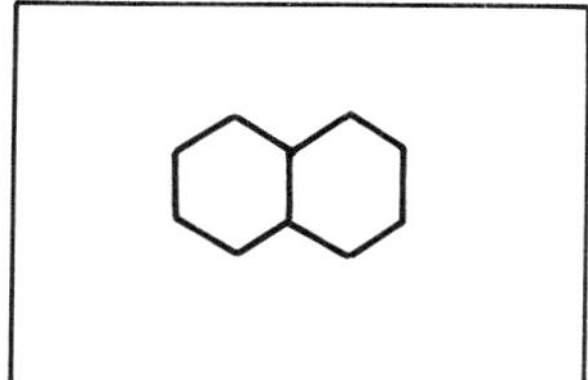

Fig. 4

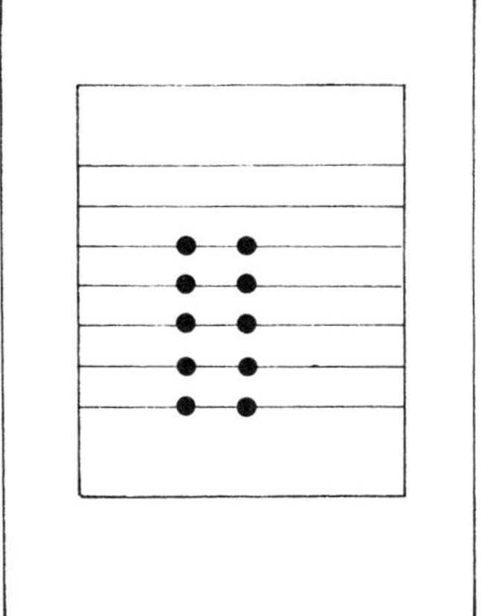

Fig.5

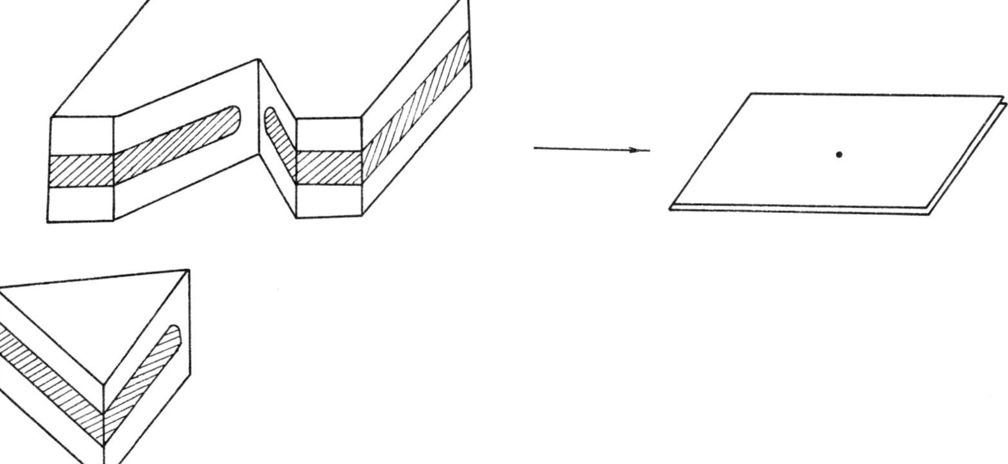

Fig.6

dU/dI
arbitrary limits
U

Fig.7

dU/dI
logarithmic
scale
arbitrary limits
ln(U)

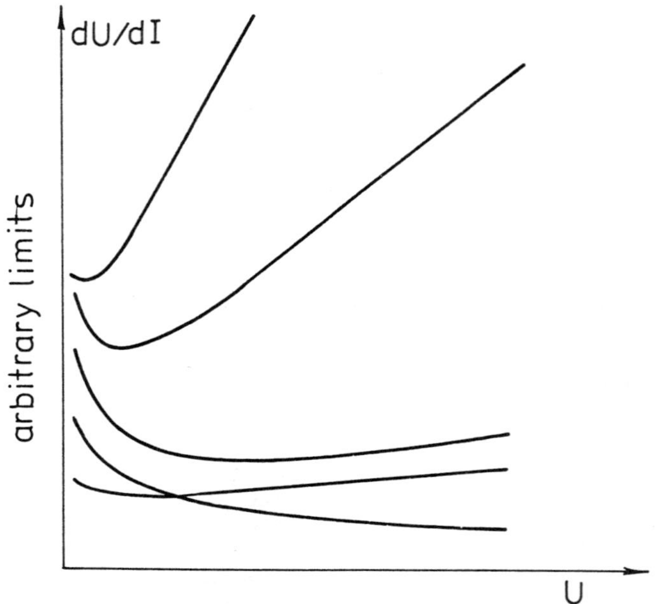

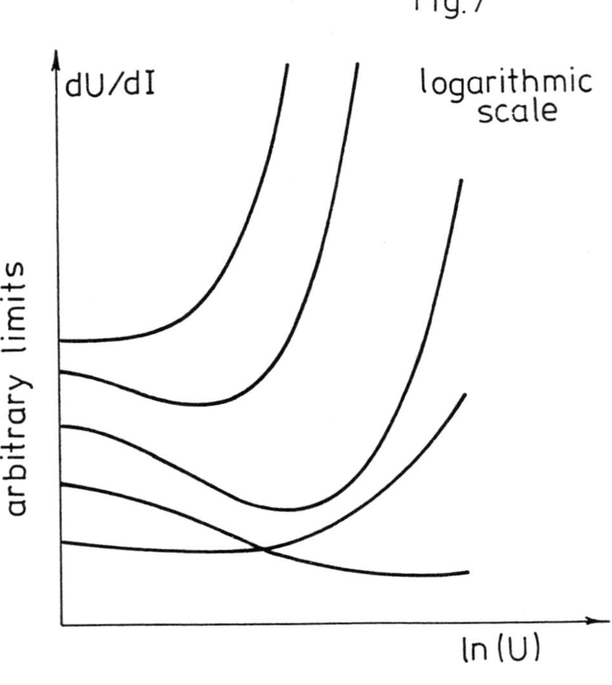

Fig.8

dU/dI [0.5mΩ/division]
0   2   4   6
Voltage U [mV]
Au + 0.03% Mg contact
R ≈ 2.1Ω , T = 1.5K

Fig. 9

dU/dI
AU + 003% Mg contact
R ≈ 21Ω ; T = 1.5K
logarithmic scale
arbitrary limits
1   voltage (mV)   10

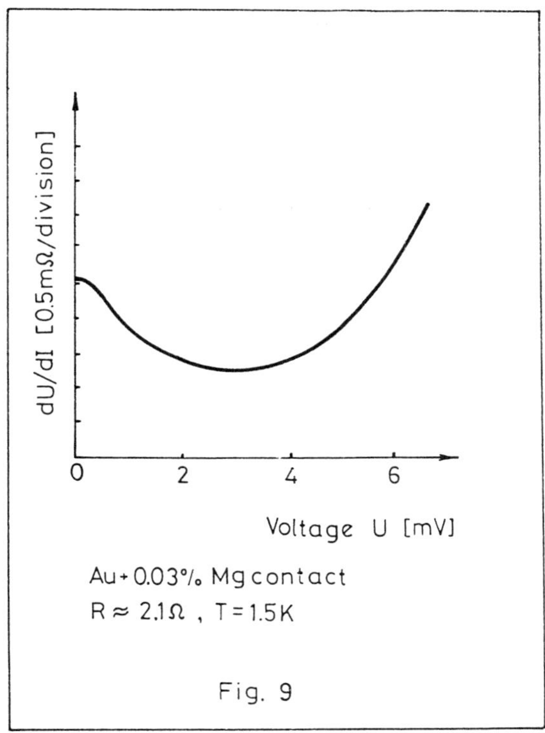

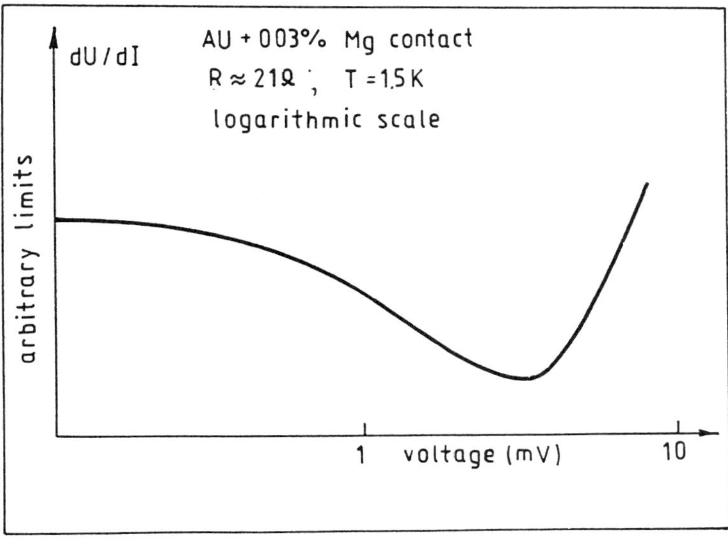

Fig. 10

High Energy Resolvent Estimates for Schrödinger Operators

ARNE JENSEN

Department of Mathematics and Computer Science
Institute for Electronic Systems
Aalborg University, Fr. Bajers 7
DK-9220 Aalborg Ø, Denmark

1 MOTIVATION. INTRODUCTION.

Let us consider a Schrödinger operator $H = -\Delta + V$ on the Hilbert space $\mathcal{H} = L^2(\mathbf{R}^m)$. We are interested in investigating the large time behavior of the evolution group e^{-itH}. Looking at the representation

$$e^{-itH} = \int_{-\infty}^{\infty} e^{-it\lambda} dE(\lambda), \tag{1}$$

where $E(\lambda)$ denotes the spectral family of H, we see that there will be two major contributions, one from the thresholds, and one from the high energy part. For the absolutely continuous part of the spectrum of H we have with $R(z) = (H - z)^{-1}$

$$E'(\lambda) = \frac{1}{2\pi i}(R(\lambda + i0) - R(\lambda - i0)) \tag{2}$$

where the boundary values exist between suitable spaces.

In this paper we consider the question of the high energy behavior. Before doing so, let us mention that there are several studies of the threshold behaviors of Schrödinger operators. A very interesting apporach is the one by Høegh-Krohn and collaborators connecting the threshold behavior with the point interactions. These results and further references can be found in the monograph [AGHH(88)]. A perturbation approach using the explicitly known kernel of the resolvent of the Laplacian has been used in [JK(79), J(80), J(84), Mu(82)]. All these results treat only potentials of very short range. For long range potentials very little is known about the threshold behavior for non-central potantials. One result in this direction is [Y82].

It follows from (1) and (2) that in order to get results on the contribution from the high energy part it suffices to study the differentiability properties of $E'(\lambda)$ between suitable spaces and estimate the large λ behavior of the derivatives. Before doing this we briefly mention some earlier results. Our main interest is in results applicable to Schrödinger operators with *long range potentials*.

2 PREVIOUS RESULTS.

We introduce the weighted L^2-spaces:

$$L^{2,s}(\mathbf{R}^m) = \{f \in L^2_{loc}(\mathbf{R}^m) \mid \|f\|_{L^{2,s}} = \|(1+x^2)^{s/2}f\|_{L^{2,s}(\mathbf{R}^m)} < \infty\}$$

for any $s \in \mathbf{R}$.

Saitō proved in [S(74)] the following result: Let $V \in C^2(\mathbf{R}^n)$ and assume for some $\varepsilon > 0$ and all multi-indices α with $|\alpha| \leq 2$

$$|(\partial/\partial x)^\alpha V(x)| \leq c_\alpha (1+|x|)^{-\varepsilon-|\alpha|}.$$

Let $H_0 = -\Delta$ and $H = H_0 + V$. Then

$$R(\lambda \pm i0) = \lim_{\varepsilon \downarrow 0} R(\lambda \pm i\varepsilon)$$

exists in the operator norm of $\mathcal{B}(L^{2,s}(\mathbf{R}^m), L^{2,-s}(\mathbf{R}^m))$ for $s > 1/2$ and all $\lambda > 0$. Furthermore, we have the estimate

$$\|R(\lambda \pm i0)\|_{\mathcal{B}(L^{2,s}(\mathbf{R}^m),L^{2,-s}(\mathbf{R}^m))} \leq c\,\lambda^{-1/2}, \quad \lambda \geq \lambda_0 > 0. \tag{3}$$

For H_0 this estimate (3) was proved in [A(75)]. Results on the differentiability of the boundary values $R(\lambda \pm i0)$ and high energy estimates for these derivatives were obtained in [JK(79), Mu(83), Mu(84)] for short range potentials.

3 RESULTS.

We obtain results for a class of generalized many-body Schrödinger operators. Let $X = \mathbf{R}^m$ and let π_j, $j = 1, 2, \ldots, N$, be orthogonal projections in X. Let $X_j = \pi_j(X)$ be the range of π_j. We consider potentials which are real-valued functions defined on X_j. Let x_j be the coordinates on X_j, and let ∇_j denote the corresponding gradient. For an integer $n \geq 2$ we make the following assumptions:

Assumption 3.1. *Let $V_j \in C^n(X_j)$, $j = 1, 2, \ldots, N$, such that*

$$\sup_{x_j \in X_j} \sum_{l=1}^{n} |((x_j \cdot \nabla_j)^l V_j)(x_j)| < \infty. \tag{4}$$

Note that we do not assume $V_j(x_j) \to 0$ as $|x_j| \to \infty$. Thus the V_j could be functions homogeneous of degree zero outside the unit ball. Furthermore, we need conditions only on a finite number of derivatives of potentials.

We introduce coupling constants $g = (g_1, g_2, \ldots, g_N) \in \mathbf{R}_+^N$ and write

$$V(g)(x) = \sum_{j=1}^{N} g_j V_j(\pi_j(x)), \quad x \in X.$$

The operator $H = H_0 + V(g)$ is the generalized many-body Schrödinger operator considered here. The resolvent is denoted $R(z) = (H(g) - z)^{-1}$. Let

$$\mu(V_j) = \max\{0, \sup_{x_j \in X_j} (V_j(x_j) + \tfrac{1}{2} x_j \cdot \nabla_j V_j(x_j))\}.$$

With these definitions we can state the main result as follows:

Theorem 3.2. *Let V_j satisfy Assumption 3.1. There exists $\lambda_0 > 0$ such that for $c_0 > 0$, $\delta > 1$, and all λ and g satisfying $\lambda \geq \lambda_0$, $\lambda \geq c_0 \max\{g_j \mid j = 1, 2, \ldots, N\}$, and $\lambda \geq \delta \sum_{j=1}^{N} g_j \mu(V_j)$ the following results hold: (i) The boundary values $R(\lambda \pm i0)$ exist in the operator norm on $\mathcal{B}(L^{2,s}, L^{2,-s})$ for $s > 1/2$. Furthermore, (ii) For $k = 0, 1, 2, \ldots, n - 2$, $s > k + (1/2)$, the map $\lambda \to R(\lambda \pm i0)$ is k times continuously differentiable from (λ_0, ∞) to $\mathcal{B}(L^{2,s}, L^{2,-s})$ and we have the estimate*

$$\|(d/d\lambda)^k R(\lambda \pm i0)\|_{\mathcal{B}(L^{2,s}, L^{2,-s})} \leq c\lambda^{-(k+1)/2},$$

where c is independent of $g \in \mathbf{R}_+^N$ satisfying the above conditions.

Method of proof: We will indicate the main ingredients of the proof.
(i) Differentiability of the boundary values has been proved using the Mourre method in [JMP(84), J(85)]. This is reduced to proving resolvent estimates for powers of the resolvent, since we have $(d/d\lambda)^k R(\lambda + i\varepsilon) = k! R(\lambda + i\varepsilon)^{k+1}$.
(ii) By a scaling argument similar to the one used in [Y86] one can reduce the problem to one at fixed energy equal to one, but with a λ-dependent potential $V(g)(\lambda^{-1/2} x)$.
(iii) The last step of the proof consists in tracking the constant dependence in the estimates used in Mourre's method. This is tedious, but straightforward. It is an essential feature of Mourre's method that it is constructive in this sense.

Let us note that the condition $\lambda \geq \delta \sum_{j=1}^{N} g_j \mu(V_j)$ implies the non-trapping condition for the corresponding classical system.

To make the constructive aspect of the Mourre method explicit, we will give the computations in an example. Let $V(x) = c|x|^{-\nu}$, $c \in \mathbf{R}$, $0 < \nu < \min\{m/2, 2\}$ and let $H = -\Delta + V(x)$. We then have the following result:

Proposition 3.3. *The operator H defined above satisfies the following estimate:*

$$\|(1 + x^2)^{-1/2}(H - \lambda - i\varepsilon)^{-1}(1 + x^2)^{-1/2}\|_{\mathcal{B}(L^2)}$$
$$\leq \frac{1}{\nu\sqrt{\lambda}}\left[1 + 6\sqrt{3}\sqrt{\nu/(2 - \nu)}[1 + 2\sqrt{\nu/(2 - \nu)} + (2 + 2\nu + m)/(\nu\sqrt{\lambda})]^{1/2}\right.$$
$$+ (2 + 2\nu + m)/(\nu\sqrt{\lambda})(1 + 2\sqrt{\nu/(2 - \nu)} + (2 + 2\nu + m)/(\nu\sqrt{\lambda})) \times$$
$$\left. \times (\chi_{(1,\infty)}(\lambda) + 4\sqrt{\lambda}\chi_{(0, 1]}(\lambda))\right]$$

for all $\lambda > 0$ and all $\varepsilon \neq 0$.

Before sketching the proof of this result we note that the estimate is independent of the constant c in the potential V. The result gives a behavior $O(\lambda^{-1/2})$ as $\lambda \to \infty$ and $O(\lambda^{-1})$ as $\lambda \to 0$.

Sketch of proof: Let $A = (-i/2)(x \cdot \nabla + \nabla \cdot x)$ be the generator of dilations. A computation yields $i[H, A] = 2H_0 + \nu V = (2 - \nu)H_0 + \nu H$. Let $\lambda > 0$ be fixed. We want to get an estimate at energy λ, so we replace H on the right hand side by λ and define $B^*B = (2 - \nu)H_0 + \nu\lambda$. It follows from [Mo(81),Proposition II.5] that the operator $G(\varepsilon) = (H - \lambda - i\mu - i\varepsilon B^*B)^{-1}$ is a bounded operator for $\mu > 0$ and $\varepsilon > 0$. Define $F(\varepsilon) = (1 + x^2)^{-1/2}G(\varepsilon)(1 + x^2)^{-1/2}$. This function is differentiable in operator norm wiht respect to ε. Computations analogous to those in [Mo(81)] give the following two estimates:

$$\|F(\varepsilon)\| \leq \frac{1}{\nu\lambda\varepsilon} \tag{5}$$

and

$$\|\frac{d}{d\varepsilon}F(\varepsilon)\| \leq \frac{2}{\sqrt{2 - \nu}}\frac{1}{\sqrt{\varepsilon}}\|F(\varepsilon)\|^{1/2} + (2 + 2\nu + m)\|F(\varepsilon)\|. \tag{6}$$

Integrate the differential inequality (6) using (5) and iterate twice to get an estimate for $\|F(\varepsilon)\|$ valid down to $\varepsilon = 0$ and thus the result of the proposition. The point here is that we can follow the constant dependence in the two iterations. To get the constants in the Proposition we used the estimate

$$-\log(s) \leq s^{1-\delta}/(\delta - 1), \quad 1 < \delta < 2, \quad 0 < s \leq 1$$

and optimized the choice of the parameter δ.

Details of the results in this section can be found in the paper [J(89)].

4 BESOV SPACE ESTIMATES.

Besov spaces were introduced into the study of spectral properties of Schrödinger operators by Agmon and Hörmander in the paper [AH(76)]. Recently, the high energy estimates for the Laplacian in Besov spaces was obtained by Agmon in [A89]. It turns out that by using the results and methods described above and a result from [AP(85)], one can prove high energy resolvent estimates in Besov spaces for the generalized many-body Schrödinger operator considered here. We shall briefly describe the results obtained.

Let A be a selfadjoint operator with domain $\mathcal{D}(A)$ in a Hilbert space $\mathcal{H}$. Its spectral family is denoted E_A. Let $I_0 = \{\lambda \in \mathbf{R} \,|\, |\lambda| \leq 1\}$, $I_j = \{\lambda \in \mathbf{R} \,|\, 2^{j-1} < |\lambda| \leq 2^j\}$,

$j = 1, 2, \ldots$, and $R_j = 2^j$, $j = 0, 1, 2, \ldots$. The abstract Besov space relative to A is defined as

$$B_A = \{\psi \in \mathcal{H} \mid \|\psi\|_{B_A} = \sum_{j=0}^{\infty} R^{1/2}\|E_A(I_j)\psi\| < \infty\}.$$

The space B_A is a Banach space with the norm $\|\cdot\|_{B_A}$. The inner product on $\mathcal{H}$ allows one to identify the dual space B_A^* with the completion of $\mathcal{H}$ in the norm

$$\|\xi\|_{B_A^*} = \sup_{j \geq 0}(R_j^{-1/2}\|E_A(I_j)\xi\|).$$

An equivalent norm is given by

$$\||\xi|\|_{B_A^*} = \sup_{R \geq 1}(R^{-1/2}\|E_A([-R, R])\xi\|).$$

If one takes $\mathcal{H} = L^2(\mathbf{R}^m)$ and $A = |x|$ (multiplication operator), then the space B_A becomes the Besov space used in [AH(76)]. Here we denote this space $B(\mathbf{R}^m)$. Note $L^{2,s}(\mathbf{R}^m) \subset B(\mathbf{R}^m) \subset L^{2,1/2}(\mathbf{R}^m)$ for all $s > 1/2$. Thus Besov space estimates are stronger than estimates in weighted L^2-spaces.

We also need a subspace of $L^\infty(\mathbf{R})$ introduced by Birman and Solomjak. Let $f \in L^\infty(\mathbf{R})$ and define for $j \geq 0$ $s_j(f) = \operatorname{ess\,sup}\{|f(\lambda)| \mid \lambda \in I_j\}$. We then define

$$\ell^2(L^\infty) = \{f \in L^\infty(\mathbf{R}) \mid \|f\|_{\ell^2(L^\infty)} = (\sum_{j=0}^{\infty} s_j(f)^2)^{1/2} < \infty\}.$$

The main result obtained is the following theorem. We use the same notation as in the previous section. In particular, A denotes the generator of dilations.

Theorem 4.1. *Let V_j, $j = 1, 2, \ldots, N$, satisfy Assumption 3.1 with $n = 2$. Then there exist $\lambda_0 > 0$ such that for $c_0 > 0$, $\delta > 1$, and all λ and g satisfying $\lambda \geq \lambda_0$, $\lambda \geq c_0 \max\{g_j \mid j = 1, 2, \ldots, N\}$, and $\lambda \geq \delta \sum_{j=1}^{N} g_j \mu(V_j)$ the following estimate*

$$\|f(A)(H - \lambda - i\varepsilon)^{-1}\psi\|_{L^2(\mathbf{R}^m)} \leq c\lambda^{-1}\|f\|_{\ell^2(L^\infty)}\|\psi\|_{B_A}$$

holds for all $f \in \ell^2(L^\infty)$, $\psi \in B_A$, and all $\varepsilon \neq 0$.

One can change from the localization given by the generator of dilations to the localization given by the multiplication operator $|x|$, by using the technique in [JP(85)] and an interpolation argument. One gets the following

Corollary 4.2. *Let H be as in the Theorem. Then there exist $d > 0$ and $\lambda_1 > \lambda_0$ such that for all $\lambda \geq \lambda_1$, $\varepsilon \neq 0$ and $R > 1$ the estimate*

$$\|\chi_R(|x|)(H - \lambda - i\varepsilon)^{-1}\psi\|_{L^2(\mathbf{R}^m)} \leq d\lambda^{-1/2}R^{1/2}\|\psi\|_{B(\mathbf{R}^m)} \tag{7}$$

holds for all $\psi \in B(\mathbf{R}^m)$. Here χ_R denotes the characteristic function of the interval $[0, R]$.

The factor $R^{1/2}$ comes from the estimate $\|\chi_R(\lambda)\|_{\ell^2(L^\infty)} \leq 2R^{1/2}$. The estimate (7) is the one obtained in [A(90)] for $H = -\Delta$ with an explicit numerical value for the constant d.

Details of the results in this section can be found in [J(90)].

REFERENCES

A(75) Agmon, S.: Spectral properties of Schrödinger operators and scattering theory. *Ann. Scuola Norm. Sup. Pisa* (4) **2** (1975), 151–218.

A(90) Agmon, S.: A representation theorem for solutions of the Helmholtz equation and resolvent estimates for the Laplacian. In: *Analysis et cetera*. Ed.: Rabinowitz, P., Zehnder, E. Academic Press, New York, 1990, p. 39–76.

AH(76) Agmon, A., Hörmander, L.: Asymptotic properties of solutions of differential equations with simple characterisitcs. *J. Analyse Math.* **30** (1976), 1–38.

AGHH(88) Albeverio, S., Gesztesy, F., Høegh-Krohn, R., Holden, H.: *Solvable Models in Quantum Mechanics*. Springer Verlag, Heidelberg 1988.

J(80) Jensen, A.: Spectral properties of Schrödinger operators and time-decay of the wave functions. Results in $L^2(\mathbf{R}^m)$, $m \geq 5$. *Duke Math. J.* **47**(1980), 57–80.

J(84) Jensen, A.: Spectral properties of Schrödinger operators and time-decay of the wave functions. Results in $L^2(\mathbf{R}^4)$. *J. Math. Anal. Appl.* **101**(1984), 491–513.

J(85) Jensen, A.: Propagagation estimates for Schrödinger-type operators. *Trans. Amer. Math. Soc.* **291** (1985), 129–144.

J(89) Jensen, A.: High energy resolvent estimates for generalized many-body Schrödinger operators. *Publ. RIMS, Kyoto Univ.* **25** (1989), 155–167.

J(90) Jensen, A.: High energy resolvent estimates for Schrödinger operators in Besov spaces. Preprint, Aalborg University, September 1990.

JK(79) Jensen., A., Kato, T.: Spectral properties of Schrödinger operators and time-

·

decay of the wave functions. *Duke Math. J.* **46** (1979), 583–611.

JMP(84) Jensen, A., Mourre, E., Perry, P.: Multiple commutator estimates and resolvent smoothness in quantum scattering theory. *Ann. Inst. H. Poincaré, Sect. A (N.S.)* **41** (1984), 207–225.

JP(85) Jensen, A, Perry, P.: Commutator methods and Besov space estimates for Schrödinger operators. *J. Operator Theory* **14** (1985), 181–188.

Mo(81) Mourre, E.: Absence of singular continuous spectrum for certain self-adjoint operators. *Commun. Math. Phys.* **78** (1981), 391–408.

Mu(82) Murata, M.: Asymptotic expansions in time for solutions of Schrödinger-type equations. *J. Funct. Anal.* **49** (1982), 10–56.

Mu(83) Murata, M.: High energy resolvent estimates, I, first order operators. *J. Math. Soc. Japan* **35** (1983), 711–733.

Mu(84) Murata, M.: High energy resolvent estimates, II, higher order elliptic operators. *J. Math. Soc. Japan* **36** (1983), 1–10.

S(74) Saitō, Y.: The principle of limiting absorption for the non-selfadjoint Schrödinger operator in $\mathbf{R}^N$, $(N \neq 2)$. *Publ. RIMS Kyoto Univ.* **9** (1974), 397–428.

Y(82) Yafaev, D. R.: Spectral properties of the Schrödinger operator with a potential having a slow falloff. *Funct. Anal. Appl.* **16** (1982), 280–286.

Y(86) Yafaev, D. R.: The eikonal approximation and the asymptotics of the total scattering cross-section for the Schrödinger equation. *Ann. Inst. H. Poincaré, Phys. Théor.* **44** (1986), 397–425.

There are a few happy cases in mathematical physics in which one can find solvable models rich enough to contain essential features of the phenomena to be studied and to serve as a starting point for gaining control of general situations by suitable approximations

(from [1])

Point interactions as solvable models

Werner Kirsch

Werner Kirsch

1   INTRODUCTION

One of the very many topics which were developed by Raphael Høegh-Krohn was the study of point interactions in quantum mechanics. He did this research as so many other things jointly with Sergio Albeverio as well as with Fritz Gesztesy and Helge Holden, a former student of his. Their common effort culminated in the book "Solvable Models in Quantum Mechanics" [1].

My own scientific contact with Raphael - at least that part that led to publications - took place in the above mentioned field. However, his influence both on my personal and scientific development goes much beyond this special interest. Among others he influenced my view of mathematics and its use in science.

I remember walking with him on the campus of Oslo University. He told me the joke about the mathematician who, on the way to his institute used to pass by a shop which displayed signs like "We wash your clothes" and " We sell used cars" etc.. One day he decided to bring his dirty clothes to this place to get them cleaned up. When he entered the shop the shopkeeper said : "Sorry, Sir, we don't wash clothes.". "But why have you got this sign in the window ? ". "Well, Sir, we just make signs". Then Raphael

261

added : "That's precisely what we mathematicians do, we just make signs". Frankly, I was embarrassed and disappointed. Mathematics as the science of making signs ? Seeing my reaction, Høegh-Krohn continued : "Making signs is an important profession. Think, how important it is to have warning signs, traffic signs etc." Raphael Høegh-Krohn has always paid a lot of attention to the use of the mathematics he developed in physics and technology. Therefore - I am sure - he didn't want to say that a mathematician should not care for what his signs mean, but rather that he has to. Signs, even if technically perfectly produced, but wrongly applied can be useless, misleading and dangerous.

In this note I will briefly survey some parts of the theory of point interactions. I will not give proofs and hardly state any theorem. Instead, I'll try to give the reader some flavor of the subject and to emphasize open problems which not only may be easier to handle in the special model but may lead to solutions for other, more realistic cases.

2   WHY POINT INTERACTIONS ?

In the point interaction model, two point particles only interact when they meet at precisely the same place, thus the potential is formally given by a Dirac function δ_{x_0}, the Hamiltonian being

$$H = -\Delta + \lambda \delta_{x_0} \tag{1}$$

Here Δ is the Laplacian, δ_{x_0} the unit point measure concentrated in $x_0 \in \mathbb{R}^d$, and λ is a suitable coupling constant. Obviously this is a rough idealization, if not a caricature of an interaction occurring in nature. Moreover, formula (1) is only a formal definition. A rigorous treatment requires some mathematical care and sophistication.

Even then, it is impossible to define a point interaction if the dimension d of the underlaying configuration space is four or higher. For dimension $d = 1$ there is a (more or less straightforward) definition of H by means of quadratic forms while for $d = 2$ or $d = 3$ the operator H can be defined only if the coupling constant λ is taken to be "infinitesimally small" in a very precise way. Given these inconveniences, why should we consider point interactions at all ?

Our motivation for this subject is that model (1) admits solutions of certain problems which are by far more complete than in other cases. For example, one can compute the spectrum of H and also of operators with many point interactions explicitly (at least up to the solution of algebraic equations). Especially in solid state physics there are very few examples for which the band structure of the Hamiltonian can be determined explicitly. In fact, in dimension $d = 3$, I know of no example but the point interaction models where this is possible.

For random potentials numerous results are proven for the Anderson model (which is a model on the lattice) but are still unknown for Schrödinger operators. Here point interactions may serve as continuous systems which share a number of features with the lattice case.

In all these cases it is of course questionable whether the results one proves are of general nature, i.e. are shared by other, more complicated (and perhaps more realistic) models. In deed, there are certain effects which are special to the specific models. We will discuss this topics in some details below. Despite of this problem, I believe, it is very convenient to have specific models where conjectures can be evaluated quantitatively.

While we will concentrate on these features of point interactions we remark that their investigation also allows for an analysis of the low energy behavior for nonsingular Schrödinger operators.

Our next goal will be to define point interactions. To me, it looks like an irony of nature that in many cases systems which can be *solved* most explicitly, are most hard to *define* rigorously.

3   WHAT IS A POINT INTERACTION ?

Taking expectation values in (1) we get

$$\langle \varphi, \, H\varphi \rangle = \langle \varphi, \, H_0\varphi \rangle + \lambda \, |\varphi(x_0)|^2.$$

Performing a Fourier transform we arrive at

$$\langle \varphi, \, \hat{H}\varphi \rangle = \langle \varphi, \, p^2\varphi \rangle + \lambda \, | \frac{1}{(2\pi)^{\frac{d}{2}}} \int \varphi(p)e^{-ipx_0}dp|^2 \; .$$

Thus the operator may be written as

$$\hat{H} = p^2 + \lambda|\phi_{x_0}\rangle\langle\phi_{x_0}| \tag{2}$$

where $\phi_{x_0} = \frac{1}{(2\pi)^{\frac{d}{2}}}e^{-ipx_0}$.

Admittedly, (2) doesn't make more sense than (1). However, from this formula we get an idea how to define $\hat{H}$ by a limiting procedure. We simply take a cutoff function $\chi_N(p) = \chi_{\{|p|\leq N\}}(p)$ and set

$$\hat{H}_N = p^2 + \lambda_N|\chi_N\phi_{x_0}\rangle\langle\phi_{x_0}\chi_N| \tag{3}.$$

As we mentioned already above we will eventually be forced to take the coupling constant λ to be infinitesimally small. To implement this we allowed the coupling constant in (3) to be N–dependent. Eventually, we will send λ_N to zero as N goes to infinity.

Next we compute the resolvent of (3). We shall write H_N for $\hat{H}_N$ and assume first that E is very negative. Since we don't make any use of the special form of the function ϕ_{x_0} we call it ψ. It is merely assumed that ψ is locally bounded. Defining $K_N\varphi = \langle \chi_N\psi, \, \varphi \rangle$, a bounded operator from L^2 to $\mathbb{R}$ we may write $|\psi\chi_N\rangle\langle\chi_N\psi|$ as $K_N^*K_N$.

Using the resolvent identity we have

$$(H_N - E)^{-1} = (H_0 - E)^{-1} - (H_0 - E)^{-1}\lambda_N K_N^* K_N (H_N - E)^{-1}$$

$$= (H_0 - E)^{-1}$$

$$- (H_0 - E)^{-1}\lambda_N K_N^* K_N (1 + (H_0 - E)^{-1}\lambda_N K_N^* K_N)^{-1}(H_0 - E)^{-1}.$$

It turns out that

$$K_N(1 + (H_0 - E)^{-1}\lambda_N K_N^* K_N)^{-1} = (1 + K_N((H_0 - E)^{-1}\lambda_N K_N^*)^{-1}K_N$$

which can be seen by use of a Neumann series:

$$K(1 + AK^*K)^{-1} = K \sum (-1)^n (AK^*K)^n = \sum (-1)^n (KAK^*)^n K = (1 + KAK^*)^{-1} K.$$

We thus obtain

$$(H_N - E)^{-1} = (H_0 - E)^{-1}$$
$$- (H_0 - E)^{-1} \lambda_N K_N^* (1 + \lambda_N K_N (H_0 - E)^{-1} K_N^*)^{-1} K_N (H_0 - E)^{-1} \quad (4).$$

While we derived (4) only for $|E|$ large it is valid for each E for which the right hand side makes sense (use analyticity, for details see e.g. [1]).

Reinserting ϕ_{x_0} into (4) we get

$$(H_N - E)^{-1} = (p^2 - E)^{-1} - (\frac{1}{\lambda_N} + \langle \chi_N \phi_{x_0}, \ (p^2 - E)^{-1} \chi_N \phi_{x_0}\rangle)^{-1}$$
$$\big| (p^2 - E)^{-1} \chi_N \phi_{x_0} \big\rangle \big\langle (p^2 - E)^{-1} \chi_N \phi_{x_0} \big| \quad (5).$$

For $E < 0$, the expressions $(p^2 - E)^{-1} \chi_N(p) e^{ipx_0}$ converge to an L^2–function when N tends to infinity iff $(p^2 - E)^{-1}$ is square integrable which happens precisely in dimension $d \leq 3$.

Thus our approach can work at most for these dimensions. This is not a limitation of the *method* above. In fact, one can prove that point interactions cannot be defined for $d \geq 4$.

So, let us suppose we have $d \leq 3$. Then the expression $\langle \chi_N \phi_{x_0}, \ (p^2 - E)^{-1} \chi_N \phi_{x_0}\rangle$ tends to a finite value only for d=1.

In this case (d=1) with $\lambda_N \equiv \lambda$ we obtain the limit

$$(H - E)^{-1} = (H_0 - E)^{-1} - (\frac{1}{\lambda} + \frac{1}{2\sqrt{|E|}})^{-1} \big| \frac{\phi_{x_0}}{p^2 - E}\big\rangle \big\langle \frac{\phi_{x_0}}{p^2 - E}\big| \quad (6)$$

(we used $\frac{1}{2\pi} \int_{-\infty}^{+\infty} \frac{1}{p^2 - E} dp = \frac{1}{2\sqrt{|E|}}$ for $E < 0$).

The right hand side of (6) has a pole for $E = -\frac{\lambda^2}{4}$, if $\lambda < 0$. In fact this value of E is a bound state for $H = -\frac{d^2}{dx^2} + \lambda \delta_0$.

For dimension d=2 or 3 the expression

$$\frac{1}{\lambda_N} + \langle \chi_N \phi_{x_0}, \ (p^2 - E)^{-1} \chi_N \phi_{x_0}\rangle = \frac{1}{\lambda_N} + \frac{1}{(2\pi)^3} \int_{|p| \leq N} \frac{1}{p^2 - E} dp \quad (7)$$

can not converge if λ_N is kept constant since $(p^2 - E)^{-1} \notin L^1(\mathbb{R}^d)$ for $d \geq 2$. Instead we choose λ_N depending on N in such a way that the divergence of the integral $\int_{|p| \leq N} \frac{1}{p^2 - E} \, dp$ is cancelled. (We should, of course, take λ_N independent of the energy E!)

Let us consider d=3 first. Then we have

$$\frac{1}{(2\pi)^3} \int_{|p| \leq N} \frac{1}{p^2 - E} \, dp = \frac{1}{2\pi^2} \int_0^N \frac{x^2}{x^2 + |E|} \, dx = \frac{1}{2\pi^2} N - \frac{1}{2\pi^2} \sqrt{|E|} \arctan\left(\frac{N}{\sqrt{|E|}}\right).$$

Thus, if we take

$$\frac{1}{\lambda_N} = \frac{-1}{2\pi^2} N - \alpha = \frac{-1}{(2\pi)^3} \int_{|p| \leq N} \frac{1}{p^2} \, dp - \alpha \tag{8}$$

the expression (7) converges to $-\alpha - \frac{\sqrt{|E|}}{4\pi}$.

Hence with the choice (8) we get as limit of the resolvent of the operators (3):

$$(H_\alpha - E)^{-1} = (p^2 - E)^{-1} + \left(\alpha + \frac{\sqrt{|E|}}{4\pi}\right)^{-1} \left|\frac{\phi_{x_0}}{p^2 - E}\right\rangle \left\langle\frac{\phi_{x_0}}{p^2 - E}\right| \tag{9}$$

This operator is in deed the resolvent of a selfadjoint operator H_α which we call the Hamiltonian with a point interaction of strength α at the point x_0. The constant α formally plays the role of a coupling constant. However, this phrase shouldn't be taken too seriously. For example, switching the point interaction off means sending α to infinity, not to zero. The expression (9) has a pole at $E = -(4\pi\alpha)^2$ if $\alpha < 0$. This pole corresponds to an eigenvalue of H_α. In *this* sense the potential is attractive if $\alpha < 0$. When α passes through zero the eigenvalue is turned into a resonance which disappears for $\alpha \to +\infty$.

Note that a small (say very negative) α leads to a slower decay of λ_N to 0. So α very negative means strongly "negative potential" while α big (e.g. positive) means weakly attractive interaction. However, we emphasize that the "potential" is negative for *every* α, since α_N always becomes negative for N large enough. That we have an eigenvalue only if the potential is negative enough is a well known fact (see e.g. Reed-Simon [2]) for $d \geq 3$. For $d \leq 2$ any negative potential no matter how small produces a negative eigenvalue.

It can be shown that the above convergence of H_N to H_α holds in the norm resolvent sense.

It is instructive to reconsider the resolvent limit (d=3) for the choice $\lambda_N \equiv \lambda < 0$. Our philosophy above might suggest that in this case H_N doesn't converge at all since

λ_N constant leads to a too strong interaction. In the contrary, H_N converges to the free Hamiltonian H_0! This puzzling fact can be explained as follows. H_N has an eigenvalue which goes to $-\infty$ thus disappears in the limit $N \to \infty$. Consequently H_N tends to H_0 only in the strong resolvent sense (for $\lambda_N \equiv \lambda$).

Let us briefly discuss the two dimensional case. Here we take

$\frac{1}{\lambda_N} = \frac{-1}{(2\pi)^2} \int_{|p|\leq N} \frac{1}{p^2+1}\, dp - \alpha$ (the simpler choice $\int_{|p|\leq N} \frac{1}{p^2}\, dp$ is divergent!).

Then

$$\frac{1}{\lambda_N} + \frac{1}{(2\pi)^2} \int_{|p|\leq N} \frac{1}{p^2 + |E|}\, dp$$

$$= -\alpha + \frac{1}{2\pi} \int_0^N \frac{p}{p^2 + |E|}\, dp - \frac{1}{2\pi} \int_0^N \frac{p}{p^2 + 1}\, dp$$

$$= -\alpha + \frac{1}{4\pi}\ln\left(N^2 + |E|\right) - \frac{1}{4\pi}\ln |E| - \frac{1}{4\pi}\ln\left(N^2 + 1\right)$$

$$\to -\alpha - \frac{\ln |E|}{4\pi}$$

Consequently (5) converges to

$$(H_\alpha - E)^{-1} = (p^2 - E)^{-1} + \left(\alpha + \frac{\ln |E|}{4\pi}\right)^{-1} \left|\frac{\phi_{x_0}}{p^2 - E}\right\rangle \left\langle \frac{\phi_{x_0}}{p^2 - E}\right| \tag{10}$$

H_α has a negative eigenvalue at $E = -e^{-4\pi\alpha}$ consequently *there is* an eigenvalue for *any* coupling constant α. The surprised reader is reminded that $\alpha = 0$ does *not* mean to switch the interaction off. Turning the interaction off is done by $\alpha \to +\infty$ in which case the eigenvalue E goes to zero as expected. As for $d = 3$ any coupling constant α corresponds to an attractive potential which for $d = 2$ always produces negative eigenvalues.

So far we have always taken the function $\phi_{x_0}(p) = (2\pi)^{-d/2} e^{-ix_0 p}$, the Fourier transform of δ_{x_0}. But nobody forbids us to take another function instead. Let's try for example with $\phi_{x_0}(p) = (2\pi)^{d/2} p_1 e^{-ix_0 p}$ (where p_1 is the first component of p). For motivation, we remark that this ϕ_{x_0} is the Fourier transform of $-i\frac{\partial}{\partial x_1}\delta_{x_0}$. Doing the above procedure we obtain (in analogy with (5)) a Hamiltonian H_N whose resolvent is given by

$$(H_N - E)^{-1} = (p^2 - E)^{-1} - \left(\frac{1}{\lambda_N} + \frac{1}{(2\pi)^d} \int_{|p|\leq N} \frac{p_1^2}{p^2 + |E|}\, dp\right)^{-1}$$

$$\left| \frac{1}{(2\pi)^{d/2}} \frac{p_1}{p^2 - |E|} \chi_N \right\rangle \left\langle \frac{1}{(2\pi)^{d/2}} \frac{p_1}{p^2 - |E|} \chi_N \right| \tag{11}$$

For the projection operator in (11) to make sense in the limit we have to require $\frac{p_1}{p^2+|E|} \in L^2(\mathbb{R}^d)$ i.e. we got to assume $d = 1$.

In this case

$$\frac{1}{2\pi} \int_{|p|\leq N} \frac{p^2}{p^2 + |E|} \mathrm{dp} = \frac{1}{\pi} \left(N - \sqrt{|E|} \, \arctan \frac{N}{\sqrt{E}} \right)$$

Thus choosing $\frac{1}{\lambda_N} = -\frac{N}{\pi} - \alpha$ we get

$$(H_\alpha - E)^{-1} = (p^2 - E)^{-1} + \left(\alpha + \frac{\sqrt{|E|}}{2} \right)^{-1}$$

$$\left| \frac{1}{(2\pi)^{1/2}} \frac{p}{p^2 + |E|} \right\rangle \left\langle \frac{1}{(2\pi)^{1/2}} \frac{p}{p^2 + |E|} \right| \tag{12}$$

H_α has an eigenvalue at $E = -(2\alpha)^2$.

We consider H_α as a (renormalized) δ'-interaction Hamiltonian. Formally δ' as a quadratic form acts as $\delta'(\varphi, \psi) = \varphi'(0)\overline{\psi'}(0) = < \hat{\varphi}, p >< p, \hat{\psi} >$, so that δ' is something like $|p > < p|$.

Let us mention that in all cases above one can compute not only eigenvalues and resolvents explicitly but also eigenfunctions, scattering operators etc. .

An important extension is the definition of many point interactions, i.e. of a potential formally given by $\sum_{y \in Y} \lambda_y \delta_y$ where Y is a finite or discrete infinite set. To do so we merely have to change the meaning of the operator K_N in formula (4). Above we set $K_N : L^2(\mathbb{R}^d) \to \mathbb{C}$, $K_N f = < \chi_N \phi_{x_0}, f >$. The proper generalisation is to define K_N as an operator from $L^2(\mathbb{R}^d)$ into $\mathbb{C}^{|Y|} \simeq l^2(Y)$ by setting

$$(K_N f)_y = < \chi_N \phi_y, f > \tag{13}$$

This operator is well defined if Y is finite while domain question arise for infinite Y. To avoid these we may take Y finite and approximate an infinite Y finally by growing finite sets Y_R. Since our discussion is formal here, we will treat also infinite (discrete) sets Y from now on without considering domain questions.

The adjoint operator to $K_N, K_N^* : l^2(Y) \to L^2(\mathbb{R})$ acts as

$$(K_N^* u)(x) = \sum_{y \in Y} u_y \phi_y(x) \tag{14}$$

The operator $K_N(H_0 - E)^{-1}K_N^*$ is now an operator on $l^2(Y)$. The term $\frac{1}{\lambda_N}$ in (5) is to be interpreted as a diagonal operator with values $\mu_N^{(y)} = \frac{-1}{\lambda_N^{(y)}}, y \in Y$ on the diagonal. The numbers $\lambda_N^{(y)}$ reflect the strength of the interaction at the point y. The operator $\left(-\mu_N^{(y)} + K_N(H_0 - E)^{-1}K_N^*\right)^{-1}$ in (4) is now meant as the inverse of the operator in brackets on the space $l^2(Y)$.

The operator $\left(-\mu_N^{(y)} + K_N(H_0 - E)^{-1}K_N^*\right)^{-1}$ is a (possible infinite) matix M with diagonal entries

$$M_{yy} = -\mu_N^{(y)} + \frac{1}{(2\pi)^{-d}} \int_{|p|\leq N} \frac{1}{p^2 + |E|} \, dp$$

and off-diagonal entries $(y \neq y')$

$$M_{yy'} = +\frac{1}{(2\pi)^{-d}} \int_{|p|\leq N} \frac{e^{ip(y-y')}}{p^2 + |E|} \, dp$$

The latter quantities happen to converge to a finite limit, namely to the kernel of $(H_0 + |E|)^{-1}(y, y')$ (which is the Fourier transform of $(p^2 + |E|)^{-1}$ for all). For $y \neq y'$ this is known to be finite.

In fact, (for $E < 0$)

$$(H_0 - E)^{-1}(x, y) = \begin{cases} \frac{1}{2\sqrt{|E|}} e^{\sqrt{|E|}|x-y|} & d = 1 \\ \frac{1}{4\pi} \frac{e^{\sqrt{|E|}|x-y|}}{|x-y|} & d = 3 \end{cases}$$

For $d = 2$ there is an explicit expression for the resolvent kernel in terms of a Hankel function.

As we have already seen the integral $\frac{1}{(2\pi)^d} \int_{|p|\leq N}(p^2 + |E|)^{-1} \, dp$ diverges for $d \geq 2$. But, if we take $\mu_N^{(y)} = \frac{1}{(2\pi)^3} \int_{|p|\leq N} \frac{1}{p^2} \, dp + \alpha_y$ for $d = 3$ (resp $\frac{1}{(2\pi)^2} \int_{|p|\leq N} \frac{1}{p^2+1} \, dp + \alpha_y$ for $d = 2$) then the divergent parts cancel and we obtain a finite limit. The same remarks apply to the δ'-interactions for $d = 1$ as well as for the one dimensional δ-interactions. In the latter case $\mu_N^{(y)}$ can be taken independently of N.

Summarizing, we were able to define point interactions in dimension $d = 1, 2, 3$ and δ'-intercations for $d = 1$. The corresponding Hamiltonians are unambiguously defined through their resolvents.

For a discrete set Y of points we define operators A_1, A_2, A_3 and B_1 on $l^2(Y)$ by:

$$A_1(y, y') = G_1^{(E)} \qquad\qquad y, y' \in \mathbb{R} \qquad\qquad (15a)$$

$$A_2(y, y') = \begin{cases} -\frac{\ln |E|}{4\pi} & \text{for } y = y' \\ G_2^{(E)}(y - y') & \text{otherwise} \end{cases} \quad y, y' \in \mathbb{R}^2 \qquad\qquad (15b)$$

$$A_3(y, y') = \begin{cases} -\frac{\sqrt{|E|}}{4\pi} & \text{for } y = y' \\ G_3^{(E)}(y - y') & \text{otherwise} \end{cases} \quad y, y' \in \mathbb{R}^3 \qquad\qquad (15c)$$

where G_1, G_2, G_3 are the kernels of the free resolvent in dimension $d = 1, 2, 3$ respectively.

$$B_3(y, y') = \begin{cases} \frac{-\sqrt{|E|}}{2} & \text{for } y = y' \\ D_1 G_1^{(E)}(y - y') & \text{otherwise} \end{cases} \quad y, y' \in \mathbb{R}^3 \qquad\qquad (15d)$$

Here $D_1 G_1$ is the derivative of the free Greens functions (= kernel of the resolvent) in dimension $d = 1$. Then the resolvent for the point interactions on the set Y in dimension $d = 1, 2, 3$ is given by

$$\left(H_0 + \sum_{y \in Y} \lambda_y \delta_y - E \right)^{-1} (x, x') = G_1^{(E)}(x, x') \qquad + \qquad \text{(for d = 1)}$$

$$\sum_{y, y' \in Y} \left(-\lambda_y^{-1} \delta_{y, y'} - A_1 \right)_{y, y'}^{-1} \cdot G_1^{(E)}(x - y) \, G_1^{(E)}(x' - y') \qquad\qquad (16a)$$

$$\left(H_0 + \sum_{y \in Y} \alpha_y \delta_y' - E \right)^{-1} (x, x') = G_d^{(E)}(x, x') \qquad + \qquad (d = 2, 3)$$

$$\sum_{y, y' \in Y} \left(\alpha_y \delta_{y, y'} - A_d \right)_{y, y'}^{-1} G_d(x - y) G_d(x' - y') \qquad\qquad (16.bc)$$

and the resolvent for δ'-interactions in $d = 1$ is :

$$\left(H_0 + \sum_{y \in Y} \alpha_y \delta_y' - E \right)^{-1} (x, x') = G_1^{(E)}(x, x') \qquad +$$

$$\sum_{y, y' \in Y} \left(\alpha_y \delta_{yy'} - B_1 \right)_{yy'}^{-1} \cdot D_1 G_1^{(E)}(x - y) \, D_1 G_1^{(E)}(x' - y') \qquad\qquad (16.d)$$

4   PERIODIC POINT INTERACTIONS

Let us denote by $H = H_0 + \sum_{i \in I} \alpha_{x_i} \delta_{x_i}$ for a discrete set $\{x_i\}_{i \in I}$ of points in $\mathrm{I\!R}^d$ the operator whose resolvent is given by formula (16).

If the set $\{x_i\}$ is finite the spectrum of H can be determined as follows. The free operator H_0 has $[0, \infty)$ as its spectrum and the spectrum is purely absolutely continuous. Since the resolvent of H can be looked upon as a finite dimensional perturbation of the resolvent of H_0, also H has $[0, \infty)$ as its essential, in fact absolutely continuous spectrum. The only possible new points in the spectrum consist of the poles of the "perturbation", i.e. of those energies E for which the finite matrix $A_d(E) = A_d$ as given in (15) has a zero eigenvalue. If N denotes the number of points $\{x_i\}$ then H has at most N eigenvalues which are less than zero, namely the values $E < 0$ for which the determinant of A vanishes.

If the index set I of $\{x_i\}$ has infinitely many elements then the determination of the spectrum is much more complicated. In fact, the essential spectrum of H will differ from $[0, \infty)$ in general. The case which can be analysed completely is a periodic point interaction.

This is classical in the one dimensional case. Virtually any text book on solid state physics will contain the "Kronig – Penney – model". This is the Hamiltonian $H_\lambda = -\frac{d^2}{dx^2} + \lambda \sum_{n=-\infty}^{+\infty} \delta(x - an)$. H_λ has purely absolutely continuous spectrum which consists of infinitely many bands $B_n = [\alpha_n, \beta_n]$, $n \in \mathrm{I\!N}$ with gaps (" forbidden zones") separating the bands.

In fact, for $\lambda > 0$ we have $\beta_n = \frac{\pi^2}{a^2} n^2 (n = 1, 2, ...)$ and $\alpha_1 > 0$, $\beta_n < \alpha_{n+1} < \beta_{n+1}$ The β_n do not depend on the coupling constant λ. This is very untypical. It comes from the fact that $f(x) = sin(\sqrt{\beta_n} x)$ (at least formally) solves the equation $H_\lambda f = \beta_n f$ for arbitrary λ. More precise information can be found in [1] III.2

The occurrence of infinitely many gaps is typical of one dimensional crystals. In higher dimensions the occurrence of one semi-infinite band above finitely many compact bands is generic. (see [8]) (" Bethe-Sommerfeld conjecture ") .

This is in deed the case for periodic δ-interactions in $d = 3$. The spectrum of $H_\alpha = -\Delta + \alpha \sum_{n \in \mathbb{Z}} \delta(x - n)$ consists of two bands, i.e. $\sigma(H_\alpha) = [E_0, E_1] \cup [E_2, \infty)$. For α

small enough we have $E_0 < 0$ while $E_1 > 0$ for arbitrary α. Moreover, $E_2 > E_1$ if α is not too big.

The situation is much less clear in $d = 3$ when there is more than one point interaction per period. In fact, the band structure above $E = 0$ is completely unknown while there are at most N (= number of point interactions per period) bands below 0 [9]. We strongly conjecture, however, that there are at most $(N + 1)$ bands on the whole energy line and precisely $(N + 1)$ bands for generic cases.

The above results are obtained, as usually for periodic potentials, by a directe integral decomposition ("Blochs theorem"). One defines so called reduced Hamiltonians $H_\alpha(k)$ which for fixed k have discrete spectrum. Their eigenvalues $E_n(k)$ move through the bands of H_α as k is varied. The direct integral decomposition can be done in our case quite explicitely.

Let us suppose that our system has the periodic structure of a lattice Λ in $\mathbb{R}^3$. Then the lattic points $n \in \Lambda$ are obtained as integral linear combinations $n = \beta_1\alpha_1 + \beta_2\alpha_2 + \beta_3\alpha_3$ of a base a_1, a_2, a_3 of $\mathbb{R}^3$, $\beta_1, \beta_2, \beta_3 \in \mathbb{Z}$. The unit cell C_o of Λ is defined by $C_o = \{\lambda_1 a_1, \lambda_2 a_2, \lambda_3 a_3 | 0 \leq \lambda_i < 1\}$ We assume we have finitely many point interactions in C_o periodically extended to R^3. Let us denote by Ξ the set of these points in C_0.

The Fourier transform on C_0 is defined by $\hat{f}(\gamma) = \frac{1}{|C_o|} \int_{C_o} f(x)e^{-ix\gamma}\,dx$ where γ runs through the so called dual lattice Γ of Λ. Γ is a lattice (in momentum space) with bases b_1, b_2, b_3. This is the the "dual base" to a_1, a_2, a_3 in the sense that $< b_i, a_j >= 2\pi\delta_{ij}$.

The Fourier transform on Γ brings us back to C_o by :

$$\hat{\phi}(k) = \sum \phi(\gamma)e^{i\gamma k}$$

It is the inverse of the Fourier transform on C_o (in L^2-sense).

Let us consider a Hamiltonian with periodic point interaction on $\Lambda + \Xi$. For $\xi \in \Xi$ we denote by α_ξ the strength of the point interaction at ξ (and - by periodicity - at $\xi + n$ for any $n \in \Lambda$.

We may uniqely write a point in (the momentum space) $\mathbb{R}^3$ as $p = \gamma + k$ for a $\gamma \in \Gamma$ and $k \in B_0$ the unit cell of the (dual) lattice Γ. B_0 is frequently called the (first) Brillouin zone.

By direct computation using the Poisson summation formula one proves that the resolvent of $H^{per}_{\{\alpha_\xi\}}$ can be decomposed into a direct integral with "reduced" resolvents $H(k)$, action on the dual lattice Γ, i.e.

$$(H - E)^{-1} = (H^{per}_{\alpha_\xi} - E)^{-1} = \int_{B_0}^{\oplus} (H(k) - E)^{-1} \, dk \tag{13}$$

with

$$(H(k) - E)^{-1}(\gamma, \gamma') = \left((k + \gamma)^2 - E\right)^{-1} \delta_{\gamma\gamma'}$$
$$+ \frac{1}{|C_0|} \sum_{\xi,\xi' \in \Xi} D(E, K)^{-1}_{\xi,\xi'} \frac{e^{i(\gamma+k)\xi}}{(\gamma+k)^2 - E} \frac{e^{i(\gamma'+k)\xi'}}{(\gamma+k)^2 - E} \tag{14}$$

where $D(E, K)$ is the operator (matrix) on $l^2(\Xi)$ given by

$$D(k, x)_{\xi\xi'} = \alpha_\xi \delta_{\xi\xi'} - \frac{1}{|C_0|} g_E(k, \xi - \xi') \tag{15}$$

$g_E(k, x)$ is the Fourier transform of $\frac{1}{|\gamma+k|^2 - E}$ with respect to the lattic Γ, i.e. (formally)

$$g_E(k, x) = \frac{1}{|C_0|} \sum_{\gamma \in \Gamma} \frac{e^{i(\gamma+k)x}}{(\gamma+k)^2 - E} \tag{16}$$

However the above some diverges for $x = 0$ (even for $x \in \Lambda$). Hence we have to renormalize again (which shouldn't be surprising after all). So, we get

$$g_E(k, 0) = \lim_{N \to \infty} \left\{ \frac{1}{|C_0|} \sum_{\substack{\gamma \in \Gamma \\ |\gamma+k| \leq N}} \frac{1}{(\gamma+k)^2 - E} - \frac{1}{2\pi^2} N \right\} \tag{17}$$

(A careful analysis can be found in [1]). It can be shown that (14) is in fact the resolvent of a selfadjoint operator which we denote by

$$H_{\alpha, \Xi}(k) = H_0(k) + \sum_{\lambda \in \Lambda} \sum_{\xi \in \Xi} \alpha_\xi \delta_{\xi + \lambda} \tag{18}$$

It can be shown that these operators have a purely discrete spectrum. Their eigenvalues are exactly given by the singularities (in E) of the resolvent (14). This expression has two ways to develop a singularity, namely in the first summand, which is just the (reduceded) free resolvent and secondly in the interaction term, i.e. at the points where the matrix $D(E, k)$ becomes singular.

The easiest case occurs if Ξ consist of only one point. By shifting we may suppose that $\Xi = \{0\}$. In this case equation (14) simply gives :

$$
(H_\alpha(k) - E)^{-1}(\gamma, \gamma') = \left(((k+\gamma)^2 - E)\right)^{-1} \delta_{\gamma\gamma'}
$$
$$
+ (\alpha g_E(k,0))^{-1} \frac{1}{(\gamma + k)^2 - E} \frac{1}{(\gamma' + k)^2 - E} \tag{19}
$$

The function $g(E) = g_E(k,0)$ has singularities at the point of the set $S_k = \{(\gamma+k)^2 | \gamma \in \Gamma\}$. The set $\mathbb{R}\backslash S_k$ consists of infinitely many open intervals, $\mathbb{R}\backslash S_k = \bigcup_{n=0}^\infty I_n$ where only I_0 is infinite. (We take, of course, I_0 to be the the lowest interval, I_1 the next one from the bottom etc.). Inside the interval $I_n = (s_{n-1}, s_n)$ the function $g(E)$ is strictly increasing, tending to $-\infty$ as $E \searrow s_{n-1}$ and to $-\infty$ as $E \nearrow s_n$.

Consequently, in each integral I_n, there is a unique E_n such that $g_{E_n}(k,0) = \alpha$. At this point the second term in (19) is singular. Thus $H(k)$ has exactly one (simple) eigenvalue in each intervall I_n.

The free resolvent (= the first summand in (19)) has singularities at the point $E \in S_k$. However, the second summand has singularities at these points, too, so that cancellations of the singularities may occur. In fact, it is not hard to see that cancellation happens, if the $\frac{1}{(\gamma+k)^2 - E}$ term has a simple singularity at a point $E \in S_k$ (i.e. if $\#\{\gamma | (\gamma + k)^2 = E\} = 1$) while $H(k)$ has an eigenvalue of multiplicity m if $\#\{\gamma | (\gamma + k)^2 = E\} = m + 1 \geq 2$.

From the result one can show by a little geometry that the spectrum of the operator $H_\alpha = H_0 + \alpha \sum_{\lambda \in \Lambda} \delta(x - \lambda)$ (Λ a lattic) has one infinite band plus a ground state band which may or may not overlap with the infinite band depending on the coupling constant α.

The above analysis was carried through in the case that $\#(\Xi) = 1$, i.e. if there is only one particle in a unit cell. However, if Ξ contains more than one element the above reasoning fails, e.g. because cancellation of singularities is much harder to investigate. Consequently, it is not known whether such an operator has finitely many or infinitely many gaps in its spectrum. This point is also a major abstract for further developments for example in the case of random operators.

5   RANDOM POINT INTERACTIONS

Let us now assume that the points $\{x_i\}$ form a lattic Λ in $\mathbb{Z}^d$ but the coupling constants α_i of the point interactions at x_i are independent, identically distributed random variables. It may be surprising that in this situation the spectra can still be calculated rather explicitly.

For $d = 1$ (and $\alpha_i(\omega) \geq 0$) Saxon and Hutner conjectured that the spectrum Σ of

$$H_\omega = \frac{d^2}{dx^2} + \sum \alpha_1(\omega)\,\delta(x - an)$$

consists of the union of the spectrum of $H(\lambda) = \frac{d^2}{dx^2} + \sum \lambda\,\delta(x - an)$ where λ runs through all the possible values the random variable is allowed to take. This conjecture was proved by Luttinger [3] and much later in a general frame work in [4]. It is, however, also true if $\alpha_i(\omega)$ is allowed to be negative or even to change sign [5]. In [5] the Saxon-Hutner conjecture is proved for the δ' - interaction as well. We emphasize that the indedendence of the upper band edges β_n from the coupling constant plays a major role in the proof. In fact, for other potential the Saxon-Hutner conjecture is wrong in its above stated form.

Also for the three-dimensional point interaction the Saxon-Hutner-conjecture is valid merely in a modified form (see [4]).

The density of states is a a much finer measure to describe the spectra of solid state systems. For one dimensional δ and δ'- interactions this quantity is well investigated. Again the upper band edges β_n play a special role. The density of states behaves there in a way known for periodic systems while the typical Lifshitz behavior occurs only at the lower band edges [6]. There seem to be no papers so far on the density of states in 2– and 3–dimensional point interaction models. However, the existence and the basic properties of the (integrated!) density of states should be not too hard to obtain. On the other hand, our poor knowledge of the multicenter periodic case may be an obstacle to the corresponding proofs.

6 LOCALIZATION VERSUS DELOCALISATION

A characteristic feature of random Schrödinger operators is the phenomenon known as Anderson localization. For one dimensional systems it is expected (and proven) that random Schrödinger operator H_ω have a pure point spectrum.

For δ and δ'- interactions this can be shown when the common distribution P_0 of the coupling constant $\alpha_i(\omega)$ has a bounded density (decreasing fast enough at infinity) (see [7]). Until now, a proof for the case of non absolutely continuous distributions does not exist, although it is probably not so hard along the lines of [10].

For the higher dimensional cases one expects pure point spectrum at the band edges and for low disorder absolutely (?) continuous spectrum away from the edges. If the disorder is increased the continuous spectrum is more and more eaten up by the point spectrum. Based on the paper [11] several authors [12] proved the occurrence of pure point spectrum near the edges of the spectrum. These works were done for the Anderson model as well as for a few continuous models.

For the 2– and 3–dimensional point interaction this was not proven yet. We strongly believe that there is pure point spectrum for low energies. It is, however, not at all clear to us that there should be point spectrum at arbitrary energies for sufficiently high disorder. In fact, there may be a threshold $E_0(= 0?)$ such that there are only extended states above E_0 regardless of the disorder parameter. (As above the locations of the point interactions are kept fixed while the coupling constants are random.)

This view is supported by preliminary results [13] which suggest that there are extended states for certain random point interactions on a 3–dimensional strip.

The existence of extended states is not proven so far for any random Schrödinger operator. It may be that these questions are to a certain extend easier for point interactions.

Virtually all results referred to so far are obtained for the specific model of randomness explained above. There are many more physically meaningful random models, such as models for amorphous materials. This is still an open field to a great extend.

7   QUANTUM CHAOS

There is a new field of research where point interactions proved to be useful as solvable models, the so called quantum chaology. This branch of research tries to find the signature of chaos found in classical systems in the corresponding quantum systems. One such signature is the level statistics, i.e. the distribution of the differences of the eigenvalues of the system. While eigenvalues of regular systems are expected to behave like independent random variables, the eigenvalues of chaotic systems should try to avoid crossing, i.e. there is level repulsion for those systems.

Albeverio and Šeba [14] found in deed level repulsion for a Hamiltonian on a torus with a point interaction. The statics they obtained does, however, not agree with the prediction of random matrix theory for large distances.

8   REFERENCES

[1] Albeverio S., Gesztesy F., Høegh–Krohn R., Holden H.: Solvable models in quantum mechanics, Springer–Verlag, Berlin–Heidelberg–New York–Tokyo, 1988.

[2] Reed M., Simon B.: Methods of Modern Mathematical Physics. IV. Analysis of Operators, Academic Press, New York–San Francisco– London, 1978.

[3] Luttinger J. M.: Wave propagation in one–dimensional structures, Philips Res.Rep.**6** (1951), 303–310.

[4] Kirsch W.: Über Spektren stochastischer Schrödingeroperatoren (in German), Ph.D. Dissertation, Ruhr–Universität Bochum, FRG, 1981.

Kirsch W., Martinelli F.: On the spectrum of Schrödinger operators with a random potential, Commun. Math. Phys. **85** (1982), 329–350.

[5] Gesztesy F., Holden H., Kirsch W.: On energy gaps in a new type of analytically solvable model in quantum mechanics, J.Math.Anal.Appl. **134** (1988), 9–29.

[6] Kirsch W., Nitzschner F.: Lifshitz–tails and non–Lifshitz–tails for one–dimensional random point interactions, Operator theory: Advances and Applications **46** (1990), 171–178.

[7] Gesztesy F., Kirsch W., Obermeit J., to be published

[8] Skriganov M.M.: Proof of the Bethe–Sommerfeld conjecture in dimension two, Soviet Math.Dokl. **20** (1979), 956–959.

Skriganov M.M.: The multidimensional Schrödinger operator with a periodic potential, Math.USSR–Izv. **22** (1984), 619–645.

Skriganov M.M.: The spectrum band structure of the three–dimensional Schrödinger operator with periodic potential, Invent.Math. **80** (1985), 107–121.

[9] Høegh–Krohn R., Holden H., Martinelli F.: The spectrum of defect periodic point interactions, Lett.Math.Phys. **7** (1983), 221–228.

[10] Carmona R., Klein A., Martinelli F.: Anderson localization for Bernouilli and other singular potentials, Commun.Math.Phys. **108** (1987), 41–66.

[11] Fröhlich J., Spencer T.: Absence of diffusion in the Anderson tight binding model for large disorder or low energy, Commun.Math.Phys. **88** (1983), 151–184.

[12] Fröhlich J., Martinelli F., Scoppola E., Spencer T.: A constructive proof of localization in Anderson tight binding model, Commun.Math.Phys. **101** (1985), 21–46.

Delyon F., Lé vy Y., Souillard B.: Approach á la Borland to multidimensional localization, Phys.Rev.Lett. **55** (1985), 618–621.

Simon B., Wolff T.: Singular continuous spectrum under rank one pertubation and localization for random Hamiltonians, Commun.Pure Appl.Math.**39** (1986), 75–90.

Holden H., Martinelli F.: A remark on the absence of diffusion near the bottom of the spectrum for a random Schrödinger operator on $L^2(\mathbb{R}^\nu)$, Commun.Math.Phys. **93** (1984), 197-217.

von Dreifus H., Klein A.: A new proof of localization in the Anderson tight binding model, Commun.Math.Phys. **124** (1989), 285–299.

[13] Goldsheid I., Kirsch W., unpublished

[14] Šeba P.: Wave chaos in singular quantum billard, Phys.Rev.Lett. **64** (1990), 1855.

Albeverio S., Šeba P.: Wave chaos in quantum systems with point interactions, Preprint.

Werner Kirsch

Institut für Mathematik

Ruhr - Universität

D-4630 Bochum, Germany

STOCHASTIC MODEL OF QUANTUM MECHANICS

V.P.Maslov

M.I.E.M., Moscow, USSR

The stochastic model of quantum mechanics given in the present paper is based on the idea that there exist deep relations between Poissonian processes and the equations of quantum mechanics. Some papers dealing with this topic appeared at the end of the 70ies when the results concerning relations between the Schrodinger equation and the Kolmogorov-Feller equations were published [1-3]. Then this relation was established for the Heisenberg picture of quantum mechanics in the paper [4], where, under certain assumptions, the Kolmogorov-Feller equation was written in the form of an equation for the Wigner function (or for the Weyl symbols of operators). This formalism was developed in the papers [5-8], where certain interesting results were obtained in these directions.

The present paper extends the idea of the article [4] to a wider class of quantum equations. The stochastic model proposed here, although it is of rather naive character, nevertheless yields the basic quantum mechanical equations, such as the Schrödinger equation, the Pauli equation and the Dirac equation.

1. In order to create a stochastic model of quantum mechanics, we shall try to use the basic concepts of quantum physics, but we shall consider them from a more abstract and, on the other hand, more deterministic point of view. First of all, we shall assume that our situation involves the Plank constant h, and we must come to classical physics as $h \to 0$ (the principle of correspondence). Further we formulate the identity principle from quite materialistic positions. Namely, we assume that we look at the world just as if we were watching a movie, observing the photographs-slides that fix the positions and momenta of N particles and appear at random times tj, $0 \le tj \le t$, j=1,...,n, n ~ 1/h. We assume that we can't know what specific particle on the next slide corresponds to a chosen specific particle on the previous slide. In other words, if we fix a particle on the j-th slide, we don't know into what specific particle on the j+1-th slide it will evolve after the time tj+1 - tj. Thus, we don't know what happens with the particles in between. Nevertheless, the picture must be continuous as $h \to 0$. Thus, we shall assume that if we number particles on the j-th and the j+1-th slide in an arbitrary way, then we get the following highly probable relation for the k-th particle

280

$$q_k^{(j)} - q_k^{(j+1)} \sim h, \quad p_k^{(j)} - p_k^{(j+1)} \sim h \qquad (1)$$

where $q_k^{(j)}, p_k^{(j)}$ are the coordinate and momenta vectors of the k-th particle related to the tj instant of time. The probability of $q_k^{(j)} - q_k^{(j+1)} \sim 1$ and $p_k^{(j)} - p_k^{(j+1)} \sim 1$ tends to zero as $h \to 0$. This means that we number the particles in an arbitrary way, but reasonably: if a particle on the j+1-th slide is very far from the given particle (~1), then we give it the number of the given particle with small probability. So, we have observable slides and the unobservable world which exists by itself. The instants of time and the positions of particles are assumed to be random and we shall operate only with averaged values which will be called completely observable values.

Our slides have two sides and thus we must take into account the orientation of a slide. Orientation is a geometrical notion connected with the physical notion of polarization. We suppose that the polarization (orientation) κ can assume two values $\pm h/2$, thus it can vary only by 0 or by $+h$ when we pass from one slide to another. From the point of view of observable values, it does not matter how much we turn the slide, it is only important to know how it is oriented. Nevertheless, slide rotation requires some energy (not completely observable) which, generally speaking, may depend on the number of the slide rotations. In order to take these facts into account, we shall consider the (not completely observable) energy on the lattice $\kappa = h(2n+1)/2, n \in \mathbb{Z}$, where $\mathbb{Z}$ is the set of integers. The completely observable values will be described by functions which are 2h-periodic in κ, which is equivalent to considering κ as a two-valued variable. Thus, we can write the value of the jump in orientation when passing from slide to slide as follows

$$\Delta\kappa = \gamma, \quad \gamma \in \{0, h\} \qquad (2)$$

Naturally, we suppose that γ assumes the values 0 and h with equal probability.

By taking the concepts (1) and (2) to be the basic ones, we obtain the following relations for increments of coordinates and momenta when we pass from slide to slide

$$\Delta q = \alpha \kappa, \quad \Delta p = \alpha'\kappa; \tag{3}$$

i.e. the coordinate and momenta increments are proportional to κ, and α, α' are random constants.

Now we shall construct a function in q, p, κ, depending on α, α', γ which is conserved when we pass from slide to slide; it will be called energy. First we consider the case $\gamma = 0$. Obviously, the following expression, which is quadratic in q, p, κ,

$$\mathcal{H}_0 = \beta[\alpha p\kappa - \alpha'q\kappa + g(\alpha,\alpha')\kappa + \psi(\alpha,\alpha')\kappa^2],$$

where β is a constant, is conserved for the increments (3)

$$\Delta\mathcal{H}_0 = \mathcal{H}_0(q+\Delta q, p+\Delta p,\kappa) - \mathcal{H}_0(q,p,\kappa) = \beta(\alpha\kappa\Delta p - \alpha'\kappa\Delta q) = \beta(\alpha\kappa^2\alpha' - \alpha'\kappa^2\alpha) = 0$$

We note that we can write for the increments Δq and Δp

$$\Delta q = \frac{1}{\beta}\frac{\partial\mathcal{H}_0}{\partial p}, \quad \Delta p = -\frac{1}{\beta}\frac{\partial\mathcal{H}_0}{\partial q} \tag{4}$$

Consider the case $\gamma \neq 0$. In this case

$$\Delta\mathcal{H}_0 = \beta[\alpha p\Delta\kappa - \alpha'q\Delta\kappa + g\Delta\kappa + \psi(\Delta\kappa)^2 + 2\psi\kappa\Delta\kappa]$$

In order to compensate $\Delta\mathcal{H}_0$, we introduce the variable φ adjoint to the orientation κ. This variable κ has the meaning of momentum and thus the adjoint variable φ plays the role of additional coordinate. In order to obtain (similarly to relations (4))

$$\Delta\kappa = -\frac{1}{\beta}\frac{\partial\mathcal{H}}{\partial\gamma}$$

and to compensate the summand $\varphi(\Delta\kappa)^2$ in $\Delta\mathcal{H}_0$, we set

$$\mathcal{H} = \beta[\alpha p\kappa - \alpha'q\kappa + g\kappa + \psi\kappa^2 - \psi\gamma\kappa - \gamma\varphi],$$

then, since $\Delta\kappa = \gamma$, we obtain

$$\Delta\mathcal{H}=\beta[(\alpha p-\alpha'q)\gamma+g\gamma+\psi\gamma^2+2\psi\gamma\kappa-\psi\gamma^2-\gamma\Delta\varphi].$$

Therefore, in order to have $\Delta\mathcal{H}=0$, it is necessary to set

$$\Delta\varphi=\alpha p-\alpha'q+g(\alpha,\alpha')+2\psi(\alpha,\alpha')\kappa$$

The additional coordinate φ is a variable adjoint to orientation κ, assuming discrete values and having the meaning of momentum. Thus, φ has the meaning of orientation angle and completely observable values must be 2π-periodic in φ. It is natural to demand that if we change the orientation to the opposite one, the increment of the orientation angle $\Delta\varphi$ will change by π. This condition is satisfied if $\psi=\dfrac{\pi}{2h}$, thus we finally have

$$\Delta\varphi=\alpha p-\alpha'q+g(\alpha,\alpha')+\frac{\pi\kappa}{h} \tag{5}$$

and for energy $\mathcal{H}$ we obtain

$$\mathcal{H}=\beta[\alpha p\kappa-\alpha'q\kappa+\kappa g(\alpha,\alpha')+\frac{\pi\kappa^2}{2h}-\frac{\pi\gamma\kappa}{2h}-\gamma\varphi]$$

It is easy to see that the increments of the variables q,p,φ,κ are defined by formulas (2),(3),(5) and the energy is conserved when we pass from slide to slide.

So, at time t, we have a set of slides made at random instants of time t_j, $t_j\in[0,t]$, which fix the position of particles in extended phase space formed by the variables q,p,φ,κ. We stress once more that we don't know what happens in between. We only know the instants of time t_j, and the increments of variables q,p,φ,κ. More precisely, we know the probability distributions of times and increments. Now let us describe these distributions.

Let the instants of time t_j, $j=1,\ldots,\nu(t)$, be uniformly distributed on the interval $[0,t]$, so that the number of these instants $\nu(t)$ is distributed according the Poisson law, i.e. the probability that $\nu(t)$ is equal to the number n is defined by the relation

$$P\{\nu(t)=n\}=\frac{(at)^n}{n!}e^{-\alpha t}$$

where a is the parameter of the Poissonian distribution. We demand that the averaged statisitcal number of points of measurement be equal to t/h. By setting this number equal to the mathematical expectation $M\nu(t)$, we obtain $M\nu(t) = at = t/h$. Thus, the parameter of Poisson distribution is $a = h^{-1}$.

The expressions (2), (3), (5) for the increments of q, p, φ, κ contain the parameters α, α', γ, which define the value of increments. Let the distribution of parameters α, α' be given by the measure $\mu(d\alpha, d\alpha')$, and suppose that γ assumes the value 0 or h with equal probability, i.e. the distribution of the parameter γ is described by the measure $[\delta_0(d\gamma) + \delta_h(d\gamma)]/2$ where $\delta_\theta(d\gamma)$ is the Dirac measure concentrated at the point θ. We note that the principle of identity for particles requires that the measure $\mu(d\alpha, d\alpha')$ and the function $g(\alpha, \alpha')$ be symmetric with respect to the vectors α, α' corresponding to the numbers of different particles.

We consider the density function of non-compensated charges $\rho(q, p, \kappa, t)$. We consider the density of non-compensated charges only, because, in contrast to the density of negative or positive charges, it can be measured, and thus, we can formulate initial conditions for ρ. In accordance with the probability model given above, we write the evolution equation

$$\frac{\partial \rho}{\partial t} = \frac{1}{2h} \int [\rho(q+\Delta q,\, p+\Delta p,\, \varphi+\Delta\varphi,\, \kappa+\Delta\kappa, t) - \rho(q, p, \varphi, \kappa, t)]\, \mu(d\alpha, d\alpha')[\delta_0(d\gamma) + \delta_h(d\gamma)] \quad (6)$$

which is the Kolmogorov-Feller equation.

The density ρ is a completely observable function; thus we must consider only two values $\kappa = \pm h/2$, and the value $\kappa + \Delta\kappa$ is taken modulo 2h. The variable φ is adjoint to a discrete variable, and thus just like as in the principle of duality known both in radiophysics and in the theory of groups [9], φ belongs to the unit circle. This means that the completely observable function ρ must be 2π-periodic in φ. It is easy to see that equation (6) does not contradict this fact. Moreover, since the variable κ assumes only two values, we shall try to restrict our considerations to the first two harmonics with respect to the adjoint variable φ, i.e. 1 and $e^{\pm i\varphi}$, which can be composed into only two real combinations, namely 1 and $\cos\varphi$. We shall show below that equation (6) is invariant on the subspace generated by these functions. First we transform equation (6). First of all, in order to compensate the second summand in the right-hand side of the equation, we consider the function $r = e^{-\beta t}\rho$, where

$$\beta = h^{-1}\!\int \mu(d\alpha,d\alpha')$$

The function r is described by the equation

$$\frac{\partial r}{\partial t} = \frac{1}{2h}\!\int r(q+\Delta q,p+\Delta p,\varphi+\Delta\varphi,\kappa+\Delta\kappa,t)\mu(d\alpha,d\alpha')[\delta_0(d\gamma)+\delta_h(d\gamma)]$$

We note that only $\Delta\kappa$ in the expression under the integral depends on γ. Since $(\kappa+\Delta\kappa)(\mathrm{mod}2h)$ is equal to κ if $\gamma=0$ and to $-\kappa$ if $\gamma=h$, we can calculate the integral with respect to the Dirac measures

$$\frac{\partial r}{\partial t} = \frac{1}{2h}\!\int [r(q+\Delta q,p+\Delta p,\varphi+\Delta\varphi,\kappa,t)+r(q+\Delta q,p+\Delta p,\varphi+\Delta\varphi,-\kappa,t)]\mu(d\alpha,d\alpha') \qquad (7)$$

Now let us average with respect to the orientation $\kappa=\pm h/2$. For this purpose we consider equation (7) for $\kappa=h/2$ and $\kappa=-h/2$. By the formulas for the increments of variables q,p,φ, we obtain

$$\frac{\partial}{\partial t}[r(q,p,\gamma,\kappa,t)+r(q,p,\gamma,-\kappa,t)]=\frac{1}{2h}\!\int \Big[r\big(q+\tfrac{h\alpha}{2},\,p+\tfrac{h\alpha'}{2},\,\varphi+\Delta\varphi\,(\tfrac{h}{2}),\tfrac{h}{2},t\big)+$$

$$+r\big(q+\tfrac{h\alpha}{2},\,p+\tfrac{h\alpha'}{2},\varphi+\Delta\varphi(\tfrac{h}{2}),\,-\tfrac{h}{2},t\big)+r\big(q-\tfrac{h\alpha}{2},\,p-\tfrac{h\alpha'}{2},\varphi-\Delta\varphi(-\tfrac{h}{2}),\,\tfrac{h}{2},t\big)+$$

$$+r\big(q-\tfrac{h\alpha}{2},\,p-\tfrac{h\alpha'}{2},\varphi+\Delta\varphi(-\tfrac{h}{2}),\,-\tfrac{h}{2},t\big)\Big]\,\mu(d\alpha,d\alpha') \qquad (8)$$

By grouping the summands in equation (8), we obtain the following equation for the function $R(q,p,\gamma,t)=\frac{1}{2}[r(q,p,\varphi,\tfrac{h}{2},t)+r(q,p,\varphi,-\tfrac{h}{2},t)]$ averaged with respect to orientation

$$\frac{\partial R}{\partial t} = \frac{1}{2h}\!\int\!\left[R\!\left(q+\frac{h\alpha}{2},\,p+\frac{h\alpha'}{2},\,\varphi+\Delta\varphi(\tfrac{h}{2}),t\right)+R\!\left(q-\frac{h\alpha}{2},p-\frac{h\alpha'}{2},\varphi-\Delta\varphi(-\tfrac{h}{2}),t\right)\right]\mu(d\alpha,d\alpha') \qquad (9)$$

Now in order to get the dependence on the additional variable φ, we shall try to represent the solution of equation (9) in the form

$$R(q,p,\varphi,t) = W(q,p,t)\,v(\varphi,t) \qquad (10)$$

$$\beta = h^{-1}\int \mu(d\alpha, d\alpha')$$

The function r is described by the equation

$$\frac{\partial r}{\partial t} = \frac{1}{2h}\int r(q+\Delta q, p+\Delta p, \varphi+\Delta\varphi, \kappa+\Delta\kappa, t)\mu(d\alpha, d\alpha')[\delta_0(d\gamma)+\delta_h(d\gamma)]$$

We note that only $\Delta\kappa$ in the expression under the integral depends on γ. Since $(\kappa+\Delta\kappa)(mod 2h)$ is equal to κ if $\gamma=0$ and to $-\kappa$ if $\gamma=h$, we can calculate the integral with respect to the Dirac measures

$$\frac{\partial r}{\partial t} = \frac{1}{2h}\int [r(q+\Delta q, p+\Delta p, \varphi+\Delta\varphi, \kappa, t)+r(q+\Delta q, p+\Delta p, \varphi+\Delta\varphi, -\kappa, t)]\mu(d\alpha, d\alpha') \quad (7)$$

Now let us average with respect to the orientation $\kappa=\pm h/2$. For this purpose we consider equation (7) for $\kappa=h/2$ and $\kappa=-h/2$. By the formulas for the increments of variables q, p, φ, we obtain

$$\frac{\partial}{\partial t}[r(q,p,\gamma,\kappa,t)+r(q,p,\gamma,-\kappa,t)]=\frac{1}{2h}\int [r(q+\frac{h\alpha}{2}, p+\frac{h\alpha'}{2}, \varphi+\Delta\varphi(\frac{h}{2}), \frac{h}{2}, t)+$$

$$+r(q+\frac{h\alpha}{2}, p+\frac{h\alpha'}{2}, \varphi+\Delta\varphi(\frac{h}{2}), -\frac{h}{2}, t)+r(q-\frac{h\alpha}{2}, p-\frac{h\alpha'}{2}, \varphi+\Delta\varphi(-\frac{h}{2}), \frac{h}{2}, t)+$$

$$+r(q-\frac{h\alpha}{2}, p-\frac{h\alpha'}{2}, \varphi+\Delta\varphi(-\frac{h}{2}), -\frac{h}{2}, t)]\,\mu(d\alpha, d\alpha') \quad (8)$$

By grouping the summands in equation (8), we obtain the following equation for the function $R(q,p,\gamma,t)=\frac{1}{2}[r(q,p,\varphi,\frac{h}{2},t)+r(q,p,\varphi,-\frac{h}{2},t)]$ averaged with respect to orientation

$$\frac{\partial R}{\partial t} = \frac{1}{2h}\int \left[R\left(q+\frac{h\alpha}{2}, p+\frac{h\alpha'}{2}, \varphi+\Delta\varphi(\frac{h}{2}),t\right)+R\left(q-\frac{h\alpha}{2}, p-\frac{h\alpha'}{2}, \varphi+\Delta\varphi(-\frac{h}{2}),t\right)\right]\mu(d\alpha, d\alpha') \quad (9)$$

Now in order to get the dependence on the additional variable φ, we shall try to represent the solution of equation (9) in the form

$$R(q,p,\varphi,t) = W(q,p,t)v(\varphi,t) \quad (10)$$

and to find the invariant subspace to which the function v belongs. And thus the multiplicative structure (10) of the solution of equation (9) is conserved. It turns out that the subspace pointed out above by using the principle of duality which is spread over the functions 1 and $\cos\varphi$ possesses the same property. Moreover, equation (9) is invariant on each of the elements 1 and $\cos\varphi$. Actually, it is evident that equation (9) is invariant with respect to representation (10) for $v=1$. In the case $v=\cos\varphi$ we shall prove the following theorem:

THEOREM 1. Let $g(\alpha,\alpha')$ be a function odd with respect to all its arguments, $g(\alpha,\alpha') = -g(-\alpha,-\alpha')$, μ be a symmetric measure such that

$$\int u(\alpha,\alpha')\mu(d\alpha,d\alpha') = \int u(-\alpha,-\alpha')\,\mu(d\alpha,d\alpha')$$

where u is an arbitrary measurable function. Then the solution of equation (9) can be represented in the form $R = W(q,p,t)\cos\varphi$.

After writing $\cos\varphi$ in the form $\frac{1}{2}(e^{i\varphi}+ e^{-i\varphi})$ we substitute $R = W\cos\varphi$ into equation (9). Since the value of increment $\Delta\varphi$ is defined by formula (5), we obtain

$$\cos\varphi\frac{\partial W(q,p,t)}{\partial t} = \frac{1}{4h}\int \{W(q+\frac{h\alpha}{2},p+\frac{h\alpha'}{2},t)[\exp(i(\varphi+p\alpha-q\alpha'+g(\alpha,\alpha')+\frac{\pi}{2})) +$$

$$+\exp(-i(\varphi+p\alpha-q\alpha'+g(\alpha,\alpha')+\frac{\pi}{2}))]+W(q-\frac{h\alpha}{2},p-\frac{h\alpha'}{2},t)[\exp(i(\varphi+p\alpha-q\alpha'+g(\alpha,\alpha')-\frac{\pi}{2}))+$$

$$\exp(-i(\varphi+p\alpha-q\alpha'+g(\alpha,\alpha')-\frac{\pi}{2}))]\}\mu(d\alpha,d\alpha') \tag{11}$$

Then we change $\alpha,\alpha' \to -\alpha,-\alpha'$ in the second group of summands under the sign of integral in the right-hand side of equation (11). Since the measure μ is symmetric and the function g is odd, we obtain the equation

$$\cos\varphi\frac{\partial W}{\partial t} = \frac{1}{4h}(e^{i\varphi}+e^{-i\varphi})\int W(q+\frac{h\alpha}{2}, p+\frac{h\alpha'}{2},t)[\exp(i(p\alpha-q\alpha'+g(\alpha,\alpha')+\frac{\pi}{2})) +$$

$$\exp(-i(p\alpha-q\alpha'+g(\alpha,\alpha')+\frac{\pi}{2}))]\mu(d\alpha,d\alpha'), \tag{12}$$

which allows to exclude $\cos\varphi$. Thus, the Theorem is proved.

We shall demand that the density of non-compensated charges (averaged with respect to the angle of orientation) be equal to zero, i.e.

$$\int_{-\pi}^{\pi} R\,(q,p,\varphi,t)\,d\varphi = 0.$$

This condition points out just the case $R(q,p,\varphi,t) = W(q,p,t)\cos\varphi$, which we shall consider below. The equation describing the function W follows directly from equation (12). We make the change $-\alpha,-\alpha' \rightarrow \alpha,\alpha'$ in the second summand in the right-hand side of equation (12), then

$$\frac{\partial W}{\partial t} = \frac{i}{2h}\int [W(q+\frac{h\alpha}{2},p+\frac{h\alpha'}{2},t) - W(q-\frac{h\alpha}{2},p-\frac{h\alpha'}{2},t)\, e^{i(p\alpha-q\alpha'+g(\alpha,\alpha'))}\,\mu(d\alpha,d\alpha') \qquad (13)$$

We note that since $\int_{-\pi}^{\pi} Rd\gamma = 0$, the value averaged with respect to the angle of orientation can be used for normalization. But since formula (10) contains $\cos\varphi$ multiplicatively, we can nevertheless speak about the distribution $W(q,p,t)$ for fixed time within the accuracy of normalization. And since the integral $\int Wdqdp$, as it is easy to see, is conserved in time, we can take it conditionally as a new normalization. This is very useful for comparing of measurements.

We note that equation (13) coincides with the equation for the Wigner function related to the Schrödinger equation with the following Weyl symbol of the Hamiltonian

$$H(q,p) = \frac{1}{2}\int e^{i(p\alpha - q\alpha' + g(\alpha,\alpha'))}\,\mu\,(d\alpha,d\alpha') \qquad (14)$$

Since the function $W(q,p,t)$ satisfies the Wigner equation related to Hamiltonian (14) , we have some properties of this function which allow to consider W as a probability distribution. So the distribution with respect to q, i.e. the density $\int Wdp$ is a value the positiveness of which is conserved. The same property holds for the value $\int F^2 Wdqdp$ if $F = F(q)$ depends only on the variable q. We can say the same about the distribution with respect to p. The positiveness of the correlation matrix for the values q and p is also conserved together with the distribution W. Thus, the function

W can be used in the simplest cases for calculating errors (dispersion). The problem of calculating dispersion in a more complicated situation can also be well posed.

We note that the measure μ and the function g can, generally speaking, depend regularly on the parameter h, thus the Hamiltonian symbol must be written more precisely in the form

$$H(q,p,h) = \frac{1}{2} \int e^{i(p\alpha - q\alpha' + g(\alpha,\alpha',h))} \, \mu(d\alpha, d\alpha', h) \tag{15}$$

The Wigner equation is transformed into the Louiville equation relation to the Hamiltonian $H(q,p,0)$ as $h \to 0$. Thus, if we know the limit picture, we can define the limit $H(q,p,h)$ as $h \to 0$ only. The whole class of Hamiltonians (before passing to the limit) is related to this limit Hamiltonian, thus we can obtain $H(q,p,h)$, using the principle of correspondence. It is assumed in quantum mechanics that the Hamiltonian symbol is independent of h, i.e. $H(q,p)$ is the same as in classical mechanics - this is the principle of correspodence. However, it is impossible to say whether there exists very weak dependence on h, so weak that it cannot be defined experimentally by modern tools. In our case not each Hamiltonian can be represented in the form (14), i.e. as the Fourier transform of the finite measure variation, but each Haniltonian can be changed very weakly (regularized) so that formula (15) will hold for it. In this case the equation for the Wigner function can be written as follows

$$\frac{\partial W}{\partial t} = \frac{i}{2h} \int [W(q+\frac{h\alpha}{2},p+\frac{h\alpha'}{2},t) - W(q-\frac{h\alpha}{2},p-\frac{h\alpha'}{2},t)] \times$$

$$\times \, e^{i(p\alpha - q\alpha' + g(\alpha,\alpha',h))} \mu(d\alpha, d\alpha', h) \tag{16}$$

and though the measure μ becomes singular as $h \to 0$, the equation is not damaged as $h \to 0$. Thus, the initial model can be also applied in this case.

We note two facts which prove that the constructed statistical model and quantum mechanical description are matched. First, the function g is odd and the measure μ is symmetric (these conditions were stated in Theorem 1) which is equivalent, as it is easy to see, to the condition that the Weyl symbol of Hamiltonian (14) is real. Secondly, if the momentum is conserved in the statistical description, it is also conserved in the quantum mechanical description. Actually, let the momentum be conserved when we pass from slide to slide, i.e. the momentum increments are equal to zero, then $\alpha'=0$ and $\mu(d\alpha,d\alpha')=\mu_0(d\alpha)\delta_0(d\alpha')$. This measure is related to the symbol of Hamiltonian (14) independent of the coordinates q, which is the condition of momentum conservation in the quantum mechanical description.

The operator form of the Wigner equation coincides with the Heisenberg equation and has the folowing form

$$\frac{\partial \hat{W}}{\partial \tau} = \frac{i}{h}\left[\hat{H},\hat{W}\right], \tag{17}$$

where $\hat{W}$ is the Wigner operator, $\hat{H}$ is the Hamilton operator, the brackets [,] denote the commutator of operators. In order to obtain equation (13) from the operator equation (17), we shall use the formula of operator composition in the Weyl calculus, according to which the symbol A of the composition $\hat{A}=\hat{A}_1\hat{A}_2$ of the operators $\hat{A}_1,\hat{A}_2$ is defined by the following expression [10]

$$A(q,p) = A_1(\overset{2}{q} + \frac{ih}{2}\overset{1}{\frac{\partial}{\partial p}}, \overset{2}{p} - \frac{ih}{2}\overset{1}{\frac{\partial}{\partial q}}) A_2(q,p) = A_2(\overset{2}{q} - \frac{ih}{2}\overset{1}{\frac{\partial}{\partial p}}, \overset{2}{p} + \frac{ih}{2}\overset{1}{\frac{\partial}{\partial q}}) A_1(q,p) \tag{18}$$

where $A_1(q,p),A_2(q,p)$ are the symbols of the operators $\hat{A}_1,\hat{A}_2$ and the numbers over the operators show the order of their action. Since the Weyl symbol of the Hamiltonian has the form (14), the first relation in (18) yields that the symbol of the operator $\frac{i}{h}\hat{H}\hat{W}$ is equal to

$$\frac{i}{2h}\int e^{i[p\alpha-q\alpha'+g]}\, e^{\frac{h\alpha}{2}\frac{\partial}{\partial q}+\frac{h\alpha'}{2}\frac{\partial}{\partial p}}\, \mu(d\alpha,d\alpha')W(q,p,t)=$$

$$=\frac{i}{2h}\int W(q+\frac{h\alpha}{2}, p+\frac{h\alpha'}{2}, t)\, e^{i[p\alpha-q\alpha'+g]}\, \mu(d\alpha,d\alpha')$$

and coincides with the second summand in the right-hand side of equation (13). The first summand is the symbol of the operator $-\frac{i}{h}\hat{W}\hat{H}$ which is the direct consequence of the second relation in fromula (18).

We note that we can pass from the Wigner equation in the operator form to the equation for symbols only with respect to certain components of the variables q,p which are somehow marked. Such example will be considered in the next section.

2. We consider our statistical model in the case of one particle when the state of the particle is described, besides coordinates q and momenta p by an additional coordinate (angle of spin) q_s and an additional momentum (momentum of spin) p_s. In this case just like as before the extended phase space is formed by the coordinates $\bar{q}$, the momenta $\bar{p}$, the orientation κ and the angle of orientation φ, $\bar{q}=(q_s,q)$, where q_s, p_s are spin variables, and q,p are ordinary coordinates and momenta. Since the spin variable q_s denotes angle, it is natural to demand that the completely observable functions be 2π-periodic in this variable. The variable p_s denotes spin momentum, which assums values multiple to half-integer Plank constant, $p_s = nh/2$, $n \in \mathbb{Z}$. As we can see, the conditions of 2π-periodicity in q_s and discreteness of the adjoint variable p_s correspond with the principle of duality.

We assume that the increments of variables q,p,κ are defined as before by formulas (2),(3), and then the increments of spin valiables and φ must be defined by the formulas

$$\Delta q_s = \alpha_s \kappa, \quad \Delta p_s = \alpha'_s \kappa$$

$$\Delta \varphi = \bar{\alpha}\,\bar{p} - \bar{\alpha}'\,q + g_s(\bar{\alpha},\bar{\alpha}') + \frac{\pi}{h}\kappa, \qquad (19)$$

where $\bar{\alpha} = (\alpha_s,\alpha)$, $\bar{\alpha}' = (\alpha'_s,\alpha)$, $g_s(\bar{\alpha},\bar{\alpha}')$ is an arbitrary function. We suppose that the spin variables vary jumpwise. If there is no jump of spin angle q_s (or spin momentum p_s), then $\alpha_s=0$ ($\alpha'_s=0$). If there are jumps of spin variables, then the values of these are fixed. It is natural to suppose that if orientation is changed by the opposite one, the value of spin angle varies by π, and the value of spin momentum varies by h. Thus, we have that if there are jumps in spin variables, then $\alpha_s = \pi/h$, and $\alpha' = 1$. We note that the variable p_s remains on the lattice $p_s = nh/2$ during these jumps. Thus, there can be four variants of jumps in spin variables: there are no jumps in both variables ($\alpha_s=0$, $\alpha'_s=0$); only one of spin variables jumps ($\alpha_s=\pi/h$, $\alpha'_s=0$ or $\alpha_s=0$, $\alpha'_s=1$); there are jumps in both spin variables ($\alpha_s=\pi/h$, $\alpha'_s=1$). Further we shall assume that these four situations take place with certain probabilities.

The additional spin variables do not change essentially the procedure proposed above, and we obtain an equation similar to equation (13)

$$\frac{\partial W_s(\bar{q},\bar{p},t)}{\partial t} = \frac{i}{2h}\int [W_s(\bar{q}+\frac{h\bar{\alpha}}{2},\bar{p}+\frac{h\bar{\alpha}'}{2},t) -$$

$$W_s(\bar{q}-\frac{h\bar{\alpha}}{2},\bar{p}-\frac{h\bar{\alpha}'}{2},t)]\, e^{i(p\bar{\alpha}\,-\,\bar{q}\bar{\alpha}'+g_s(\bar{\alpha},\bar{\alpha}'))}\,\mu_s(d\bar{\alpha},d\bar{\alpha}') \qquad (20)$$

The measure μ_s can be written as follows (by taking into account the assumptions about spin jumps made above)

$$\mu_s(d\bar{\alpha},d\bar{\alpha}') = \mu_0(d\alpha,d\alpha')\delta_0(d\alpha_s)\delta_0(d\alpha'_s)+\mu_1(d\alpha,d\alpha')\delta_{\frac{\pi}{h}}(d\alpha_s)\delta_0(d\alpha'_s) +$$

$$+\mu_2(d\alpha,d\alpha')\delta_{\frac{\pi}{h}}(d\alpha_s)\delta_1(d\alpha'_s)+\mu_3(d\alpha,d\alpha')\delta_0(d\alpha_s)\delta_1(d\alpha'_s). \qquad (21)$$

We can integrate the right-hand side of equation (20) with respect to the Dirac measures which are written in (21). As the result we obtain

$$\frac{\partial W_s}{\partial t} = +\frac{i}{2h}\int e^{i(p\alpha-q\alpha')}\{[W_s^+(q_s,p_s) - W_s^-(q_s,p_s)]e^{ig_0}\mu_0(d\alpha,d\alpha')+$$

$$[W_s^+(q_s+\frac{\pi}{2},p_s)-W_s^-(q_s-\frac{\pi}{2},p_s)]e^{i(p_s\frac{\pi}{h}+g_1)}\mu_1(d\alpha,d\alpha')+$$

$$+[W_s^+(q_s+\frac{\pi}{2},p_s+\frac{h}{2})-W_s^-(q_s-\frac{\pi}{2},p_s-\frac{h}{2})]e^{i(p_s\frac{\pi}{h}-q_s+g_2)}\mu_2(d\alpha,d\alpha')+$$

$$[W_s^+(q_s,p_s+\frac{h}{2}) - W_s^-(q_s,p_s-\frac{h}{2})]e^{i(-q_s+g_3)}\mu_3(d\alpha,d\alpha'), \qquad (22)$$

Here

$$W_s^{\pm}(q_s,p_s) = W_s(q_s,q\pm\frac{h\alpha}{2},p_s,p\pm\frac{h\alpha'}{2},t);\ g_i=g_i(\alpha,\alpha');\ g_0(\alpha,\alpha')=g_s(0,\alpha,0,\alpha'),$$

$$g_1(\alpha,\alpha')=g_s(\frac{\pi}{h},\alpha,0,\alpha'),\ g_2(\alpha,\alpha')=g_s(\frac{\pi}{h},\alpha,1,\alpha'),\ g_3(\alpha,\alpha')=g_s(0,\alpha,1,\alpha'),$$

and we assume that the function $g_s(\alpha_s,\alpha,\alpha'_s,\alpha')$ is even with respect to spin variables.

We assume that the comletely observable function W_s is 2π-periodic in spin angle q_s, and since p_s denotes the spin momentum, it is 2h-periodic in p_s. It is easy to see that equation (22) does not contradict to this assumption. Moreover, equation (22) refllects the principle of duality. A variable adjoint to the variable, with respect to which completely observable functions is periodic, must be discrete with step of lattice equal to $\pi h/T$, where T is the period. Actually, the form of equation (22) allows to consider the function W_s on the lattice $p_s = nh/2$, $n \in \mathbb{Z}$, with respect to the variable p_s, and to consider it on a lattice with step $\pi/2$ with respect to the variable q_s.

Equation (22) is the Wigner equation with the Weyl symbol of Hamiltonian

$$H(q_s,q,p_s,p) = H_0(q,p)+H_1(q,p)e^{i\frac{\pi p_s}{h}}+H_2(q,p)e^{i\frac{\pi p_s}{h}}e^{-iq_s}+H_3(q,p)e^{-iq_s}, \qquad (23)$$

where

$$H_j(q,p)=\frac{1}{2}\int e^{i(p\alpha-q\alpha'+g_j(\alpha,\alpha'))}\mu_j(d\alpha,d\alpha'). \qquad (24)$$

We perform the Weil quantization in spin variables only changing p_s by the operator $-ih\frac{\partial^2}{\partial q_s}$ and q_s by the operator $\frac{q_s^1+q_s^3}{2}$

$$\hat{H}(q,p)=H(\frac{q_s^1+q_s^3}{2},q,-ih\frac{\partial^2}{\partial q_s},p)=H_0+H_1e^{\pi\frac{\partial}{\partial q_s}}+iH_2e^{-iq_s}e^{\pi\frac{\partial}{\partial q_s}}+H_3e^{-iq_s}. \qquad (25)$$

Operator (25) structure allows to consider it over the functions on a lattice with step π, in particular, over 2π-periodic functions on the lattice $q_s = \pi n$, $n \in \mathbb{Z}$, which can be defined by two values at points 0 and π, i.e. by two-component vectors. In this case the operator $\hat{H}$ turns into a matrix of the second order

$$H_s(q,p) = \sum_{i=0}^{3} H_i(q,p)\sigma_i, \qquad (26)$$

where $\sigma_0 = I$ is the unit matrix of the second order, and σ_j (j = 1,2,3) is the Pauli matrix

$$\sigma_1 = \begin{pmatrix} 0 & 1 \\ 1 & 0 \end{pmatrix}, \quad \sigma_2 = \begin{pmatrix} 0 & -i \\ i & 0 \end{pmatrix}, \quad \sigma_3 = \begin{pmatrix} 1 & 0 \\ 0 & -1 \end{pmatrix} \tag{27}$$

Now we pass from the Weyl symbol with respect to spin variables $W_s(q_s, q, p_s, p, t)$ to the operator $\hat{W}_s(q, p, t) = W_s\left(\frac{\overset{1}{q_s}+\overset{3}{q_s}}{2}, q, -ih\frac{\overset{2}{\partial}}{\partial q_s}, p, t\right)$. Since the symbol W_s is 2h-periodic in p_s, the operator $\hat{W}_s$ is invariant over functions defined on a lattice with step π. By considering $\hat{W}_s$ over the subspace of 2π-periodic functions given on the lattice $q_s = n\pi$, $n \in \mathbb{Z}$, we pass from the operator $\hat{W}_s(q, p, t)$ to a matrix of the second order $\mathcal{W}(q, p, t)$. Equations (22) yields the equation defining the matrix function $\mathcal{W}$.

THEOREM 2. Let 2h-periodic in the variable p_s function W_s satisfy equation (22), then the corresponding matrix function $\mathcal{W}$ satisfies the equation

$$\frac{\partial \mathcal{W}}{\partial t} = \frac{i}{2h} \sum_{j=0}^{3} \int [\sigma_j \mathcal{W}(q + \frac{h\alpha}{2}, p + \frac{h\alpha'}{2}, t) -$$
$$- \mathcal{W}(q - \frac{h\alpha}{2}, p - \frac{h\alpha'}{2}, t) \sigma_j] e^{i(p\alpha - q\alpha' + g_j[\alpha,\alpha'])} \mu_j(d\alpha, d\alpha'). \tag{28}$$

Actually we write equation (22) in the operator form with respect to spin variables only. The composition formula (18) yields

$$\frac{\partial \hat{W}_s}{\partial t} = \frac{i}{2h} \int e^{i(p\alpha - q\alpha')} \{ [\hat{W}_s^+ - \hat{W}_s^-] e^{ig_0} \mu_0(d\alpha, d\alpha') +$$

$$+ [e^{\pi\frac{\partial}{\partial q_s}}\hat{W}_s^+ - \hat{W}_s^- e^{\pi\frac{\partial}{\partial q_s}}] e^{ig_1} \mu_1(d\alpha, d\alpha') - \tag{29}$$

$$- [i e^{-iq_s} e^{\pi\frac{\partial}{\partial q_s}} \hat{W}_s^+ - i\hat{W}_s^- e^{-iq_s} e^{\pi\frac{\partial}{\partial q_s}}] e^{ig_2} \mu_2(d\alpha, d\alpha') +$$

$$+ [e^{-iq_s}\hat{W}_s^+ - \hat{W}_s^- e^{-iq_s}] e^{ig_3} \mu_3(d\alpha, d\alpha')\}$$

where

$$\hat{W}_s^{\pm} = W_s\left(\frac{\overset{1}{q_s}+\overset{2}{q_s}}{2},\; q\pm\frac{h\alpha}{2},\; -ih\frac{\overset{3}{\partial}}{\partial q_s},\; p^{\pm}\frac{h\alpha'}{2},\; t\right).$$

. The restriction of the operator equation (29) to the subspace of 2π-periodix functions defined at the points $q_s=\{0,\pi\}$ by their values yields straightforward equation (28). The operators contained in equation (29) can be restricted to functions defined on a lattice with step π, since their symbols are 2h-periodic with respect to the variable p_s.

The matrix equation (28) is an equation for the Wigner matrix related to the Pauli equation

$$ih\frac{\partial\psi(q,t)}{\partial t} = \sum_{j=0}^{3} \hat{H}_j\, \sigma_j\, \psi(q,t)$$

where $\psi(q,t)$ are two-component vector-functions, and the Weyl symbols of the operators H_j are defined by expressions (24). Thus, we see that the classical statistical model considered here leads to the quantum mechanical Pauli equation.

3. From the point of view of our model we shall consider the relativistic case. Let the state of a particle be described besides the coordinates q, the momenta p, the orientation κ and the spin variables q_s, p_s by a pair of additional relativistic variables q_r, p_r. We shall call q_r and p_r the proper time and energy of a particle repectively. We assume that the proper energy of a particle may assum only discrete values multiple to $\pi h/T$, where T is a constant, possessing the dimension of time. According to the principle of duality, the fact that p_s is discrete implies the periodicity of completely observable functions with respect to the adjoint variable q_r. And since $p_r=\pi hn/T$, $n \in \mathbb{Z}$, the period is equal to T.

Just like as before we set for the increments

$$\Delta q_r = \alpha_r \kappa, \qquad \Delta p_r = \alpha'_r \kappa,$$

$$\Delta\varphi = \tilde{\alpha}\tilde{p} - \tilde{\alpha}''\tilde{q} + g_r(\tilde{\alpha},\tilde{\alpha}') + \frac{\pi}{h}\kappa, \qquad (30)$$

where $\tilde{q} = (q_r,q_s,q)$, $\tilde{p} = (p_r,p_s,p)$, $\tilde{\alpha} = (\alpha_r,\alpha_s,\alpha)$, $\tilde{\alpha}' = (\alpha'_r,\alpha'_s,\alpha')$, $g_r(\tilde{\alpha},\tilde{\alpha}')$ is an arbitrary function. The increments in the variables q_s, q, p_s, p, κ are defined as before.

Acting in the same way as in the previous section, we obtain the equation similar to equation (28)

$$\frac{\partial \mathcal{W}_r}{\partial t} = \frac{i}{2h} \sum_{j=0}^{3} \int [\sigma_j \mathcal{W}_r^{+} - \mathcal{W}_r^{-}\sigma_j] \exp(i(p_r\alpha_r + p\alpha - q_r\alpha'_r - q\alpha' + g_{rj}(\alpha_r,\alpha,\alpha'_r,\alpha')))] \times$$

$$\times \mu_{rj}(d\alpha_r, d\alpha, d\alpha'_r, d\alpha'), \qquad (31)$$

$\mathcal{W}_r^{\pm} = \mathcal{W}_r(q_r \pm \frac{h\alpha_r}{2}, q \pm \frac{h\alpha}{2}, p_r \pm \frac{h\alpha'_r}{2}, p \pm \frac{h\alpha'}{2}, t)$, $\mathcal{W}_r = \mathcal{W}_r(q_r, q, p_r, p, t)$, where $\mathcal{W}_r$ is a matrix of the second order, $g_{r0} = g_r\big|_{\alpha_s = \alpha'_s = 0}$, $g_{r1} = g_r\big|_{\alpha_s = \pi/h, \alpha'_s = 0}$, $g_{r2} = g_r\big|_{\alpha_s = \pi/h, \alpha'_s = 1}$, $g_{r3} = g_r\big|_{\alpha_s = 0, \alpha'_s = 1}$.

Let us consider the measures μ_{rj} more precisely. We assume that the values of jumps in relativistic variables satisfy the quantization condition of Bohr-Zommerfeld type

$$\Delta q_r \Delta p_r = \frac{\pi h}{2}(m+\frac{1}{2})$$

where m is integer. In order to satisfy this condition and to stay on the lattice $p_r = \pi h n/T$ with respect to the variable p_r, it is necessary to set $\Delta p_r = (2k+1)\pi h/T$, where k is integer. Then $\Delta q_r = (2l+1)T/4$, where l is integer. It is easy to see that solutions of equation (31) with initial conditions T-periodic in q_r will be T-periodic for such values of jumps in the variable p_r. We note that it is sufficient to restrict oneself to the jumps $\Delta q_r = (2l+1)T/4$, $l = 0,-1$ due to the T-periodicity. This fact corresponds to the idea that q_r is the proper time and hence can jump into the nearest possible positions only.

So, by formulas (30) and $\kappa = \pm h/2$, we get that $\alpha_r = T/2h$, $\alpha'_r = 2\pi/T(2k+1)$ if we have jumps in relativistic variables. Further, we shall assume that the jumps of proper energy are bounded by a certain value, i.e. $k = 0, 1, \dots, K$. And the cases of different k have equal possibilities. Let us state the following rules forbidding certain jumps in relativistic variables:

1. We have a jump of energy ($\Delta p_r \neq 0$) if jumps of proper time are forbidden ($\Delta q_r = 0$).

2. Jumps of spin variables are forbidden ($\Delta q_r = \Delta p_r = 0$) iff jumps of proper time are forbidden ($\Delta q_r = 0$).

The first rule means that a jump of proper energy must be instantaneous when the proper time does not vary. The second rule means that in order to measure each of spin variables in the scale of proper time it is necessary to consider two instants of time.

Thus the measures μ_{rj} can be writtem inthe form

$$\mu_{r0}(d\alpha_r, d\alpha, d\alpha'_r, d\alpha') = \frac{\mu'_{r0}(d\alpha, d\alpha')}{K+1} \sum_{k=0}^{K} \delta_{\frac{2\pi}{T}(2k+1)}(d\alpha'_r)\delta_0(d\alpha_r),$$

$$\mu_{ri}(d\alpha_r, d\alpha, d\alpha'_r, d\alpha') = \mu'_{ri}(d\alpha, d\alpha')\delta_0(d\alpha'_r)\delta_{\frac{T}{2h}}(d\alpha_r), \qquad i = 1,2,3. \qquad (32)$$

Equations (31) with measures (32) can be written as follows

$$\frac{\partial W_r}{\partial t} = \frac{i}{2h} \left\{ \frac{1}{K+1} \sum_{k=0}^{K} \int \exp[i(p\alpha - q\alpha' + g_{r0}^{(k)}(\alpha,\alpha') - \frac{2\pi}{T}(2k+1)q_r)][W_r^+(q_r, p_r + \frac{\pi h}{T}(2k+1)) - \right.$$

$$\left. - W_r^-(q_r, p_r - \frac{\pi h}{T}(2k+1))]\, \mu'_{r0}(d\alpha, d\alpha') + \sum_{j=1}^{3} \int \exp[i(p\alpha - q\alpha' + g'_{rj}(\alpha,\alpha') + \frac{Tp_r}{2h})]\times \right.$$

$$\left. \times[\sigma_j W_r^+(q_r + \frac{T}{4}, p_r) - W_r^-(q_r - \frac{T}{4}, p_r)\sigma_j]\mu'_{rj}(d\alpha, d\alpha') \right\}, \qquad (33)$$

where

$$W_r^{\pm}(q_r, p_r) = W_r(q_r, q \pm \frac{h\alpha}{2}, p_r, p \pm \frac{h\alpha'}{2}, t), \quad g_{r0}^{(k)}(\alpha,\alpha') = g_{r0}(0, \alpha, \frac{2\pi(2k+1)}{T}, \alpha'),$$

$$g'_{rj}(\alpha,\alpha') = g_{rj}(\frac{T}{2h}, \alpha, 0, \alpha'), \quad j = 1,2,3.$$

We shall assume that the proper time q_r assumes discrete values. The equation shows that the step of lattice with respect to the variable q_r is equal to $T/4$. Then the principle of duality yields that the completely observable function W_r must be $4h\pi/T$-periodic in the variable p_r. It is easy to see that equation (33) does not contradict to the assumption about $4h\pi/T$-periodicity of the fynction W_r in the variable p_r.

Then we perform the Weyl quantization with respect to the relativistic variables. In this case the function W_r will be the operator

$$\hat{\hat{\mathcal{W}}}_r = \mathcal{W}_r(\frac{\overset{3}{q_r}+\overset{1}{q_r}}{2},q,-ih\frac{\overset{2}{\partial}}{\partial q_r},p,t),$$

and equation (33) is written in the form

$$\frac{\partial \hat{\hat{\mathcal{W}}}_r}{\partial t} = \frac{i}{2h}\{\frac{1}{K+1}\sum_{k=0}^{K}\int \exp[i(p\alpha - q\alpha' + g_{r0}^{(k)})\times$$

$$\times\{\exp[-i\frac{2\pi}{T}(2k+1)q_r]\hat{\hat{\mathcal{W}}}_r^{+} - \hat{\hat{\mathcal{W}}}_r^{-}\exp[-i\frac{2\pi}{T}(2k+1)q_r]\}\mu_{r0}'(d\alpha,d\alpha')+ \qquad (34)$$

$$+\sum_{j=1}^{3}\int \exp[i(p\alpha - q\alpha' + g_{rj}')][\exp(\frac{T}{2}\frac{\partial}{\partial q_r})\sigma_j\hat{\hat{\mathcal{W}}}_r^{+} - \hat{\hat{\mathcal{W}}}_r^{-}\sigma_j\exp(\frac{T}{2}\frac{\partial}{\partial q_r})]\mu_{rj}'(d\alpha,d\alpha')$$

where

$$\hat{\hat{\mathcal{W}}}_r^{\pm} = \mathcal{W}_r(\frac{\overset{3}{q_r}+\overset{1}{q_r}}{2},q\pm\frac{h\alpha}{2},-ih\frac{\overset{2}{\partial}}{\partial q_r},p\pm\frac{h\alpha'}{2},t).$$

Since $\mathcal{W}_r$ is periodic in the variable p_r, the operator $\hat{\hat{\mathcal{W}}}_r$ and equation (34) can be reduced to the subspace of functions defined on the lattice with step $T/2$. Considering the operator $\hat{\hat{\mathcal{W}}}_r$ (the symbol of which is a matrix of second order) on a subspace of T-periodic functions defined on the lattice $q_r = nT/2$, $n \in \mathbb{Z}$, we passfrom the operator $\hat{\hat{\mathcal{W}}}_r$ to a matrix of the fourth order $\mathcal{W}$. Equation for the matrix function $\mathcal{W}(q,p,t)$ can be obtained straightforward from equation (34) by reducing it to the subspace of T-periodic functions defined by their values at two points $q_r = \{0, T/2\}$.

THEOREM 3. Let $g_{r0}^{(k)}(\alpha,\alpha') = g_{r0}'(\alpha,\alpha')$ for any $k = 0,1,\ldots,K$. If $\mathcal{W}_r$ satisfies equation (33), the matrix function of the fourth order $\mathcal{W}(q,p,t)$, related to it, satisfies the equation

$$\frac{\partial \mathcal{W}}{\partial t} = \frac{i}{2h}\sum_{j=0}^{3}\int[\gamma_j\mathcal{W}(q+\frac{h\alpha}{2},p+\frac{h\alpha'}{2},t) -$$

$$\mathcal{W}(q-\frac{h\alpha}{2},p-\frac{h\alpha'}{2},t)\gamma_j]\exp[i(p\alpha - q\alpha' + g_{rj}'(\alpha,\alpha'))]\mu_{rj}'(d\alpha,d\alpha'), \qquad (35)$$

where γ_j is the Dirac matrix

$$\gamma_0 = \begin{pmatrix} I & 0 \\ 0 & -I \end{pmatrix}, \quad \gamma_i = \begin{pmatrix} 0 & \sigma_i \\ \sigma_i & 0 \end{pmatrix}, \quad i=1,2,3.$$

Equation (35) is just the equation for the Wigner matrix function related to the Dirac equation with the Hamiltonian

$$\hat{H}_r = \sum_{i=0}^{3} \hat{H}_{\eta i} \gamma_i$$

where the Weyl symbols of the operators $\hat{H}_{\eta j}$ have the form

$$\hat{H}_{\eta j}(q,p) = \frac{1}{2} \int \exp[i(p\alpha - q\alpha' + g'_{\eta j}(\alpha,\alpha'))] \mu'_{\eta j}(\alpha,\alpha')$$

The matrix $\mathbb{W}(q,p,t)$ plays the role of the Wigner matrix function.

LITERATURE

1. V.P.Maslov, A.M.Chebotarev. On the Monte-Carlo calculation of the Feynmans paths integrals in the P-representation. Proc. of the Second IMACS International symposium on computer methods for partial differential equations. Lehigh Univ.. Bethlehem. Pa(USA). June 22-24,1977.
2. V.P.Maslov, A.M.Chebotarev. Jump processes and their applications to quantum mechanics. VINITI, Itogi Nauki. v.15 (1978) (in Russian).
3. V.P.Maslov, A.M.Chebotarev. Process a sauts et leurs applications dans la mecanique quantique, in: Feynman Path Integral Proceedings. Marseille 1978. Leture Notes in Physics. 106 (Springer Verlag. Berlin, Heidelberg. New York. 1979).
4. V.P.Maslov. Kolmogorov-Feller equation and stochastic model of quantum mechanics. Itogi nauki i tekhniki, VINITI, 1982. v.19. p.55-84.
5. Ph.Combe. R.Hoegh-Krohn. R.Rodriguez, M.Sirugue, M.Sirugue-Collin. Poisson processes on groups and Feynman path integral. Commun. Math. Phys., v.77 (1980), p.268-288.
6. S.Albeverio. Ph.Blauchard, R.Hoegh-Krohn. M.Sirugue. Local relativistic invariant flows for quantum fields. Commun. Math. Phys., v.90 (1983), p.329-351.

7. Ph.Combe, F.Guerra, R.Rodrigues, M.Sirugue, M.Sirugue-Collin. Quantum dynamical time evolution of stochastic flows in phase space. Physica, 124A, 561-574 (1984).
8. Ph.Blauchard, M.Sirugue. Large deviations from classical paths. Comm. Math. Phys., v.101, p.173-135 (1985).
9. V.G.Danilov, V.P.Maslov. Pontryagin duality principle for calculating Cherenkov-type effects in crystals and difference schemes.I. Trudy MIAN SSSR, 1984, v.166, p.130-160.
10. M.V.Karasev, V.P.Maslov. Asymptotic and geometric quantization. Uspekhi Mathem. Nauk, 1984, v.39, N6, p.115-173.

Maximality of Infinite Dimensional Dirichlet Forms and
Høegh-Krohn's Model of Quantum Fields

by

Sergio Albeverio [*,***,#] and Shigeo Kusuoka [**,***]

Abstract

We study Dirichlet forms on infinite dimensional Banach spaces. We
give a characterization of the maximal Dirichlet form obtained by
extension from a form defined on smooth cylinder functions. We also
obtain criteria for its equality with other closed extensions. In
the case of Dirichlet forms given by log concave probability measures
we obtain further characterizations of the associated self-adjoint
operators. We also study the limits of sequences of such Dirichlet
forms and apply all results to the case of Høegh-Krohn's model of
quantum fields, obtaining in particular a characterization of the
physical Hamiltonian as the self-adjoint operator associated with the
maximal Dirichlet form.

Facultät für Mathematik, Ruhr-Universität Bochum,
D-4630 Bochum 1 (FRG)

[*] Research Institute Mathematical Sciences,
Kyoto University, Kyoto (Japan)

[**] SFB 237 Bochum-Essen-Düsseldorf

BiBos Research Centre

0. Introduction

The theory of Dirichlet forms is intimately connected with the theory of symmetric Markov processes, see e.g. [12], [13], [21]. In the case where the state space is locally compact this theory has close connections with classical potential theory as well as with axiomatic potential theory, see e.g. the above references and [20]. In the case where the state space is infinite dimensional (non locally compact) the theory was developed in [3], [4], [5], in the framework of rigged Hilbert spaces resp. suitable compactifications, extended in [18] for Banach spaces (with a complete construction of the associated process). The work [19] should also be quoted for a corresponding analytic capacity theory, as well as the work [1], [8], [11], [16] for other, partly more abstract frameworks. In [6] the rigged Hilbert space and Banach space approaches [3], [4], [5], [18] were pursued further in the context of Suslin spaces. Whereas this work yields sufficient and necessary conditions for existence of the forms and processes involved, it leaves open the corresponding question of characterizing the different Dirichlet forms extending a minimally defined one. In the case of finite dimensional state space a strong result on uiqueness of Dirichlet forms extending the minimal one has been given by Wielens [23], see also [6], [13], [22].

In the present paper we shall study the family of Dirichlet forms extending a minimally defined one and give a precise characterization of the maximal Dirichlet form as well as criteria for its equality with other Dirichlet forms. The content of the paper is as follows.

In Section 1 we introduce basic definitions, in particular the type of Dirichlet forms on Banach spaces we shall study. They are given in terms of probability measures μ on a Banach space B by

$\frac{1}{2} \int_B (Du(z), Dv(z))_H \, \mu(dz)$, with H a Hilbert space densely and continuously contained in B, D being a stochastic derivative. A basic "integration by parts" Lemma is also given.

In Section 2 it is shown (Theorem(2.5)) that the Dirichlet forms defined in Section 1 are maximal in a natural sense (roughly speaking they correspond in the finite dimensional case, to the maximal Markovian extensions of Dirichlet forms defined on smooth functions of compact support in a bounded region given by Neumann boundary conditions (cfr. [12] Ch. 2, §2.3) in the case of the form $\sum_{i=1}^{d} \int_\Lambda \frac{\partial u}{\partial x_i} \cdot \frac{\partial v}{\partial x_i} \, dx$, Λ a domain in a natural Sobolev space in our infinite dimensional setting. The proof of the above maximality involves martingale methods and stochastic calculus in the L^2-setting of the theory of infinite dimensional Dirichlet forms.

In Section 3 we consider Dirichlet forms given by probability measure μ on Banach spaces which have the additional property of being log concave (in a sense related to the log-concave functions used by Brascamp-Lieb in [9]). We show that if μ has this property and is differentiable in a sense described in Section 1 then the self-adjoint operator associated with the maximal form described in Section 1 can be obtained as strong limit of self-adjoint operators associated with finite dimensional subspaces (here we use the finite dimensional uniqueness result by Wielens).

In Section 4 we study the limits of the Dirichlet forms associated with a sequence of log-concave probability measures of the type discussed in Section 3.

In Section 5 we show that interesting examples for log concave probability measures on Banach spaces are provided by models of quantum fields with exponential interactions (Høegh-Krohn's model). Using the fact that for these models the global Markov property has been proven [14], [24], we show that the Dirichlet form uniquely associated with the physical Hamiltonian coincides with the maximal form discussed in this paper, given by the probability measure ("vacuum measure") associated with the sharp time fields of this model.

Acknowledgements

We are very grateful to Prof. Dr. Ludwig Streit resp. Prof. Dr. David Elworthy for kind invitations at ZiF (Project No.2)(1984), where this work was begun resp. at the University of Warwick (1987), where it was continued. We thank Prof. Dr. Michael Röckner for stimulating and very interesting discussions.

1. Preliminary Results

Let B be a separable real Banach space and H be a separable rea Hilbert space for which H is continuously and densely embedded in B by $i:H \to B$. For simplicity of notation, we identify $i(H)$ with H. Also, we identify the dual space H^* of H with H itself. Since we may regard the dual space B^* of B as a subspace of H^*, we have the triple $B^* \subset H \subset B$.

Let K be a dense vector subspace of B^* and μ be a probability measure on B. For each $u \in B$, we define $S(u):B \to B$ by $S(u)z = z + u$.

(1.1)**Definition.** We say that μ is *K-quasi-invariant*, if the image measure $S(k)\mu$ is absolutely continuous relative to μ for all $k \in K$.

We write $L^2(d\mu)$ for $L^2(B, \mathscr{B}(B), d\mu)$, where $\mathscr{B}(B)$ is the Borel σ-algebra of B. If μ is K-quasi-invariant, we can introduce a famil of unitary operators $\{V(k); k \in K\}$ in $L^2(d\mu)$ given by

$$(1.2) \quad (V(k)f)(z) = \left(\frac{dS(k)\mu}{d\mu}\right)(z)^{1/2} f(z+k), \quad f \in L^2(d\mu).$$

Then it is obvious that the map $k \to V(k)$ gives a unitary representation of the Abelian group K. Since $\{V(tk); t \in \mathbb{R}\}$ is C_0-continuous (i.e. strongly continuous) in t (c.f. Araki[7],

Hegerfeldt[15]), there is a self-adjoint operator $\pi(k)$ for each $k \in K$ such that

(1.3) $V(tk) = \exp(it\pi(k))$, $t \in \mathbb{R}$.

(1.4)**Definition.** We say that μ is K-*differentiable*, if μ is K-quasi-invariant and $1 \in \bigcap_{k \in K} \mathcal{D}om(\pi(k))$.

For each $k \in K$ with $k \neq 0$, let σ_k be a measure on B defined by

(1.5) $\sigma_k(A) = \int_{\mathbb{R}} dt \ (S(tk)\mu)(A)$, $A \in \mathcal{B}(B)$.

(1.6)**Definition.** We say that μ is K-*strictly-positive*, if there is a measurable function $f_k : B \to \mathbb{R}$ for each $k \in K$ such that

1) $\mu(dz) = f_k(z)\sigma_k(dz)$ and

2) ess.inf$\{ f_k(z+tk); -T<t<T \} > 0$ for any $z \in B$ and $T > 0$.

Remark. Recently it has been shown that a K-differentiable probability measure always satisfies some kind of strictly positivity namely "Hamza's condition" [6]).

From now on, we assume that μ is K-differentable and K-strictly-positive. Let

(1.8) $\mathcal{F}C_b^\infty \underset{\text{def}}{\equiv} \{ f : B \to \mathbb{R}; \text{ there are } k_1, \ldots, k_n \in K \text{ and } \tilde{f} \in C_b^\infty(\mathbb{R}^n; \mathbb{R})$

$\text{such that } f(z) = \tilde{f}(\langle z, k_1 \rangle, \ldots, \langle z, k_n \rangle), \ z \in B \}.$

Here $\langle \ , \ \rangle$ is the dualization between B and B^*. Let us define $\beta : K \to L^2(d\mu)$ by $\beta(k) = 2i\pi(k)1$, $k \in K$.

The following is due to Albeverio-Høegh-Krohn [4].

(1.9)**Proposition.** (1) $\beta : K \to L^2(d\mu)$ *is linear.*

2) $\mathcal{F}C_b^\infty \subset \bigcap_{k \in K} \mathcal{D}om(\pi(k))$ *and*

(1.10) $(\pi(k)u)(z) = \frac{1}{i}((k, \nabla u(z))_H + \frac{1}{2} \beta(k)(z)u(z))$, $u \in \mathcal{F}C_b^\infty$.

Here $(\ , \ )_H$ *is the scalar product in H and* $\nabla u : B \to H$ *is given by*

$(h, \nabla u(z))_H = \frac{d}{dt} u(z+th)\big|_{t=0}.$

(1.11)**Definition.** We define a dense sub-vector space $\mathcal{D}(\tilde{\mathcal{E}})$ in $L^2(d\mu)$ by the following. $u \in \mathcal{D}(\tilde{\mathcal{E}})$, if

1) $u \in L^2(d\mu)$,

(1) for each $k \in K$, there is a measurable $\tilde{u}_k : B \to \mathbb{R}$ satisfying

  (i) $\tilde{u}_k(z) = u(z)$ μ-a.e.z   and

  (ii) $\tilde{u}_k(z+tk)$ is absolutely continuous in t for any $z \in B$,

and

(2) there is a measurable $Du : B \to H$ such that

$$\int_B \|Du(z)\|_H^2 \mu(dz) < \infty \text{ and}$$

$$\mu(\{ z \in B;\ \tfrac{1}{t} |\ u(z+th) - u(z) - t(k,Du(z))_H\ | > \varepsilon \}) \to 0, \ t \to 0,$$

for any $k \in K$ and $\varepsilon > 0$.

(1.12)*Remark.*   (1) $Du(z)$ is uniquely determined for μ-a.e.z.

(2) If $u \in \mathcal{F}C_b^\infty$, then $u \in \mathcal{D}(\tilde{\mathcal{E}})$ and $Du(z) = \nabla u(z)$ μ-a.e.z.

   We define a symmetric bilinear form $\tilde{\mathcal{E}} : \mathcal{D}(\tilde{\mathcal{E}}) \times \mathcal{D}(\tilde{\mathcal{E}}) \to \mathbb{R}$ by

$$(1.13) \quad \tilde{\mathcal{E}}(u,v) = \frac{1}{2} \int_B (Du(z),Dv(z))_H \mu(dz).$$

We call $(\tilde{\mathcal{E}},\mathcal{D}(\tilde{\mathcal{E}}))$ the *Sobolev Dirichlet form* associated with μ.

Then we have the following (c.f. Kusuoka[18]).

(1.14)**Proposition.** $(\tilde{\mathcal{E}},\mathcal{D}(\tilde{\mathcal{E}}))$ *is a Dirichlet form in* $L^2(d\mu)$, *i.e.*,

$(\tilde{\mathcal{E}},\mathcal{D}(\tilde{\mathcal{E}}))$ *is a closed form in* $L^2(d\mu)$ *and for any* $u \in \mathcal{D}(\tilde{\mathcal{E}})$ *and*

$g \in C^1(\mathbb{R};\mathbb{R})$ *with* $g(0) = 0$ *and* $|g'| \leq 1$, $g \circ u \in \mathcal{D}(\tilde{\mathcal{E}})$ *and* $\tilde{\mathcal{E}}(g \circ u, g \circ u) \leq$
$\tilde{\mathcal{E}}(u,u)$.

   We will use the following lemma frequently.

(1.15)**Lemma.**   *For any* $u \in \mathcal{D}(\tilde{\mathcal{E}})$, $v \in \mathcal{F}C_b^\infty$ *and* $k \in K$,

$$\int_B (k,Du)_H v\, d\mu = - \int_B u(k,\nabla v)\, d\mu - \int_B uv\beta(k)\, d\mu .$$

Proof.   We may assume that $\|k\|_H = 1$.   It is sufficient to show this

lemma in the case where u is bounded.   Observe that

$$(1.16) \quad \int_B (k,Du(z))_H v(z)\, \mu(dz)$$

$$= \int_B (k,Du(z))_H v(z) f_k(z)\, \sigma_k(dz)$$

$$= \int_B \mu(dz) \int_\mathbb{R} ds\ (k,Du(z+sk))_H v(z+sk) f_k(z+sk)$$

$$= \int_B \mu(dz) \int_\mathbb{R} ds\ \frac{d}{ds} \tilde{u}_k(z+sk) \cdot v(z+sk) f_k(z+sk), \text{ and}$$

$$(1.17) \quad - \int_B u(k,\nabla v)_H\, d\mu - \int_B uv\beta(k)\, d\mu$$

$$= - \int_B m(dz) \int_{\mathbb{R}} ds \; u(z+sk)\{ \frac{d}{ds} v(z+sk) + v(z+sk)\beta(k)(z+sk)\} f_k(z+sk).$$

By virtue of Proposition(1.9), we see that $v \in \mathcal{D}om(\pi(k))$. Combining (1.10) with the fact that $(\pi(k)1,v)_{L^2(d\mu)} = (1,\pi(k)v)_{L^2(d\mu)}$, we see that our lemma is true if $u = 1$. Thus we have

$$\int_B \mu(dz) \int_{\mathbb{R}} ds \; \frac{d}{ds} v(z+sk) \cdot f_k(z+sk)$$

$$= - \int_B \mu(dz) \int_{\mathbb{R}} ds \; v(z+sk)\beta(k) f_k(z+sk) \quad \text{for all } v \in \mathcal{F}C_b^{\infty}.$$

Let P be the orthogonal projection from H onto $\mathbb{R}k$. Then P is extensible to a bounded operator $P:B \to K$. Let $v_0 \in C_0^{\infty}(\mathbb{R};\mathbb{R})$ and $v_1 \in \mathcal{F}C_b^{\infty}$ and let $v \in \mathcal{F}C_b^{\infty}$ be given by $v(z) = v_0(\langle z,k \rangle)v_1(z-Pz)$, $z \in B$. Then we have

$$\int_B \mu(dz)v_1(z-Pz) \int_{\mathbb{R}} ds \; v_0'(\langle z,k \rangle +s) f_k(z+sk)$$

$$= - \int_B \mu(dz)v_1(z-Pz) \int_{\mathbb{R}} ds \; v_0(\langle z,k \rangle +s)\beta(k)(z+sk) f_k(z+sk).$$

Then there is a measurable subset Ω in B with $\mu(\Omega) = 1$ such that

$$\int_{\mathbb{R}} ds \; v_0'(\langle z,k \rangle +s) f_k(z+sk)$$

$$= - \int_{\mathbb{R}} ds \; v_0(\langle z,k \rangle +s)\beta(k)(z+sk) f_k(z+sk)$$

for all $v_0 \in C_0^{\infty}(\mathbb{R};\mathbb{R})$ and $z \in \Omega$, because both sides are measurable with respect to the sub-σ-algebra generated by $(I_B-P)^{-1}A$, $A \in \mathcal{B}(B)$. Here I_B denotes the identity map in B. Thus we see that as a distribution in s

$$\frac{d}{ds} f_k(z+sk) = - \beta(k)(z+sk) \cdot f_k(z+sk), \quad z \in \Omega.$$

Therefore we see that $f_k(z+sk)$ has a nice version in s for each $z \in \Omega$. Then we see that

$$\frac{d}{ds}(\tilde{u}_k(z+sk) f_k(z+sk))$$

$$= (\frac{d}{ds} \tilde{u}_k(z+sk) + \tilde{u}_k(z+sk)\beta(k)(z+sk)) f_k(z+sk)$$

is a distribution in s for $z \in \Omega$. Combining this with (1.16) and (1.17), we have our lemma.   Q.E.D.

2. Maximality of $(\tilde{\mathcal{E}}, \mathcal{D}(\tilde{\mathcal{E}}))$.

As in the previous section, we assume that μ is K-differentiable and K-strictly-positive.

Let $u \in \mathcal{F}C_b^\infty$. Then there is a finite dimensional subspace E in such that $\nabla u(z) \in E$ for all $z \in B$. Let us define $L_0^E u \in L^2(d\mu)$ by

$$(2.1) \quad L_0^E u(z) = \frac{1}{2} \sum_{j=1}^{n} \left(\frac{d^2}{dt^2} u(z+te_j)\big|_{t=0} + (e_j, \nabla u(z))_H \cdot \beta(e_j)(z) \right),$$

where $\{e_j\}_{j=1}^{n}$ is an orthonormal basis of E.

Then by Lemma(1.15), we have the following.

$$(2.2) \quad (L_0^E u, v)_{L^2(d\mu)} = -\frac{1}{2} \int_B (\nabla u, \nabla v)_H \, d\mu \quad \text{for any } v \in \mathcal{F}C_b^\infty .$$

Since $\mathcal{F}C_b^\infty$ is dense in $L^2(d\mu)$, we see that $L_0^E u$ is independent of the choice of E. We denote $L_0^E u$ by $L_0 u$. Then L_0 is a linear operator from $\mathcal{F}C_b^\infty$ into $L^2(d\mu)$. Moreover, by Lemma(1.15), we see that

$$(2.3) \quad (L_0 u, v)_{L^2(d\mu)} = -\tilde{\mathcal{E}}(u,v) \quad \text{for any } u \in \mathcal{F}C_b^\infty \text{ and } v \in \mathcal{D}(\tilde{\mathcal{E}}) .$$

(2.4)**Definition.** We say that a Dirichlet form $(\mathcal{E}, \mathcal{D}om(\mathcal{E}))$ in $L^2(d\mu)$ is an extension of L_0, if $\mathcal{F}C_b^\infty \subset \mathcal{D}om(\mathcal{E})$ and $(L_0 u, v) = -\mathcal{E}(u,v)$ for an $u \in \mathcal{F}C_b^\infty$ and $v \in \mathcal{D}om(\mathcal{E})$.

Obviously $(\tilde{\mathcal{E}}, \mathcal{D}(\tilde{\mathcal{E}}))$ is an extension of L_0. The following is the main result of this section.

(2.5)**Theorem.** *If a Dirichlet form $(\mathcal{E}, \mathcal{D}om(\mathcal{E}))$ in $L^2(d\mu)$ is an extension of L_0, then*

(1) $\mathcal{D}om(\mathcal{E}) \subset \mathcal{D}(\tilde{\mathcal{E}})$, and

(2) $\tilde{\mathcal{E}}(u,u) \leq \mathcal{E}(u,u)$ for all $u \in \mathcal{D}om(\mathcal{E})$.

(2.6)**Corollary.** *$(\tilde{\mathcal{E}}, \mathcal{D}om(\tilde{\mathcal{E}}))$ is the unique extension of L_0, if and only if $\mathcal{F}C_b^\infty$ is dense in $\mathcal{D}om(\tilde{\mathcal{E}})$ with respect to the inner product $\tilde{\mathcal{E}}_1$. Here $\tilde{\mathcal{E}}_1(u,v) = (u,v)_{L^2(d\mu)} + \tilde{\mathcal{E}}(u,v)$, $u,v \in \mathcal{D}om(\tilde{\mathcal{E}})$.*

In order to prove Theorem(2.5), we make some preparations. Le us assume that a Dirichlet form $(\mathcal{E}, \mathcal{D}om(\mathcal{E}))$ is an extension of L_0 throughout this section.

Let $\{P_t\}_{t \geq 0}$ be the associated Markov semigroup in $L^2(d\mu)$ with the Dirichlet form $(\mathscr{E},\mathscr{D}om(\mathscr{E}))$ and let L be the infinitesimal generator of this semigroup. Let $T = \{ \frac{k}{2^n};$ k and n are non-ngative integers $\}$, $\Omega = B^T$ and $\tilde{\mathscr{B}}$ be the Borel algebra of Ω. We define a probability measure P on Ω by

(2.7) $P[\omega(r_1) \in A_1, \ldots, \omega(r_n) \in A_n]$

$= (1, \chi_{A_n} P_{r_n - r_{n-1}} \chi_{A_{n-1}} P_{r_{n-1} - r_{n-2}} \cdots P_{r_2 - r_1} \chi_{A_1} 1)_{L^2(d\mu)}$

for any $r_1, \ldots, r_n \in T$ with $r_1 \leq r_2 \leq \cdots \leq r_n$ and $A_1, \ldots, A_n \in \mathscr{B}(B)$. By virtue of Kolmogorov's extension theorem, such a probability measure exists and is unique. Let $(\Omega, \mathscr{B}, P)$ the completion of $(\Omega, \tilde{\mathscr{B}}, P)$.

(2.8)**Proposition.** *For any* $f \in L^2(d\mu)$, *there is a stochastic process* $\{\tilde{f}_t\}_{t \geq 0}$ *such that*

1) $\tilde{f}_r(\omega) = f(w(r))$ *P-a.s.ω for all* $r \in T$, *and*

2) *a map* $t \to \tilde{f}_t$ *from* $[0, \infty)$ *into* $L^2(\Omega, \mathscr{B}, P)$ *is continuous.*

Proof. This follows from the fact that

$E[| f(\omega(r)) - f(\omega(r'))|^2] = 2 (f, f - P_{|r-r'|} f)_{L^2(d\mu)} \to 0$

as $|r-r'| \to 0$.   Q.E.D.

Let $\tilde{\mathscr{F}}_r = \sigma\{ \omega(r'); r' \in T, r' < r\} \vee \{ A \in \mathscr{B}; P(A) = 0 \text{ or } 1\}$, $\in T$, and let $\mathscr{F}_t = \bigcap_{\substack{r \in T \\ r > t}} \tilde{\mathscr{F}}_r$, $t \geq 0$. Then $\{\mathscr{F}_t\}_{t \geq 0}$ is a right continuous increasing σ-algebras. Obviously $\tilde{f}_t$ is $\mathscr{F}_t$-measurable.

For each $u \in \mathscr{D}om(L)$, let

$\tilde{M}_t^{[u]} = \tilde{u}_t - \tilde{u}_0 - \int_0^t \tilde{Lu}_s \, ds, \quad t \geq 0.$

(2.9)**Proposition.** $\tilde{M}_t^{[u]}$ *is an* $\mathscr{F}_t$-*martingale for each* $u \in \mathscr{D}om(L)$.

Proof. This follows from the fact that if $t > s > r$, $r \in T$,

$E[(\tilde{M}_t^{[u]} - \tilde{M}_s^{[u]}) f(\omega(r))]$

$= E[(\tilde{u}_t - \tilde{u}_s) f(\omega(r))] - \lim_{n \to \infty} E[\frac{1}{2^n} \sum_{k=[2^n s]}^{[2^n t]} Lu(\omega(\frac{k}{2^n})) \cdot f(\omega(r))]$

$= (P_{t-r} u - P_{s-r} u - \int_{s-r}^{t-r} P_\tau Lu \, d\tau, f)_{L^2(d\mu)}$

$= 0$ for any $f \in L^{\infty}(d\mu)$.   Q.E.D.

By Proposition(2.9), we see that $\tilde{M}_t^{[u]}$, $u \in \mathcal{D}om(L)$, has a good version $M_t^{[u]}$ such that $t \to M_t^{[u]}$ is right continuous and has a left limit. $M_t^{[u]}$ is decomposed to the continuous part $M_t^{[u]c}$ and the discontinuous part $M_t^{[u]d}$ (c.f. Dellacherie-Meyer[5]).

(2.10)**Proposition.** *For any* $u,v \in \mathcal{D}om(L)$,

$$E[M_t^{[u]} M_t^{[v]}] = E[\langle M^{[u]}, M^{[v]} \rangle_t] = 2t\, \mathcal{E}(u,v).$$

Here $\langle M \rangle_t$ *denotes the quadratic variation process corresponding to a martingale* M_t *(see e.g. Dellacherie-Meyer[5], Ikeda-Watanabe[8,p.53]*

Proof.   This follows from

$$E[(M_t^{[u]})^2]$$

$$= \sum_{k=1}^{n} E[(M_{kt/n}^{[u]} - M_{(k-1)t/n}^{[u]})^2]$$

$$= \sum_{k=1}^{n} E[(\tilde{u}_{kt/n} - \tilde{u}_{(k-1)t/n} - \int_{(k-1)t/n}^{kt/n} \tilde{L u}_s\, ds)^2]$$

$$= 2n \int_B (u^2 - u P_{t/n} u + \frac{1}{2} \int_0^{t/n} d\tau \int_0^{t/n} ds\, Lu \cdot P_{|\tau - s|} Lu) \, d\mu$$

$$\to 2t\, \mathcal{E}(u,u) \quad \text{as } n \to \infty. \hspace{3cm} \text{Q.E.D.}$$

By virtue of Proposition(2.10), we can define a martingale $M_t^{[u]}$ for each $u \in \mathcal{D}om(\mathcal{E})$ as a limit of, $M_t^{[v_n]}$ if $\{v_n\}_1^{\infty} \subset \mathcal{D}om(L)$ and $\|u - v_n\|_{L^2(d\mu)}^2 + \mathcal{E}(u - v_n, u - v_n) \to 0$, $n \to \infty$.   Also, we have

(2.11) $E[M_t^{[u]} M_t^{[v]}] = E[\langle M^{[u]}, M^{[v]} \rangle_t] = 2t\, \mathcal{E}(u,v)$

for any $u,v \in \mathcal{D}om(\mathcal{E})$ (c.f. Fukushima[12] Chapter 5).

(2.12)**Proposition.** *For each* $u \in \mathcal{F}C_b^{\infty} \subset \mathcal{D}om(L)$, $M_t^{[u]}$ *is a continuous martingale.*

Proof.   Note that $u^n \in \mathcal{F}C_b^{\infty} \subset \mathcal{D}om(L)$, $n = 1,2,\dots$.   Let $g(t) = E[|\tilde{u}_t - \tilde{u}_0|^4]$.   Then

$$g(t) = \int_B (2u^4 - 8u^3 P_t u + 6u^2 P_t(u^2))\, d\mu.$$

Therefore,

$$g'(t) = -8 \int_B L(u^3) P_t u\, d\mu + 6 \int_B L(u^2) P_t(u^2)\, d\mu \ , \quad \text{and}$$

$$g''(t) = -8 \int_B L(u^3) P_t Lu \, d\mu + 6 \int_B L(u^2) P_t L(u^2) \, d\mu.$$

Hence we have by (2.2)

$$g(0) = 0,$$

$$g'(0) = -8 \int_B (\nabla(u^3), \nabla u)_H \, d\mu + 6 \int_B (\nabla(u^2), \nabla(u^2))_H \, d\mu = 0 \quad \text{and}$$

$$|g''(t)| \leq C = 8\|L(u^3)\|_{L^2(d\mu)} \|Lu\|_{L^2(d\mu)} + 6\|L(u^2)\|_{L^2(d\mu)}^2.$$

So we obtain that

$$E[|\tilde{u}_t - \tilde{u}_s|^4] = g(t-s) \leq C |t-s|^2.$$

By virtue of this and Kolmogorov's theorem, we see that $\tilde{u}_t$ has a
path-continuous version. Clearly $\int_0^t \widetilde{Lu}_s \, ds$ has a path-continuous
version. Thus $M_t^{[u]}$ has a path-continuous version.   Q.E.D.

2.13)**Proposition.** *For any* $u,v \in \mathcal{F}C_b^\infty$,

$$\langle M^{[u]}, M^{[v]} \rangle_t = \int_0^t \tilde{f}_s \, ds \quad P\text{-}a.s., \quad \text{where } f(z) = (\nabla u(z), \nabla v(z))_H .$$

Proof. Note that $L(u^2) = L_0(u^2) = 2uLu + (\nabla u, \nabla u)_H$ μ-a.e. Therefore
$\tilde{u^2}_t - 2 \int_0^t \tilde{u}_s \widetilde{Lu}_s \, ds - \int_0^t \tilde{g}_s \, ds$ is a martingale, where $g = (\nabla u, \nabla u)_H$.
By Proposition(2.12) and Ito's formula, we also see that
$\tilde{u}_t^2 - 2 \int_0^t \tilde{u}_s \widetilde{Lu}_s \, ds - \langle M^{[u]}, M^{[u]} \rangle_t$ is a martingale. Therefore we
have $\langle M^{[u]}, M^{[u]} \rangle_t = \int_0^t \tilde{g}_s \, ds$. This proves our assertion.   Q.E.D.

Now we define $T_k^{[u]} : \mathcal{F}C_b^\infty \to \mathbb{R}$, $u \in L^2(d\mu)$ and $k \in K$, by

$$T_k^{[u]}(g) = -\int_B u \cdot (k, \nabla g)_H \, d\mu - \int_B ug \cdot \beta(k) \, d\mu.$$

is obvious $T_k^{[u]}(g)$ is linear in u,g and k.

2.14)**Proposition.** *Let* $\bar{g} \in C_b^\infty(\mathbb{R}; \mathbb{R})$ *and* $\bar{h} \in C_b^\infty(\mathbb{R}^n; \mathbb{R})$ *with*
$\in C_0^\infty(\mathbb{R}; \mathbb{R})$, *and let* $k, k_1, \ldots, k_n \in K$ *with* $\|k\|_H = 1$ *and* $(k, k_i)_H = 0$,
$= 1, \ldots, n$. *Let* $g, h, f \in \mathcal{F}C_b^\infty$ *be given by* $g(z) = \bar{g}(\langle z, k \rangle)$, $h(z) =$
$\bar{h}(\langle z, k_1 \rangle, \ldots, \langle z, k_n \rangle)$ *and* $f(z) = \bar{g}'(\langle z, k \rangle) \bar{h}(\langle z, k_1 \rangle, \ldots, \langle z, k_n \rangle)$, $z \in B$.
Then we have

$$T_k^{[u]}(f) = E[\, M_1^{[u]} \cdot \int_0^1 \tilde{h}_s \, dM_s^{[g]}\,] \quad for \ u \in \mathcal{D}om(\mathcal{E}).$$

Proof. It is sufficient to show this in the case where $u \in \mathcal{D}om(L)$.

It is obvious that

$$T_k^{[u]}(f)$$

$$= -2\,(uh, L_0 g)_{L^2(d\mu)}$$

$$= \lim_{t \to 0} \frac{1}{t}\, E[((\widetilde{uh})_t - (\widetilde{uh})_0)(\tilde{g}_t - \tilde{g}_0)]$$

$$= \lim_{n \to \infty} E[\sum_{k=1}^n ((\widetilde{uh})_{k/n} - (\widetilde{uh})_{(k-1)/n})(\tilde{g}_{k/n} - \tilde{g}_{(k-1)/n})]\;.$$

Since $M_t^{[h]}$ is a continuous martingale by Proposition(2.12), we have $[M^{[u]}, M^{[h]}] = \langle M^{[u]c}, M^{[h]} \rangle$. Therefore by Ito's formula, we have

$$\tilde{u}_t \tilde{h}_t - \tilde{u}_0 \tilde{h}_0$$

$$= \int_0^t \tilde{h}_s \, dM_s^{[u]} + \int_0^t \tilde{u}_s \, dM_s^{[h]} + \int_0^t (\tilde{h}_s L\tilde{u}_s + \tilde{u}_s L\tilde{h}_s)\, ds + \langle M^{[u]c}, M^{[h]} \rangle$$

Because $\tilde{g}_t$ is bounded and $\tilde{g}_t - \tilde{g}_0 = M_t^{[g]} + \int_0^t L\tilde{g}_s \, ds$, we have

$$T_k^{[u]}(f) = E[\, \int_0^1 \tilde{h}_s \, d\langle M^{[u]c}, M^{[g]} \rangle_s + \int_0^1 \tilde{u}_s \, d\langle M^{[h]}, M^{[g]} \rangle_s\,].$$

But by Proposition(2.13) and the assumption, we see that $\langle M^{[h]}, M^{[g]} \rangle_t = 0$. Thus we have

$$T_k^{[u]}(f) = E[\, \int_0^1 \tilde{h}_s \, d\langle M^{[u]}, M^{[g]} \rangle_s\,] = E[\, M_1^{[u]} \int_0^1 \tilde{h}_s \, dM_s^{[g]}\,].$$

Q.E.D.

Take a $\varphi \in C_0^\infty(\mathbb{R};\mathbb{R})$ with $\varphi(t) = t$, $|t| \leq 1$. For each $k \in K$, let $g_n^{(k)} \in \mathcal{F}C_b^\infty$, $n \geq 1$, be given by $g_n^{(k)}(z) = n \cdot \varphi(\frac{1}{n}\langle z, k \rangle)$, $z \in B$. Then it is easy to see that $\mathcal{E}(g_n^{(k)} - g_m^{(k)}, g_n^{(k)} - g_m^{(k)}) \to 0$ as $n, m \to \infty$. Therefore, by Proposition(2.10) and Doob's theorem, there is a continuous martingale such that

$$E[\, |M_t^{(k)} - M_t^{(g_n^{(k)})}|^2\,] \to 0, \ n \to \infty, \ t \geq 0.$$

It is easy to see the following.

(2.15)**Proposition.** (1) $\langle M^{(k)}, M^{(k')} \rangle_t = t(k, k')_H$, $k, k' \in K$.

(2) *Let* $k \in K$ *and* $\psi \in C_b^\infty(\mathbb{R};\mathbb{R})$. *Let* $u, u' \in \mathcal{F}C_b^\infty$ *be given by* $u(z) =$

$\psi(\langle z,k\rangle)$ *and* $u'(z) = \psi'(\langle z,k\rangle)$, $z \in B$. *Then*

$$M_t^{[u]} = \int_0^t \tilde{u}'_s \, dM_s^{(k)} \quad P\text{-}a.s., \quad t \geq 0.$$

The following is an easy consequence of Propositions (2.14) and (2.15).

(2.16)**Lemma.** *For any* $k \in K$, $u \in \mathcal{D}om(\mathcal{E})$ *and* $g \in \mathcal{F}C_b^\infty$,

$$T_k^{[u]}(g) = E[\, M_1^{[u]} \int_0^1 \tilde{g}_t \, dM_t^{(k)} \,].$$

(2.17)**Corollary.** (1) $|T_k^{[u]}(g)| \leq 2\|k\|_H \|g\|_{L^2(d\mu)} \mathcal{E}(u,u)^{1/2}$

for any $u \in \mathcal{D}om(\mathcal{E})$, $k \in K$ *and* $g \in \mathcal{F}C_b^\infty$.

(2) *If* $\{k_i\}_{i=1}^\infty \subset K$ *is an orthonormal basis in* H, *then*

$$\sum_{i=1}^\infty |T_{k_i}^{\{u\}}(g_i)| \leq (\sum_{i=1}^\infty \|g_i\|_{L^2(d\mu)}^2)^{1/2} (2\mathcal{E}(u,u))^{1/2}$$

for any $u \in \mathcal{D}om(\mathcal{E})$ *and* $\{g_i\}_{i=1}^\infty \subset L^2(d\mu)$.

Proof. The assertion (1) follows from

$$|T_k^{[u]}(g)|^2$$

$$\leq E[(\int_0^1 \tilde{g}_s \, dM_s^{(k)})^2] E[M_1^{[u]2}]$$

$$= 2\|g\|_{L^2(d\mu)}^2 \|k\|_H^2 \mathcal{E}(u,u).$$

To show the assertion (2), we set $\varepsilon_i = \mathrm{sign}(T_{k_i}^{[u]}(g_i))$. Then

$$\sum_{i=1}^N |T_{k_i}^{\{u\}}(g_i)|$$

$$= \sum_{i=1}^N T_{k_i}^{\{u\}}(\varepsilon_i g_i)$$

$$= E[\, M_1^{[u]} \sum_{i=1}^N \int_0^1 \varepsilon_i \tilde{g}_{is} \, dM_s^{(k_i)} \,]$$

$$\leq E[(\sum_{i=1}^N \int_0^1 \varepsilon_i \tilde{g}_{is} \, dM_s^{(k_i)})^2]^{1/2} E[M_1^{[u]2}]^{1/2}$$

$$= (\sum_{i=1}^N \|g_i\|_{L^2(d\mu)}^2)^{1/2} (2\mathcal{E}(u,u))^{1/2}.$$

Letting $N \to \infty$, we obtain our assertion. Q.E.D.

Now, let us prove Theorem(2.5). Let $u \in \mathcal{D}om(\mathcal{E})$ and u be bounded. Let $\{k_i\}_{i=1}^\infty \subset K$ be a complete orthonormal basis of H. Then by Corollary(2.17), we see that there is a $g_i \in L^2(d\mu)$ for each $i \geq 1$

such that $T_{k_i}^{[u]}(f) = \int_B g_i f \, d\mu$ for any $f \in L^2(d\mu)$. Noting that

$$\sum_{i=1}^{N} |T_{k_i}^{[u]}(g_i)| \leq \left(\sum_{i=1}^{N} \|g_i\|_{L^2(d\mu)}^2 \right)^{1/2} \mathcal{E}(u,u)^{1/2}, \quad N \geq 1,$$

we have $\sum_{i=1}^{\infty} \|g_i\|_{L^2(d\mu)}^2 \leq 2\mathcal{E}(u,u)$.

Let $F(z) = \sum_{i=1}^{\infty} g_i(z)k_i$, $z \in B$. Then $F \in L^2(B \to H; d\mu)$ and

$$(2.18) \quad \int_B \|F(z)\|_H^2 \, \mu(dz) \leq 2 \, \mathcal{E}(u,u).$$

Also, by Corollary(2.17), we see that

$$(2.19) \quad T_k^{[u]}(g) = \int_B (k, F(z))_H g(z) \, \mu(dz), \quad k \in K \text{ and } g \in L^2(d\mu).$$

Take a $k \in K$ with $\|k\|_H = 1$ and fix it. Let $P: B \to H$ be a map given by $Pz = \langle z, k \rangle$, $z \in B$. Then $P|_H : H \to H$ is the orthogonal projection onto $\mathbb{R}k$. Let $g_0 \in C_0^{\infty}(\mathbb{R}; \mathbb{R})$ and $g_1 \in \mathcal{F}C_b^{\infty}$, and let $g \in \mathcal{F}C_b^{\infty}$ be given by $g(z) = g_0(\langle z, k \rangle)g_1(z - Pz)$, $z \in B$. Then

$$T_k^{[u]}(g)$$
$$= -\int_B u \cdot (k, \nabla g)_H \, d\mu - \int_B u \cdot g \cdot \beta(k) \, d\mu$$
$$= -\int_B \mu(dz) \, g_1(z-Pz) \int_{\mathbb{R}} ds \, \{ \, u(z+sk) \cdot g_0'(\langle z, k \rangle + s)$$
$$+ u(z+sk) \cdot g_0(z+sk) \cdot \beta(k)(z+sk) \, \} f_k(z+sk)$$

and we have by (2.19)

$$T_k^{[u]}(g)$$
$$= \int_B \mu(dz) \, g_1(z-Pz) \int_{\mathbb{R}} ds \, (k, F(z+sk))_H g_0(\langle z, k \rangle + s) f_k(z+sk),$$

where f_k is the one in Definition(1.6).

Therefore there is a measurable subset Ω_1 in B with $\mu(\Omega_1) = 1$ such that

$$(2.20) \quad \int_{\mathbb{R}} (k, F(z+sk))_H g_0(\langle z, k \rangle + s) f_k(z+sk) \, ds$$
$$= -\int_{\mathbb{R}} \{ \, u(z+sk)g_0'(\langle z, k \rangle + s)$$
$$+ u(z+sk)g_0(\langle z, k \rangle + s)\beta(k)(z+sk) \, \} \, f_k(z+sk) \, ds$$

for all $g_0 \in C_0^{\infty}(\mathbb{R}; \mathbb{R})$ and $z \in \Omega_1$, because both sides of (2.20) are $\sigma\{(I_B - P)z\}$ measurable. Thus we see that for any $z \in \Omega_1$ as a distribution in s

$$(2.21) \quad \frac{d}{ds} \left(u(z+sk)f_k(z+sk) \right)$$
$$= (k,F(z+sk))_H f_k(z+sk) + u(z+sk)\cdot\beta(k)(z+sk)f_k(z+sk).$$

As is shown in the proof of Lemma(1.15), there is a measurable subset Ω_2 in B with $\mu(\Omega_2) = 1$ such that for each $z \in \Omega_2$

$$\frac{d}{ds} f_k(z+sk) = \beta(k)(z+sk)f_k(z+sk)$$

as a distribution in s. Thus, using the assumption on f_k in Definition(1.6), we have

$$\frac{d}{ds} \left(1/f_k(z+sk) \right) = -\beta(k)(z+sk)/f_k(z+sk)$$

as a distribution in s for all $z \in \Omega_2$. Therefore we see that

$$(2.22) \quad \frac{d}{ds} u(z+sk) = (k,F(z+sk))_H$$

as a distribution in s for all $z \in \Omega_1 \cap \Omega_2$.

Take a $\psi \in C_0^\infty(\mathbb{R};\mathbb{R})$ with $\int_{\mathbb{R}} \psi(t)\, dt = 1$, and let

$$u_n(z) = n\cdot\int_{\mathbb{R}} u(z+sk)\cdot\psi(ns)\, ds, \quad z \in B.$$

Since u is bounded, we see that

$$\int_B |u(z)- u_n(z)|\, \mu(dz)$$
$$= \int_B \mu(dz) \int_{\mathbb{R}} |u(z+sk)- u_n(z+sk)|\cdot f_k(z+sk)\, ds \to 0, \quad n \to\infty.$$

So, taking a subsequence if necessary, we may think that there is a measurable subset Ω_3 with $\mu(\Omega_3) = 1$ such that

$$(2.23) \quad u_n(z) \to u(z), \quad n \to \infty, \quad \text{for each } z \in \Omega_3.$$

Take a σ-compact set Ω such that $\Omega \subset \Omega_1 \cap \Omega_2 \cap \Omega_3$ and $\mu(\Omega) = 1$. Then for any $z \in \Omega$

$$\frac{d}{ds} u_n(z+sk)$$
$$= n\cdot\int_{\mathbb{R}} \frac{d}{d\tau} u(z+sk+\tau k)\, \psi(n\tau)\, d\tau$$
$$= n\cdot\int_{\mathbb{R}} (k,F(z+sk+\tau k))_H \psi(n\tau)\, d\tau,$$

and so

$$u_n(z+tk) - u_n(z)$$
$$= \int_0^t ds \int_{\mathbb{R}} (k,F(z+sk+\tau k))_H\, n\cdot\psi(n\tau)\, d\tau$$

$$\rightarrow \int_0^t ds \ (k,F(z+sk))_H \qquad \text{as } n \rightarrow \infty.$$

Therefore $u_n(z)$ converges for any $z \in \Omega + \mathbb{R}k$. Let us define $\tilde{u}_k : B \rightarrow \mathbb{R}$ by

$$\tilde{u}_k(z) = \begin{cases} \lim_{n \rightarrow \infty} u_n(z) & \text{if } z \in \Omega + \mathbb{R}k \\ \\ 0 & \text{otherwise} \end{cases}$$

Then $u(z) = \tilde{u}_k(z)$, $z \in \Omega$, and

$$\tilde{u}_k(z+tk) - \tilde{u}_k(z) = \begin{cases} \int_0^t (k,F(z+sk))_H \ ds & \text{if } z \in \Omega + \mathbb{R}k \\ \\ 0 & \text{otherwise} \end{cases}$$

Therefore $\tilde{u}_k(z+tk)$ is absolutely continuous in t for all $z \in B$.

Let $A(t) = \{ z \in B; \ \frac{1}{t} |u(z+tk) - u(z) - t(k,F(z))_H| > \varepsilon \}$. Then $\mu(A(t))$

$$= \int_B \mu(dz) \int_{\{s \in \mathbb{R}; \ z+sk \in A(t)\}} f_k(z+sk) \ ds \rightarrow 0, \ t \rightarrow 0.$$

Thus we see that $u \in \mathcal{D}(\tilde{\mathcal{E}})$ and $Du(z) = F(z)$ μ-a.e.z. Combining this with (2.18), we obtain our theorem.

This completes the proof of Theorem(2.5).

Now we study the relation between our Dirichlet form and the Dirichlet form defined in Albeverio-Høegh-Krohn[4]. Let $\mathcal{P}(K)$ denote the set of all orthogonal projections in H such that their range is finite dimensional subspace of K. Then each P in $\mathcal{P}(K)$ can be extended to a bounded linear operator $\tilde{P}: B \rightarrow \text{Image}(P)$.

(2.24)**Definition.** We say that a K-quasi-invariant probability measu μ is K-*locally-Lipschitz* if for each $P \in \mathcal{P}(K)$, there is a measurable function $\rho_P : \text{Image}(P) \times B \rightarrow \mathbb{R}$ such that

(1) $\rho(\cdot,z): \text{Image}(P) \rightarrow \mathbb{R}$ is locally Lipschitz continuous for each $z \in B$, and

(2) $\int_A \rho(x,z) \ m(dx) = P^\mu [\ \tilde{P}^{-1}(A) | \mathcal{F}_{P^\perp}](z)$ μ-a.e.z for each Borel se A in $\text{Image}(P)$. Here $\mathcal{F}_{P^\perp} = \sigma\{(I_B - \tilde{P})z\}$ and m is the Lebesgue measure on $\text{Image}(P)$.

Let μ be an K-differentiable probability measure on B. For eac

$P \in \mathscr{P}(K)$, we define a bilinear form $\tilde{\mathscr{E}}_P$ in $\mathscr{F}C_b^\infty$ by

$$\tilde{\mathscr{E}}_P(u,v) = \frac{1}{2} \int_B (P\nabla u(z), \nabla v(z))_H \, \mu(dz), \quad u,v \in \mathscr{F}C_b^\infty.$$

Then by virtue of Albeverio-Høegh Krohn[4], we see that $\tilde{\mathscr{E}}_P$ is closable in $L^2(d\mu)$ and has the Markov property. Let $(\mathscr{E}_P, \mathscr{D}om(\mathscr{E}_P))$ be the smallest closed extension of $(\tilde{\mathscr{E}}_P, \mathscr{F}C_b^\infty)$ in $L^2(d\mu)$. Let H_P be the self-adjoint operator in $L^2(d\mu)$ associated with $(\mathscr{E}_P, \mathscr{D}om(\mathscr{E}_P))$, i.e., $\mathscr{D}om(H_P) \subset \mathscr{D}om(\mathscr{E}_P)$ and $\mathscr{E}_P(u,v) = (H_P u, v)_{L^2(d\mu)}$, $u \in \mathscr{D}om(H_P)$ and $v \in \mathscr{D}om(\mathscr{E}_P)$. We call the self-adjoint operator H_P the *partial Hamiltonian operator along* P *associated to* μ.

By virtue of the result of Wielens [23], we have the following.

(2.25)**Lemma.** *If* μ *is K-locally-Lipschitz, then* $H_P|_{\mathscr{F}C_b^\infty} : \mathscr{F}C_b^\infty \to L^2(d\mu)$ *is essentially self-adjoint for any* $P \in \mathscr{P}(K)$.

In [2], it is shown that there is a self-adjoint operator H in $L^2(d\mu)$ such that $H_P \to H$ strongly as $P \uparrow$ Identity, $P \in \mathscr{P}(K)$. Then we have the following.

(2.26)**Theorem.** *If* μ *is K-locally-Lipschitz and K-strictly-positive, then* H *is the self-adjoint operator associated with the Sobolev Dirichlet form* $(\tilde{\mathscr{E}}, \mathscr{D}(\tilde{\mathscr{E}}))$ *associated to* μ.

Proof. By Lemma(2.25), we see that $\mathscr{D}om(\mathscr{E}_P) \supset \mathscr{D}(\tilde{\mathscr{E}})$, $P \in \mathscr{P}(K)$. In [2] it is shown that $\mathscr{D}om(H^{1/2}) = \bigcap_{P \in \mathscr{P}(K)} \mathscr{D}om(\mathscr{E}_P)$ and $(H^{1/2}u, H^{1/2}v) = \lim_{\uparrow} \mathscr{E}_P(u,v)$, $u,v \in \mathscr{D}om(H^{1/2})$, $H|_{\mathscr{F}C_b^\infty} = L_0|_{\mathscr{F}C_b^\infty}$, and that H is Markov. Thus we have $\mathscr{D}om(H^{1/2}) \supset \mathscr{D}(\tilde{\mathscr{E}})$ and $(H^{1/2}u, H^{1/2}v)_{L^2(d\mu)} = \tilde{\mathscr{E}}(u,v)$, $u,v \in (\tilde{\mathscr{E}})$. Thus Theorem(2.5) implies our assertion.

We call the self-adjoint operator H the *maximal Hamiltonian operator associated with* μ.

3. Log Concave Probability Measure

(3.1)**Definition.** We say that a function $\mathbb{R}^N \to [0,\infty)$ is *log concave* if $f(\lambda x+(1-\lambda)y) \geq f(x)^\lambda f(y)^{1-\lambda}$ for any $x,y \in \mathbb{R}^N$ and $\lambda \in (0,1)$.

The following is due to Brascamp-Lieb[9].

(3.2)**Theorem.** (1) *If* $F:\mathbb{R}^{N+M} \to [0,\infty)$ *is log concave, and if* $f(x) = \int_{\mathbb{R}^M} F(x,y)\,dy < \infty$, $x\in\mathbb{R}^N$, *then* $f:\mathbb{R}^N \to [0,\infty)$ *is log concave.*

(2) *If* $f:\mathbb{R}^N \to [0,\infty)$ *and* $g:\mathbb{R}^N \to [0,\infty)$ *are log concave and* $f*g(x) = \int_{\mathbb{R}^N} f(x-y)g(y)\,dy < \infty$, $x\in\mathbb{R}^N$, *then* $f*g:\mathbb{R}^N \to [0,\infty)$ *is log concave.*

(3.3)**Definition.** We say that a probability measure μ is *strongly log concave*, if there is a log concave function $f:\mathbb{R}^N \to [0,\infty)$ such that $\mu(dx) = f(x)dx$.

Let $g_\varepsilon(x) = (\frac{1}{2\pi\varepsilon})^{N/2} \exp(- \frac{|x|^2}{2\varepsilon})$, $x\in\mathbb{R}^N$. Then we have the following.

(3.4)**Lemma.** *A probability measure μ on $\mathbb{R}^N$ is strongly log concave if and only if* (1) μ *is absolutely continuous, and* (2) *the function* $x \to \int_{\mathbb{R}^N} g_\varepsilon(x-y)\mu(dy)$ *from* $\mathbb{R}^N$ *into* $[0,\infty)$ *is log concave for all* $\varepsilon > 0$.

Proof. 'only if' part is an immediate consequence of Theorem(3.2). Let us show 'if' part. Since the Radon-Nikodym derivative $\frac{d\mu}{dx} \in L^1(\mathbb{R}^N)$, there is a sequence $\{\varepsilon_n\}_1^\infty$ such that $\varepsilon_n \downarrow 0$ and $g_{\varepsilon_n}*\frac{d\mu}{dx}(x) \to \frac{d\mu}{dx}(x)$ a.e.x as $n\to\infty$. Set $f(x) = \lim_{n\to\infty} \int_{\mathbb{R}^N} g_{\varepsilon_n}(x-y)\mu(dy)$, $x\in\mathbb{R}^N$. Then it is easy to see that f is log concave.   Q.E.D.

(3.5)**Definition.** We say that a probability measure μ on $\mathbb{R}^N$ is log concave if the function $x \to \int_{\mathbb{R}^N} g_\varepsilon(x-y)\mu(dy)$ from $\mathbb{R}^N$ into $[0,\infty)$ is log ccncave for all $\varepsilon > 0$.

The following is an easy consequence of Definition(3.5).

(3.6)**Proposition.** *Suppose that* μ, μ_n, $n \geq 1$, *are probabilty measures on* $\mathbb{R}^N$ *and that* $\mu_n \to \mu$ *weakly as* $n \to \infty$. *If* μ_n, $n \geq 1$, *are* *log concave, then μ is also log concave.*

(3.7)**Definition.** We say that a probability measure μ on the Banach space B is *K-log-concave* if for any $P \in \mathcal{P}(K)$, the image measure $\mu \circ \tilde{P}^{-1}$ is a log concave probability measure on Image(P).

(3.8)**Proposition.** *If a probability measure μ on B is K-quasi-invariant and K-log-concave, then for each $P \in \mathcal{P}(K)$, the image measure $\mu \circ \tilde{P}^{-1}$ has a Radon-Nikodym density ρ relative to the Lebesgue measure on Image(P) such that log ρ :Image(P) → $\mathbb{R}$ is a concave and locally Lipschitz continuous function.*

Proof. Since μ is K-quasi-invariant, $\mu \circ \tilde{P}^{-1}$ is mutually absolutely continuous relative to the Lebesgue measure m on Image(P). Therefore from Lemma(3.4), $\mu \circ \tilde{P}^{-1}$ has a log concave density function ρ relative to m. Since $0 < \rho(x) < \infty$ a.e.x, we see that $0 < \rho(x) < \infty$ for all $x \in \mathbb{R}^N$. Thus log ρ:Image(P)→$\mathbb{R}$ is concave, and so is locally Lipschitz continuous. Q.E.D.

(3.9)**Lemma.** *Suppose that μ is a K-quasi-invariant and K-log-concave probability measure. Then for each $P \in \mathcal{P}(K)$, there is a function ρ_P:Image(P)×B→$\mathbb{R}$ such that*

(1) $\rho(\cdot, z)$:Image(P)→$\mathbb{R}$ is log concave and locally Lipschitz continuous for each $z \in B$, and

(2) $\int_A \rho(x,z) \, m(dx) = P^\mu[\tilde{P}^{-1}(A) | \mathcal{F}_P^\perp](z)$ μ-a.e.z   for each Borel set A in Image(P). Here $\mathcal{F}_P^\perp = \sigma\{(I_B - \tilde{P})z\}$ and m is the Lebesgue measure on Image(P).

In particular, μ is K-locally-Lipschitz.

Proof. We can take a sequence $\{Q_n\}_1^\infty$ in $\mathcal{P}(K)$ such that Image(P) $\subset$ Image(Q_1) $\subset$ Image(Q_2) $\subset \ldots$ and $\cup_n$ Image(Q_n) is dense in B^* with respect to weak[*] topology. Let $\mathcal{F}_n = \sigma\{(\tilde{Q}_n - \tilde{P})z\}$, $n \geq 1$. Then we see that $\vee_n \mathcal{F}_n = \mathcal{B}(B)$ up to μ-null set. Let $\rho_n(x,z) = \dfrac{d\mu \circ \tilde{Q}_n^{-1}}{dm_n}(x + (\tilde{Q}_n - \tilde{P})z),$

$x \in \text{Image}(P)$ and $z \in B$, where m_n is the Lebesgue measure on $\text{Image}(Q_n)$.
By Proposition(3.8), we may assume that $\log \rho_n(\cdot, z)$ is concave for each $z \in B$. Since $\int_A \rho_n(x,z)\, m(dx) = P^\mu[\, \tilde{P}^{-1}(A)\,|\,\mathcal{F}_n\,](z)$ μ-a.e.z for each Borel set A in Image(P), by virtue of Doob's Theorem on the convergence of martingales we see that $\rho_n(x,z)m(dx)$ converges to a probability measure $\tilde{\rho}(dx,z)$ on Image(P) weakly for μ-a.e.z and that $\int_A \tilde{\rho}(dx,z) = P^\mu[\, \tilde{P}^{-1}(A)\,|\,\mathcal{F}_P{}^\perp\,](z)$ for each Borel set A. By Proposition (3.6) and K-quasi-invariance, we see that $\tilde{\rho}(dx,z)$ is log concave and mutually absolutely continuous relative to m for μ-a.e.z. Thus similarly to the proof of Proposition(3.8), we have our assertion.   Q.E.D.

By Lemma(2.5) and Theorem(2.6), we have the following.

(3.10)**Proposition.** *If μ is K-differentiable and K-log-concave, then $H_P|_{\mathcal{F}C_b^\infty} : \mathcal{F}C_b^\infty \to L^2(d\mu)$ is essentially self-adjoint for any $P \in \mathcal{P}(K)$. Here H_P is the partial Hamiltonian operator along P associated with μ.*

(3.11)**Proposition.** *Let H denote the maximal Hamiltonian operator associated with μ. If μ is K-differentiable and K-log-concave, then H is the self-adjoint operator associated with the Sobolev Dirichlet form $(\tilde{\mathcal{E}}, \mathcal{D}(\tilde{\mathcal{E}}))$.*

The log concavity is essential in the following Lemma.

(3.12)**Lemma.** *Let $f : \mathbb{R}^N \to \mathbb{R}$ be a concave function with $\int_{\mathbb{R}^N} e^{f(x)}\, dx = 1$ and $\int_{\mathbb{R}^N} |\nabla f(x)|^2 e^{f(x)} dx < \infty$. Then for any $u \in C_b^\infty(\mathbb{R}^N)$ and integers n, with $1 \le n < m \le N$,*

$$(3.13) \qquad \int_{\mathbb{R}^N} \Big| \sum_{i=1}^{n} \Big(\frac{\partial^2 u}{\partial x_i^2}(x) - \frac{\partial f}{\partial x_i}(x) \frac{\partial u}{\partial x_i}(x) \Big) \Big|^2 e^{f(x)}\, dx$$

$$\le \int_{\mathbb{R}^N} \Big| \sum_{i=1}^{m} \Big(\frac{\partial^2 u}{\partial x_i^2}(x) - \frac{\partial f}{\partial x_i}(x) \frac{\partial u}{\partial x_i}(x) \Big) \Big|^2 e^{f(x)}\, dx\ .$$

Proof. It is sufficient to prove (3.13) for $u \in C_0^\infty(\mathbb{R}^N)$. Let $\rho = e^{f}$

and $\rho_\varepsilon = g_\varepsilon * \rho$, $\varepsilon > 0$. Then $\log \rho_\varepsilon : \mathbb{R}^N \to \mathbb{R}$ is a smooth concave function and so $\left(- \dfrac{\partial^2}{\partial x_i \partial x_j} \log \rho_\varepsilon(x) \right)_{1 \le i, j \le N}$ is a non-negative definite symmetric matrix for each $x \in \mathbb{R}^N$.

Let $L_n^{(\varepsilon)} = \displaystyle\sum_{i=1}^{n} \left(\dfrac{\partial^2}{\partial x_i^2} - \left(\dfrac{\partial}{\partial x_i} \log \rho_\varepsilon \right) \cdot \dfrac{\partial}{\partial x_i} \right)$. Then we have

$$\int_{\mathbb{R}^N} (L_n^{(\varepsilon)} u(x))^2 \, \rho_\varepsilon(x) \, dx$$

$$= - \sum_{i=1}^{n} \int_{\mathbb{R}^N} \frac{\partial u}{\partial x_i}(x) \cdot \frac{\partial}{\partial x_i} (L_n^{(\varepsilon)} u)(x) \, \rho_\varepsilon(x) \, dx$$

$$= - \sum_{i=1}^{n} \int_{\mathbb{R}^N} \frac{\partial u}{\partial x_i}(x) \cdot (L_n^{(\varepsilon)} \frac{\partial}{\partial x_i} u)(x) \, \rho_\varepsilon(x) \, dx$$

$$\qquad - \sum_{i,j=1}^{n} \int_{\mathbb{R}^N} \left(\frac{\partial^2}{\partial x_i \partial x_j} \log \rho_\varepsilon \right)(x) \, \frac{\partial u}{\partial x_i}(x) \, \frac{\partial u}{\partial x_j}(x) \, \rho_\varepsilon(x) \, dx$$

$$= \sum_{i,j=1}^{n} \int_{\mathbb{R}^N} \left(\frac{\partial^2 u}{\partial x_i \partial x_j}(x) \right)^2 \rho_\varepsilon(x) \, dx$$

$$\qquad + \sum_{i,j=1}^{n} \int_{\mathbb{R}^N} \left(- \frac{\partial^2}{\partial x_i \partial x_j} \log \rho_\varepsilon \right)(x) \, \frac{\partial u}{\partial x_i}(x) \, \frac{\partial u}{\partial x_j}(x) \, \rho_\varepsilon(x) \, dx.$$

Thus we have

$$\int_{\mathbb{R}^N} (L_n^{(\varepsilon)} u(x))^2 \, \rho_\varepsilon(x) \, dx \;\le\; \int_{\mathbb{R}^N} (L_m^{(\varepsilon)} u(x))^2 \, \rho_\varepsilon(x) \, dx$$

for $1 \le n < m \le N$ and $u \in C_0^\infty(\mathbb{R}^N)$. Letting $\varepsilon \downarrow 0$, we have (3.13).

Q.E.D.

The following are main results in this section and play a key role in the next section.

3.14) **Lemma.** *Suppose that μ is K-differentiable and K-log-concave. Then for any $P, Q \in \mathcal{P}(K)$ with* $\mathrm{Image}(P) \subset \mathrm{Image}(Q)$,

1) $\mathcal{D}om(H_Q) \subset \mathcal{D}om(H_P)$, *and*

2) $\| H_P u \|_{L^2(d\mu)} \le \| H_Q u \|_{L^2(d\mu)}$ *for $u \in \mathcal{D}om(H_Q)$* .

Proof. By Lemma(3.9) and Proposition(3.11), we see that

3.15) $\| H_P u \|_{L^2(d\mu)} \le \| H_Q u \|_{L^2(d\mu)}$ for $u \in \mathcal{F}C_b^\infty$.

Since $H_Q |_{\mathcal{F}C_b^\infty}$ is essentially self-adjoint by Proposition(3.10), $\mathcal{F}C_b^\infty$ is dense in $\mathcal{D}om(H_Q)$ with respect to the graph norm. Therefore (3.15) implies our assertion.

(3.16)Theorem. *Let* $\{P_n\}_{n=1}^{\infty} \subset \mathcal{P}(K)$ *with* $P_n \uparrow$ *Identity as* $n \to \infty$. *If* μ *is K-differentiable and K-log-concave, then*

$$\mathcal{D}om(H) = \{\, u \in \bigcap_{n=1}^{\infty} \mathcal{D}om(H_{P_n}); \; \sup_n \|H_{P_n} u\|_{L^2(d\mu)} < \infty \,\}, \; and$$

$$H_{P_n} u \to Hu \; strongly \; in \; L^2(d\mu) \; as \; n \to \infty \; for \; each \; u \in \mathcal{D}om(H).$$

Proof. First, note that $(\lambda I + H_{P_n})^{-1} \to (\lambda I + H)^{-1}$ strongly, $n \to \infty$, for any $\lambda > 0$. Then by Lemma(3.4), for any $u \in L^2(d\mu)$ and $v \in \mathcal{D}om(H_{P_n})$, $n \geq 1$,

$$|(H_{P_n} v, (\lambda I + H)^{-1} u)_{L^2(d\mu)}|$$

$$= \lim_{m \to \infty} |(H_{P_n} v, (\lambda I + H_{P_m})^{-1} u)_{L^2(d\mu)}|$$

$$= \lim_{m \to \infty} |(v, H_{P_n}(\lambda I + H_{P_m})^{-1} u)_{L^2(d\mu)}|$$

$$\leq \|u\|_{L^2(d\mu)} \|v\|_{L^2(d\mu)} .$$

Therefore we see that

$$(3.17) \quad \mathcal{D}om(H) \subset \mathcal{D}om(H_{P_n}) \, , \; n \geq 1, \; and$$

$$(3.18) \quad \|H_{P_n} u\|_{L^2(d\mu)} \leq \|Hu\|_{L^2(d\mu)} \, , \; u \in \mathcal{D}om(H) \; and \; n \geq 1.$$

Now suppose that $u \in \bigcap_{n=1}^{\infty} \mathcal{D}om(H_{P_n})$ and $\sup_n \|H_{P_n} u\|_{L^2(d\mu)} < \infty$. Then $\{H_{P_n} u\}_1^{\infty}$ is relative compact with respect to the weak topology in $L^2(d\mu)$. Observe that for any $v \in L^2(d\mu)$,

$$(H(I+H)^{-1} v, u)_{L^2(d\mu)}$$

$$= (v - (I+H)^{-1} v, u)_{L^2(d\mu)}$$

$$= \lim_{n \to \infty} (v - (I + H_{P_n})^{-1} v, u)_{L^2(d\mu)}$$

$$= \lim_{n \to \infty} ((I + H_{P_n})^{-1} v, H_{P_n} u)_{L^2(d\mu)}$$

$$= \lim_{n \to \infty} ((I+H)^{-1} v, H_{P_n} u)_{L^2(d\mu)} .$$

Therefore we see that $u \in \mathcal{D}om(H)$ and $H_{P_n} u \to Hu$ weakly, $n \to \infty$. Thus (3.17) and (3.18) imply that $\lim_{n \to \infty} \|H_{P_n} u\|_{L^2(d\mu)} = \|Hu\|_{L^2(d\mu)}$ and so $H_{P_n} u \to Hu$ strongly as $n \to \infty$. Thus we have our assertion.

4.   The Limit of Semigroups

Let B, H, K be as in Section 1, and let μ_n, n = 1,2,...,∞, be K-differentiable and K-log-concave probability measures on B. We denote by $(\tilde{\mathcal{E}}^{(n)}, \mathcal{D}(\tilde{\mathcal{E}}^{(n)}))$, n = 1,2,...,$\infty$, the Sobolev Dirichlet forms in $L^2(d\mu_n)$ as is defined in Definition(1.11) by taking $\mu = \mu_n$. Also, we denote by $\beta^{(n)}$ the map $\beta : K \to L^2(d\mu)$ defined in Section 1 for $\mu = \mu_n$. So β_n is a linear map from K into $L^2(d\mu_n)$. Let $H^{(n)}$ denote the non-negative self-adjoint operator associated with the Sobolev Dirichlet form $(\tilde{\mathcal{E}}^{(n)}, \mathcal{D}(\tilde{\mathcal{E}}^{(n)}))$.

Then we have the following.

(4.1)**Theorem.** *Suppose that $\mu_n \to \mu_\infty$ weakly as n $\to \infty$, and there is a Markov semigroup $\{P_t\}_{t \geq 0}$ in $L^2(d\mu_\infty)$ satisfying*

(4.2) $(\exp(-tH^{(n)})f,g)_{L^2(d\mu_n)} \to (P_t f,g)_{L^2(d\mu_\infty)}$, n $\to \infty$,

for any f,g $\in \mathcal{F}C_b^\infty$ and t $\geq$ 0, and suppose moreover that

(4.3) $\sup_n \|\beta^{(n)}(k)\|_{L^2(d\mu_n)} < \infty$ *for any k $\in$ K.*

Then $P_t = \exp(-tH^{(\infty)})$, t $\geq$ 0.

Proof. Take a sequence $\{P_m\}_1^\infty \subset \mathcal{P}(K)$ such that $P_m \uparrow$ Identity as m $\to \infty$. Let $H_m^{(n)}$ denote the partial Hamiltonian operator along P_m associated to μ_n given in Section 3. By the assumption, we see that $\int_B f\, d\mu_n \to \int_B f\, d\mu_\infty$, n $\to \infty$, for any f $\in \mathcal{F}C_b^\infty$. Therefore for any f,g $\in \mathcal{F}C_b^\infty$,

(4.4) $(H_m^{(n)} f,g)_{L^2(d\mu_n)}$

$$= \int_B (P_m \nabla f(z), \nabla g(z))_H\, \mu_n(dz)$$

$$\to (H_m^{(\infty)} f,g)_{L^2(d\mu_\infty)} \quad \text{as n} \to \infty, \text{ and}$$

(4.5) $\|\exp(-tH^{(n)})f - g\|_{L^2(d\mu_n)}^2$

$$= (\exp(-2tH^{(n)})f,f)_{L^2(d\mu_n)} - 2(\exp(-tH^{(n)})f,g)_{L^2(d\mu_n)}$$

$$+ (g,g)_{L^2(d\mu_n)}$$

$$\to \|P_t f - g\|_{L^2(d\mu_\infty)}^2 \quad \text{as n} \to \infty.$$

Therefore for any $f, g, h \in \mathcal{F}C_b^\infty$,

$$\varlimsup_{n\to\infty} |(\exp(-tH^{(n)})f, H_m^{(n)}g)_{L^2(d\mu_n)} - (P_t f, H_m^{(\infty)}g)_{L^2(d\mu_\infty)}|$$

$$\leq \varlimsup_{n\to\infty} |(\exp(-tH^{(n)})f-h, H_m^{(n)}g)_{L^2(d\mu_n)} - (P_t f-h, H_m^{(\infty)}g)_{L^2(d\mu_\infty)}|$$

$$+ \varlimsup_{n\to\infty} |(h, H_m^{(n)}g)_{L^2(d\mu_n)} - (h, H_m^{(\infty)}g)_{L^2(d\mu_\infty)}|$$

$$\leq \|P_t f-h\|_{L^2(d\mu_\infty)} \cdot (\sup_n \|H_m^{(n)}g\|_{L^2(d\mu_\infty)} + \|H_m^{(\infty)}g\|_{L^2(d\mu_\infty)}).$$

Since $\mathcal{F}C_b^\infty$ is dense in $L^2(d\mu_\infty)$ and $\sup_n \|H_m^{(n)}g\|_{L^2(d\mu_n)} < \infty$ by the

assumption (4.3), we see that

$$(4.6) \quad (\exp(-tH^{(n)})f, H_m^{(n)}g)_{L^2(d\mu_n)} \to (P_t f, H_m^{(\infty)}g)_{L^2(d\mu_\infty)}, \quad n \to \infty,$$

for any $f, g \in \mathcal{F}C_b^\infty$ and $m \geq 1$.

   Let $-\tilde{H}$ be the infinitesimal generator of the semigroup $\{P_t\}_{t\geq 0}$

Then by (4.6), we have

$$(4.7) \quad ((I+H^{(n)})^{-1}f, H_m^{(n)}g)_{L^2(d\mu_n)} \to ((I+\tilde{H})^{-1}f, H_m^{(\infty)}g)_{L^2(d\mu_\infty)}, \quad n \to \infty,$$

for any $f, g \in \mathcal{F}C_b^\infty$ and $m \geq 1$. By Theorem(3.16) (or (3.18)), we hav

$$|((I+H^{(n)})^{-1}f, H_m^{(n)}g)_{L^2(d\mu_n)}| \leq \|f\|_{L^2(d\mu_n)}\|g\|_{L^2(d\mu_n)}.$$

Thus we see that

$$|((I+\tilde{H})^{-1}f, H_m^{(\infty)}g)_{L^2(d\mu_\infty)}| \leq \|f\|_{L^2(d\mu_\infty)}\|g\|_{L^2(d\mu_\infty)}$$

for any $f, g \in \mathcal{F}C_b^\infty$ and $m \geq 1$. Thus, again by Theorem(3.16), we see

that

$$(4.8) \quad (I+\tilde{H})^{-1}f \in \mathcal{D}om(H^{(\infty)}) \quad \text{for any } f \in \mathcal{F}C_b^\infty, \text{ and}$$

$$(4.9) \quad \|H^{(\infty)}(I+\tilde{H})^{-1}f\|_{L^2(d\mu_\infty)} \leq \|f\|_{L^2(d\mu)}, \quad f \in \mathcal{F}C_b^\infty.$$

These implies that $\mathcal{D}om(\tilde{H}) \subset \mathcal{D}om(H^{(\infty)})$.

   Since $\mathcal{F}C_b^\infty \subset \mathcal{D}om(H^{(n)})$, we have

$$(4.10) \quad \|\exp(-tH^{(n)})f-f\|_{L^2(d\mu_n)} \leq t\|H^{(n)}f\|_{L^2(d\mu_n)}$$

for any $f \in \mathcal{F}C_b^\infty$, $t \geq 0$ and $n \geq 1$. Thus by (4.5), we have

$$(4.11) \quad \|P_t f-f\|_{L^2(d\mu_\infty)} \leq t \sup_n \|H^{(n)}f\|_{L^2(d\mu_n)}$$

for any $f \in \mathcal{F}C_b^\infty$. By our assumption (4.3), we see that

$\sup_{n} \|H^{(n)}f\|_{L^2(d\mu_n)} < \infty$.   Therefore (4.11) implies that $\mathcal{F}C_b^\infty \subset \mathcal{D}om(\tilde{H})$.

On the other hand, we see that

$$\left| (\exp(-tH^{(n)})f-f,g)_{L^2(d\mu_n)} - \frac{t}{2}\int_B (\nabla f,\nabla g)_H \, d\mu_n \right|$$

$$= \left| ((\exp(-tH^{(n)}) - I - tH^{(n)})f,g)_{L^2(d\mu_n)} \right|$$

$$\leq t^2 \, \|H^{(n)}f\|_{L^2(d\mu_n)} \|H^{(n)}g\|_{L^2(d\mu_n)}$$

for any $f,g \in \mathcal{F}C_b^\infty$, $t > 0$ and $n \geq 1$.   Thus again by (4.5), we have

$$\left| (P_t f-f,g)_{L^2(d\mu_\infty)} - \frac{t}{2}\int_B (\nabla f,\nabla g)_H \, d\mu_\infty \right|$$

$$\leq t^2 \sup_{n} \|H^{(n)}f\|_{L^2(d\mu_n)} \|H^{(n)}g\|_{L^2(d\mu_n)}$$

for any $f,g \in \mathcal{F}C_b^\infty$ and $t > 0$.   This implies that

$$(4.12) \quad (\tilde{H}f,g)_{L^2(d\mu_\infty)} = \frac{1}{2}\int_B (\nabla f,\nabla g)_H \, d\mu_\infty \,, \quad f,g \in \mathcal{F}C_b^\infty \,,$$

and so we have

$$(4.13) \quad \tilde{H}f = H^{(\infty)}f, \quad f \in \mathcal{F}C_b^\infty \,.$$

Then, for any $u \in \mathcal{D}om(\tilde{H}) \subset \mathcal{D}om(H^{(\infty)})$ and $f \in \mathcal{F}C_b^\infty$, we have

$$(\tilde{H}u,f)_{L^2(d\mu_\infty)} = (u,\tilde{H}f)_{L^2(d\mu_\infty)} = (u,H^{(\infty)}f)_{L^2(d\mu_\infty)} = (H^{(\infty)}u,f)_{L^2(d\mu_\infty)}.$$

However, since both $\tilde{H}$ and $H^{(\infty)}$ are self-adjoint, we have $\tilde{H} = H^{(\infty)}$.

Q.E.D.

Combining Theorem(4.1) with the results in Section 3, we have the following.

(4.14)**Corollary.** *Under the assumptions of Theorem(4.1), and calling* $-\tilde{H} = -H^{(\infty)}$ *the infinitesimal generator of the Markov semigroup* $\{P_t\}_{t\geq0}$, *we have*

$$\mathcal{D}om(\tilde{H}) = \{ u \in \bigcap_{n=1}^{\infty} \mathcal{D}om(H_{P_n}^{(\infty)}); \ \sup_{n}\|H_{P_n}^{(\infty)}u\|_{L^2(d\mu)} < \infty \} \quad and$$

$$H_{P_n}^{(\infty)}u \to \tilde{H}u \ strongly \ in \ L^2(d\mu) \ as \ n \to \infty, \ for \ each \ u \in \mathcal{D}om(\tilde{H}).$$

Moreover $\mathcal{D}om(\tilde{H}^{1/2}) = \mathcal{D}(\tilde{\mathcal{E}})$, *where* $\mathcal{D}(\tilde{\mathcal{E}})$ *is given by Definition(1.1).*

5. Application to Scalar Field with Interactions of Exponential type

In this section we apply our results to the scalar fields with interactions of exponential type (Høegh-Krohn's models), discussed in [2].

Let $B = \{u \in \mathscr{S}'(\mathbb{R}); \ (1-(\frac{d}{dx})^2)^{-1}u \in L^2_{loc}(\mathbb{R}),$
$\int_{\mathbb{R}} |(1-(\frac{d}{dx})^2)^{-1}u(x)|^2 \frac{dx}{1+|x|^2} < \infty\}$, and let $H = L^2(\mathbb{R})$ and $K = \mathscr{S}(\mathbb{R})$.
Let μ_0 be a probability measure in B such that

$$\int_B \exp(\sqrt{-1}_{\mathscr{S}}, <\phi,u>_{\mathscr{S}}) \ \mu_0(d\phi) = \exp(-\frac{1}{2} \int_{\mathbb{R}} u(x)\cdot(1-(\frac{d}{dx})^2)^{-1/2}u(x)dx \ )$$

for any $u \in \mathscr{S}(\mathbb{R})$. Then μ_0 is the sharp time field of the free scalar field.

Let ν be a bounded positive measure in $\mathbb{R}$ whose support is in the open interval $(-\sqrt{2\pi},\sqrt{2\pi})$ and let $V(s) = \int \exp(\alpha s)\ \nu(d\alpha)$. Then by virtue of Albeverio-Høegh-Krohn[1], we see that for each $r > 0$, we can define a measurable function V_r in B by

$$V_r(\phi) = \int_{|x|<r} :V(\phi(x)): \ dx$$
$$= \underset{\varepsilon\downarrow 0}{l.i.m.} \ V_{r,\varepsilon}(\phi)$$

where $V_{r,\varepsilon}(\phi) = \int_{|x|<r} dx \int_{\mathbb{R}} \exp(\alpha(1-(\frac{d}{dx})^2)^{-\varepsilon}\phi(x) - \alpha^2 C_\varepsilon)\nu(d\alpha)$ and

$C_\varepsilon = \frac{1}{2} \int_{\mathbb{R}} (1+p^2)^{-(1/2+2\varepsilon)}\ dp$. So we see that $V_r(\phi) \geq 0 \ \mu_0$-a.e.ϕ.

Let $-H_0$ be the generator associated with Dirichlet form $(\tilde{\mathscr{E}},\mathscr{D}om(\tilde{\mathscr{E}}))$ in $L^2(d\mu_0)$ given in Definition (1.11) with $\mu = \mu_0$. $H_r = H_0 + V_r(\phi)$ is a well-defined self-adjoint operator in $L^2(d\mu_0)$, see Albeverio-Høegh-Krohn[2]. Let $\Omega_r \in L^2(d\mu_0)$, with norm 1, be such that $H_r\Omega_r = E_r\Omega_r$, with E_r the infimum of the spectrum of H_r. Let $d\mu_r = |\Omega_r|^2 d\mu_0$. From [2], we have

(5.1) There is a probability measure μ in $\mathscr{S}'(\mathbb{R})$ such that the sequence of probability measures μ_r converges to μ weakly as probability measures in $\mathscr{S}'(\mathbb{R})$.

We shall call μ the sharp time field measure (vacuum measure) of the Høegh-Krohn's model of quantum fields.

Remark.   The existence of this measure and its properties extend to the case where supp $\nu \subset (-\sqrt{4\pi},\sqrt{4\pi})$ (see [2]).

Let μ^E be the Euclidean (path space) measure of Høegh-Krohn's model of quantum fields, as defined in [2] (see also Gielerak[14] and Zegarliński[24]). μ coincides with the restriction of μ^E to the σ-algebra of time zero fields. μ^E is, as shown in [2], the weak limit for $r\to\infty$ of the measures

$$d\mu_r^E(\tilde{\phi}) \equiv \tilde{Z}_r^{-1} \exp(- V_r^E(\tilde{\phi}))d\mu_0^E(\tilde{\phi}),$$

with $V_r^E(\tilde{\phi}) = \int_{|x|\leq r} :V(\tilde{\phi}(x)): d^2x$, $\tilde{Z}_r \equiv \int_{\mathcal{G}'(\mathbb{R}^2)} \exp(-V_r^E(\tilde{\phi}))d\mu_0^E(\tilde{\phi})$,

and μ_0^E is Nelson's free field measure, i.e.,

$$\int_{\mathcal{G}'(\mathbb{R}^2)} \exp(\sqrt{-1}\ _{\mathcal{G}'}\langle\tilde{\phi},u\rangle_{\mathcal{G}})\ \mu_0^E(d\tilde{\phi})$$

$$= \exp(- \frac{1}{2}\int_{\mathbb{R}^2} u(x)((1-\Delta)^{-1}u)(x)\ dx\ ).$$

One has $V_r^E(\tilde{\phi}) = \mathrm{l.i.m.}_{\varepsilon\downarrow 0} V_{r,\varepsilon}(\tilde{\phi})$, with $V_{r,\varepsilon}(\tilde{\phi})$ defined similarly as $V_{r,\varepsilon}(\phi)$, which shows that $V_r^E(\tilde{\phi}) \geq 0\ \mu_0^E$-a.e.$\tilde{\phi}$. Moreover $\mu_{r,\varepsilon}^E$ converges weakly to μ_r^E, where $\mu_{r,\varepsilon}^E$ is defined corespondingly to μ_r^E with $V_{r,\varepsilon}$ replacing V_r. Since $\mu_{r,\varepsilon}^E$ is K-log-concave, we see by Proposition(3.6) that μ_r^E, as well as μ^E and μ, are K-log-concave.

Remark.   We also remark incidentally that also the probability measure $\tilde{\mu}_r(d\phi) = Z_r^{-1}\exp(-V_r(\phi))\mu_0(d\phi)$, where $Z_r= \int_B \exp(-V_r(\phi))\mu_0(d\phi)$ is, by Proposition(3.6), K-log-concave, as shown by representing $\tilde{\mu}_r$ as the weak limit of correspomding "ε-regularized" measure $\tilde{\mu}_{r,\varepsilon}$, constructed by replacing V_r by $V_{r,\varepsilon}$.

(4.2) There are constants $C < \infty$ and $m > 0$ such that

$$\int_{\mathcal{G}'}\ _{\mathcal{G}'}\langle\phi,u\rangle_{\mathcal{G}}^2\ \mu_r^E(d\tilde{\phi}) \leqq C \int_{\mathbb{R}^2}(\int_{\mathbb{R}^2} \exp(\sqrt{-1}px)u(x)dx)^2 (m^2+p^2)^{-1}dp$$

for any $u \in \mathcal{G}(\mathbb{R})$ and $r > 0$ ([2]).

Let $(\tilde{\mathscr{E}}^{(r)}, \mathscr{D}(\tilde{\mathscr{E}}^{(r)}))$ be the Dirichlet form in $L^2(d\mu_r)$ as is defined in Definition(1.11) by taking $\mu = \mu_r$. Also, we denote by $\beta^{(r)}$ the map $\beta: K \to L^2(d\mu)$ defined in Section 1 for $\mu = \mu_r$. Let $\{P_t^{(r)}\}_{t \geq 0}$ denote the Markov semigroup associated to the Dirichlet form $(\tilde{\mathscr{E}}^{(r)}, \mathscr{D}(\tilde{\mathscr{E}}^{(r)}))$.

(5.3) There is a Markov semigroup $\{P_t\}_{t \geq 0}$ in $L^2(d\mu)$ such that

$$(P_t^{(r)} f, g)_{L^2(d\mu_r)} \to (P_t f, g)_{L^2(d\mu_\infty)} \, , \, r \to \infty,$$

for any $f, g \in \mathcal{F}C_b^\infty$ and $t \geq 0$ ([2],[14],[24]).

(5.4) $\sup_r \| \beta^{(r)}(k) \|_{L^2(d\mu_n)} < \infty$ for any $k \in K$ ([4]).

Therefore we have the following from Theorem(4.1) and Corollary(4.14).

(5.5) **Theorem.**   *Let μ be the sharp time field of the scalar field with exponential type interaction in two dimensional space time. Let $\{P_t\}_{t \geq 0}$ be the Markov semigroup in $L^2(d\mu)$ induced by the scalar field. Let $-H$ be the infinitesimal generator of the Markov semigroup $\{P_t\}_{t \geq 0}$. Then $\{P_t\}_{t \geq 0}$ is the Markov semigroup associated to the Dirichlet form $(\tilde{\mathscr{E}}, \mathscr{D}(\tilde{\mathscr{E}}))$ given in Definition(1.1). Therefore $H = \tilde{H}$ $\tilde{H}$ as in (4.14), $\mathscr{D}om(H^{1/2}) = \mathscr{D}(\tilde{\mathscr{E}})$. Moreover, for any $\{P_n\}_{n=1}^\infty \subset \mathcal{P}(K)$ with $P_n \uparrow I_H$ as $n \uparrow \infty$,*

$$\mathscr{D}om(H) = \{ u \in \bigcap_{n=1}^\infty \mathscr{D}om(H_{P_n}) ; \sup_n \| H_{P_n}^{(\infty)} u \|_{L^2(d\mu)} < \infty \} \text{ and}$$

$H_{P_n}^{(\infty)} u \to \tilde{H}u$ strongly in $L^2(d\mu)$ as $n \to \infty$, for each $u \in \mathscr{D}om(\tilde{H})$. Here H_P denotes the partial Hamiltonian operator along $P \in \mathcal{P}(K)$ associate to μ.

References

[1] Albeverio, S., J.E. Fenstad, R. Høegh-Krohn, and T. Lindstrøm, Nonstandard methods in stochastic analysis and mathematical physics

Academic Press 1986, Orlando.

[2] Albeverio, S., and R. Høegh-Krohn, The Wightman axioms and the mass gap for strong interactions of exponential type in two dimensional space-time, J. Funct. Anal. 16(1974), 39-78.

[3] Albeverio, S., and R. Høegh-Krohn, Quasi invariant measures, symmetric diffusion processes and quantum fields, pp. 11-59 in Colloques Int. CNRS, No. 248, Edts. CNRS (1976).

[4] Albeverio, S., and R. Høegh-Krohn, Dirichlet forms and diffusion processes on rigged Hilbert spaces, Z. Wahrscheinlichkeitstheorie verw. Gebiete 40(1977), 1-57.

[5] Albeverio, S., and R. Høegh-Krohn, Hunt processes and analytic potential theory on rigged Hilbert spaces, Ann. Inst. H. Poincaré B 1977), 269-291.

[6] Albeverio, S., and M. Röckner, Classical Dirichlet forms on topological vector spaces - closability and a Cameron-Martin formula, to appear in J. Funct. Anal.

[7] Araki, H., Hamiltonian formalism and the canonical commutation relations in quantum field theory, J. Math. Phys. 1(1960), 492-504.

[8] Bouleau, N., and F. Hirsch, Formes de Dirichlet générales et densité des variables aléatoires réelles sur l'espace de Wiener, J. Funct. Anal. 69(1986), 229-259.

[9] Brascamp, H., and E. Lieb, Some inequalities for Gaussian measures, in 'Functional integration and its applications' ed. by A.M. Arthur, Clarendon Press 1975, Oxford.

[10] Dellacherie, C., and P.A. Meyer, Probabilities and potential B, North Holland 1982, Amsterdam-New York-Oxford.

[11] Dynkin, E.B., Green's and Dirichlet spaces for a symmetric Markov transition function, Lect. Notes London Math. Soc. (1982).

[12] Fukushima, M., Dirichlet forms and Markov processes, North Holland/Kodansha 1980, Amsterdam-Oxford-New York/Tokyo.

[13] Fukushima, M., Energy forms and diffusion processes, pp. 65-97 in "Mathematics + Physics", vol. 1, ed. L. Streit, World Scientific 1985, Singapore.

[14] Gielerak, R., Verification of the global Markov property in some class of strongly coupled exponential interactions, J. Math. Phys. 24(1983), 347-355.

[15] Hegerfeldt, G.C., On canonical commutation relations and infinite-dimensional measures, J. Math. Phys. 13(1972), 45-50.

[16] Hida, T., J. Potthof, and L. Streit, Dirichlet forms and white noise analysis, Commun. Math. Phys. 116(1988), 235-245.

[17] Ikeda, N., and S. Watanabe, Stochastic differential equations a diffusion processes, North Holland/Kodansha 1981, Amsterdam-Oxford-New York/Tokyo.

[18] Kusuoka, S., Dirichlet forms and diffusion processes on Banach spaces, J. Fac. Sci. Univ. Tokyo Sec. IA 29(1982), 79-95.

[19] Paclet, Ph., Espace de Dirichlet en dimension infinie, CR Ac. Sci., Paris, Ser. A 288(1979), 981-983.

[20] Röckner, M., Self-adjoint harmonic spaces and Dirichlet forms, Hiroshima Math. J. 14(1984), 55-66.

[21] Silverstein, M.L., Boundary theory for symmetric Markov processes, Lect. Notes Math. vol.516, Springer 1976, Berlin.

[22] Takeda, M., On the uniqueness of Markovian self-adjoint extensions of diffusion operators on infinite dimensional spaces, Osaka J. Math. 22(1985), 733-742.

[23] Wielens, N., On the essential self-adjointness of generalized Schrödinger operators, J. Funct. Anal. 61(1985), 98-115.

[24] Zegarliński, B., Uniqueness and the global Markov property for Euclidean fields: The case of general exponential interactions, Com Math. Phys. 96(1984), 195-221.

GLOBAL MARKOV PROPERTY
IN QUANTUM FIELD THEORY
AND STATISTICAL MECHANICS:
A REVIEW ON RESULTS AND PROBLEMS

by

Sergio Albeverio[*,**] , Boguslaw Zegarliński [*]

Fakultät für Mathematik
Ruhr–Universität
D4630 Bochum 1
Germany

Abstract

We shortly review the problem of proving the global Markov property for homogenous (generalized) random fields describing models of relativistic quantum scalar fields or translation invariant models of classical statistical mechanics.

0. Global Markov Property: Results and Problems

The problem of the global Markov property was one of the hardest issues in euclidean field theory and statistical mechanics during the past twenty years. In solving this problem Raphael Høegh-Krohn has played a prominent role. The purpose of our paper is to give a brief historical review and to show the present understanding of the problem.

---

[*] SFB 237 Essen – Bochum – Düsseldorf
[**] BiBoS Research Centre, 4800 Bielefeld 1;
     CERFIM, Locarno (Switzerland)

331

As a starting point stands Nelson's work [N1,2], which provided a bridge from a quantum field theory in euclidean region given by a linear process $\{\varphi(f) : f \in \mathcal{S}(I\!\!R^d)\}$ on a probability space $(\mathcal{S}'(I\!\!R^d), \mathcal{B}, \mu)$, to a corresponding field theory (satisfying Wightman axioms [Wig] plus some regularity conditions) in Minkowski region. A significant role in Nelson's theory is played by the <u>Markov Axiom</u> i. e. the assumption that for any function $F(\varphi)$ measurable with respect to σ-algebra $\Sigma_{[0,\infty)\times I\!\!R^{d-1}}$ of "events in the future", its conditional expectation associated with the probability measure μ satisfies

$$E_\mu(F|\Sigma_{[0,\infty)\times I\!\!R^{d-1}}) = E_\mu(F|\Sigma_{\{0\}\times I\!\!R^{d-1}}), \qquad \mu - \text{a. e.} \tag{0.1}$$

where $\Sigma_{\{0\}\times I\!\!R^{d-1}}$ denotes the σ-algebra of "present events" ("time zero" σ-algebra).
The fundamental example of a probability measure on the space of tempered distributions, defining a free field theory satisfying all Nelson's axioms, is given by a Gaussian measure μ_0 of mean zero and covariance

$$G \equiv (-\Delta + m^2)^{-1} \tag{0.2}$$

where Δ is d-dimensional Laplacian and $m^2 \geq 0$ $(m^2 > 0$ for $d \leq 2)$ a constant, see [N2]. For discussions of Nelson's axioms and other reconstruction theorems giving the passage from euclidean fields to relativistic fields and using conditions weaker than the Markov property, we refer to [N3], [Si1], [Si2], [Ch], [He], [O'CO], [OS], [Gl], [Kl], [JSt].
First nontrivial examples of a field theory satisfying a weaker set of axioms than those of Nelson, namely those of [OS], [Gl] given in euclidean region by a probability measure μ, has been constructed in two dimensions, see [GJS1], [AHK1] (see also [GJ], [Si2] and refs. therein). However at that time the Markov Axiom of Nelson had not been verified. It was only shown in [New] that the measure μ of a model with polynomial interaction satisfies the following property, called by the author the <u>Local Markov Property</u> (**LMP**):
For any measurable function $F(\varphi)$ localized in a bounded (regular) region $\Lambda \subset I\!\!R^2$ one has

$$E_\mu(F(\varphi)|\Sigma_{\Lambda^c}) = E_\mu(F(\varphi)|\Sigma_{\partial\Lambda}), \qquad \mu - \text{a. e.} \tag{0.3}$$

where E_μ is the conditional expectation of the measure μ with respect to the σ-algebra Σ_{Λ^c}, respectively $\Sigma_{\partial\Lambda}$ of events localized in the complement Λ^c, respectively boundary $\partial\Lambda$, of the bounded region Λ,(see also [GRS1]).
By contradistinction the equality (0.1) has got the name of <u>Global Markov Property</u> (**GMP**)

Remark : In fact more precisely this is the Markov property with respect to the σ-algebra associated with the $\{x_0 = 0\}$ hyperplane, when we denote a point $x \in I\!\!R^d$ by $(x^0, \overrightarrow{x}) \in I\!\!R \times I\!\!R^{d-1}$. In the case of fields which are rotation invariant (isotropic) (in the probabilistic sense, with finite dimensional distributions being rotation invariant) this property is equivalent with the corresponding property with respect to any $d-1$-dimensional hyperplane through the origin and, if the fields are also translation invariant, also with respect to arbitrary d-1 dimensional hyperplanes. In the literature often global Markov property is understood in the more general sense where $\{x_0 = 0\}$ is replaced by the σ-algebra of any boundary $\partial\Lambda$ of a measurable set $\Lambda \subset I\!\!R^d$, cfr [AHK2] . It is also called sharp global Markov property to distinguish it from other Markov properties, where the σ-algebra associated with $\partial\Lambda$ is defined differently. See Remark 2.1 at the end of Section 0. In the following , however, for simplicity we will understand as global Markov property the one expressed by (0.1) .

Due to its intrinsic connection with euclidean field theory as with the question of existence and construction of homogeneous Markov fields the problem of proving the global Markov Property, i. e. the question under what additional assumptions the local Markov property implies **GMP**, became a subject of intensive studies. It appeared that a natural context to consider this problem was provided by the theory of Gibbs measures introduced before for statistical mechanical systems by R. L. Dobrushin [Do1-3] and O. E. Landford III and D. Ruelle [LR].

In this theory (formalized and developed further in [Fö1], see also [Av], [Pre], [Ge] and references therein) we have a family $\mathcal{E} \equiv \{E_\Lambda\}_{\Lambda\in\mathcal{F}}$, given a priori and called <u>a local specification</u>, consisting of probability kernels E_Λ defined on a standard Borel space (Ω, Σ). The index set $\mathcal{F}$ is chosen to be a family of bounded (regular) subsets of a noncompact (metric) space Γ, partially ordered by the inclusion relation. It is assumed that $\mathcal{F}$ includes a subfamily $\mathcal{F}_0 \equiv \{\Lambda_n \in \mathcal{F}\}_{n\in I\!N}$, called a countable base, which is increasing, absorbing (i.e. for any $\Lambda \in \mathcal{F}$ there is $n \in I\!N$ such that $\Lambda \subseteq \Lambda_n$) and invading all the space Γ (i. e. $\bigcup_{\Lambda_n\in\mathcal{F}_0} \Lambda_n = \Gamma$).

To any set $\Lambda \in \mathcal{F}$ and its complement $\Lambda^c \equiv \Gamma \setminus \Lambda$ there are associated sub-σ-algebras Σ_Λ and respectively Σ_{Λ^c}, so that their union $\Sigma_\Lambda \cup \Sigma_{\Lambda^c}$ generates Σ. A local specification $\mathcal{E} \equiv \{E_\Lambda\}_{\Lambda\in\mathcal{F}}$, satisfies by definition the following requirements:

$(\mathcal{E}^i)$ *For any* $\omega \in \Omega$ *and* $\Lambda \in \mathcal{F}$, E_Λ^ω *is a probability measure on* (Ω, Σ). *Moreover for any real measurable function* F *on* (Ω, Σ) *the function*

$$\Omega \ni \omega \longrightarrow E_\Lambda^\omega(F) \tag{0.4}$$

is Σ_{Λ^c} *measurable and if* F *is actually* Σ_{Λ^c} *measurable then*

$$E_\Lambda^\omega(F) = \delta_\omega(F), \tag{0.5}$$

δ_ω *being the point measure concentrated at* ω.

$(\mathcal{E}^{ii})$ <u>*Compatibility condition:*</u> *If* $\Lambda_1, \Lambda_2 \in \mathcal{F}$, $\Lambda_1 \subset \Lambda_2$ *then*

$$E_{\Lambda_2}E_{\Lambda_1} = E_{\Lambda_2} \tag{0.6}$$

The set of <u>Gibbs measures</u> $\mathcal{G}(\mathcal{E})$ for a local specification $\mathcal{E}$ consists of all probability measures μ on (Ω, Σ) which satisfy

(DLR) $\mu E_\Lambda F = \mu F$ (0.7)

for all $\Lambda \in \mathcal{F}$ and measurable functions F , where μF denotes the integral $\int F d\mu$ of a function F against μ. I. e. the elements of the family $\mathcal{E}$ coincide with (a good version of) the conditional expectations associated to μ. As shown in [Fö1] any Gibbs measure admits a unique convex linear representation in terms of elements ν of the set $\partial \mathcal{G}(\mathcal{E})$ of <u>extremal Gibbs measures</u> which are characterized by the requirement that any event belonging to the <u>σ-algebra at infinity</u>

$$\Sigma_\infty \equiv \bigcap_{\Lambda \in \mathcal{F}} \Sigma_{\Lambda^c}, \tag{0.8}$$

has corresponding ν-measure 0 or 1. In physical terminology the above description corresponds to associating to any thermodynamical system in a finite volume Λ interacting with a fixed external configuration ω_{Λ^c}, a probability measure E_Λ^ω which can be written in a Gibbsian form

$$E_\Lambda^\omega(\cdot) = \delta_\omega \frac{\mu_{0|\Lambda}\left(e^{-U_\Lambda} \cdot\right)}{\mu_{0|\Lambda}\left(e^{-U_\Lambda}\right)} \tag{0.9}$$

with $\mu_{0|\Lambda}$ being some "free measure" restricted to the volume Λ (corresponding to a system without interaction) and U_Λ some interaction functional. The condition ($\mathcal{E}_{ii}$) allows us to identify a system in a volume $\Lambda_1 \in \mathcal{F}$ as a subsystem of a larger one in $\Lambda_2 \in \mathcal{F}$, if $\Lambda_1 \subset \Lambda_2$. Any extremal Gibbs measure μ describes a pure phase of the infinite system in Γ and, although it can not be written in the Gibbsian form (0.9), we have

$$\mu = \lim_{\mathcal{F}_0} E_\Lambda^\omega \tag{0.10}$$

for μ-a. e. $\omega \in \Omega$, see [Föl]* If there exists more than one extremal Gibbs measure one speaks about the existence of a <u>phase transition</u> (of the first order). For the structure of Gibbs states , in particular in reation with translation invariance , see also e.g. [Ai], [Hi], [Me] and [Co], [Pa], [Kes], [Sl], [Spi], [Win] .

For our purposes it is also useful to introduce the concept a local <u>Markov specification</u> $\mathcal{E} \equiv \{E_\Lambda\}_{\Lambda \in \mathcal{F}}$, describing a system with interaction of finite range $0 \leq r < \infty$, defined by requiring that for any $\Lambda \in \mathcal{F}$ and any Σ_Λ-measurable function F (shortly $F \in \Sigma_\Lambda$) we have

$$E_\Lambda^{\cdot}(f) \in \Sigma_{\partial_r \Lambda} \tag{0.11}$$

where

$$\partial_r \Lambda \equiv \{x \in \Lambda^c : d(x, \Lambda) \leq r\} \tag{0.12}$$

is the boundary of the set Λ, relative to the finite range of the interaction (region where the interaction acts). We see that by its very definition any Gibbs measure μ for a local Markov specification $\mathcal{E}$ has the local Markov property in the sense (generalizing (0.3)) that

$$E_\mu(F|\Sigma_{\Lambda^c}) = E_\mu(F|\Sigma_{\partial_r \Lambda}), \qquad \mu - \text{a. e.} \tag{0.13}$$

---

* Sometimes given a nonextremal Gibbs measure μ one may find a configuration $\omega_0 \in \Omega$ such that

$$\mu = \lim_{\mathcal{F}_0} E_\Lambda^{\omega_0}$$

We do not know whether this is a general property i. e. if it holds for any (nonextremal) Gibbs measure.

for any bounded set $\Lambda \in \mathcal{F}$ and any function $F \in \Sigma_\Lambda$. In general the local Markov property does not implies the global Markov property (i. e. the property (0.13) for unbounded domains Λ). To show this it is enough to consider a situation where for a given unbounded region $Q \subset \Gamma$ there exist two Gibbs measures μ_1, μ_2 for a Markov specification $\mathcal{E}$, whose restrictions to $\Sigma_{\partial_r Q}$ coincide but which are singular to each other on Σ_Q. Then the measure

$$\mu \equiv \frac{1}{2}(\mu_1 + \mu_2)$$

is locally Markov, but has not the global Markov property (with respect to $\partial_r Q$). Explicit examples based on this idea can be found in [Go] (where the three dimensional Ising model at sufficiently low temperature is considered) and in [Ze1] (where the Gibbs measure on $\mathcal{D}'$ corresponding to the local specification of the free euclidean field in $\mathbb{R}^d$ is considered,see also [GRS1], [HoSt]). A natural conjecture appears to be the following one:

(H)   **GMP** *is implied by* **LMP** *when a local specification satisfies some strong uniqueness condition*

To consider this more closely let us note that for any unbounded set $Q \subset \Gamma$ and any Gibbs measure μ we have

$$E_\mu(F|\Sigma_{Q^c})(\omega) \equiv E_Q^\omega(F) = \lim_{\mathcal{F}_0} E_{(\Lambda \cap Q)}^\omega(F), \tag{0.13a}$$

μ-a. e. (for some countable base $\mathcal{F}_0$).
Suppose now that some kind of "patching together" of configurations in Ω is possible, so that for any $\omega, \omega_o \in \Omega$ we can define a configuration $\tilde{\omega} \equiv \tilde{\omega}(\omega, \omega_0)$ coinciding on Q^c with ω and on $Q \setminus \partial_r Q$ with ω_0, in the sense that for all $\Lambda \in \mathcal{F}$

$$E_{(\Lambda \cap Q)}^{\tilde{\omega}}(F) = \delta_\omega(F) \tag{0.14a}$$

if $F \in \Sigma_{Q^c}$ and

$$E_{(\Lambda \cap Q)}^{\tilde{\omega}}(F) = \delta_{\omega_0}(F) \tag{0.14b}$$

if $F \in \Sigma_{\Lambda^c \cap Q \setminus \partial_r Q}$. Then we see that the **GMP** for a Gibbs measure μ of a Markov specification $\mathcal{E}$ holds, if for any $F \in \Sigma$ and for some configuration $\omega_0 \in \Omega$

$$\lim_{\mathcal{F}_0} E_{(\Lambda \cap Q)}^\omega(F) = \lim_{\mathcal{F}_0} E_{(\Lambda \cap Q)}^{\tilde{\omega}}(F), \qquad \mu - \text{ a. e.} \tag{0.15}$$

with some countable base $\mathcal{F}_0$ independent of F and ω.
One may treat this as a more general case of the uniqueness problem for a restricted local specification $\mathcal{E}_{Q,\omega} \equiv \{E_\Lambda^{\omega'}\}_{\Lambda \in \mathcal{F}, \omega'_{Q^c} = \omega_{Q^c}}$ (for related investigations of such problems in statistical mechanics see e. g. [Do1-5], [Sh]). A first important example for statistical mechanics and field theory which illustrates the above situation is given by taking Γ as a countable set (called a lattice) and the space Ω to be S^Γ for some standard Borel space S (called a single spin space). In particular if S is a finite set resp. a compact Riemannian manifold than one can take the free measure μ_0 to be the product of the

uniform probability measures on S resp. of a normalized Riemannian volume on S. Then taking a <u>Gibbsian interaction</u>

$$\Phi : \mathcal{F} \mapsto \mathcal{C}(\Omega, I\!R)$$

satisfying

$$\|\Phi\| \equiv \sup_{i \in \Gamma} \sum_{\substack{X \in \mathcal{F} \\ i \in X}} \|\Phi_X\| < \infty \tag{0.16}$$

we may define an interaction functional

$$U_\Lambda \equiv \sum_{X \cap \Lambda \neq \phi} \Phi_X \tag{0.17}$$

Using these data we construct a local specification $\mathcal{E} \equiv \{E_\Lambda\}_{\Lambda \in \mathcal{F}}$ via (0.9), with $\mu_{0|\Lambda}$ being conditional expectation of μ_0 with respect to σ-algebra Σ_{Λ^c} generated by the coordinate functions $\{\omega_i : \Omega \to S\}_{i \in \Lambda^c}$. Additionally if Γ is equiped with a metric ρ and the interaction Φ is of finite range $0 \leq r < \infty$, i. e.

$$\Phi_X = 0$$

for all $X \in \mathcal{F}$ with diam $X \equiv \sup_{i,j \in X} \rho(i,j) > r$, then our local specification is a Markov specification. If the norm (0.16) of interaction Φ is sufficiently small then using a high temperature expansion method of statistical mechanics as in [AHKO] or Dobrushin uniqueness method as in [Fö2], one may show that (0.15) is true for any unbounded set $Q \subset \Gamma$ and for all $\omega \in \Omega$, with any $\tilde{\omega}, \tilde{\omega}_i \equiv \omega_i$ for $i \in Q^c$. In this case then one has the global Markov property.(See also [Hi1].)

The other particular case we want to mention is <u>a scalar field theory on the lattice</u> $\Gamma = \mathbb{Z}^d$. In this case $S \equiv I\!R$. For a finite set $\Lambda \in \mathcal{F}$ we define the <u>hamiltonian function</u>.

$$H_\Lambda \equiv -\frac{1}{2} \sum_{i \in \Lambda} \varphi_i(-\Delta\varphi)_i + \sum_{i \in \Lambda} V(\varphi_i) \tag{0.18}$$

where φ_i denotes the ith coordinate function on $\Omega = I\!R^\Gamma$, V is a real valued function which is bounded from below and measurable and Δ is difference Laplacian defined by

$$(-\Delta\varphi)_i := \sum_{i'}(\varphi_{i'} - \varphi_i) \tag{0.19}$$

with summation going over the nearest neighbors of the point $i \in \Gamma$.

With this notation we define a local Markov specification $\mathcal{E} = \{E_\Lambda\}_{\Lambda \in \mathcal{F}}$ by setting

$$E_\Lambda^\omega(\cdot) := \delta_\omega \frac{\int d\varphi_\Lambda \left(e^{-H_\Lambda} \cdot\right)}{\int d\varphi_\Lambda e^{-H_\Lambda}} \tag{0.20}$$

where $d\varphi_\Lambda$ denotes the product of Lebesgue measures $d\varphi_i, i \in \Lambda$.

(For a corresponding definition of field theory on a simplicial complex, necessary for a natural approximation of quantum fields on a Riemannian manifold, see [AZ1,2].) Restricting our local specification to a set $\mathcal{S}'_\Gamma$ consisting of tempered sequences in Ω, one may extend Dobrushin's uniqueness method and show (0.15) for all configurations $\omega \in \mathcal{S}'_\Gamma$ and any unbounded $Q \subset \Gamma$ in the case where the local interaction V is a convex function growing sufficiently fast at infinity, see [BP] and [Ro]. In this case we have also a unique Gibbs measure (see also [COPP]). (Note that the restriction to $\mathcal{S}'_\Gamma$ is important and in general we can get nonuniqueness working in Ω, see e. g. [Ro] for a corresponding example of Gaussian fields). Let us mention that one can check (0.15) directly, i. e. investigating the kernels $E_{\Lambda \cap Q}$, also in some models of euclidean fields on $I\!\!R^2$ (with weak trigonometric [AHK2] or exponential interactions [Ze1]). We will describe these results in more details in the next section.

Let us stress that in all above mentioned examples the corresponding set of Gibbs measures consisted of a unique point, confirming the conjecture (**H**).

Another interesting sufficient condition for **GMP**, extending (0.15), has been given in [Go]. By simple arguments it is shown there that:

If a Gibbs measure μ for a Markov specification $\mathcal{E}$ has a representation of the form
(Ci)

$$\mu = \lim_{F_0} E_\Lambda^{\omega_0} \tag{0.21}$$

for some $\omega_0 \in \Omega$, and, moreover, for an unbounded set $Q \subset \Gamma$ the sequence of measures

$$\mu_{Q,\Lambda,\omega_0}(\cdot) \equiv \mu E_{\Lambda \cap Q}^{\tilde{\omega}}(\cdot) \tag{0.22}$$

where $\tilde{\omega}_{|Q} = \omega_{0|Q}$ and $\Lambda \in \mathcal{F}$, satisfies

(Cii)

$$\lim_{\mathcal{F}_0} \mu_{Q,\Lambda,\omega_0} = \mu \tag{0.23}$$

*then μ has **GMP** (with respect to ∂Q).*

Using this condition one of the authors was able to shown **GMP** also in the interesting case of systems with attractive interactions where more than one extremal Gibbs measure exist [Ze5].

More precisely such a situation can be described as follows (see [Pre]): We take an ordered single spin space S. Then the configuration space $\Omega \equiv S^\Gamma$ possess an induced (measurable) partial order relation $\leq$ and one can define a set $\mathcal{A}^\uparrow$ of increasing measurable functions on (Ω, Σ). For such a system one may consider <u>an attractive local specification</u> $\mathcal{E} = \{E_\Lambda\}_{\Lambda \in \mathcal{F}}$ distinquished by the requirement that for any $\Lambda \in \mathcal{F}$

$$E_\Lambda(\mathcal{A}^\uparrow) \subset \mathcal{A}^\uparrow \tag{0.24}$$

Let us recall two important examples of local Markov specifications which are attractive:
<u>Example 1:</u> The local specification (0.15) of an Ising model on the space $\Omega = \{-1, +1\}^{\mathbb{Z}^d}$ defined with μ_0 being a product of uniform probability measures on $\{-1, +1\}$ and an interaction Φ defined by setting, with a constant $J > 0$

$$\Phi_X = -J \omega_i \omega_j \tag{0.25}$$

if $X = \{i, j\}$ and $i, j \in \mathbb{Z}^d$ are nearest neighbors, and zero for other sets $X \in \mathcal{F}$.

Example 2: The local specification of a euclidean (scalar) field on a lattice given in (0.20). In [Go] the author considers first a system on a lattice Γ with a compact single spin space S, so that there are maximal configurations $\omega_+, \omega_- \in \Omega$ such that

$$\omega_- \leq \omega \leq \omega_+ \tag{0.26}$$

for any $\omega \in \Omega$. Then it is known (see e. g. [Pre]) that for any attractive local specification $\mathcal{E}$ on (Ω, Σ) there exist two (not necessarily different) extremal Gibbs measures $\mu_+, \mu_- \in \partial\mathcal{G}(\mathcal{E})$

$$\mu_\pm = \lim_{\mathcal{F}_0} E_\Lambda^{\omega_\pm} \tag{0.27}$$

which are FKG-maximal in the sense that

$$\mu_-(F) \leq \mu(F) \leq \mu_+(F) \tag{0.28}$$

for any $F \in \mathcal{A}^\uparrow$. The latter relation defines the so called FKG order in the space of measures and it is shortly denoted by

$$\mu_- \underset{FKG}{\leq} \mu \underset{FKG}{\leq} \mu_+ \tag{0.29}$$

The author of [Go] shows that the $\mu_\pm$ states satisfy the condition (C) and therefore they have the global Markov property. This gives another proof of a result first obtained in [Fö2](see also [AFHKL]).

Let us mention that in [Go] it is also shown that this method applies as well to some systems in the continuum (Widom-Rollinson model in $\mathbb{R}^d$) .

An extension of the mentioned results for attractive Markov specifications to noncompact single spin space S, important for euclidean field theory on a lattice, has been worked out in [Ze4].(That such an extension should exist has been earlier observed in [BHK].) In this situation, instead of the maximal configuration ω_+ (0.26), one has to take, for a given Gibbs measure μ, configurations which are "dominating at infinity" with probability one, in the sense that the set

$$\bigcup_{\Lambda_n \in \mathcal{F}_0} \{\omega \in \Omega : \forall \Lambda \in \mathcal{F}_0, \Lambda_n \subset \Lambda \quad \omega_{|\partial\Lambda} \leq \omega_{|\partial\Lambda}^+\} \tag{0.30}$$

has μ-measure one (and similarly for a configuration ω^-). The idea of this extension has been applied afterwards in [AHKZ2] for measures of euclidean fields on the space of (real) tempered distributions $\mathcal{S}'(\mathbb{R}^2)$.

In [Ze4] also other conditions for extremality and **GMP** of a Gibbs measure have been proposed. This resulted in [Ze5] in proving these properties for non-FKG maximal measures of an attractive Markov specification of a spin system on the lattice $\mathbb{Z}^2$. There the following condition (which implies (C)) has been used, with $\mathcal{E} = \{E_\Lambda\}_{\Lambda \in \mathcal{F}}$ being a specification which is both attractive and Markov:

There is $\omega_0 \in \Omega$ such that

$$\mu = \lim_{\mathcal{F}_0} E_\Lambda^{\omega_0} \qquad (0.31)$$

(C') and

$$\lim_{\mathcal{F}_0} \mu E_\Lambda^{\omega_0 \wedge \omega} = \lim_{\mathcal{F}_0} \mu E_\Lambda^{\omega_0 \vee \omega} = \mu \qquad (0.32)$$

where ω denotes integration variable and $\omega_0 \wedge \omega \equiv inf\,(\omega_0, \omega)$ resp $\omega_0 \vee \omega \equiv \sup(\omega_0, \omega)$

Explicit examples of applications of (C') have been obtained by choosing a system on two-dimensional lattice with a ferromagnetic interaction satisfying the assumptions of Pirogov-Sinai theory ([PS], [Sin], [Za]) and ω_0 being a ground state corresponding to a non FKG maximum Gibbs measure. The idea behind (0.32) is that for a pure phase μ associated to a ground state ω_0 the typical configurations $\omega \in \Omega$ are "the small" fluctuations around the ground state and so are the corresponding configurations $\omega \vee \omega_0$ and $\omega \wedge \omega_0$.
(In two dimensions it should be possible to show the result for any pure state of the Pirogov-Sinai theory with arbitrary, finite range interaction. This is probably not true when $d \neq 3$).

Taking into account the fact that the property (Ci) is satisfied by all extremal Gibbs states as well as by all examples considered in [Go] (in fact in all examples known at that time) Goldstein made the following conjecture.

$(\tilde{\mathbf{H}})$
Any state of the form (Ci) and in particular any extremal Gibbs state $\mu \in \partial\mathcal{G}(\mathcal{E})$

*of a Markov specification has the Global Markov Property *.*

Let us note that as shown in [D-NR], generalizing the results of [Spi] and [RoY], this statement is true in one dimensions i. e. for $\Gamma = \mathbb{Z}$ or $\mathbb{R}$.
Despite the large number of examples confirming conjecture $(\tilde{\mathbf{H}})$ it turned out not to be true in general.
Counter examples are discussed:
- in [Ke1-2], where some examples of (± 1)-spins ferromagnetic models on an (essentially) one dimensional lattice were first given. These examples involve "infinite energies" or potentials of unbounded norm, which are nontranslation invariant (see also [AFHKL]);
- in [Is], some explicit models of multicomponent spin systems were exhibited on the three dimensional lattice (with ferromagnetic potentials of bounded norm; in one of the examples translation invariance holds).
(In connection with this discussion see also [W1-2], [Ge].) In both cases the violation of **GMP** for an extremal Gibbs measure appeared together with existence of more than one pure phase. On this basis in [Is] the following (open) question was raised:
Does the global Markov property hold whenever the Gibbs state for a potential Φ is unique?
It seems that the answer to this question is also negative and it should be possible to construct corresponding examples for some models of disordered spin systems with frustration or Czech type models ([Sh]) on a $d \geq 3$ dimensional lattice.

---

* The related more general conjecture that local Markov property plus triviality of the σ-algebra at infinity (tail field) implies **GMP** was disproved by a counterexample of [W1].

Not only the connection of the attractivity property with **GMP**, but also its relation with translation invariance is of big interest. This is so in particular because the property of translation invariance is an intrinsic feature of models of quantum fields. From an example in [Is] we know that even in the presence of attractivity and translation invariance of the interaction an extremal Gibbs measure can suffer from the lack of **GMP**. In the papers [Ku1-3] (see also [Sch]) it has been shown that under some additional conditions (as invertibility of the transfer matrix and reflection invariance of the interaction) one may have **GMP** with respect to hyperplanes for any extremal Gibbs measure of a lattice model. The Potts model is one of the examples covered by this result. These results have been extended in [GoKS]. The proof of these results is not based directly on the local Markov property but uses rather the variational principle which states that equilibrium states minimize the free energy (cfr. [Sch]). The relation between the global Markov property and the absence of some types of symmetry breaking is also discussed in [GoKS]. It would be interesting to continue the investigations in this direction including continuum models **

Up to now we were mainly concerned with the **GMP** problem for pure states. Let us consider also this problem for nonextremal Gibbs measures. For this let us remind that the very first examples of violation of **GMP** given in [Go] used non-pure Gibbs states. In general taking a mixture of extremal global Markov Gibbs measures $\{\mu^{(\alpha)}\}$ we can again obtain a state with **GMP** with respect to some unbounded set $Q \subset \Gamma$ whenever the restrictions $\mu^{(\alpha)}|\Sigma_{\partial_r Q}$ as well as $\mu|\Sigma_Q{}^{(\alpha)}$ are mutually singular for different α's. A particular example with a mixture of FKG-maximal measures for the 3 dimensional Ising model can be found in [Hi1]. For general structural results on Gibbs measures satisfying **GMP** see [Mi] and [Ke3].
Let us mention shortly the corresponding results for Gibbs measures of euclidean scalar fields in two dimensions (a more detailed description will be given in the next section.)
The first proof of extremality and **GMP** has been obtained in [AHK2] for a model with weak trigonometric interactions (see also [GiZ] for corresponding result in a statistical mechanical model). Afterwards the same problem has been solved for the exponential interaction (Høegh-Krohn) model in some particular cases in [Gi1] and the general case in [Ze1] and for strong trigonometric interaction in [Gi4]. The much harder case of polynomial interactions have been considered in [AHKZ2], where a general approach to the problem was presented and an explicit proof was given for the $: \varphi^4 :_2$ model. (For the corresponding extremality problem see also [AHKZ1] and [Gi3].)
There are still a lot of interesting mathematical problems connected with two dimensional models of scalar euclidean fields as far as the structure of Gibbs states and the Markov property are concerned. Let us also mention that a probabilistic approach, in particular the study of **GMP** as well as the extremality problem for higher dimensional scalar field theories, e. g. $: \varphi^4 :_3$ theory, are completely open.

Before closing this section let us add a few

** This has also been stressed by Michael Aizenmann during a colloquium lecture in Princeton by one of the authors. We are grateful to him for this remark.

Remarks

1) Importance of GMP

The importance of the global Markov property (more precisely the Markov property with respect to $d-1$-dimensional hyperplanes) for applications in classical statistical mechanics as well as in quantum field theory is great, for several reasons.

For simplicity let us consider only the latter case, the former being entirely similar. Let Σ_0 be the σ-algebra associated with the "time zero" hyperplane $\{x^0 = 0\}$ in $I\!\!R^d$, writing

$$x = (x^0, \overrightarrow{x}) \in I\!\!R \times I\!\!R^{d-1} \ .$$

The **GMP** of a Gibbs measure μ permits to obtain a symmetric Markov semigroup P_t, $t \geq 0$ in the Hilbert space $L^2(\nu_0)$, where ν_0 is the restriction of μ to Σ_0. We have

$$P_t = E_0 U_t E_0 \ ,$$

where $E_0 \equiv E(\cdot | \Sigma_0)$ is the conditional expectation with respect to μ and Σ_0, looked upon as a projection in $L^2(\mu)$; U_t is the strongly continuous unitary group in $L^2(\mu)$ representing "translations by t", i.e. $(x^0, \overrightarrow{x}) \to (x^0 + t, \overrightarrow{x})$; μ being invariant under the transformation induced on path space by this operation. Let H be the self-adjoint generator of P_t i.e.

$$P_t = e^{-tH} \ , t \geq 0.$$

We have $P_t 1 = 1$ and P_t is a strongly continuous contraction semigroup in all $L^p(\nu_0)$, $1 \leq p < \infty$. By the general theory of Dirichlet forms and Markov semigroups, valid in an abstract measure theoretical setting, see e.g. [ABrR], we know that

$$(u, v) \longrightarrow (H^{\frac{1}{2}} u, \ H^{\frac{1}{2}} v) \ ,$$

for u, v in the domain $D(H^{\frac{1}{2}})$ of $H^{\frac{1}{2}}$ is a Dirichlet form (i.e. a closed bilinear form with a suitable contraction property) (on the right hand side $(\ , \ )$ is the scalar product in $L^2(\mu)$.

Entirely similar results hold in the discrete case, in the latter case e^{-H} is called the transfer matrix, see e.g. [MinS] for a study of spectral properties of this operator (whose existence often was only postulated).

In the case of the Gibbs measures for quantum fields the importance of above connections has been stressed originally as a constructive tool for relativistic quantum fields by Nelson [N1,2] (see also e.g. [Si1,2], [He], [Ch] for further discussions). This is based on the fact that P_t has an analytic continuation to purely imaginary t yielding the unitary time evolution semigroup of time zero quantum fields and yielding for H the interpretation of Hamiltonian for those quantum fields. One has in this case the strongest possible form of causality which can be present in a relativistic theory, the Hilbert space being $L^2(\nu_0)$ (i.e. generated by "time zero fields"), a property called "cyclicity of time zero fields".

The proofs of the Markov property in the situations we mentioned, cfr. also Sect. 1, permit the verification of Nelson's axioms in these situations [AHK1], [Ze1], [AHKZ2].

It should also be mentioned that H is a truly infinite dimensional elliptic operator in the following sense. The above Dirichlet form associated with H can be realized as Dirichlet form $\mathcal{E}$ on the space $\mathcal{S}'(\mathbb{R}^{d-1})$, for which the general theory of Dirichlet forms on Souslin's spaces applies [AMRö]. By construction, the restriction of $\mathcal{E}$ to the space FC_b^∞ of smooth cylinder functions in $L^2(\nu_0)$ coincides with the restriction to FC_b^∞ of the classical Dirichlet form $\mathcal{E}_0$ given by ν_0 in the sense of [ARö1-5]. In fact ν_0 satisfies all assumptions of the theory of classical Dirichlet forms, as initiated in [AHK6] and pursued in [AHK6-13]–[ARö1-5] (and references therein; see also e.g. [Gro1,2], [Ze8], [StZe] for connection with hypercontractive semigroups). In the interaction free case $\mathcal{E}$ even coincides with $\mathcal{E}_0$ on their full domain. In the case of Høegh–Krohn's model (exponential interaction) $\mathcal{E}$ coincides with a maximal Dirichlet form, as described in [AKu] (it is still open whether $\mathcal{E} = \mathcal{E}_0$ in this case, see however [RZh], for this type of uniqueness result for "space–cut–off interactions", see also [ARZh] for some consequences of this, and [Wie], [Tak], [KonT1,2] for other results around this topics).

As a consequence of **GMP** the time zero fields operators are realized as the operator multiplication by the coordinate function, $X(\varphi), \varphi \in \mathcal{S}(\mathbb{R}^{d-1})$, the conjugate momentum $\Pi(\varphi)$ is defined by the natural generator of translations in the directions of $\varphi(\mathbb{R}^{d-1})$ in $L^2(\nu_0)$, and $X(\varphi)$, $\Pi(\varphi)$ satisfy canonical commutation relations (on a dense domain, or in the Weyl form). $X(\varphi), \Pi(\varphi)$ are essentially self–adjoint on smooth cylinder functions, one has a strongly continuous representation of the Poincare Group in $L^2(\nu_0)$. All generators of the Lorentz group have representations in terms of Dirichlet forms. Equations of motion hold in $L^2(\nu_0)$. Furthermore canonical commutation relations hold for the fields and conjugate momenta, and one has a representation of the Lie algebra of the Lorentz group in terms of Dirichlet forms. See [AHK4-13] and especially [AHKR] for details.

As well known in the study of quantum fields (and statistical mechanics) the Markov property is often replaced by some more general property like T-positivity [He] or reflection positivity [Gl], [OS], [FrILS]. In this case the interpretation of the Hilbert space as the space of time zero fields is lost.

2) Other types of Markov fields

All the homogenous Markov fields, on $\mathbb{Z}^d$ or $\mathbb{R}^d$, ordinary or generalized, we have been discussing, were of Gibbs type and scalar and the Markov property we had in mind was the global Markov property, understood in the sense of (0.1) (or in the generalized sense where Σ_0 is replaced by $\Sigma_{\partial\Lambda}$ for any measurable $\Lambda \subset \mathbb{Z}^d$ resp. $\Lambda \subset \mathbb{R}^d$). Let us now make a few comments about work concerned with other types of fields and/or other Markov properties, starting from fields having Markovian properties which are weaker than the global sharp Markov property discussed above.

2.1 Fields with other Markovian properties

The interest in Markovian fields was initiated by Ottaviani (discrete case) and P. Lévy [Lé]. Already in the 40's and 50's it was realized, by Gnedenko and Jadrenko [Ja], that ordinary random fields over $\mathbb{R}^d$ which are homogeneous with respect to the Euclidean group cannot have a sharp Markov property unless they are deterministic. This was discussed in details by Wong in the Gaussian case see e.g. [Wo], [WoH].

Various extensions of the sharp Markov property, have been discussed by Yu. Rozanov in '67 ("L- Markov property") and '71 [Roz1-6] (a "stable Markov Property" with respect to enlargements of state space, also discussed by Mandrekar).

To the best of our knowledge, Urbanik first introduced generalized Markov fields in 1962 [U]. Here in general, σ-algebras have to be defined in a not sharp sense. McKean in 1963 introduced in connection with Lévy Brownian motion [Lé] the concept of ε-Markov property with σ-algebras to $\partial\Lambda$ defined by intersections of σ-algebras of an "enlargement" $\partial\Lambda_\varepsilon$ of $\partial\Lambda$. Kusuoka [Kus2] introduced the concept of zero Markov property and discussed its relation with other concepts, see also [Iw1,2]. Relations between different concepts of Markov property were discussed in several publications, see e.g. [Ac], [Kol], [Roz1-6], [At], [Bu], [Dy], [Es], [KaM], [KoP], [KoO], [Ok],[Kot], [Pi1,2], [Kü], .

Let us now shortly describe another type of homogeneous random fields over $I\!\!R^d$ having a global sharp Markov field with respect to certain regions (not including above hyperplanes however) and with values in the integers (thus an extension of Ising models to the continuum), describing results from a paper by Høegh-Krohn , with Surgailis and myself, which was initiated already in the late fifties but finished only recently ([AHKS]). The Markov fields $\xi(x)(\gamma)$, $x \in I\!\!R^d$, $d \geq 2$ are associated with classical Gibbs states $\nu = \lim_{\Lambda \uparrow I\!\!R^d} \nu_\Lambda$, with

$$dv_\Lambda(\gamma) = Z_\Lambda^{-1} z^{|\gamma|} e^{-\beta V(\gamma)} d\lambda(\gamma) \ ,$$

a Gibbs measure on the set $\mathcal{B}_\Lambda$ of all finite subsets $\gamma \subset \Lambda$, with pair potential

$$V(\gamma) = \sum_{\{x_i,x_j\}\subset\gamma} \Phi(x_i - x_j) \ ,$$

Φ being a real–valued function (of stable type) and $z > 0$ being sufficiently small:

$$Z_\Lambda \equiv \int e^{-\beta V(\gamma)} d\lambda(\gamma) \ .$$

The random field $\xi(x)(\gamma)$ is defined by

$$\xi(x)(\gamma) = \sum_{y\in\gamma} \chi\left(|x - y| \leq r\right) \ ,$$

χ being the indicator function and $r > 0$ a parameter. In particular if Φ has an hard core of radius $0 < r_0 \leq r$, then $\xi(x)(\cdot)$ takes a.s. a finite number $m(r,r_0,d)$ of values. For $r = r_0$ we have $m = 2$ and then $\xi(x)(\cdot)$ is a 0–1–valued (Ising like) field over $I\!\!R^d$. In all cases ξ has the global Markov property in the following sense: for any disjoint closed sets $\Lambda_1, \Lambda_2 \subset I\!\!R^d$ separated by $\Lambda_0 = I\!\!R^d - (\Lambda_1 \cup \Lambda_2)$, the σ–algebras Σ_{Λ_i} $(= \sigma\{\xi(x), \ x \in \Lambda_i\})$ associated with the Λ_i, $i = 1,2$ are conditionally independent given the σ–algebra Σ_{Λ_0}. In the proof the global Markov property of the Gibbs measure ν is used. For $d \geq 3$, ξ has also the sharp global Markov property with respect to a large class of d–1–dimensional hypersurfaces, in the sense that Σ_{Λ_1}, Σ_{Λ_2} are conditionally independent given Σ_{Λ_0} for

$\Lambda_0 = \partial\Lambda_i$ such a $d-1$–dimensional hypersurface. Here however the case where Λ_0 is a hyperplane is excluded, Λ_0 should rather be "spherically irregular" in the sense of being "non flat, non spherical in any neighborhood of its points". The proof uses essentially geometrical facts, see [AHKS] for details.

Let us also mention that a somewhat related class of "geometrical Markov fields" with polygonal realizations was constructed in [ArS] (see also [Ar]) for $d = 2$.

Very different techniques are requested to study non homogeneous Markov fields associated with hyperbolic partial differential equations (wave equations e.g.) and multiparameter martingales and stochastic integrals. Work on this has its roots in investigations by Mc Kean and has been pursued by a number of mathematicians, including Cairoli, Carnal, Dozzi, Guyon, Hajek, Korezioglu, Lefort, Maziotto, Merzbach, Nualart, Pitt, Prum, Rozanov, Russo, Sanz, Walsh, Wong and Zakai, see e.g. [As], [Doz], [DozW], [MeN], [Roz4,5], [Ru1,2], [Fö3], [Ha], [WoZ] and references therein.

2.2 Homogeneous nonscalar Markov fields

The reader should observe that the whole discussion until now, and in fact in all this paper, has concerned scalar- valued fields* . In reality there is a literature also concerning the study of vector - valued Markov fields, although not quite so extensive and systematic (see however [BrFL], [Lev]) . Let us limit ourselves here to mention some results for the case of vector fields over $I\!\!R^d$. Gaussian Markov vector generalized fields have been studied particularly in connection with questions of quantum field theory, especially concerning electromagnetic fields and Proca fields. The proof of the Markov property for latter fields was given by L. Gross [Gro3], who also raised the question of the similar property for free electromagnetic fields. For the latter case the question is delicate because of gauge invariance, the Markov property being sensibly dependent on the choice of test functions and of gauge; see [AIK2], and for previous results [Gu1], [GeP], [Lö], [Ek] This discussion can be looked upon as a particular case of the discussion of when for multicomponent fields, X, Y and a linear constant coefficients differential operator L one has that if X solves $LX = Y$ and Y has a certain Markov property then X has also the same Markov property. This has been discussed by Okabe [Ok] and Kusuoka [Kus2], under certain assumptions, which exclude degeneracy of L. In the degenerate case, which is the one in particular for the free Euclidean electromagnetic field or more generally of the fields we discuss in 1974, the answer has been found by K. Iwata [Iw1,2] (some particular cases had been discussed before , with a proof of a weaker form of Markov property ,in [Su1,2]) . The latter results allow in particular to prove that the class of proper Euclidean generalized random vector fields of non linear electromagnetic type $I\!\!R^4$, first discussed in [AHK14,15], [AHKI], [AIK1,2,4], do indeed possess the global Markov property (with respect to 3-dimensional hyperplanes). This is indeed acknowledge at present the only known class of (non Gaussian, non independent at every point) homogenous global Markov fields. The proof of these results involves essentially the fact that L is of first order, (similarly as in the proof of the Markov property of the free Nelson's field), together with the stability of the test function space under a natural module operation (corresponding to splitting the space according to the complementary regions for which the Markov property is considered), for

---

* We also did not discuss quantum statistical mechanical problems , see e.g. [Ma], [Wic] .

more details we refer to [AIK2], [AIK4]. (See also [Osi1] for related results for fields over $I\!R^8$ and [Osi2], [AIK3] for fields $I\!R^2$.) In the study of additive functionals of these Markov fields, cfr. [AHKI], possibly similar techniques as the one discussed in this paper will have to be used.

Finally let us mention that in connection with gauge fields or more generally random fields associated with extended geometrical objects rather than points other interesting problems arise, which have been discussed in recent years.

In 2 dimensions one can look upon gauge fields as particular cases of Markov cosurfaces [AHKH1-4] , objects which have a natural (sharp, global) Markov property. In fact, the stochastic holonomy associated with the stochastic connection 1-form A of the gauge field, through $\int_\gamma A$, γ being a loop $\gamma = \partial B$, is determined by $\eta(B)$, where η is a stochastic measure with values in the gauge group G. η is a multiplicative noise, hence a fortiori global Markov. A similar situation arises for the Markov cosurfaces associated to $d-1$ dimensional hypersurfaces $d \geq 3$, for G abelian. Also in the non abelian case one can study multiplicative stochastic measures and associated generalised Markov semi- groups. All these quantities have, by construction, global Markov properties. See e.g. [AHKH1-4].

1. Extremality and the Global Markov Property for Euclidean Scalar Fields.

In this section we would like to describe in more details the results and ideas connected to the **GMP** problem of euclidean scalar fields defined by probability measures on the (real) Schwartz space $\mathcal{S}'(I\!R^d)$ (or $D'(I\!R^d)$). Let us note that investigation of such theories in the context of Gibbs measure theory is natural and does not restrict the generality. This is because, as shown in [Kuz], any probability measure on a nuclear space is a Gibbs measure for some local specification.

We will restrict our description mainly to the fields over a two-dimensional space, where the **GMP** and extremality problems are already nontrivial, and where actually we have a quite complete picture. Let us mention that the Gibbsian description of these fields has been proposed first in [GRS1] and [DoM1], [DoM2].

As a starting point we consider a free field theory, given in euclidean space $I\!R^d$ by a Gaussian measure μ_0 with mean zero and covariance operator

$$G \equiv (-\Delta + m^2)^{-1} \tag{1.1}$$

with Δ being the Laplacian and the mass $0 \leq m^2 < \infty$ (assumed to be strictly positive if dimension $d \leq 2$), see [N2] [Si2], [GJ] .

Then is known, see [N2,3] (see also, [Roz1,2], [R1]) that μ_0 has (global) Markov property in the sense that for any open set $\Lambda \subset I\!R^d$ and any $\Sigma_{\bar{\Lambda}}$ - measurable function F we have

$$E_{\mu_0}(F|\Sigma_{\Lambda^c}) = E_{\mu_0}(F|\Sigma_{\partial\Lambda}), \quad \mu_0 - \text{a.e.}, \tag{1.2}$$

where Σ_Λ (resp. Σ_{Λ^c}) are the σ-algebras generated by the coordinate functions $\{\varphi(f) : f \in \mathcal{D}, \text{ supp } f \subset \Lambda\}$, (resp. supp $f \subset \Lambda^c$), completed by μ_0-measure, zero sets, and

$$\Sigma_{\partial\Lambda} := \left\{ \bigcap \Sigma_{\mathcal{O}} : \mathcal{O} \subset I\!\!R^d \text{ open }, \partial\Lambda \subset \mathcal{O} \right\}$$

where $\partial\Lambda$ is the (topological) boundary of Λ.

Moreover the measure μ_0 is an extremal Gibbs measure for a local specification $\mathcal{E}_0 \equiv \{E_\Lambda\}_{\Lambda \in \mathcal{F}}$ defined by

$$E^\eta_\Lambda(F(\varphi)) = \mu_0^{\partial\Lambda}(F(\varphi + \psi_\eta^{\partial\Lambda})) \tag{1.4}$$

where $\mu_0^{\partial\Lambda}$ is a Gaussian measure (with Dirichlet boundary conditions on $\partial\Lambda$) of mean zero and covariance

$$G^{\partial\Lambda} \equiv (-\Delta^{\partial\Lambda} + m^2)^{-1} \tag{1.5}$$

with $\Delta^{\partial\Lambda}$ denoting the Laplace operator with Dirichlet boundary conditions on $\partial\Lambda$, and where $\psi_\eta^{\partial\Lambda}$ denotes a (generalized) solution of the following elliptic Dirichlet problem.

$$\left. \begin{array}{ll} (-\Delta + m^2)\psi_\eta^{\partial\Lambda}(x) = 0 & \text{in } \Lambda \\[2mm] \psi_\eta^{\partial\Lambda}(x) = \eta(x) & \text{"on } \partial\Lambda\text{"} \end{array} \right\} \tag{1.6}$$

A solution $\psi_\eta^{\partial\Lambda}$ of the problem (1.6) has been constructed for the bounded open sets Λ with piecewise smooth boundary in [DoM3] on some Banach space $\mathcal{T}'$ dual to a space $\mathcal{T} \subset H_1$ containing a Poisson kernel of the corresponding problem. In [AHK2] a definition of $\psi_\eta^{\partial\Lambda}$ has been given for $\mu -$ a.a. $\eta \in \mathcal{S}'$, with μ being a (weakly) regular probability measure in the sense that

$$\mu\varphi(f)^2 \leq C\|f\|^2_{-1} \tag{1.7}$$

for some constant $0 < C < \infty$ (independent of $f \in \mathcal{D}$ but possibly dependent on supp f) and

$$\|f\|^2_{-1} \equiv \|G^{\frac{1}{2}}f\|_{L_2}. \tag{1.8}$$

The authors solve first the corresponding classical Dirichlet problem with boundary condition given by

$$\eta_\kappa(x) \equiv \eta(h_\kappa(\cdot - x)) \tag{1.9}$$

for some $h_\kappa \in \mathcal{C}^\infty h_\kappa(x) \to \delta(x)$ weakly as $\kappa \to \infty$. Then they use condition (1.7) to show that the sequence $\psi_{\eta_\kappa}^{\partial\Lambda}$ converges in $L_2(\mu)$ as $\kappa \to \infty$. The generalization of this construction using potential theoretic methods can be found in [R1] (see also [RZ] for corresponding quasilinear Dirichlet problem with boundary data given by a distribution). A complete description of the set of Gibbs measures for $\mathcal{E}_0$ has been given in [HoSt] (see also [R2]) where it was shown that any $\mu \in \partial\mathcal{G}(\mathcal{E}_0)$ has the form

$$\mu F(\varphi) = \mu_0 F(\varphi + \psi) \tag{1.10}$$

with ψ being a harmonic function on R^d corresponding to the elliptic operator $(-\Delta + m^2)$.

To study the fields with nontrivial local interaction we define a local specification $\mathcal{E}_U$ by perturbing $\mathcal{E}_0$ using a multiplicative functional. In two dimensions one may do that as follows : restricting to the bounded "regular" sets $\Lambda \in \mathcal{F}$ we introduce an additive functional

$$U_\Lambda(\varphi) := \int_\Lambda \, : U(\varphi) : (x) d_2 x \qquad (1.11)$$

where U can be a function of various types :

- polynomial interactions (as e.g. in [GJ], [Si2], [DoM1,2], [GRS1])

$$U(x) \equiv \lambda P(x) = \lambda \sum_{k=1}^{2n} a_k x^k; \qquad (1.12)$$

- trigonometric interactions (as e.g. in [AHK3], [Fr1], [FrP], [FrSe], [AHK2])

$$U(x) = \lambda \int dv(\alpha) cos(\alpha x + \vartheta_\alpha) \qquad (1.13)$$

with dv a measure of bounded variation supported in $(-2\sqrt{\pi}, 2\sqrt{\pi})$ (for interactions with $|\alpha| > 2\sqrt{\pi}$ see [Ni]);

- exponential interactions (as e.g. in [AHK1] , [FrP], [Ze1])

$$U(x) = \lambda \int d\nu(\alpha) e^{\alpha x} \qquad (1.14)$$

with a probability measure $d\nu(\alpha)$ supported in $(-2\sqrt{\pi}, 2\sqrt{\pi})$, (for $\sqrt{8\pi} > |\alpha| \geq 2\sqrt{\pi}$ see [Kus]; for $|\alpha| > \sqrt{8\pi}$ see [AHK16], [AGHK], [Osi3]);

- F interactions [E] constructed as in (1.13) but with some more general measures $d\nu(\alpha)$.

The coupling constant λ is assumed to be finite and nonnegative. The double dots : : in (1.11) denote the normal ordering with respect to the free measure μ_0. To construct $\mathcal{E}_U$ we shall consider a probability measure μ on $(\mathcal{D}', \Sigma)$, regular in the sense (more restrictive than (1.7)) that

$$\mu e^{\varphi(f)} \leq e^{C_\Lambda(\|f\|_{-1}^2 + \|f\|_{-1}^p)} \qquad (1.15)$$

for any open bounded $\Lambda \in \mathcal{F}$ with a constant $0 < C_\Lambda < \infty$ independent of $f \in \mathcal{D}$, supp $f \subset \Lambda$. Let us note that, as shown in [AHKZ1], this class contains all the measures associated to the interactions mentioned above. One may see that for suitable subfamily $\mathcal{F}'$ of "regular" sets $\Lambda \in \mathcal{F}$ we have

$$U_\Lambda(\varphi + \psi_\eta^{\partial\Lambda}) \in L_2(\mu_0^{\partial\Lambda} \otimes \mu) \qquad (1.16)$$

and

$$\mu_0^{\partial\Lambda} \left(e^{-U_\Lambda(\varphi + \psi_\eta^{\partial\Lambda})} \right) < \infty \qquad (1.17)$$

for μ - a. a. $\eta \in \mathcal{D}'$. Using this we define $\mathcal{E}_U = \{E_\Lambda\}_{\Lambda \in \mathcal{F}'}$ formally by setting

$$E_\Lambda^\eta(F) = \frac{\mu_0^{\partial\Lambda}\left(e^{-U_\Lambda(\varphi + \psi_\eta^{\partial\Lambda})} F(\varphi + \psi_\eta^{\partial\Lambda})\right)}{\mu_0^{\partial\Lambda}\left(e^{-U_\Lambda(\varphi + \psi_\eta^{\partial\Lambda})}\right)} \tag{1.18}$$

A rigorous construction of the local specification $\mathcal{E}_U$ for polynomial interactions (which extends also to the other cases) has been given in [R2] (see also [AHK2], [Ze1]).

By construction the specification $\mathcal{E}_U$ is a Markov specification. Let us note that this specification is also attractive in the following sense :

For any function of the form

$$F(\varphi) = \hat{F}(\varphi(f_1), ..., \varphi(f_n)) \tag{1.19}$$

with $f_i \in \mathcal{D}$, $f_i \geq 0$, $i = 1, ..., n, (n \in N)$ and $\hat{F}$ a function on $\mathbb{R}^n$ increasing with respect to each coordinate, or any limit of such functions,

(Ai)   If $\omega, \tilde{\omega} \in \mathcal{D}'$, $\omega - \tilde{\omega} \geq 0$ then

$$E_\Lambda^{\tilde{\omega}} F \geq E_\Lambda^\omega F \tag{1.20}$$

for all $\Lambda \in \mathcal{F}$.

(Aii)   Moreover for any $\Lambda \in \mathcal{F}$ the function

$$\mathcal{D}' \ni \eta \longmapsto E_\Lambda^\eta(F) \tag{1.21}$$

can be represented as a limit of a sequence of increasing cylinder functions.

This fact has been shown for the first time in [AHKZ2], where a (converging) lattice approximation was used. The property (Ai) may be viewed as corresponding to "second quantization" of the maximum principle for classical quasilinear partial differential operators. Note that E_Λ^η satisfies the following formal analog of a quasilinear differential equation

$$E_\Lambda^\eta\left((-\Delta + m^2)\varphi(x) + U'(\varphi)(x)\right) = 0 \tag{1.22}$$

Using this idea and the fact that the measure $\mu_0^{\partial\Lambda}$ (defining E_Λ^η) satisfies FKG inequalities [FKG], i. e. inequalities of type

(FKG) $\mu(F, G) \geq 0$ (1.23)

with $\mu(F, G) \equiv \mu(F, G) - \mu(F)\mu(G)$, for any increasing cylinder functions F and G (or limits of such functions), it is possible to show (A) directly in the continuum without invoking the lattice approximation [Ze7]. (For other connections of Gibbs measures and partial differential equations see [Schr], [Ze2] and [Ze3]).

Let us note that the attractivity property (Ai) for the lattice specifications of euclidean fields is sufficient for the (**FKG**) property (see e. g. [GJ]). The usual proof of **FKG** for measures of euclidean fields in the continuum goes via lattice approximation. It would be very interesting to find a proof of **FKG** for these measures directly in the continuum.

Let us leave now for a while the analysis of **FKG** structure (we will come back to this later on), and consider the problem of compactness of a local specification $\mathcal{E}$. This is the question whether the set $\mathcal{G}(\mathcal{E})$ is nonempty and compact. A sufficient condition for compactness of $\mathcal{E} = \{E_\Lambda\}_{\Lambda \in \mathcal{F}}$ on a measurable space $(\mathcal{S}', \mathcal{B})$ similarly as in [BHK] (see also e. g. [Sin]), is the existence of a functional $A_\Lambda(\eta, f)$ such that for any $\Lambda \in \mathcal{F}$ and $f \in \mathcal{D}$

$$E_\Lambda^\eta e^{\varphi(f)} \le e^{A_\Lambda(\eta, f)} \tag{1.24}$$

and for some countable base $\mathcal{F}_0$

$$\lim_{\mathcal{F}_0} |A_\Lambda(\eta, f)| < C(f) < \infty \quad \mu - \text{a. e.} \tag{1.25}$$

with some functional $0 < C(f) < \infty$ independent of $\eta \in \mathcal{S}'$ and for any regular measure μ. A simple example of a compact local specification is given by the free specification $\mathcal{E}_0$ on $(\mathcal{S}', \mathcal{B})$. In this case we have obviously for $E_\Lambda^\eta \in \mathcal{E}_0$ that

$$E_\Lambda^\eta e^{\varphi(f)} = e^{\frac{1}{2}\|f\|^2_{-1,\partial\Lambda} + \psi_\eta^{\partial\Lambda}(f)} \le e^{\frac{1}{2}\|f\|^2_{-1} + \psi_\eta^{\partial\Lambda}(f)} \tag{1.26}$$

where $\|\cdot\|_{-1,\partial\Lambda}$ denotes H_{-1} norm defined with $G^{\partial\Lambda}$ the covariance with Dirichlet boundary conditions on $\partial\Lambda$. (Note however that the specification given by the same formula but on $(\mathcal{D}', \Sigma)$ is noncompact.)

Let us mention that the same compactness estimate remains true for the local specifications corresponding to exponential interactions in two dimensions, as can be seen by a simple application of Brascamp - Lieb inequalities [BrL].

It is also not very hard to get a compactness estimate for the weak coupling trigonometric interactions considered in [AHK2]. (One should expect to have it also for any coupling constant).

It would be interesting to complete this picture by proving the compactness of local specifications also for polynomial interactions.

Let us now review briefly some results connected to the existence, extremality and global Markov property of Gibbs measures for the local specifications which interest us. When we discussed the examples of interactions, we already mentioned some works where (among other things) the problem of existence of the thermodynamic limit of the moments or of the generating functional of finite volume measures μ_Λ with various "classical" boundary conditions [GRS2] is considered.

We recall that the first proof of existence of the infinite volume Euclidean measure has been given in [New], where it was shown, using the cluster expansion of [GJS1], that the restrictions $\mu_{\Lambda|\Sigma_{\tilde\Lambda}}$ of the finite volume measures μ_Λ for a weak coupling $P(\varphi)_2$ interaction to local σ-algebra $\Sigma_{\tilde\Lambda}$, $\tilde\Lambda \in \mathcal{F}$, are absolutely continuous with respect to $\mu_{0|\Sigma_{\tilde\Lambda}}$, μ_0 being a free measure. Moreover the related densities $\dfrac{d\mu_{\Lambda|\Sigma_{\tilde\Lambda}}}{d\mu_{0|\Sigma_{\tilde\Lambda}}}$ were shown to converge in $L_1(\mu_0)$ (see also [AHK], [AHK3] for related work in a regularized model). These results were obtained before the Gibbs theory entered into the domain of euclidean fields, what happend in [AHK], [AHK3], [GRS1] and [DoM1] (see also [DoM2]). A very general result has been given in [FrSi] where it has been shown that any measure μ constructed with a polynomial interaction (via cluster expansion, monotonicity method based on GKS inequalities, and **FKG** inequalities by switching off a large extremal magnetic field with arbitrary "classical" boundary conditions) is <u>ultraregular</u> in the following sense:

For any rectangle $\Lambda \in \mathcal{F}$

$$\mu_{|\Sigma_\Lambda}(\quad \cdot \quad) = \mu_{0|\Sigma_\Lambda} \left(\rho^{\partial\Lambda} e^{-\alpha_\infty |\Lambda|} e^{-U_\Lambda} \quad \cdot \quad \right) \tag{1.27}$$

where α_∞ is the corresponding infinite volume pressure ([GRS1]) and the boundary density $\rho^{\partial\Lambda} \in \Sigma_{\partial\Lambda}$ is a strictly positive function satisfying for any $1 \le p < \infty$

$$\|\rho^{\partial\Lambda}\|_{L_p(\mu_0)} \le e^{c|\partial\Lambda|} \tag{1.28}$$

with some constant $0 < c < \infty$ which can be taken equal to zero if $1 \le p < \frac{4}{3}$.

These results extend also to the other interactions listed earlier. Let us mention also the existence theorem for Gibbs measures given in [R2] with a proof based on application of a general compactness criterion of Ch. Preston (unpublished).

The literature concerning the properties of constructed probability measures defining field theories in two dimensional euclidean space is very rich (see e.g. an extensive list of references in [GJ] and [AHK5]). We would like to mention some of the results. First of all let us mention that it was possible to understand relatively fast and well how to construct the translation invariant ergodic measures by applying various generalizations of methods from statistical mechanics. In particular the construction of a unique measure in a class of translation invariant probability measures was made, either by use of cluster expansion as in [GJS1], [Sp] for polynomial interactions, [FrSe] for trigonometric interactions, (extending early work in [AHK], [AHK3], which first pointed out the connection between Euclidean methods and statistical mechanical methods in this model) or by use of various correlations inequalities as in e.g. [AHK1] [Si2], and application of Lee - Yang type arguments for φ_2^4 models in [GrSi].

For arbitrary polynomial interactions Fröhlich and Simon [FrSi] constructed the states $\mu_\pm$ formally analogous to **FKG** - maximal Gibbs measures of ferromagnetic lattice spin systems (which by results of [LePr], [BHK] [Ze4] are the extremal Gibbs measures).

Additionaly in the same paper it was shown (extending the results of [Si3] and [Gu2]) that for an interaction $P(\varphi) + h_0\varphi$ there exists a unique probability measure which is ergodic with respect to translations, if the corresponding infinite volume pressure $\alpha_\infty(P(\varphi) + h\varphi)$ is differentiable with respect to h at the point $h \equiv h_0$. Let us note that the analogous classical result of [LePr] implies uniqueness of Gibbs measure.

Results concerning the phase transitions in the sense of existence of more that one translation invariant Gibbs measure for a given interaction can be found e.g. in [GJS2], [G] [Fr2], [Im] (see also references therein). The most complete description in this case has been given in [Im], where the Pirogov - Sinai theory of phase transition ([PS], [Sin], [Za]) for lattice spin systems has been essentially extended to the case of the measures of euclidean field theory. (Let us however note that in Pirogov - Sinai theory one can show that all the measures corresponding to "stable" ground states are extremal Gibbs measures. This should be also true for the measures constructed in [Im], however up to now no general proof of this fact exists.)

For a long time there were no results concerning extremality and global Markov property of the constructed Gibbs measures. The first results and important break through has been obtained for models with weak trigonometric interactions in [AHK2]. The authors considered the interactions

$$U_\Lambda(\varphi) = \lambda \int_\Lambda \int d\nu(\alpha) : \cos(\alpha\varphi + \vartheta_\alpha) :_\circ (x) d_2 x \qquad (1.29)$$

with $0 < \lambda < \infty$ small and, $0 \le \vartheta_\lambda < 2\pi$ a ν-measurable function, ν a measure of bounded variation supported in $(-\sqrt{2\pi}, \sqrt{2\pi})$ (and $: \quad :_\circ$ Wick ordering with respect to the free field). Due to the special character of interaction it was possible to apply the cluster expansion of ([GJS]) [FrSe] directly to the conditional measures E_Λ^η. The most important technical ingredients were given by the facts that the cluster expansion converges uniformly with respect to boundary conditions η and that the related solution $\psi_\eta^{\partial\Lambda}$ of the Dirichlet problem, in $\Lambda \in \mathcal{F}$ decays exponentially, locally uniformly to zero i.e. for any $\Lambda_0 \in \mathcal{F}$

$$\lim_{\mathcal{F}_0} \sup_{x \in \Lambda_0} |\psi^{\partial\Lambda}(x)| e^{md(\Lambda_0, \partial\Lambda)} = 0$$

for some $0 < m < 2m_0$ and $\mu - $ a.a. $\eta \in \mathcal{S}', \mu$ being a regular measure (in the weak sense with a constant coefficient of regularity, see below).

To formulate precisely the results of [AHK2] let us denote by $\mathcal{M}_0$ the space of regular measures μ satisfying by definition the inequality

$$\mu\varphi(f)^2 \le c\|f\|_{-1}^2 \qquad (1.30)$$

with a constant $0 < c < \infty$ independent of $f \in \mathcal{S}$.

Theorem 1 :

Let $\mathcal{E}_U \equiv \{E_\Lambda^\eta\}$ be a local specification defined for trigonometric interaction (1.29) by (1.18) with a fixed bare mass $0 < m_0 < \infty$ of the measures $\mu_0^{\partial\Lambda}, \Lambda \in \mathcal{F}$. There is $0 < \lambda_0 < \infty$, dependent only on m_0 and the variation $\int |\nu(d\alpha)|$ of the measure ν, such that for any $0 < \lambda < \lambda_0$ the set $\mathcal{M}_0 \cap \mathcal{G}(\xi)$ consists of a unique extremal Gibbs measure $\mu \in \partial\mathcal{G}(\xi)$. Moreover this measure has the global Markov property with respect to any unbounded (sufficiently regular) sets $Q \subset \mathbb{R}^2$ with piecewise $\mathcal{C}^1$ boundary.

○

Remark: Sufficiently regular means that, for any $\Lambda \in \mathcal{F}$, $U_{\Lambda \cap Q}$ is well defined so that (1.16) and (1.17) are satisfied.

○

The second result concerning extremality and **GMP** for euclidean fields has been presented in [Gi1] for a special class of exponential interactions. The proof of the following result for general exponential interactions has been given in [Ze1].

Theorem 2:

For any exponential interactions

$$U_\Lambda(\varphi) \equiv \lambda \int_\Lambda d_2 x \int \nu(d\alpha) : e^{\alpha \varphi} :_\circ (x) \quad , \Lambda \in \mathcal{F} \tag{1.31}$$

*defined with $0 < \lambda < \infty$ and a probability measure $\nu(d\alpha)$ supported in $(-2\sqrt{\pi}, 2\sqrt{\pi})$, there exists a unique extremal Gibbs measure μ which is regular. Moreover $\mu \in$ **GMP** (in the sense as in Theorem 1).*

○

Analogously as for trigonometric interactions the proof has been obtained by investigating directly the measures E_Λ^η. The basic technical facts used in the proofs are the cluster property of the measures E_Λ^η independent of $\eta \in \mathcal{S}'$, $\Lambda \in \mathcal{F}$, dominated by that of corresponding free measure, and exponential bounds. The first property, which is due to the convexity of interactions, has been obtained by applying GKS inequalities [Gri], [KS]. To obtain the exponential bounds the author applied FKG inequalities [FKG].
(Actually one can give a simpler proof by invoking to Brascamp - Lieb inequalities [BrL] in order to get a simpler exponential bounds, see [Ze6]). The above results should be also true for the case of the exponential interactions considered in [Kus1].
Let us mention that for the trigonometric and for some exponential interactions one may also construct nontempered Gibbs measures. In fact one can show that the structure of the set of Gibbs states on $(\mathcal{D}', \Sigma)$ for trigonometric interactions is the same as that of corresponding free theory. In the case of exponential interactions defined with the measure $d\nu(\alpha)$ supported in $(0, 2\sqrt{\pi})$ (or $(-2\sqrt{\pi}, 0)$) for any nonpositive (resp. nonnegative) harmonic function of the operator $(-\Delta + m_0^2)$, m_0 being a bare mass, one may find an extremal Gibbs measure possessing the global Markov property, see [Ze3].
The most complicated and hardest case to deal with, is the case of polynomial interactions. It has taken a long time to get the corresponding extremality (uniqueness) or **GMP** results for this interactions. Let us first mention the result concerning extremality and uniqueness for weakly coupled polynomial interactions obtained in [AHKZ1]. To formulate it we denote by $\mathcal{M}_{ur}$ a set of ultraregular measures in the sense of (1.27) and (1.28).

Theorem 3:

Let $\mathcal{E} = \{E_\Lambda\}_{\Lambda \in \mathcal{F}}$ be a local specification for polynomial interaction

$$U_\Lambda(\varphi) = \lambda \int_\Lambda : P(\varphi) :_\circ (x) d_2 x \tag{1.32}$$

with $0 < \lambda < \infty$ and a semibounded polynomial P. There is $0 < \lambda_0 < \infty$ such that for any $\lambda \in [0, \lambda_0)$ the set $\mathcal{G}(\mathcal{E}) \cap \mathcal{M}_{ur}$ of ultraregular Gibbs measures for $\mathcal{E}$ consists of a unique extremal Gibbs measure

$$\mu \equiv \lim_{\mathcal{F}_0} E_\Lambda^\circ, \tag{1.33}$$

with a Fisher sequence $\mathcal{F}_0$ and

$$E_\Lambda^\circ \equiv E_{\Lambda|\eta=0}^\eta.$$

$\circ$

The idea of the proof is to consider the quantities

$$\mu |E_\Lambda^\eta F - E_\Lambda^\circ F| \tag{1.34}$$

for $\Lambda \in \mathcal{F}_0$ and a polynomial function F localized in some set $\tilde{\Lambda} \in \mathcal{F}, \tilde{\Lambda} \subset \Lambda$. Using ultraregularity of μ and definition (1.18) of E_Λ we represent it as follows

$$\mu |E_\Lambda^\eta F - E_\Lambda^\circ F| = \tag{1.35}$$

$$= \mu_{0|\Sigma_\Lambda} \left(\rho^{\partial\Lambda} \frac{e^{-\alpha_\infty |\Lambda|}}{Z_\Lambda} \left| \mu_0^{\partial\Lambda} \otimes \tilde{\mu}_0^{\partial\Lambda} e^{-U_\Lambda(\varphi + \psi_\eta^{\partial\Lambda}) - U_\Lambda(\tilde{\varphi})} \left(F(\varphi + \psi_\eta^{\partial\Lambda}) - F(\tilde{\varphi}) \right) \right| \right)$$

with

$$Z_\Lambda \equiv \mu_0^{\partial\Lambda} e^{-U_\Lambda(\varphi)} \tag{1.36}$$

and $\tilde{\mu}_0^{\partial\Lambda}$ being an isomorphic copy of $\mu_0^{\partial\Lambda}$.
Taking a smooth function χ equal to one on $\Lambda_1 \equiv \{x \in \Lambda : d(x, \partial\Lambda) \geq 3\}$ and supported in $\Lambda_2 \equiv \{x \in \Lambda : d(x, \partial\Lambda) \geq 2\}$ we perform a change of integration variables in the measure $\mu_0^{\partial\Lambda}$ by

$$\varphi \longmapsto \varphi' = \varphi + \chi \psi_\eta^{\partial\Lambda} \tag{1.37}$$

Then we apply the cluster expansion to the nonnormalized measure in the absolute value. We observe that all terms in this expansion defined with a Dirichlet countour Γ contained in Λ_2 are equal to zero (due to symmetry of the measure $\mu_0^\Gamma \otimes \tilde{\mu}_0^\Gamma$ and the interaction $U_X(\varphi + (1 - \chi)\psi_\eta^{\partial\Lambda}) + U_X(\tilde{\varphi})$ for $X \subset \Lambda_2$ and antisymmetry of the integrand $F(\varphi) - F(\tilde{\varphi})$. This allows to get the estimates showing that (1.34) converges to zero as $\Lambda \uparrow \mathbb{R}^2$, what implies uniqueness and extremality results.
In the paper [AHKZ2] we developed a general method to study the uniqueness, extremality and **GMP** for Gibbs measures of euclidean fields. Our arguments involve the attractivity property of local specification described earlier and a uniform continuity of local specification in the following sense.

Definition : *A local specification $\mathcal{E} = \{E_\Lambda\}_{\Lambda \in \mathcal{F}}$ is <u>uniformly continuous</u> iff there is a sequence $\{\varepsilon \equiv \varepsilon(\Lambda) : \Lambda \in \mathcal{F}_0\}$ such that*

$$\lim_{\mathcal{F}_0} \left| E_\Lambda^{\psi_\eta^{\partial \Lambda_\varepsilon}} F - E_\Lambda^\eta F \right| = 0, \quad \mu - \text{a.e.} \tag{1.38}$$

for any regular measure μ and any local function F, where

$$\Lambda_\varepsilon \equiv \{x \in \mathbb{R}^2 : d(x, \Lambda) \leq \varepsilon(\Lambda)\} \tag{1.39}$$

o

Let us note that all local specification introduced before satisfy the requirements of this definition.

Another important observation is the fact that for any regular probability measure μ there is a nonnegative sufficiently fastly increasing function ω which <u>dominates at infinity</u> the samples $\eta \in \mathcal{S}'$ with probability one in the sense that the set

$$\Omega_{\omega,\varepsilon} \equiv \bigcup_{\Lambda \in \mathcal{F}_0} \Omega_{\Lambda,\omega,\varepsilon} \equiv \bigcup_{\tilde{\Lambda} \in \mathcal{F}_0} \left\{ \eta \in \mathcal{S}' : \forall \Lambda \in \mathcal{F}_0 \quad \tilde{\Lambda} \subset \Lambda \sup_{x \in \partial \Lambda} \left| \psi_\eta^{\partial \Lambda_\varepsilon}(x) \right| \leq \omega_{|\partial \Lambda} \right\} \tag{1.40}$$

is of μ-measure one.

Let us note that in fact ω and ε do not depend on μ directly, but only through the corresponding parameters c_Λ from definition (1.15). Using these facts, we extended the arguments of [Ze4] for lattice systems and proven the following result

Theorem 4:

Suppose $\mu \in \mathcal{G}(\mathcal{E})$ for an attractive and uniformly continuous specification $\mathcal{E} = \{E_\Lambda\}_{\Lambda \in \mathcal{F}}$ can be represented as

$$\mu = \lim_{\mathcal{F}_0} E_\Lambda^\omega \tag{1.41a}$$

or

$$\mu = \lim_{\mathcal{F}_0} E_\Lambda^{-\omega} \tag{1.41b}$$

with ω dominating at infinity $\mu - $ a.e. Then $\mu \in \partial \mathcal{G}(\mathcal{E})$. If in addition 1.41a) and b) hold simultanously, then μ is the unique regular Gibbs measure for $\mathcal{E}$.

o

To give an idea of the proof let us take a nonnegative nondecreasing function F and consider the case 1.41a); the second case is similar. Then for any $\eta \in \Omega_{\Lambda,\omega,\varepsilon}$, using attractivity of $\mathcal{E}$, we get

$$E_\Lambda^\eta F = E_\Lambda^{\psi_\eta^{\partial \Lambda_\varepsilon}} F + \left(E_\Lambda^\eta F - E_\Lambda^{\psi_\eta^{\partial \Lambda_\varepsilon}} F \right) \leq E_\Lambda^\omega F + \left(E_\Lambda^\eta F - E_\Lambda^{\psi_\eta^{\partial \Lambda_\varepsilon}} F \right) \tag{1.42}$$

Hence by (1.41a) with ω dominating at infinity and by the use of the uniform continuity of the local specification we conclude that

$$\lim_{\mathcal{F}_0} E_\Lambda^\eta F \leq \lim_{\mathcal{F}_0} E_\Lambda^\omega F = \mu F, \quad \mu - \text{a.e} \tag{1.43}$$

which implies extremality of μ. Similar considerations yield also the uniqueness result if 1.41a) and b) are simultanously satisfied.

One may show that assumptions (1.41a) and b) are satisfied for weak trigonometric and exponential interactions. Additionally in [AHKZ2] we have shown that it is possible to apply the above arguments to the φ_2^4 models and by this to show the uniqueness (and extremality) in the weak coupling region and extremality of the (FKG maximal) measures $\mu_\pm$ in the region of phase transition. In this proof we use GHS inequality [GrHSh] to show the existence of a configuration ω dominating at infinity. Moreover, assuming the measures for an interaction $P(\varphi)_2 \pm h_\infty\varphi$ for some sufficiently big $0 < h_\infty < \infty$ (constructed by the cluster expansion [Sp] and proven to be unique and extremal in [AHKZ1]) have the representation 1.41a) ,b)resp., we are able to identify the measures $\mu_\pm$ of $P(\varphi)_2$ interaction constructed in [FrSi] with FKG maximal extremal Gibbs measures. This in turn (using arguments of [FrSi]) implies that at the points of differentiability of the pressure $\alpha_\infty(P(\varphi)_2 + h\varphi)$ with respect to h, we have a unique (and so extremal) Gibbs measure similarly as in the classical picture for lattice fields ([LePr], [BHK], [Pre]). Essentially the same ingredients as before are sufficient to settle the **GMP** problem by arguments generalizing [Ze4], based on an application of Goldstein [Go] criterion for **GMP**. Let us first formulate the following general theorem

Theorem 5:
Suppose $\mathcal{E} = \{E_\Lambda\}_{\Lambda \in \mathcal{F}}$ is an attractive uniformly continuous specification, and let ω be a configuration dominating at infinity μ - a.e. for a unique regular Gibbs measure

$$\mu = \lim_{\mathcal{F}_0} E_\Lambda^{\pm\omega}. \tag{1.44}$$

Suppose additionally that for an unbounded (regular) set $Q \subset \mathbb{R}^2$ and some $|\tilde\omega| \leq \omega$ the sequence

$$\left\{ E_{(\Lambda \cap Q)}^{\eta_{|\partial Q} + \tilde\omega_{|\partial\Lambda}} \right\}_{\Lambda \in \mathcal{F}_0} \qquad \text{converges} \qquad \mu - \text{a.e.}$$

Then μ has the global Markov property with respect to ∂Q.

$$\circ$$

To show the above statement it is sufficient to follow [Go] and prove that a sequence of measures formally defined by

$$\mu_{Q,\Lambda} \equiv \mu E_{(\Lambda \cap Q)}^{\eta_{|\partial Q} + \tilde\omega_{|\partial\Lambda}} \tag{1.45}$$

with some $|\tilde\omega| \leq \omega$, converges to μ as $\Lambda \uparrow \mathbb{R}^2$ i. e.

$$\lim_{\mathcal{F}_0} \mu_{Q,\Lambda} = \mu \tag{1.46}$$

To do that we consider two nondecreasing, bounded and nonnegative functions $F \in \Sigma_Q$ and $G \in \Sigma_{Q^c}$ and proceed as follows. First we use the Gibbs property to write

$$\mu_{Q,\Lambda} GF \equiv \mu\left(G E_{(\Lambda \cap Q)}^{\eta_{|\partial Q} + \tilde{\omega}_{|\partial \Lambda}}(F)\right) = \mu E_{\Lambda}^{\eta}\left(G(\eta') E_{\Lambda \cap Q}^{\eta'_{|\partial Q} + \tilde{\omega}_{|\partial \Lambda}}(F)\right) \tag{1.47}$$

Then we continue by representing the right hand side of (1.47) as

$$\mu E_{\Lambda}^{\psi_{\Lambda}^{\partial \Lambda_\epsilon}}\left(G E_{\Lambda \cap Q}^{\eta'_{|\partial Q} + \tilde{\omega}_{|\partial \Lambda}}(F)\right) + \mu\left(\left(E_{\Lambda}^{\eta} - E_{\Lambda}^{\psi_{\eta}^{\partial \Lambda_\epsilon}}\right)\left(G E_{(\Lambda \cap Q)}^{\eta'_{|\partial Q} + \tilde{\omega}_{|\partial \Lambda}}(F)\right)\right) \tag{1.48}$$

Under the assumption that $E_{(\Lambda \cap Q)}^{\eta'_{|\partial Q} + \tilde{\omega}_{|\partial \Lambda}}(F)$ converges we may use the uniform continuity of the specification to show that the second term in (1.48) converges to zero as $\Lambda \uparrow I\!\!R^2$. To treat the first term in (1.48) we use the arguments based on attractivity of the specification and the fact that by our assumption ω is dominating at infinity μ-a.e. For any $\Lambda \in \mathcal{F}_0$ we have

$$\mu \chi_{\Omega_{\Lambda,\omega,\epsilon}} E_{\Lambda}^{\psi_{\eta}^{\partial \Lambda_\epsilon}}\left(G E_{(\Lambda \cap Q)}^{\eta'_{|\partial Q} + \omega_{|\partial \Lambda}}(F)\right) \leq E_{\Lambda}^{\omega}\left(G E_{(\Lambda \cap Q)}^{\eta'_{|\partial Q} + \tilde{\omega}_{|\partial \Lambda}}(F)\right)$$

$$\leq E_{\Lambda}^{\omega}\left(G E_{(\Lambda \cap Q)}^{\eta'_{|\partial Q} + \omega_{|\partial \Lambda}}(F)\right) = E_{\Lambda}^{\omega}(GF), \tag{1.49}$$

where $\chi_{\Omega_{\Lambda,\omega,\epsilon}}$ is the characteristic function of $\Omega_{\Lambda,\omega,\epsilon}$. We used here (Aii), and in the last step, the compatibility condition of our specification. This by use of (1.44) implies that

$$\lim_{\mathcal{F}_0} \mu_{Q,\Lambda} \underset{\text{FKG}}{\leq} \mu. \tag{1.50}$$

Applying similar arguments with $E_{\Lambda}^{-\omega}$ we obtain also that

$$\mu \underset{\text{FKG}}{\leq} \lim_{\mathcal{F}_0} \mu_{Q,\Lambda}. \tag{1.51}$$

This implies (1.46) and so $\mu \in \mathbf{GMP}$ (with respect to Q).

One may see that the above arguments can be applied "directly" to weak trigonometric and exponential interactions. In [AHKZ2] we considered also weakly coupled symmetric φ_2^4 model and by use of GHS inequalities [GHS] we have shown that (1.44) is true. Then using ultraregularity of the measure μ we have shown by some involved estimates that

$$\lim_{\mathcal{F}_0} \mu\left|\left(E_{\Lambda}^{\eta} - E_{\Lambda}^{\psi_{\eta}^{\partial \Lambda_\epsilon}}\right)\left(G(\eta') E_{\Lambda \cap Q}^{\eta'_{|\partial Q}}(F)\right)\right| = 0 \tag{1.52}$$

By this and similar arguments as before we got that

$$\lim_{\mathcal{F}_0} \mu_{Q,\Lambda} \equiv \lim_{\mathcal{F}_0} \mu E_{(\Lambda \cap Q)}^{\eta_{|\partial Q}} = \mu \tag{1.53}$$

what implies $\mu \in \mathbf{GMP}$ in the case we consider. Essentially the same arguments used to prove (1.52) extend also to the other polynomial models, see [AHKZ2] .

The verification of (1.44) for the other models with polynomial interactions besides the above mentioned φ_2^4 - case, is still an open problem.

Acknowledgements

We are greatly indebted to Raphael Høegh–Krohn with whom we started the whole program of study of Markov fields.

Our understanding of the problems have been also influenced by a number of collaborations over many years, in particular with Jens Erik Fenstad, Roman Gielerak, Christoph Kessler, Shigeo Kusuoka, Tom Lindstrøm, Helge Holden, Koichiro Iwata, Torbjørn Kolsrud, Michael Röckner, Yuri Rozanov, Francesco Russo to whom we are also grateful.

We thank Martin Jarrath, Brigitte Richter, Ursula Weber, Bożena Zegarliński for their skilful help with the setting of the manuscript.

References

[ABrR] Albeverio S., Brasche J., Röckner M.: *Dirichlet forms and generalized "Schrödinger operators"*, pp. 1–42 in Proc. Sønderborg Conf. "Schrödinger operators", Edts. H. Holden, A. Jensen, Lect. Notes Phys. **345**, Springer, Berlin (1989)

[Ac] Accardi L.: *Local Perturbations of Conditional Expectations*, J. Math. Anal. and Appl. **72** (1979), 34–69

[Ai] Aizenmann M.: *Translation invariance and instability of phase coexistence in two–dimensional Ising system*, Commun. Math. Phys. **73**, (1989), 83–94

[AFHKL] Albeverio S., Fenstad J.E., Høegh-Krohn R., Lindstrøm T.: *Nonstandard methods in stochastic analysis and mathematical physics*, Academic Press, New York (1986)

[AGHK] Albeverio S., Gallavotti G., Høegh-Krohn R.: *Some results for the exponential interaction in 2 or more dimensions*, Comm. Math. Phys. **70**, 187–192 (1979)

[AHK] Albeverio S., Høegh-Krohn R.: *Homogeneous random fields and statistical mechanics*, J. Funct. Anal. **19** (1975) 242-272

[AHK1] Albeverio S., Høegh-Krohn R.: *The Wightman Axioms and the Mass Gap for a Strong Interaction of Exponential Type in Two-Dimensional Space-Time*, J. Funct. Anal. **16** (1974), 39–82

[AHK2] Albeverio S., Høegh-Krohn R.: *Uniqueness and the Global Markov Property for Euclidean Fields. The Case of Trigonometric Interactions*, Commun. Math. Phys. **68** (1979), 95–128

[AHK3] Albeverio S., Høegh-Krohn R.: *Uniqueness of the physical vacuum and the Wightman functions in the infinite volume limit for some nonpolynomial interactions*, Commun. Math. Phys. **30** (1973), 171–200

[AHK4] Albeverio S., Høegh-Krohn R.: *Local and global Markoff fields*, **19**, 225–248 (1984)

[AHK5] Albeverio S., Høegh-Krohn R.: *Diffusion fields, quantum fields, and fields with values in Lie groups*, pp 1–98 in "Stochastic Analysis and Applications", Ed. M. A. Pinsky, M. Dekker, New York (1984)

[AHK6] Albeverio S., Høegh-Krohn R.: *Quasi invariant measures, symmetric diffusion processes and quantum fields*, pp. 11–59 in "Proceedings of the International Colloquium on Mathematical Methods of Quantum Field Theory", Editions du CNRS. 1976, (Colloques Internationaux du Centre National de la Recherche Scientifique, No. 248)

[AHK7] Albeverio S., Høegh-Krohn R.: *Dirichlets forms and diffusion processes on rigged Hilbert spaces*, Zeitschrift für Wahrscheinlichkeitstheorie und verwandte Gebiete **40**, 1–57 (1977)

[AHK8] Albeverio S., Høegh-Krohn R.: *Hunt processes and analytic potential theory on rigged Hilbert spaces*, Ann. Inst. H. Poincaré (Probability Theory) **B13**, 269–291 (1977)

[AHK9] Albeverio S., Høegh-Krohn R.: *The global Markov property for Euclidean and lattice fields*, Phys. Letters **84B**, 89–90 (1979)

[AHK10] Albeverio S., Høegh-Krohn R.: *Uniqueness and global Markov Property for Euclidean fields and lattice Systems*, pp.303–329 in "Quantum Fields-Algebras, Processes", Proc. Bielefeld (Ed. L. Streit), Springer-Verlag, Wien 1980.

[AHK11] Albeverio S., Høegh-Krohn R.: *Markov processes and Markov fields in quantum theory, group theory, hydrodynamics and C^*–algebras*, (Lectures given at the London

Math. Soc. Symposium, Durham, July 1980), pp.497–540, in "Stochastic integrals", Ed. D. Williams Lecture Notes in Mathematics 851, Springer Verlag Berlin, 1980

[AHK12] Albeverio S., Høegh-Krohn R.: *Uniqueness of Gibbs states and global Markov property for Euclidean fields, pp. 37–64 in*, Proc. Colloquium "Random fields – rigorous results in statistical mechanics and quantum fields theory", Esztertgom 1979, Edts. J. Fritz. J.L. Lebowitz, D. Szász, Colloquia Mathematica Societatis Janos Bolyai **27**, North Holland (1981)

[AHK13] Albeverio S., Høegh-Krohn R.: *Stochastic methods in quantum theory and hydrodynamics*, (Lectures given at les Houches Workshop Stochastic Equations and Physics, June 1980), pp.193–214 in "New Stochastic Methods in Physics", Ed. C. De Witt-Morette, K.D. Elworthy, Physics Reports **77** (1981)

[AHK14] Albeverio S., Høegh-Krohn R.: *Euclidean Markov fields and relativistic quantum fields from stochastic partial differential equations in four dimensions*, Phys. Letts. bf B177, 175–179 (1986)

[AHK15] Albeverio S., Høegh-Krohn R.: *Construction of interacting local relativistic quantum fields in four space-time dimensions*, Phys. Letts. **B200**, 108–114 (1988),**B 202** 621 Err. Phys. Letts. (1988)

[AHK16] Albeverio S., Høegh-Krohn R.: *Martingale convergence and the exponential interaction in $I\!R^n$*, pp.331–335 in "Quantum Fields-Algebras, Processes", Proc. Bielefeld Int. Colloquium, (Ed. L. Streit), Springer Verlag, Wien, 1980

[AHKH1] Albeverio S., Høegh-Krohn R., Holden H., *Markov processes on infinite dimensional spaces, Markov fields and Markov cosurfaces*, pp. 11–40 in "Stochastic Space–Time Models and Limit Theorems", Ed. L. Arnold, P. Kotelenez, D. Reidel, Dordrecht (1985)

[AHKH2] Albeverio S., Høegh-Krohn R., Holden H., Kolsrud T., *Markov cosurfaces and gauge fields*, pp. 211–231 in Proc. Schladming 84, 23. Intern. Universitätswochen für Kernphysik, Acta Phys. Austr. Suppl. XXVI (1984): "Stochastic Methods and Computer Techniques in Quantum Dynamics, Edts. H. Miller, L. Pittner

[AHKH3] Albeverio S., Høegh-Krohn R., Holden H., Kolsrud T., *Stochastic multiplicative measures, generalized Markov semigroups and group valued stochastic processes and fields*, J. Funct. Anal. **78**, 154–184 (1988)

[AHKH4] Albeverio S., Høegh-Krohn R., Holden H., Kolsrud T., *Representation and construction of multiplicative noise*, J. Funct. Anal. **87**, (1989) 250–272

[AHKH5] S.Albeverio, R. Høegh-Krohn, H. Holden, T. Kolsrud, *Construction of quantised Higgs-like fields in two dimensions*, Phys. Letts. **B 222**, 263–268 (1989)

[AHKI] Albeverio S., Høegh–Krohn R., Iwata K.: *Covariant Markovian random fields in four space–time dimensions with nonlinear electromagnetic interaction*, pp. 69–83 in "Applications of Self–Adjoint Extensions in Quantum Physics" , in Proc. Dubna Conf. 1987, Edts. P. Exner, P. Seba, Lec. Notes in Physics **324**, Springer, Berlin (1989)

[AHKO] Albeverio S., Høegh-Krohn R., Olsen G.: *The Global Markov Property for Lattice Systems*, J. Multiv. Anal. **11** (1981), 599–607

[AHKR] Albeverio S., Høegh-Krohn R., Röckner M.: *Canonical quantum fields and infinite dimensional Dirichlet processes*, in preparation

[AHKS] Albeverio S., Høegh-Krohn R., Surgailis D.: *Some euclidean integer–valued random fields with Markov properties*, in this volume

[AHKZ1] Albeverio S., Høegh-Krohn R., Zegarlinski B.: *Uniqueness of Gibbs States for General $P(\varphi)_2$–weak Coupling Models by Cluster Expansion*, Commun. Math. Phys. **121** (1989), 683–697

[AHKZ2] Albeverio S., Høegh-Krohn R., Zegarlinski B.: *Uniqueness and Global Markov Property for Euclidean Fields: The Case of General Polynomial Interactions*, Commun. Math. Phys. **123** (1989), 377–424

[AIK1] Albeverio S., Iwata K., Kolsrud T.: *Random fields as solution of the homogeneous Cauchy–Riemann equation. I; Invariance and analytic continuation*, Commun. Math. Phys. **132** (1990), 555–580

[AIK2] Albeverio S., Iwata K., Kolsrud T.: *Homogeneous Markov generalized vector fields and quantum fields over 4–dimensional space–time*, SFB 237–Preprint, to appear in Proc. Trento Conf. "Stochastic ppartial differential equations", Edts. G. Da Prato, Tubaro, (1991)

[AIK3] Albeverio S., Iwata K., Kolsrud T.: *Conformally invariant and reflection positive random fields in two dimensions*, Bochum–Stockholm Preprint (1990), to appear in volume dedicated to M. Zakai, Edt. E. Merzbach, Acad. Press, New York (1991)

[AIK4] Albeverio S., Iwata K., Kolsrud T.: *A model of four space–time dimensional gauge fields: reflection positivity for coupled loop variables*, BiBoS–Preprint , to appear in Proc. Liblice Conf. Edts. Dittrich, Exner, World Scient. Singapore (1991)

[AKu] Albeverio S., Kusuoka S.: *Maximality of infinite dimensional Dirichlet forms and R. Høegh–Krohn's model of quantum fields* ,in this volume

[AMRö] Albeverio S., Ma Z. M., Röckner M.: *Dirichlet forms and Markov fields – A report on recent developments*, to appear in M. Pinsky Ed. "Diffusion Processes and Related Problems in Analysis", Birkhäuser, New York (1991)

[ARö1] Albeverio S., Röckner M.: *Classical Dirichlet forms on topological vector spaces - the construction of the associated diffusion process*, Prob. Theory and Rel. Fields **83** (1989) 405-438.

[ARö2] Albeverio S., Röckner M.: *New developments in theory and applications of Dirichlet forms*, in: Stochastic Processes, Physics and Geometry, (eds. S. Albeverio, G. Casati, U. Cattaneo, D. Merlini, R. Moresi) , World Scientific, Singapore, 1990, pp. 27-76.

[ARö3] Albeverio S., Röckner M.: *Classical Dirichlet forms on topological vector spaces - closability and a Cameron-Martin formula*, J. Func. Anal. **88** (1990) 395-436.

[ARö4] Albeverio S., Röckner M.: *Stochastic differential equations in infinite dimensions: solutions via Dirichlet forms*, Prob. Th. Rel. Fields (1991) to appear.

[ARö5] Albeverio S., Röckner M.: *New developments in theory and applications of Dirichlet forms*, pp. in "Stochastic Processes, Physics and Geometry", Proc. 2nd Int. Conf. Ascona–Locarno–Como 1988, Eds. S. Albeverio , G. Casati, U. Cattaneo, D. Merlini, R. Moresi, World Scient. Singapore (1990)

[ARZh] Albeverio S., Röckner M., Zhang T.S.: *Girsanov transform for symmetric diffusions with infinite dimensional space state space* , Bonn Prerint (1990)

[Ar] Arak S.: *A class of Markov fields with finite range*, pp.994–999 in Proc ICM, Berkeley , California , AMS , Providence (1986)

[AS] Arak T., Surgailis D.: *Markov fields with polygonal realizations*, Prob. Th. Rel. Fields **80** (1989), 543–579

[At] Atkinson B.: *On Dynkin's Markov property of random fields associated with symmetric Markow processes*, Stoch. Proc. Appl. **15** (1983), 193–201

[As] Asao Y.: *On random fields on hyperbolic spaces*, Mem. Fac. Lit. Sci., Shimane Univ. Nat. Sci., Matsue, **10** (1976) , 35–42

[Av] Averintsev M. B.:*Description of Markovian random fields by Gibbsian conditional probabilities*, Th. Prob. Appl. **17**, (1972), 20-33

[AZ1] Albeverio S., Zegarlinski B.: *Construction of Convergent Simplicial Approximations of Quantum Fields on Riemannian Manifolds*, Commun. Math. Phys. **132** (1990), 39–71

[AZ2] Albeverio S., Zegarlinski B.: *Some stochastic techniques in quantization, new developments in Markov fields and quantum fields*, in: Stochastic Quantization, (eds. P.H. Damgaard, H. Hüffel, A. Rosenblum), Plenum Press, 1990, pp. 1–19.

[BHK] Bellissard J., Høegh-Krohn R.: *Compactness and the maximal Gibbs State for Random Gibbs Fields on a Lattice*, Commun. Math. Phys. **84** (1982), 297–327

[BP] Bellissard J., Picco P.: *Lattice Quantum Fields: Uniqueness and Markov Property*, CPT–CNRS Marseille, Preprint 1976

[BrFL] Bricmont J., Fontaine J. R., Landau L. J.: *Absence of symmetry breakdown and uniqueness of the vacuum for multicomponent field theories*, Commun. Math. Phys. **64** (1979), 49–72

[BrL] Brascamp H.J., Lieb E.H.: *On extensions of the Brunn-Minkowski and Prékopa-Leindler theorem, including inequalities for log-concave functions and with applications to the diffusion equations*, J. Funct. Anal. **22** (1976), 366–389

[Bu] Buljichev V. A.: *Markov property of the stochastic evolution equation*, Th. Prob. Appl. **31** (1986), 373–378

[Ch] Challifour J.: *Euclidean Field Theory II : Remarks on Embedding the Relativistic Hilbert Space*, J. Math. Phys. **17** (1976), 1889–1892

[Co] Cox T.: *An example of phase transition in countable one–dimensional Markov random fields*, I. Appl. Prob. **14**, (1977), 205-211

[COPP] Cassandro M., Olivieri E., Pellegrinotti A., Presutti E.: *Existence and uniqueness of DLR equations for unbounded spin systems*,Z. Wahr. und ver. Geb. , **41**, (1978), 313-334

[D-NR] Dang-Ngoc N., Royer G.: *Markov Property of Extremal Local Fields*, Proc. Ann. Math. Soc. **70** (1978), 185–188

[Do1] Dobrushin R.L.: *The description of a Random Field by means of Conditional Probabilities and Conditions of its Regularity*, Theor. Prob. Appl. **13** (1968), 201–229 (in Russian) (transl. **13** (1968), 197–224)

[Do2] Dobrushin R.L.: *The Problem of Uniqueness of a Gibbsian Random Field and the Problem of Phase Transition*, Funct. Anal. Appl. **2** (1968), 44–57 (in Russian) (transl. **2**, 302–312, (1968)

[Do3] Dobrushin R.L.: *Prescribing a System of Random Variables by Conditional Distributions*, Theor. Prob. Appl. **XV** (1970), 469–497, (resp. **15**, 3 (1970), 485–486)

[Do4] Dobrushin R.L.: *Gibbs States, Describing Coexistence of Phases of the Three - Dimensional Ising Model*, Usp. Mat. Nauk **XVII** (1972), 619–639 (in Russian) (trans. **17**, (1972), 382-600

[Do5] Dobrushin R.L.: *An Investigation of Gibbsian States for Three-Dimensional Lattice Systems*, Theor. Prob. Appl. **XVIII** (1973), 261–279 (in Russian), (resp. 253–271)

[DoM1] Dobrushin R.L., Minlos R.A.: *Construction of one dimensional quantum field via a continuous Markov field*, Funct. Anal. and its Appl. **7** (1973), 324–325

[DoM2] Dobrushin R.L., Minlos R.A.: *The theory of Markov random fields and its applications to quantum field theory*, in Functional and Probabilistic Methods in Quantum Field Theory **I**, 23–49, Jancewicz B., ed., Acta Univ. Wratisl. **368**, Wroclaw 1976

[DoM3] Dobrushin R.L., Minlos R.A.: *Investigation of the Properties of Generalized Gaussian Random Fields*, in The Problems of Mechanics and Mathematical Physics, Moscow (1976), 117–165 (in Russian), (Sel. Math. Sov. **1** (1981), 215–263)

[Doz] Dozzi M.: *Propriétés markoviennes de processus sur $I\!R^2$*, Ann. Inst. H. Poincaré **BXIX** (1983), 209–221

[DozW] Dozzi M., Walsh J. B.: *On the Markov property of stochastic partial ditterential equations*, Bern Preprint (1990)

[Dy] Dynkin E. B.: *Markov processes and random fields*, Bull. Am. Math. Soc. **3** (1980), 975–999

[E] English H.: *Remarks on Mc Bryan's Convergence proof for $(\alpha^{-4}(\cos(\alpha\varphi) - 1 + \frac{1}{2}\alpha^2\varphi^2))_2$- Quantum Fields*, Reports on Math. Phys. **18** (1980), 387–397

[Ek] Ekhaguere G. O. S.: *A class of Gaussian fields of markov type* , Physica **99A** (1979), 545–568

[Es] Estigneev I. V.: *Random sets and Markov fields*, pp 339–344 in Proc. 1st World Congr. Bernouilli Soc., VNU Science Prep. Utrecht (1988)

[Fö1] Föllmer H.: *Phase Transition and Martin Boundary*, in Sem. de Probabilités IX, Strasbourg 1975, 305–317, Lect. Notes in Math. 465

[Fö2] Föllmer H.: *On the Global Markov Property*, in Quantum Fields - Algebras, Processes, Ed. Streit L., Springer, Wien 1980

[Fö3] Föllmer H.: *Von der Brownschen Bewegung zum Brownschen Blatt: einige neuere Richtungen in der Theorie der stochastischen Prozesse*, in: Perspective in Mathematics, (eds. W. Jäger, J. Moser, R. Rempert), Birkhäuser, Basel 1984, pp. 159-190.

[FKG] Fortuin C.M., Kasteleyn P.W., Ginibre J.: *Correlation Inequalities on Some Partially Ordered Sets*, Commun. Math. Phys. **22** (1971), 89–103

[Fr1] Fröhlich J.: *Classical and quantum statistical mechanics in one and two dimensions: two component Yukawa and Coulomb systems*, Commun. Math. Phys. **47** (1976), 233–268

[Fr2] Fröhlich J.: *Phase transitions, Goldstone boson and topological superselection rules*, Acta. Phys. Austriace Suppl. **XV** (1976), 133–269

[FrILS] Fröhlich J., Israel R., Lieb E., Simon B.: *Phase transition and reflection positivity*, I, Commun. Math. Phys. **62** (1978) 1–34; II, J. Stat. Phys. **22** (1980), 297–347

[FrP] Fröhlich J., Park Y.M.: *Remarks on exponential interactions and quantum sine - Gordon equation in two space-time dimensions*, Helv. Phys. Acta. **50** (1977), 315–329

[FrSe] Fröhlich J., Seiler E.: *The massive Thirring-Schwinger model QED_2. Convergence of perturbation theory and particle structure*, Helv. Phys. Acta. **49** (1976), 889–924

[FrSi] Fröhlich J., Simon B.: *Pure states for general $P(\varphi)_2$ theories; construction, regularity and variational inequality*, Ann. of Math. **105** (1977), 493–526

[G] Gawedzki K.: *Existence of Three Phases for a $P(\varphi)_2$ Model of Quantum Field*, Commun. Math. Phys. **59** (1978), 117–142

[GaW] Gårding L., Wightman A.S.: *Fields as Operator-Valued Distributions in Relativistic Quantum Theory*, Arkiv för Fysik **26** (1964), 129–184

[Ge] Georgii H.-O.: *Gibbs Measures and Phase Transitions*, W. de Gruyter, Berlin, New York 1988

[GeP] Georgescu V., Purice R.: *On the Markov property for the free Euclidean electromagnetic field*, Lett. Math. Phys. **4** (1980), 465–667

[Gi1] Gielerak R.: *Verification of the Global Markov Property in some class of strongly Coupled Exponential Interactions*, J. Math. Phys. **24** (1983), 347–355

[Gi2] Gielerak R.: *Trigonometric perturbation of the Gaussian generalized fields. Small coupling results.*, Ann. Inst. H. Poincaré **48** (1988), 205–228

[Gi3] Gielerak R.: *On the DLR Equation for the $(\lambda : \varphi^4 : +b : \varphi^2 : +\mu\varphi, \ \mu \neq 0)_2$ Euclidean Quantum Field Theory: The Uniqueness Theorem*, Ann. Phys. **189** (1989), 1–28

[Gi4] Gielerak R.: *On the DLR equation for the two-dimensional sine-Gordon model* , J. Math. Phys. **27** (1986), 2892–2902

[GiZ] Gielerak R., Zegarlinski B.: *Uniqueness and Almost Global Markov Property for Regularized Yukawa Gases*, Fortschr. der Phys. **32** (1984), 1–24

[Gl] Glaser V.: *On the equivalence of the Euclidean and Wightman formulation of field theory*, Commun. Math. Phys. **37** (1974), 257–272

[GJ] Glimm J., Jaffe A.: *Quantum Physics: A Functional Integral Point of View*, 2nd ed. Springer 1986

[GJS1] Glimm J., Jaffe A., Spencer T.: *The particle structure of the weakly coupled $P(\varphi)_2$ model and other applications of high temperature expansions*, in: Constructive quantum field Theory (Erice 1973), Velo G. and Wightman A.S. (eds.), Lect. Notes Phys. **25** Springer 1973

[GJS2] Glimm J., Jaffe A., Spencer T.: *A convergent expansion about mean field theory*, I Ann. Phys **101** (1976), 610–630, II Ann. Phys. **101** (1976), 631–669

[Go] Goldstein S.: *Remark on the global Markov property*, Commun. Math. Phys. **74** (1980), 223–234

[GoKS] Goldstein S., Kuick R., Schlijper A.G.: *Entropy and Global Markov Property*, Commun. Math. Phys. **126** (1990), 469–482

[Gri] Griffiths R.B.: *Correlation in Ising ferromagnet*,
I. J. Math. Phys. **8** (1967), 478–483
II. J. Math. Phys. **8** (1967), 484–489
III. Commun. Math. Phys. **6** (1967), 121–127

[Gro1] Gross L.: *Logarithmic Sobolev inequalities*, Amer. J. Math. **97** (1975), 1061–1083

[Gro2] Gross L.: *Logarithmic Sobolev inequalities on loop groups*, Preprint, Cornell (1991)

[Gro3] Gross L.: *The free Euclidean Proca and electromagnetic fields*, pp. 69-82 in Ed. Arthurs, "Functional integration and its applications"

[GrHSh] Griffiths R.B., Hurst C., Shermann S.: *Concavity of magnetization of an Ising ferromagnet in a positive external field*, J. Math. Phys. **11** (1970), 790–795

[GrSi] Griffiths R.B., Simon B.: *The $(\varphi)_2$ field theory as a classical Ising model*, Commun. Math. Phys. **33** (1973), 145–164

[Gu1] Guerra F.: *Local algebras in Euclidean quantum field theory* , pp 13–26, Symp. Math. XX, Academic Press, New York (1976)

[Gu2] Guerra F.: *Exponential Bounds in Lattice Field Theory*, pp. 185–206 in Les Méthodes Mathématiques De La Théorie Quantique Des Champs, (Marseille 1975), Guerra F., Robinson D.W. and Store R. (Eds.) Paris 1976

[GRS1] Guerra F., Rosen L., Simon B.: *The $P(\varphi)_2$ Euclidean quantum field theory as classical statistical mechanics*, Ann. Math. **101** (1975), 111–259

[GRS2] Guerra F., Rosen L., Simon B.: *Boundary conditions for the $P(\varphi)_2$ Euclidean Field Theory*, Ann. Inst. H. Poincare **XXV** (1976) 231–334

[Ha] Hajek B.: *Markov properties of stochastic differential equation*, Proc. J. Ant. Control Conf. 1980, Vol II

[He] Hegerfeldt G.: *From Euclidean to Relativistic Fields and on the notion of Markoff fields*, Commun. Math. Phys. **35** (1974), 155–171

[Hi1] Higuchi Y.: *A Remark on the Global Markov Property for the d-Dimensional Ising Model*, Proc. Japan Acad. **60A** (1984), 49–52

[Hi2] Higuchi Y.: *On the absence of nontranslation invariant Gibbs states for the two-dimensional Ising model*, Proc. Esztergom, Fritz J. et al. (eds.) Coll. Math. Soc. Balyai J., Amsterdam, North Holland 1979

[HoSt] Holley R., Stroock D.W.: *The DLR conditions for translation invariant Gaussian measures on $S'(I\!R^d)$*, 2. Wahr. Verw. Geb. **53** (1980), 293–304

[Im] Imbrie J.Z.: *Phase diagrams and cluster expansion for low-temperature $P(\varphi)_2$ models. I. The phase diagram, II. The Schwinger function*, Commun. Math. Phys. **82** (1981), 261–304, 305– 343

[Is] Israel R.B.: *Some examples concerning the global Markov property*, Commun. Math. Phys. **105** (1986), 669–673

[Iw1] Iwata K.: *On linear maps preserving Markov properties and applications to multi-component generalized random fields*, Ph. D. Thesis, Bochum (1990)

[Iw2] Iwata K.: *The inverse of a local operator preserves the Markov property*, SFB 237– Preprint, Bochum (1991)

[Iw3] Iwata K.: *Reversible measures of a $P(\Phi)_1$–time evolution*, pp. 195–209, Taniguchi Symp. PMMP, Katata 1985

[Ja] Jadrenko M. I.: *Isotropic Gaussian random fields of Markov type on a sphere*, Dokl. Akad. Nauk. URSR. (1959), 231-236, (Ukr.)

[JSt] Jakóbczyk L., Strocchi F.: *Euclidean Formulation of Quantum Field Theory without Positivity*, Commun. Math. Phys. **119** (1988), 529–541

[KaM] Kallianpur G., Mandrekar V.: *Markov Property for Generalized Gaussian Random Fields*, Ann. Inst. Fourier (Grenoble) **24** (1974), 143

[KS] Kelley D.G., Sherman S.: *General Griffith's inequalities on correlations in Ising ferromagnets*, J. Math. Phys. **9** (1969), 466–484

[Ke1] Kessler Ch.: *Attractiveness of interactions for binary lattice systems and the global Markov property*, Stoch. Proc. Appl. **24** (1987), 309–313

[Ke2] Kessler Ch.: *Examples of extremal lattice fields without the global Markov property*, Publ. RIMS **21** (1985), 877–888

[Ke3] Kessler Ch.: *Nonstandard conditions for the global Markov property for lattice fields*, Acta Appl. Math. **7** (1986), 225-256

[Ke4] Kessler Ch.: *Markov type properties for mixtures of probability measures*, Prob. Th. Rel. Fields **78** (1988), 253–259

[Kes] Kesten H.: *Existence and uniqueness of countable one dimensional random fields*, Ann. Prob. **4** (1976), 557–569

[Kl] Klein A.: *Stochastic process associated with quantum systems*, pp. 329–338 in "Stochastic Methods in Physics", Edts. C. De Witt–Morette, K. D. Elworthy, Phys. Repts. **77** (1981)

[Kol] Kolsrud T.: *On the Markov property for certain Gaussian random fields*, Prob. Th. Rel. Fields **74** (1986), 393–402

[KonT1] Kondrat'ev Jv. G., Tsycalenko T. V.: *Infinite–dimensional hyperbolic equations corresponding to Dirichlet operators*, Sov. Math. Dokl. **33** (1986),775-778

[KonT2] Kondrat'ev Jv. G., Tsycalenko T. V.: *Infinite–dimensional Dirichlet operators*, Bielefeld/Kiev Preprint, in preparation

[KoO] Kotani S., Okabe Y.: *On a Markovian property of stationary Gaussian processes with a multidimensional parameter*, pp. 153–163 in "Hyperfunctions and Pseudodifferential Equations", Lect. Notes Maths. **287**, Springer (1975)

[KoP] Kotecky R., Preiss D.: *Markoff property of generalized random fields*, 7th Winter School on Abstract Analysis, Math. Inst. of Cz. Acad. of Sciences, Praha 1979

[Kot] Kotani S.: *On a Markov property for stationary Gaussian processes with a multidimensional parameter*, pp. 239–250, In: Proc. 2nd Japan-USSR Symp., Lec. Notes Math. **330**, Springer 1974

[Ku1] Kuik R.: *The Global Markov Property for Equilibrium States which are Determined by Correlations in a Strip*, Commun. Math. Phys. **115** (1988), 177–189

[Ku2] Kuik R.: *Markov and Stability Properties of Equilibrium States for Nearest-Neighbour Interactions*, Commun. Math. Phys. **115** (1988), 529–552

[Ku3] Kuik R.: *On the q–state Potts-model by means of noncomutative algebras*, Groningen Ph. D. Thesis (1986)

[Kus1] Kusuoka S.: in this volume

[Kus2] Kusuoka S.: *Markov fields and local operators*, J. Fac. Sci. Univ. Tokyo Sec IA 26 (1979), 199–212

[Kü] Künsch H.: *Gaussian Markov random fields*, J. of the Fac. of Sciences Univ. of Tokyo Sec. IA, **26** (1979), 53–73

[Kuz] Kuznetzov C.E.: *Construction of Consistent Conditional Distributions*, Sec. I, 4.3 in Rozanov Yu.A., Markov random fields, Springer 1982

[LR] Lanford III O.E., Ruelle D.: *Observable at infinity and states with short range correlations in statistical mechanics*, Commun. Math. Phys. **13** (1969), 194–215

[LePr] Lebowitz J.L., Presutti E.: *Statistical Mechanics of Systems of unbounded Spins*, Commun. Math. Phys. **50** (1976), 195–218

[Lev] Levin S. L.: *Application of Dobrushin's uniqueness theorem to N-vector models*, Commun. Math. Phys. **78** (1980), 65–74

[Lé] Lévy P.: *A special problem of Brownian motion, and a general theory of Gaussian random functions*, Proc. 3. Berkeley Symp. Math. Stat. Prob. **2** (1956), 357–378

[Lö] Löffelholz J., *The Markov property of the free Euclidean electromagnetic field*, Karl Marx Universität (1982)

[Ma] Matsui T.: *Uniqueness of the translationally invariant ground state in quantum spin systems*, Commun. Math. Phys. **126** (1990), 453–467

[Me] Merlini D.: *Boundary conditions and cluster property in two–dimensional Ising ferromagnets*, J. Stat. Phys. **21** (1979), 739–745

[MeN] Merzbach E., Nualart D.: *Markov properties for point process on the plane*, Ann. Prob. **18** (1990), 342–358

[Mi] Miyamoto M.: *Phase transition in one dimensional Ising models with spatially in homogenous potentials*, J. Math. Kyoto Univ. **24**, (1984), 679–688

[MinS] Minlos R. A., Sinai Ya. G.: *Investigation of the specification of stochastic operators arising in lattice models of a gas*, Theor. Math. Phys. **2** (1970), 167–176

[Mo] Molchan G.M.:*Characterization of Gaussian fields with Markovian property*, Dokl. Akad. Nauk. SSSR **197** 784–787, Translation: Sov. Math. Dokl. **12** (1971), 563–567

[N1] Nelson E.: *The construction of quantum fields from Markov fields*, J. Funct. Anal. **12** (1973), 97–112

[N2] Nelson E.: *The free Markov field*, J. Funct. Anal. **12** (1973), 211-217

[N3] Nelson E.: *Quantum fields and Markov fields*, in "Partial Differential Equations", Ed. D. C. Spencer, Providence (1973)

[New] Newman C.: *The construction of stationary two-dimensional Markoff fields with an application to quantum field theory*, J. Funct. Anal. **14** (1973) 44–61

[Ni] Nicolò F.: *On the massive sine-Gordon equation in the higher regions of collapse*, Commun. Math. Phys. **88** (1983), 581–600

[O'CO] O'Corroll M., Otterson P.: *Some aspects of Markov and Euclidean field theories*, Commun. Math. Phys. **36**, (1974), 37–58

[Ok] Okabe Y.: *On the germ field of stationary Gaussian processes with Markovian property*, J. Math. Soc. Japan **28** (1976), 86–95

[Osi1] Osipow E.P.: *Markovian properties of solutions of stochastic partial differential equations in a finite volume*, this volume

[Osi2] Osipow E.P.: *Two-dimensional random fields as solutions of stochastic differential equations*, Bochum Preprint (1989)

[Osi3] Osipow E.P.: *On triviality of the :* $\exp \lambda\phi$ *:$_4$ quantum field theory in a finite volume*, Rep. Math. Phys. **20** (1984) 111-116

[OS] Osterwalder K., Schrader R.: *Axioms for Euclidean Green's functions I,II*, Commun. Math. Phys. **31** (1973), 83–112, **42** (1975), 281–305

[Pa] Papangelou F.: *On the absence of phase transition in continuous one–dimensional Gibbs systems with no hard core*, Probab. Th. rel Fields **74** (1987), 485–496

[Pi1] Pitt L. D.: *A Markov property for Gaussian processes with a multidimensional parameter*, Arch. Rat. Mech. Anal. **43** (1971), 367–391

[Pi2] Pitt L. D.: *Stationary Gaussian–Markov fields on $\mathbb{R}^d$ with deterministic component*, J. Math. Anal. **5** (1975), 300–313

[PrK] Preis D., Kotecky R.: *Markov property of generalized random fields*, VII Winter School, Czechoslovakia (1979)

[PS] Pirogov S. A., Sinai Ya. G.: *Phase diagrams of classical lattice systems*, (Transl.: Theor. Math. Phys. **25** (1975), 1185–1192) II.: Theor. Mat. Fiz. **26** (1976), 61–76 (Transl.: Theor. Math. Phys. **26** (1976), 39–49)

[Pre] Preston C.: *Random fields*, Lec. Notes in Math. 534, Springer 1976

[R1] Röckner M.: *Markov Property of Generalized Fields and Axiomatic Potential Theory*, Math. Ann. **264** (1983), 153–177

[R2] Röckner M.: *Specifications and Martin Boundaries for $P(\varphi)_2$-Random Fields*, Commun. Math. Phys. **106** (1986), 105–135

[R3] Röckner M.: *Generalized Markov fields and Dirichlet forms*, Acta Appl. Math. **3** (1985), 285–311

[R4] Röckner M.: *Dirichlet problem for distributions and specifications for random fields*, Mem. AMS **324** (1985), 76 pp.

[R5] Röckner M.: *Specifications and Martin boundary for $P(\phi)_2$ random fields*, Comm. Math. Phys. **106** (1986) 105-135.

[R6] Röckner M.: *Traces of harmonic functions and a new path space for free quantum fields*, J. Funct. Anal. **79** (1988), 211–247

[RZ] Röckner M., Zegarlinski B.: *The Dirichlet problem for quasilinear partial differential operators with boundary data given by a distribution*, In Stochastic Processes and their Applications in Mathematics and Physics, Edts. S. Albeverio, Ph. Blanchard, L. Streit; Dortrecht: Reidel 1990

[RZh] Röckner M., Zhang I. S.: *On uniqueness of generalized Schrödinger operators and applications*, J. Funct. Anal. (1991)

[Ro] Royer G.: *Etude des champs euclidiens sur un réseau $\mathbb{Z}^\nu$*, J. Math. Pures Appl. **56** (1977), 455–478

[RoY] Royer G., Yor M.: *Représentation intégrale de certaines mesures quasi-invariantes sur $C(\mathbb{R})$; measures extrémales et propriété de Markov*, Ann. Inst. Fourier **26** (1976), 7–24

[Roz1] Rozanov Yu. A.: *Markov random fields and stochastic partial differential equations*, Math. USSR - Sb. **32** (1977), 515–534

[Roz2] Rozanov Yu. A.: *Stochastic Markovian fields*, Developments in statistics, Vol.2, pp. 204–234, Academic Press 1979

[Roz3] Rozanov Yu. A.: *Markov random fields*, Springer, New York (1982)

[Roz4] Rozanov Yu. A.: *Boundary problems for stochastic partial differential equations*, pp. 233–267 in "Stochastic Processes – Mathematics and Physics II, Eds. S. Albeverio, Ph. Blanchard. L. Streit; Lect. Notes Math. **1250**, Springer, Berlin (1987)

[Roz5] Rozanov Yu. A.: *Some functional models for random fields*, Chapel Hill Preprint (1988)

[Roz6] Rozanov Yu. A.: *On the paper "Markov random fields and stochastic partial differential equations"*, Math. Sb. ß5 (1979), 157-164

[Ru1] Russo F.: *Champs markoviens et prediction*, Thèse, EFP Lausanne (1987)

[Ru2] Russo F.: *Linear extrapolation concerning Hilbert valued planar functions*, SFB 237 – Preprint (1990)

[Sch] Schlijper A. G.: *Exact variational methods and cluster–variation approximations*, J. Stat. Phys. **35**, (1984), 285–301

[Schr] Schrader R.: *Nonlinear boundary value problems in quantum field theory* Berlin Preprint (1988)

[Sh] Shlosman S.B.: *Uniqueness and half–space nonuniqueness of Gibbs states in Czech models*, Theor. Math. Phys. **66** (1989), 248–293

[Sin] Sinai Ya.G.: *Theory of Phase transitions: rigorous results*, Pergamon Press, Oxford 1982

[Si1] Simon B.: *Positivity of the Hamiltonian Semigroup and the Construction of Euclidean Region Fields*, Helv. Phys. Acta **46** (1973), 686–696

[Si2] Simon B.: *The $P(\varphi)_2$ Euclidean (Quantum) field theory*, Princeton Univ. (1974)

[Si3] Simon B.: *Correlation inequalities and the mass gap in $P(\varphi)_2$, II. Uniqueness of the vacuum for a class of strongly coupled theories*, Ann. of Math. **101** (1975), 260–267

[Sl] Slawny J.: *Low-temperature Properties of Classical Lattice Systems: Phase Transitions and Phase Diagrams*, pp 127–205 In: Phase Transitions Vol. 11, Domb D., Lebowitz J.L. (eds.) Academic Press 1985

[Sp] Spencer T.: *The mass gap for the $P(\varphi)_2$ quantum field model with a strong external field*, Commun. Math. Phys. **39** (1974), 63–76

[Spi] Spitzer F.: *Phase transition in one-dimensional nearest neighbour systems*, J. Funct. Anal. **20** (1975), 240 254

[Su1] Surgailis D.:*On covariant stochastic differential equations and Markov property of their solution*, Ist. Fisico Univ. Roma, Preprint (1979)

[Su2] Surgailis D.: *On the Markov property of a class of linear infinitely divisible fields*, Z. Wahr. verv. Geb. **49** (1979), 293–311

[SZ1] Stroock D.W., Zegarlinski B.: *The Logarithmic Sobolev Inequality for Continuous Spin Systems on a Lattice* , J.Func.Anal. (1991) to appear

[SZ2] Stroock D.W., Zegarlinski B.: *The Equivalence of the Logarithmic Sobolev Inequality and the Dobrushin – Shlosman Mixing Condition*, Commum. Math.Phys. (1991) to appear

[Tak] Takeda M: *On the uniqueness of Markov self–adjoint expansions of diffusion operators on infinite dimensional spaces*, Osaka J. Math. **22** (1983), 733–742

[U] Urbanik K.: *Generalized stationary process of Markovian character*, Studia Math. **3** (1962) 261–282

[W1] Weizsäcker V.H.: *A simple example concerning the global Markov Property of lattice random fields*, 6th Winter School on Abstract Analysis Math. Inst. of the Cz. Acad. of Sciences, Praha 1980

[W2] Weizsäcker V.H.: *Exchanging the order of taking suprema and countable intersections of σ-algebras*, Ann. Inst. H. Poincaré **BXIX** (1983), 91–100

[Wic] Wick D.: *Convergence to equilibrium of the stochastic Heisenberg model*, Commun. Math. Phys. **81** (1981), 361–377

[Wie] Wielens N.: *The essential self–adjointness of generalized Schrödinger operators*, J. Funct. Anal. **61**, 98–115 (1985)

[Wig] Wightman A.S.: *Quantum Field Theory in Terms of Vacuum Expectation Values*, Phys. Rev. **101** (1956), 860–866

[Win] Winkler G.: *The number of phases of homogenous Markov fields with finite spaces on N and Z and their behaviour at infinity*, Math. Nachr. **104** (1981), 101–112

[Wo] Wong E.: *Homogenous Gauss–Markov random fields*, Ann. Math. Stat. **40** (1969), 1625-1634

[WoH] Wong E., Hajek B.: *Stochastic Processes in Engineering Systems*, Springer (1985)

[WoZ] Wong E., Zakai M.: *Isotropic Gauss Markov current*, Prob. Th. Rel. Fields **82** (1989) 137–154

[Za] Zahradnik M.: *An Alternative Version of Pirogov-Sinai Theory*, Commun. Math. Phys. **93** (1984), 559–581

[Ze1] Zegarlinski B.: *Uniqueness and the Global Markov Property for Euclidean Fields: The Case of General Exponential Interaction* , Commun. Math. Phys. **96** (1984), 195–221

[Ze2] Zegarlinski B.: *The Gibbs measure and partial differential equations*, I. Commun. Math. Phys. **107** (1986), 411–429

[Ze3] Zegarlinski B.: *The Gibbs measure and partial differential equations II. The global aspects*, (unpublished).

[Ze4] Zegarlinski B.: *Extremality and the global Markov property I: The Euclidean Fields on Lattice*, J. Multiv. Anal. **21** (1987), 158–167

[Ze5] Zegarlinski B.: *Extremality and the global Markov property II: The Global Markov Property for Non-FKG maximal Gibbs Measures*, J. Stat. Phys. **107** (1986), 570–588

[Ze6] Zegarlinski B.: *On extremality and GMP for exponential models*, unpublished manuscript (1988)

[Ze7] Zegarlinski B.: unpublished manuscript (1989)

[Ze8] Zegarlinski B.: *Dobrushin Uniqueness Theorem and Logarithmic Sobolev Inequalities*, J.Funct.Anal. (1991) to appear

Hypercontractivity:
A Bibliographic Review

E.Brian Davies,[a] Leonard Gross[b] and Barry Simon[c]

a. Department of Mathematics, King's College, The Strand, London WC2R 2LS, England

b. Department of Mathematics, White Hall, Cornell University, Ithaca, NY 14853

c. Division of Physics, Mathematics, and Astronomy, California Institute of Technology, 253-37, Pasadena, CA 91125; Research partially funded under NSF grant number DMS-8801918.

370

§1. Overview and Self-Adjointness

For the past twenty years, a set of ideas known as hypercontractivity has played a continuing role in analysis with ramifications in quantum field theory, self-adjointness of Schrödinger operators, best constants in classical inequalities and bounds on semigroup kernels. A significant role was played by a paper of Raphael Hoegh-Krohn and one of us [98] which codified previous work and coined the term "hypercontractive." Our goal here is to give a brief historical review and a rather complete bibliography.

In looking at the history, one must bear in mind that for several papers there were lengthy delays between submission and publication, roughly two years for Simon and Hoegh-Krohn [98] and Gross [48].

The theory of hypercontractive semigroups was introduced in a fundamental paper of Nelson [74], who also discovered the simplest and most basic example.

<u>Definition</u> Let (Ω, μ_0) be a probability measure space. $H_0 \geq 0$ is a self-adjoint operator on $L^2(\Omega, \mu)$. We say that e^{-tH_0} ($t \geq 0$) is a <u>hypercontractive</u> <u>semigroup</u> if and only if:

(a) e^{-tH_0} is a contraction on $L^2(\Omega, \mu_0)$ for all $t > 0$.

(b)   For some T, e^{-TH_0} is a bounded map from $L^2(\Omega, \mu_0)$ to $L^4(\Omega, \mu_0)$.

General principles (interpolation and duality) imply that e^{-tH_0} is then a contraction from any L^p to itself and bounded from any L^p to any L^q ($1 < p, q < \infty$) if t is sufficiently large (depending on p,q).

<u>Example (Nelson[74])</u> Let $H = -\frac{1}{2}\frac{d^2}{dx^2} + \frac{1}{2}x^2 - \frac{1}{2}$, the harmonic oscillator on $L^2(\mathbb{R}, dx)$. Let $\Omega_0 = (\pi)^{-1/4} \exp\left(-\frac{1}{2}x^2\right)$ be the unique positive unit vector with $H\Omega_0 = 0$ and

consider

$$H_0 = \Omega_0^{-1} \, H\Omega_0 = \tfrac{1}{2}\left(-\frac{d}{dx^2} + 2x\frac{d}{dx}\right).$$ Then e^{-tH_0} is hypercontractive on $L^2(\mathbb{R}, \Omega_0^2 \, dx)$.

Nelson applied this result to deduce semiboundedness of certain cutoff quantum field Hamiltonians. Glimm [46] made an important observation: Suppose that e^{-tH_0} is hypercontractive. If $H_0 1 = 0$, zero is a simple eigenvalue and for some $m > 0$: $\sigma(H_0) \subset \{0\} \cup [m, \infty)$, then e^{-TH_0} is actually a contraction from L^2 to L^4 for T sufficiently large. This allows one to extend Nelson's example to the second quantization of any strictly positive operator (see [98]).

Nelson implemented his semiboundedness proof by making extensive use of path integrals. An alternative proof of semiboundedness was given by P. Federbush [44] based on differentiating the already established hypercontractive inequality $\|e^{-tH}f\|_{p(t)} \leq \|f\|_2$ with respect to t at $t = 0$. This yielded the first log Sobolev inequalities, providing a precursor to the work we will discuss in §2.

Segal [90,91,92] studied an abstract version of the theory, and in particular showed that essential self-adjointness followed from the same L^p properties that Nelson used. Simon - Hoegh-Krohn [98] codified and extended this work, and, in particular show that :

<u>Theorem 1</u>: Suppose that e^{-tH_0} is a hypercontractive semigroup on $L^2(\Omega,\mu)$. Let V be a function on Ω (and also the associated multiplication operator). Suppose that $e^{-V} \in \bigcap_{p<\infty} L^p$ and $V \in L^p$ for some $p>2$. Then $H_0 + V$ is bounded from below and essentially self-adjoint on $D(H_0) \cap D(V)$. The same result is true if $V \geq 0$ and $V \in L^2$.

At the time of this work, self-adjointness of the spatially cut off quantum field Hamiltonian was important in the construction of infinite volume asymptotics of the dynamics. It was first proven for $(\phi^4)_2$ field theories by Glimm-Jaffe [47] using different methods and then by Rosen [83] and Segal [91] for general $P(\phi)_2$ theories. Segal used variants of the above theorem. Hoegh-Krohn [54] applied it to the $: \exp(\alpha\phi) :$ interaction.

A second example of some historical significance is in the Simon-Hoegh-Krohn paper [98]. Let $V \geq 0$ be in $L^2(\mathbf{R}^d, e^{-x^2}dx)$. Then, by the above theorem and a small additional argument, $-\Delta + x^2 + V(x)$ is essentially self-adjoint on $C_0^\infty(\mathbf{R}^d)$. This is of interest because prior to this, all such theorems had required V to be locally L^p with $p \geq d/2$ if $d \geq 4$. Simon [96] showed how to get rid of the x^2 and prove $-\Delta + V$ essentially self adjoint on C_0^∞ if $V \geq 0$ in $L^2(\mathbf{R}^d, e^{-ax^2}dx)$ some $a>0$. Motivated by this, in a celebrated work, Kato [62], using different methods, showed that $V \in L^2_{loc}(\mathbf{R}^d)$ suffices if $V \geq 0$.

It is interesting that the self-adjointness result is now mainly of historical interest. Kato's work has replaced any application to Schrödinger operators. And, because of the Euclidean revolution in quantum field theory, self-adjointness became irrelevant, although the L^p estimates associated to hypercontractivity are still significant; see Guerra et al. [52,53].

§2 <u>Logarithmic</u> <u>Sobolev</u> <u>Inequalities</u> <u>and</u> <u>hypercontactivity</u>

Some aspects of these concepts are most easily understood in finite dimensions. Let $d\mu(x) = (2\pi)^{-n/2} \exp[-\|x\|^2/2]dx$ denote Gauss measure on $\mathbf{R}^n$. The inequality

$$\int_{\mathbf{R}^n} |f(x)|^2 \ell n|f(x)|d\mu(x) \leq c \int_{\mathbf{R}^n} |\nabla f(x)|^2 d\mu(x) + \|f\|^2_{L^2(\mu)} \ell n\|f\|_{L^2(\mu)} \qquad (2.1)$$

is the prototype of logarithmic Sobolev inequalities. If one defines an operator N on $L^2(\mathbf{R}^n, \mu)$ by $(Nf,g)_{L^2(\mu)} = \int_{\mathbf{R}^n} \nabla f(x) \cdot \nabla \bar{g}(x)d\mu(x)$ then (2.1) reads

$$\int_{\mathbf{R}^n} |f(x)|^2 \ell n|f(x)|d\mu(x) \leq c(Nf,f)_{L^2(\mu)} + \|f\|^2_{L^2(\mu)} \ell n\|f\|_{L^2(\mu)}. \qquad (2.2)$$

In P. Federbush's proof [44] of semiboundedness of H_0+V, he first differentiated the hypercontractivity inequality for e^{-tH_0} at $t=0$ to obtain the inequality (2.2) for H_0 (with $\mathbf{R}^n$ replaced by an infinite dimensional space and N replaced by H_0). He then showed that the inequality (2.2) was itself sufficient to prove semiboundedness. L. Gross [48] later showed that, conversely, one could recover hypercontractivity of e^{-tN} from (2.1) so that hypercontractivity and logarithmic Sobolev inequalities were actually equivalent for Dirichlet form operators (such as N). The techniques in Gross' argument for the direction "log. Sobolev for H_0 implies bounds on e^{-tH_0}" has yielded tremendous advances recently in the understanding of heat kernels. This will be discussed in the next section. A direct proof of (2.1) with the best constant $c = 1$ was also given by Gross [48] using a central limit theorem argument applied to a logarithmic Sobolev-like inequality for an operator on L^2 of a two point measure space. The "two point inequality " was an outgrowth of his earlier work [49] on hypercontractivity for Fermions but was also anticipated by Bonami [14] in his work on harmonic analysis on finite groups.

There are now many proofs of these two types of inequalities for Gauss measure. Some prove hypercontractivity directly [9, 10, 19, 46, 74, 75, 76, 77, 91] while others prove the logarithmic Sobolev inequality directly [2, 5, 13, 43, 48, 84]. The most elementary direct proof of hypercontractivity with best constants is that of E. Nelson [76] while the most elementary and simplest direct proof of the logarithmic Sobolev inequality with best constants ($c = 1$) is is that of O. Rothaus [84].

Both kinds of inequalities were developed in various directions in the 1970's. Inequality (2.2) can be interpreted as saying that $(N + 1)^{-1/2}$ is a bounded operator from $L^2(\mu)$ to the Orlicz space $L^2 \ell n\, L$. G. Feissner showed more generally that $(N + 1)^{-k/2}$ is a bounded operator from $L^p(\mu)$ to $L^p \ell n^k\, L$. See also [8]. Furthermore one may ask whether, given a measure ν on $\mathbb{R}^n$ with a reasonable density, its Dirichlet form operator satisfies a logarithmic Sobolev inequality. There is a procedure by which Dirichlet form operators arise naturally in quantum mechanics and quantum field theory; if V is a suitable real valued function on $\mathbb{R}^n$ then the operator $H := -\Delta + V$ is a self-adjoint operator with a unique lowest eigenvector ψ of unit norm which may be taken strictly positive. If $d\nu(x) = \psi(x)^2 dx$ and $\lambda = \inf$ spectrum H, then the unitary operator $U : f \longrightarrow f/\psi$ from $L^2(\mathbb{R}^n, dx)$ to $L^2(\mathbb{R}^n, d\nu)$ converts $(H - \lambda)$ into an operator $\hat{H} := U(H - \lambda)U^{-1}$ on $L^2(\mathbb{R}^n, d\nu)$ which turns out to be a Dirichlet operator for ν [40]. Hence, by Gross' theorem, hypercontractivity and the logarithmic Sobolev inequality are equivalent for $\hat{H}$. Conditions on V which assure that both hold were obtained by J. P. Eckmann [40], R. Carmona [25], J. Rosen [82], J. Hooton [58]. Moreover J. Rosen [82] showed that if the density of ν decreases very quickly at ∞, then for $1 < p < \infty$, $\|e^{-t\hat{H}}\|_{L^2 \to L^p} < \infty$ for all $t > 0$. This is strictly stronger than hypercontractivity and was called supercontractivity. For a review of studies of $\hat{H}$, see B. Simon [94]. An even stronger notion, ultracontractivity, will be discussed in §3.

An application of hypercontractive ideas in yet another direction was made by W.

Beckner [10] who used extensions of the above-mentioned two point inequality to get the exact bounds in the Hausdorff-Young inequality. The study of e^{-zN}, for complex z, from the point of view of hypercontractivity has recently been completed in a definitive manner by J. Epperson [41]. One may ask whether the Laplace-Beltrami operator on a manifold other than $\mathbb{R}^n$ generates a logarithmic Sobolev inequality. This has been addressed in a number of works [32, 33, 34, 35, 87, 89] with startlingly complete results for S^n in [106]. Finally we mention that applications of logarithmic Sobolev inequalities in infinite dimensions to statistical mechanics were made by Holley and Stroock [55, 56, 57]. The bibliography contains many more works which touch on hypercontractivity or logarithmic Sobolev inequalities in one way or another and not described here or below.

§3 Ultracontractivity

It was shown in [28] that if $H = -\dfrac{d^2}{dx^2} + V$ on $L^2(a, b)$ where $V \in L^1$ and $-\infty < a < b < \infty$, then the semigroup $e^{-t\hat{H}}$ is bounded from $L^2(d\nu)$ to $L^\infty(d\nu)$ for all $t > 0$. This property is called ultracontractivity and is equivalent to $e^{-t\hat{H}}$ having a pointwise bounded integral kernel $K(t, x, y)$ for all $t > 0$.

A more general investigation of such inequalities from the point of view of logarithmic Sobolev inequalities was initiated by Davies and Simon in [38]. If (Ω, μ) is a measure space then there is a very close relationship between bounds of the form

$$\| e^{-tH} \|_{L^2 \to L^\infty} \leq c(t) < \infty \tag{3.1}$$

for all $t > 0$, and

$$\int_\Omega | f |^2 \ln | f | \, d\mu \leq \epsilon \, \| H^{1/2} f \|_2^2$$

$$+ \, \beta(\epsilon) \, \| f \|_2^2 + \| f \|_2^2 \log \| f \|_2 \tag{3.2}$$

for all $f \in \mathrm{Dom}\,(H^{1/2})$ and all $\epsilon > 0$. In particular one has an equivalence between (3.1) and (3.2) in the case where

$$c(t) = c\,t^{-N/4} \tag{3.3}$$

and

$$\beta(\epsilon) = a - \frac{N}{4} \ln \epsilon.$$

Although theorems of this type are not applicable to the harmonic oscillator, they are important for some other Schrödinger operators $H = -\Delta + V$ on $L^2(\mathbf{R}^n)$, For example if $\lambda > 0$ and

$$V(x) = (a^2 + |x|^2)^\lambda$$

then $e^{-t\hat{H}}$ is ultracontractive if and only if $\lambda > 1$, with constants $c(t)$ which diverge as $t \to 0$ much more rapidly than (3.3). The harmonic oscillator therefore stands at the borderline between ultracontractive behavior and the absence of any $L^2 \to L^p$ smoothing properties.

It was subsequently shown by Varopoulos [102] and Fabes and Stroock [42] that less singular problems of this type, for which (3.3) is valid, could be handled by the use of ordinary Sobolev inequalities or what were called Nash inequalities. We remark that the paper of Nash [73] involves "entropy inequalities" which antedate the introduction of logarithmic Sobolev inequalities by a decade.

The paper [38] has spawned a substantial literature concerning pointwise bounds on the heat kernels of second order elliptic operators in divergence form; see[34] for a comprehensive survey. One can easily handle uniformly elliptic operators with measurable coefficients on $\mathbf{R}^n$ and on regions in $\mathbf{R}^n$ subject to Dirichlet or Neumann boundary conditions. The use of logarithmic Sobolev inequalities also allows one to obtain pointwise bounds on the heat kernels of many singular elliptic operators of second order [32,78].

When one turns to the study of the Laplace-Beltrami operator on a Riemannian manifold, many different techniques can be used to obtain pointwise bounds on the heat kernel [34]. Some [26, 71, 36] make little or no use of logarithmic Sobolev inequalities, while others [32, 33, 35] depend essentially upon such a use.

Suppose one has a uniform bound on the heat kernel of the form (3.1), or equivalently

$$0 \leq K(t, x, y) \leq a(t) < \infty.$$

By proving a logarithmic Sobolev inequality for the non-self-adjoint operator $\varphi\, e^{-Ht}\, \varphi^{-1}$, where φ is a suitable weight, it is often possible to show that

$$0 \leq K(t, x, y) \leq c_\delta(t) \exp\left[-d(x, y)^2/(4 + \delta)\, t\right] \tag{3.4}$$

for any $\delta > 0$, where $d(x, y)$ is a Riemannian metric constructed directly from the operator

[30, 33, 35].   See also [31, 42, 21, 34], and [33] for an analogous result for certain second order hypoelliptic operators.   By comparison with earlier literature [3, 81] the advantage of (3.4) is that one has obtained the sharp constant 4 in the exponential.

The possibility of obtaining sharp constants is a recurring feature of the use of logarithmic Sobolev inequalities, and demonstrates their deep significance.   Many developments of the above applications are being currently investigated, and one particularly looks forward to the proof of lower bounds on heat kernels which are of comparable accuracy to the upper bounds mentioned above.

References

[1] R. A. Adams, General logarithmic Sobolev inequalities and Orlicz imbeddings, J. Funct. Anal. $\underline{34}$ (1979), 292-303.

[2] ___________and F. H. Clarke, Gross's logarithmic Sobolev inequality: a simple proof, Amer. J. Math. $\underline{101}$ (1979), 1265-1270.

[3] D. G. Aronson, Non-negative solutions of linear parabolic equations, Ann. Sci. Norm. Sup. Pisa (3) $\underline{22}$ (1968), 607-694.

[4] Joel Avrim, Perturbation of Logarithmic Sobolev generators by electric and magnetic potentials, Berkeley Thesis 1982.

[5] D. Bakry, Une remarque sur les diffusions hypercontractives, Preprint Nov. 1987, 3 pp.

[6] D. Bakry and M. Emery, Hypercontractivité de semi-groupes de diffusion, Compte-Rendus $\underline{299}$ (1984), 775-778.

[7] _______________, Diffusions hypercontractives, pp. 177-207 in Sem. de Probabilites XIX, Lecture Notes in Mathematics #1123, eds. J. Azema and M. Yor, Springer- Verlag, New York, 1985.

[8] D. Bakry and P. A. Meyer, Sur les inequalities de Sobolev logarithmique I & II Preprints, Univ. de Strasbourg 1980/81.

[9] W. Beckner, Inequalities in Fourier Analysis on $\mathbf{R}^n$, Proc. Nat. Acad. Sci. U.S.A. $\underline{72}$ (1975), 638-641.

[10] _______________, Inequalities in Fourier Analysis, Ann. of Math. $\underline{102}$ (1975), 159-182.

[11] _______________, S. Janson and D. Jerison, Convolution inequalities on the circle, Conference on harmonic analysis in honor of Antoni Zygmund, Wadsworth, Belmont 1983, vol. 1, 32-43.

[12] Iwo Bialynicki-Birula and Jerzy Mycielski, Uncertainty relations for information entropy in

wave mechanics, Comm. Math. Phys. 44 (1975), 129-132.

[13] ______________, Wave equations with logarithmic nonlinearities, Bull. Acad. Polon. Sci. Cl. III, 23 (1975), 461.

[14] A. Bonami, Étude des coefficients de Fourier des fonctions de $L^P(G)$, Ann. Inst. Fourier 20:2 (1970), 335-402 [Contains "two point" inequalities, see p. 376-380].

[15] Christer Borell, Positivity improving operators and hypercontractivity, Math. Z. 180 (1982), 225-234.

[16] ______________, A Gaussian correlation inequality for certain bodies in R^n , Math. Ann. 256 (1981), 569-573.

[17] ______________, Convexity in Gauss space, Les aspects statistiques et les aspects physiques des processus gaussiens, CNRS, Paris, No. 307, (1981), 27-37.

[18] ______________, On polynomial chaos and integrability, Preprint 1980, Univ. of Göteborg, Prob. & Math. Statistics.

[19] H. J. Brascamp and E. H. Lieb, Best constants in Young's inequality, its converse, and its generalization to more than three functions, Advances in Math. 20 (1976), 151-173.

[20] Henri Buchwalter et chen Dian Jie, Une liaison entre une inegalité logarithmique de Weissler sur le cercle et une question d'extremum dans le probleme des moments de Stieltjes, Preprint U.E.R. de Mathematiques, Univ. Claude-Bernard-Lyon 1, 43, bd du 11 Nov. 1918; 69622 Villeurbanne Cedex.

[21] Eric Carlen, S. Kusuoka and D.W. Stroock, Upper bounds for symmetric Markov transition functions, Ann. Inst. H. Poncairé 23 (1987), 245-287.

[22] Eric Carlen and Daniel Stroock, An application of the Bakry-Emery criterion to infinite dimensional diffusions, Strasbourg Sem. de Probabilite XX, 1986, Ed. Azima and Yor.

[23] ______________, Ultracontractivity, Sobolev inequalities, and all that, M.I.T. preprint,

June 1985.

[24] Eric Carlen and Michael Loss, Extremals of functionals with competing symmetries, Princeton preprint, 1988.

[25] R. Carmona, Regularity properties of Schrödinger and Dirichlet semigroups, J. Funct. Anal. $\underline{33}$ (1979), 259-296.

[26] I. Chavel, Eigenvalues in Riemannian geometry, Academic Press (1984).

[27] E. B. Davies, Spectral properties of metastable Markov semigroups, J. Funct. Anal. $\underline{52}$ (1983), 315-329 (and other references to Davies therein).

[28] _______________, Hypercontractive and related bounds for double well Schrödinger operators, Quart. J. Math. Oxford Ser. (2) $\underline{34}$ (1983), 407-421.

[29] _______________, Perturbations of ultracontractive semigroups, Quart. J. Math. Oxford (2) $\underline{37}$ (1986), 167-176.

[30] _______________ , Explicit constants for Gaussian upper bounds on heat kernels, Amer. J. Math. $\underline{109}$ (1987), 319-334.

[31] _______________, The equivalence of certain heat kernel and Green function bounds, J. Funct. Anal. $\underline{71}$ (1987), 88-103.

[32] _______________, Heat kernel bounds for second order elliptic operators on Riemannian manifolds, Amer. J. of Math. $\underline{109}$ (1987), 545-570.

[33] _______________, Gaussian upper bounds for heat kernels of some second order operators on Riemannian manifolds, J Funct. Anal. $\underline{80}$ (1988), 16-32.

[34] _______________, Heat kernels and spectral theory, Cambridge University Press 1989, Cambridge, England.

[35] E. B. Davies and N. Mandouvalos, Heat kernel bounds on manifolds with cusps, J. Funct. Anal. $\underline{75}$ (1987), 311-322.

[36] _______________, Heat kernel bounds on hyperbolic space and Kleinian groups, Proc.

London Math. Soc. (3) $\underline{57}$ (1988), 182-208.

[37] E. B. Davies and B. Simon, Ultracontractive semigroups and some problems in analysis, in "Aspects of Mathematics and its Applications", North-Holland, Amsterdam 1984. 18 pp.

[38] __________, Ultracontractivity and the heat kernel for Schrödinger operators and Dirichlet Laplacians, J. Funct. Anal. $\underline{59}$ (1984), 335-395.

[39] __________, L^1 properties of intrinsic Schrödinger semigroups, J. Funct. Anal. $\underline{65}$ (1986), 126-146.

[40] J. P. Eckmann, Hypercontractivity for anharmonic oscillators, J. Funct. Anal. $\underline{16}$ (1974), 388-406.

[41] J. Epperson, The hypercontractive approach to exactly bounding an operator with complex Gaussian kernel, J. Funct. Anal. $\underline{86}$ (1989).

[42] E. B. Fabes and D. W. Stroock, A new proof of Moser's parabolic Harnack inequality via the old ideas of Nash, Arch. Rat. Mech. Anal. $\underline{96}$ (1986), 327-338.

[43] W. Faris, Product spaces and Nelson's inequality, Helv. Phys. Acta, $\underline{48}$ (1975), 721-730.

[44] P. Federbush, A partially alternate derivation of a result of Nelson, J. Math. Phys. $\underline{10}$ (1969), 50-52.

[45] G. Feissner, Hypercontractive semigroups and Sobolev's inequality, T.A.M.S. $\underline{210}$ (1975), 51-62.

[46] J. Glimm, Boson fields with nonlinear self-interaction in two dimensions, Comm. Math. Phys. $\underline{8}$ (1968), 12-25.

[47] __________ and A. Jaffe, A $\lambda\phi^4$ quantum field theory without cutoffs, I, Phys. Rev. $\underline{176}$ (1968), 1945-1951.

[48] L. Gross, Logarithmic Sobolev inequalities, Amer. J. of Math. $\underline{97}$ (1975), 1061-1083.

[49] __________, Existence and uniqueness of physical ground states, J. Funct. Anal. $\underline{10}$

(1972), 52-109.

[50] ______________ , Hypercontractivity and logarithmic Sobolev inequalities for the Clifford-Dirichlet form, Duke Math. J. $\underline{42}$ (1975), 383-396.

[51] O. Guennoun, Inegalités logarithmiques de Gross-Sobolev et hypercontractivité , Thèse 3ème cycle, Université Lyon 1 (1980), 90 pages.

[52] F. Guerra, L. Rosen and B. Simon, The $P(\phi)_2$ Euclidean quantum field theory as classical statistical mechanics, Ann. Math. $\underline{101}$ (1975), 111-259.

[53] ______________, Boundary conditions for the $P(\phi)_2$ Euclidean quantum field theory, Ann. Inst. H. Poincaré, $\underline{25A}$ (1976), 231-334.

[54] R. Hoegh-Krohn, A general class of quantum fields without cutoffs in two space-time dimensions, Commun. Math. Phys. $\underline{21}$ (1971), 224-255.

[55] R. Holley and D. Stroock, Diffusions on an infinite dimensional torus, J. Funct. Anal. $\underline{42}$ (1981), 29-63.

[56] ______________, Logarithmic Sobolev inequalities and stochastic Ising models, J. of Statistical Physics $\underline{46}$ (1987), 1159-1194.

[57] ______________, Uniform and L^2 convergence in one dimensional stochastic Ising models, M.I.T. preprint, Sept. 1988, 8 pp.

[58] J. G. Hooton, Dirichlet forms associated with hypercontractive semigroups, Trans. A.M.S. $\underline{253}$ (1979), 237-256.

[59] ______________, Dirichlet semigroups on bounded domains, Rocky Mountain J. Math. $\underline{12}$ (1982), 283-297.

[60] ______________, Compact Sobolev imbeddings on finite measure spaces, J. Math. Anal. and Applications $\underline{83}$ (1981), 570-581.

[61] S. Janson, On hypercontractivity for multipliers on orthogonal polynomials [Best constant for Laplacian on S^2] Ark. Mat. $\underline{21}$ (1983), 97-110.

[62] T. Kato, Schrödinger operators with singular potentials, Amer. J. Math. <u>13</u> (1972), 135-148.

[63] Abel Klein and Bernard Russo, Sharp inequalities for Weyl operators and Heisenberg groups, Math. Ann. <u>235</u> (1978), 175-194.

[64] Andrzej Korzeniowski, On hypercontractive estimates for the Laguerre semigroup, Preprint, U. of Texas, Arlington, Feb. 1984.

[65] __________, On logarithmic Sobolev constant for diffusion semigroups, J. Funct. Anal. <u>71</u> (1987), 363-370.

[66] Andrzej Korzeniowski and D. W. Stroock, An example in the theory of hypercontractive semigroups, Proc. A.M.S. <u>94</u> (1985), 87-90.

[67] W. Krakowiak and J. Szulga, Hypercontraction principle and random multilinear forms, Wroclaw Univ. Preprint #34 (1985), to appear in Prob. Theory and Related Fields.

[68] S. Kusuoka and D. Stroock, Some boundedness properties of certain stationary diffusion semigroups, J. Funct. Anal. <u>60</u> (1985), 243-264.

[69] S. Kwapien and J. Szulga, Hypercontraction methods for comparison of moments of random series in normed spaces, Case Western Reserve Preprint, Sept. 1, 1988, 13 pages.

[70] M. Ledoux, Isopérimétrie et inégalités de Sobolev logarithmiques Gaussiennes, C. R. Acad. Sci. Paris, ser. 1, v. 306 (1988), 79-82.

[71] P. Li and S. T. Yau, On the parabolic kernel of the Schrödinger operator, Acta Math. <u>156</u> (1986), 153-201.

[72] Paul Meyer, Some analytical results on the Ornstein-Uhlenbeck semigroup in infinitely many dimensions, in IFIP Working Conference on Theory and Applications of Random Fields, Bangalore, Jan. 4-9, 1982 preprint (10 pages)

[73] J. Nash, Continuity of solutions of parabolic and elliptic equations, Amer. J. Math. 80 (1958), 931-954.

[74] E. Nelson, A quartic interaction in two dimensions, in Mathematical Theory of Elementary Particles, 69-73. (R. Goodman and I. Segal, eds.) M.I.T. Press, Cambridge, Mass. 1966.

[75] ———————, Probability theory and Euclidean field theory, in Constructive quantum field theory. Ed. G. Velo and A. Wightman, Springer Lecture Notes in Physics, Vol. 25 (1973).

[76] ———————, The free Markoff field, J. Funct. Anal. 12 (1973), 211-227.

[77] J. Neveu, Sur l'espérance conditionnelle par rapport à un mouvement brownien, Ann. Inst. H. Poincare 12 (1976), 105-109. [Contains a probabilistic proof of hypercontractivity]

[78] M. M. H. Pang, L^1 properties of singular second order elliptic operators, J. London Math. Soc. (1987), to appear.

[79] J. Peetre, A class of kernels related to the inequalities of Beckner and Nelson, "A tribute to Ake Pleijel" Uppsala 1980, 171-210.

[80] Ross Pinsky, A new approach to a class of logarithmic Sobolev inequalities motivated by large deviations theory, Technion preprint, July 1986.

[81] F. O. Porper and S. D. Eidel'man, Two sided estimates of fundamental solutions of second order parabolic equations and some applications, Russian Math. Surveys 39 (3) (1984), 119-178.

[82] J. Rosen, Sobolev inequalities for weight spaces and supercontractivity, Trans. A.M.S. 222 (1976), 367-376.

[83] L. Rosen, A $\lambda\phi^{2n}$ field theory without cutoffs, Commun. Math. Phys. 16 (1970), 157-183.

[84] O. S. Rothaus, Lower bounds for eigenvalues of regular Sturm-Liouville operators and the

logarithmic Sobolev inequality, Duke Math. Journal $\underline{45}$ (1978), 351-362.

[85] _______________, Logarithmic Sobolev inequalities and the spectrum of Sturm-Liouville operators, J. Funct. Anal. $\underline{39}$ (1980), 42-56.

[86] _______________, Logarithmic Sobolev inequalities and the spectrum of Schrödinger operators, J. Funct. Anal. $\underline{42}$ (1981), 110-120.

[87] _______________, Diffusion on compact Riemannian manifolds and logarithmic Sobolev inequalities, J. Funct. Anal. $\underline{42}$ (1981), 102-109.

[88] _______________, Analytic inequalities, isoperimetric inequalities and logarithmic Sobolev inequalities, J. Funct. Anal. $\underline{64}$ (1985), 296-313.

[89] _______________, Hypercontractivity and the Bakry-Emery criterion for compact Lie groups, J. Funct. Anal. $\underline{65}$ (1986), 358-367.

[90] I. E. Segal, Notes towards the construction of nonlinear relativistic quantum fields. III. Properties of the C^*-dynamics for a certain class of interactions, Bull. A.M.S. $\underline{75}$ (1969), 1390-1395.

[91] _______________, Construction of nonlinear local quantum processes, I, Ann. of Math. $\underline{92}$ (1970), 462-481.

[92] _______________, Construction of nonlinear local quantum processes, II, Inventiones Math. $\underline{14}$ (1971), 211-241.

[93] I. Shigekawa, Existence of invariant measures of diffusions on an abstract Wiener space, Osaka J. Math. $\underline{24}$ (1987), 37-59.

[94] B. Simon, Schrödinger semigroups, Bull. A.M.S. $\underline{7}$ (1982), 447-526.

[95] _______________, A remark on Nelson's best hypercontractive estimates, Proc. Amer. Math. Soc. $\underline{55}$ (1976), 376-378.

[96] _______________, Essential self-adjointness of Schrödinger operators with positive potentials, Math. Ann. $\underline{201}$ (1973), 211-220.

[97] ———————, The $P(\phi)_2$ Euclidean (Quantum) field theory, Princeton Univ. Press, Princeton, N.J. 1974.

[98] B. Simon and R. Hoegh-Krohn, Hypercontractive semi-groups and two dimensional self-coupled Bose fields, J. Funct. Anal. 9 (1972), 121-180.

[99] D. W. Stroock, An introduction to the theory of large deviations, "Universitext", Springer-Verlag 1984, 196 pages.

[100] Jerzy Szulga, On hypercontractivity of α-stable random variables, $0 < \alpha < 2$, Preprint, July 1987, U. of North Carolina Statistics Dept.

[101] N. Th. Varopoulos, Iteration de J. Moser, Perturbation de semigroups sous-markoviens, C. R. Acad. Sci. Paris Ser. 1, t.300 (1985), 617-620.

[102] ———————, Hardy-Littlewood theory for semigroups, J. Funct. Anal. 63 (1985), 240-260.

[103] F. Weissler, Logarithmic Sobolev inequalities for the heat- diffusion semigroup, Trans. A.M.S. 237 (1978), 255-269.

[104] ———————, Two-point inequalities, the Hermite semigroup, and the Gauss-Weierstrass semigroup, J. Funct. Anal. 32 (1979), 102-121.

[105] ———————, Logarithmic Sobolev inequalities and hypercontractive estimates on the circle, J. Funct. Anal. 37 (1980), 218-234.

[106] ——————— and C. E. Mueller, Hypercontractivity for the heat semigroup for ultraspherical polynomials and on the n-sphere, J. Funct. Anal. 48 (1982), 252-283.

[107] I. Wilde, Hypercontractivity for Fermions, J. Math. Phys. 14 (1973), 791-792.

[108] J.M. Lindsay, Gaussian hypercontractivity revisited. Preprint, King's College, London. 1989

[109] A. Klein, Self-adjointness of the locally correct generators of Lorentz transformations for $P(\varphi)_2$, in Mathematics of Contemporary Physics, ed R. F. Streater, Academic Press, London (1972), 227–236.

[110] ――――, Quadratic expressions of a free Boson field, Trans. Amer. Math. Soc. $\underline{181}$ (1973), 439–456.

[111] A. Klein and L. J. Landau, Construction of a unique self-adjoint generator for a symmetric local semigroup, J. Funct. Anal. $\underline{44}$ (1981), 121–137.

[112] ――――, Singular perturbations of positivity preserving semigroups via path space techniques, J. Funct. Anal. $\underline{20}$ (1975), 44–82.

[113] A. Klein, L. J. Landau and D. Shucker, Decoupling inequalities for stationary Gaussian processes, Annals of Probability $\underline{10}$ (1982), 702–708.

Progress and Problems in Algebraic Quantum Field Theory

S. DOPLICHER

Dipartimento di Matematica,
Università di Roma "La Sapienza", 00185 Rome, ITALY

1. INTRODUCTION

Algebraic Quantum Field Theory studies the general features of QFT which are grounded on few basic principles involving only local observables.

One does not assume the existence of any a priori given Lagrangian, and the main emphasis is on the locality principle and its rigid consequences.

This analysis on one side leads to a new view on structures which are generally accepted, but historically ad hoc since they are not expressed in terms of observable quantities: to a large extent, Algebraic Quantum Field Theory legitimates these structures as mathematical constructs that one can derive in a canonical way from the observables, and one learns in some cases how they could possibly generalize. On the other side this analysis is a challenge to the basic principles: it probes their power of generating detailed structures or possibly their excessive generality.

This kind of analysis was carried out so far mostly in view of applications to purely massive theories (but see [22,40,41] and References therein). For such theories one gets a fairly complete analysis of superselection structure, particle statistics, gauge invariance of the first kind and a canonical derivation of a field algebra with normal (Bose/Fermi type) commutation properties at spacelike separations [1]. These results apply to theories on a spacetime with at least three space dimensions. The theory modifies quite non trivially to apply to models in two or three spacetime dimensions, and despite a good deal of recent progress the theory is not complete in those cases. This problem received much attention both because it is mathematically attractive and related to developments on knot invariants and quantum groups, and in view of applications to classifying conformal theories and indirectly to string theory.

390

However there are several open important problems for QFT in the four dimensional spacetime related to the above mentioned subject or to the analysis of local aspects of superselection rules, aiming to a full Quantum Noether theorem. We shortly review the progress on superselection structure and emphasize some of the several open problems.

2. GENERAL ASSUMPTIONS

The theory is specified by the local observables acting in their vacuum representation on the (separable) Hilbert space $\mathcal{H}_0$. Namely, there is an *inclusion preserving* map from the set $\mathcal{K}$ of double cones in Minkowski space to von Neumann algebras acting on $\mathcal{H}_0$

$$\mathcal{O} \in \mathcal{K} \to \mathfrak{A}(\mathcal{O}) \subset B(\mathcal{H}_0) \tag{1}$$

where $\mathfrak{A}(\mathcal{O})$ is generated by the observables which can be measured within $\mathcal{O}$ [2]. The most significant first principle requires that, as a manifestation of Einstein causality, the $\mathfrak{A}(\mathcal{O})$'s commute elementwise for spacelike separated double cones (*locality*). The Hilbert space $\mathcal{H}_0$ should be *minimal* i.e. the C^*–algebra $\mathfrak{A}$ generated by $\mathfrak{A}(\mathcal{O})$, $\mathcal{O} \in \mathcal{K}$ is *irreducible* and the defining representation of $\mathfrak{A}$ is associated with just one sector, *the vacuum superselection sector*. The C^*–subalgebra of $\mathfrak{A}$ generated by all the $\mathfrak{A}(\mathcal{O}_1)$, $\mathcal{O}_1 \in \mathcal{K}$ included in the spacelike complement $\mathcal{O}'$ of $\mathcal{O}$, is denoted by $\mathfrak{A}(\mathcal{O}')$ and locality says that

$$\mathfrak{A}(\mathcal{O}) \subset \mathfrak{A}(\mathcal{O}')' \equiv \mathfrak{A}^{\mathrm{d}}(\mathcal{O}) . \tag{2}$$

Whenever the local algebras $\mathfrak{A}(\mathcal{O})$ can be generated by Wightman fields, $\mathfrak{A}^{\mathrm{d}}$ will be also a *local* net [3] and hence fulfill *essential duality* [4]:

$$\mathfrak{A}^{\mathrm{d}}(\mathcal{O}) = \mathfrak{A}^{\mathrm{dd}}(\mathcal{O}) , \quad \mathcal{O} \in \mathcal{K} . \tag{3}$$

A stronger condition, requiring that the local algebras are maximal, is *Haag duality*

$$\mathfrak{A}^{\mathrm{d}}(\mathcal{O}) = \mathfrak{A}(\mathcal{O}) , \quad \mathcal{O} \in \mathcal{K} . \tag{4}$$

The situation where duality fails but essential duality holds is typical of theories with spontaneously broken gauge symmetries [5]. We will not discuss here this topic and will assume duality (4).

In interesting situations there will be, of course, more specific properties of the inclusion preserving map (1) related to translation (or Poincaré) covariance and to the spectrum condition, characterizing the *vacuum state* ω_0, induced by a unit vector $\Omega \in \mathcal{H}_0$, as the ground state in the vacuum sector. Covariance and spectral properties, however, play no direct role in most of the analysis of superselection structure, except for an algebraic consequence [6] called property B: if $\mathcal{O}_1, \mathcal{O}_2 \in \mathcal{K}$ and $\mathcal{N}$ is a neighbourhood of 0 s.t. $\mathcal{O}_1 + \mathcal{N} \subset \mathcal{O}_2$, each non zero projection in $\mathfrak{A}(\mathcal{O}_1)$ is equivalent to I mod $\mathfrak{A}(\mathcal{O}_2)$, i.e.

$$\mathcal{O} \neq E = E^* E \in \mathfrak{A}(\mathcal{O}_1) \Rightarrow E = WW^*, W^*W = I \, , \, W \in \mathfrak{A}(\mathcal{O}_2) \, . \quad (5)$$

All that is used in the analysis of superselection structure and of internal symmetry dual to it is that the net defined by the inclusion preserving map (1) is irreducible and fulfills property B and duality (essential duality would suffice [4,1]); the covariance and spectral properties of the vacuum sector, when present, are merely spectators (cf. also the comments at the end of the next section).

3. SUPERSELECTION STRUCTURE, NORMAL FIELD SYSTEMS AND GAUGE SYMMETRY

Superselection sectors are described by the unitary equivalence classes of some irreducible representations of the C^*–algebra $\mathfrak{A}$ generated by local observables [2]. The relevant representations should be generated (by the GNS construction) by states ω which describe elementary excitations of the vacuum state ω_0.

For theories with short range forces, this was taken in [7,8] to mean that the overall deviation of ω from ω_0 in the spacelike complement $\mathcal{O}'$ of $\mathcal{O} \in \mathcal{K}$ becomes negligible when $\mathcal{O}$ increases to the whole space, i.e.

$$\|(\omega - \omega_0)|_{\mathfrak{A}(\mathcal{O}')}\| @ >> \mathcal{O} \uparrow \mathbb{R}^4 > 0 \, . \quad (6)$$

The criterion (6) leads (cf. [7, Appendix]) to an important class of sectors, those carrying "localizable charges", which have a neat mathematical structure. Up to unitary equivalence, the relevant representations are obtained composing the vacuum representation π_0 of $\mathfrak{A}$ with a semigroup of endomorphisms of $\mathfrak{A}$, *the localized morphisms*: namely, an endomorphism ρ is in this class if it is *localized* i.e. there is $\mathcal{O}_\rho \in \mathcal{K}$ s.t. ρ leaves $\mathfrak{A}(\mathcal{O}'_\rho)$ pointwise fixed, and is *localizable everywhere*, i.e. for each $\mathcal{O} \in \mathcal{K}$ ρ is unitarily equivalent to an endomorphism localized in $\mathcal{O}$.

We can view such morphisms ρ, ρ' as representations acting on the fixed

Hilbert space $\mathcal{H}_0$ and the intertwining operators (ρ, ρ') for these representations lie in $\mathfrak{A}$ as a consequence of duality (4). Therefore the superselection structure is described by the full subcategory of End $\mathfrak{A}$ with objects the localized morphisms. As a consequence of property B this category has direct sums and subobjects. It is a *strict monoidal* category with the operation $\rho, \rho' \to \rho\rho'$ and $T \in (\rho_1, \rho_2)$, $T' \in (\rho'_1, \rho'_2) \to T \times T' \in (\rho_1\rho'_1, \rho_2\rho'_2)$, $T \times T' = T\rho_1(T')$. Each ρ has a *statistics* ([7]; cf. also [42]) described by the unitary representations of the permutation groups specified by assigning to the flip $(r+1, r+2)$ the element

$$\rho^r(\varepsilon(\rho, \rho)) \in (\rho^{r+2}, \rho^{r+2}) =$$
$$= \rho^{r+2}(\mathfrak{A})' \, ,$$

where $\varepsilon(\rho, \rho)$ is the restriction to the diagonal of the unique *permutation symmetry* ε [9] for our category such that $\varepsilon(\rho, \sigma) = I$ whenever ρ, σ are localized in spacelike separated double cones (so that, as easily proved, ρ and σ commute [7]) (a permutation symmetry is specified assigning to each pair of objects ρ, σ a unitary $\varepsilon(\rho, \sigma) \in (\rho\sigma, \sigma\rho)$ having the formal properties of the permutation flip of tensor products $x \otimes y \to y \otimes x$ in a category of vector spaces where $\otimes$ is strictly associative [9,12]).

The statistics of ρ is either infinite, i.e. all irreducible representations of the permutation groups $\mathbb{P}(n)$ occur, or, if ρ is irreducible, either paraBose or paraFermi of finite order d_ρ, i.e. for each n, up to infinite multiplicity, we find either the unitary representation which permutes the factors in the tensor product of n d_ρ–dimensional Hilbert spaces (paraBose case) or the same multiplied by the sign of the permutation (paraFermi case).

For the sake of discussing the superselection structure of massive theories on the physical four dimensional spacetime, the infinite statistics can be disregarded ([10,11]; see also [8, Appendix] for an earlier partial result), and we will denote by Δ the subset of localized morphisms with *finite statistics*.

This subset is a subsemigroup and each $\rho \in \Delta$ is the *finite* direct sum of irreducible elements in Δ [7].

The equivalence classes of elements in Δ form an abelian semigroup $\Delta/\mathcal{I}$ ($\mathcal{I} = \Delta \cap$ Inner $\mathfrak{A}$) with a *unique* involution $\xi \to \bar{\xi}$ commuting with direct sums such that $\xi\bar{\xi}$ contains as a direct summand the class of the identity automorphism ι i.e. the vacuum sector (particle–antiparticle conjugation of superselection sectors [7,8]). Two morphisms $\rho, \bar{\rho} \in \Delta$ are representatives of conjugate classes if and only if there are intertwiners $R, \overline{R}$ such that

$$R \in (\iota, \overline{\rho}\rho) , \quad \overline{R} \in (\iota, \rho\overline{\rho})$$
$$\overline{R} = \hat{\varepsilon}(\overline{\rho}, \rho)R \tag{7}$$

$$\overline{R}^{*}\rho(R) = I = R^{*}\overline{\rho}(\overline{R}) . \tag{8}$$

Here $\hat{\varepsilon}(\rho, \sigma)$ denotes the *bosonized* symmetry which, if ρ, σ are irreducible, is just $\pm\varepsilon(\rho, \sigma)$, the — sign occurring only if both ρ, σ are paraFermi. The use of $\hat{\varepsilon}$ in place of ε in (7) makes relations (8) possible in that form, which is the exact analogue of the conjugate relations for complex conjugate finite dimensional unitary representations of a group [12].

These facts can be summed up by saying that the full subcategory $\mathcal{T}$ of End $\mathfrak{A}$ with objects Δ describes superselection structure (equivalence in the category means unitary equivalence of representations and classes of irreducible objects describe superselection sectors) and $(\mathcal{T}, \hat{\varepsilon})$ is a strict symmetric monoidal C^{*}–category with conjugates [12]. Moreover, $\mathcal{T}$ has subobjects and direct sums and the selfintertwiners of the monoidal unit (i.e. the identity automorphism ι associated with the vacuum representation itself, which is irreducible) reduce to $\mathbb{C}$.

By the main Theorem of [12], $\mathcal{T}$ is then the dual of a unique compact group G. This group can be viewed more concretely as the stabilizer of $\mathfrak{A}$ in the automorphism group of the cross product $\mathfrak{A} \times \mathcal{T}$ defined in [9] (cf. [32] for a survey).

This is a canonical C^{*}–algebra which contains $\mathfrak{A}$ as a subalgebra with trivial relative commutant, and is generated by Hilbert spaces H_{ρ} with support I which induce ρ on $\mathfrak{A}$, $\rho \in \Delta$. The operator product of the H_{ρ}'s is just a strictly *associative* realization of the *tensor product* and the corresponding permutation flip $\vartheta_{H_{\rho}, H_{\sigma}}$ s.t.

$$\vartheta_{H_{\rho}, H_{\sigma}}\psi\psi' = \psi'\psi , \quad \psi \in H_{\rho} , \ \psi' \in H_{\sigma} \quad \rho, \sigma \in \Delta \tag{9}$$

has to agree with the bosonized symmetry $\hat{\varepsilon}_{\rho, \sigma}$.

The induced representation π of $\mathfrak{A} \times \mathcal{T}$, determined by the vacuum representation π_{0} of $\mathfrak{A}$ and by the mean $m : \mathfrak{A} \times \mathcal{T} \longrightarrow \mathfrak{A}$ over the compact group G, provides the vacuum representation of the field algebra, where G is implemented by a strongly continuous unitary representation leaving $\mathcal{H}_{0}$ pointwise fixed.

The *field net* can be defined assigning to each $\mathcal{O} \in \mathcal{K}$ the von Neumann algebra $F(\mathcal{O})$ generated by $\pi(H_\rho), \rho \in \Delta$ localized in $\mathcal{O}$.

By definition (9) and by the condition $\hat{\varepsilon} = \vartheta$, if ρ, σ are irreducible and localized in mutually spacelike double cones (so that we have $\varepsilon(\rho, \sigma) = I$) we get for each $\psi \in \pi(H_\rho)$ $\psi' \in \pi(H_\sigma)$,

$$\psi'\psi = \pi(\vartheta_{H_\rho, H_\sigma})\psi\psi' = \pi(\hat{\varepsilon}(\rho, \sigma))\psi\psi'$$
$$= \pm\pi(\varepsilon(\rho, \sigma))\psi\psi' =$$
$$= \pm\psi\psi'$$

and our field net is generated by field operators normally commuting or anti-commuting at spacelike separations, and connecting the vacuum subspace $\mathcal{H}_0$ to subspaces associated with every superselection sector. We can summarize the result of the construction by saying that it is a complete, normal field system with gauge symmetry.

A *field system* with gauge symmetry is a triple $\{\pi, G, F\}$ where π is a representation of $\mathfrak{A}$ on a Hilbert space $\mathcal{H}$ containing $\mathcal{H}_0$ as a subspace reducing π to π_0, G is a strongly compact unitary group on $\mathcal{H}$ leaving $\mathcal{H}_0$ pointwise fixed, and $\mathcal{O} \to F(\mathcal{O})$ is an inclusion preserving map form $\mathcal{K}$ to von Neumann algebras acting on $\mathcal{H}$ with irreducible union, so that the $g \in G$ induce automorphisms α_g of each $F(\mathcal{O})$ with fixed points $\pi(\mathfrak{A}(\mathcal{O}))$, $\mathcal{O} \in \mathcal{K}$, and $F(\mathcal{O})\mathcal{H}_0$ is dense in $\mathcal{H}$ for every $\mathcal{O} \in \mathcal{K}$. Moreover the nets F and $\pi(\mathfrak{A})$ should commute locally with one another.

Such a system is *normal* if there is a central involutive element $k \in G$ s.t. the net F obeys graded local commutativity for the $\mathbb{Z}_2$ grading defined by k; it is *complete* if every superselection sector (i.e., up to equivalence, each irreducible $\rho \in \Delta$) appears as a subrepresentation of π.

The main result of [1, Section 3] can be then summarized as

THEOREM. *There is a complete, normal field system with gauge symmetry and this system is unique up to equivalence.*

This applies to localizable charges on a three or more dimensional space–time. In a two dimensional world the analysis of statistics modifies to give rise to representations of the braid groups ([13,18]; see also [19,36,37]); the invariants for the statistics are related to Jones polynomials [14,15]; the reconstruction theorem has not been extended to these situations and would need an extension of the duality theory in [12] to an abstract duality theory for *braided monoidal categories* and "Quantum groups" (cf. comments in [1,12]). A similar problem arises in Conformal Field Theory (cf. e.g. [39]). In this context it is to be noted that, independently from [12], a parallel result has been established by P. Deligne [16], providing an abstract characterization of the category of representations of an algebraic group on finite dimensional vector spaces over a field of a characteristic zero.

But do massive particles in the four dimensional spacetime necessarily carry only localizable charges? Does the existence of a massive one particle state associated with an isolated point in the mass spectrum in a given sector imply sufficiently strong localization properties? Buchholz and Fredenhagen, motivated also by work of Swieca on charge screening, found a full answer, positive for the second question [10] and negative to the first one ([10,17] see also [18]). If the representation π describes a sector with the above properties, π is unitarily equivalent to a vacuum representation π_0 when restricted to subalgebra of $\mathfrak{A}$ associated with the spacelike complement of any arbitrary *spacelike cone* $\mathcal{C}$, i.e.

$$\pi|_{\mathfrak{A}(\mathcal{C}')} \cong \pi_0|_{\mathfrak{A}(\mathcal{C}')} . \tag{10}$$

The possible localization regions are thus not necessarily bounded, but rather a suitable neighbourhood of a string connecting a given point to spacelike infinity.

The analysis of statistics [7] applies also to these more general "quantum topological charge" provided spacetime is at least four dimensional ([10]; see also [1, Sect. 4]). In a three dimensional spacetime quantum topological charges might give rise to braid statistics [18,19,36,37].

The construction of fields yields, in the four dimensional spacetime, a unique complete normal *extended* field system with gauge symmetry, which assigns von Neumann algebras of field operators to *spacelike cones*. Every irreducible representation fulfilling (10) with finite statistics will appear on an appropriate subspace of $\mathcal{H}$.

The field operators associated with *bounded* spacetime regions will no longer generate the whole Hilbert space $\mathcal{H}$ when acting on the vacuum Hilbert space $\mathcal{H}_0$, but all the sectors carrying *localizable charges* will be reached in this way [1, Sect. 5].

The unique associated compact group G of internal symmetries has now a closed normal subgroup N such that a sector carries a "topological charge" if and only if it corresponds to an irreducible representation of G which is non trivial on N. The gauge group associated with the subclass of localizable charges is the quotient group G/N [1].

If the vacuum sector is translation covariant, it is natural to restrict attention to representations fulfilling (10) with finite statistics which are translation covariant. This should not be a proper subclass in reasonable theories. The spectrum condition then holds everywhere if it holds in the vacuum sector and there is a unique associated covariant, normal, extended field system with gauge symmetry. Similar results apply to Poincaré covariant theories [1, Sect. 6].

4. TRENDS AND PROBLEMS

The developments described in Section 3 dealt with purely massive theories hence apply to theories where both electroweak and gravitational forces are neglected.

If one would take into account the quantum aspects of gravitation, not only the basic locality principle, but the very notion of spacetime as a classical underlying geometry had to be abandoned. A less radical but physically important scheme would deal with quantum theory in an external gravitational field, and there is interesting progress along this line [20,21].

The principle of locality should however be a central feature of theories with gauge symmetries of the second kind, without gravitation.

What are the distinguishing aspects of such theories, in terms of the local algebras of observables? This problem, long proposed by Rudolf Haag, is still open. The theory of net cohomology [4] of John Roberts was a sound approach to this question; a final answer is not yet known.

In our opinion a better understanding will be possible if we will be able to extend the algebraic picture of superselection structure and statistics to theories *with* massless particles. There the long proposed picture should be as follows. In massive theories, we may think of the labels distinguishing among superselection sectors (i.e. among unitary equivalence classes of relevant irreducible representations) as "charge" quantum numbers (such as baryon number, strangeness,...). In presence of massless particles, on the contrary, there will be several superselection sectors which describe the same "charge" values and differ by infrared waves of infinitely many soft massless particles. There should be a looser criterion than those of Section 3 which selects the relevant representations, and an equivalence relation looser than unitary equivalence that groups together sectors associated with the same charge values.

By [22,10] a plausible selection criterion is covariance, spectrum condition and the validity of (10) for *some* spacelike cones, those whose shadow at spacelike infinity contains a given open region. The appropriate charge equivalence notion is unknown in general. For the important case of QED (or of theories where only massive particles carry charges) there is an interesting proposal [22] which can be phrased as follows (see also [23,24]): $\pi \sim \pi'$ if π and π' are unitarily equivalent when restricted to $\mathfrak{A}(V)$, where V is any future or past light cone.

If we change the localization region of π from a spacelike cone $\mathcal{C}_1$ to $\mathcal{C}_2$ in the spacelike complement of $\mathcal{C}_1$, e.g. by a rotation, it is to be expected that π changes not by a unitary equivalence but within its charge class.

Problem. Can we still analyze on these grounds the general notion of statistics of a charge class? Are again ordinary paraBose or paraFermi statistics the only possibilities? Is braid group statistics excluded a priori on the four dimensional spacetime?

Understanding these problems might shed some light on the intrinsic features of gauge theories.

So far the only sound manifestation of gauge theories at the local algebraic level is the possible appearance of "quantum topological charges", associated with representations fulfilling (10) but not (6). Therefore, even if we consider purely massive theories, a problem related to the local algebraic understanding of gauge theories is the following.

Problem. Can we characterize by algebraic properties of the net of local observables $\mathcal{O} \to \mathfrak{A}(\mathcal{O})$ the theories where quantum topological charges occur, i.e. the normal subgroup N of G is not trivial (cf. Sect. 3)?

In general, can we say anything more about N? (e.g. is it abelian? (K. Fredenhagen)).

A borderline area between massless and massive theories is spontaneous breakdown of symmetries. For some progress in this context we refer to a forthcoming paper [25], and here we turn to questions related to localizable charges. We mentioned already in Section 3 the important mathematical problems posed by understanding superselection structure and internal symmetries in the two dimensional world, and we will limit ourselves to mention problems on theories in ordinary Minkowski space. Understanding the first of them might also help on the way of studying gauge theories.

A theory fulfills the *split property* if for each pair of double cones $\mathcal{O}_1$, $\mathcal{O}_2$ such that $\mathcal{O}_1 + \mathcal{N} \subset \mathcal{O}_2$ for some neighbourhood $\mathcal{N}$ of the origin, there is a type I factor $\mathcal{N}$ between $F(O_1)$ and $F(O_2)$, the net F being as in the theorem of Section 3. This assumption is related to basic spectral properties of localizable states [26] and to existence of equilibrium states [27]. It rules out many undesirable models [28] and plays the role of the assumption that there is an underlying lagrangian, at least as far as the structure of symmetries is concerned. Indeed, in theories with the split property we have a weak form of Noether Theorem [29,30,31] providing local objects, analogous to the integrals of a regularized current density over a finite space volume, forming "local current algebras", associated to Lie groups of global symmetries.

Problem. Can we derive conserved current vector fields as pointwise limits of the local current algebras?

This problem should provide a full Quantum Noether Theorem and a general mechanism for the appearance of Schwinger terms [29,32].

We could also consider a local subnet generated by (bounded functions of operators obtained spreading out with smooth functions) those currents, and take its gauge invariant part (the currents associated with the gauge group would not be observables unless the gauge group is abelian). For instance, we could apply this procedure starting with the spacetime translation group and the gauge group, and our subnet would be generated by the energy momentum tensor and by invariant combinations of the gauge currents. The case of a non connected or discrete gauge group can also be handled, since these cases are covered by our weak form of Quantum Noether Theorem.

Problem. Characterize those theories where the subnet generated by those Noether currents coincides with the given net of observables.

Such a characterization would answer a fundamental question: local observables are "defined" by locality, but also charged Bose fields (the subnet of F commuting with k, cf. Section 3) obey that property. Is there an intrinsic notion of observables? Coincidence of nets as mentioned in the above problem would be a reasonable characterization of local observables which are *intrinsically* observables. A priori only the spacetime localization of each $A \in \mathfrak{A}$ should matter, irrespectively of our specific interpretation. However, the weak form of Noether Theorem provides *local observables with a specific physical interpretation*, and the above problems, if solved, would provide *Wightman fields* with a specific physical interpretation, possibly generating all local observables.

A necessary condition for the required coincidence of nets is that the originally given net of local observables should have itself no nontrivial gauge group of the first kind. Is this also a sufficient condition [35]?

The split property refers to the complete normal field system with gauge symmetry associated to our net $\mathcal{O} \to \mathfrak{A}(\mathcal{O})$; it would be desirable to characterize those properties of the net itself which are equivalent to the split property of the field net (which is known to imply the split property for $\mathfrak{A}$, while the converse is not known [29]). Recent work on modular nuclearity should be relevant to this question [38].

Eventually we would like to mention the problem of stability of superselection structure and of its invariants (an old problem, but stated in print only in [33]). As suggested there, the order $d(\rho)$ of parastatistics of a localized morphism $\rho \in \Delta$ can be viewed as a non commutative version of the index of a Fredholm operator, where invariance under perturbations by inner elements of Δ takes the role of invariance under compact perturbations.

Problem. Are there stability properties of this index under deformations of the theory and can we identify it with "topological" invariants? Are there wider stability properties of the Bose/Fermi character of parastatistics under perturbations of dynamics?

In other words we ask whether there is a non commutative version of the Atiyah Singer index theorem which applies to (localized) endomorphisms. The analytical index should correspond to the order of parastatistics while a "topological" definition of the index is not known.

The recent work by R. Longo on index of infinite subfactors and classification of braid statistics in low dimensional theories has shown remarkable connections with the Jones theory of inclusions, showing in particular that

$$\mathrm{d}(\rho)^2 = \mathrm{ind}\left(\rho(\mathfrak{A}(\mathcal{O}))\,,\,\mathfrak{A}(\mathcal{O})\right) \tag{11}$$

where $\mathcal{O}$ is any double cone where ρ is localized, and the index is defined by appropriately extending the notion of Jones index [34] to purely infinite von Neumann algebras [14,15].

Relation (11) applies also to situations where $\mathrm{d}(\rho)$ is not necessarily integer (theories on low dimensional spacetimes) and hence goes beyond our discussion here; it gives striking evidence that $\mathrm{d}(\rho)$ has to be thought of as an index.

REFERENCES

[1] S. Doplicher and J.E. Roberts, *Why there is a field algebra with a compact gauge group describing the superselection structure in particle physic*, Commun. Math. Phys **131** (1990) 51–107.

[2] R. Haag and D. Kastler, *An algebraic approach to Quantum Field Theory*, J. Math. Phys. **5** (1964) 848–861.

[3] J.J. Bisognano, E.H. Wichmann, *On the Duality Condition for Quantum Fields*, J. Math. Phys. **17** (1976) 303–321.

[4] J.E. Roberts, *Net Cohomology and its Applications to Field Theory* in "Quantum Fields — Algebras, Processes", L. Streit ed. Wien, New York, Springer–Verlag (1980).

[5] J.E. Roberts, *Spontaneously Broken Gauge Symmetries and Superselection Rules*, Proceedings of the International School of Mathematical Physics, Camerino 1974. G. Gallavotti ed., Università di Camerino, 1976.

[6] H.J. Borchers, *A Remark on a Theorem of B. Misra*, Commun. Math. Phys. **4** (1967) 315–323.

[7] S. Doplicher, R. Haag and J.E. Roberts, *Local Observable and Particle Statistics I* Commun. Math. Phys. **23** (1971) 199–230.

[8] S. Doplicher, R. Haag and J.E. Roberts, *Local Observables and Particle Statistics II*, Commun. Math. Phys. **35** (1974) 49–85.

[9] S. Doplicher and J.E. Roberts, *Endomorphism of C*-Algebras, Cross Products and Duality for Compact Groups*, Annals of Mathematics **130** (1989) 75–119.

[10] D. Buchholz and K. Fredenhagen, *Locality and the Structure of Particle States*, Commun. Math. Phys. **84** (1982) 1–54.

[11] K. Fredenhagen, *On the existence of antiparticles*, Commun. Math. Phys. **79** (1981) 141–151.

[12] S. Doplicher and J.E. Roberts, *A New Duality Theory for Compact Groups*, Inventiones Mathematicae **89** (1989) 157–218.

[13] K. Fredenhagen, K.H. Rehren, B. Schroer, *Superselection Sectors with Braid Group Statistics and Exchange Algebras*, Commun. Math. Phys. **125** (1989) 201–226.

[14] R. Longo, *Index of subfactors and statistics of Quantum Fields*, I. Commun. Math. Phys. **26** (1989) 217–247.

[15] R. Longo, *Index of subfactors and statistics of Quantum Fields*, II Commun. Math. Phys. to appear.

[16] P. Deligne, *Categories Tannakiennes*, Grothendieck Festschrift, Birkhäuser 1990.

[17] K. Fredenhagen and M. Marcu, *Charged States in Z_2 Gauge Theories*, Commun. Math. Phys. **92** (1983) 81–119.

[18] J. Froehlich, P.A. Marchetti, Commun. Math. Phys. **121** (1989) 177–224; J. Froehlich, F. Gabbiani, P.A. Marchetti in *Algebraic Theory of Superselection Sectors*, D. Kastler ed., World Sci. P. Co 1990.

[19] K. Fredenhagen, *Quantum Field Theory in Low Dimensional Space Time*, to appear in the Proceedings of the Schladming Winter School 1990, Springer Proceedings in Physics.

[20] R. Haag, H. Narnhofer, U. Stein, Commun. Math. Phys. **94** (1984) 219.

[21] K. Fredenhagen, R. Haag, Commun. Math. Phys. **108** (1987), 91 and **127** (1990) 273.

[22] D. Buchholz, *The Physical State Space of Quantum Electrodynamics*, Commun. Math. Phys. **85** (1982) 49–71.

[23] S. Doplicher, M. Spera, Commun. Math. Phys. **89** (1985) 19–25; S. Doplicher, F. Figliolini, D. Guido, Ann. Inst. H. Poincaré **41** (1984) 49–62.

[24] D. Buchholz, S. Doplicher, Ann. Inst., H. Poincaré **40** (1984) 175–184.

[25] D. Buchholz, S. Doplicher, R. Longo and J.E. Roberts, *Broken Symmetries and Degeneracy of the Vacuum in Quantum Field Theory*, preprint in preparation.

[26] D. Buchholz and E.H. Wichmann, *Causal Independence and the Energy–level Density of Localized States in Quantum Field Theory*, Commun. Math. Phys. **106** (1986) 321.

[27] D. Buchholz and P. Junglas, *On the Existence of Equilibrium States in Local Quantum Theory*, Commun. Math. Phys. **121** (1989) 255–270.

[28] S. Doplicher and R. Longo, *Standard and Split Inclusions of von Neumann Algebras*, Inventiones Math. **75** (1984) 493–536.

[29] S. Doplicher, *Local Aspects of Superselection Rules*, Commun. Math. Phys. **85** (1982) 73–86.

[30] S. Doplicher and R. Longo, *Local Aspects of Superselection Rules II* Commun. Math. Phys. **88** (1983) 399–409.

[31] D. Buchholz, S. Doplicher and R. Longo, *On Noether's Theorem in Quantum Field Theory* Ann. Phys. **170** (1986) 1–17.

[32] S. Doplicher, *Local Observables and the Structure of Quantum Field Theory* in "Algebraic Theory of Superselection Sectors", D. Kastler ed., World Sci. P. Co. 1990.

[33] S. Doplicher, *Problems in Quantum Field Theory and Operator Algebras* in "Operator Algebras and their connections with Ergodic Theory and Topology", H. Araki, C.C. Moore, S. Stratila and D. Voiculescu eds., Springer L.N. in Math. 1132.

[34] V.F.R. Jones, *Index for Subfactors* Inventiones Math. **72** (1983) 1–25.

[35] R. Conti, S. Doplicher, work in progress.

[36] J. Fröhlich, *Quantum Statistics and locality*, preprint ETH (1990).

[37] Contributions of D. Kastler, M. Mebkhout and K.H. Rehren; K.H. Rehren; G. Mack and V. Schomerus in "Algebraic Theory of Superselection Sectors", D. Kastler ed., World Sci. P. Co. 1990.

[38] D. Buchholz, C. D'Antoni, R. Longo, *Nuclear Maps and modular structures I*, J. Functional Analysis **88** (1990) 233–250; II, Commun. Math. Phys. **129** (1990) 115–138.

[39] S. Majid, *Reconstruction theorems and Rational Conformal Theories*, Preprint DAMTP, Cambridge (1990).

[40] D. Buchholz, *On Particles, Infraparticles and the Problem of Asymptotic completeness*, Proc. of the Marseille Conference on Mathematical Physics, M. Mebkhout, R. Seneor eds., World Sci. P. Co. (1987).

[41] D. Buchholz, *Harmonic analysis of local operators*, Desy preprint, (1989).

[42] J.E. Roberts, *Lectures on Algebraic Quantum Field Theory*, in the Algebraic Theory of Superselection Sectors D. Kastler ed., World Sci. P. Co. (1990).

$$\text{H\o egh-Krohn's Model of Quantum Fields and}$$
$$\text{the Absolute Continuity of Measures}$$

by

Shigeo Kusuoka

0. Introduction.

Let μ be a Gaussian measure in $\mathscr{S}'(\mathbb{R}^2)$ such that

$$\int_{\mathscr{S}'} \exp(\sqrt{-1}\cdot {}_{\mathscr{S}'}\langle\phi,u\rangle_{\mathscr{S}})\,\mu(d\phi) = \exp\left(-\tfrac{1}{2}((1-\Delta)^{-1}u,u)_{L^2(\mathbb{R}^2)}\right), \quad u \in \mathscr{S}(\mathbb{R}^2).$$

For each $h \in \mathscr{S}'(\mathbb{R}^2)$, let $S_h:\mathscr{S}'(\mathbb{R}^2)\to\mathscr{S}'(\mathbb{R}^2)$ be a map given by $S_h(\phi) = \phi + h$. Then by Cameron-Martin's theorem we see that the probability measure $\mu\circ S_h^{-1}$ is absulutely continuous relative to μ, if and only if $(1-\Delta)^{1/2}h \in L^2(\mathbb{R}^2,dx)$. Also, we see that the Radon-Nikodym derivative is given by

$$(0.1) \quad \frac{d\mu\circ S_h^{-1}}{d\mu}(\phi) = \exp\left(-\langle\phi,(1-\Delta)h\rangle - \tfrac{1}{2}\left(\int_{\mathscr{S}'}\langle\tilde\phi,(1-\Delta)h\rangle^2\mu(d\tilde\phi)\right)\right).$$

Let $p(t,x) = \dfrac{1}{4\pi t}\exp\left(-\dfrac{|x|^2}{4t}\right) = e^{t\Delta}(0,x)$, $t > 0$ and $x \in \mathbb{R}^2$ (the heat kernel), and let $g(x) = \displaystyle\int_0^\infty e^{-t}p(t,x)dt = (1-\Delta)^{-1}(0,x)$, $x \in \mathbb{R}^2$. Also, let $g_x(y) = g(y-x)$, $x,y \in \mathbb{R}^2$. Then $g_x \in \mathscr{S}'(\mathbb{R}^2)$ for each $x \in \mathbb{R}^2$. Since $(1-\Delta)^{1/2}g_x$ does not belong to $L^2(\mathbb{R}^2)$ for any $x \in \mathbb{R}^2$, $\mu\circ S_{\alpha g_x}^{-1}$ is not absolutely continuous relative to μ unless $\alpha = 0$. However, it is not clear whether the measure $\displaystyle\int_{[0,1]^2} \mu\circ S_{\alpha g_x}^{-1}\,dx$ is absolutely continuous relative to μ.

Formally we have

$$\int_{[0,1]^2} dx\,\mu\circ S_{\alpha g_x}^{-1}(d\phi)$$

$$= \left(\int_{[0,1]^2} \exp\left(-\alpha\cdot\phi(x) - \frac{\alpha^2}{2}\int_{\mathscr{S}'}\tilde\phi(0)^2\mu(d\tilde\phi)\right)dx\right)\mu(d\phi),$$

and the right hand side is exactly the potential function which was
first studied by Høegh-Krohn [7] with the relation to the quantum
field theory. By his result we have the following (cf. [2]).

(0.2)**Theorem.** *If* $\alpha^2 < 4\pi$, *then the measure* $\displaystyle\int_{[0,1]^2} dx\ \mu\circ S_{\alpha g_x}^{-1}$ *is*
absolutely continuous relative to μ *and the Radon-Nikodym derivative*
belongs to $L^2(\mathscr{S}'(\mathbb{R}^2),d\mu)$.

After that, this subject was studied by several authors, because
it has the relations not only with the quantum field theory but also
with the energy representation of gauge groups. And the following
has been proven (cf. Albeverio-Gallavotti-Høegh-Krohn [1], Ossipov
[9] and Wallach [10]).

(0.3)**Theorem.** *If* $\alpha^2 > 8\pi$, *then the measure* $\displaystyle\int_{[0,1]^2} dx\ \mu\circ S_{\alpha g_x}^{-1}$ *is*
singular relative to μ.

However, it was unkown what happens in the case $\alpha^2 \in [4\pi,8\pi]$. In
the present paper we prove the following.

(0.4)**Theorem.** *If* $\alpha^2 < 8\pi$, *then the measure* $\displaystyle\int_{[0,1]^2} dx\ \mu\circ S_{\alpha g_x}^{-1}$ *is*
absolutely continuous relative to μ *and the Radon-Nikodym derivative*
belongs to $L^p(\mathscr{S}'(\mathbb{R}^2),d\mu)$, $p \in (1,(\frac{8\pi}{\alpha^2})\wedge 2)$.

The case where $\alpha^2 = 8\pi$ remains an open problem.

We also prove the similar results for the Gaussian random
distribution with a parameter in compact Riemannian manifolds as
follows.

Let M be a compact sooth Riemannian manifold, and let $p_M(t,x,y)$,
$t > 0$, $x,y \in M$, be the heat kernel. Let $g_M(x,y) =$
$\displaystyle\int_0^\infty (p_M(t,x,y) - \mathrm{vol}(M)^{-1})\ dt$, $x,y \in M$. Let μ_M be a Gaussian measure
in $\mathscr{D}'(M)$ such that

$$\int_{\mathscr{D}'} \exp(\ \sqrt{-1}\ {}_{\mathscr{D}'}\langle\phi,u\rangle_{\mathscr{D}}\ )\ \mu(d\phi) = \exp(\ -\tfrac{1}{2}\int_{M\times M} g_M(x,y)u(x)u(y)dx\otimes dy\ ),$$

$u \in \mathcal{D}(M)$. Let $S_h : \mathcal{D}'(M) \to \mathcal{D}'(M)$, $h \in \mathcal{D}'(M)$, be a map given by $S_h \phi = \phi + h$. Let $g_x \in \mathcal{D}'(M)$, $x \in M$, be given by $g_x(y) = g_M(y,x)$.

Then we have the following.

(0.5) **Theorem.** *If $\alpha^2 < 8\pi$, then the measure $\int_M \mu \circ S_{\alpha g_x}^{-1} dx$ in $\mathcal{D}'(M)$ is absolutely continuous relative to μ and the Radon–Nikodym derivative belongs to $L^p(\mathcal{D}'(M), d\mu)$, $p \in (1, (\frac{8\pi}{\alpha^2}) \wedge 2)$.*

Theorem(0.5) may be useful for the representation thery of Gauge gruoups or the relative string theory.

Acknowledment. This work was done when the author stayed in Rhur University, Bochum in the summer in 1988. He thanks Prof. Albeverio for kind invitation, for suggesting the problem and for the stimulating discussion. He also thanks Prof. S. Paycha and Prof. S. Scarlatti for useful discussion.

After finishing this work, the author was informed by Professor H.Sato that Theorems (0.4) and (0.5) follow from the results of J.P.Kahane, Sur le chaos multiplicatif, Ann. Sci. Math. Québec, 9(2), 1985, 105-150. Moreover, his results imply that the measure $\int_{[0,1]^2} \mu \circ S_{\alpha g_x}^{-1} dx$ is singular relative to μ if $\alpha^2 = 8\pi$.

1. Preliminary Results.

(1.1) **Lemma.** *Let Z_k, $k = 1, 2, \ldots$, be non-negative random variables. Then we have*

$$P\left[\sum_{i=1}^{2^m} Z_i \geq A \right]$$
$$\leq P\left[\max_{1 \leq i \leq 2^m} Z_i \geq (1-\varepsilon)^m A \right] + \varepsilon^{-1}(1-\varepsilon)^{-2m+1} A^{-2} \sum_{1 \leq i < j \leq 2^m} E[Z_i \cdot Z_j]$$

for any $A > 0$, $m \geq 0$ and $\varepsilon \in (0, 1/2]$.

408   S. Kusuoka

Proof. We prove this by induction in m. for m = 0, this is obvious.
So we assume that this statement is true for m = n. Then we have

$$P[\sum_{i=1}^{2^{n+1}} Z_i \geq A]$$

$$= P[\sum_{i=1}^{2^n} (Z_i + Z_{2^n+i}) \geq A]$$

$$\leq P[\max_{1 \leq i \leq 2^n} (Z_i + Z_{2^n+i}) \geq (1-\varepsilon)^n A]$$

$$+ \varepsilon^{-1}(1-\varepsilon)^{-2n+1} A^{-2} \sum_{1 \leq i < j \leq 2^n} E[(Z_i + Z_{2^n+i})(Z_j + Z_{2^n+j})]$$

Note that

$$P[\max_{1 \leq i \leq 2^n} (Z_i + Z_{2^n+i}) \geq (1-\varepsilon)^n A]$$

$$= P[Z_i + Z_{2^n+i} \geq (1-\varepsilon)^n A \quad \text{for some } i \text{ with } 1 \leq i \leq 2^n]$$

$$\leq P[Z_i + Z_{2^n+i} \geq (1-\varepsilon)^n A \quad \text{and} \quad \min\{Z_i, Z_{2^n+i}\} < \varepsilon(1-\varepsilon)^n A$$
$$\text{for some } i \text{ with } 1 \leq i \leq 2^n]$$

$$+ P[Z_i + Z_{2^n+i} \geq (1-\varepsilon)^n A \quad \text{and} \quad \min\{Z_i, Z_{2^n+i}\} \geq \varepsilon(1-\varepsilon)^n A$$
$$\text{for some } i \text{ with } 1 \leq i \leq 2^n]$$

$$\leq P[\max_{1 \leq i \leq 2^{n+1}} Z_i \geq (1-\varepsilon)^{n+1} A] + \sum_{i=1}^{2^n} P[Z_i \cdot Z_{2^n+i} \geq \varepsilon(1-\varepsilon)^{2n+1} A],$$

and

$$\varepsilon^{-1}(1-\varepsilon)^{-2n+1} A^{-2} \sum_{1 \leq i < j \leq 2^n} E[(Z_i + Z_{2^n+i})(Z_j + Z_{2^n+j})]$$

$$+ \sum_{i=1}^{2^n} P[Z_i \cdot Z_{2^n+i} \geq \varepsilon(1-\varepsilon)^{2n+1} A]$$

$$\leq \varepsilon^{-1}(1-\varepsilon)^{-2n-1} A^{-2} \sum_{1 \leq i < j \leq 2^{n+1}} E[Z_i \cdot Z_j].$$

These implies our assertion for m = n+1.   Q.E.D.

(1.2)**Corollary.** *If* Z_k, *k = 1,2,...., are non-negative random*
variables, then

$$E\left[\ \left(\sum_{i=1}^{2^m} Z_i\right)^p\ \right]$$

$$\leq\ (1-\varepsilon)^{-mp}\sum_{i=1}^{2^m} E[\ Z_i^{\ p}] + \frac{2}{2-p}\ \{\ \varepsilon^{-1}(1-\varepsilon)^{-2m+1}\sum_{1\leq i<j\leq 2^m} E[Z_iZ_j]\}^{p/2}$$

for any $m \geq 0$, $p \in (1,2)$ *and* $\varepsilon \in (0,1/2]$.

Proof. Note that

$$E[\ Z^p\ ] = \int_0^\infty pA^{p-1}P[\ Z \geq A\ ]\ dA\ .$$

Therefore by Lemma(1,1), we have

$$E[\ \left(\sum_{i=1}^{2^m} Z_i\right)^p\ ]$$

$$\leq\ p\cdot\int_0^\infty A^{p-1}P[\ \sum_{i=1}^{2^m} Z_i \geq A\ ]\ dA$$

$$\leq\ p\cdot(1-\varepsilon)^{-mp}\int_0^\infty A^{p-1}P[\ \max_{1\leq i\leq 2^m} Z_i \geq A\ ]\ dA$$

$$+\ p\cdot\int_0^\infty A^{p-1}\cdot\ (((\varepsilon^{-1}(1-\varepsilon)^{-2m+1}\sum_{1\leq i<j\leq 2^m} E[Z_iZ_j])A^{-2})\wedge 1)\ dA$$

$$=\ (1-\varepsilon)^{-mp} E[(\ \max_{1\leq i\leq 2^m} Z_i)^p]$$

$$+\ \frac{2}{2-p}\ \{\ \varepsilon^{-1}(1-\varepsilon)^{-2m+1}\sum_{1\leq i<j\leq 2^m} E[Z_iZ_j]\}^{p/2}\ .$$

This implies our assertion.

(1.3)**Lemma.** *Let* M *be a smooth manifold and* $\{X(x);x\in M\}$ *and* $\{Y(x);x\in M\}$ *be mean zero Gaussian random fields with smooth paths.* *Suppose that* $E[X(x)X(y)] - E[Y(x)Y(y)]$, $x,y \in M$, *is nonnegative definite.* *Then for any finite measure* ν *on* M *and* $p \in [1,\infty)$

$$E[\{\int_M \exp(\ Y(x) - \frac{1}{2} E[Y(x)^2]\ )\nu(dx)\}^p\ ]$$

$$\leq\ E[\{\int_M \exp(\ X(x) - \frac{1}{2} E[X(x)^2]\ )\nu(dx)\}^p\ ]\ .$$

Proof. Let $\{Z(x);x\in M\}$ be a mean zero Gaussian random fields with smooth paths such that $\{Y(x);x\in M\}$ and $\{Z(x);x\in M\}$ are independent and

$$E[Z(x)Z(y)] = E[X(x)X(y)] - E[Y(x)Y(y)],\ x,y \in M.$$

Then we have

$$E[\{\int_M \exp(\ X(x)\ -\ \tfrac{1}{2}\ E[X(x)^2]\ )\nu(dx)\}^p\ ]$$

$$=\ E[\{\int_M \exp(\ (Y(x)+Z(x))\ -\ \tfrac{1}{2}\ E[Y(x)^2+Z(x)^2]\ )\nu(dx)\}^p\ ]$$

$$\geq\ E[\{E[\int_M \exp(\ (Y(x)+Z(x))\ -\ \tfrac{1}{2}\ E[(Y(x)+Z(x))^2]\ )\nu(dx)|Y]\}^p\ ]$$

$$=\ E[\{\int_M \exp(\ Y(x)\ -\ \tfrac{1}{2}\ E[Y(x)^2]\ )\nu(dx)\}^p\ ]\ .$$

This completes the proof.

(1.4)**Lemma.** *Let* M *be a smooth manifold and* $\{X_n(x);x\in M\}$, $n \in M$, *be mean zero Gaussian random fields with smooth paths. Also let* $R:M{\times}M{\to}\mathbb{R}$ *be a nonnegative definite smooth function. Now let* $\{Y_n(x);x\in M\}$, $n\in\mathbb{N}$, *be mean zero Gaussian random fields with smooth paths such that*

$$E[Y_n(x)Y_n(y)]\ =\ E[X_n(x)X_n(y)]\ +\ R(x,y),\quad x,y \in M.$$

Then for any $p \in [1,\infty)$ *and finite measure* ν *on* M *whose support is compact, there is a constant* $C < \infty$ *such that*

$$\sup_n\ E[\{\int_M \exp(\ Y_n(x)\ -\ \tfrac{1}{2}\ E[Y_n(x)^2]\ )\nu(dx)\}^p\ ]$$

$$\leq\ C{\cdot}\sup_n\ E[\{\int_M \exp(\ X_n(x)\ -\ \tfrac{1}{2}\ E[X_n(x)^2]\ )\nu(dx)\}^p\ ]\ .$$

Proof.   Let $\{W(x);x\in M\}$ be a mean zero Gaussian random fields with smooth paths such that $\{W(x);x\in M\}$ is independent from $\{X_n(x);x\in M, n\in\mathbb{N}\}$ and that $E[W(x)W(y)] = R(x,y)$, $x,y \in M$. Let K the support of the measure ν. Then K is compact. Then by Landau-Shepp-Fernique's theorem we see that $C = E[\ \exp(p{\cdot}\sup\{|W(x)|;x\in K\})\ ] < \infty$.

Then we have

$$E[\{\int_M \exp(\ Y_n(x)\ -\ \tfrac{1}{2}\ E[Y_n(x)^2]\ )\nu(dx)\}^p\ ]$$

$$=\ E[\{\int_M \exp((X_n(x)+W(x))-\ \tfrac{1}{2}\ E[X_n(x)^2+W(x)^2]\ )\nu(dx)\}^p\ ]$$

$$\leq\ C{\cdot}E[\{\int_M \exp(\ X_n(x)\ -\ \tfrac{1}{2}\ E[X_n(x)^2]\ )\nu(dx)\}^p\ ]\ .$$

This completes the proof.

(1.5)**Lemma.**   *Let* $\{U_n\}_{n=1}^{\infty} \subset C_0^{\infty}(\mathbb{R}^d{\times}\mathbb{R}^d;\mathbb{R})$ *satisfy the following.*

(1.6) $\sup\{(1+|x|+|y|)^{2d+1}|\dfrac{\partial^{\beta+\gamma}}{\partial x^{\beta}\partial y^{\gamma}}\ U_n(x,y)|;\ n \in \mathbb{N},\ x,y \in \mathbb{R}^d\ \} < \infty$

for any multi-indice β and γ.

Then there is a V ∈ C^∞(ℝ^d;ℝ) such that

$$(1.7) \quad \left| \int_{\mathbb{R}^d \times \mathbb{R}^d} U_n(x,y)\varphi(x)\varphi(y) \, dxdy \right| \leqq \int_{\mathbb{R}^d \times \mathbb{R}^d} V(x-y)\varphi(x)\varphi(y) \, dxdy$$

for any n ∈ ℕ and φ ∈ C_0^∞(ℝ^d;ℝ).

Proof. Let $W_n(\xi,\eta) = \int_{\mathbb{R}^d \times \mathbb{R}^d} e^{-i(x \cdot \xi + y \cdot \eta)} U_n(x,y) \, dxdy$ and let $\rho(\xi) = \sup_n \{ \int_{\mathbb{R}^d} (|W_n(\xi,\eta)| + |W_n(\eta,\xi)|) d\eta \}$. Then by the assumption (4.2), we see that $\sup\{(1+|\xi|)^{2n}\rho(\xi); \xi \in \mathbb{R}^d\} < \infty$ for any n ∈ ℕ. Observe that

$$\left| \int_{\mathbb{R}^d \times \mathbb{R}^d} U_n(x,y)\varphi(x)\varphi(y) \, dxdy \right|$$

$$= (2\pi)^{-2d} \left| \int_{\mathbb{R}^d \times \mathbb{R}^d} W_n(\xi,\eta)\hat{\varphi}(\xi)\hat{\varphi}(\eta) \, d\xi d\eta \right|$$

$$\leqq (2\pi)^{-2d} \left(\int_{\mathbb{R}^d \times \mathbb{R}^d} |W_n(\xi,\eta)||\hat{\varphi}(\xi)|^2 \, d\xi d\eta \right)^{1/2}$$

$$\times \left(\int_{\mathbb{R}^d \times \mathbb{R}^d} |W_n(\xi,\eta)||\hat{\varphi}(\eta)|^2 d\xi d\eta \right)^{1/2}$$

$$\leqq (2\pi)^{-2d} \left(\int_{\mathbb{R}^d} \rho(\xi)|\hat{\varphi}(\xi)|^2 d\xi d\eta \right).$$

Thus letting $V(x) = (2\pi)^{-d} \int_{\mathbb{R}^d} e^{ix \cdot \xi} \rho(\xi) \, d\xi$, we obtain our assertion.

2. Basic Theorem.

Let $p(t,x) = \frac{1}{4\pi t} \cdot \exp\left(- \frac{|x|^2}{4t}\right)$, t > 0 and x ∈ ℝ^2, as in Introduction. The folowing is the main result in this section.

(2.1) **Theorem.** *Let* $\{\Phi_n(x); x \in \mathbb{R}^2\}$, n ∈ ℕ, *be a Gaussian random field such that*

(1) E[$\Phi_n(x)$] = 0, x ∈ ℝ^2,

(2) E[$\Phi_n(x)\Phi_n(y)$] = $\int_{2^{-2n}}^{1} p(t,x-y) \, dt$, x,y ∈ ℝ^2,

and

(3) $\Phi_n(\cdot):\mathbb{R}^2 \to \mathbb{R}$ *is smooth with probability one.*

Then, if p > 1 and $\alpha^2 p < 8\pi$,

(2.2) $\sup_{n} E[\{ \int_{[0,1]^2} \exp(\alpha \cdot \Phi_n(x) - \frac{1}{2} \alpha^2 E[\Phi_n(0)^2]) \, dx \}^p] < \infty.$

(2.3) *Remark.* Since $\int_{2^{-2n}}^{1} p(t,x-y) \, dt$ is a positive definite smooth

function in x and y, there exists a Gausian random field satisfying

(1), (2) and (3).

We make some preparations to prove Theorem(2.1). Let $R(x) = \int_{1}^{4} p(t,x) \, dt$, $x \in \mathbb{R}$. Then we have

(2.4) $R(0) = \frac{1}{2\pi} \log 2,$

and

(2.5) $R(2^n x) = \int_{2^{-2n}}^{2^{-2n+2}} p(t,x) \, dt$, $x \in \mathbb{R}^2.$

Now let Y_k, $\tilde{Y}_k$, $k \in \mathbb{Z}$, be i.i.d. Gaussian random fields over $\mathbb{R}^2$ with

smooth paths such that

(2.6) $E[Y_0(x)] = 0$ and $E[Y_0(x)Y_0(y)] = R(x-y)$, $x,y \in \mathbb{R}^2.$

Since $R:\mathbb{R}^2 \to \mathbb{R}$ is smooth and positive definite, the existence of such

random fields are guaranteed. Let

(2.7) $X_n(x) = X_n^{(K)}(x) \underset{\text{def}}{=} \sum_{k=-K}^{n-K} Y_k(2^k x)$, $x \in \mathbb{R}^2$, $n \in \mathbb{N}$ and $K \in \mathbb{Z}$,

and

(2.8) $\tilde{X}_m(x) = \sum_{k=0}^{m-1} \tilde{Y}_k(2^k x)$, $x \in \mathbb{R}^2$ and $m \in \mathbb{N}.$

Then we see that $X_n(\cdot)$ is a mean zero stationary Gaussian random

field such that

(2.9) $E[X_n(x)X_n(y)] = \int_{2^{-2n+2K}}^{2^{2K+2}} p(t,x-y) \, dt$, $x,y \in \mathbb{R}^2.$

Therefore our goal is to prove (2.2) for $\Phi_n(x) = X_n^{(-1)}(x).$

(2.10) **Proposition.** *The random fields* $\{X_{n+m}(x); x \in \mathbb{R}^2\}$ *and*

$\{X_n(2^m x)+\tilde{X}_m(2^{-K}x); x \in \mathbb{R}^2\}$, *have the same probability law.*

Proof. This follows from

$\qquad E[X_{n+m}(x)X_{n+m}(y)]$

$$= \int_{2^{-2m-2n+2K}}^{2^{-2m+2K+2}} p(t,x-y) \, dt + \int_{2^{-2m+2K+2}}^{2^{2K+2}} p(t,x-y) \, dt$$

$$= \int_{2^{-2n+2K}}^{2^{2K+2}} p(t,2^m(x-y)) \, dt + \int_{2^{-2m+2}}^{4} p(t,2^{-K}(x-y)) \, dt$$

$$= E[\, X_n(2^m x) X_n(2^m y) \,] + E[\, \tilde{X}_m(2^{-K} x) \tilde{X}_m(2^{-K} y) \,] \, . \qquad\qquad \text{Q.E.D.}$$

Since $\tilde{X}_m(x)$ is smooth in x with probability one, by Landau-Schepp-Fernique's theorem, we see that for any $m \in \mathbb{N}$, there is some $\beta_m \in (0,\infty)$ such that

$$(2.11) \quad E[\, \exp(\, \beta_m \cdot \sup_{x,y\in[0,1]^2} (\frac{|\tilde{X}_m(x)-\tilde{X}_m(y)|}{|x-y|})^2 \,) \,] \leqq 2 \, .$$

Let $B_{k,\ell} = B_{k,\ell}^{(m)} = [2^{-m}(k-1), 2^{-m}k] \times [2^{-m}(\ell-1), 2^{-m}\ell] \subset \mathbb{R}^2$, $1 \leqq k,\ell \leqq 2^m$. Then we have by Corollary(1.2)

$$(2.12) \quad E[\, \{ \int_{[0,1]^2} \exp(\, \alpha \, X_{n+m}(x) - \frac{1}{2} \alpha^2 \, E[\, X_{n+m}(0)^2 \,]) \, dx \,\}^p \,]$$

$$\leqq \{E[\{ \sum_{k+\ell=odd} \int_{B_{k,\ell}} \exp(\alpha X_{n+m}(x) - \frac{1}{2}\alpha^2 \, E[\, X_{n+m}(0)^2 \,]) dx \,\}^p \,]^{1/p}$$

$$+ E[\{ \sum_{k+\ell=even} \int_{B_{k,\ell}} \exp(\alpha X_{n+m}(x) - \frac{1}{2}\alpha^2 \, E[\, X_{n+m}(0)^2 \,]) dx \,\}^p \,]^{1/p}\}^p$$

$$\leqq 2^p \cdot E[\{ \sum_{k+\ell=even} \int_{B_{k,\ell}} \exp(\alpha X_{n+m}(x) - \frac{1}{2}\alpha^2 \, E[\, X_{n+m}(0)^2 \,]) dx \,\}^p \,]$$

$$\leqq 2^p (1-\varepsilon)^{-(2m-1)} \cdot 2^{2m-1} E[\{ \int_{B_{1,1}} \exp(\alpha X_{n+m}(x) - \frac{1}{2}\alpha^2 \, E[\, X_{n+m}(0)^2 \,]) dx \,\}^p]$$

$$+ \frac{2^{p+1}}{2-p} \{\varepsilon^{-1}(1-\varepsilon)^{-(4m-3)} \cdot \sum_{\substack{(k,\ell)\neq(i,j) \\ k+\ell=even \\ i+j=even}} E[\int_{B_{k,\ell}} dx \int_{B_{i,j}} dy$$

$$\exp(\alpha X_{n+m}(x) + \alpha X_{n+m} - \alpha^2 E[X_{n+m}(0)^2])]\}^{p/2} \, .$$

(2.13)**Proposition.**

$$\sup_{n\geqq 0} E[\int_{B_{k,\ell}} dx \int_{B_{i,j}} dy \, \exp(\alpha X_n(x) + \alpha X_n(y) - \alpha^2 E[X_n(0)^2])] < \infty$$

for any (k,ℓ), $(i,j) \in \{1,2,\ldots,2^m\}^2$ *satisfying* $(k,\ell)\neq(i,j)$, $k+\ell =$ *even and* $i+j =$ *even.*

Proof. Note that

$$E[X_n(x) X_n(y)]$$

$$\leq \int_0^{2^{2K+2}} p(t, x-y)\, dt$$

$$\leq \int_0^{2^{2K+2}|x|^{-2}} \frac{1}{4\pi t}\, e^{-1/4t}\, dt$$

$$\leq \int_0^{2^{2K+2}} \frac{1}{4\pi t}\, e^{-1/4t}\, dt + \int_1^{2^{2K+2}} (|x|^{-1}\vee 1)^2\, \frac{dt}{4\pi t}$$

$$= \frac{1}{2\pi} \log_+(\frac{1}{|x|}) + C(K) ,$$

where $C(K) = \dfrac{K+1}{2\pi} \log 2 + \displaystyle\int_0^{2^{2K+2}} \frac{1}{4\pi t}\, e^{-1/4t}\, dt$.

Then we have

$$E[\int_{B_{k,\ell}} dx \int_{B_{i,j}} dy\, \exp(\alpha X_n(x)+\alpha X_n(y)- \alpha^2 E[X_n(0)^2]))]$$

$$= \int_{B_{k,\ell}} dx \int_{B_{i,j}} dy\, \exp(\, \alpha^2 E[X_n(x)X_n(y)]\,)$$

$$\leq \exp(\, \alpha^2 C(K))\cdot \int_{B_{k,\ell}} dx \int_{B_{i,j}} dy\, (\frac{1}{|x-y|}\vee 1)^{\alpha^2/2\pi} .$$

Therefore (2.13) is obvious if $B_{k,\ell}\cap B_{i,j} = \phi$.   If $B_{k,\ell}\cap B_{i,j}\neq \phi$, then we have

$$\int_{B_{k,\ell}} dx \int_{B_{i,j}} dy\, (\frac{1}{|x-y|}\vee 1)^{\alpha^2/2\pi}$$

$$\leq \int_{[-1,0]^2} dx \int_{[0,1]^2} dy\, (\frac{1}{|x-y|}\vee 1)^{\alpha^2/2\pi}$$

$$\leq \int_{[0,1]^2} dx \int_{[0,1]^2} dy\, (\frac{1}{|x+y|}\vee 1)^{\alpha^2/2\pi}$$

$$\leq \int_{[0,2]^2} (\frac{1}{|x|}\vee 1)^{\alpha^2/2\pi}\cdot |x|^2 dx < \infty,$$

if $\alpha^2/2\pi < 4$.

This completes the proof.

(2.14)**Lemma.**   *Let $p > 1$ and $\alpha \in \mathbb{R}^2$ with $\alpha^2 p < 8\pi$.   Then there are constants $\gamma(m,K,\alpha,p) < \infty$ and $C(m,K,\alpha,p,\varepsilon) < \infty$ such that*

$$(2.15)\quad E[\, \{ \int_{[0,1]^2} \exp(\, \alpha\, X_{n+m}(x) - \frac{1}{2}\, \alpha^2\, E[\, X_{n+m}(0)^2]) \, dx \}^p \,]$$

$$\leq \gamma(m,K,\alpha,p)\cdot 2^{-(2m-1)(p-1)} (1-\varepsilon)^{-(2m-1)p}$$

$$\times \; E[\, \{ \int_{[0,1]^2} \exp(\, \alpha\, X_n(x) - \tfrac{1}{2}\alpha^2\, E[\, X_n(0)^2\,]\,)\; dx \}^p \,]$$

$$+\; C(m,K,\alpha,\varepsilon)$$

for any $m,n \in \mathbb{N}$, $K \in \mathbb{Z}$ *and* $\varepsilon \in (0,1/2]$,

and that

$$(2.16)\quad \lim_{K\to\infty} \gamma(m,K,\alpha,p) = \exp(\, \frac{\alpha^2}{4\pi}\, mp(p-1)\log 2\,), \quad m \in \mathbb{N}.$$

Here $\gamma(m,K,\alpha,p)$ *depends only on* m, K, α *and* p, *and* $C(m,K,\alpha,\varepsilon)$ *depends only on* m, K, α, p *and* ε. *In particular, both are independent from* n.

Proof. By Proposition(2.10), we have

$$E[\, \{ \int_{B_{1,1}} \exp(\alpha X_{n+m}(x) - \tfrac{1}{2}\alpha^2\, E[\, X_{n+m}(0)^2\,])\,dx \}^p \,]$$

$$= E[\, \{ \int_{B_{1,1}} \exp(\alpha X_n(2^m x) - \tfrac{1}{2}\alpha^2 E[X_n(0)^2])$$

$$\times \exp(\alpha \tilde{X}_m(2^{-K}x) - \tfrac{1}{2}\alpha^2 E[\tilde{X}_m(0)^2])\; dx \}^p \,]$$

$$\leqq E[\, \exp(\, \alpha p \sup_{x\in B_{1,1}} \tilde{X}_m(2^{-K}x) - \tfrac{1}{2}\alpha^2 pE[\tilde{X}_m(0)^2])\,]$$

$$\times\; E[\, \{ 2^{-2m}\cdot \int_{[0,1]^2} \exp(\, \alpha\, X_n(x) - \tfrac{1}{2}\, \alpha^2\, E[\, X_n(0)^2\,])\; dx \}^p \,]\; .$$

Let

$$\gamma(m,K,\alpha,p) = E[\, \exp(\, \alpha p \sup_{x\in B_{1,1}} \tilde{X}_m(2^{-K}x) - \tfrac{1}{2}\alpha^2 pE[\tilde{X}_m(0)^2])\,],$$

and

$$C(m,K,\alpha,p,\varepsilon)$$

$$= \frac{2^{p+1}}{2-p}\{\varepsilon^{-1}(1-\varepsilon)^{-(4m-3)}\cdot \sum_{\substack{(k,\ell)\neq(i,j)\\ k+\ell=\text{even}\\ i+j=\text{even}}} \sup_{n\geqq 0} E[\int_{B_{k,\ell}} dx \int_{B_{i,j}} dy$$

$$\exp(\alpha X_{n+m}(x)+\alpha X_{n+m} - \alpha^2 E[X_{n+m}(0)^2])]\}^{p/2}\; .$$

Then by (2.11) and Proposition(2.13), we see that $\gamma(m,K,\alpha,p) < \infty$ and $C(m,K,\alpha,p,\varepsilon) < \infty$. Also (2.15) is obvious from (2.12). By virtue of (2.4) and (2.12), we have

$$\lim_{K\to\infty} \gamma(m,K,\alpha,p)$$

$$= E[\ \exp(\ \alpha p\ \tilde{X}_m(0)\ -\ \tfrac{1}{2}\alpha^2 pE[\tilde{X}_m(0)^2])\ ]$$

$$= \exp(\ \tfrac{1}{2}\ \alpha^2 p(p-1)E[\tilde{X}_m(0)^2])$$

$$= \exp(\ \tfrac{1}{2}\ \alpha^2 p(p-1)m\cdot R(0)\ )$$

$$= \exp(\ \frac{\alpha^2}{4\pi}\ mp(p-1)\log 2\ ).$$

This proves (2.16).   Q.E.D.

(2.17)**Lemma.**   *If* $\alpha^2 p < 8\pi$, *then*

$$(2.18)\quad \sup_{n\in\mathbb{N}}\ E[\{\ \int_{[0,1]^2}\ \exp(\ \alpha\ X_n^{(K)}(x)\ -\ \tfrac{1}{2}\ \alpha^2\ E[\ X_n^{(K)}(0)^2])\ dx\}^p]\ <\ \infty$$

for any $K \in \mathbb{Z}$.

Proof. First we show that there is a $K_0 \in \mathbb{Z}$ such that (2.18) holds for any $K \geq K_0$. Let

$$M_n^{(K)}\ =\ E[\{\ \int_{[0,1]^2}\ \exp(\ \alpha\ X_n^{(K)}(x)\ -\ \tfrac{1}{2}\ \alpha^2\ E[\ X_n^{(K)}(0)^2])\ dx\}^p]\ .$$

Then by Lemma(2.14) we see that

$$M_{n+m}^{(K)}\ \leqq\ \gamma(m,K,\alpha,p)2^{-(2m-1)(p-1)}(1-\varepsilon)^{-(2m-1)p}M_n^{(K)}\ +\ C(m,K,\alpha,p,\varepsilon)\ .$$

Note that

$$\lim_{m\to\infty}\ (\ \frac{1}{m}\cdot\lim_{K\to\infty}\ \log(\ \gamma(m,K,\alpha,p)2^{-(2m-1)(p-1)}(1-\varepsilon)^{-(2m-1)p})$$

$$=\ -\ (2\ -\ \frac{\alpha^2 p}{4\pi})(p-1)\cdot\log 2\ -\ 2p\cdot\log(1-\varepsilon)\ .$$

Therefore there are $K_0 \in \mathbb{Z}$, $m \in \mathbb{N}$ and $\varepsilon \in (0,1/2]$ such that

$$\gamma\ =\ \sup_{K\geqq K_0}\ \gamma(m,K,\alpha)2^{-(2m-1)(p-1)}(1-\varepsilon)^{-(2m-1)}\ <\ 1.$$

Then we have

$$\overline{\lim_{n\to\infty}}\ M_n^{(K)}\ \leqq\ (1-\gamma)^{-1}C(m,K,\alpha,p,\varepsilon)\ <\ \infty\quad \text{for any } K\ \geqq\ K_0\ .$$

This implies (2.18) for $K \geqq K_0$.

Note that

$$E[X_{n+K_0}^{(K_0)}(x)X_{n+K_0}^{(K_0)}(y)]\ =\ E[X_{n+K}^{(K)}(x)X_{n+K}^{(K)}(y)]\ +\ \int_{2^{2K+2}}^{2^{2K_0+2}}\ p(t,x-y)\ dt$$

for $K < K_0$. Thus we have (2.19) for $K < K_0$ from Lemma(1.3).

    This completes the proof.

Theorem(2.1) follows from Lemma(2.17) immediately.

3. The Proof of Theorem(0.4).

Let $g_{n,x}(y) = \int_{2^{-2n-1}}^{\infty} e^{-t} p(t,y-x)\, dt$, $x,y \in \mathbb{R}^2$ and $n \in \mathbb{N}$. Then we see that $(1-\Delta)^{1/2} g_{n,x} \in L^2(\mathbb{R}^2)$ and $((1-\Delta) g_{n,x})(y) = p(2^{-2n-1}, y-x)$, $y \in \mathbb{R}^2$, for any $x \in \mathbb{R}^2$ and $n \in \mathbb{N}$. Let ν_n be a probability measure given by

$$(3.1)\quad \nu_n = \int_{[0,1]^2} \mu \circ S_{\alpha g_{n,x}}\, dx\ ,\quad n \in \mathbb{N}.$$

Then we see that ν_n is absolutely continuous relative to μ and the Radon-Nikodym derivative $f_n : \mathcal{S}'(\mathbb{R}^2) \to \mathbb{R}$ is given by

$$(3.2)\quad f_n(\phi)$$
$$= \int_{[0,1]^2} \exp\left(- \alpha \langle \phi, p(2^{-2n-1}, \cdot - x) \rangle - \frac{\alpha^2}{2} \cdot \int_{\mathcal{S}'} \langle \tilde{\phi}, p(2^{-2n-1}, \cdot) \rangle^2 \mu(d\tilde{\phi}) \right)\, dx$$

for μ-a.s. ϕ.

Then we have the following.

(3.3)**Lemma.** *For any $\alpha \in \mathbb{R}$ and $p \in (1,2)$ with $\alpha^2 p < 8\pi$,*

$$\sup_n \int_{\mathcal{S}'} f_n(\phi)^p\, \mu(d\phi) < \infty.$$

Proof. Let $X_n(x;\phi) = \langle \phi, p(2^{-2n-1}, \cdot - x) \rangle$, $x \in \mathbb{R}^2$ and $\phi \in \mathcal{S}'(\mathbb{R}^2)$. Then $\{X_n(x;\phi); x \in \mathbb{R}^2\}$ is a mean zero Gaussinan stationary measure under $\mu(d\phi)$ such that

$$(3.4)\quad \int_{\mathcal{S}'} X_n(x;\phi) X_n(y;\phi)\, \mu(d\phi) = \int_{2^{-2n}}^{\infty} e^{-t} p(t, x-y)\, dt.$$

Now let $\{Y_n(x); x \in \mathbb{R}^2\}$, $n \in \mathbb{N}$, are mean zero Gaussian Random fields with smooth paths such that

$$(3.5)\quad E[Y_n(x) Y_n(y)] = \int_{2^{-2n}}^{1} e^{-t} p(t, x, y)\, dt\ ,\quad x,y \in \mathbb{R}^2.$$

Then by Lemma(1.3) and Theorem(2.1), we have

$$\sup_n E\left[\left\{ \int_{[0,1]^2} \exp\left(\alpha Y_n(x) - \frac{\alpha^2}{2} E[Y_n(0)^2] \right)\, dx \right\}^p \right] < \infty.$$

Therefore Lemma(1.4), (3.4) and (3.5) imply

$$\sup_n \int_{\mathscr{S}'} \{ \int_{[0,1]^2} \exp(\ \alpha(X_n(x;\phi))\ -\ \frac{\alpha^2}{2}\ E[X_n(0;\phi)^2])\ dx\}^p]\ \mu(d\phi)\ <\ \infty.$$

This completes the proof.

Now let us prove Theorem(0.4). Note that

$$\nu_n \to \nu = \int_{[0,1]^2} \mu \circ S_{\alpha g_x}\ dx\quad \text{weakly as}\ n \to \infty.$$

On the other hand, by Lemma(3.3) there is a subsequence $\{n_k\}_1^\infty$ and $f \in L^p(\mathscr{S}'(\mathbb{R}^2),d\mu)$ such that $f_{n_k} \to f$ weakly in $L^p(\mathscr{S}'(\mathbb{R}^2),d\mu)$. Thus we see

that $d\nu = f \cdot d\mu$. This completes the proof.

4. The Proof of Theorem(0.5).

Let B_R denote $\{\ x \in \mathbb{R}^2;\ |x| < R\ \},\ R > 0$.

(4.1)**Lemma.** *Let* $p_R(t,x,y)$ *be the heat kernel on* B_R *with Dirichlet boundary condition. Let* $\{X_n(x);x \in B_{R/2}\}$, $n \in \mathbb{N}$, *be mean zero Gaussian random fields such that*

$$E[X_n(x)X_n(y)] = \int_{2^{-2n}}^1 p_R(t,x,y)\ dt,\quad x,y \in B_{R/2}\ .$$

Then for any $\alpha \in \mathbb{R}$ *and* $p > 1$ *with* $\alpha^2 p < 8\pi$,

$$\sup_n E[\ \{\int_{B_{R/2}} \exp(\ \alpha \cdot X_n(x)\ -\ \frac{\alpha^2}{2} \cdot E[X_n(x)^2]\ )\ dx\}^p\ ]\ <\ \infty.$$

Proof.   It is well known that

$$\sup\{|\frac{\partial^{\beta+\gamma}}{\partial x^\beta \partial y^\gamma}(p_R(t,x,y)-p(t,x-y)|;t \in (0,1],x,y \in B_{4R/5}\}\ <\ \infty$$

for any multi-indice β and γ (c.f. [9]). Thus there is $\{U_n\}_{n=1}^\infty \subset C_0^\infty(\mathbb{R}^2 \times \mathbb{R}^2;\mathbb{R})$ satisfying (1.6) and

$$U_n(x,y) = \int_{2^{-2n}}^1 (p_R(t,x,y)-p(t,x,y))\ dt,\quad x,y \in B_{3R/5}.$$

Then by Lemma(1.5), we see that there is a $V \in C^\infty(\mathbb{R}^2;\mathbb{R})$ satisfying

(1.7). This implies that $V(x-y) + \int_{2^{-2n}}^1 p(t,x,y)dt$

$$- \int_{2^{-2n}}^{1} p_R(t,x,y)dt, \quad x,y \in B_{R/2}, \quad \text{is nonnegative definite.}$$ Then our assertion follows from Lemmas (1.3) and (1.4).

(4.2)**Lemma.** *Let* $a^{ij} \in C^{\infty}(\mathbb{R}^2;\mathbb{R})$, $i,j = 1,2$, *satisfy*

$$\sum_{i,j=1}^{2} a^{ij}(x)\xi_i\xi_j \geq |\xi|^2, \quad x,\xi \in \mathbb{R}^2. \text{ Let } H_0 \text{ be the selfadjoint}$$

operator in $L^2(B_R,dx)$ *wich is the unique extension of*

$$\sum_{i,j=1}^{2} \frac{\partial}{\partial x^i}(a^{ij}(x)\frac{\partial}{\partial x^j}) \text{ on } \{u \in C^2(\overline{B_R};\mathbb{R}); \ u|_{\partial B_R} = 0\}. \text{ Also let } q(t,x,y),$$

$(t,x,y) \in (0,\infty) \times B_R \times B_R$, *be the kernel of* $\exp(tH_0)$ *in* $L^2(B_R,dx)$. *Let* $a > 0$ *and let* $\{X_n(x); x \in B_{R/2}\}$, $n \in \mathbb{N}$, *be mean zero Gaussian random fields with smooth paths such that*

$$E[X_n(x)X_n(y)] = \int_{2^{-2n}}^{\infty} e^{-at}q(t,x,y) \ dt, \quad x,y \in B_{R/2}.$$

Then for any $\alpha \in \mathbb{R}$ *and* $p > 1$ *with* $\alpha^2 p < 8\pi$

$$\sup_n E[\int_{B_{R/2}} \exp(\alpha \cdot X_n(x) - \frac{\alpha^2}{2} \cdot E[X_n(x)^2]) \ dx \ ] < \infty.$$

Proof. Let $\{Y_n(x); x \in B_{R/2}\}$, $n \in \mathbb{N}$, be mean zero Gaussian random fields with smooth paths such that

$$E[Y_n(x)Y_n(y)] = \int_{2^{-2n}}^{\infty} e^{-at}p_R(t,x,y) \ dt, \quad x,y \in B_{R/2}.$$

Then by Lemmas (1.3), (1.4) and (4.1), we have

$$\sup_n E[\int_{B_{R/2}} \exp(\alpha \cdot Y_n(x) - \frac{\alpha^2}{2} \cdot E[Y_n(x)^2]) \ dx \ ] < \infty.$$

Let Δ_R be the Laplacian operator in $L^2(B_R,dx)$ with Dirichlet boundary condition on ∂B_R. Then $p_R(t,x,y)$ is the kernel of $\exp(t\Delta_R)$. From the assumption we see that $-\Delta_R \leqq -H_0$, and so $(a-H_0)^{-1} \leqq (a-\Delta_R)^{-1}$. Therefore we have

$$(4.3) \quad \exp(2^{-2m-1}\Delta_R)(a-H_0)^{-1}\exp(2^{-2m-1}\Delta_R) \leqq \int_{2^{-m}}^{\infty} e^{-at}\exp(t\Delta_R) \ dt.$$

Let $R_{n,m}(x,y)$

$$= \int_{B_R} dz_1 \int_{B_R} dz_2 \int_{2^{-2n}}^{\infty} dt \; p_R(2^{-2m-1},x,z_1)q(t,z_1,z_2)p_R(2^{-2m-1},z_2,y),$$

$x,y \in B_{R/2}$. Then by (4.3), we see that $\int_{2^{-2n}}^{\infty} p_R(t,x,y)dt - R_{n,m}(x,y)$,

$x,y \in B_{R/2}$, is nonnegative definite. Let $\{Z_{n,m}(x); x \in B_{R/2}\}$, $n,m \in \mathbb{N}$,

be mean zero Gaussian random fields with smooth paths such that

$$E[Z_{n,m}(x)Z_{n,m}(y)] = R_{n,m}(x,y), \quad x,y \in B_{R/2}.$$

Then we have by Lemma(1.3),

$$\sup_n E\left[\int_{B_{R/2}} \exp\left(\alpha \cdot X_n(x) - \frac{\alpha^2}{2} \cdot E[X_n(x)^2] \right) dx \right]$$

$$= \sup_n \lim_{m \to \infty} E\left[\int_{B_{R/2}} \exp\left(\alpha \cdot Z_{n,m}(x) - \frac{\alpha^2}{2} \cdot E[Z_{n,m}(x)^2] \right) dx \right]$$

$$\leqq \sup_m E\left[\int_{B_{R/2}} \exp\left(\alpha \cdot Y_m(x) - \frac{\alpha^2}{2} \cdot E[Y_m(x)^2] \right) dx \right]$$

$$< \infty.$$

This proves our assertion.

(4.4)**Lemma**. *Let* $a^{ij} \in C^{\infty}(\mathbb{R}^2;\mathbb{R})$ *and* H_0 *be as in Lemma(4.2). Let* $V \in C^{\infty}(\mathbb{R}^2;\mathbb{R})$ *and let* H *be the selfadjoint operator in* $L^2(B_R,dx)$ *given by* $H = H_0+V$. *Let* $q^V(t,x,y)$, $(t,x,y) \in (0,\infty) \times B_R \times B_R$, *be the smooth kernel of* $\exp(tH)$ *in* $L^2(B_R,dx)$. *Let* $a > \sup\{V(x); x \in B_R\}$ *and let* $\{X_n(x); x \in B_{R/2}\}$, $n \in \mathbb{N}$, *be mean zero Gaussian random fields with smooth paths such that*

$$E[X_n(x)X_n(y)] = \int_{2^{-2n}}^{\infty} e^{-at}q^V(t,x,y)\,dt, \quad x,y \in B_{R/2}.$$

Then for any $\alpha \in \mathbb{R}$ *and* $p > 1$ *with* $\alpha^2 p < 8\pi$

$$\sup_n E\left[\int_{B_{R/2}} \exp\left(\alpha \cdot X_n(x) - \frac{\alpha^2}{2} \cdot E[X_n(x)^2] \right) dx \right] < \infty .$$

Proof. Let $a_0 = \sup\{V(x); x \in B_R\}$. Then we have $- H_0 - a_0 \leqq - H$.

Therefore we get our assertion similarly to the proof of Lemma(4.2).

(4.5)**Corollary**. *Let* $q^V(t,x,y)$, $(t,x,y) \in (0,\infty) \times B_R \times B_R$ *be as in Lemma(4.4). Let* $\{X_n(x); x \in B_{R/2}\}$, $n \in \mathbb{N}$, *be mean zero Gaussian random* *fields with smooth paths such that*

$$E[X_n(x)X_n(y)] = \int_{2^{-2n}}^{1} q^V(t,x,y)\,dt, \quad x,y \in B_{R/2}.$$

Then for any $\alpha \in \mathbb{R}$ and $p > 1$ with $\alpha^2 p < 8\pi$

$$\sup_n E\left[\int_{B_{R/2}} \exp\left(\alpha \cdot X_n(x) - \frac{\alpha^2}{2}\cdot E[X_n(x)^2]\right) dx\right] < \infty.$$

Proof.   Let $a = \sup\{V(x); x\in B_R\} + 1$ and β be a

real number such that $|\beta| > |\alpha|$ and $\beta^2 p < 8\pi$.   Now let m be a natural

number such that $\exp(-2^{-2m}a)\beta^2 > \alpha^2$.   Then we see that

$$\beta^2\int_{2^{-n}}^{\infty} e^{-at}q^V(t,x,y)\,dt - \alpha^2\int_{2^{-n}}^{2^{-m}} q^V(t,x,y)\,dt, \quad x,y \in B_R, \text{ is nonnegative}$$

definite for $n \geq m$.   Let $\{Y_n(x); x\in B_{R/2}\}$ be a mean zero Gaussian

random fields such that

$$E[Y_n(x)Y_n(y)] = \int_{2^{-n}}^{2^{-m}} q^V(t,x,y)\,dt, \quad x,y \in B_R.$$

Then by Lemmas (1.3) and (4.2), we have

$$(4.6) \quad \sup_n E\left[\int_{B_{R/2}} \exp\left(\alpha \cdot Y_n(x) - \frac{\alpha^2}{2}\cdot E[Y_n(x)^2]\right) dx\right] < \infty.$$

Since $\int_{2^{-m}}^{1} q^V(t,x,y)\,dt$, $x,y \in B_R$, is smooth, we have our assertion

from Lemma(1.4) and (4.6).

(4.7)**Lemma.**   *Let $a^{ij} \in C^{\infty}(\mathbb{R}^2;\mathbb{R})$, $i,j = 1,2$, satisfying*

$\sum\limits_{i,j=1}^{2} a^{ij}(x)\xi_i\xi_j \geq |\xi|^2$, $x,\xi \in \mathbb{R}^2$, *and let $\rho \in C^{\infty}(\mathbb{R}^2;\mathbb{R})$ satisfying*

$\rho(x) > 0$, $x \in \mathbb{R}^2$. *Let L be the selfadjoint operator in $L^2(B_R,\rho(x)dx)$*

which is the unique extension of $\rho(x)^{-1} \cdot \sum\limits_{i,j=1}^{2} \dfrac{\partial}{\partial x^i}\left(\rho(x)a^{ij}(x)\dfrac{\partial}{\partial x^j}\right)$ on

$\{u\in C^2(\overline{B_R};\mathbb{R}); u|_{\partial B_R} = 0\}$. *Let $r(t,x,y)$, $t > 0$, $x,y \in B_R$, be the smooth*

kernel of $\exp(tL)$ in $L^2(B_R,\rho(x)dx)$.   Let $\{Y_n(x); x\in B_{R/2}\}$, $n \in \mathbb{N}$, be

mean zero Gaussian random fields with smooth paths such that

$$E[Y_n(x)Y_n(y)] = \int_{2^{-2n}}^{1} r(t,x,y)\,dt, \quad x,y \in B_{R/2}.$$

Then for any $\alpha \in \mathbb{R}$ *and* $p > 1$ *with* $\alpha^2 p < 8\pi \cdot \inf\{\rho(x); x \in B_R\}$

$$\sup_n E\left[\int_{B_{R/2}} \exp(\alpha \cdot Y_n(x) - \frac{\alpha^2}{2} \cdot E[Y_n(x)^2])\,dx\right] < \infty.$$

Proof. Let $\varphi(x) = \rho(x)^{1/2}$ and let $V(x) =$

$-\frac{1}{2} \sum_{i,j=1}^{2} \frac{\partial}{\partial x^i}(a^{ij}(x)\rho(x)^{-1/2}\frac{\partial}{\partial x^j}\rho(x))$. Then we see that the map $u \rightarrow$ φu is a unitary map from $L^2(B_R, \rho(x)dx)$ onto $L^2(B_R, dx)$. Now let H be the selfadjoint operator in $L^2(B_R, dx)$, q^V be the kernel of e^{tH} in $L^2(B_R, dx)$ and $\{X_n(x); x \in B_{R/2}\}$, $n \in \mathbb{N}$, be the Gaussian random fields as in Corollary(4.5). Then we see that $\varphi L\varphi^{-1} = H$, and so $\varphi e^{tL}\varphi^{-1} = e^{tH}$, $t > 0$. Thus we see that

$$\varphi(x)\int_{B_R} r(t,x,y)\varphi(y)^{-1}u(y)\rho(y)dy = \int_{B_R} q^V(t,x,y)u(y)dy, \quad u \in L^2(B_R, dx).$$

Thus we see that $r(t,x,y) = \varphi(x)^{-1}q^V(t,x,y)\varphi(y)^{-1}$, $x,y \in B_{R/2}$. So $\{\varphi(x)^{-1}X_n(x); x \in B_{R/2}\}$ and $\{Y_n(x); x \in B_{R/2}\}$ have the same law. Thus we have our assertion from Lemma(1.3) and Corollary(4.5).

(4.8)**Lemma**. *Let M be a compact smooth Riemannian manifold of two dimensions, and* $p_M(t,x,y)$, $(t,x,y) \in (0,\infty) \times M \times M$, *be the heat kernel on M. Let* $\{X_n(x); x \in M\}$, $n \in \mathbb{N}$, *be mean zero Gaussian random fields such that*

$$E[X_n(x)X_n(y)] = \int_{2^{-2n}}^{1} p_M(t,x,y)\,dt, \quad x,y \in M.$$

Then for any $\alpha \in \mathbb{R}$ *and* $p > 1$ *with* $\alpha^2 p < 8\pi$

$$\sup_n E\left[\int_M \exp(\alpha \cdot X_n(x) - \frac{\alpha^2}{2} \cdot E[X_n(x)^2])\,m(dx)\right] < \infty.$$

Here m is the Riemannian volume on M.

Proof. It is sufficient to show that for any $x_0 \in M$, there is an open neighborhood U of x_0 such that

$$(4.9) \quad \sup_n E\left[\int_U \exp(\alpha \cdot X_n(x) - \frac{\alpha^2}{2} \cdot E[X_n(x)^2])\,dx\right] < \infty.$$

Fix a $x_0 \in M$ and a orthonormal basis $\{e_1, e_2\}$ in $T_{x_0}(M)$. Let $\Phi : \mathbb{R}^2 \to M$ be a map given by $\Phi(x) = \exp_{x_0}(x^1 e_1 + x^2 e_2)$. Then there is an $\varepsilon > 0$ such that $\Phi|_{B_{2\varepsilon}} : B_{2\varepsilon} \to \Phi(B_{2\varepsilon})$ is a diffeomorphism. Then we see that there are $a^{ij} \in C^\infty(\mathbb{R}^2 ; \mathbb{R})$, $i,j = 1,2$, and $\rho \in C^\infty(\mathbb{R}^2 ; \mathbb{R})$ with $\rho(x) > 0$, $x \in \mathbb{R}^2$, such that $\Delta_M(u \circ \Phi^{-1})(\Phi(x)) = \rho(x)^{-1} \cdot \sum_{i,j=1}^{2} \frac{\partial}{\partial x^i}(\rho(x) a^{ij}(x) \frac{\partial}{\partial x^j} u)(x)$ for any $u \in C_0^\infty(B_\varepsilon ; \mathbb{R})$, $m \circ \Phi(dx) = \rho(x)dx$ on B_ε, $\rho(0) = 1$ and $a^{ij}(0) = \delta_{ij}$, $i,j = 1,2$. Therefore letting R be a sufficienyly small positive number with $R < \varepsilon$ and letting $r(t,x,y)$, $(t,x,y) \in (0,\infty) \times B_R \times B_R$, and $\{Y_n(x) ; x \in B_{R/2}\}$, $n \in \mathbb{N}$, be as in Lemma(4.7), we have

$$\sup_n E[\int_{B_{R/2}} \exp(\alpha \cdot Y_n(x) - \frac{\alpha^2}{2} \cdot E[Y_n(x)^2]) \, \rho(x)dx] < \infty .$$

Note that

$$\sup\{ | \frac{\partial^{\beta+\gamma}}{\partial x^\beta \partial y^\gamma}(p_M(t,\Phi(x),\Phi(y)) - r(t,x,y)) | ; \ (t,x,y) \in (0,1] \times B_{2R/3} \times B_{2R/3} \}$$

$$< \infty$$

for any multi-indice β and γ. Therefore similarly to the proof of Lemma(4.1), we have (4.9) for $U = \Phi(B_{R/2})$.

Now the proof of Theorem(0.5) is similar to the proof of Theorem(0.4).

References.

[1] Albeverio, S., G.Galavotti, and R. Høegh-Krohn, Some results for the exponential interaction in two or more dimensions, Comm. Math. Phys. 70(1979), 187-192.

[2] Albeverio, S., and R. Høegh-Krohn, Then Wightman axioms and the mass gap for strong interactions of exponential type in two-dimensional space-time, J. Funct. Anal. 16(1974), 39-82.

[3] Albeverio, S., R. Høegh-Krohn, S. Paycha, and S. Scarlatti, Path space measure for the Liouville quantum field theory and the

construction of relativistic strings, Ruhr-University Preprint (1986).

[4] Albeverio, S., R. Høegh-Krohn, and D. Testrand, Irreducibility
and reducibility for the energy representation of the group of
mappings of a Riemannian manifold into a compact semi-simple Lie
group, J. Funct. Anal. 41(1981), 378-396.

[5] Gelfand, I.M., M.I.Graev and A. Vershik, Representations of the
group of smooth mappings of a manifold X into compact Lie group,
Commpositio Math. 35(1977), 299-334.

[6] Gelfand, I.M., M.I.Graev and A. Vershik, Representations of the
group of functions taking values in a compact Lie group,
Commpositio Math. 42(1982), 217-243.

[7] Høegh-Krohn R., A general class of quantum fields without cutoffs
in two space-time dimensions, Comm. Math. Phys. 21(1971), 244-255.

[8] Ismagilov, R., On unitary representations of the group $C_0(X,G)$,
G=SU(2), Mat. Sb. 100(1976), 117-131. Math. USSR Sb. 29(1976),
105-117.

[9] Kusuoka, S., and D.W. Stroock, Applications of the Malliavin
calculus, Part II, J. Fac. Sci. Univ. Tokyo Sec. IA, 32(1985),1-76.

[10] Ossipov, Y., On the triviality of the quantum field theory in a
finite volume, Novosibirsk Preprint (1979).

[11] Wallach, N.R., On the irreducibility and in equivalence of
unitary representations of gauge groups, Compositio Math. 64(1987),
3-29.

Research Institute for Mathematical Science

Kyoto University, Kyoto, Japan

BARYON STATES IN LATTICE QCD

R.A. MINLOS, E.A. ZHINZHINA

1 INTRODUCTION This paper is a continuation of our works Minlos, Zhinzina (a),(b) , where the lower meson states were constructed in the lattice QCD (for small κ and β, see below). In this note we construct the lower baryon (and antibaryon) states for this model. We show that there are two one-particle baryon invariant spaces of Hamiltonian with the energy $3\ln\kappa + 0(1)$:   1) the *octet* of the baryons with the spin 1/2 and 2) the *decaplet* of the baryons with the spin 3/2. There are the similar spaces for the antibarions.

2. We consider the model of the quark field with the gauge field interaction (QCD model) on the lattice $\tilde{Z}^4$ (see Fröhlich, King (1988)).

The complete action for the model in Λ is:

$$S_\Lambda = \sum_{x\in\Lambda} \sum_{n=1}^{2} \sum_{i=1}^{3} \sum_{\alpha=1}^{4} \bar{\psi}_{n,i}^{\alpha}(x)\psi_{n,i}^{\alpha}(x)$$

$$+ \kappa \sum_{b=<xy>\in\Lambda} \sum_{n=1}^{2} \sum_{i,j=1}^{3} \sum_{\alpha,\beta=1}^{4} \bar{\psi}_{n,i}^{\alpha}(x)\Gamma^{\alpha\beta}(b)\,U^{ij}(b)\,\psi_{n,j}^{\beta}(y) \tag{1}$$

$$+ \beta \sum_{P\subset\Lambda} Re\,tr(U(\partial P)),$$

where $\Lambda \subset \tilde{Z}^4$ is the finite subset of the lattice $\tilde{Z}^4$ obtained from the lattice Z^4 as a result of the shift by vector $(\frac{1}{2},0,0,0)$. The quark field is represented by anticommuting variables $\{\bar{\psi}_{n,i}^{\alpha}(x),\ \psi_{n,i}^{\alpha}(x)\}$ (generators of the Grassmann field algebra L).
We denote by $\alpha = 1,2,3,4$ the spinor index, by $n = 1,2,3$ the flavor index and by $i = 1,2,3$ the color index.

425

Then $U = \{U(b), b = \, <xy> \,$ is an oriented lattice bond$\}$ is the gauge field defined on the oriented lattice bonds b and with values in the color group $SU(3)$, such that $U(-b) = U^*(b)$. We consider the gauge field U in the radial gauge (i.e. $U(b) \equiv e \in SU(3)$ for all b in the time direction). For any oriented plaquette $P \subset \check{Z}^4 : U(\partial P) = \prod_{b \in \partial P} U(b)$ is the ordered product along the boundary ∂P of P.
The spin matrices $\Gamma(b)$ are:

$$
\Gamma(b) = \begin{cases} \Gamma_{+\mu} = 1 + \gamma_\mu, & \text{if } b \text{ in } +\mu \text{ direction,} \\ \Gamma_{-\mu} = 1 - \gamma_\mu, & \text{if } b \text{ in } -\mu \text{ direction,} \end{cases}
$$

$\gamma_\mu, \mu = 1, 2, 3, 4$ are the 4×4 Euclidean Dirac matrices, satisfying the relations $\gamma_\mu \gamma_{\mu'} + \gamma_{\mu'} \gamma_\mu = 2\delta_{\mu,\mu'} E$.

Let A be the algebra of the bounded continuous functions from the gauge field configurations, then the action (1) defines the Gibbs quasistate $< \cdot >$ on the field algebra $\boldsymbol{A} = A \otimes L$ as the thermodynamic limit of the local quasistates $< \cdot >_\Lambda$ on the algebra $\boldsymbol{A}_\Lambda = A_\Lambda \otimes L_\Lambda$ defined by the formula:

$$
<F>_\Lambda = \frac{1}{Z_\Lambda} \int_{\Omega_\Lambda} \int_{L_\lambda} F e^{-S_\Lambda} \prod_{b \in \Lambda} dU(b) \prod_{\substack{m \in M \\ x \in \Lambda}} d\psi_m(x) d\bar{\psi}_m(x).
$$

Here Ω_Λ is the space of gauge field configurations in Λ, $dU(b)$ is the normed Haar measure on the group $SU(3)$, $F \in A_\Lambda \otimes L_\Lambda$, $\int_{L_\Lambda} \cdot \prod_{m \in M, \, x \in \Lambda} d\psi_m(x) d\bar{\psi}_m(x)$ is the Beresin integral, $m = (\alpha, n, i)$ is a multiindex, M is the set of all multiindexes, Z_Λ is the normalizing factor.

For small κ and β the limit $\lim <F>_\Lambda$ does exist and defines the Gibbs field on $\boldsymbol{A}$. The antilinear timeinvolution θ is defined on the generators of the algebra $\boldsymbol{A}$ by the formulae:

$$
\theta\psi(x) = \bar{\psi}(\phi x)\gamma_0, \quad \theta\bar{\psi}(x) = \gamma_0\psi(\phi x), \quad \theta U(b) = \bar{U}(\phi b),
$$

where $\phi_x = (-x_0, \bar{x})$, if $x = (x_0, \bar{x})$, $x_0 \in Z_1$, $\bar{x} \in Z_3$, $\phi b = \, <\phi_x, \phi_y>$, if $b = \, <x, y>$. The involution θ can be extended on the algebra $\boldsymbol{A}$ by the relation:

$$
\theta(AB) = \theta B \cdot \theta A, \quad A, B \in \boldsymbol{A}.
$$

Then the field $< \cdot >$ is invariant with respect to $\theta : <\theta F> = <\bar{F}>$ and it prossesses the property of OS-positivity (see Seiler (1982)): $<F\theta F> \geq 0$ for any $F \in \boldsymbol{A}_+$, where $\boldsymbol{A}_+ \subset \boldsymbol{A}$ is the algebra of the "future" (see Seiler (1982)). For this field we can define the physical Hilbert space H (by means of the Osterwalder-Schrader construction) with the scalar product $(f, g) = \, <f\theta g>$, $f, g \in H$; the semigroup of time shifts

$\{T_t,\ t \in \mathbf{N}\}$, and the group of space shifts $\{U_w,\ w \in \mathbf{Z}^3\}$ acting in H. It is easy to see that the quasistate $< \cdot >$ is invariant with respect to the homeomorphisms of the algebra L, which are produced by the action of $O(\mathbf{Z}^3)$ (the full rotation group $\mathbf{Z}^3$ acting on the bispinor indexes) and $SU(3)$ (the global flavor group).

Note that this action defines the unitary representations of $O(\mathbf{Z}^3)$ and $SU(3)$ in H:

$$g \to T_g,\ \ g \in O(\mathbf{Z}^3),$$
$$q \to V_q,\ \ q \in SU(3).$$

Let $H^{(i)} \subset H$ be the subspace of all functionals from H, invariant with respect to action of the local gauge transformation:

$$\psi(x) \ \to \ \ell_x^{-1}\psi(x),$$
$$\bar{\psi}(x) \ \to \ \bar{\psi}(x)\ell_x,$$
$$U(b) \ \to \ \ell_x^{-1}U(b)\ell_y,\ \ \ell_x,\ \ell_y \in SU(3),\ \ b =< xy > .$$

The subspace $H^{(i)}$ is invariant with respect to operators T_n, U_w, T_g, V_q, and further we shall consider the restrictions of these operators on $H^{(i)}$, for which we retain the same notations. The operator T (transfer-matrix) commutes with the operators U_w, T_g, V_q.

The space $H^{(i)}$ can be decomposed into the direct sum of two orthogonal subspaces:

$$H^{(i)} \ = \ H^{(i)}_{\text{odd}} \otimes H^{(i)}_{\text{even}},$$

where $H^{(i)}_{\text{odd(even)}}$ consists of the odd (even) elements of the algebra $\mathbf{A}$. Both spaces are invariant with respect to the transfer-matrix T. The baryon states described here are the leading branches of the spectrum of the transfer-matrix in $H^{(i)}_{\text{odd}}$.

3. From the general physical point of view the baryons are formed by three quarks localized on one point of the lattice $\mathbf{Z}^3$. It means that the corresponding space consists of the forms

$$\phi = \sum_{x \in \mathbf{Z}^3} \sum_{i,j,k=1}^{3} \sum_{\ell,m,n=1}^{3} \sum_{\alpha\beta,\delta=1}^{4} A_{i,j,k;\ell,m,n;\alpha,\beta,\delta}(x)\psi_{\ell,i}^{\alpha}(x)\psi_{m,j}^{\beta}(x)\psi_{n,k}^{\delta}(x). \tag{2}$$

These forms are gauge invariant if and only if

$$A_{i,j,k;\ell,m,n;\alpha,\beta,\delta}(x) \ = \ \varepsilon_{ijk} B_{(\ell\alpha),(m\beta),(n\delta)}(x),$$

where ε_{ijk} is the antisymmetrical tensor:

$$\varepsilon_{ijk} \ = \ \begin{cases} 0, & \text{if some two indices are equal,} \\ (-1)^{\pi}, & \text{if all indices are different, and } \pi \\ & \text{is the parity of its permutation.} \end{cases}$$

The coefficients $B_{(\ell\alpha),(m\beta),(n\delta)}(x)$ are invariant with respect to the permutations of the pairs of the indices $(\ell\alpha),(m\beta),(n\delta)$.

The calculations of the matrix elements of the transfer-matrix $(T\phi,\phi)$ on the forms (2) and the analysis of symmetries of these forms based on the representation theory show that there are only two main subspaces of the forms (2): $L^{(oct,1/2)}$ and $L^{(dec,3/2)}$. These subspaces consists of the forms (2) where the spinor indices $\alpha,\beta,\delta = 1,2$. These subspaces are invariant with respect to the actions T_g, V_q of the groups $SU(3)$ (the flavor group) and $O(\mathbf{Z}^3)$ (the full rotation group), and the translations U_w, $w \in \mathbf{Z}^3$.

The representations of $SU(3)$ and $O(\mathbf{Z}^3)$ are (multiplet) irreducible ones on each of the spaces $L^{(oct,1/2)}$ and $L^{(dec,3/2)}$: on $L^{(oct,1/2)}$ these representations coincide with eight-dimensional irreducible representation of $SU(3)$ (octet) and two-dimensional irreducible one of $O(\mathbf{Z}^3)$ (spin $1/2$); on $L^{(dec,3/2)}$ these representations coincide with ten-dimensional irreducible representation of $SU(3)$ (decaplet) and four-dimensional irreducible one of $O(\mathbf{Z}^3)$ (spin $3/2$).

But the spaces $L^{(oct,1/2)}$, $L^{(dec,3/2)}$ are not invariant with respect to the transfer-matrix T. In this note we show that there exist small disturbances of these spaces invariant with respect to the translations U_w, the rotations of $\mathbf{Z}^3$ and the flavor symmetry.

4 Theorem. Let κ and β be small and $\kappa > 0$. Then there exist two mutually orthogonal subspaces $H^{(oct,1/2)}$ and $H^{(dec,3/2)}$ invariant with respect to the operators T, U_w, $w \in \mathbf{Z}^3$, T_g, $g \in O(\mathbf{Z}^3)$, and V_q, $q \in SU(3)$. In $H^{(oct,1/2)}$ and $H^{(dec,3/2)}$ there are the orthonormal bases:

$$\{h_{a,r}^{(oct,1/2)}(x), \quad a = 1,...,8; \quad r = \pm 1/2, \quad x \in \mathbf{Z}^3\}$$

and

$$\{h_{a,r}^{(dec,3/2)}(x), \quad a = 1,...,10; \quad r = \pm 1/2, \pm 3/2, \quad x \in \mathbf{Z}^3\},$$

in which the operators U_w, T_g and V_q act by the formulae:

$$
\begin{aligned}
U_w h_{a,r}^{\tau}(x) &= h_{a,r}^{\tau}(x + w), \\
T_g h_{a,r}^{\tau}(x) &= \sum_{r'} b_{r,r'}^{(s)}(g) h_{a,r'}^{\tau}(g^{-1}x), \\
V_q h_{a,r}^{\tau}(x) &= \sum_{a'} m_{a,a'}^{(f)}(q) h_{a',r}^{\tau}(x),
\end{aligned}
$$

where $\tau = (oct,1/2)$ or $(dec,3/2)$; $f = oct, dec$; $s = 1/2, 3/2$, $\tau = (f,s)$.

Here $B^{(s)}(g) = \{b_{r,r'}^{(s)}(g)\}$ and $M^{(f)}(q) = \{m_{a,a'}^{(f)}(q)\}$ are the matrices of correspondent irreducible representations of $SU(3)$ and $O(\mathbf{Z}^3)$ in some canonical bases (for example the Gelfand-Zetlin basis).

The actions of the transfer-matrix T in these bases are:

$$Th^{\tau}_{a,r}(x) = \sum_{y \in \mathbf{Z}^3 ; r'} a^{\tau}_{r,r'}(x-y)h^{\tau}_{a,r'}(y),$$

and the matrices $A^{\tau}(\xi) = \{a^{\tau}_{r,r'}(\xi)\}$ satisfy the equations:

$$B^{(s)}(g)\, A^{\tau}(\xi)\, B^{(s)}(g^{-1}) = A^{\tau}(g^{-1}\xi),$$
$$\tau = (f,s),\ \ \xi \in \mathbf{Z}^3,\ \ g \in O(\mathbf{Z}^3).$$

If $\beta << \kappa$, then the Fourier transforms of the matrices $A^{(oct,1/2)}$ and $A^{(dec,3/2)}$ are:

$$\tilde{a}^{(oct,1/2)}_{r,r'}(\lambda) = 8\kappa^3\delta_{r,r'} + 16\kappa^5\delta_{r,r'} + 32\kappa^6\sum_{\mu=1}^{3}\cos\lambda_\mu \cdot \delta_{r,r'} + O(\kappa^6);$$

$$r,r' = \pm\frac{1}{2}, \tag{3}$$

$$\tilde{a}^{(dec,3/2)}_{r,r'}(\lambda) = 8\kappa^3\delta_{r,r'} + 16\kappa^5\delta_{r,r'} + 32\kappa^6\sum_{\mu=1}^{3}\cos\lambda_\mu \cdot \delta_{r,r'} + O(\kappa^6);$$

$$r,r' = \pm\frac{1}{2}, \pm\frac{3}{2},\ \lambda \in T^3,\ T^3$$

is three-dimensional torus.

The proof of this theorem is obtained with the help of the cluster decompose of the transfer-matrix T in the special multiplicative basis in $H^{(i)}_{\text{odd}}$ (see Minlos, Zhizhina (b)) and by the exact calculation of some leading matrix elements of τ in this basis. More detailed to obtaine the decomposition of the functions $\tilde{a}^{\tau}_{r,r'}(\lambda)$ up to κ^6 it is sufficient to calculate the matrix elements of T on two forms (2) (contributing to the main order κ^3), and the matrix elements of T on the forms (2) and the gauge-invariant form:

$$\sum_{x \in \mathbf{Z}^3}\sum_{\mu=1}^{3}\sum_{i,j,k,k'=1}^{3}\sum_{\ell,m,n=1}^{3}\sum_{\alpha,\beta,\delta=1}^{4} \varepsilon_{ijk} B_{(\ell\alpha),(m\beta),(n\delta)}(x,\mu) \cdot$$
$$\cdot \psi^{\alpha}_{\ell,i}(x)\psi^{\beta}_{m,j}(x)U^{k,k'}(x,x \pm e_\mu)\psi^{\delta}_{n,k'}(x \pm \ell_\mu),$$

where $\alpha,\beta = 1,2,\ \delta = 1,2,3,4$ and the coefficients $B_{(\ell\alpha),(m\beta),(n\delta)}(x)$ are symmetric ones with respect to the pairs $(\ell\alpha)$, $(m\beta)$ and $(n\delta)$.

The cluster decomposition of T implies that the contribution of the rest matrix elements of T has no influence on the coefficients in (3).

5 REFERENCES

Fröhlich, J., King, C., (1988), *Meson masses and the U(1) problem in lattice QCD,* Nucl. Phys. B, **290**, no.2, 157-187.

Seiler, E., (1982), *Gauge theories*, Springer-Verlag.

Minlos, R.A., Zhizhina, E.A., (a) *Meson states in lattice QCD*, Submitted to Comm. Math. Phys.

Minlos, R.A., Zhizhina, E.A., (b) *The main and perturbed meson states in lattice QCD*, In press.

THE ENERGY OF A WEYL SOLITON

R.F. Streater

Department of Mathematics

King's College

Strand

London WC2R 2LS

Abstract

We show that the energy created by a time-dependent external chiral field $V(x,t)$ interacting with a left-moving Weyl quantized Fermion field in $1 + 1$ dimensions by

$$E = \frac{e^2}{4\pi} \int dx \, (\int \frac{\partial V}{\partial x}(x + t, t)dt)^2$$

when the initial state is the vacuum; for a couloured multiplet with scattering operator $S(x)$ acting locally, it is

$$E = \int \frac{dx}{4\pi} Tr(\frac{\partial S^\star(x)}{\partial x} \frac{\partial S(x)}{\partial x})$$

where the trace is over the colour group.

States of fractional charge appear as condensates of minimum energy; this provides an unconventional model of quarks.

431

1 INTRODUCTION

The occurrence of anomalies in quantum field theory [1] is usually taken as a bad sign,
and ingenious, even desperate, measures are taken to avoid or cancel then. However,
from another point of view [2] anomalies are useful because they enable us to implement
(so far, for the free field) the deep proposal of Haag and Kastler [3], [4] that the
observable fields of a theory ought to be enough to determine the charged states of the
theory. This is "modern" version of an idea of Skyrme [5] that Fermions should be
exitations of a Boson field (which could be observable).

It is not necessary, in the modern version, for the observable field to be a canonical
Boson field; but it should be Boson-like, in that the observables commute at space-like
separated points. In this paper we discuss the Haag-Kastler programme for the free
field.

Consider a multiplet $\{\psi^A : A = 1, ..., N\}$ of free Dirac fields of mass $m \geq 0$, on which
the group $U(N)$ acts (as a "rigid" group, not a local gauge group). Let $F = c^*$algebra
generated by the fields ψ, and let $A \subseteq F$ be the subalgebra of operators left invariant
by the action of $U(N)$ on F. Recent deep work [6] suggests that this typical of the
general set-up.

The "vacuum sector" is the set of states obtained by acting with A on the vacuum state
ψ_0 of the relativistic free field; this action gives the vacuum representation π_0 of A.

Similarly, by acting with A on a state of charge q we get all the states of charge
q, and the action defines the representation π_q of A. The idea of [3] is that these
representations of A are irreducible and inequivalent to each other; and, naturally, the
full set of representations of A is determined in principle by the algebra A itself. We say
that the theory determines its own superselection rules [7]. In fact, A has far too many
representations, and we should limit attention to those representations π, acting on a
Hilbert space H_π which carries a unitary representation U of the Poincaré group with
positive energy, where $U(a, \Lambda)$ implements the space-time translations a, and Lorentz
transformations Λ, of the fields ψ^A. Such a representation π of A is said to be <u>covariant</u>.
(In $d =$ four dimensions with massless fields we would drop the requirement of Lorentz
covariance). One may ask, how can one construct the charged representations π_q,
knowing only A or possibly π_0, and even more, how can one construct ψ, the Fermi
field, out of the observables in A, as desired by Skyrme's programme? Borchers [8]
contributed some insight by making use of the local structure of A: Let $A(\Omega)$ denote
the c^*-algebra generated by elements of A localized in the space-time region $\Omega \subseteq R^d$;
that is $U(N)-$ invariant combinations of fields $\psi^A(f) = \int \psi^A(\vec{x}, t) f(\vec{x}, t) dx dt$ with
$f(\vec{x}, t) = 0$ unless $(\vec{x}, t) \in \Omega$. We call $A(\Omega)$ a local algebra of observables. From
the field property of ψ, we have $A = c^*$-alg $\{A(\Omega) : \Omega \subseteq R^d\}$. The net $\{A(\Omega)\}$

obeys the axioms of Haag and Kastler [3]. Now suppose $\Omega \subseteq \mathbf{R}^d$ is such that its space-like compliment, Ω' is non-empty and open; here $\Omega = \{x \in \mathbf{R}^d; x - y$ is space-like $\forall y \in \bar{\Omega}\}$. Now let f be a test-function zero outside Ω'; then the normal commutation relations between observables and spinor fields requires $A\psi(f) = \psi(f)A$ for all $A \in A(\Omega)$.

More precisely, acting on states of charge q

$$\pi_{q-1}(A)\,\psi(f) = \psi(f)\pi_q(A), \quad A \in A(\Omega) \tag{1}$$

since A, acting on states of charge q, is in the representation π_q, and $\psi(f)$ lowers the charge by 1. This shows that $\psi(f)$ is an intertwining operator between the representations π_{q-1} and π_q regarded as representations of $A(\Omega)$. (No intertwiners except 0 exist between π_{q-1} and π_q as representations of the whole A, since these are irreducible and inequivalent - this is Schur's lemma). Borchers suggested that there should be a unitary operator u doing the work of $\psi(f)$ as a charge-raising operator. Then (1) would say $\pi_{q-1}(A) = u\pi_q(A)\bar{u}^1$ that is, π_q and π_{q-1} are locally equivalent. Like $\psi(f)$, u is localized in Ω'. As we enlarge Ω, we must vary u: the same u will not do for all Ω.

The construction of the charged sectors is clarified by using the general concept of a state on A: a state is a positive linear form i.e. a linear map $\rho : A \to C$ obeying 1) $\rho(1) = 1$, 2) $\rho(A^*A) > 0$ for all $A \in A$. We interpret $\rho(A)$ as the expectation of A in repeated measurements of A when the state is ρ. If π is a representation of A on a Hilbert space H_π, then a vector $\psi \in H_\pi$ define a state, written ρ_ψ, by

$$\rho_\psi(A) = <\psi, \pi(A)\psi>, \quad A \in A.$$

Such a state is called a vector state belonging to π. Every state ρ is a vector state of some representation π_ρ (the GNS representation belonging to the state), and a vector state of one irreducible representation π_1 cannot be a vector state of an inequivalent one, π_2. As a result of this, the expectation values of A in a state of charge q determine the value of q.

The other idea that we need is that of automorphism, the abstract version of a canonical transformation. An automorphism τ of A is a linear bijection $\tau : A \to A$ obeying $\tau(AB) = \tau(A)\tau(B)$ and $\tau(A^*) = \tau(A)^*$ for all $A, B \in A$. We write $\tau \in AutA$.

An example of an automorphism is given by conjugation with a unitary operator U in $A : \tau(A) = UAU^{-1}$. Such an automorphism is called inner. Not all automorphisms are inner. If τ is an automorphism of A and π is a representation of A on H, we say that τ is spatial in π if there exists a unitary operator V on H such that $\pi(\tau(A)) = V\pi(A)V^{-1}$ for all $A \in A$. Thus an inner automorphism is spatial in any representation. If π is a representation of A on H, then we get a new representation of A on the same Hilbert space H by $\pi \circ \tau$, that is $(\tau \circ \tau)(A) \overset{def}{=} \pi(\tau(A))$. Cleary, π and $\pi \circ \tau$ are equivalent if and only if τ is spatial in π. Thus, π and $\pi \circ \tau$ can be inequivalent (as actions on H) even though they act on the same space H (whose vectors give different states in π and $\pi \circ \tau$). We recover the various representations of A from (π, H) by letting τ run over $Aut A$. Borchers' idea, then, can be summarised by saying that the states of charge 1 of A can be obtained from the states of charge 0 by a non-spatial automorphism τ, which is locally spatial: to each Ω with a non-empty open spatial complement, there is a unitary operator u such that $(\pi(\tau(A)) = u\pi(A)u^{-1}$ for all $A \in A(\Omega)$. The unitary operator u is then localized in Ω' and is the charge-raising operator we are looking for; vector states in $\pi \circ \tau^n$ then have charge n, $n \in \mathbf{Z}$.

It is generally agreed that Borchers did not fully succeed in his aims, and a systematic study was made by Doplicher, Haag and Roberts [9]. They extend the method for getting new representations, by allowing morphisms (for which inverses might not exists) as well as automorphisms. Then the sectors may obey parastatistics, which for parafermions means that the observable algebra A is the quotient of a quark Fermion algebra J by a compact Lie group (like $U(N)$). Our set-up takes account of this generalization.

In [10] we found the charge sectors of the (gradient of) the free massless Boson field in $1 + 1$ dimensions. States of non-integral charge correspond to particles that obey statistics intermediate between Boson and Fermion. The automorphishms are given by displacement operators $\phi(x,t) \to \phi(x,t) + \eta(x,t)$ where $\circ \eta = 0$ and η is a c-number source. In [11] we found local covariant quasi-free representations of the free massless Dirac field. These are obtained by local automorphisms

$$\psi(x) \to e^{i\alpha(x) + i\beta(x)\gamma^5}\psi(x) \tag{2}$$

and also lead to states of non-integral charge. In fact, the observable algebras of these two models are isomorphic, and all the representations are, one to one, unitary equivalent. This is the theorem of bosonization [12], [13].

The automorphisms of [10] and [11] map each local algebra $A(\Omega)$ to itself. There is no need for this property, in the general theory of [9]. In fact, if the mass is > 0, or if the dimensions of space-time is > 2, local gauge transformations such as (2) lead to states of infinite energy. Such states could not be made in the laboratory - they would be confined. For this reason, we proposed [2] that the sectors of a free Fermi field ought to be created by scattering automorphisms caused by external gauge fields (for axial charge creation) and external axial gauge fields (for charge creation). This is in line with Roberts' insight [14] that the operators needed to measure charge should be dual to the operators needed to create charge. Scattering automorphisms have been studied by Palmer [15]. We follow the formulation of I.E. Segal (see Bongaarts [16]). Thus we consider the c-number Dirac equation with time-dependent potentials A^0, A^1

$$[i(\partial - ieA\mu)\gamma\mu + m]\psi(x,t) = 0 \tag{3}$$

We choose $A^\mu(x,t)$ to be smooth, and zero outside a compact interval in time.

Here, $\psi(.,t) \in L^2(\mathbf{R},\mathbf{C}^2) = K$ for each time.

Let S be the classical scattering operator for these dynamics, S is unitary, $K \to K$. The scattering automorphism of the Fermi algebra F is the defined by duality. Thus, if $\hat{\psi}_\alpha$ is the Fermi field, the smeared fields are $\psi(f) = \sum_\alpha \int \hat{\psi}_\alpha(x)f_\alpha(x)dx$, $f \in K$; then define $\tau_s(\hat{\psi}(f)) = \hat{\psi}(Sf)$, and define τ_s on polynomials in $\hat{\psi}$ by the multiplicative property of τ_s. Then τ_s extends to an automorphism of F. Bellisard [17] has shown that this automorphism is the same (for a limited class of A^μ) as that given by the ususal physicists' second quantized S operator obtained by solving the Dirac equation (3) for $\hat{\psi}$ instead of ψ.

An automorphism τ can be thrown, by duality, onto the states, and defines the linear transformation τ^* (defined by $\tau^*\omega(A) = \omega(\tau(A))$ for each state ω, $(A \in A)$). It might be thought that the scattering automorphism τ_s would not change the charge of a state, as the formal Hamiltonian is gauge invariant (or seems to be). But the presence of anomalies enables ut to use τ_s^* to create charge (or axial charge if $m = 0$ if the winding number of A^1, A^0 are non-zero [18].

The usual [18] interpretation of $\tau_s^* \omega_0$, where ω_0 is the free vacuum as the "dressed vacuum". When its charge q is non-zero, however, we claim that it is unnatural to interpret it as a vacuum; if $q = \pm 1$ we should interpret $\tau_s^* \omega_0$ as a dressed one-particle state, called a soliton.

In this note we develop a method, used by Fujiwara and Ohnuki [19] for the index, for finding the energy of the state $\tau_s^* \omega_0$ generated from the vacuum by the scattering automorphism τ_s.

We deal with a $U(N)$-multiplet of massless right-going Weyl spinors in $1 + 1$ dimensions. We obtain, §2, the usual form for the index, which is the charge of $\tau_s^* \omega_0$:

$$n = \frac{i}{2\pi} \int dx\, Tr(\frac{\partial S^*(x)}{\partial x} S(x))$$

and for the energy we get (§3)

$$E = \int \frac{dx}{4\pi}\, Tr\left(\frac{\partial S^*(x)}{\partial x}\frac{\partial S(x)}{\partial x}\right)$$

where the trace is over the colour group. The energy is given by a Dirichlet form on the space of maps $S : \mathbf{R} \to U(N)$. We might then consider (as e.g. the most stable configuration) the map S which minimises E, subject to a fixed value of n. This problem is of the same type as that studied by Brézis, Coron and Lieb [20], namely, one of finding "harmonic forms with defects". But, as $\mathbf{R}$ is not compact, the minimum is not achieved.

We find the configuration of minimum energy and charge 1 and located in the interval [0,L]. We argue that this is a possible explanation for the success of the quark picture.

2 THE INDEX

Most of the technical results on the equivalence, or otherwise, of representations of local fields concern the field algebra F rather than the observable algebra A. For Fermion fields we then have available the criterion of Shale and Stinespring [21] (for the Clifford algebra) and Berezin [22] (for the CAR algebra). One should mention the papers of Baker and Powers [23] on the question of representations of a quotient algebra. The ususal criteria concern the class of quasi-free representations. In [14], [25] we classify the non-Fock quasi-free covariant representations of the free Boson and Fermi field and we give some local examples in [11]. But the Fock representations of F already contains various charges states belonging to inequivalent representations of A, the gauge-invariant sub-algebra of A, the gauge-invariant sub-algebra of F. If the mass in zero, and the space-time dimension is 2, then we have, in addition, the representations of non-integral charge [11].

In [26], Labonte considered quasi-free automorphisms of the Dirac field that are spatial in the relativistic free representations. He remarked that when the "determinant" of the transformation is zero (we would now say, when there are zero modes) then the automorphism can create a net charge. Thus, let π be the automorphism of F which on the field $\hat{\Psi}$ has the effect

$$\pi\,\hat{\Psi}(f) = \hat{\Psi}(Sf)$$

where S is unitary on the one-particle space K. Let P_+ be the projection onto the positive-energy solutions of the Dirac equation, so that

$$S = \begin{pmatrix} S_+ & S_\pm \\ S_\mp & S_- \end{pmatrix} = \begin{pmatrix} P_+SP_+ & P_+SP_- \\ P_-SP_+ & P_-SP_- \end{pmatrix}$$

is the decomposition of S in the direct sum $K = P_+ K \oplus P_- K$. For τ to be spatial $S_\pm$ and $S_\mp$ must be in the Hilbert-Schmidt class which implies that S_+ and S_- are Fredholm operators. Labonté showed that the charge created by the automorphism τ is the Fredholm index, Ind S_+ of S_+. That this can be non-zero in $1 + 1$ dimensions, where S is the scattering from a magnetic field, is known as the Klein paradox. It is also known as "polarization of the vacuum", wrongly I think, since according to Maxwell's ideas polarization means the splitting of the aether into equal and opposite charges, with overall conservation. Attempts have been made to derive the paradox in four dimensions [27], but this now seems unlikely [28]. We can understand this now: the phenomenon is associated with the axial anomaly; in $1 + 1$ dimensions the space and time components of the current j^μ are dual, in that $\varepsilon^{\mu\nu} j_\nu$ is the axial current. In four dimensions it is the axial current, the Hodge dual of γ^μ in the Dirac exterior algebra, that creates the charge, not magnetic part of $\gamma^\mu A^\mu$.

It is known that Ind S_+ is the spectral flow of the family of Hamiltonians $\{H(t) = H_0 + V(t), \ -\infty < t < \infty\}$ at least, if space is the torus [29]. When the space is $\boldsymbol{R}$, so that the momentum spectrum is continuous, we cannot count the number of energy-levels passing through 0; we now show that the appropriate generalization of spectral flow is the cyclic cocycle [30].

Let us take Labonté's case, where τ is spatial (as a transformation of F) and S_+ is Fredholm. Since S_+ is partially isometric, its Fredholm index is given by

$$Ind\, S_+ = Tr_+(1 - S_+^* S_+) - Tr_+(1 - S_+ S_+^*)$$

where the trace is over $K_+ = P_+ K$, and 1 is the identity on K_+. Thus

$$Ind\, S_+ = -Tr(S_+^* S_+ - S_+ S_+^*)$$

Note that $S_+^* S_+$ and $S_+ S_+^*$ are not separately of trace class, so we cannot use the cyclicity of the trace to arrive at the incorrect conclusion

$$Ind\, S_+ = -Tr_+(S_+^* S_+) + Tr_+(S_+ S_+^*) = 0$$

Indeed, $Ind\, S_+ = -Tr(P_+ S^* P_+ S P_+ - P_+ S P_+ S^* P_+)$ where the trace is now over K; this is common form of cyclic cocyle constructed on a Fredholm module [30].

This is related to the heuristic method of Fujiwara and Ohnuki [19] as follows. We consider the more general case where the one-particle space has N components instead of one; thus $K = L^2(\boldsymbol{R}, \boldsymbol{C}^N)$, the space of N Fermions of positive chirality. The Dirac energy operator is α. $\nabla = i\partial/\partial x$, and $P_+ = F^{-1}\theta(p)F$ where $F = $ Fourier transform and $\theta(p) = 1, \ p > 0$ and $\theta(p) = 0, \ p < 0$. Let $f \to Sf = f'$ be the unitary local classical scattering operator, $K \to K$. Thus

$$f'_j(x) = S_{jk}(x) f_k(x)$$

Let us adopt the physicists' convention, and take the same transformation

$$\hat{\Psi}'_j(x) = S_{jk}(x) \hat{\Psi}_k(x)$$

for the second-quantized Dirac fiels $\hat{\Psi}_j(x)$ (instead of the transpose of this, as is implied by the definition of τ_S). In terms of the Fourier transforms

$$\hat{\tilde{\Psi}}(p) = \frac{1}{\sqrt{2\pi}} \int e^{-ipx} \hat{\Psi}(x) dx$$

$$\tilde{S}(q) = \frac{1}{\sqrt{2\pi}} \int e^{-iqx} S(x) dx$$

the transformation becomes

$$\hat{\tilde{\Psi}}'(p) = \frac{1}{\sqrt{2\pi}} \int_{-\infty}^{\infty} du\, \tilde{S}_{jk}(p - u) \hat{\tilde{\Psi}}_k(u)$$

and its adjoint

$$\hat{\tilde{\Psi}}'^*(p) = \frac{1}{\sqrt{2\pi}} \int_{-\infty}^{\infty} du\, \hat{\tilde{\Psi}}_k^*(u)\, \tilde{\bar{S}}_{jk}(p - u)$$

We note that $\widetilde{S^*}(q) = \tilde{S}^*(-q)$.

We interpret $\hat{\tilde{\Psi}}(p)$ as a creation operator if $p < 0$ and as an annihilation operator if $p > 0$.

The electron number, $=$ minus the charge, has the formal expression

$$N = \int_0^{\infty} dp\, \hat{\tilde{\Psi}}^*(p) \hat{\tilde{\Psi}}(p) - \int_{-\infty}^0 dp\, \hat{\tilde{\Psi}}(p)\, \hat{\tilde{\Psi}}^*(p)$$

This is not rigorous yet, as $\hat{\tilde{\Psi}}(p)$ is a distribution and should not be multiplied by itself at the same point. Also, the convergence of the infinite integral needs to be examined. But, since N is Wick-ordered, it makes sense as a quadratic form in Fock space. The method of Fujiwara and Ohnuki assumes that the automorphism τ can be extended from A, where it is defined, to singular functions of $\hat{\Psi}(x)$ such as N. Thus Fujiwara and Ohnuki define $\tau(N)$ and assume that the mean number of particles in the state

$\tau^* \omega_0$ is given by

$$\tau^* \omega_0(N) = \; < \int_0^\infty dp\, \tilde{\hat{\Psi}}^*(p)\tilde{\hat{\Psi}}'(p) - \int_{-\infty}^0 dp\, \tilde{\hat{\Psi}}'(p)\tilde{\hat{\Psi}}^*{}'(p) >$$

where $<>$ means expectation in the relativistic vacuum Ψ_0. This gives

$$\tau^* \omega_0(N) = \; < \int_0^\infty dudu' \frac{dp}{2\pi}\, \tilde{\hat{\Psi}}_\beta^*(u')\, \bar{\tilde{S}}_{\alpha\beta}(p-u')\, \tilde{S}_{\alpha\gamma}(p-u)\, \tilde{\hat{\Psi}}_\gamma(u)$$

$$- \int_{-\infty}^0 \frac{dp}{2\pi} dudu'\, \tilde{S}_{\alpha\beta}(p-u)\, \tilde{\hat{\Psi}}_\beta(u)\tilde{\hat{\Psi}}_\gamma^*(u')\tilde{\hat{\Psi}}_{\alpha\gamma}(p-u') >$$

Now, $< \tilde{\hat{\Psi}}_\beta^*(u')\, \tilde{\hat{\Psi}}_\gamma(u) > = \delta(u'-u)\delta_{\beta\gamma}$ if $u < 0$, $< \tilde{\hat{\Psi}}_\beta(u)\tilde{\hat{\Psi}}_\gamma^*(u') > = \delta(u'-u)\delta_{\beta\gamma}$ if $u' > 0$ and zero otherwise. Hence, using $\bar{\tilde{S}}_{\alpha\beta}(p) = \tilde{S}^*_{\beta\alpha}(-p)$,

$$\tau^* \omega_0(N)$$
$$= \int_0^\infty \frac{dp}{2\pi} \int_{-\infty}^0 \tilde{S}^*_{\beta\alpha}(u-p)\, \tilde{S}_{\alpha\beta}(p-u)du - \int_{-\infty}^0 \frac{dp}{2\pi} \int_0^\infty du\, \tilde{S}_{\alpha\beta}(p-u)\, \tilde{S}^*_{\beta\alpha}(u-p)$$
$$= Tr(S^* P_+ S P_- - P_- S P_+ S^*) = Tr(S^*(1-P_-)SP_- - P_- S(1-P_-)S^*) \quad (4)$$
$$= -Tr(S^* P_- S P_- - P_- S P_- S^*)$$

using $S^* S = 1 = SS^* = -Tr_-(S^*_- S_- - S_- S^*_-)$
$= Tr_-[(1_{K_-} - S^* S_-) - (1_{K_-} - S_- S^*_-)] = \operatorname{Ind} S_- = -\operatorname{Ind} S_+ = -\text{charge}.$

Thus, the method of Fujiwara and Ohnuki gives the correct answer, in spite of its heuristic nature.

The scattering operator S is local (i.e. a unitary multiplication operator on $L^2(\mathbf{R}, \mathbf{C}^N)$) in the massless case. Thus the equation

$$(\partial_x + \partial_t)\Psi_j + iV^{jk}(x,t)\Psi_k(x,t) = 0$$

has the S operator

$$S(x) = \mathrm{T}\exp[-1 \int_{-\infty}^\infty dt\, V(x+t,t)]$$

The potential, V, has values in the defining representation of the Lie algebra of $U(N)$. If $N = 1$, the time-ordering, T, can be omitted. To evaluate the index we note the following kernels of $P_\pm$:

$$P_\pm(x,y) = \mp \frac{1}{2\pi i} \frac{1}{x - y \pm i\varepsilon}.$$

Moreover, $P_+S^*P_-S$ and $P_+SP_-S^*$ are separately of trace-class, so we may permute the operators cyclically to get

$$\mathrm{Ind}\,S_+ \;=\; Tr(P_-S^*P_+S \,-\, P_+S^*P_-S) \;=\; TrK.$$

The operator $K = P_-S^*P_+S - P_+S^*P_-S$ has kernel

$$K(x,y) = \int dx\,\frac{1}{2\pi i}\frac{1}{x-z-i\varepsilon}S^*(z)\,(\frac{-1}{2\pi i})\,\frac{1}{x-z+i\varepsilon}S(y)$$
$$-(\frac{-1}{2\pi i})\,\frac{1}{x-z+i\varepsilon}S^*(z)\frac{1}{2\pi i}\frac{S(y)}{z-y-i\varepsilon}\}$$

Putting $x = y$ and integrating over x gives

$$TrK = \int \frac{dx}{2\pi}\frac{d\xi}{2\pi}\,Tr\,S^*(\xi)S(x)\,(\frac{-1}{(\xi-x+i\varepsilon)^2} + \frac{1}{(\xi-x-i\varepsilon)^2})$$
$$= \int \frac{dx}{2\pi}\frac{d\xi}{2\pi}\,Tr\,S^*(\xi)S(x)\frac{\partial}{\partial x}(\frac{1}{x-\xi-i\varepsilon} - \frac{1}{x-\xi+i\varepsilon})$$
$$= \int \frac{dx}{2\pi}\frac{d\xi}{2\pi}\,Tr(S^*(\xi)S(x))\,2\pi i\frac{\partial}{\partial x}\,\delta(x-\xi)$$
$$= -i\int \frac{dxd\xi}{2\pi}\,Tr(S^*(\xi)\frac{\partial S(x)}{\partial x})\,\delta(x-z)$$

if $\frac{\partial S}{\partial x}$ is continuous. This gives

$$\mathrm{Ind}\,S_+ = -i\int \frac{dx}{2\pi}\,Tr(S^*(x)\frac{\partial S(x)}{\partial x})$$

(the usual form for the index of a Toeplitz operator [31]), if $S(x)$ is constant at ∞, which gives the usual formula

$$\frac{1}{2\pi}\int dx\int dt\,V(x+t,t)\quad\text{in the abelian case.}$$

3 THE ENERGY

The new representations $\pi_0 \circ \tau$ are covariant, and so there exists a positive self-adjoint energy operator H generating the free dynamics. It does not follow from this that the state $\tau^*\omega_0$ has finite energy - technically it might not be in the form domain of H. This question has to be examined in each case.

The second quantized energy operator corresponding to the (first quantized) Dirac operator $-i\gamma_5\frac{\partial}{\partial x} = \gamma_5 p$ is

$$H_0 = \int_{p>0} pdp\,\tilde{\tilde{\Psi}}^*(p)\tilde{\tilde{\Psi}}(p) - \int_{p<0} pdp\,\tilde{\tilde{\Psi}}(p)\tilde{\tilde{\Psi}}^*(p)$$

since $\gamma_5 =$ on right-movers and -1 on left-movers. Then in the spirit of [19] the energy of the state $\tau^* \omega_0$, if finite, is given by

$$\tau^* \omega_0(H_0) \;=\; < H > \;=\; < \int_{p>0} p \, dp \, \tilde{\tilde{\Psi}}^{\,\prime *}(p) \tilde{\tilde{\Psi}}^{\,\prime}(p) \;-\; \int_{p<0} p \, dp \, \tilde{\tilde{\Psi}}^{\,\prime}(p) \tilde{\tilde{\Psi}}^{\,\prime *}(p) \;>$$

if finite.

Lundberg [32] has considered generators, X, of one-parameter groups, but chooses to define the zero-point such that $< X > = 0$. While the energies in inequivalent representations of A are incomperable in that the representations act on different Hilbert spaces, in our cases we know that the representations $\pi_0 \circ \tau$ gives us the family of free Fermion states, for which there is a well-established convention for energies. To put it zero is throwing the baby out with the bathwater.

The only difference between $< N' >$ and $< H' >$ is a factor p. Hence, proceeding as in §2, we get instead of (4)

$$\tau^* \omega_0(H) \;=\; \int_0^\infty \frac{dp}{2\pi} \int_{-\infty}^0 du \, \tilde{S}_{\beta\alpha}(u-p) \, p \tilde{S}_{\alpha\beta}(p-u)$$

$$-\int_{-\infty}^0 \frac{dp}{2\pi} \int_0^\infty du \, \tilde{S}^*_{\beta\alpha}(u-p) p \tilde{S}_{\alpha\beta}(p-u).$$

This is equal to $Tr(P_- S^* P_+ p \, S - P_+ S^* P_- p S)$ a cyclic cocycle written entirely in terms of operators on the one-particle space. Compared with the kernel of K, this has an extra $p = -i\nabla$; since the kernel of $\frac{d}{dx} A$ is $\frac{\partial A}{\partial x}(x, y)$, when $A(x, y)$ is the kernel of an operator A, we get for the energy the trace of the operator with kernel

$$\int d\xi \, \{ \frac{1}{2\pi i} \frac{1}{x - \xi - i\varepsilon} S^*(\xi) \frac{\partial}{\partial \xi} (\frac{-1}{2\pi i} \frac{1}{\xi - y + i\varepsilon}) S(y)$$

$$-\frac{-1}{2\pi i} \frac{1}{x - \xi + i\varepsilon} S^*(\xi) \frac{\partial}{\partial \xi} (\frac{1}{2\pi i} \frac{S(y)}{\xi - y - i\varepsilon}) \}$$

Putting $x = y$, tracing over colour and integrating over x gives

$$< H > = i \int \frac{dx \, d\xi}{2\pi^2} Tr S^*(\xi) S(x) \{ \frac{-1}{(x - \xi - i\varepsilon)^3} + \frac{1}{(x - \xi + i\varepsilon)^3} \}$$

$$= -i \int \frac{dx \, d\xi}{8\pi^2} Tr S^*(\xi) S(x) \frac{\partial}{\partial x} \frac{\partial}{\partial \xi} \{ \frac{1}{x - \xi - i\varepsilon} - \frac{1}{x - \xi + i\varepsilon} \}$$

$$= -i \int \frac{dx \, d\xi}{8\pi^2} Tr S^*(\xi) S(x) \frac{\partial}{\partial x} \frac{\partial}{\partial \xi} 2\pi i \, \delta(x - \xi)$$

so

$$E = Tr \int \frac{dx}{4\pi} \frac{\partial S^*}{\partial x}(x) \frac{\partial S}{\partial x}(x) \tag{5}$$

provided S is constant at ∞.

Luckily, it is non-negative. This expression for the energy is a Dirichlet form for the class of smooth maps $S : \mathbf{R} \to U(N)$ that are constant at ∞.

We might expect that the most stable states, among the class $\tau^* \omega_0$, would be those for which $< H >$ is a minimum. Such minimizing maps are known as harmonic forms [20]. We are thus led to the classical variational problem of minimizing, among maps $S : \mathbf{R} \to U(N)$, the expression (5) subject to a fixed index

$$n = \frac{-i}{2\pi} \int dx \, Tr \, (\frac{\partial S^*}{\partial x}(x)S(x)).$$

The solutions to such problems are known as harmonic maps with defects [20], and are used to describe the equilibrium states of classical spin systems with a given dislocation.

The minimum to be problem at hand is not achieved, since the space $\mathbf{R}$ is not compact. To make progress, let us seek the minimum of E and the minimizing function, among maps in the <u>form domain</u> of the Dirichlet form, and such that $S(x)$ is constant outside the specific set $[0, L] \subseteq \mathbf{R}$ (and with given index). Since the index is the winding number of det $S(x)$, only the diagonal part of S contributes to the index, whereas the $SU(N)$ part contributes to the energy. Hence the map giving the minimum energy is of the form

$$S(x) = \begin{pmatrix} e^{i f_1(x)} & & & 0 \\ & \cdot & & \\ & & \cdot & \\ & & & \cdot \\ 0 & & & e^{i f_N(x)} \end{pmatrix}$$

with energy

$$E = \frac{1}{4\pi} \sum_{j=1}^{N} \int_0^L |f_j'(x)|^2 \, dx$$

subject to

$$\frac{1}{2\pi} \sum_{1}^{n} \int_0^L f_j'(x) \, dx = n$$

The function

$$
f_j(x) = \begin{cases} 0 & x < 0 \\ \frac{n_j}{L}x & 0 \le x \le L \\ \frac{n_j}{L} & x \ge L \end{cases}
$$

is in the form domain of the Dirichlet form; this is the set of f such that $p\tilde{f}(p) \in L^2$. Also, such an f_j clearly minimises

$$
\int_0^L | f_j'(x) |^2 \ dx
$$

subject to $f_j(L) - f_j(0) = n_j$.

The energy of such a choice for f_j is

$$
E = \frac{1}{4\pi} \sum_{j=1}^{n} \frac{n_j^2}{L}
$$

The minimum of this, subject to $\sum n_j = n$ is achieved when all n_j are equal, to n/N. The energy is thus $E = \frac{n^2}{4\pi N L}$.

In particular, the state of charge one in the $U(N)$ model, of minimum energy and localized in $[0, L]$, contains N identical states of the N fermions $\hat{\Psi}_j, ..., \hat{\Psi}_N$; each state has charge $\frac{1}{N}$, and is neither a Boson nor a Fermion but is a twisted collective mode of the massless field $\hat{\Psi}_j$, described in [33], [34]. Thus, at least in $1 + 1$ dimensions, the dressed one-particle state of charge 1 would appear to be composed of massless "quarks" of charge $1/N$. However, the model is not satisfactory; the quarks of different colour do not interact, in that the energy is unchanged when the quarks are pulled apart, e.g. if $\hat{\Psi}_1$ is twisted in $[0, L]$, $\hat{\Psi}_2$ is twisted in $[L, 2L]$ etc. This model is related by bosonization to N copies of the free massless field ϕ, whose sectors where found in [10]. Thus, the c^*-algebra A (for the Dirac field with $N = 1$) is isomorphic to the CCR algebra generated by the current $\partial_\mu \phi$, $\varepsilon^{\mu\nu}\partial_\nu \phi$ and moreover, the sectors of A of charges q, q_5 also are unitarily equivalent to the corresponding Boson sectors [13]. The "vacuum" energy $\frac{1}{4\pi} \int | f'(x) |^2 \ dx$, is just the energy of the coherent state of the corresponding representation of the Boson field. However, with $N > 1$, Bosonization is more involved [35], and it is difficult to obtain the energy rigorously from Boson theory.

One can show that these states are eigenstates of charge [36]. In other respects the model is unrealistic, since all states of the field have zero mass.

5 REFERENCES

1 S. Adler; Axial-vector Vertex in Spinor Electrodynamics. Phys. Rev. **177**, 2426-2438 (1969)

R. Jackiw and J.S. Nuovo Cimento **51**, 47 (1969).

2 R.F. Streater; Charge, Anomalies and Index Theory, in *Fundamental Aspects of Quantum Theory* pp. 253-266. V. Gorini and A. Frigerio (Eds.). Plenum Press, 1986.

3 R. Haag and D. Kastler; An Algebraic Approach to Quantrum Field Theory. J. Math Phys. **5**, 848-861 (1964). R. Haag in [4].

4 R.F. Streater (Ed.) *Mathematics of Contemporary Physics*, Academic Press, 1972.

5 T.H.R. Skyrme; A Non-Linear Field Theory I Proc. Royal Soc. **A247**, 260 (1958); II, ibid, **A252**, 236 (1959). III, ibid, **A260**, 127 (1961). Particle States of a Quantized Meson Field; ibid, **A262**, 237 (1961).

6 S. Doplicher and J.E. Roberts, Endomorphisms of C^*-algebras, Cross Products and Duality for Compact Group. Warwick preprint.

7 J.L. Bonnard and R.F. Streater; Local Gauge Models Predicting their own Superselection Rules; Proceedings of the Moscow Conference in Mathematical Physics, 1972. Trudy Mat. Inst. Steklov, 135 (1975). Helv. Phys. Acta **49**, 259-267 (1976).

8 H.J. Borchers; Local Rings and the Connection of Spin with Statistics. Commun. Math. Phys. **1**, 291-307 (1965).

9 S. Doplicher, R. Haag and J.E. Roberts, Field, Observables and Gauge Transformations. I. Commun. Math. Phys. **13**, 1-23 (1969). II: ibid, **15**, 173-200 (1969); Local Observables and Particle Statistics I; ibid, **23**, 199 (1971).

10 R.F. Streater and I.F. Wilde; Fermion States of a Boson Field. Nucl. Phys. **B24**, 561-575 (1970).

11 Z. Hermaszewski and R.F. Streater. Some Local Covariant Representations of the Canonical Anticommutation Relations. J. Phys. A. Math. Gen. **16**, 2801-2810 (1983).

12 R.F. Streater; Gauge Theory of Superselection Rules. Lectures at Schladming, 1973. p. 317 (Springer, Ed. R. Urban); Acta Phys. Aust. Suppl. XI.

13 R.F. Streater and A.S. Wightman. PCT, Spin and Statistics and all That. (2nd Ed., Appendix). Benjamin-Cummings, 1978.

14 J.E. Roberts. Localization in Algebraic Field Theory. Commun. Math. Phys. **85**, 87-98 (1982).

15 J. Palmer; Scattering Automorphisms of the Dirac Field; Jour. Math. Anal. And Appl. **64**, 189-215 (1976).

16 P.J.M. Bongaarts; The Electron-Positron Field, coupled to External Electromagnetic Potentials, as an Elementary C^*-Algebra Theory. Ann. Phys. **56**, 108-139 (1970). See also, Linear Fields According to I.E. Segal, in Ref [4].

17 J. Bellisard, Quantized Fields in Interaction with External Fields. I. Exact Solutions and Perturbation expansion. Commun. Math. Phys. **41**, 235-266 (1975). II. Existence Theorems. Ibid, **46**, 53-74 (1976).

18 A.K. Raina and G. Wanders: The Gauge Transformations of the Schwinger Model. Ann. of Phys. **132**, 404-426 (1981). M. Klaus and G. Scharf, Helv. Phys. Acta **50**, 803 (1977).

19 T. Fukiwara and Y. Ohnuki, preprint of Nagoya Univ. DPNU - 86-07 (1986). Anomalies and Weyl Modes in Chiral Gauge Theories I, II.

20 H. Brézis, J.M. Coron and E.H. Lieb. Harmonic Maps with Defects. Commun. Math. Phys. **107**, 649-705 (1986).

21 D. Shale and W.F. Stinespring. States of the Clifford algebra; Ann. of Math. **80**, 365-381 (1964); J. Math and Mech. **14**, 315 (1965).

22 F. Berezin. *The Method of Second Quantization.* Academic Press, New York, 1966.

23 B.M. Baker and R.T. Powers. Product States of the Gauge Invariant and Rotationally Invariant CAR Algebras. J. Operator Theory **10**, 365-393 (1983) and papers cited therein. See also T. Matsui: On Quasi-equivalence of Quasifree States of Gauge Invariant CAR-algebras. J. Operator Theory, **17**, 281-290 (1987).

24 K. Kraus and R.F. Streater. Some Covariant Representations of Massless Fermi Fields. J. Phys. Math. Gen. **A14**, 2467-2478 (1981). A.L. Carey and D.M.O. Brien. Absence of Vacuum Polarization in Fock space; Lett. Math. Phys. **6** 335-340 (1982).

25 L. Polley, G. Reents and R.F. Streater. Some covariant representations of Massless Boson Fields. J. Phys. A. Math. Gen. **A14**, 2479-2488 (1981).

26 G. Labonté. On the Nature of "Strong Bogoliubov Transformations for Fermions. Commun. Math. Phys. **36**, 59-72 (1974).

27 A.I. Akhiezer and V.B. Berestetsky; Quantum Electrodynamics; Dept. of Commerce, (USA) Wash. D.C. (1953).

28 G. Scharf and H.P. Seipp; Charged Vacuum, Spontaneous Position Production and All That. Phys. Lett. **B108**, 196-198 (1982). A.L. Carey and D.M. O'Brien. Absence of Vacuum Polarisation in Fock Space. Lett. Math. Phys. **6**, 335-340 (1982).

29 N.H. Christ. Conservation Law Violation by Anomalies. Phys. Rev. **D31**, 1591-1602 (1980).

30 A. Connes. Non-commutative Differential Geometry. Publ. Math. IHES, **62**, 257-360 (1985).

31 R.G. Douglas. Banach Algebra Techniques in the Theory of Toeplitz Operators. Regional Conf. Series In Mathematics, No. 15. (Amer. Math. Soc., Providence, R.I.).

32 L.E. Lundberg; Quasi-free Quantization; Commun. Math. Phys. **50**, 103-112 (1976).

33 J. Frøhlich and P. Marchetti; Bosonization, Topological Solitons and Fractional Charges in Two-Dimensional Quantum Field Theory Commun. Math. Phys. **116**, 127-173 (1988).

34 F. Gallone, A. Sparzani, R.F. Streater and C. Ubertone; Twisted Condensates of Quantized Fields. J. Phys. A Math. Gen. **19**, 241-258 (1986).

35 E. Witten; Non-abelian Bosonization in Two Dimensions. Commun. Math. Phys. **92**, 455-472 (1984).

36 D. Prorok; Some Remarks on the Dressed Vacuum in a Two-Dimensional Model. Kings preprint.

MORPHOLOGY AND CLASSIFICATION
OF GALAXIES
A STOCHASTIC MODEL

Sergio ALBEVERIO [*,#], Philippe BLANCHARD [**,#]
Daniel GANDOLFO [***,##], Raphael HØEGH-KROHN
Mohamed MEBKHOUT [****,##]

Abstract: We propose a probabilistic approach to the old problem of classification of galaxies. We show that under suitable dynamical hypothesis, there exist equilibrium measures associated with stochastic processes modeling the motion of typical galactic objects in R^3, whose maxima match all the observed galactic morphologies, as classified by Hubble.

Subject headings: Stochastic Process. Newtonian diffusion Process. Classification of galaxies.

*   Fakultät für Mathematik, RUHR-UNIVERSITAT D-4630 BOCHUM1 (GERMANY) SFB 237 (ESSEN-BOCHUM-DUSSELDORF).

**   Fakultat für Physik, UNIVERSITAT BIELEFELD, D-4800 BIELEFELD (GERMANY).

***   Département de Mathématiques, UNIVERSITE DE TOULON ET DU VAR (FRANCE).

****   Département de Physique, UNIVERSITE D'AIX-MARSEILLE II, LUMINY (FRANCE).

\#   BIBOS, Bielefeld-Bochum Stokastik.

\#\#   Centre de Physique Théorique C.N.R.S. LUMINY Case 907 MARSEILLE.

447

I. INTRODUCTION

The most recent works, both theoretical [Pf1] and computational [LC], concerned with the study of galactic structures, reveal the fact that up to now, a satisfactory explanation of the observed shapes of galaxies is still lacking.

Nevertheless, the hypothesis of existence and primordial influence of the so-called "dark matter", seems well accepted by the astrophysical community as a main ingredient involved in the mechanism of early stage formation of galaxies [Ko].

The constituents of this dark population, whose total mass is large enough to play a significant role in galactic dynamics, as we shall see [San], comprise among others: degenerate stars, neutron stars and black holes [BM]. These objects are considered to be remains of old stars which, under gravitational collapse have produced this massive and widely dispersed dark component. The adjective "dark" comes from the fact that, due to a very intense gravitational field, these objects emit little light, consequently it is only through indirect channels, such as strongly energetic perturbative gravitational effects that they can be detected, or via spinning magnetic fields, like in the case of neutron stars.

Let us turn now to galaxy shapes. Concerning elliptical galaxies, it is believed that the initial gas has turned into stars quickly enough due to a low velocity dispersion of the black matter. Then because stars interact seldomly with each others, this could explain the more or less ellipsoïdal structure of this type of galaxies. Alternatively, if the star formation process is slower, corresponding to a high velocity dispersion of the black matter, then the gas has got enough time to aggregate under gravity, and collapse into a disk; only a part of it contributes to form a spheroïdal central component, the overall process leading to a disk galaxy.
The total mass of the galaxy is thought to be the origin for the value of the dark matter velocity dispersion, although this hypothesis is still controversal. The most massive objects are thought to give rise to elliptical galaxies.

Disk galaxies separate into two major classes according to their central structure. One of these, the spiral galaxies, shows a central spheroïdal concentration of stars connected to two spiral arms. We know that arms formation aroused latter in the galactic evolution, through some density wave process. In the second class one finds galaxies presenting in their centre, a more or less rectilinear bar of stars. These galaxies are called for this reason: barred galaxies. In fact, contrary to what was believed earlier, the proportion of barred galaxies among disk galaxies is not negligible at all. The origin of mechanisms that allow formation of this bar is not yet totally understood. A pointer to the right explanation would be as follows[Pf1]: the perturbation of a star circular orbit leads to an elongation of this orbit in the direction perpendicular to the direction of the perturbation. This perturbation could come for example from the well known phenomenon of differential rotation of the galaxy [BM] [SS1] [SS2], arising from the fact that stars angular speed is a function of the radial distance, then resonances between orbits frequency and bar rotation frequency increases the perturbation phenomenon. These instabilities lead to the fact that the stars orbits cannot be periodic any longer, but rather quasi-periodic, their superposition tending more and more to a bar shaped structure. These considerations are confirmed by numerical computations on supercomputers[Pf2], but analytical confirmations are still missing.

The aim of this paper is to show that through a suitable probabilistic approach, it is possible to compute and exhibit the observed galactic structures, classified in the famous, still up to date, Hubble's sequence [Hub].

It has been shown [Wie] that one could explain the observed increase of velocity dispersion of stars, by the existance of local fluctuations of the galactic gravitational field, the effect of wich can be described by a diffusion process. The resulting diffusion of stellar orbits prevents any determination of stellar birth places, and has strong implications for stellar dynamics of galaxies in general.

Actually binary stars collisions cannot account alone for these fluctuations. Spitzer and Schwarzschild recognized ([SS1], [SS2]) that encounters with interstellar cloud complexes could explain the large variations of kinetic energy experienced by stars, and they were able to derive a quantitative description for stellar velocity dispersion. Another well known destabilizing phenomenon is the propagation of M.H.D. waves from the galactic center [Sof], whose influence on arm formation in galaxies seems to be well established.

Our approach uses the hypothesis that a typical object, at an early stage of development of the larger structure it belongs to, moves in an external force field according to a suitable stochastic (mean) Newton law, performing a diffusion process (at least in a certain approximation). We have already mentioned how stochasticity enters in the description of the motion of this typical object. Mathematically we choose to model such a motion by an equation of the type of the one of stochastic mechanics (see [Ne], [BCZh] and, for applications to model various physical phenomena, e.g. [ABH1], [ABH2], [B], [ABH3]).

The position at time t of the typical object being denoted by X_t, we have that X_t solves a stochastic differential equation of the form:

$$dX_t = b(X_t, t)\, dt + \sqrt{\sigma(X_t)}\, dW_t$$

where W_t is a Brownian motion, σ is, in general, a diffusion matrix, and $b(X_t, t)$ a drift vector. The physical picture behind such an equation is that X_t performs approximately (for small values of σ) a classical motion solving $\frac{dX_t}{dt} = b(X_t, t)$, perturbed by a stochastic term $\sqrt{\sigma(X_t)}\frac{dW_t}{dt}$, where $\frac{dW_t}{dt}$ is a "white noise" (completely uncorrelated noise). For general references on the mathematical theory of stochastic differential equations see e.g. [Ar], [Gr] and for some discussion in connections to physics see e.g. [K], [DeE]. The specific form of the above equation in stochastic mechanics is determined by the assumption that a stochastic Newton equation (Newton equation in the mean) holds (see Sect. III below for details). This dynamical input then essentially determines X_t, and in particular its probability distribution as well as the drift $b(X_t, t)$. One calls such processes X_t, Newtonian diffusions (see e.g. [ABH3], [BCZh]). It is not the purpose of this article to give a full dynamical justification of such a model, see however e.g. [ABH1], [BCZh] for some plausibility arguments and methods, and [Dü] for investigations in related models. In this paper we assume the validity of such a model for the motion of a typical object in its environment (at least at an early stage in the evolution of the larger structure it belongs) and deduce some consequences.

The probability distribution density ρ of the process at time t, is such that the probability that X_t is in A is given by $\int_A \rho(x, t)\, dx$ for any Borel measurable subset A of the configuration space in which X_t moves. The basic observation we exploit is that for any Newtonian diffusion, the nodal set $N_\rho \equiv \{(x, t) \,|\, \rho(x, t) = 0\}$ acts as an impenetrable barrier, i.e. with probability one, the process does not get through the barrier. If the nodal set is non trivial, it can be shown that there is a decomposition of the configuration space into invariant regions separated by N_ρ; one speaks in this case of "trapping" or "confinement" or "non tunneling" since the process X_t does not cross N_ρ with probability one. We refer to [AH], [Na], [AFK], [F], [C], [ZhM], [BGo] for a mathematical discussion of this confinement effect (which has been applied in the last few years to many phenomena revealing regular patterns [ABH1], [ABH2], [B], [ABH3], [ABCRSC], [BCZh], [Na]).

In Section II we give a short description of the main goals of stochastic mechanics and explain in an heuristic way the mechanism for the formation of impenetrable barriers for Newtonian diffusions. Basic mathematical notions about Newtonian diffusions are given in Section III. In Section IV we give an application of the confinement mechanism for Newtonian diffusion to the problem of morphology and classification of galaxy structures.

This paper was begun during a visit in Oslo following an invitation by Raphael, few months before his death. The other authors have completed the paper after his departure, trying to preserve the spirit of this work, according to the discussions we had with him.

II. TRAPPING FOR NEWTONIAN DIFFUSIONS
THE PHYSICAL MECHANISM

Let us first briefly explain the goals of stochastic mechanics. For a particle of mass m, moving in Euclidean space R^N under the influence of a force F derived from a (smooth) potential function V, i.e.

$$F = -\nabla V \qquad (2.1)$$

the dynamics is given by Newton's law:

$$m\ddot{x}(t) = -\nabla V(x(t)) \qquad (2.2)$$

From this it follows that the possible trajectories of the particle are smooth curves. The physical case is of course $N = 3$, but for model building it is also useful to consider other values of N.

Stochastic Mechanics can be viewed as a minimal randomization of Classical Mechanics in the following sense. The kinematical law asserts that the paths of the particles are given by the trajectories of a stochastic (strong) Markov process with continuous paths. The Markov property (past and future being independant given the present) reflects the fact that due to high frequency of random changes coming from local irregularities and collisions, the particle keeps memory of the past only through its present state. In other words the trajectories are given by the solutions of a stochastic differential equation of the type :

$$dX_t = b(X_t, t)\, dt + \sqrt{\sigma(X_t)}\, dW_t \qquad (2.3)$$

where $\frac{dW_t}{dt}$ (W_t being the standard brownian motion in R^N), is a stochastic translation invariant and isotropic perturbation (white noise), b is a drift vector and σ a diffusion matrix. It is well known that roughly speaking dW_t is proportional to $\sqrt{dt}$, so that the paths of W_t, hence of X_t are nowhere differentiable with probability one (for these well known facts see e.g. [Ar], [K]).

Due to the non differentiability of X_t it is necessary to define, following Nelson [Ne] (see Section III), a mean stochastic acceleration a, to generalize Newton's law in this stochastic framework:

$$m\, a(X_t) = -\nabla V(X_t) \qquad (2.4)$$

This (stochastic) dynamical law plays in this context the role of a constraint allowing to determine the drift vector field b and the probability density $\rho(t, x)$ of the process X_t. For example the probability for the process X_t to be in the subset A of R^N is given by:

$$P[X_t \in A] = \int_A \rho(t, x)\, dx \qquad (2.5)$$

The drift field b contains a term of the form $\frac{\nabla \rho}{\rho}$ and is therefore singular on the nodal surface N_ρ of the density ρ, defined by:

$$N_\rho = \{x \in R^N \,|\, \rho(t, x) = 0\} \qquad (2.6)$$

The trapping property for the process X_t results from the fact that the nodal surface N_ρ can never be reached. This property can be explained in an heuristic way. Indeed the gradient of a function pointing in the direction where the function increases, points always from N_ρ and typical paths of the process X_t are repelled by the nodal surface (see e.g. [AH1], [AFK], [BGo]).

III. BASIC MATHEMATICAL FACTS ABOUT NEWTONIAN DIFFUSIONS

We come now to a short discussion of some mathematical concepts and results. For a careful account of the theory see [Ne], [BCZh]. Let Δ be the associated Laplace operator in N dimensional Euclidean space R^N. We consider infinitesimal generators of the form:

$$L = \frac{\sigma}{2}\Delta + b \cdot \nabla \tag{3.1}$$

$$b = (b_1, ..., b_N) \quad , \quad b \cdot \nabla = \sum_{i=1}^{N} b_i \frac{\partial}{\partial x_i}$$

where b is a (non random) smooth vector field (the drift) (with components b_i), which might depand explicitly on the time t. $\sigma(x)$ is a $N \times N$ strictly positive definite matrix (the diffusion matrix) depanding smoothly on x.

Let X_t be a diffusion process running on R^N which is analytically described by its infinitesimal generator L. A purely probabilistic description of the process is given by the solution of the stochastic differential equation (2.3). In addition, if we take an initial condition or distribution, then under suitable conditions the process will be well defined for all t, see [Ar], [Gr].

Let dx be the volume element on R^N. Since b and a are smooth, we know that there exists a smooth density $\rho(t, x)$ of the probability law P of X_t, with respect to dx, i.e.

$$P(X_t \in dx) = \rho(t, x)\, dx$$

Let f be a smooth real valued function on R^N, then:

$$E(f(X_t)) = \int_{R^N} f(x)\, \rho(t, x)\, dx \tag{3.2}$$

and

$$\frac{d}{dt} E(f(Xt)) = \int_{R^N} (Lf)(x)\, \rho(t, x)\, dx \tag{3.3}$$

$E(.)$ denoting the expectation (average) with respect to the process X_t.

By taking the adjoint in the last formula we obtain the Fokker-Planck equation:

$$\frac{\partial}{\partial t}\rho = \frac{\sigma^2}{2}\Delta\rho - \operatorname{div}(b \cdot \rho) \tag{3.4}$$

The connection between the drift b and the process X_t is that $b(X_t)$ is the (mean) forward derivative $D_t X_t$ of X_t at time t in the sense that:

$$b^i(x, t) = \lim_{\Delta t \downarrow 0} \frac{E\left[X_{t+\Delta t}^i - X_t^i \,|\, X_t = x\right]}{\Delta t} \tag{3.5}$$

where $E(\,.\,|\,X_t = x\,)$ means conditional expectation with respect to $X_t = x$ (for these concepts see e.g. [Ar], [Ne], [BCZh]). In other words the average is taken under the condition that we know that the process is at time t in point x. $b(X_t, t)$ can be interpreted as the mean (forward) velocity of the particles living the point x at time t.

Let $\hat{X}_t \equiv X_{-t}$ be the time reversed process. It is again a Markov process with infinitesimal generator:

$$\hat{L} = \frac{\sigma^2}{2}\Delta - b^* \cdot \nabla \tag{3.6}$$

b^* being the backward drift defined by:

$$b^* = b - \frac{\sigma^2}{2}\nabla \ln \rho = \lim_{\Delta t \downarrow 0} \frac{E\left[X_t - X_{t-\Delta t}\mid X_t = x\right]}{\Delta t} = D_- X_t \tag{3.7}$$

in other words, b^* is the backward derivative of X_t. It can be interpreted as the mean velocity of the particles entering x at time t.

By the same procedure as above, one is lead to Fokker-Planck equation for the time reversed process:

$$-\frac{\partial}{\partial t}\rho = \frac{\sigma^2}{2}\Delta\rho - \operatorname{div}\left(b^* \cdot \rho\right) \tag{3.8}$$

Let $v = \frac{b+b^*}{2}$ and assume that v is a gradient i.e. there exists a smooth function S, such that $v(X_t) = \nabla S(t, \hat{X}_t)$.

The current velocity v satisfies the continuity equation:

$$\frac{\partial \rho}{\partial t} + \operatorname{div}\left(\rho v\right) = 0 \tag{3.9}$$

We require that the stochastic Newton law:

$$m\, a(X_t) = -\nabla V(X_t) \tag{3.10}$$

holds for some function V on R^N, with:

$$a(X_t) = \frac{1}{2}(D_+ b^* + D_- b)(X_t) = \frac{1}{2}(D_+ D_- + D_- D_+)(X_t) \tag{3.11}$$

so that a is the mean acceleration. Setting now:

$$\psi(t, x) = \sqrt{\rho(t, x)}\, e^{i\frac{S(t, x)}{\sigma}} \tag{3.12}$$

it can be shown ([Ne], [BCZh]) that ψ satisfies the following Schrödinger like partial differential equation:

$$i\frac{\partial \psi}{\partial t} = H\psi \tag{3.13}$$

with $H = -\frac{\sigma}{2}\Delta + \frac{1}{\mu\sigma}V$. This generalization of the classical dynamical law appropriate to diffusion motion, gives a dynamical meaning to the drift b. Conversely, if ψ satisfies this equation, and if we define ρ and S as above, then setting $v = \frac{b+b^*}{2} = -\nabla S$, and $u = \frac{b-b^*}{2} = \frac{\sigma^2}{2}\nabla \ln \rho$, we obtain b, which yields L, and hence a process X_t with probability distribution $\rho(t, x)$ for all t. Moreover the process X_t satisfies the Newton's law in the mean (3.10).

The stationary probability distributions, i.e. $\rho(t, x) = \rho(x)$ indepandent of t, are given by $\rho(x) = |\psi(x)|^2$, with ψ solution of the eigenvalue problem (stationary Schrödinger like equation):

$$H\psi = E\psi, \quad \text{with} \quad \psi \in L^2(R^N, dx). \tag{3.14}$$

$L^2(R^N, dx)$ being the space of square integrable functions over R^N. For more details see [ABH2], [ABH3], [BCZh].

In the case where the nodal set $N_\rho = \{x \in M \mid \rho(x) = 0\}$ is non trivial, it has been shown (see e.g. [AH1], [AH2], [ABH2], [B]) that there is a decomposition of R^N into invariant regions C_i separated by N_ρ. One speak in this situation of "trapping" or "confinement" in C_i since the process started in C_i does not leave C_i with probability 1, the nodal set N_ρ acting as impenetrable barrier. This is the mathematical description of the physical mechanism we discussed in Section II.

Remark : if the Euclidean space R^N is replaced by a Riemannian manifold M, all results discussed in this section still hold. The Laplace operator on R^N is replaced by the Laplace-Beltrami operator on M and the Lebesgue measure dx by a Riemannian volume element.

IV. APPLICATION TO THE CLASSIFICATION OF GALAXIES

IV.1. Some remarks about the choice of stationary distributions

We can now specify the physical parameters related with stellar diffusion in galaxies, according to the above specified mathematical framework. Let us consider a stellar object of mass μ in the gravitational field of a massive central kernel of mass M. The choice of this generic object depands clearly of the scale of the considered physical phenomenon. For our purpose, we will consider that this object is a typical constituent of the galaxy at an early stage formation of it. In this application, the diffusion process X_t describes the stochastic motion of this typical diffusing object. According to our modeling, the motion of this object is governed by the following stochastic differential equation:

$$dX_t = b(X_t, t)\, dt + \sqrt{\sigma(X_t)}\, dW_t \tag{4.1}$$

where b is the drift field, σ the diffusion matrix, and W_t a standard Brownian motion in R^3. Assuming, as explained in Sections II and III, the Newton's law in the mean:

$$\mu a(X_t) = -\nabla V(X_t), \tag{4.2}$$

a being the stochastic mean acceleration defined in Section III, and $V(X_t) = -\frac{GM\mu}{|x|}$, the Newton potential (G is the Newton's gravitational constant), we proceed by applying the results of Section III to determine the stationary distributions of our process. For simplicity, we assume σ to be a constant multiple of the unit matrix.

These stationary distributions depand only on the radial variable $r = |x|$ and on the angles θ and ϕ, the potential being central.

We consider invariant measures of the form:

$$\rho(x)dx = \left| \sum_i \alpha_i\, \psi_{n_i, l_i, m_i}(r, \theta, \phi) \right|^2 dx = \left| \sum_i \alpha_i\, R_{n_i, l_i}(r)\, Y_{l_i}^{m_i}(\theta, \phi) \right|^2 dx \tag{4.3}$$

the $Y_{l_i}^{m_i}(\theta, \phi)$ being the spherical harmonics of indices l_i and m_i; $R_{n_i, l_i}(r)$ is a radial function. The natural numbers n_i account for the different eigenvalues of the radial part of $H = -\frac{\sigma}{2}\Delta + \frac{1}{\mu\sigma}V$. The diffusion matrix σ is, in all these discussions, a free adjustable parameter of the model. The maxima of the density in each trapping region (one of those into which $\rho = 0$ divides R^3), can be interpreted as domains of large matter concentration. Between two maxima, the model predicts volumes of space, which are quite empty.

What this work will show is that, for suitable choices of parameters, the maxima of ρ match the observed galactic patterns referenced in Hubble's classification scheme.

The results shown at the end of this paper, have been obtained for small values of n_i, typically $n_i = 1$ to 4, so for the least energetic configurations, (smallest eigenvalues of H), henceforth the most stable (in the sense of stochastic mechanics).

In the search for suitable solutions, we have been guided by symmetry considerations as well as structural ones. The solutions $\rho(x)$ were choosen such that, in a spherical coordinate system (r, θ, ϕ), the probability density $\rho(r, \theta, \phi)$ be a maximum near the plane $\theta = \frac{\pi}{2}$. This choice allows us to look for solutions that could fit in the class of disk galaxies. Symmetry with respect to the origin of the coordinate system (more or less the center of mass of the galactic kernel), has also been a selecting criterium for acceptable solutions. Finally, the angular depandance of spherical harmonics, of the form $e^{im\phi}$, has been a deciding factor concerning the arm structure of disk galaxies.

Up to now the diffusion σ was considered as a free adjustable parameter of the model. If σ is assumed to be a constant diffusion coefficient, then its value can be evaluated in the following way.

The radial depandence of ρ satisfies the following differential equation:

$$-\frac{\sigma^2}{2}\frac{d^2}{dr^2}(rR_{n_i,l_i}(r)) + \frac{\sigma^2}{2}\frac{l_i(l_i+1)}{r^2}(rR_{n_i,l_i}(r)) - V(r)\,rR_{n_i,l_i}(r) = -\frac{G^2M^2\mu^2}{2\sigma^2}R_{n_i,l_i}(r)$$

This radial depandence is given by a Laguerre polynomial times an exponential of the radial variable: $e^{-n\alpha r}$, where $\alpha = \sqrt{\frac{2E}{\mu\sigma^4}} > 0$. From this it is possible to get an estimate for the diameter of the structures we identify with galaxies. Knowing the energy E of the configuration: $E = \frac{\mu}{2}\left(\frac{GM}{n\sigma^2}\right)^2$, and typical values of galaxy parameters, such that the kernel mass M, and its galactic diameter, one can deduce a value for the diffusion coefficient σ.

One finds $\sigma \simeq 10^{22}\,m^2\,s^{-1}$, to be compared to known values of observed stellar diffusion.

IV.2. Numerical results

In the pictures shown at the end of this paper, we present the results of numerical computations and plottings of various solutions of our eigenvalue problem, corresponding to different galactic morphologies.

The graphics shown here are density plots representations. The brightest parts of each plot reflect the locii of the maximum of the probability density ρ. We have choosen this representation for its obvious relationship with camera made pictures of galaxies, and also because it gives clear informations about the nodal set N_ρ. In all examples, $\rho(x)$ contains as component, the eigenfunction $\psi_{1,0,0}(r)$, wich is a purely spherical contribution, whose main role has been to elaborate the galactic kernel.

Figures 1 to 8 exhibit known standard galaxy structures, going from the most typical ones, elliptical and lenticular, to several types of spiral galaxies (normal and barred). Figures 9 to 11 show that this model can also give quite interesting patterns connected with evolutionary structures (it is possible to built an animation of these pictures, giving rise to a very suggestive dynamical transfert effect!). Figure 12 is remarquable in the sense that it shows that our model can exhibit some very fondamental elementary patterns involved in galaxy structure. As already mentioned, we have only used eigenfunctions $\psi_{n,l,m}(x)$ with $n \leq 4$, this fact insuring that these structures are among the most stable one.

All computation and graphics where performed on Mathematica software (Wolfram Research).

IV.3. Conclusions and possible extensions

What comes out from this analysis is that there is still a long way before we can give satisfactory answers to all the questions asked to us by galactic world. Many subjects are in constant theoretical development (dark matter physics, etc...), and computer simulations are of primary help by suggesting possible scenarios for evolution phenomena.

Our stochastic model allows, through some very natural hypothesis, to recover, at least phenomenologically, the Hubble's classification scheme of galaxies, by solving an eigenvalue problem, through stochastic analysis of trajectories of stellar objects. A possible extension of this work would be to study the time depandent eigenvalue problem in order to find new hints concerning dynamical properties for the evolution of galactic structures.

To conclude let us remark that our mathematical framework is a reinterpretation of Nelson's stochastic mechanics, which originally was conceived to understand quantum mechanics from classical mechanics (see e.g. [Ne], [BCZh]) and provides a general description of a class of classical dynamical systems randomly perturbed. It is fascinating that the model seems to indicate a striking connection between phenomena in the very large and phenomena in the very small. This question has already been investigated by several authors, in related contexts (see e.g. [ABHKM], [Ca], [Ch], [ChRo], [Co1], [CoTi], [D1], [G], [So]).

Acknowledgements

We are indebted to J.M. Souriau and R. Triay for helpfull discussions.
Various stays at the Université d'Aix-Marseille II, Faculté des Sciences de Luminy as well as the partial support of the Volkswagenstiftung (Project BiBoS) are gratefully acknowledged.

References

[ABCRSC] S. Albéverio, Ph. Blanchard, Ph. Combe, R. Rodriguez, M. Sirugue, M. Sirugue Collin, Trapping in stochastic mechanics and applications to covers of clouds and radiation belts, in "Quantum Probability and Applications II", Edts. L. Accardi, W. von Waldenfels, Lect. Notes Math. 1136, Springer (1985)

[ABHl] S. Albéverio, Ph. Blanchard, R. Hoegh-Krohn, A stochastic model for the orbits of planets and satellites. An interpretation of Titius-Bode-law, Exp. Math. 4, 365-373 (1983)

[ABH2] S.Albéverio, Ph. Blanchard, R. Hoegh-Krohn, Diffusions sur une variete riemanienne. Barrieres infranchissables et applications, in "Colloque en l'Honneur de Laurent Schwartz", Asterisque, Societe Mathematique de France, 132, 181-201 (1985)

[ABH3] S. Albéverio, Ph. Blanchard, R. Hoegh-Krohn, Newtonian diffusions and planets, with a remark on non-standard Dirichlet forms and polymers, pp. 1-24, "Stochastic Analysis and Applications", Edts. A. Truman, D. Williams, Lect. Notes Math. 1095 Springer, Berlin (1984)

[ABHKM] S. Albéverio, Ph. Blanchard, R. Hoegh-Krohn, M. Mebkhout, Strata and voids in galactic structures. A probabilistic approach. In preparation.

[AFK]S. Albéverio, M. Fukushima, W. Karwowski, L. Streit, Capacity and quantum mechanical tunneling, Commun. Math. Phys.81, 501-513 (1985)

[AH1] S. Albéverio, R. Hoegh-Krohn, A remark on the connection between stochastic mechanics and the heat equation, J. Math. Phys. 15 1745-1747 (1974)

[AH2] S. Albéverio, R. Hoegh-Krohn, Dirichlet forms and diffusion processes on rigged Hilbert spaces, Zeitsch. für Wahrscheinlichkeitstheorie und verwandte Gebiete 40, 1-57 (1977)

[Ar] L. Arnold, Stochastic Differential Equations, J. Wiley (1974)

[B] Ph. Blanchard, Trapping for Newtonian diffusion processes, Acta Physica Austriaca, Suppl. XXVI 185-209 (1984)

[BCZh] Ph. Blanchard, Ph. Combe, W.A. Zheng, Mathematical and Physical Aspects of Stochastic Mechanics. Lect. Notes in Physics, Springer Verlag, (1986)

[BGo] Ph. Blanchard, S. Golin, Global existence for diffusion processes with singular drifts. Commun. Math. Phys. 109, 421-435 (1987)

[BM] Binney, Mihalas. Galactic Astronomy, Structure and Kinematics. Freeman and Co. 1981.

[C] E. Carlen, Existence and sample path properties of the diffusion in Nelson's stochastic mechanics in "Stochastic processes: Mathematics and Physics", Edts. S. Albéverio, Ph.Blanchard, L. Streit, Lect. Notes Math. 1158, Springer, Berlin (Proc. BiBoSSymp.I)

[Ca] J.C. Carvalho, On large scale quantization, Lett. Nuovo Cim. 44, 337-342 (1985)

[Ch] G.Chincarini, Redshifts and large scale structures, pp. 159-165 in G.O. Abell, G. Chincarini, Early evolution of the universe and its present structure, IAU Symp. No. 104, D. Reidel (1983)

[ChRo] G. Chincarini, H.J. Rood, The cosmic tapestry, Sky and Telescope, May 1980

[Col] W.J. Cocke, A theoretical framework for quantized redshifts and uncertainty in cosmology, Astrophys. Letters 23, 239-249 (1983)

[Co2] W.J. Cocke, Theory and interpretation of quantized extragalactic redshiflts, Astrophys. J. 288, 22-28 (1985)

[CoTi] W.J. Cocke, W.G. Tifft, Redshift quantization and the discrete groups of galaxies, Astrophys. J. 268, 56-59 (1983)

[Dl] M. Desarkissian, Does wave-particle duality apply to galaxies?, Lett. Nuovo Cim. 40, 390-394 (1984)

[D2] M. Desarkissian, Possible evidence for gravitational Bohr orbits in double galaxies, Lett. Nuovo Cim. 44, 629-636 (1985)

[Dü] D. Dürr, V. Naroditsky, N. Zanghi, On the motion of an impurity in a harmonic crystal, pp. 187-236 in "Stochastic Processes in Classical and Quantum Systems", Proc. Ascona 198, Edts. S. Albéverio, G. Casati, D. Merlini, Lect. Notes Phys. 262, Springer (1986)

[DeE] C. De Witt-Morette, D. Elworthy, Edts., New stochastic methods in physics, Phys. Repts. 77, 121-382 (1981)

[F] M. Fukushima, Energy Forms and Diffusion Processes, pp. 65-97, in "Mathematics and Physics", Lecture on Recent Results, Vol. I (L. Streit, ed.), World Scientific Publishing, Singapore (1985)

[G] D.W. Greenberger, Quantization in the large, Found. Phys. 13, 903-951 (1983)

[Gr] Th. C. Gard, Introduction to stochastic differential equations, M. Dekker, New York(1988)

[Gre] S.A. Gregory, L.A. Thompson "Superclusters and voids in the Distribution of Galaxies", Scientific American, March 1982

[Hub] E. P. Hubble, The realm of the nebulae, Yale University Press. 1936.

[K] N. G. v. Kampen, Stochastic Processes in Physics and Chemistry, North-Holland, Amsterdam (1981)

[Ko] J. Kormendy, G.R. Knapp, Edts., Dark matter in the universe, IAU Sympos. No. 117, D. Reidel, Dordrecht (1987)

[LC] G. Lake, R.G. Carlberg, Astr. J. N 96, 1988.

[Na] M. Nagasawa, Segregation of a population in an environment, J. Math. Biology 9, 213-235 (1980)

[Ne] E. Nelson, Quantum fluctuations, Princeton University Press(1985)

[Pf1] D. Pfenniger, The Structure of Barred Galaxies, Europhysics News. N 21, 1990.

[Pf2] D. Pfenniger, Astr. Astrophys. N 141, 1984.

[Sa] H. Sato, Voids in the expanding universe, pp. 289-312 in B. Bertott et al., Edts., General relativity and gravitation, D. Reidel, Dordrecht (1984)

[San] A. Sandage, Astr. Astroph., N161, 1988.

[Sof] Y. Sofue, Propagation of M.H.D. waves from the galactic center. Astr. Astroph. N60,1977.

[So] J.M. Souriau, Un modèle d'univers confronté aux observations, pp. 114-160, in "Dynamics and Processes", Edts. Ph. Blanchard, L. Streit, Lect. Notes Math. 1031, Springer (1983)

[Th] T.X. Thuan, "La formation de l'Univers", La Recherche 174, Fevrier 1986

[SS1] L. Spitzer, M. Schwarzschild, Astrophys. J. N114, pp 385, 1951.

[SS2] L. Spitzer, M. Schwarzschild, Astrophys. J. N118, pp 106, 1953.

[Wie] R. Wielen, The diffusion of stellar orbits derived from the observed age depandence of the velocity dispersion. Astr. Astrophys. N60, 1977.

[ZhM] W.A. Zheng, P.A. Meyer, Construction de processus de Nelson reversibles, pp. 12-26 in Sem. Prob. XIX, Edts. J. Azema, M. Yor, Lect. Notes Math. 1123, Springer, Berlin 1985

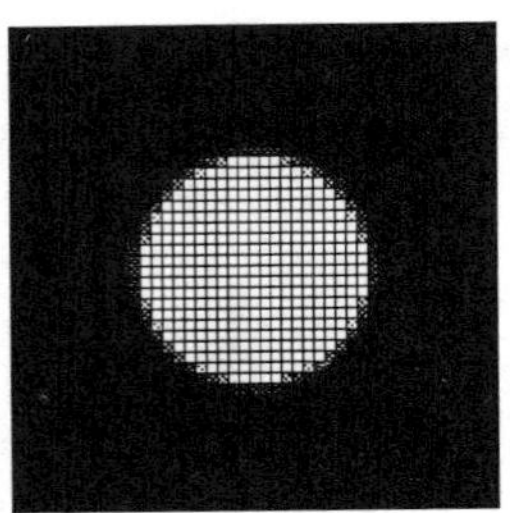

Fig. 1 : E0 galaxy (elliptical)

$$\rho(r,\theta=\tfrac{\pi}{2},\phi) = \|\ \alpha\ \Psi_{1,0,0}\ \|^2$$

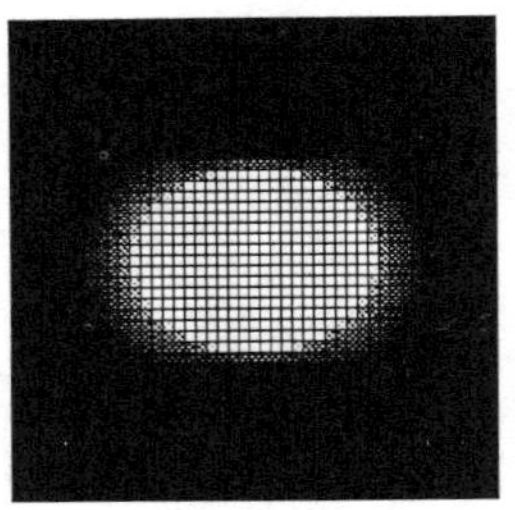

Fig. 2 : E7 galaxy (elliptical)

$$\rho(r,\theta=\tfrac{\pi}{2},\phi) = \|\ \alpha\ \Psi_{1,0,0} + \beta\ \Psi_{3,2,0} + \gamma\ \Psi_{3,2,2}\ \|^2$$

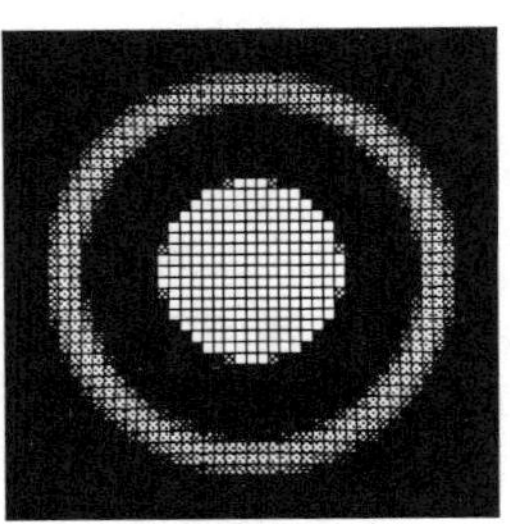

Fig. 3 : S0 galaxy (lenticular)

$$\rho(r,\theta=\tfrac{\pi}{2},\phi) = \|\ \alpha\ \Psi_{1,0,0} + \beta\ \Psi_{3,0,0}\ \|^2$$

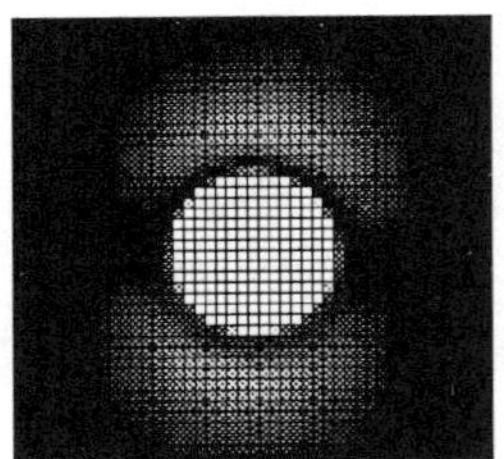

Fig. 4 : SB0 galaxy (lenticular)

$$\rho(r,\theta=\tfrac{\pi}{2},\phi) = \|\ \alpha\ \Psi_{1,0,0} + \beta\ \Psi_{3,1,1} + \gamma\ \Psi_{3,1,-1}\ \|^2$$

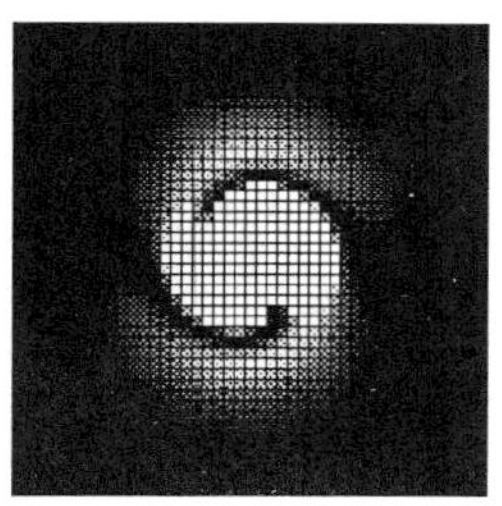

Fig. 5 : Sa galaxy (normal spiral)

$$\rho(r,\theta=\tfrac{\pi}{2},\phi) =$$
$$\| \, \alpha\Psi_{1,0,0} + \beta_{\pm}\,\Psi_{2,1,\pm1} + \gamma_{\pm}\,\Psi_{3,1,\pm1} \, \|^{2}$$

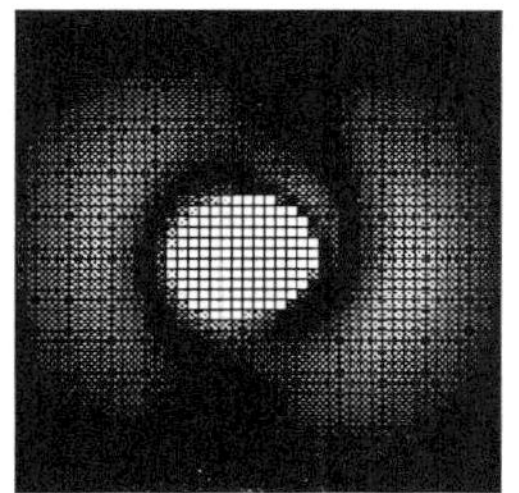

Fig. 6 : Sb galaxy (normal spiral)

$$\rho(r,\theta=\tfrac{\pi}{2},\phi) =$$
$$\| \, \alpha\Psi_{1,0,0} + \beta\Psi_{2,1,-1} + \gamma\Psi_{3,2,2} + \delta_{\pm}\,\Psi_{4,1,\pm1} \, \|^{2}$$

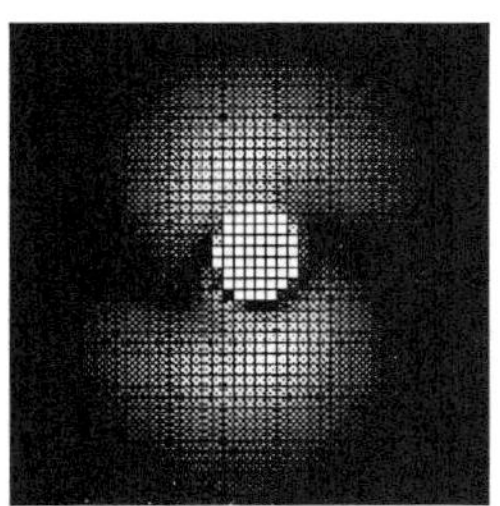

Fig. 7 : Sc galaxy (normal spiral)

$$\rho(r,\theta=\tfrac{\pi}{2},\phi) =$$
$$\| \, \alpha\Psi_{1,0,0} + \beta_{\pm}\,\Psi_{2,1,\pm1} + \gamma_{\pm}\,\Psi_{3,1,\pm1} + \delta_{\pm}\,\Psi_{4,1,\pm1} \, \|^{2}$$

Fig. 8 : SB galaxy (barred spiral)

$$\rho(r,\theta=\tfrac{\pi}{2},\phi) =$$
$$\| \, \alpha\Psi_{1,0,0} + \beta\Psi_{3,2,0} + \gamma\Psi_{3,2,2} + \delta_{\pm}\,\Psi_{4,2,\pm1} \, \|^{2}$$

Phenomena of mass transfer

$$\rho = \|\alpha\Psi_{1,0,0}+\beta\Psi_{3,1,\_1}+\gamma\Psi_{3,2,2}+\delta\Psi_{3,2,0}+\varepsilon_{\pm}\Psi_{4,1,\pm1}\|^2$$

Fig. 9

Pictures: 9,10,11, describe a phenomena of mass transfer between the central body of a galactic strucure and a   arm .

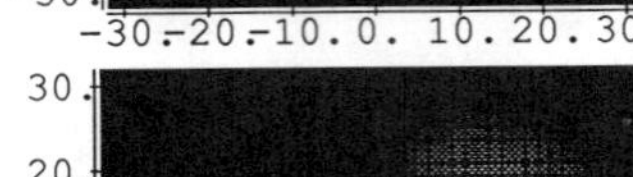

Fig. 10

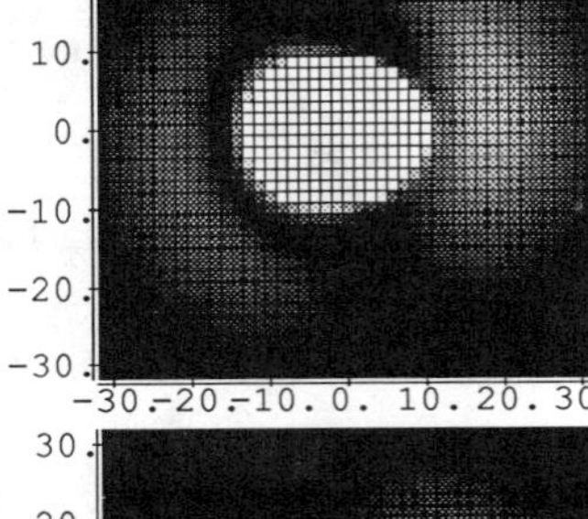

Fig. 11

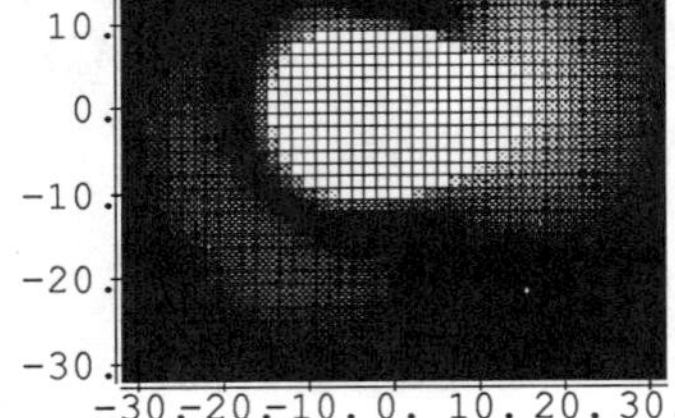

Fig. 12: NGC 4303: primary spectral component.

This shows a pattern similar to the one giving the primary spectral component of NGC4303, through spectral analysis. (Internal Kinematics and Dynamics of Galaxies, Symposium n°100, International Astronomical Union, 1982)

$$\rho = \|\alpha\Psi_{1,0,0}+\beta_{\pm}\Psi_{3,1,\pm1}+\gamma_{\pm}\Psi_{3,1,\pm1}+\delta\Psi_{4,1,\pm1}\|^2$$

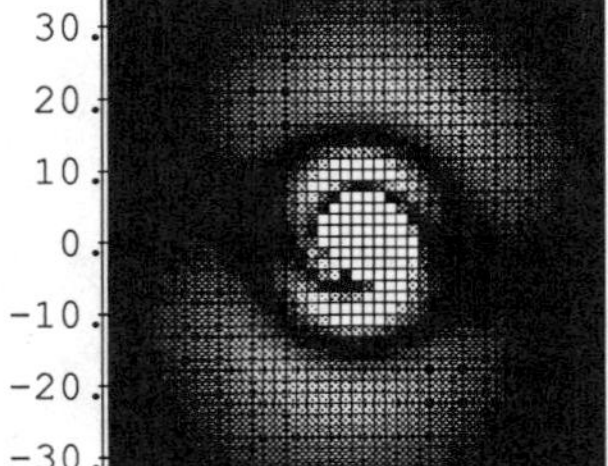

Thermodynamic Inequalities for the Surface Tension and the Geometry of the Wulff Construction

R.L. DOBRUSHIN, S.B. SHLOSMAN

Institute for the Problems of Information Transmission of the Acamdemy of Sciences, Moscow, USSR

Abstract

We study positive functions F on $S^{k-1} \subset \mathbf{R}^n$, which satisfy some natural inequalities, inspired by the statistical mechanics considerations. We call them Weak Simplex Inequalities (WSI) and Strong Simplex Inequalities (SSI). It turns out that WSI select the subset of functions, for which the Wulf construction (resulting in a convex shape W_F when applied to F) is one-to-one transformation. The SSI is the only obstruction to smoothness: if F satisfies WSI and is smooth, then the boundary ∂W_F is smooth iff F satisfies SSI.

1 INTRODUCTION.

In a recent paper [DKS] the validity of the well-known Wulff construction for the shape of a droplet of a phase floating in another phase was rigorously proven for two-dimensional ferromagnets in the canonical ensemble formalism.

The main ingredient required for carrying out the Wulff construction is the surface tension function $\tau_\beta(n), n \in S^1$ (where S^1 is the unit circle) defined defined as a limit

$$\tau_\beta(n) = -1/\beta \lim_{N \to \infty} (\ell_N(n))^{-1} \ln[Z_N^\beta(n)/Z_N^\beta(+)]. \tag{1.1}$$

Here $\ell_N(n)$ is the length of the intersection of a straight line $s(n)$ in $\mathbf{R}^2$, passing through the origin in the direction orthogonal to n, with a square

$$\{x = (x^1, x^2) \in \mathbf{R}^2 : |x_i| \leq N, \quad i = 1, 2\},$$

$Z_N^\beta(n)$ and $Z_N^\beta(+)$ are partition functions for two-dimensional ferromagnetic finite range model with inverse temperature β in a square volume

$$V_N = \{t = (t^1, t^2) \in \mathbf{Z}^2 : |t^i| \leq N, i = 1, 2\}$$

461

with boundary conditions

$$\bar{\sigma}_n(t) = \begin{cases} 1 & (n,t) \geq 0, t \in \mathbf{Z}^2 \\ -1 & (n,t) < 0, t \in \mathbf{Z}^2 \end{cases} \tag{1.2}$$

and

$$\bar{\sigma}_+(t) \equiv 1, \qquad t \in \mathbf{Z}^2. \tag{1.3}$$

The partition function $\mathbf{Z}_N^\beta(\bar{\sigma})$ in a volume V_N at inverse temperature β and with boundary conditions $\bar{\sigma}(t), t \in \mathbf{Z}^2$, where $\bar{\sigma}(t) = \pm 1$, is defined by

$$Z_N^\beta(\bar{\sigma}) = \sum_\sigma \exp\{-\beta(\sum_{\{s,t\} \subset V_N} U(s-t)\sigma_s\sigma_t + \sum_{\substack{s \in V_N, \\ t \in \mathbf{Z}^2 \setminus V_N}} U(s-t)\sigma_s\bar{\sigma}_t)\}, \tag{1.4}$$

where the summation is taken over all configurations $\sigma = (\sigma_t, t \in V_N) \in \{-1,1\}^{V_N}$. Finally, finite-range ferromagnetic potential $U(t), t \in \mathbf{Z}^2$ is a non-positive function such that $U(t) = 0$ if $|t| > R$, where $R < \infty$ is the range of interaction. Of course, $\tau^\beta(n)$ is an even function of n.

In [DKS] it is proven among other results, that for large enough β the limit (1.1) exists for any $n \in S^1$, is analytic in n and satisfies the inequality

$$\tau_\beta(n) + \tau_\beta''(n) > 0, \qquad n \in S^1. \tag{1.5}$$

The quantity $\tau_\beta + \tau_\beta''$ is known in physical literature as a stiffness coefficient, [H], [FFW].

It is easy to check that (1.5) implies by integration the following inequality (see Sect. 3). Let A, B, C be three points in general position in $\mathbf{R}^2$; n_{AB}, n_{BC}, n_{AC} be unit vectors, orthogonal to the sides of the triangle ABC, and $|AB|, |BC|, |AC|$ are the lengths of its sides. Then for any such triangle ABC

$$|AB|\tau_\beta(n_{AB}) + |BC|\tau_\beta(n_{BC}) > |AC|\tau_\beta(n_{AC}). \tag{1.6}$$

We shall call (1.6) as Strong Triangle Inequality and (1.5) - as a differential variant of Strong Triangle Inequality.

In some cases only weak variants of these inequalities hold. We shall give the name Weak Triangle Inequality to the inequality

$$|AB|\tau_\beta(n_{AB}) + |BC|\tau_\beta(n_{BC}) \geq |AC|\tau_\beta(n_{AC}) \tag{1.7}$$

and we shall refer to the inequality

$$\tau_\beta(n) + \tau_\beta''(n) \geq 0, \qquad n \in S^1 \tag{1.8}$$

as to differential variant of Weak Triangle Inequality. It is easy to see that if we consider a natural "conical" extension of the surface tension function to $\mathbf{R}^2$, given by

$$\tau_\beta(x) = |x|\tau_\beta\left(\frac{x}{|x|}\right), x \neq 0, \quad \tau_\beta(0) = 0, \tag{1.9}$$

then Weak Triangle Inequality is equivalent to the convexity of the extended function, and Strong Triangle Inequality - to its strict convexity.

To give an example, let us consider a ground state surface tension

$$\tau_{\beta=\infty}(n) = \lim_{\beta \to \infty} \tau_\beta(n). \tag{1.10}$$

It is easy to find that for ferromagnetic models

$$\tau_\infty(n) = a|(n, e_1)| + b|(n, e_2)|, \tag{1.11}$$

where $e_1, e_2 \in S^1$ are coordinate lattice vectors, and a, b are constants, depending on the potential. This function satisfies only the Weak Triangle Inequality (1.7): the inequality (1.7) becomes an equality if all three normal vectors n_{AB}, n_{BC}, n_{AC} lie in the same quadrant. The inequality (1.8) becomes an equality in all points where $\tau_\infty(n)$ is differentiable.

The rigorous proof of the inequality (1.6) given in [DKS] is quite complicated, so we shall give here only main ideas of it. Let $A = (t_A^1, t_A^2)$ and $C = (t_C^1, t_C^2)$ be points in the dual lattice $\mathbf{Z}^{2*}$ such that $t_A^1 < t_C^1$ and the set

$$V_{AC} = \{t = (t^1, t^2) \in \mathbf{Z}^2, t_A^1 < t^1 < t_C^1; |t^2| \le N\},$$

where N and the difference $t_C^1 - t_A^1$ are large enough. We consider the Gibbs ensemble in V_{AC} with boundary conditions of the type (1.2) where $n = n_{AC}$ is orthogonal to AC. For any configuration σ in V_{AC} there exists a unique contour $\Gamma(\sigma)$ (i.e., a connected line formed by the links of the dual lattice which separate different values of σ) which connects the points A and C (see Fig. 1). This contour is a (random) separation line between the two phases.

FIGURE 1

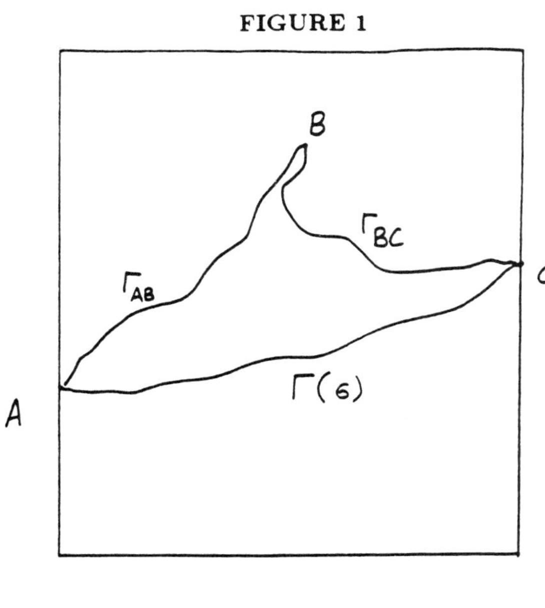

The Gibbs ensemble in V_{AC} defines the probability distribution for $\Gamma(\sigma)$:

$$Prob(\Gamma(\sigma) = \Gamma) = Z_{AC}^{-1} \times \sum_{\sigma:\Gamma(\sigma)=\Gamma} \exp\{-\beta(\sum_{(s,t)\in V_{AC}} U(s-t)\sigma_s\sigma_t + \sum_{s\in V_{AC}, t\in V_{AC}^C} U(s-t)\sigma_s\bar{\sigma}_t)\} \tag{1.12}$$

where Z_{AC} is the partition function in V_{AC} with the boundary conditions (1.2), where $n = n_{AC}$. Let also a point $B = (t_B^1, t_B^2)$ in the dual lattice be given, such that $t_A^1 < t_B^1 < t_C^1$ and $|t_B^2| < N$. We suppose that it is far enough from the boundary of V_{AC}. We consider a subset $G(B)$ of the set G of all contours Γ, which pass through B. Each contour $\Gamma \in G(B)$ consists of two parts, $\Gamma = \Gamma_{AB} \cup \Gamma_{BC}$; Γ_{AB} connects A and B and Γ_{BC} connects B and C. To find a probability of the event $\Gamma \in G(B)$ we have to take the sum of the probabilities (1.12) over all $\Gamma \in G(B)$, and we can sum separately over Γ_{AB} and Γ_{BC}. If one neglects the interaction between Γ_{AB} and Γ_{BC} and normalize, one gets

$$Prob(\Gamma(G) \in G(B)) \approx \left(\frac{Z_{AC}}{Z_{AC}(+)}\right)^{-1} \frac{Z_{AB}}{Z_{AB}(+)} \frac{Z_{BC}}{Z_{BC}(+)}, \tag{1.13}$$

where $Z_{AC}(+)$ is the partition function in the volume V_{AC} with boundary condition (1.3) and partition functions $Z_{AC}, Z_{BC}, Z_{AB}(+), Z_{BC}(+)$ are defined similarly to $Z_{AC}, Z_{AC}(+)$. (Note that a rigorous justification of (1.13) is far from trivial.) Using the definition (1.1) we can rewrite (1.13) as

$$Prob(\Gamma(G) \in G(B)) \approx \exp\{-\beta(|AB|\tau_\beta(n_{AB})) + |BC|\tau_\beta(n_{BC}) - |AC|\tau_\beta(n_{AC})\}. \tag{1.14}$$

Now Weak Triangle Inequality (1.7) follows from the fact that the probability $Prob(\Gamma(G) \in G(B)) \leq 1$. To derive from (1.13) the Strong Triangle Inequality (1.6) we need to know the existence of a positive constant $\alpha > 0$ (depending on the triangle ABC), such that for large $|AC|$

$$Prob(\Gamma(\sigma) \in \sigma(B)) \approx \exp\{-\alpha|AC|\}. \tag{1.15}$$

To prove this statement is much more difficult.

The idea is that a typical contour $\Gamma \in G$ deviates from the segment AC by only $const.|AC|^{1/2}$, and the event $\Gamma \in G(B)$ means that a very large fluctuation of the position of that contour takes place. The chapter of the probability theory called the theory of large deviations studies the asymptotic behavior of probabilities of large fluctuations of sums of a large number of independent or weakly dependent random variables. For large $|AC|$ the height of Γ at the point with the abscissa t_B^1 can be represented as a sum of increments of the heights of subcontours in which we can subdivide Γ. It is possible to show that these increments are weakly dependent and so it is possible to apply the ideas of the theory of large deviations to prove the estimate (1.15). It is possible also to give a probabilistic interpretation to the statement (1.5) about positivity of the stiffness coefficient. In [DKS] it is proven that the variance of the height of Γ at a middle points of the segment AC is asymptotically proportional to

$$|AC|(\tau_\beta(n) + \tau_\beta''(n)). \tag{1.16}$$

So if the stiffness is not positive, then the fluctuations of the boundary of phases cannot be described by a Gaussian probability distribution with the usual normalization.

It seems that similar ideas can be applied to a very general class of two-dimensional interactions which need not be ferromagnetic. So we expect that the Weak Triangle Inequality is valid in a very general situation, but the Strong Triangle Inequality holds only under some additional restrictions and may be violated in general.

As it was shown in [DKS], the Triangle Inequality has some important implications for the geometry of Wulff Curve. We remind the reader this construction for the two-dimensional case. Let $F(n), n \in S^1$ be a continuous strictly positive even function and γ is a self avoiding curve in $\mathbf{R}^2$ which is supposed to be piecewise continuously differentiable. Let

$$\varphi_F(\gamma) = \int_\gamma F(n(s))ds, \tag{1.17}$$

where $n(s) \in S^1$ is a normal to γ at the point s and ds is the length differential. We are looking for the minimizer of the integral (1.17) in a class of closed curves γ such that the area inside γ equals one. This variational problem has a unique solution and the resulting curve γ_F is called Wulff Curve corresponding to the function F. This Wulff Curve is given by the following Wulff Construction. Let $R_F(n), n \in S^1$ be a half-plane $\{x \in \mathbf{R}^2 : (x, n) \le F(n)\}$. Let

$$\tilde{W}(F) = \cap_{n \in S^1} R_F(n),$$

and $W(F)$ is homothetic to $\tilde{W}(F)$ and of a unit area. Then the boundary of $W(F)$ is Wulff Curve. The main result of [DKS] states that the shape of the droplet in the ferromagnetic model asymptotically coincides with Wulff Curve defined by the function $\tau_\beta(n)$.

It is natural to consider an inverse problem of a reconstruction of the function F from its Wulff Curve γ_F. In general this problem has many solutions. But it was noted in [DKS] that if we restrict ourselves to the class of functions F for which Weak Triangle Inequality holds, then for any curve γ which enclose a symmetric convex domain there exists a unique function F in this class such that $\gamma_F = \gamma$. This fact is a reformulation of a well-known Minkovsky Theorem [M] (see also [BF] or [V]), which gives conditions for a function to be a support function of a convex set.

In [DKS] the following implication of Strong Triangle Inequality is noted. In general Wulff Curve can have angles. For example in case $F = \tau_\infty$ (see (1.11)) the corresponding Wulff Curve is a rectangle. However, Strong Triangle Inequality for a function F implies the smoothness of the corresponding Wulff Curve.

The main aim of the present paper is to suggest a natural multidimensional generalization of the Triangle Inequalities. In this introductory section we shall discuss for simplicity only the physically interesting three-dimensional case. Let $F(n), n \in S^2$ be a continuous strictly positive even function. We say that a function F satisfies Strong Simplex Inequality, if for any points A, B, C, D in $\mathbf{R}^3$ in general position

$$F(n(ABC))|ABC| + F(n(ABD))|ABD| + F(n(ACD))|ACD| > F(n(BCD))|BCD|, \tag{1.18}$$

where $n(ABC) \in S^2$ is the normal to the plane, defined by $A, B, C, |ABC|$ is the area of the triangle ABC and where other quantities in (1.18) are defined similarly. We say that a function F satisfies Weak Simplex Inequality if for any four points $A, B, C, D \in \mathbf{R}^3$ in general position

$$F(n(ABC))|ABC| + F(n(ABD))|ABD| + F(n(ACD))|ACD| \ge F(n(BCD))|BCD|. \tag{1.19}$$

Together with Simplex Inequalities it is natural to consider also Triangle Inequalities for function $F(n), n \in S^2$. These inequalities are defined again by the relations (1.6), (1.7), where now we use triangles ABC in $\mathbf{R}^3$, and the normals n_{AB}, n_{BC}, n_{AC} are supposed now to lie in the plane ABC. It is proven in Sect. 3 of the present paper that Weak Triangle Inequality and Weak Simplex Inequality are equivalent to each other and also equivalent to the convexity of the extended surface tension. Strong Triangle Inequality implies Strong Simplex Inequality but not vice versa. A differential form of this inequality can be given in terms of the Hessian of F. It turns out (see Sect. 3) that Weak Triangle Inequality corresponds to the case when both eigenvalues $\lambda_1, \lambda_2 (\lambda_1 \geq \lambda_2)$ of the Hessian are such that $\lambda_1 \geq -F(n), \lambda_2 \geq -F(n)$, Strong Simplex Inequality corresponds to the case when $\lambda_1 > -F(n), \lambda_2 \geq -F(n)$, and Strong Triangle Inequality - to the case when $\lambda_1 > F(n), \lambda_2 > -F(n)$. The last property is an analog of (1.5). The Theorem by Minkovsky mentioned above implies that in a class of functions satisfying Weak Triangle Inequality and Weak Simplex Inequality the surface tension is uniquely defined by its Wulff shape. For completeness we give the proof of this fact in Sect. 2. In Sect. 4 we study the connection between Strong Simplex Inequality and the absence of angle points in Wulff Shape, and also between Strong Triangle Inequality and the absence of angle points and angle edges (which is equivalent to the smoothness of Wulff Shape). In the last Section 5 we consider the case when the condition (1.5) holds for all $n \in S^2$. In this case if the function F is ℓ times differentiable, then Wulff Shape is $(\ell - 1)$ times differentiable, and if F is a holomorphic function, then the corresponding Wulff Shape has the same property. If (1.5) does not hold at some point, then Wulff Shape has points of infinite curvature despite any smoothness properties of F.

As in two-dimensional case, the problem of proving the validity of Strong Triangle Inequality or Strong Simplex Inequality for three-dimensional systems is connected with a study of large deviations of phase separation surface. In case of Simplex Inequality it is necessary to consider a volume which is a cylinder having a triangle base, and in case of Triangle Inequality - a cylinder, which base is a two dimensional strip (see Fig. 2). This problem is open and seems to be difficult, as well as other problems concerning the statistics of surfaces.

FIGURE 2

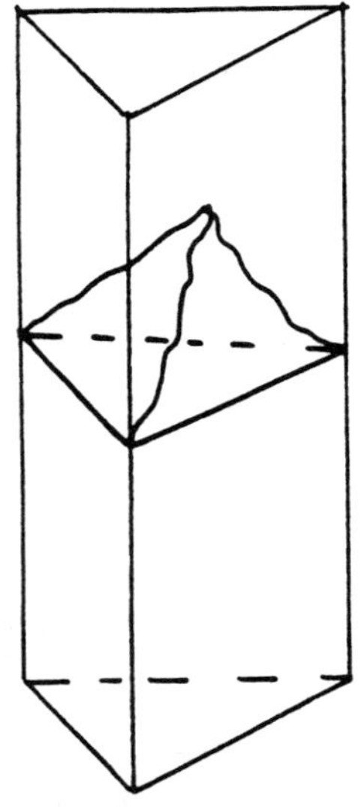
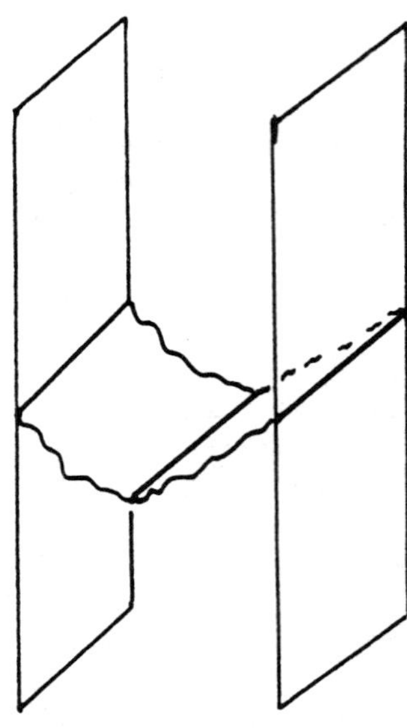

Following the report about the results of this paper, given by one of the authors in Marseille, Messager, Miracle-Sole and Ruis [MMR] proved Weak Simplex Inequality by the help of correlation inequalities for a large class of ferromagnets.

The authors were surprised by the absence of any mentioning of Simplex Inequality in the literature, despite the fact the inequalities seem to be so natural from both physical and geometrical points of view. Still, the authors are not sure that their knowledge of literature is complete enough and that the ideas presented here are very original. They hope, though, that the systematic exposition of them can stimulate further studies of analytical properties of the surface tension and the corresponding Wulff Shapes.

In the main part of the paper, which is independent of this introduction, we consider geometrical questions only, which are treated in a general multidimensional situation.

2 Convex sets.

We start by reviewing some known facts about convex sets in finite-dimensional spaces (see for example [V]).

Let $V \subset \mathbf{R}^d$ be a closed convex set and ∂V denotes its boundary. A $(d-1)$-dimensional hyperplane $L \subset \mathbf{R}^d$ is called a support plane to V corresponding to the direction $n \in S^{d-1}$ (where $S^{d-1} = \{n \in \mathbf{R}^d; |n| = 1\}$), if L is orthogonal to n and

$$L \cap V \neq \phi, \qquad L \cap V \subset \partial V. \tag{2.1}$$

It is well-known (see [V]) that for each $x \in \partial V$ there exists a support plane to V containing x. The set of all directions $n \in S^{d-1}$, such that the corresponding support plane contains x, we denote by $\mathcal{L}_V(x)$. A set $\mathcal{L}_V(x)$ is an intersection of a closed convex cone with S^{d-1}. A point $x \in \partial V$ is called an angle point of rank $k(k = 0, 1, \cdots, d-1)$ if the manifold $\mathcal{L}_V(x)$ is k-dimensional. The support plane to V in a given direction $n \in S^{d-1}$ is unique and it exists for all $n \in S^{d-1}$ if the set V is bounded. We denote such support plane by $L_V(n)$.

We suppose in the following that the origin $0 \in V$. A function

$$h_V(n) = \begin{cases} |n| dist(0, L_V(\frac{n}{|n|})), & n \neq 0 \\ 0, & n = 0 \end{cases} \tag{2.2}$$

will be called a support function of a bounded closed convex set V.

Let $F(n), n \in S^{d-1}$ be a non-negative function. We let

$$R_F(n) = \{x \in \mathbf{R}^d, (x, n) \leq F(n)\} \tag{2.3}$$

and will call the intersection

$$W_F = \cap_{n \in S^{d-1}} R_F(n) \tag{2.4}$$

the Wulff Shape defined by F.

The following result is due to Minkovsky [M] (see also [V]). For completeness we give here the proof.

THEOREM 2.1. *A nonnegative function $h(n), n \in \mathbf{R}^d$ is a support function of a bounded closed convex set $V \subset \mathbf{R}^d$ which contains an origin iff it satisfies the following two conditions:*

i) *(semilinearity)*

$$h(\lambda n) = \lambda h(n), \quad n \in \mathbf{R}^d, \quad \lambda \geq 0, \tag{2.5}$$

ii) *(convexity)*

$$h(\lambda_1 n_1 + \lambda_2 n_2) \geq \lambda_1 h(n_1) + \lambda_2 h(n_2), \lambda_1, \lambda_2 \geq 0. \tag{2.6}$$

The set V coincides with the Wulff Shape defined by the restriction of its support function $h_V(n)$ to S^{d-1}.

Proof: Let V be a bounded closed convex set and $0 \in V$. It is evident from the definition (2.2) that $h_V(n) \geq 0$ and that the semilinearity property for $h_V(n)$ is valid. From the definition (2.2) follows also that for any $n \in S^{d-1}, n_1, n_2 \in \mathbf{R}^d$ there exists a point $x \in \partial V$ such that

$$(x, n) = h_V(n) \tag{2.7}$$

and that

$$(x, n_1) \leq h_V(n_1), \qquad (x, n_2) \leq h_V(n_2). \tag{2.8}$$

If for some $\lambda_1, \lambda_2 \geq 0$

$$n = \lambda_1 n_1 + \lambda_2 n_2, \tag{2.9}$$

then

$$h_V(n) = (x, n) = \lambda_1(x, n_1) + \lambda_2(x, n_2) \leq \lambda_1 h_V(n_1) + \lambda_2 h_V(n_2). \tag{2.10}$$

So we have checked (2.6) for $|n| = 1$. The restriction $|n| = 1$ is completely innocuous because of a semilinearity property of $h_V(n)$. So we have proven the necessity part of the conditions (2.5), (2.6).

Note that the Wulff Shape W_h defined by a nonnegative function $h(n)$ is always a bounded closed convex set such that its support function

$$h_{W_h}(n) \leq h(n). \tag{2.11}$$

So to prove the sufficiency statement of the Theorem it is enough to check that under conditions i) and ii) for any $n \in S^{d-1}$ there exists a point $x_n \in W_h$ such that

$$(x_n, n) = h(n). \tag{2.12}$$

But the condition $x \in W_h$ is equivalent to the inequality

$$(x, \tilde{n}) \leq h(\tilde{n}) \text{ for all } \tilde{n} \in S^{d-1}. \tag{2.13}$$

We consider now a dual cone which is a subset $G_h \subset \mathbf{R}^{d+1} = \mathbf{R}^d \times \mathbf{R}^1$, such that a point $z = (n; y), n \in \mathbf{R}^d, y \in \mathbf{R}^1$ is in G_h iff

$$y \geq h(n). \tag{2.14}$$

It follows from the conditions (2.5), (2.6) that G_h is a closed convex cone. With any point $x \in \mathbf{R}^d, x \neq 0$ let us associate the d-dimensional hyperplane

$$M_x = \{z = (n; y) \in \mathbf{R}^{d+1} : (n, x) = y\} \tag{2.15}$$

in $\mathbf{R}^{d+1}$. Evidently, any d-dimensional hyperplane M in $\mathbf{R}^{d+1}$, which does not contain the line $\ell = \{(0; y), 0 \in \mathbf{R}^d, y \in \mathbf{R}^1\}$, is of the form M_x for some $x \in \mathbf{R}^d$. Note that the line ℓ is inside G_h. From (2.13) it follows that $x \in W_h$ iff the hyperplane M_x does not contain inner points of G_h. Let us now consider the support hyperplane $M(n)$ to G, passing through the point $(n, h(n))$. Because $\ell \not\subset M(n)$, we have $M(n) = M_{x_n}$ for some $x_n \in \mathbf{R}^d$. This point satisfies the equation (2.12). $\qquad\square$

3   Simplex inequalities.

Let $F(n), n \in S^{d-1}$ be a nonnegative function. Let $\{A_0, A_1, \cdots, A_k\} \subset \mathbf{R}^d, k = 2, 3, \cdots, d$ be any set of $(k+1)$ points in general position, and Δ be the simplex defined by these points. For any $i = 0, \cdots, k$ define

$$\Delta_i = \Delta(A_0, \cdots, A_{i-1}, \hat{A}_i, A_{i+1}, \cdots, A_k) \tag{3.1}$$

to be the corresponding $(k-1)$ dimensional simplex defined by the same set of points, except A_i. Let $|\Delta_i|$ be the $(k-1)$ dimensional volume of Δ_i and L_i be the $(k-1)$-dimensional hyperplane containing Δ_i. Let a family of unit vectors $n_1, \cdots, n_k \in S^{d-1}$ be in general position, and a unit vector n_0 is their linear combination. We call the family $n_0, n_1, \cdots, n_k$ direction consistent, if

$$n_0 = \sum_{i=1}^{k} \lambda_i n_i \tag{3.2}$$

with all $\lambda_i > 0, i = 1, \cdots, k$. It is easy to see that there are exactly two direction consistent families $n_0, n_1, \cdots, n_k$, belonging to the plane of the simplex Δ, such that the vectors n_i are orthogonal to L_i for all $i = 0, \cdots, k$. One family corresponds to the case when n_0 points outside Δ, while $n_1, \cdots, n_k$ point inside Δ, the other is obtained by changing signs of all $n_0, \cdots, n_k$.

We say that a function $F(n)$ satisfies Weak Simplex Inequality of rank $k(k = 2, \cdots, d)$, if for any set $\{A_0, \cdots, A_k\}$ of $(k+1)$ points in general position and any direction consistent family $n_0, \cdots, n_k$ of vectors orthogonal to Δ_i and lying in the plane of Δ,

$$\sum_{i=1}^{k} F(n_i)|\Delta_i| \geq F(n_0)|\Delta_0|. \tag{3.3}$$

A function $F(n)$ satisfies Strong Simplex Inequality of rank $k(k = 2, \cdots, d)$ if for any such families $\{A_i\}, \{n_i\}$,

$$\sum_{i=1}^{k} F(n_i)|\Delta_i| > F(n_0)|\Delta_0|. \tag{3.4}$$

It is easy to see that a function $F(n)$ satisfies Weak Simplex Inequality of rank 2 iff its conical extension

$$F'(n) = \begin{cases} |n|\, F(n/|n|), & n \neq 0, n \in \mathbf{R}^d \\ 0, & n = 0 \end{cases} \tag{3.5}$$

satisfies the conditions i) and ii) of the Theorem 2.1. If a function F is even,

$$F(n) = F(-n) \tag{3.6}$$

(which is natural for the surface tension), then the direction consistency property of the family $n_0, n_1, \cdots, n_k$ is redundant.

                                               R. Dobrushin & S. Shlosman

THEOREM 3.1. *For any nonnegative function $F(n), n \in S^{d-1}$ and any $k, k' = 2, \cdots, d$ the Weak Simplex Inequality of rank k for F implies the Weak Simplex Inequality of rank k' for F.*

(For the case $k' < k$ the proof can be easily obtained via suitable limiting procedure, see Figure 3.)

FIGURE 3

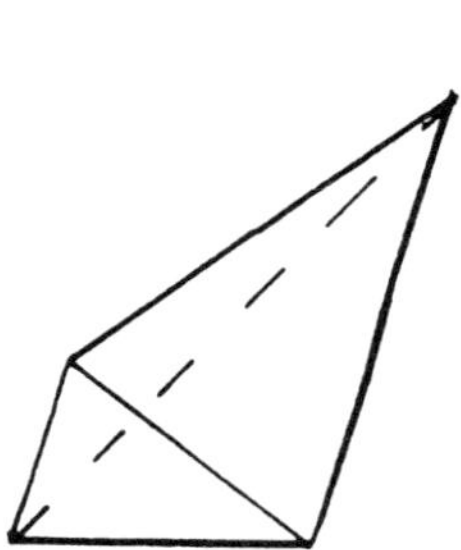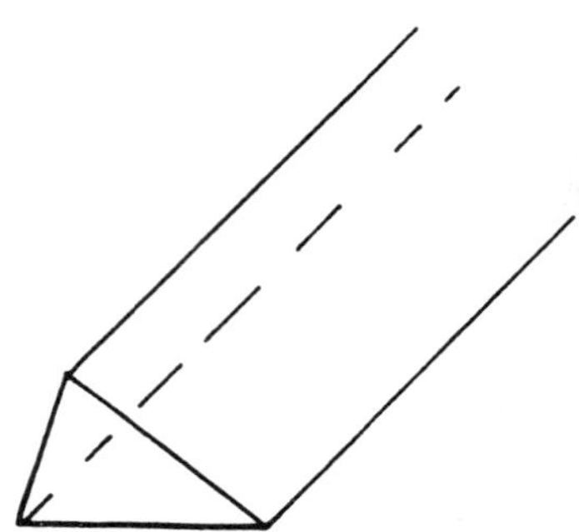

We start the proof with three lemmas. Let a half-space $R_F(n)$ be defined as in (2.3) and a hyperplane

$$L_F(n) = \{x \in \mathbf{R}^d : (x, n) = F(n)\} \tag{3.7}$$

be its boundary. Let $n_0, n_1, \cdots, n_d \in S^{d-1}$ be direction consistent family of vectors. Consider the d-faced angle

$$R_F(n_1, \cdots, n_d) = \cap_{i=1}^d R_F(n_i). \tag{3.8}$$

Let a_F be the vertex of this angle, i.e., a point of the intersection of the planes $L_F(n_1), \cdots, L_F(n_d)$ (which is unique because the vectors $n_1, \cdots, n_d$ are in general position).

LEMMA 3.2. *If the point a_F lies in $L_F(n_0)$ and $n_0, n_1, \cdots, n_d$ is a direction consistent family, then the intersection*

$$R_F(n_1, \cdots, n_d) \cap L_F(n_0) \tag{3.9}$$

consists of this single point a_F.
On the other hand, if for all $x \in \mathbf{R}^d$ the intersection

$$R_F(n_1, \cdots, n_d) \cap [L_F(n_0) + x] \tag{3.10}$$

is bounded, then one of the families: $n_0, n_1, \cdots, n_d$ or $-n_0, n_1, \cdots, n_d$ is direction consistent.

Proof: Suppose $y \in R_F(n_1, \cdots, n_d) \cap L_F(n_0), y \neq a_F$. Then

$$(n_i, y) \leq F(n_i), i = 1, \cdots, d, (n_0, y) = F(n_0), \tag{3.11}$$

and for some $i_0 = 1, \cdots, d$ one of the above inequalities is strict: $(n_{i_0}, y) < F(n_{i_0})$. But $\lambda_{i_0} > 0$ in (3.2), hence

$$F(n_0) = (n_0, y) = \sum_{i=1}^{d} \lambda_i(n_i, y) < \sum_{i=1}^{d} \lambda_i F(n_i) = \sum_{i=1}^{d} \lambda_i(n_i, a_F) = (n_0, a_F) = F(n_0).$$

The contradiction proves the first half of the statement.

To prove the second half consider the decomposition

$$n_0 = \lambda_1 n_1 + \cdots + \lambda_d n_d. \tag{3.12}$$

If some $\lambda_i = 0$, then for some x_0

$$\cap_{i \neq i_0} L_F(n_i) \subset L_F(n_0) + x_0$$

so the intersection (3.10) is infinite for $x = x_0$. Hence all λ_I are nonzero. Suppose that $\lambda_1 > 0, \cdots, \lambda_k > 0, \lambda_{k+1} < 0 \cdots, \lambda_d < 0, 1 \leq k < d$. Then the family $n_0, n_1, \cdots, n_k, -n_{k+1}, \cdots, -n_d$ is direction consistent. On the other hand, the intersection

$$R_F(\pm n_1, \cdots, \pm n_d) \cap [L_F(n_0) + x]$$

is bounded iff all the signs are the same, hence the intersection

$$R_F(n_1, \cdots, n_k, -n_{k+1}, \cdots, -n_d) \cap [L_F(n_0) + x]$$

is unbounded for all x. Without loss of generality we can suppose that

$$a_F \in R_F(n_1, \cdots, n_k, -n_{k+1}, \cdots, -n_d) \cap L_F(n_0).$$

This, together with the above unboundedness, contradicts to already proven part of the Lemma. $\qquad \Box$

LEMMA 3.3. *A function F satisfies Weak Simplex Inequality of rank d iff for any direction consistent family $n_0, n_1, \cdots, n_d$ the intersection*

$$I_F(n_0; n_1, \cdots, n_d) \equiv R_F(n_1, \cdots, n_d) \cap L_F(n_0) \neq \phi. \tag{3.13}$$

The function F satisfies Strong Simplex Inequality of rank d iff for any direction consistent family $n_0, n_1, \cdots, n_d$ the above intersection is a non-degenerate $(d-1)$-dimensional simplex.

Proof: From the above lemma it follows that the intersection $I_F(n_0; n_1, \cdots, n_d)$ is necessarily bounded. So there are only three possibilities (see Fig. 4):

$$I_F(n_0, n_1, \cdots, n_d) = \begin{cases} \phi, \\ a_F, \\ \text{a nondegenerate simplex.} \end{cases} \tag{3.14}$$

FIGURE 4

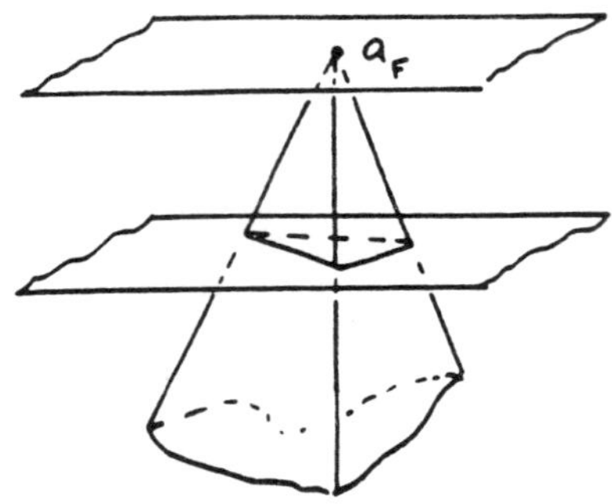

Let the points $A_0, \cdots, A_d$ be intersection points,

$$A_i = \cap_{j:j=0,1,\cdots,d,j\neq i} L_F(n_j), \qquad i = 0, \cdots, d, \tag{3.15}$$

simplexes Δ_i be defined by (3.1), and let $|\Delta_i|$ be their volumes. Our claim is that to the above three possibilities correspond the following three relations:

$$\sum_{i=1}^d |\Delta_i| F(n_i) \begin{cases} > & |\Delta_0|\, F(n_0), \\ = & |\Delta_0|\, F(n_0), \\ < & |\Delta_0|\, F(n_0). \end{cases} \tag{3.16}$$

To check it we consider the simplexes

$$\bar{\Delta}_i = \Delta(A_0, \cdots, A_{i-1}, \hat{A}_i, A_{i+1}, \cdots, A_d, 0), \quad i = 1, \cdots, d, \tag{3.17}$$

defined by the origin 0 and by all the points $A_0, \cdots, A_d$ except the point A_i. It is easy to see that to the three possibilities in (3.14) the following three relations correspond:

$$\sum_{i=1}^d |\bar{\Delta}_i| \begin{cases} > & |\bar{\Delta}_0|, \\ = & |\bar{\Delta}_0|, \\ < & |\bar{\Delta}_0|. \end{cases} \tag{3.18}$$

(Note that in the second case we have $|\bar{\Delta}_i| \equiv 0, i = 0, \cdots, d$). Since

$$|\bar{\Delta}_i| = (d)^{-1} F(n_i) |\Delta_i|, \quad i = 0, \cdots, d, \tag{3.19}$$

the statement (3.16) holds, and that implies the lemma. $\qquad\square$

NOTE 3.4. *It follows from the definitions that Weak Simplex Inequality or Strong Simplex Inequality of rank k for a function F are valid iff they are valid for restrictions of F to any $(k-1)$-dimensional subsphere of S^{d-1}. By a $(k-1)$-dimensional subsphere of S^{d-1} we mean an intersection of S^{d-1} with a k-dimensional subspace of $\mathbf{R}^d$.*

LEMMA 3.5. *A non-negative function $F(n), n \in S^{d-1}$ satisfies Weak Simplex Inequality of rank k iff for any direction consistent family $n_0, n_1, \cdots, n_k \in S^{d-1}$*

$$F(n_0) \leq \sum_{i=1}^{k} \lambda_i F(n_i). \tag{3.20}$$

It satisfies Strong Simplex Inequality of rank k iff for any such family

$$F(n_0) < \sum_{i=1}^{k} \lambda_i F(n_i). \tag{3.21}$$

Proof: From the last Note it follows that it is enough to consider only the case $k = d$. Evidently, the point $x \in R_F(n_1, \cdots, n_d)$ iff

$$(x, n_i) \leq F(n_i) \text{ for all } i = 1, \cdots, d, \tag{3.22}$$

and x is an inner point of $R_F(n_1, \cdots, n_d)$ iff

$$(x, n_i) < F(n_i) \text{ for all } i = 1, \cdots, d. \tag{3.23}$$

From Weak Simplex Inequality and the last Lemma 3.3 it follows the existence of a point x such that:

i)

$$(x, n_0) = F(n_0) \tag{3.24}$$

ii) The condition (3.22) holds.

Hence

$$F(n_0) = (x, n_0) = \sum_{i=1}^{d} \lambda_i(x, n_i) \leq \sum_{i=1}^{d} \lambda_i F(n_i), \tag{3.25}$$

which implies (3.20). To derive from (3.20) the validity of Weak Simplex Inequality, consider the vertex $a_F = a_F(n_1, \cdots, n_d)$, defined after (3.8). By definition,

$$(a_F, n_i) = F(n_i), \tag{3.26}$$

which implies

$$\sum \lambda_i(a_F,, n_i) = (a_F, n_0) \geq F(n_0). \tag{3.27}$$

Now, the origin $0 \in R_F(n_1, \cdots, n_d)$, and $(0, n_0) = 0 \leq F(n_0)$. So the interval $[0, a]$ contains a point x such that $(x, n_0) = F(n_0)$. This point $x \in I_F(n_0; n_1, \cdots, n_d)$. In view of the Lemma 3.3 that proves Weak Simplex Inequality. The proofs for the Strong Simplex Inequality case are similar. $\qquad\square$

Proof of the theorem 3.1. Taking limits in Lemma 3.5 we find that Weak Simplex Inequality of rank k for F is equivalent to the inequality (3.20) for any $n_0, n_1, \cdots, n_k \in S^{d-1}$ with non-negative λ_i in (3.2) (instead of $n_1, \cdots, n_k$ in general position and positive λ_i of Lemma 3.5). Passing to the conical extension (3.5) we obtain that Weak Simplex Inequality is equivalent to the inequality

$$F'(\sum_{i=1}^{k} \lambda_i n_i) \geq \sum_{i=1}^{k} \lambda_i F'(n_i) \tag{3.28}$$

for any $n_1, \cdots, n_k \in \mathbf{R}^d, \lambda_1, \cdots, \lambda_k \geq 0$. Evidently, conditions (3.26) are equivalent for different k. $\qquad\square$

From now on we can, in view of the last Theorem 3.1, adopt the following terminology: we shall say that a function satisfies Weak Simplex Inequality if it satisfies Weak Simplex Inequality of some rank (all ranks) $k = 2, \cdots, d$.

COROLLARY 3.6. *If a function F satisfies Weak Simplex Inequality, then it is a support function of its Wulff Shape.*

Proof: Using Weak Simplex Inequality of rank 2 we see that conditions of the Theorem 2.1 hold for the conical extension of F. So our statement follows from this Theorem. $\qquad\square$

COROLLARY 3.7. *If functions F and G satisfy Weak Simplex Inequality, and their Wulff Shapes coincide, then $F = G$.*

Proof: It follows from the previous Corollary because the support function of a set is uniquely defined by it. $\qquad\square$

Further formulations of Weak Simplex Inequality and Strong Simplex Inequality can be given in terms of second derivatives of F.

PROPOSITION 3.8. *Let $S^1 \subset S^{d-1}$ be a one-dimensional subsphere. Let*

$$\varphi = \arccos(r, r_0), \quad r \in S^1, \tag{3.29}$$

where $r_0 \in S^1$ is fixed. Consider the restriction of F to S^1, which we shall treat as a function $F(\varphi)$ of φ.

If F satisfies Weak Simplex Inequality, and for some φ_0 the function $F(\varphi)$ is twice differentiable at φ_0, then

$$\frac{d^2}{d\varphi^2}F(\varphi_0) + F(\varphi_0) \geq 0. \tag{3.30}$$

If F is in $C^2(S^{d-1})$ and (3.30) is valid for all one-dimensional subspheres and all points φ_0 on them, then F satisfies Weak Simplex Inequality.

Proof: Because of Note 3.4 it is enough to consider only the case $d = 2$. Let $F'(x), x \in \mathbf{R}^2$ be a conical extension of F. Because Weak Simplex Inequality is equivalent to the inequality (3.28), the function F' is convex. We consider the usual orthogonal xy coordinates in $\mathbf{R}^2$ together with polar coordinates φ, ρ, where we measure φ from the x axis. We suppose also that the angle φ_0 is zero. For F' we then have:

$$F'(x,y) = F(arctg\frac{y}{x})(x^2 + y^2)^{1/2}. \tag{3.31}$$

Taking derivatives, we find that

$$\frac{\partial^2}{\partial y^2}F'(x,y)\big|_{\substack{y=0 \\ x=x_0}} = \frac{1}{x_0}[\frac{d^2F'}{d\varphi^2}(0) + F'(0)], \tag{3.32}$$

$$\frac{\partial^2}{\partial x^2}F'(x,y)\big|_{\substack{y=0 \\ x=x_0}} = \frac{\partial^2}{\partial x\partial y}F'(x,y)\big|_{\substack{y=0 \\ x=x_0}} = 0.$$

Because F' is convex, the form

$$\frac{\partial^2}{\partial y^2}F'(x_0,0)dy^2 + 2\frac{\partial^2}{\partial x\partial y}F'(x_0,0)dxdy + \frac{\partial^2}{\partial x^2}F'(x_0,0)dx^2 \tag{3.33}$$

has to be non-negatively defined, which in our situation is equivalent to (3.28). That proves the first statement.

To prove the second half, note that the smoothness of F implies that of F' everywhere except the origin. Since the form (3.33) is non-negatively defined, it implies the usual convexity

$$F'(\lambda_1 x_1 + \lambda_2 x_2) \geq \lambda_1 F'(x_1) + \lambda_2 F'(x_2), \quad \lambda_1 + \lambda_2 = 1, \quad \lambda_1, \lambda_2 \geq 0 \tag{3.34}$$

PROPOSITION 3.9. *Let for some $n_0 \in S^1 \subset S^{d-1}$ the inequality (3.3) does not hold for some*

$$\varphi_0 = \varphi(n_0) = \arccos(n_0, r_0).$$

Then for any $k \geq 2$ and for some direction consistent family $n_0, n_1, \cdots, n_k$ the Weak Simplex Inequality (3.3) of rank k is violated.

Proof: We are going to find first a direction consistent family with $k = 2$. Let $\Delta\varphi > 0$; we consider two vectors n_1, n_2 is S^1 such that

$$\varphi(n_1) = \varphi(n_0) + \Delta\varphi, \quad \varphi(n_2) = \varphi(n_0) - \Delta\varphi,$$

and all three vectors n_0, n_1, n_2 are in the same semicircle of the circle S^1 defined by two points: $r_0, -r_0$. This is always possible provided r_0 is such that $\varphi(n_0) \neq 0, \pi$ and $\Delta\varphi$ is small enough.

Consider the hyperplanes $L_F(n_i), i = 0, 1, 2$, and let A_0, A_1, A_2 be the triangle defined by them in the two dimensional plane, which contains the circle S^1. By an elementary calculation it is easy to see that

$$F(n_1) \cdot |A_0 A_2| + F(n_2)|A_0 A_1| - F(n_0)|A_1 A_2| = [\frac{d^2}{d\varphi^2} F(\varphi(n_0)) + F(\varphi(n_0))](\Delta\varphi)^2 + o((\Delta\varphi)^2).$$

$$(3.35)$$

Hence when $\Delta\varphi$ is small enough, the triangle inequality is violated, according to our hypothesis.

The case of general k is done by induction. We will present here only the first step, leaving the rest to the reader.

Consider the 2-faced angle

$$R_F(n_1, n_2) = \cap_{i=1,2} R_F(n_i).$$

By Lemma 3.3 we know that because F does not satisfy Weak Simplex Inequality of rank k,

$$I_F(n_0; n_1, n_2) = R_F(n_1, n_2) \cap L_F(n_0) = \phi.$$

Let n_3 be a unit vector in a general position with n_0, n_1, n_2. Then of course

$$I_F(n_0; n_1, n_2, n_3) = R_F(n_1, n_2, n_3) \cap L_F(n_0) = \phi, \qquad (3.36)$$

where

$$R_F(n_1, n_2, n_3) = \cap_{i=1}^3 R_F(n_i).$$

If n_0, n_1, n_2, n_3 would be direction consistent, then Lemma 3.3 would imply that rank 3 Weak Simplex Inequality for F is violated. However in the linear decomposition

$$n_0 = \lambda_1 n_1 + \lambda_2 n_2 + \lambda_3 n_3$$

we have $\lambda_1, \lambda_2 > 0$ while $\lambda_3 = 0$. We are going to show that one can rotate the vectors n_1, n_2 slightly, so that for the resulting vectors n_1', n_2' the intersection $I_F(n_0; n_1', n_2', n_3)$ is again empty, while the coefficients in the decomposition

$$n_0 = \lambda_1' n_1 + \lambda_2' n_2 + \lambda_3' n_3$$

are now all positive.

Without loss of generality we can suppose now that $d = 3$. Let the line ℓ be parallel to the hyperplane $L_F(n_0)$ and orthogonal to the line $L_F(n_1) \cap L_F(n_2)$. Denote by O_κ the orthogonal rotation around the axis ℓ on the angle κ. It easy to see that for ϵ small enough and for all rotations O_κ with $\kappa \in [0, \epsilon]$ or with $\kappa \in [-\epsilon, 0]$ the intersection

$$[O_\kappa(R_F(n_1, n_2)) \cap R_F(n_3)] \cap L_F(n_0) = \phi.$$

From the continuity of F it follows then that for some κ

$$R_F(O_\kappa n_1, O_\kappa n_2, n_3) \cap L_F(n_0) = \phi \tag{3.37}$$

because

$$R_F(O_\kappa n_1, O_\kappa n_2, n_3) = R_F(O_\kappa n_1, O_\kappa n_2) \cap R_F(n_3),$$

and because the 2-faced angle $R_F(O_\kappa n_1, O_\kappa n_2)$ is a shift of the angle $O_\kappa(R_F(n_1, n_2))$, while the size of that shift can be made arbitrary small provided κ is small.

Lemma 3.2 implies now that the family n_0, n_1, n_2, n_3 is direction consistent, while Lemma 3.3 says that the Weak Simplex Inequality of rank 3 is violated for F at n_0. $\qquad\square$

Now we shall do the same for Strong Simplex Inequality of rank k. Because of Note 3.4 it is enough to consider only the case $k = d$.

PROPOSITION 3.10. *Let F satisfies Weak Simplex Inequality. It does not satisfy Strong Simplex Inequality of rank k iff there exists an open set $G \subset S^{d-1}$ such that for any $r \in G$ and any one-dimensional subsphere $S^1 \subset S^{d-1}, r \in S^1$*

$$\frac{d^2 F}{d\varphi^2}(\varphi_0) + F(\varphi_0) = 0 \tag{3.38}$$

where $F(\varphi)$ is defined as in Proposition 3.8 and φ_0 is the polar coordinate of r on S^1.

Proof: The relation (3.21) shows that Strong Simplex Inequality of rank d is violated iff for some direction consistent family $n_0, n_1, \cdots, n_d \in S^{d-1}, n_0 = \sum_{i=1}^{d} \lambda_i n_i$

$$F(n_0) = \sum_{i=1}^{d} \lambda_i F(n_i). \tag{3.39}$$

Let $L(n_0)$ be a tangent plane to S^{d-1} at the point n_0. Define the coefficients $\mu_i, i = 1, \cdots, d$ by the property: the endpoints of all the vectors $n'_i = \mu_i n_i$ are in $L(n_0)$. Let $\lambda'_i = \lambda_i / \mu$. Then

$$F'(n_0) = \sum_{i=1}^{d} \lambda'_i F'(n'_i), \tag{3.40}$$

$$n_0 = \sum_{i=1}^{d} \lambda'_i n'_i, \tag{3.41}$$

and

$$\sum_{i=1}^{d} \lambda'_i = 1. \tag{3.42}$$

Now, for any convex function f the equality

$$f(\sum_{i=1}^{d} \lambda_i x_i) = \sum_{i=1}^{d} \lambda_i f(x_i) \tag{3.43}$$

valid for some points $x_1, \cdots, x_d$ in general position and for some positive $\lambda_1, \cdots, \lambda_d$ with $\lambda_1 + \cdots + \lambda_d = 1$ implies that the restriction of f to the simplex defined by vertices $x_1, \cdots, x_d$ is linear. The restriction of F' to $L(n_0)$ is convex because Weak Simplex Inequality is valid for F. So Strong Simplex Inequality of rank d is violated iff for some $n_0 \in S^{d-1}$ the restriction of the function F' to some neighborhood of the point n_0 in the plane $L(n_0)$ is linear. Let $S^1 \subset S^{d-1}$ be a one-dimension subsphere containing n_0. The restriction of F' to the tangent line to S^1 which passes through n_0 is connected with the restriction of F to S^1 via (3.31). The linearity of F' around n_0 implies that for some $a, b \in \mathbf{R}^1$ and for small y

$$ay + b = F(arctg\, y)(y^2 + 1)^{1/2}. \tag{3.44}$$

This easily implies that

$$F(\varphi) = a \sin \varphi + b \cos \varphi, \tag{3.45}$$

which is a general form of the solution of (3.38). $\qquad\square$

Proposition 3.10 and Note 3.4 imply the following

COROLLARY 3.11. *For any k, k', where $2 \leq k < k' \leq d$ Strong Simple Inequality of rank k implies Strong Simplex Inequality of rank k'.* $\qquad\square$

The results of the following Section 4 imply that the previous statement is not valid for $k > k'$.

It is also instructive to describe Weak Simplex Inequality and Strong Simplex Inequality in terms of the Hessian of F. Let again $L(n_0)$ be a tangent hyperplane to S^{d-1} at n_0. We

The results of the following Section 4 imply that the previous statement is not valid for $k > k'$.

It is also instructive to describe Weak Simplex Inequality and Strong Simplex Inequality in terms of the Hessian of F. Let again $L(n_0)$ be a tangent hyperplane to S^{d-1} at n_0. We consider a usual orthonormal coordinates $y_1, \cdots, y_{d-1}$ in $L(n_0)$ and we introduce the angular coordinates $\varphi_1, \cdots, \varphi_{d-1}$ in a neighborhood of n_0 in S^d in a following way: if $x \in S^{d-1}$, and a ray from 0, containing x, intersects the tangent plane $L(n_0)$ in a point $y = (y_1, \cdots, y_{d-1})$, then we define $\varphi_i = (sign\, y_i) arctg|y_i|$.

COROLLARY 3.12. *Let F be in $C^2(S^{d-1})$, and $A(n_0) = \|a_{ij}\|$ with*

$$a_{ij} = \frac{\partial^2 F(\varphi_1, \cdots, \varphi_{d-1})}{\partial \varphi_i \partial \varphi_i}\Big|_{\varphi_1 = \cdots = \varphi_{d-1} = 0} \tag{3.46}$$

be the matrix of it second derivatives at n_0. If for all n_0 all the eigenvalues $\lambda_1, \cdots, \lambda_{d-1}$ of the matrix $A(n_0)$ are not less than $[-F(n_0)]$, then F satisfies Weak Simplex Inequality. If, in addition, for all n_0 the number of eigenvalues which equal $[-F(n_0)]$ is smaller than $k-1$, then F satisfies Strong Simplex Inequality of rank k.

Proof: It is enough to note that for any one-dimensional subsphere $S^1 = \{\varphi \in [0, 2\pi)\}$ of S^{d-1} the second derivative $d^2/d\varphi^2\, F(\varphi)$ with respect to φ given by (3.29) equals to the second derivative of F in coordinates $\varphi_1, \cdots, \varphi_{d-1}$ in the corresponding direction and then use the Propositions 3.8 and 3.9. To see this equality one has to compare two different systems of angular coordinates $\varphi_1, \cdots, \varphi_{d-1}$ and $\psi_1, \cdots, \psi_{d-1}$ centered at the same point n_0. Then the Jacobian of this change of coordinates at n_0 is just an orthogonal rotation, and all the second derivatives $\partial^2/\partial \varphi_i \partial \varphi_j(\psi_k)|_{\varphi_1 = \cdots = \varphi_{d-1} = 0} = 0, i, j, k = 1, \cdots, d-1$. The claim now follows from the formula for the derivatives of the composition of two functions.

In what follows we shall adopt the following terminology. We say that stiffness is positive at a point n_0 if F is twice differentiable at n_0 and all eigenvalues of $A(n_0)$ are greater than $[-F(n_0)]$. If stiffness is positive everywhere, then F satisfies Strong Simplex Inequality of rank 2. However, this last property is compatible with violation of the stiffness positivity in a discrete set of points of S^{d-1}.

4   Angles of the Wulff Shape.

THEOREM 4.1. *Let F satisfies Weak Simplex Inequality. Then the Wulff Shape has no angle points of rank $k - 1(k = 2, \cdots, d)$ and higher iff F satisfies Strong Simplex Inequality of rank k.*

Proof for the case $k = d$. Let W_F has an angle point x of rank $d - 1$. Then there exists a direction consistent family $n_0, \cdots, n_d$ of vectors lying in $\mathcal{L}_{W_F}(x)$ (see Sect. 2). From Weak Simplex Inequality and Corollary 3.6 it follows that any hyperplane $L_F(n)$ is a support plane of W_F. So the angle $R_F(n_1, \cdots, n_d)$ has the point x as its vertex and Lemma 3.2 implies that the intersection

$$R_F(n_1, \cdots, n_d) \cap L_F(n_0) = \{x\}. \tag{4.1}$$

But then Lemma 3.3 tells us that Strong Simplex Inequality of rank d cannot be valid.

Suppose now that W_F has no angle points of rank $d - 1$. Then for any d vectors $n_1, \cdots, n_d$ in general position the vertex a_F of the angle $R_F(n_\lambda, \cdots, n_d)$ does not belong to W_F. Suppose now that the family $n_0, n_1, \cdots, n_d$ is direction consistent. Because W_F lies inside $R_F(n_1, \cdots, n_d)$, the hyperplane $L_F(n_0)$ has non-empty intersection with $R_F(n_1, \cdots, n_d)$. The hyperplane $L_F(n_0)$ is a support plane, and so in view of the Lemma 3.3 it cannot contain the vertex a_F. Hence the intersection $I_F(n_0; n_1, \cdots, n_d)$ is a non-degenerate simplex and Lemma 3.3 implies Strong Simplex Inequality of rank d.

The proof in general case is based on the following

LEMMA 4.2. *Let K be a k-dimension subspace of $\mathbf{R}^d$, π_K be the orthogonal projection on K and $\pi_K(W_F)$ be the image of W_F under it. Then a point $x \in \partial[\pi_K(W_F)]$ is an angle point of rank j $(j = 1, \cdots, k - 1)$ iff each point $x' \in \pi_K^{-1}(x) \cap W_F$ is an angle point of rank j or higher. The projection $\pi_K(W_F)$ is the Wulff Shape defined by the restriction $F|_{S^{d-1} \cap K}$.*

The proof follows immediately from the following observation. The set of all support hyperplanes to $\pi_K(W_F)$ (in K) coincides with the set of intersections with K of all support hyperplanes to W_F (in R^d), which are orthogonal to K.   □

Proof of the Theorem for arbitrary k. Let x be an angle point of rank j. Consider the (j-dimensional) manifold $\mathcal{L}_{W_F}(x)$, defined in the beginning of Sect. 2, and let K' be the intersection of all support hyperplanes from $\mathcal{L}_{W_F}(x)$. The dimension of K' is $d - 1 - j$. Let K be the orthogonal complement to K'. Then the projection $\pi_K(x)$ is an angle point of $\pi_K(W_F)$ of rank j, while the dimension of K is $j + 1$. The Theorem follows now from the above Lemma 4.2, Note 3.4 and from the proven part of the Theorem.   □

COROLLARY 4.3. *Wulff Shape W_F is smooth (i.e., has unique tangent plane at each point) iff rank 2 Strong Simplex Inequality holds for F.*

5   Special points of a smooth Wulff Shape.

Throughout this Section we suppose that F satisfies Strong Simplex Inequality of rank 2, so Wulff Shape has smooth boundary. We shall describe a set of special points of ∂W_F which can exist despite of smoothness or even analyticity of F.

Let W be convex, and the origin 0 is an inner point of W. Then for each $r \in S^{d-1}$ there exists exactly one positive real number $X(r)$ such that $X(r)r \in \partial W$. It is convenient to describe a surface ∂W_F by the help of the corresponding function X, which we denote by X_F. Suppose now that $W = W_F$ with F satisfying Strong Simplex Inequality of rank 2. Then for each point x in ∂W_F there exists a unique tangent plane to ∂W_F containing x. Consider a map $R_F(r) : S^{d-1} \to S^{d-1}$, where $R_F(r)$ is a vector orthogonal to a tangent plane to ∂W_F at $X_F(r)$. It is easy to see that

$$X_F(R) = F(R_F(r))/(R, R_F(r)) \tag{5.1}$$

(compare [Z], f-la (3.6)), where $(r, R_F(r))$ is a scalar product, so the analytic properties of $X_F(r)$ and $R_F(r)$ are closely connected.

THEOREM 5.1. *Suppose a positive function F satisfy stiffness positivity for all $n \in S^{d-1}$. If F is in $C^{\ell}(S^{d-1}), \ell \geq 2$, then X_F is in $C^{\ell-1}(S^{d-1})$. If F is holomorphic then so is X_F.*

(A function on S^{d-1} is called holomorphic if it can be extended to a holomorphic function on a neighborhood of S^{d-1} in $S_{\mathbf{C}}^{d-1}$ where $S_{\mathbf{C}}^{d-1}$ is given by the equation $z_1^2 + \cdots + z_d^2 = 1$ in the d-dimensional complex space.)

Proof: By definition, $(r, R_F(r)) \neq 0$. So because of (5.1) it is enough to prove the claim of the Theorem for the function R_F. It is geometrically evident that the ratio

$$F(n)/(n, r) \tag{5.2}$$

as a function of n has exactly one minimum, which is attained at the point $n = R_F(r)$. Because F is continuous, so is R_F. Now let $r_0 \in S^{d-1}, n_0 = R_F(r_0)$ and $\varphi_1, \cdots, \varphi_{d-1}, \psi_1, \cdots, \psi_{d-1}$ be the systems of angular coordinates centered at r_0, resp. n_0, introduced in Sect. 3. Let $r = (\varphi_1, \cdots, \varphi_{d-1}), n = (\psi_1, \cdots, \psi_{d-1})$ be two points in the vicinity of r_0, n_0 respectively. It is easy to see that

$$\begin{aligned}
(n, r) &= (1 + \sum_{i=1}^{d-1} \varphi_i^2)^{-1/2}(1 + \sum_{j=1}^{d-1} \psi_j^2)^{-1/2} \\
&\quad \times (A + \sum_{i=1}^{d-1} A_i \varphi_i + \sum_{j=1}^{d-1} B_j \psi_j + \sum_{i,j=1}^{d-1} C_{ij} \varphi_i \psi_j),
\end{aligned} \tag{5.3}$$

where

$$A = (n_0, r_0), A_i = (n_0, e_i^r), B_j = (e_j^n, r_0), C_{ij} = (e_j^n, e_i^r) \tag{5.4}$$

and $e_i^r, e_j^n, i, j = 1, \cdots, d-1$ are orthonormal vectors in the tangent planes $L(r_0), L(n_0)$, used in the construction of the angular coordinates. If F is in $C^1(S^{d-1})$, then the necessary conditions for the extremum imply that the points $r = (\varphi_1, \cdots, \varphi_{d-1})$ and $R_F(r) = n = (\psi_1, \cdots, \psi_{d-1})$ are connected by the relations

$$H_j(n,r) = 0, \qquad j = 1, \cdots, d-1, \tag{5.5}$$

where

$$H_j(n,r) = \frac{\partial}{\partial \psi_j} F(n)/(n,r) = \frac{\partial F(n)}{\partial \psi_j}(n,r)^{-1} - F(n)(n,r)^{-2}\frac{\partial}{\partial \psi_j}(n,r). \tag{5.6}$$

It follows from (5.5) and (5.6) that

$$\frac{\partial}{\partial \psi_j} F(n) \cdot A^{-1} - F(n)A^{-2}B_j = 0. \tag{5.7}$$

The sequence $H = (H_1, \cdots, H_{d-1})$ defines a map from a neighborhood of a point (n_0, r_0) in $\mathbf{R}^{d-1} \times \mathbf{R}^{d-1}$ into $\mathbf{R}^{d-1}$, and we are going to use the well known Implicit Function Theorem. If F is in C^2, then H is in C^1. To apply the Implicit Function Theorem we need to check only that the differential of H as a function of n is nondegenerate at the point (r_0, n_0). From (5.6), (5.3) and (5.7) it follows, that

$$\begin{aligned}
\frac{\partial}{\partial \psi_k} H|_{n_0, r_0.} &= \frac{\partial^2}{\partial \psi_j \partial \psi_k} F(n_0)A^{-1} - \frac{\partial}{\partial \psi_j} F(n_0)A^{-2}B_k \\
&\quad - \frac{\partial}{\partial \psi_k} F(n_0)A^{-2}B_j + F(n_0)A^{-1}\delta_{jk} + 2F(n_0)A^{-3}B_k B_j \\
&= \frac{\partial^2}{\partial \psi_j \partial \psi_k} F(n_0)A^{-1} + F(n_0)A^{-1}\delta_{jk},
\end{aligned} \tag{5.8}$$

where δ_{jk} is Kronecker δ. Now stiffness positivity implies that the matrix with matrix elements given by (5.8) is nondegenerate. So Implicit Function Theorem implies that the equations (5.5) define a function $R_F(r)$ which is in $C^1(S^{d-1})$. Using formulas for higher derivatives of R_F we find that R_F is in $C^{\ell-1}$ if F is in C^ℓ. When F is holomorphic, H can be treated as a transformation from a neighborhood of a point (n_0, r_0) in $\mathbf{C}^{d-1} \times \mathbf{C}^{d-1}$ into $\mathbf{C}^{d-1}$. The same reasoning shows that the function R is differentiable in the corresponding region, which implies that it is actually holomorphic.

The stiffness positivity condition of Theorem 5.1 is necessary. Without it the surface $X(r)$ can have angles, as we have seen above. But even if it is smooth, but stiffness positivity does not hold for F in C^2 at some r_0 (i.e., equality (3.38) holds), then at the corresponding point n_0 the surface $X(n_0)$ either has no second derivatives or has infinite curvature in some direction and so is not in C^2. Note first that because of the second statement of the Lemma 4.2 it is enough to consider only the case $d = 2$. Indeed, if the boundary of the projection $\pi_K(W_F)$ has infinite curvature at some point p along some direction, then the same is true for any point $q \in \partial W_F$, such that $\pi_K(q) = p$. In our case we consider the two-dimensional plane K, which contains the circle, along which the derivative in (3.38) is taken. So let the curvature of the curve $X(n)$ exists and is bounded at the point n_0. We are going to show first that there exists a disc $B(n_0) \subset W_F$ such that $X(n_0) \in B(n_0)$. After changing the coordinate system we can suppose that

$$W = \{(x,y) \in \mathbf{R}^2 : x \in [a,b], f(x) \le y \le g(x)\},$$

where $a < 0 < b, f$ is convex, g is concave, $f \ge 0$ and $X(n_0) = (0,0)$. Because $f''(0) < C$ for some C, we have $f(x) \le Cx^2/2$ in some neighborhood of 0. Evidently, one can find $c < 0 < d$ in this neighborhood such that $Cc^2/2 \le g(c), Cd^2/2 \le g(d)$. Hence, the segment of the parabola $Cx^2/2$ lying below the interval $[(c, \frac{Cc^2}{2}), (d, \frac{Cd^2}{2})]$, belongs to W_F, which evidently proves the existence statement.

Let p be the radius of $B(n_0), m(n_0)$ be its center and let $G = F - \rho/2$. Then it is geometrically obvious that the line $L_G(n_0)$ is a support line to W_G (the disc $b(n_0)$ centered at $m(n_0)$ with radius $\rho/2$ is in W_G and so the intersection $L_G(n_0) \cap W_G$ contains the point of intersection $L_G(n_0) \cap b(n_0)$ and is thus nonempty). On the other hand stiffness is negative for G at the point r_0, so because of Proposition 3.9 Weak Triangle Inequality is violated and the line $L_G(n_0)$ cannot be a support line.

References

[BF]   Bonnesen T., Fencheln. Theorie der Konvexen Körper, Verlag van J. Springer, Berlin, (1934).

[DKS]   Dobrushin R.L., Kotecky R., Shlosman S.B.: Wulff construction: global shape from local interactions. To appear.

[H]   Herring C. Phys. Rev. **82**, 87 (1951).

[MMR]   Messager A., Miracle-Sole S., Ruiz J. Private communication.

[M]   Minkowski H. Theorie der Konvexen Korper, ins besondere Begründung ihres Oberflächebegriffs, Ges. Abhande. Leipzig-Berlin, vol. 2, 131-229, 1911.

[V]   Valentine F.A. Convex sets. McGraw Hill Book Co., 1964.

[Z]   Zia R.K.P. Anisotropic surface tension and equilibrium crystal shapes, in: Progress in Statistical Mechanics, ed. by C.K. Hu, World Scientific, 1988, pp. 303-357.

Verification of the global Markov property for strongly coupled trigonometric interactions

by

Roman Gielerak*

* Institute of Theoretical Physics, University of Wroclaw, 50-205 Wroclaw, Poland
BiBoS Research Centre, D-4800 Bielefeld 1, FRG

Abstract

We consider a class of local specifications fulfilling certain certain compactness conditions. It is shown that the corresponding FKG maximal Gibbs states are extremal and possess the global Markov property. A special class of trigonometric interactions leads to examples of local specifications fulfilling all our assumptions and conclusions.

1.Introduction.

The present volume contains an exhaustive and deep review of the problem of the global Markov property in Euclidean Field Theory written by Albeverio and Zegarlinski [1]. We refer to this contribution for general background, history, literature and the present status of this problem. Because of this we will be rather brief in this article. The text below is a summary of work [2], where all details of proofs are contained.

The present contribution gives a method of verifying the global Markow property for a class of interactions which leads to local specifications obeying certain compactness conditions. Extremality is known not to be sufficient for the global Markov property to hold, whereas uniqueness of the Gibbs state is not known to always imply the global Markow property, see [1]. In the case of uniqueness, strong uniqueness is sufficient [1]. However attractivity properties and certain compactness conditions are sufficient to prove that the uniqueness of the Gibbs state implies the global Markow property. The proof is obtained by adapting and completing suitably ideas of the papers [3,4]. Alternatively it is possible to use the Goldstein criterion [5] (as was used in [12] and [2]). Strongly coupled trigonometric interactions will serve as an example of interactions which lead to the local specification obeying the requirements of our scheme.

2.Gibbsian approach of the verification of the global Markov property

Let us denote by FR the sets of all PBC measures μ on $(\mathcal{D}'(R^2), \mathcal{B})$ which fulfil the following regularity condition:

$$\mu(e^{\varphi(f)}) \leq \exp\{c_1\|f\|_{-1}^2 + c_2\|S * f\|_1 + c_3\|S * f\|_p^p\} \tag{2.1}$$

484

for some constants $c_1, c_2, c_3 \geq 0$, $p > 2$, where $S = (-\Delta + 1)^{-1}$.

The set of admissible (by this we mean the polynomial, trigonometric or exponential like) interactions as described in [6,7 and ref. therein] shall be denoted by $\mathcal{A}$.

Let μ_C^0 be a Gaussian measure on $(\mathcal{D}'(R^2), \mathcal{B})$ with mean equal to zero and covariance given by C. The indexing space for the free scalar field μ_S^0 is choosen to be $H_{-1}(R^2)$ and the σ–algebras $\Sigma(\Lambda)$ are generated by the closed subspace $H_{-1}(\Lambda)$ of the space $H_{-1}(R^2)$ consisting of those $f \in H_{-1}(R^2)$ for which $\operatorname{supp} f \subset \Lambda$. From the general theory it is known that the conditional expectation values for μ_S^0 are given by the following formula:

$$E_{\mu_S^0}\{F(\varphi)|\Sigma(\Lambda^c)\}(\eta) = \mu_{S^{\partial\Lambda}}^0(F(\varphi + \Pi_{\Lambda^c}^*(\eta))) \tag{2.2}$$

where $S^{\partial\Lambda} \equiv (-\Delta^{\partial\Lambda} + 1)^{-1}$ and $\Pi_{\Lambda^c}^*$ is dual to the orthogonal projector $\Pi_{\Lambda^c} : H_{-1}(R^2) \rightarrow H_{-1}(\Lambda^c)$. More precisely it is known that there exists a Borel subset $\Xi_\Lambda^0 \subset \mathcal{D}'(R^2)$ with full μ_S^0–measure such that for any $\eta \in \Xi_\Lambda^0$ the following holds:

$$\forall_{f \in H_{-1}(R^2)} : < \eta, \Pi_{\Lambda^c}(f) > = < \Pi_{\Lambda^c}^*(\eta), f > \tag{2.3}$$

Taking $U \in \mathcal{A}$ and $\Lambda \subset R^2$ and defining the new measure

$$\mu_\Lambda(d\varphi) = \frac{1}{Z_\Lambda} \cdot \exp U_\Lambda(\varphi) \cdot \mu_S^0(d\varphi) \tag{2.4}$$

we have the following expression for the corresponding conditional expectation values (with $\eta \in \Xi_\Lambda^0$):

$$E_\Lambda^\eta(F) = \frac{\mu_{S^{\partial\Lambda}}^0(F(\varphi + \Pi_{\Lambda^c}^*(\eta)) \exp U_\Lambda(\varphi + \Pi_{\Lambda^c}^*(\eta))}{\mu_{S^{\partial\Lambda}}^0(\exp U_\Lambda(\varphi + \Pi_{\Lambda^c}^*(\eta)))} \tag{2.5}$$

In order to apply general ideas from the works [3,4,5] we have to select a class $\mathcal{R}_b$ of bounded subsets of R^2 and a sufficiently large, Borel subset Ξ of $\mathcal{D}'(R^2)$ such the system of probability kernels $\varepsilon \equiv \{E_\Lambda^\eta\}_{\Lambda \in \mathcal{R}_b, \eta \in \Xi}$ forms a local specification in the following sense

(c1) $\forall \Lambda \in \mathcal{R}_b, \forall \eta \in \Xi$; E_Λ^η is PBC measure on $(\mathcal{D}'(R^2), \mathcal{B})$.

(c2) $\forall \Lambda \in \mathcal{R}_b$; $E_\Lambda^\eta(F)$ is $\Sigma(\Lambda^c)$–measurable.

(c3) $\forall \Lambda_1, \Lambda_2 \in \mathcal{R}_b : \Lambda_1 \subset \Lambda_2 \Rightarrow E_{\Lambda_2}(FG) = E_{\Lambda_2}(FE_{\Lambda_1}(G))$ provided F is $\Sigma(\Lambda_1^c)$–measurable.

The problem of construction of local specifications in this sense is known to have a satisfactory solution [1,2,8].

Proposition 2.1 Let:

$$\mathcal{R}_b \equiv \{\Lambda \subset R^2 \mid |\Lambda| < \infty \text{ and } \partial\Lambda \text{ is piecewise } C^\infty \text{ and } \log\text{–normal}\},$$

$$\Xi \equiv \{\eta \in \mathcal{D}'(R^2) \mid \Pi_\Lambda^*(\eta) \in \bigcap_{\alpha < \frac{1}{2}} H_\alpha(\Lambda) \text{ for any } \Lambda \in \mathcal{R}_b\}$$

Then for any $U \in \mathcal{A}$ the system of probabilistic kernels (2.5) gives $(\mathcal{R}_b, \Xi)$–local specification $\varepsilon(U)$.

For a given $U \in \mathcal{A}$, the set of all Gibbs measures coresponding to the local specification $\varepsilon(U)$ will be denoted by $\mathcal{G}(U)$ and its intersection with the set FR by $\mathcal{G}_{FR}(U)$. The set of their extremal points (i.e. the set of those Gibbs measures for $\varepsilon(U)$ which have no nontrivial convex decompositions in $\mathcal{G}(U)$) is denoted by $\partial\mathcal{G}(U)$. From the general theory of local specifications it follows that $\mathcal{G}(U)$ is uniquely described by its skeleton $s\mathcal{G}(U)$ which is defined as the set of all thermodynamic limits $\mu_\infty^\eta \equiv \lim_\alpha E_{\Lambda_\alpha}^\eta$ as η varies over Ξ. In [2] we have proved that for any $\mu \in FR : \mu(\Xi) = 1$. The local decay properties of $\Pi_{\Lambda^c}^*(\eta)$ (see i.e. [1]) combined with the Stone–Weierstrass theorem gives then $\lim_\alpha E_{\Lambda_\alpha}^\eta = \lim_\alpha \mu_{\Lambda_\alpha}^\eta$ for any $\eta \in \Xi$, where the conditioned, finite volume measure μ_Λ^η is given by

$$\mu_\Lambda^\eta(d\varphi) = (Z_\Lambda^\eta)^{-1} \exp(U_\Lambda(\varphi + \Pi_{\Lambda^c}^*(\eta)))\mu_{S^{\partial\Lambda}}^0(d\varphi) \tag{2.6}$$

For the case of an arbitrary $U \in \mathcal{A}$ a suitable adaption of Röckner arguments from [8] gives rise to:

Poposition 2.2   For any $U \in \mathcal{A}$ the set $\mathcal{G}_{FR}(U)$ is nonempty.

The question we would like to adress in the present contribution is which Gibbs measures from $\mathcal{G}_{FR}(U)$ have global Markov property with respect to half–spaces. Extremality is known to be sufficient for the global Markov property to hold. Uniqueness of the Gibbs state is not known to always imply the global Markoy property, however strong uniqueness is sufficient, see [1]. This route has been used before in [9,10,11,12] for veryfying global Markov property in some cases. However the strong uniqueness result is hard to obtain even in the case where the uniqueness of (a suitably regular) Gibbs state is known as i.e. [12,13]. Attractivity properties of the corresponding local specifications substantially simplify the problem [3,4,5,11]. In the next section we shall introduce the FKG order into the set $s\mathcal{G}(U)$ and then we avoid the difficult strong uniqueness problem by adapting suitably arguments of [3,5].

The effective application of the method of the FKG ordering requires selection of special interactions $C\mathcal{A} \subset \mathcal{A}$ which lead to the (weakly) compact local specifications $\varepsilon(U)$.

Definition:

An interaction $U \in \mathcal{A}$ fulfills the first compactness condition iff

$$\text{(C1)} \qquad \mu_\Lambda^\eta(e^{\varphi(f)}) \leq \exp\{c_1'\|f\|_{-1}^2 + c_2'\|f\|_1 + c_3'\|f\|_p^p\} \tag{2.7}$$

uniformly in $\Lambda \in \mathcal{R}_b$ and $\eta \in \Xi$.

Trigonometric and exponential interactions provides examples of interactions fulfilling the first compactness condition (C1). Whether the polynomial interactions fulfills (C1) needs to be checked.

3. Ferromagnetic ordering in $s\mathcal{G}(U)$

Let us denote by Δ^+ the set of regularising sequences for the δ-distribution, i.e.
$\Delta^+ = \{(\chi_n)_n \subset C^\infty(R^2) | \chi_n \geq 0 \text{ and } \omega\text{-}\lim_{n\uparrow\infty} \chi_n = \delta\}$. The following binary relation in
$\mathcal{D}'(R^2)$: $\eta \bowtie \eta'$ if for any $(\chi_n)_n \in \Delta_0^+$, any $\Lambda \in \mathcal{R}_b$ and any $\alpha > 1$

$$\chi_{\partial\Lambda} \cdot (-\Delta + \alpha)(\eta * \chi_n)(x) \leq \chi_{\partial\Lambda} \cdot (-\Delta + \alpha)(\eta' * \chi_n)(x) \tag{3.1}$$

defines a measurable, partial order in the space $\mathcal{D}'(R^2)$. It is our aim to transfer this
semiorder into the set $s\mathcal{G}(U)$. For this goal let us define $S_\alpha^{\partial\Lambda} = (-\Delta + 1 + \alpha\chi_{\bar{\Lambda}^c})^{-1}$ (for
$\alpha \geq 0$ and on the space $L_2(\Lambda)$) and then the regularized kernel

$$E_{\Lambda,\alpha}^{\eta_n}(F) = \frac{\mu_{S_\alpha^{\partial\Lambda}}^0(F(\varphi + \Pi_{\Lambda^c}^*(\eta * \chi_n)) \exp U_\Lambda(\varphi + \Pi_{\Lambda^c}^*(\eta * \chi_n)))}{\mu_{S_\alpha^{\partial\Lambda}}^0(\exp U_\Lambda(\varphi + \Pi_{\Lambda^c}^*(\eta * \chi_n)))} \tag{3.2}$$

The simple shift transformation then shows

$$E_{\Lambda,\alpha}^{\eta_n}(F) = \frac{\mu_{S_\alpha^{\partial\Lambda}}^0(F \exp U_\Lambda(\varphi) \exp \varphi(\Gamma_{\alpha,n}^\Lambda(\eta)))}{\mu_{S_\alpha^{\partial\Lambda}}^0(\exp U_\Lambda(\varphi) \exp \varphi(\Gamma_{\alpha,n}^\Lambda(\eta)))} \tag{3.3}$$

where

$$\Gamma_{\alpha,n}^\Lambda(\eta) = \begin{cases} 0 & \text{for } x \in \text{Int}\Lambda \\ (-\Lambda + 1 + \alpha)(\eta * \chi_n)(x) & \text{for } x \in \partial\Lambda \\ 0 & \text{for } x \in \bar{\Lambda}^c \end{cases} \tag{3.4}$$

Using the Wiener integral representation for the kernel $S_\alpha^{\partial\Lambda}$ it follows that $S_\alpha^{\partial\Lambda} \to S^{\partial\Lambda}$
strongly on the space $L^2(\Lambda)$ as $\alpha \uparrow \infty$. This fact and well known estimates ([7,6]) enables
us to show the following convergence result.

Proposition 3.1 Let $U \in \mathcal{A}$. Then for any $\Lambda \in \mathcal{R}_b$, any $\eta \in \Xi$:

$$\lim_{n\uparrow\infty} \lim_{\alpha\uparrow\infty} E_{\Lambda,\alpha}^{\eta_n} = \lim_{\alpha\uparrow\infty} \lim_{n\uparrow\infty} E_{\Lambda,\alpha}^{\eta_n}$$
$$= E_\Lambda^\eta \tag{3.5}$$

On account of this convergence one can conclude easily that $\eta \bowtie \eta'$ yields $E_\Lambda^\eta \overset{FKG}{\leq} E_\Lambda^{\eta'}$,
where $\overset{FKG}{\leq}$ denotes the familiar FKG semi-order relation. Let μ_∞^η and $\mu_\infty^{\eta'}$ be any accumu-
lation points of the probabilistic kernels $(E_\Lambda^\eta)_\Lambda$, resp. $(E_\Lambda^{\eta'})_\Lambda$. If $\eta \bowtie \eta'$ then $\mu_\infty^\eta \overset{FKG}{\leq} \mu_\infty^{\eta'}$.
To see this let us take the regularized versions of E_Λ^η and $E_\Lambda^{\eta'}$ as given by (3.2) and let
assume that G fulfills the conditions for which FKG correlation inequalities are valid. By
the fundamental theorem of calculus:

$$(E_{\Lambda,\alpha}^{\eta'_n} - E_{\Lambda,\alpha}^{\eta_n})(G) = \int_0^1 ds \frac{d}{ds}(E_{\Lambda,\alpha}^{(s\eta'+(1-s)\eta)*\chi_n}(G))$$
$$= \int_0^1 ds E_{\Lambda,\alpha}^{(s\eta'+(1-s)\eta)*\chi_n}(G; \varphi(\Gamma_{\alpha,n}^\Lambda(\eta'-\eta))^T \geq 0 \tag{3.6}$$

This uses the definition of our regularisation and the very definition of FKG correlation inequalities. Then from Proposition (3.1) it follows that $E_\Lambda^{\eta'}(G) \leq E_\Lambda^\eta(G)$ and this holds for any $\Lambda \in \mathcal{R}_b$ and therefore can be transferred into the thermodynamic limit. The existence of the accumulation points μ_∞^η is guaranteed by the first compactness condition (C1) in our scheme. Thus we have transferred the FKG order into the skeleton $s\mathcal{G}(U)$, provided $U \in C\mathcal{A}$.

From the compactness asumption (C1) it follows that for any FKG ordered net $(\mu_\alpha)_\alpha$ in the set $s\mathcal{G}_{FR}(U)$ there exist maximal elements. From the Kuratowski–Zorn lemma it follows the existence of maximal elements in the set $s\mathcal{G}_{FR}(U)$. In fact one can show the existence of a unique, extremal (with respect to the semiorder $\overset{FKG}{\leq}$) elements μ_+ and μ_- in the set $s\mathcal{G}_{FR}(U)$, provided $U \in C\mathcal{A}$. Those extremal elements μ_+ and μ_- are automatically Euclidean invariant and reflection positive. The equality $\mu_+ = \mu_-$ shows that $\# s\mathcal{G}_{FR}(U) = 1$ and therefore also $\# \mathcal{G}_{FR}(U) = 1$.

An useful criterion for the equality $\mu_+ = \mu_-$ can be described conveniently in terms of the conditioned free energy densities $p_\Lambda^\eta(U)$ which are defined by

$$p_\Lambda^\eta(U) = -\frac{1}{|\Lambda|} \ln \mu_{S^{\partial\Lambda}}^0(\exp U_\Lambda(\varphi + \Pi_{\Lambda^c}^*(\eta))) \tag{3.7}$$

Proposition 3.2 ([14]) Let $U \in \mathcal{A}$ and let $\mu \in FR$. Then for any sequence $(\Lambda_n)_n \subset \mathcal{R}_b$ such that $\Lambda_n \uparrow R^2$ monotonously and such that $\overline{\lim}_{n\uparrow\infty} \frac{|\partial\Lambda_n|^{1+p}}{|\Lambda_n|} = 0$, for some $p > \frac{1}{4}$ the unique limit $p_\infty^\eta(U) \equiv \lim_{n\uparrow\infty} p_{\Lambda_n}^\eta(U)$ exists for μ–a.e. $\eta \in \mathcal{D}'(R^2)$ and moreover

$$p_\infty^\eta(U) = p_\infty^{\eta\equiv 0}(U) \equiv p_\infty^{H.D}(U) \quad (\mu\text{–a.e}) \tag{3.8}$$

For $h \in R$ and $U \in \mathcal{A}$ let $U_\Lambda^h(\varphi) = U_\Lambda(\varphi) + h \int_\Lambda \varphi(x)dx$ and let $p_\infty^{H.D}(h) = p_\infty^{H.D}(U^h)$. Clearly $p_\infty^{H.D}(h)$ is a concave function of h (and therefore an almost everywhere differentiable function of h).

Proposition 3.3 (see also [1]) Let $U^h \in C\mathcal{A}$ for any $h \in R$ and let h_0 be a point of differentiability of the corresponding free energy density $p_\infty^{H.D}(U^h)$. Then the set $\mathcal{G}_{FR}(U^h)$ consists of an exactly one element.

We outline the proof. Let $\mu \in s\mathcal{G}_{FR}(U^h)$ be of the form $\mu = \lim_\alpha E_{\Lambda_\alpha}^\eta$ where $(\Lambda_\alpha)_\alpha \subset \mathcal{R}_b$ is any net fulfilling the assumptions of Proposition 3.2. From the assumed regularity of h_0 it follows that:

$$\begin{aligned}
\frac{\partial p_\Lambda^\eta(U^h)}{\partial h}\Big|_{h=h_0} &= \lim_\alpha \frac{\partial p_\Lambda^\eta(U^h)}{\partial h^\alpha}\Big|_{h=h_0} \\
&= -\lim_\alpha \frac{1}{|\Lambda_\alpha|} \int_{\Lambda_\alpha} dx\, \mu_{\Lambda_\alpha}^\eta(\varphi(x) + \Pi_{\Lambda_\alpha^c}^*(\eta)) \\
&= \frac{\partial p_\infty^{H.D}(U^h)}{\partial h}\Big|_{h=h_0} \\
&= \mu_\infty^{\eta\equiv 0}(\varphi(0))
\end{aligned} \tag{3.9}$$

In particular taking $\mu = \mu_{\pm}$ we obtain the equality $\mu_+(\varphi(0)) = \mu_-(\varphi(0))$ from which $\mu_+ = \mu_-$ it follows (see i.e. [15]).

4. From uniqueness to the global Markov property.

For concreteness we choose $\Gamma = \{x \in R^2 | x_1 = 0\}$ and let $R^2_{\pm} = \{x \in R^2 | x_1 > (<)0\}$. Then we have $R^2 - \Gamma = R^2_+ \cup R^2_-$. Let us denote by $\mathcal{R}^{\pm}_b$ the family of bounded subsets Λ of $R^2_{\pm}$ which have piecewise C^∞–boundaries and which are log–normal. On the inductive systems $\{\mathcal{D}'(\bar{R}^2_{\pm}), \Sigma(\Lambda)_{\Lambda \in \mathcal{R}^{\pm}_b}, \Sigma(R^{\pm})\}$ we define the following probabilistic kernels:

$$E^{\eta \vee \eta'}_{R^2_{\pm},\Lambda}(F) = \frac{\mu^0_{S^{\Gamma \cup \partial \Lambda}}(F(\varphi + \Pi^*_{\Gamma \cup \partial \Lambda}(\eta,\eta')\exp U_\Lambda(\varphi + \Pi^*_{\Gamma \cup \partial \Lambda}(\eta,\eta')))}{\mu^0_{S^{\Gamma \cup \partial \Lambda}}(\exp U_\Lambda(\varphi + \Pi^*_{\Gamma \cup \partial \Lambda}(\eta,\eta')))} \tag{4.1}$$

where $\Pi^*_{\Gamma \cup \partial \Lambda}(\eta,\eta')$ is defined (informally!) as the solution of the following Dirichlet problem:

$$\left\{ \begin{array}{l} (-\Delta + 1)\Pi^*_{\Gamma \cup \partial \Lambda}(\eta,\eta')(x) = 0 \text{ for } x \in \mathrm{Int}(\Lambda \cap R^2_{\pm}) \\[2mm] \Pi^*_{\Gamma \cup \partial \Lambda}(\eta,\eta')(x) = \left\{ \begin{array}{l} \eta(x) \text{ for } x \in \Gamma \\[1mm] \eta'(x) \text{ for } x \in \partial \Lambda - \Gamma \end{array} \right\} \end{array} \right\} \tag{4.2}$$

Assuming $\mu \in FR$ and taking $U \in \mathcal{A}$ one can show that there exist Borel subspaces $\Xi^{\pm} \subset \mathcal{D}'(\bar{R}^2_{\pm})$ of full μ–measure such that the system of probabilistic kernels as defined in (4.1) gives $(\mathcal{R}^{\pm}_b, \Xi^{\pm})$–local specifications on $\mathcal{D}'(\bar{R}^2_{\pm}), \Sigma(\bar{R}^2_{\pm}))$.

For any $\alpha > 0$ and $(\chi_n)_n \in \Delta^+$ the regularized kernels $E^{\eta_n \vee \eta'_n}_{R^2_{\pm},\Lambda,\alpha}$ are given by formulae like (4.1), but the Gaussian measure $\mu^0_{S^{\Gamma \cup \partial \Lambda}}$ should be replaced by the Gaussian measure $\mu^0_{S^{\Gamma \cup \partial \Lambda}_\alpha}$, where the covariance $S^{\Gamma \cup \partial \Lambda}_\alpha$ is given by $S^{\Gamma \cup \partial \Lambda}_\alpha = (-\Delta^\Gamma + 1 + \alpha\chi_{\overline{(R^2_{\pm}-\Lambda)^c}})^{-1}$ and η (resp. η') is replaced by $\eta * \chi_n$ (resp. $\eta' * \chi_n$). Using the Wiener integral representation it follows that $S^{\Gamma \cup \partial \Lambda}_\alpha \to S^{\Gamma \cup \partial \Lambda}$ strongly in $L^2(\Lambda)$ as $\alpha \uparrow \infty$. This enables us to show, that for any $\eta' \in \Xi^{\pm}$, any $\eta \in \Xi$ the following convergence takes place:

$$\lim_{n \uparrow \infty} \lim_{\alpha \uparrow \infty} E^{\eta_n \vee \eta'_n}_{R^2_{\pm},\Lambda,\alpha} = E^{\eta \vee \eta'}_{R^2_{\pm},\Lambda} \tag{4.3}$$

In complete analogy with the previous section the FKG order can be introduced into the skeletons of the sets of the Gibbs states $\mathcal{G}^{\eta}_{\pm}(U)$ corresponding to the local specifications $\varepsilon^{\eta}_{\pm} = ((E^{\eta \vee \eta'}_{R^2_{\pm},\Lambda})_{\Lambda,\eta'})$ provided they exists. To assure the existence of Gibbs states in $\mathcal{G}^{\eta}_{\pm}(U)$ we now impose the second compactness condition. Owing to the local decay properties of $\Pi^*_{\Gamma \cup \partial \Lambda}(\eta,\eta')$ as $\Lambda \uparrow R^2_{\pm}$ (see as an example Thm. 2.2 in [9]) and referring to the determining properties of the set $\{\exp i\varphi(f)| \mathrm{supp}\, f\text{–compact}\}$ it is enough to guarrant the existence of weak limits of the corresponding conditioned measures $\mu^{\eta \vee \eta'}_{\pm,\Lambda}$ which are defined by:

$$\mu^{\eta \vee \eta'}_{\pm,\Lambda}(F) = \frac{\mu^0_{S^{\Gamma \cup \partial \Lambda}}(F(\varphi)\exp U_\Lambda(\varphi + \Pi^*_{\Gamma \cup \partial \Lambda}(\eta,\eta')))}{\mu^0_{S^{\Gamma \cup \partial \Lambda}}(\exp U_\Lambda(\varphi + \Pi^*_{\Gamma \cup \partial \Lambda}(\eta,\eta')))} \tag{4.4}$$

Second Compactness Condition (C2) We say that the conditional local specifications $\varepsilon_{\pm}^{\eta}$ fulfill the second compactness condition iff there exists a continuous norm $||| - |||$ on the space $\mathcal{D}(R^2)$ such that the following estimate

$$\mu_{\pm,\Lambda}^{\eta\vee\eta'}(e^{\varphi(f)}) \leq e^{O_\eta(1)|||f|||} \tag{4.5}$$

holds uniformly in $\eta' \in \Xi^{\pm}$ and $\Lambda \in \mathcal{R}_b^{\pm}$.

From now on we will assume that the interaction $U \in \mathcal{A}$ leads to the local specifications fulfilling the first (C1) and the second (C2) compactness condition. From the second compactness condition it follows that for any $\mu \in \mathcal{G}_{FR}(U)_0$ the set $s\mathcal{G}_+^\eta(U)$ includes maximal (with respect to FKG semiorder) elements $\mathcal{M}_+^\eta$ and $\mathcal{M}_-^\eta$ for μ a.a. $\eta \in \mathcal{D}'(R^2)$. Moreover it is due to noncompactness of the corresponding Ξ–spaces that there exists $\eta^* \in \Xi$ such that $\mu_+ = \lim_{\Lambda \uparrow R^2} E_\Lambda^{\eta^*}$ and $\lim_{\Lambda \uparrow \bar{R}^2} E_{R_+^2,\Lambda}^{\eta\vee\eta^*} = \mathcal{M}_+^\eta$. It is an important remark that the measure $\mathcal{M}_+^\eta$ is an extremal point of $\mathcal{G}_+^\eta(U)$ from which it follows that the map $\Xi \ni \eta \to \mathcal{M}_+^\eta$ is $\Sigma(\Gamma)$ measurable. In other words it means that the corresponding σ–algebra "at infinity" for the measure is nontrivial only on the $\Sigma(\Gamma)$ piece. Alternatively the weaker condition (C) of Goldstein [5] or some simple modification of it can be used as well [2].

Now let us take $G \in \Sigma(R_-^2)$ and $F \in \Sigma(R_+^2)$ and let $(\Lambda_n)_n \subset R_b$ be such that $\Lambda_n \uparrow R^2$ monotonously and by inclusion. Then we have

$$\begin{aligned}
\mu_+(FG) &= \lim_{n\uparrow\infty} E_{\Lambda_n}^{\eta^*}(FG) \\
&= \lim_{n\uparrow\infty} E_{\Lambda_n}^{\eta^*}(E_{\Lambda_n\cap R_+^2}^{\cdot}(FG)) \\
&= \lim_{n\uparrow\infty} E_{\Lambda_n}^{\eta^*}(GE_{R_+^2,\Lambda_n^+}^{\cdot\vee\eta^*}(F)) \\
&= \mu_+(G\mathcal{M}_+^{\cdot}(F))
\end{aligned} \tag{4.6}$$

where $\Lambda_n^{\pm} = \Lambda_n \cap R_{\pm}^2$ and the DLR condition $(c3)$ has been used.
From the very definition of the conditional expectation value it follows that:

$$E_{\mu_+}\{F|\Sigma(R_-^2)\}(\cdot) = \lim_{n\uparrow\infty} E_{R_+^2,\Lambda_n^+}^{\cdot\vee\eta^*}(F) \tag{4.7}$$

which is $\Sigma(\Gamma)$ measurable, due to the remarks above. Thus we have showed that the weak compactness of the conditional local specifications $\varepsilon_{\pm}^\eta$ is a sufficient condition for the FKG–extremal states $\mu_{\pm} \in s\mathcal{G}_{FR}(U)$ to have the global Markov property with respect to the hyperplane Γ. In particular it follows from the uniqueness condition that this unique Gibbs measure $\mu \in \mathcal{G}_{FR}(U)$ is automatically a globally markovian, generalized random field providing the conditions (C1) and (C2) are fulfilled.

5. Strongly coupled trigonometric interactions.

The two–dimensional sine–Gordon theory constructed and discussed in [16,17,18,19] will serve as an example which leads to the local specification fulfilling the first and the second compactness conditions (C1) and (C2). In the pioneering paper [9] the global Markov property of the corresponding weakly coupled Gibbs measure was proved. Essentially the boundness of perturbations was one of the most important ingredients of the proof of the uniform (in the boundary data) convergence of the corresponding cluster expansions which yields the strong uniqueness result. For an additive functional U_Λ of the form:

$$U_\Lambda(\varphi) = U_\Lambda^z(\varphi) = z \int_\Lambda \; : \cos\alpha\varphi : (x)dx, \quad z \geq 0, \quad |\alpha| < \left(\frac{2}{1 - \frac{1}{2\pi}} \right)^{\frac{1}{2}} \tag{5.1}$$

we extend the result of [9].

Theorem 5.1.   Let U_Λ be given by (5.1). Then for any regular value of $z_0 > 0$ the set $\mathcal{G}_{FR}(z_0)$ consists of exactly one element $\mu_\infty(z_0)$ which has the global Markov property. Outline of the proof:

The following correlation inequalities of the Ginibre type play an important role in our proof. The finite–volume, with "empty b.c" Gibbs measure μ_Λ is defined by:

$$\mu_\Lambda(d\varphi) = Z_\Lambda^{-1} \cdot \exp U_\Lambda(\varphi) \cdot \mu_S^0(d\varphi). \tag{5.2}$$

(CI 1) Let $(f_i)_{i=1,\ldots,n}$ be some real measurable functions. Then we have:

$$\left| \mu_\Lambda\left(\prod_{i=1}^n \; : \cos\alpha_i(\varphi(x_i) + f_i(x_i)) : \right) \right| \leq \mu_\Lambda\left(\prod_{i=1}^n \; : \cos\alpha_i\varphi(x_i) : \right) \tag{5.3}$$

(CI 2) For any integers n, m

$$\mu_\Lambda\left(\prod_{i=1}^n \; : \cos\alpha\varphi(x_i) : \; ; \; \prod_{j=1}^m \; : \cos\beta\varphi(y_j) : \right)^T \geq 0 \tag{5.4}$$

(CI 3) $\cos\alpha\varphi$–bound:

$$\bigvee_{f \in \mathcal{D}(R^2)} \; \mu_\Lambda\left(e^{:\cos\alpha\varphi:(f)} \right) \leq e^{c_1 \|f*S\|_1 + c_2 \|f*S\|_p^p} \tag{5.5}$$

$$\text{for } p > \frac{1}{1 - \frac{\alpha^2}{4\pi}} \text{ and uniformly in } \Lambda$$

Using (CI1), (CI2) and (CI3) it follows easily

$$\mu_\Lambda^\eta\left(e^{\varphi(f)} \right) \leq \exp\frac{1}{2}\|f\|_{-1}^2 \exp z(\|S^{\partial\Lambda} * f\|_1 + \|S^{\partial\Lambda} * f\|_p^p). \tag{5.6}$$

Therefore the first compactness condition (C1) is fulfilled. Similarly we are checking the validity of the second compactness condition (C2):

$$\mu_{R_{\pm}^2,\Lambda}^{\eta\vee\eta'}(e^{\varphi(f)}) \le \exp\{\frac{1}{2}\|f\|_{-1}^2 + z(\|S^{\Gamma\cup\partial\Lambda} * f\|_1 + \|S^{\Gamma\cup\partial\Lambda} * f\|_p^p)\}$$

$$\le \exp\frac{1}{2}\|f\|_{-1}^2 \exp z \cdot O(1)(\|S * f\|_1 + \|S * f\|_p^p). \tag{5.7}$$

The following correlation inequalities (isolated in [18]) together with (CI2) give the existence of a unique thermodynamic limit $\mu_\infty = \lim_\Lambda \mu_\Lambda$.

(CI 4)

$$\mu_\Lambda(e^{t\varphi(f)} \; ; \; : \cos\alpha\varphi(x) :)^T \le 0 \tag{5.8}$$

(CI 5)

$$\mu_\Lambda(\varphi^2(f) \; ; \; : \cos\alpha\varphi(x) :)^T \le 0 \tag{5.9}$$

It follows easily that μ_∞ fulfills:

$$\mu_\infty(e^{\varphi(f)}) \le e^{\frac{1}{2}\|f\|_{-1}^2} \tag{5.10}$$

and thus μ_∞ is completely regular Gibbs measure.

The following correlation inequality (isolated originally by Pfister in [20] in a similar context) plays an important role in the proof of uniqueness:

(CI 6)

$$(\mu_\Lambda^{\eta=0} \otimes \mu_\Lambda^\eta)((\prod_{i=1}^n : \cos\alpha\varphi(x_i) : - \prod_{i=1}^n : \cos\alpha\varphi'(x_i) :)$$

$$\exp\delta \int_\Lambda : \cos\alpha\varphi(x) : : \cos\alpha\varphi'(x) : dx) \ge 0 \tag{5.11}$$

for any $\delta \in R$.

As we have demonstrated in [19] correlation inequality (CI6) leads to the following bootstrap principle:

$$\text{if } \lim_\Lambda \mu_\Lambda^\eta(: \cos\alpha\varphi(x) :) = \lim_\Lambda \mu_\Lambda^{\eta=0}(: \cos\alpha\varphi(x) :) \tag{5.12}$$

and moreover μ_∞^η is translationally invariant, then $\mu_\infty^\eta = \lim_\Lambda \mu_\Lambda^{\eta=0} \equiv \mu_\infty^{H.D}$ for any $\eta \in \Xi$.

Application of Proposition 3.2 and the assumed regularity of z_0 yield:

$$\mu_\infty^\eta(: \cos\alpha\varphi(0) :)(z_0) = \mu_\infty^{H.D}(: \cos\alpha\varphi(0) :)(z_0)$$

provided μ_∞^η is translationally invariant. Taking into account that the extremal Gibbs measures $\mu_\pm(z_0)$ from $s\mathcal{G}_{FR}(z_0)$ are translationally invariant we obtain

$$\mu_\pm(: \cos\alpha\varphi : (0))(z_0) = \mu_\infty^{H.D}(: \cos\alpha\varphi : (0))(z_0) \tag{5.13}$$

Integration by parts gives

$$\mu_\pm(\varphi(0))(z_0) = \mu_\infty^{H.D}(\varphi(0))(z_0)$$

from which it follows (with the help of FKG ordering arguments) $\mu_+ = \mu_-$. Adapting arguments of Röckner [8] it follows that $\mu_\infty \in \mathcal{G}_{FR}(z)$ and therefore $\mathcal{G}_{FR}(z_0) = \{\mu_\infty(z_0)\}$.

Since the second compactness (C2) is fulfilled we can apply arguments of section 4 to conclude that $\mu_\infty(z_0)$ has global Markov property with respect to Γ.

Remarks. A weaker uniqueness theorem has been provided by us in [19]. The optimal value for α in Theorem 5.1 should be $|\alpha| < 2\sqrt{\pi}$ but at the moment we do not know how to relax (5.1). From the correlation inequality (C1) it follows that for any $z > 0$ the Gibbs state $\mu_\infty(z)$ is an extremal point of the set $\mathcal{G}_{FR}(z)$. It is natural to conjecture that the global Markow property holds for any $z \geq 0$.

Acknowledgements.

I would like to express my deep gratitude to Prof. S. Albeverio whose important remarks significantly improved the presentation of the material contained in this article. The presented research has been begun during my participation in the SFB–273 program in summer 1987.

References

1. Albeverio, S., Zegarlinski, B., in this volume
2. Gielerak, R.:Towards the verification of the global Markow property in the Euclidean (Quantum) Field Theory, to appear as BiBoS preprint.
3. Föllmer, H.: in Streit, L. (ed.) Quantum fields–algebras–processes, Springer (1975).
4. Bellissard, J., Høegh–Krohn, R.: Comm. Math. Phys. __84__, 297–327 (1982)
5. Goldstein, Sh.: Comm. Math. Phys. __74__, 223–234, (1980)
6. Simon, B.: The $\mathcal{P}(\varphi)_2$ Euclidean (Quantum) Field Theory. Princeton, N.J. Princeton Univ. (1974)
7. Glimm, J., Jaffe, A.: Quantum Physics, 2^{nd}edition, Springer–Verlag (1986)
8. Röckner, M.: Comm. Math. Phys. __106__, 105–135 (1986)
9. Albeverio, S., Høegh–Krohn, R.: Comm. Math. Phys. __68__, 95–128 (1979)
10. Gielerak, R.: Journ. Math. Phys. __24__ 347–355 (1983)
11. Zegarlinski, B.: Comm. Math. Phys. __96__, 195–221 (1984)
12. Albeverio, S., Høegh–Krohn, R., Zegarlinski, B.: Comm. Math. Phys. __121__, 683–697 (1989)
13. Gielerak, R.: Ann. of Phys. 189 (1), 1–28 (1989)
14. Gielerak, R,: Rept. Math. Phys. __24__, 145–154 (1986), and the addendum recently submitted.
15. Fröhlich, J., Simon, B.: Ann. Math. __105__, 493–526 (1977)
16. Fröhlich, J.: Comm. Math. Phys. __47__, 233–268 (1976)
17. Fröhlich, J., Seiler, G.: Helv. Phys. Acta __49__, 889–924 (1976)
18. Fröhlich, J., Park, Y.M.: Comm. Math. Phys. __59__, 235–265 (1978)
19. Gielerak, R.: Journ. Math. Phys. 27 (12) 2892–2902 (1986)
20. Pfister, Ch–E.: Comm. Math. Phys. __86__, 375–390 (1982)

DECOMPOSITION OF ISING MODELS AND THE MAYER EXPANSION

C. M. Newman
Department of Mathematics
University of Arizona
Tucson, AZ 85721 USA

George A. Baker, Jr.
Theoretical Division
Los Alamos National Laboratory
University of California
Los Alamos, NM 87545 USA

Abstract: It was shown by Bassalygo and Dobrushin (for high temperature T) and by Hajek and Berger (in general) that a spin-1/2 Ising model can be represented as a noninteracting system of spins in some (correlated) random external field of constant strength h provided h exceeds a threshold value $h_c(T)$. For the standard Ising ferromagnet or antiferromagnet on $\mathbb{Z}^d$ with zero external field, we show that $\exp(-h_c(T))$ coincides with $R(T)$, the radius of convergence of the classical Mayer series for the *antiferromagnet*. This yields an identity for the entropy theoretic quantity denoted δ_c by Hajek and Berger: $\delta_c = R/(1 + R)$. For $d = 2$, we evaluate $R(T)$ numerically by a series analysis.

1. Background and Results

Consider a ± 1 valued random field $(S_i, i \in V)$, also denoted (S_i) or simply S, such as a spin-1/2 Ising model, where the index set V might be $\mathbb{Z}^d$ or some finite subset of $\mathbb{Z}^d$. We will define a quantity denoted δ_c by Hajek and Berger [HB] in three equivalent ways.

First, following [HB], we say that $D = (D_i)$ with each D_i in $[0,1/2)$ is *extractable* from S if there is *some* ± 1 valued random field W such that (S_i) has the same joint distribution as $\left(W_i U_i(D_i) \right)$ where the $U_i(D_i)$'s are *independent* random variables with $P\left(U_i(D_i) = -1 \right) = D_i = P\left(U_i(D_i) = +1 \right)$ and (W_i) is independent of $\left(U_i(D_i) \right)$. I.e., D is extractable from S when the distribution of S can be decomposed as a convolution (in the sense of the product group of $\{+1,-1\}$'s) of the distribution of $\left(U_i(D_i) \right)$ and some other distribution (that of W). The quantity δ_c is defined to be the supremum of δ's in $[0,1/2)$ such that $(D_i \equiv \delta)$ is extractable from S. It is an information theoretic quantity relevant in statistical communications theory when one transmits binary messages $(s_i, i \in V)$ chosen acoording to the distribution of S.

A second definition of δ_c may be obtained by considering the correlations of S,

$$\phi_S(A) = \left\langle \prod_{i \in A} S_i \right\rangle, \qquad A \subseteq V, \quad A \text{ finite,}$$

where $\langle \cdot \rangle$ denotes expectation with respect to the distribution of the random field S. Then δ_c may be defined as the supremum of δ's in $[0,1/2)$ such that

$$\phi_S(A)/(1 - 2\delta)^{|A|}, \qquad A \subseteq V, \quad A \text{ finite,}$$

are the correlations $\phi_W(A)$ of some ± 1 valued random field W; here $|A|$ denotes the number of sites in A.

Finally, another equivalent definition of δ_c is

$$\delta_c = e^{-h_c}/(e^{h_c} + e^{-h_c}),$$

where h_c is defined as follows. For a given (inhomogeneous) magnetic field $(H_i, i \in V)$, denote by $\nu_{(H_i)}$ the distribution of uncoupled spins in an external field (H_i); i.e., the $\nu_{(H_i)}$-probability of a spin configuration $(s_i = \pm 1, i \in A \subseteq V)$ is

$$Z_A^{-1} \exp\left(\sum_{i \in A} H_i s_i\right), \qquad Z_A = \prod_{i \in A}\left(e^{H_i} + e^{-H_i}\right).$$

Then h_c is the minimum $h \geq 0$ such that the distribution of S can be expressed as a (quenched) average of such $\nu_{(H_i)}$'s over (H_i)'s with fixed $|H_i| = h$; i.e., such that the distribution of S has the form

$$\int \nu_{(hw_i)} d\mu(w)$$

for some probability measure μ on sign configurations $w = (w_i = \pm 1, i \in V)$ of a *random* external field (hW_i) of constant magnitude h.

Hajek and Berger showed [HB] that for a large class of Ising models (with finite range interactions), δ_c can be bounded away from zero uniformly in the size of the region V; Bassalygo and Dobrushin had earlier shown this for high temperature [BD]. Newman subsequently discovered a relation between δ_c and the location of partition function zeros which yielded a quite different proof of the Hajek-Berger result with somewhat improved bounds [N]. The main result of this paper, based on this latter approach, is an identity for δ_c in terms of the radius of convergence R of the Mayer series: $\delta_c = R/(1 + R)$.

For a finite region V, the Mayer series is the power series expansion, in powers of $z = e^{-2h}$, of $\ln \Lambda (z)$ where

$$\ln \Lambda (z) = \frac{1}{|V|}\ln\left(Q_S(z)/Q_S(0)\right), \quad Q_S(z) = \left\langle \exp\left(h\sum_{i \in V}(S_i - 1)\right)\right\rangle. \tag{1.1}$$

(In case $Q_S(0) = 0$ so that $\ln \Lambda (z)$ is undefined, our convention will be to set $R = 0$.) For an infinite V one takes a limit of finite volume approximations to (1.1) with appropriate boundary conditions; we will always take free boundary conditions.

Our identity for δ_c is valid for Ising models with pair antiferromagnetic interactions and negative or zero external field; i.e., ones whose distribution is of the form

$$Z^{-1}\exp\left(\sum_{\{i,j\}} K_{ij}s_i s_j + \sum_i H_i s_i\right) \tag{1.2}$$

with $K_{ij} \leq 0$ and $H_i \leq 0$ for all i, j. (The temperature has been absorbed into the K_{ij}'s and H_i's.) It is easy to see that δ_c is unchanged by arbitrary gauge transformations - i.e., when (S_i) is replaced by $(s_i S_i)$ with (s_i) an arbitrary configuration of ± 1's. This allows one to extend the above results to antiferromagnet models with positive external field $(H_i \geq 0$ for all $i)$ and to certain ferromagnetic $(K_{ij} \geq 0$ for all i, $j)$ models, e.g., to the nearest neighbor ferromagnet on the cubic lattice $\mathbb{Z}^d$ (but not on a triangular lattice) with zero external field (or with a staggered external field). Note in this context that the resulting identity relates the δ_c of the ferromagnet to the radius of convergence R of the antiferromagnet obtained after a gauge transformation.

In the next section of the paper, we derive the identity $\delta_c = R/(1 + R)$ by combining the approach of [N] with known results about the Mayer series [Gr] and about zeros of partition functions [LS]. In the third section, we restrict attention to the standard nearest neighbor planar Ising model with zero external field; here $\Lambda = \mathbb{Z}^2$, $K_{ij} = K$ $(= J/kT)$ for $\{i,j\}$ a nearest neighbor pair and zero otherwise and $H_i \equiv 0$. Although, unlike the one-dimensional case [G], there is no explicit solution for R and δ_c as a function of K here, we can and do evaluate these functions numerically.

2. *Theory*

We begin by restricting attention to finite V. An elementary calculation based on the matrix identity,

$$\begin{bmatrix} 1 - \delta & \delta \\ \delta & 1 - \delta \end{bmatrix}^{-1} = \text{const.} \begin{bmatrix} 1 & z \\ z & 1 \end{bmatrix}, \quad \text{with} \quad z = \frac{\delta}{1-\delta}, \tag{2.1}$$

shows that when D is extractable from S, the distribution of the other factor W is given by

$$P(W = w) = \text{const.} \left\langle \prod_i z_i^{(1-w_i S_i)/2} \right\rangle, \quad \text{with} \quad z_i = \frac{D_i}{1 - D_i}. \tag{2.2}$$

Setting $D_i \equiv \delta$, noting that probabilities must be non-negative and recalling the definition (1.1) of $Q_S(z)$, we have the following.

<u>Proposition 1</u>. Suppose V is finite. For any choice $w = (w_i, i \in V)$ of ± 1's, define λ_w as the minimum λ in $[0,\infty)$ such that $-\lambda$ is a zero of $Q_{wS}(z)$; here wS is the random field $(w_i S_i, i \in V)$ and $\lambda_w = +\infty$ if Q_{wS} doesn't vanish on $(-\infty,0]$. Let $\bar{\lambda} = \min_w \lambda_w$; then $\delta_c/(1 - \delta_c) = \bar{\lambda}$ or equivalently $\delta_c = \bar{\lambda}/(1 + \bar{\lambda})$.

<u>Proof</u>. This result follows immediately from the discussion preceding the proposition and from the remark that if (S_i) is equidistributed with $\left(\tilde{W}_i U_i(\delta) \right)$ with $0 < \delta < \delta_c$, then $P(\tilde{W} = w) > 0$ for each w.

To obtain the identity $\delta_c = R/(1 + R)$ for antiferromagnets, we need to identify $\bar{\lambda}$ with R. Since $Q_S(z)$ is a polynomial and hence analytic, the previous definition of R as a radius of convergence of $\ln\left(Q_S(z)/Q_S(0) \right)$ is equivalent to $R = R_{(1)} \left(= R_{(w_i \equiv 1)} \right)$, where

$$R_w = \min\{|z| : Q_{wS}(z) = 0\}. \tag{2.3}$$

When S has a distribution of the form (1.2), $Q_{wS}(0) = P(S = w) > 0$ for each w, and hence R_w is also the radius of convergence of $\ln Q_{wS}(z)$. The next two theorems will immediately yield the desired identification of $\bar{\lambda}$ with R_0. The first is part of Theorem 2 of [Gr] (see also Prop. 2 of [BF]). The second is an immediate corollary of Theorem 3.2 of [LS].

$\underline{\text{Theorem}}$ $\underline{2}$ [Gr]. Suppose V is finite and S has the distribution (1.2) with $K_{ij} \leq 0$ for all $\{i,j\}$ (and H_i arbitrary). Then all the multi-Taylor coefficients $\left(\text{at}\ (z_i \equiv 0)\right)$ of $-\ln Q_S\left(-(z_i)\right)$, where

$$Q_S\left((z_i)\right) = \left\langle \prod_i z_i^{(1-S_i)/2} \right\rangle, \tag{2.4}$$

are non-negative. Hence so are the Taylor coefficients of $-\ln Q_S(-z)$ and thus $\lambda_{(1)} = R_{(1)}$.

$\underline{\text{Theorem}}$ $\underline{3}$. [LS]. Suppose V is finite, S has the distribution (1.2) and S' has a similar distribution with parameters K'_{ij} and H'_i. Suppose that $(\rho_i, i \in V)$ are numbers such that

$$Q_S\left((z_i)\right) \neq 0 \ \text{ when } \ |z_i| < \rho_i \ \text{ for each } i;$$

if $K_{ij} \leq K'_{ij}$ and $H_i \leq H'_i$ for all i, j, then $Q_{S'}\left((z_i)\right)$ does not vanish in the same region. Hence if $K_{ij} \leq 0$ and $H_i \leq 0$ for all i, j, it follows that for every w, $R_{(1)} \leq R_w$.

$\underline{\text{Proof}}$. For the first part of the theorem, it suffices by induction to consider the case where S' and S differ in a single K_{ij} or a single H_i. The latter case is trivial because it amounts to replacement of z_i by $\exp\left(-2(H'_i - H_i)z_i\right)$; the former case follows from Theorem 3.2 of [LS]. To obtain the last part of the theorem from the first part, simply note that changing S to $S' = wS$ corresponds to changing K_{ij} to $w_i w_j K_{ij}$ and H_i to $w_i H_i$.

Our main result now follows.

<u>Theorem 4</u>. Suppose V is finite and S has the distribution (1.2) with $K_{ij} \leq 0$ and $H_i \leq 0$ for all i, j. Then $\delta_c = R/(1 + R)$ where $R = R_{(1)}$, the radius of convergence of the Taylor series of

$$\ln\left\langle \prod_i z^{(1-S_i)/2} \right\rangle.$$

<u>Proof</u>. By Prop. 1, it suffices to prove that $\bar{\lambda} \equiv \min_w \lambda_w = R_{(1)}$. Now from the definitions of λ_w and R_w, we see that $\lambda_w \geq R_w$ so $\bar{\lambda} \geq \min_w R_w$. On the other hand, by Theorems 2 and 3, $\lambda_{(1)} = R_{(1)} = \min_w R_w$ so that $\bar{\lambda} \leq \lambda_{(1)} = R_{(1)} = \min_w R_w$. Thus $\bar{\lambda} = R_{(1)}$ as desired.

<u>Remark</u>. The above arguments yield, under the hypotheses of Theorem 4, the following characterization of $\mathcal{D} \equiv \left\{ D : D \text{ is extractable from } S \right\}$. Define $\mathcal{T}$ to be the closure of $\left\{ (t_i) : t_i \geq 0 \right.$ for each i and the multi-Taylor series for $\ln Q_S\big((z_i)\big)$ is absolutely convergent at $\left. (z_i) = (t_i) \right\}$. Then $\mathcal{D} = \mathcal{T}/(1 + \mathcal{T})$; i.e.,

$$\mathcal{D} = \left\{ \left(\frac{t_i}{1+t_i} \right) : (t_i) \in \mathcal{T} \right\}.$$

The multivariate generalization of the radius of convergence R is $\mathcal{R}$, the "boundary" of $\mathcal{T}$, which we identify as

$$\mathcal{R} = \left\{ (R_i) : R_i \geq 0, \ Q_S\big((-R_i)\big) = 0, \ Q_S\big((-\alpha R_i)\big) \neq 0 \text{ for all } \alpha \in (0,1) \right\}.$$

The remainder of this section is devoted to extending Theorem 4 to infinite V. We suppose that S has the Gibbs distributrion (1.2) with free boundary conditions; i.e., we suppose that there is a fixed collection of couplings K_{ij}, H_i and an increasing sequence of finite sets V_n converging to V such that the sequence $\left(S_i^{(n)} : i \in V_n \right)$ with distribution (1.2) converges to S in the sense that for each configuration s and each finite $\tilde{V}$,

$$P\left(S_i^{(n)} = s_i, i \in \tilde{V} \right) \rightarrow P\left(S_i = s_i, i \in \tilde{V} \right). \tag{2.5}$$

We further assume that

$$\ln \Lambda(z) = \lim_{n \to \infty} \ln \Lambda_n(z) \equiv \lim_{n \to \infty} \frac{1}{|V_n|} \ln \left(Q_{S^{(n)}}(z) / Q_{S^{(n)}}(0) \right) \tag{2.6}$$

exists and is finite for all $z > 0$.

$\underline{\text{Theorem } 5}$. Suppose V is infinite and S has the Gibbs distribution (1.2) with free boundary conditions. Let $R^{(n)}$ denote the radius of convergence of the Taylor series of $\ln \left(Q_{S^{(n)}}(z) \right)$. If $K_{ij} \leq 0$ and $H_i \leq 0$ for all i, j, then $R^{(n)}$ is decreasing in n and

$$\delta_c = R/(1 + R), \quad \text{where} \quad R = \lim_{n \to \infty} R^{(n)}; \tag{2.7}$$

if $R > 0$, then R equals the radius of convergence of the Taylor series for $\ln \Lambda (z)$.

$\underline{\text{Proof}}$. For $m \leq n$, $Q_{S^{(n)}}$ is converted into $Q_{S^{(m)}}$ times some other polynomial by setting $K_{ij} = 0$ for all $\{i,j\}$ with $i \in V_m$ and $j \in V_n \backslash V_m$. Thus, for an antiferromagnetic model, Theorem 3 implies that $R^{(n)}$ is decreasing in n.

Now suppose $\delta = R'/(1 + R')$ with $0 < R' < R = \lim R^{(n)}$. Then by Theorem 4, for each n, $\left(D_i \equiv \delta, i \in V_n \right)$ is extractable from $\left(S_i^{(n)}, i \in V_n \right)$ and so for $\tilde{V} \subseteq V_n$, $\left(D_i \equiv \delta, i \in \tilde{V} \right)$ is extractable from $\left(S_i^{(n)}, i \in \tilde{V} \right)$. Letting $n \to \infty$ for fixed finite $\tilde{V}$, we see that for each finite $\tilde{V} \subset V$, $\left(D_i \equiv \delta, i \in \tilde{V} \right)$ is extractable from $\left(S_i, i \in \tilde{V} \right)$ which implies that $\left(D_i \equiv \delta, i \in V \right)$ is extractable from $\left(S_i, i \in V \right)$. Thus $\delta_c \geq R/(1 + R)$.

Next suppose $\delta > R/(1 + R)$. Let us assume that $(D_i \equiv \delta, i \in V)$ is extractable from $\left(S_i, i \in V \right)$ and search for a contradiction. The assumption implies that for any finite $\tilde{V} \subset V$, $\left(D_i \equiv \delta, i \in \tilde{V} \right)$ is extractable from $\left(S_i, i \in \tilde{V} \right)$. Since $\left(S_i, i \in \tilde{V} \right) = \lim_{n \to \infty} \left(S_i^{(n)}, i \in \tilde{V} \right)$, this further implies that for any $0 < \hat{\delta} < \delta$, $\left(D_i \equiv \hat{\delta}, i \in \tilde{V} \right)$ is extractable from $\left(S_i^{(n)}, i \in \tilde{V} \right)$ for all large n. Choose

$\hat{\delta} > R/(1 + R)$ and then choose $\tilde{V} = V_m$ with m sufficiently large so that $\hat{\delta} > R^{(m)}/(1 + R^{(m)})$ $\left(\text{or equivalently } R^{(m)} < \hat{\delta}/(1 - \hat{\delta})\right)$. Define

$$Q_{m,n}(z) = \left\langle \prod_{i \in V_m} z^{(1-S_i^{(n)})/2} \right\rangle;$$

by Theorem 4 (or more accurately, by the remark following Theorem 4) $\ln Q_{m,n}(z)$ must have radius of convergence $R_{m,n} \geq \hat{\delta}/(1 - \hat{\delta})$. But $Q_{S^{(m)}}(z)$ may be obtained from $Q_{m,n}(z)$ by setting $K_{ij} = 0$ for all $i \in V_m$ and $j \in V_n \backslash V_m$; hence by Theorem 3, $R^{(m)} \geq R_{m,n} \geq \hat{\delta}/(1 - \hat{\delta})$ which contradicts a previous inequality. This contradiction implies that $\delta_c \leq R/(1 + R)$.

The remaining task is to show that a positive R equals $\hat{R}$, the radius of convergence of $\ln \Lambda(z)$. Now for each n, $Q_{S^{(n)}}(z)$ has no zeros in $\{z : |z| < R\}$ and further $|Q_{S^{(n)}}(re^{i\theta})| \leq |Q_{S^{(n)}}(r)|$; thus

$$\Lambda_n(z) \equiv \exp\left(\frac{1}{|V_n|} \ln(Q_{S^{(n)}}(z)/Q_{S^{(n)}}(0))\right)$$

is analytic and nonvanishing in $\{z : |z| < R\}$ and bounded by its values on $\{z : 0 \leq z < R\}$. By standard analytic function arguments it follows that the limit of $\ln \Lambda_n$, defined initially $\left(\text{by } (2.6)\right)$ for $z \geq 0$ extends to an analytic function on $|z| < R$. This shows that $\hat{R} \geq R$. Now the Taylor coefficients, U_k, of $-\ln \Lambda(-z)$, are the limits of the corresponding coefficients $U_{k,n}$ of $-\ln \Lambda_n(-z)$, which are all non-negative. The final inequality, $\hat{R} \leq R$, follows from the inequalities [Gr],

$$U_{k,n} \leq U_k. \tag{2.8}$$

The proof of Theorem 5 is now complete.

We conclude with a corollary which applies Theorem 5 to nearest neighbor Ising ferromagnets or antiferromagnets on $\mathbb{Z}^d$. Here $\Lambda = \mathbb{Z}^d$, $K_{ij} = \pm K/T$ when i, j are nearest neighbors and zero otherwise with $T > 0$ the temperature and K a fixed positive constant; the plus sign is the ferromagnetic case and the minus sign is the antiferromagnetic case. We suppose that there is zero external field; i.e., $H_i \equiv 0$. Taking a free boundary condition infinite volume limit with V_n equal to (say) $\{-n, \cdots, n - 1, n\}^d$, we obtain the infinite volume ferromagnet $\left(S_i^F, i \in \mathbb{Z}^d\right)$ or anitferromagnet $\left(S_i^A, i \in \mathbb{Z}^d\right)$ and their corresponding (infinite volume) "pressures", $\ln \Lambda^F(z)$ and $\ln \Lambda^A(z)$, as a function of activity z. (These differ slightly from the usual definition of pressure.) In either case, the limiting radius of convergence is positive, e.g., by the arguments of [R], and hence these functions are analytic in some T-dependent disk of (maximal) radius $R^F(T)$ or $R^A(T)$.

$\underline{\text{Corollary } \underline{6}}$. For a standard nearest neighbor Ising ferromagnet S^F or antiferromagnet S^A on $\mathbb{Z}^d$ at temperature T, with zero external field and periodic boundary conditions, the quantities $\delta_c^F(T)$ and $\delta_c^A(T)$ are related to the radius of convergence $R^A(T)$ of the Mayer series for the antiferromagnet $\left(\text{i.e., the Taylor series of } \ln \Lambda^A(z)\right)$ by:

$$\delta_c^F(T) = \delta_c^A(T) = R^A(T)/\left(1 + R^A(T)\right). \tag{2.9}$$

$\underline{\text{Proof}}$. For the antiferromagnet, this is an immediate consequence of Theorem 5 and the discussion preceding the corollary. The ferromagnetic case then follows since the ferromagnet is obtained from the antiferromagnet by a gauge transformation in which all variables are flipped on an alternating sublattice. As mentioned in the introduction, gauge transformations do not affect δ_c.

3. *Numerics*

In this section we restrict attention to the quantities $\delta_c(T)$ $\left(= \delta_c^F(T) = \delta_c^A(T)\right)$ for the standard d-dimensional Ising ferromagnet or antiferromagnet with zero field and the radius of convergence $R(T) = R^A(T)$ for the Mayer series of the antiferromagnet, as defined at the end of the last section. (Note that for the ferromagnet, $R^F(T) \equiv 1$ by the Lee-Yang circle theorem [LY]).

These are related by $\delta_c = R/(1 + R)$ according to the results of the last section.

For $d = 1$, δ_c was derived in [G] to be

$$\delta_c(T) = \tfrac{1}{2}\left(1 - \left(1 - e^{-4K/T}\right)^{1/2}\right). \tag{3.1}$$

For $d \geq 2$, it was shown in [N] that

$$\delta_c(T) \geq \tfrac{1}{2}\left(1 - \tanh\left[d \, \cosh^{-1}\left(e^{2K/T}\right)\right]\right). \tag{3.2}$$

Let us define a variable $u = \exp(4K/T)$, so that using the identity $R = \delta/(1 - \delta)$, we may rewrite (3.2) as

$$R(T) \geq \left[u^{1/2} + (u - 1)^{1/2}\right]^{-2d}. \tag{3.3}$$

These lower bounds behave asymptotically like:

$$R \gtrsim 1 - 2d(u - 1)^{1/2}, \qquad \delta_c \gtrsim \tfrac{1}{2} - d(K/T)^{1/2} \quad \text{as } u \to 1, \quad T \to \infty; \tag{3.4}$$

$$R \gtrsim 2^{-2d}u^{-d}, \qquad \delta_c \gtrsim 2^{-2d}e^{-4dK/T} \quad \text{as } u \to \infty, \quad T \to 0. \tag{3.5}$$

In the remainder of this section, we present and discuss numerical results on R as a function of u for $d = 2$. These results utilize a series analysis based on the fact that

$$\ln \Lambda^A(z) = \sum_{s=1}^{\infty} L_s(u)z^s$$

where each $L_S(u)$ is a finite polynomial in u. These polynomials have been explicitly tabulated for $d = 2$ through order $s = 15$ [SGMEE]. For example,

$$L_1 = u^2, \quad L_2 = 2u^3 - (5/2)u^4, \quad L_3 = 6u^4 - 16u^5 - (31/3)u^6.$$

To obtain a numerical value of R at a given u based on these 15 coefficients, we proceed as follows.

Both the pressure, $\ln \Lambda^A$, and the (reduced) magnetization per spin,

$$M^A(z) = 1 - 2z\frac{d}{dz}\ln \Lambda\,(z),$$

are analytic in $|z| < R$ and both diverge as z approaches $-R$ along the negative real axis. We postulate that to leading order, the divergence of M^A has the form,

$$M^A(z) \sim (z + R)^{-\delta},$$

for some $\delta > 0$. Taking a logarithmic derivative yields, to leading order, a simple pole,

$$f(z) \equiv d \ln M^A/dz \sim - \frac{\delta}{z + R}. \tag{3.6}$$

We remark that each $M_n^A(z) \equiv 1 - 2z\frac{d}{dz}\ln \Lambda_n(z)$, and hence $M^A(z)$, does not vanish for $|z| < R$ so that $f(z)$ has no singularities closer to the origin than the one at $-R$. To see this, express the finite volume magnetization M_n^A as:

$$M_n^A(z) = \frac{1}{|V_n|} \sum_{j=1}^{|V_n|} \frac{1+z/z_j}{1-z/z_j}$$

where the z_j's are the zeros of $Q_{S(n)}(z)$. Then notice that for $|z| < \min(|z_j|) = R$, $\mathrm{Re}(M_n^A(z)) > 0$, so M_n^A does not vanish for $|z| < R$.

The coefficients through order z^{14} in the Taylor series of f may be obtained from the coefficients $L_1, \cdots, L_{15}$ of $\ln \Lambda^A$. To estimate numerically the R (and δ) of (3.6), we employ the method of Padé approximants [B;BG-M]. Briefly, a Padé approximant to a power series, $f(z) = \sum_{j=0}^{\infty} f_j z^j$, is defined by

$$Q_M(z)f(z) - \mathcal{P}_L(z) = O\left(z^{L+M+1}\right) \quad \text{as} \quad z \to 0, \tag{3.7}$$

where Q_M and $\mathcal{P}_L$ are polynomials of degree M and L respectively with $Q_M(0) = 1$. The coefficients of Q_M and $\mathcal{P}_L$ are determined from the f_j's (with $j \leq L + M$) by the system of linear equations obtained from (3.7). The Padé approximant $\left(\text{of order } (L,M)\right)$ is denoted $[L/M]$ and defined as the rational function

$$[L/M] \equiv \mathcal{P}_L(z)/Q_M(z). \tag{3.8}$$

The "best" approximants for a given number of series coefficients f_j are usually the "diagonal" $[N/N]$ or "near diagonal" $[N\pm 1/N]$ ones. We have computed these for a variety of values of u (and $N \leq 7$) and then determined the location (and residue) of the resulting pole closest to the origin. This pole was always on the negative real axis, consistent with the fact that f has its closest singularity to the origin at $z = -R$. The results are tabulated in Table 1. The values of R and δ given there are those obtained from the $[7/7]$ approximant while the quoted errors in the last decimal place are apparent errors estimated from the variation in R or δ values obtained from analysis of the $[7/7]$, $[7/6]$, $[6/7]$ and $[6/6]$ approximants. Several decimal places of accuracy are typically obtained by this method of analysis. Here, some care must be exercised to retain sufficient numerical accuracy as in some of the cases reported, more than 12 decimal places were lost in the course of the computations.

For $u = 1$ $(T = \infty)$, $M^A(z) = (1 - z)/(1 + z)$ and $f(z)$ has a pole at $z = -R = -1$ with residue $-\delta = -1$ as well as a pole at $z = +1$ with residue $+1$ resulting from the vanishing of M^A at $z = 1$. Table 1 covers values of u ranging from 1.001 to 20.0. The slow variation of δ with u over this range is consistent with a u-independent value, $\delta = 0.25 \pm 0.05$ for all $u > 1$.

Figure 1 plots the estimated R vs. $(u - 1)^{1/2}$. It is plausible that this variable correctly describes the variation of δ_c and R with temperature near $u = 1$ $(T = \infty)$, as was the case for the exact one-dimensional solution (3.1) of [G] and the lower bound (3.2)-(3.4) of [R;N] for general d. We estimate

$$R \sim 1 - C(u - 1)^{1/2}, \quad \delta_c \sim \tfrac{1}{2} - \tfrac{C}{2}(K/T)^{1/2}, \quad C = 3.4 \pm 0.1, \quad \text{as } u \to 1, \ T \to \infty,$$

compared to the previous lower bound (3.4) with C replaced by 4 (for $d = 2$).

Figure 2 plots the estimated $u^2 R$ vs. $1/u$. The estimated intercept at $1/u = 0$ has a value 0.119 ± 0.001, so

$$R \sim C'/u^2, \quad \delta_c \sim C' e^{-8K/T}, \quad C' = 0.119 \pm 0.001, \quad \text{as } u \to \infty, \ T \to 0.$$

This is consistent with the $\text{const.}/u^d = \text{const.} e^{-4dK/T}$ behavior of the exact one-dimensional solution (3.1) and the d-dimensional lower bound (3.5) (which has C' replaced by $1/16$ for $d = 2$).

Acknowledgments. This work was supported in part by NSF grant DMS85-14834 to C. M. N. and AFOSR Universities Research Initiative Contract No. F49620-86-C0130 to the Arizona Center for Mathematical Sciences, and also in part by the U. S. DOE (Contract W-7405-ENG-36) .

[B]   Baker, Jr., G. A. (1975). *Essentials of Padé Approximants*, Academic, N. Y.

[BD]   Bassalygo, L. A. and Dobrushin, R. L. (1985). Rate-distortion function of the Gibbs field. *Problems Inform. Trans.* 23, 3-15.

[BF]   Brydges, D. and Federbush, P. (1978). A new form of the Mayer expansion in classical statistical mechanics. *J. Math. Phys.* 19, 2064-2067.

[BG-M]   Baker, Jr., G. A. and Graves-Morris, P. R. (1981). *Padé Approximants I. Basic Theory* and *II. Extensions and Applications*. Addison-Wesley, Reading, MA.

[G]   Gray, R. M. (1970). Information rates of autoregressive processes. *IEEE Trans. Inform. Theory* IT-16, 412-421.

[Gr]   Groeneveld, J. (1962). Two theorems on classical many-particle systems. *Phys. Lett.* 3, 50-51.

[HB]   Hajek, B. and Berger, T. (1987). A decomposition theorem for binary Markov random fields. *Ann. Probab.* 15, 1112-1125.

[LS]   Lieb, E. H. And Sokal, A. D. (1981). A general Lee-Yang theorem for one-component and multi-component ferromagnets. *Comm. Math. Phys.* 80, 153-179.

[LY]   Lee, T. D. and Yang, C. N. (1952). Statistical theory of equations of state and phase transition.II. Lattice gas and Ising model. *Phys. Rev.* 87, 410-419.

[N]   Newman, C. M. (1987). Decomposition of binary random fields and zeros of partition functions. *Ann. Probab.* 15. 1126-1130.

[R]   Ruelle, D. (1973). Some remarks on the location of zeros of the partition function for lattice systems. *Comm. Math. Phys.* 31, 265-277.

[SGMEE]   Sykes, M. F., Gaunt, D. S., Mattingly, S. R., Essam, J. W. and Elliott, C. J. (1973). Derivation of low-temperature expansions for Ising model.III. Two-dimensional lattices - field grouping. *J. Math. Phys.* 14, 1066-1070.

Table 1. Numerical estimates for the closest singularity to the origin of the antiferromagnetic spin-1/2 plane square lattice Ising model magnetization as a function of the activity variable z: $M^A(u,z) \sim (z + R)^{-\delta}$, $u = e^{4K/T}$. The estimates are obtained from the closest pole at $-R$, with residue $-\delta$, of the [7/7] Padé approximant to $d \log M^A/dz$. The error brackets listed are only apparent errors.

u	R	δ	u	R	δ
1	1	1	1.5	0.1045 ± 3	0.24
1.001	0.900 ± 1	0.30 ± 3	1.6	0.0860 ± 1	0.24
1.005	0.7906 ± 5	0.30 ± 1	1.7	0.0722 ± 1	0.23
1.010	0.7168 ± 8	0.29 ± 2	1.8	0.06164 ± 5	0.23
1.015	0.6652 ± 5	0.29 ± 2	1.9	$0.0533 + 1$	0.23
1.02	0.625 ± 1	0.29 ± 2	2.0	0.0465 ± 1	0.23 ± 1
1.04	0.5136 ± 5	0.27 ± 2	3.0	0.01732 ± 1	0.22
1.06	0.4428 ± 3	0.26 ± 2	4.0	$(9.03 \pm 1) \times 10^{-3}$	0.22
1.08	0.3910 ± 2	0.26 ± 1	5.0	$(5.543 \pm 1) \times 10^{-3}$	0.222 ± 2
1.1	0.351 ± 1	0.26 ± 1	6.6	$(3.0610 \pm 5) \times 10^{-3}$	0.221 ± 1
1.2	0.230 ± 1	0.25	10.0	$(1.2819 \pm 5) \times 10^{-3}$	0.218 ± 4
1.3	0.168 ± 1	0.24	20.0	$(3.089 \pm 5) \times 10^{-4}$	0.21 ± 1
1.4	0.130 ± 1	0.24			

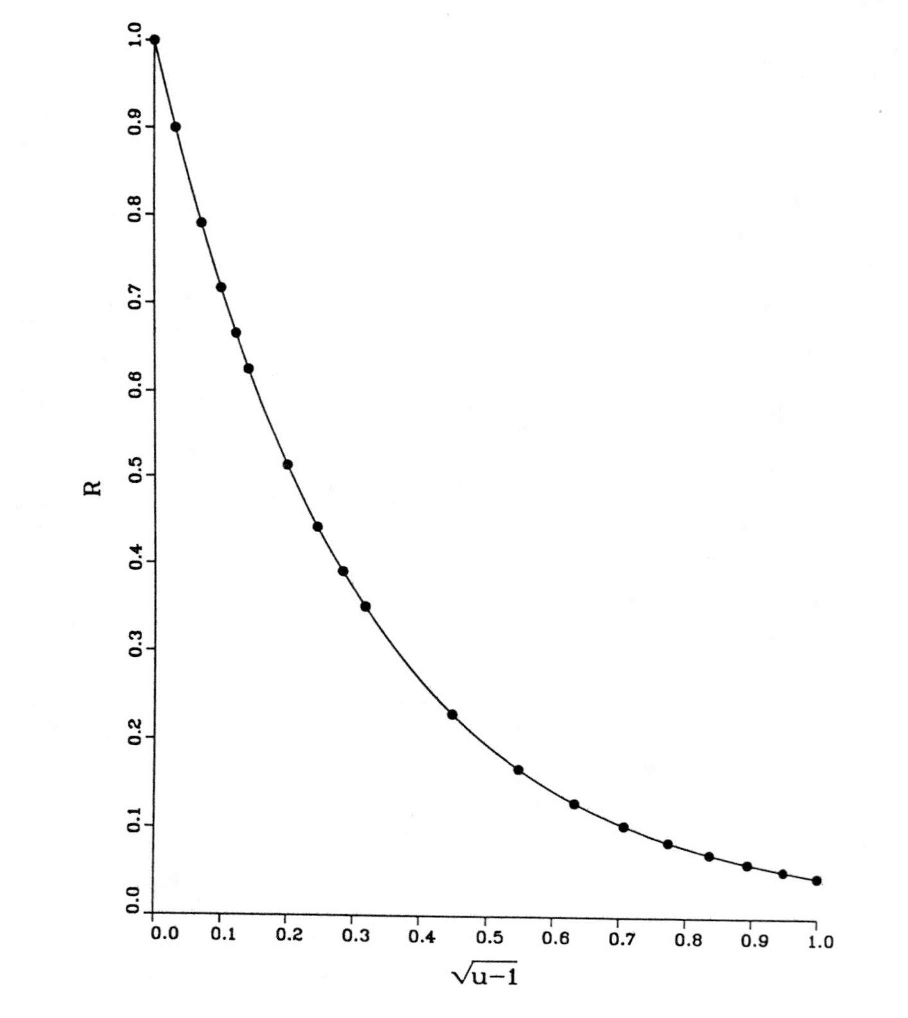

Figure 1. A plot of R vs. $(u - 1)^{1/2}$. R is the radius of convergence of the Mayer series in the activity variable z for the plane-square spin - 1/2 Ising antiferromagnet with nearest neighbor coupling $J/kT = -K/T$; $u = e^{4K/T}$. The data points are numerical estimates obtained from Table 1.

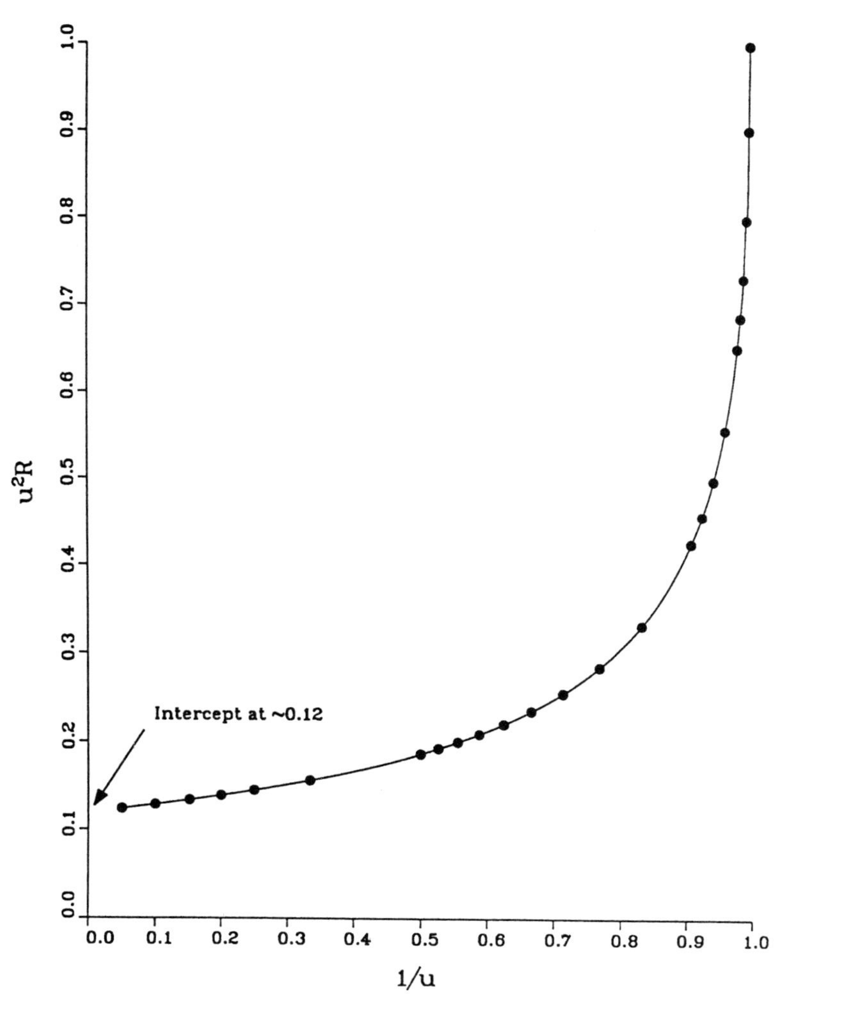

Figure 2. A plot of u^2R vs. $1/u$ for the plane-square spin - 1/2 Ising antiferromagnet. R and u are as in Fig. 1. The critical temperature corresponds to $1/u = (1 + \sqrt{2})^{-2} \approx 0.17$.

Towards Time - Dynamics for Bosonic Systems
in Quantum Statistical Mechanics

by

A.G. Shuhov[1], Yu.M. Suhov[1,2], and A.V. Teslenko[1]

Abstract:

Consider a one-dimensional lattice boson system with the Hamiltonian in a finite box Λ, $H_\Lambda = K_\Lambda + U_\Lambda$. Here K_Λ is the kinetic energy (the discrete Laplacian) and U_Λ is the potential energy corresponding to a finite-range pair interaction. For a class of states $\mathcal{S}$ of the infinite system, we prove the existence of the limit $\mathcal{S}_t(A) = \lim_{\Lambda \nearrow \mathbf{Z}^1} \mathcal{S}(e^{itH_\Lambda} A e^{-itH_\Lambda})$ for any $t \in \mathbf{R}^1$ and any local observable A. Thereby a family $\{\mathcal{S}_t, t \in \mathbf{R}^1\}$ of locally normal states is determined which describes the time-evolution of the initial state $\mathcal{S}$.

Key words and phrases: one-dimensional lattice boson system, diagonal state, time evolution, path integrals.

1. Introduction

The problem of constructing time-dynamics for an infinitely extended system is a major one in statistical mechanics, both classical and quantum. So far, in quantum statistical mechanics this problem has been solved in a satisfactory way for a particular class of systems only, namely, for quantum spin systems. Here Robinson's theorem [1] (see also [2]) asserts that, given a "reasonable" interaction potential of a general form, there exists a corresponding strongly continuous *-automorphism group of the quasilocal C^*-algebra. This provides, in particular, an elegant definition of an equilibrium state via the KMS boundary condition.

However, for other types of quantum systems of interest, e.g., for interacting particle systems in a Euclidean space, the problem remains open (some results, both of a positive and a negative character, are available for free systems; see, e.g., [3], [4]). A wide-spread opinion is that the "traditional" C^*-algebras (the CAR, the CCR and the quasilocal algebras) are not appropriate for this purpose; by analogy with the classical case, it is believed that trouble may arise from singularities which occur in a system with infinitely many degrees of freedom, such as uncontrolled "collapses" or "accelerations" of particles.

From this point of view, it seems natural to consider the time-evolution of a state rather

[1] Institute for Problems of Information Transmission, USSR Academy of Sciences, 19 Yermolova St., GSP - 4 Moscow, 101447 USSR

[2] School of Theoretical Physics, Dublin Institute for Advanced Studies, 10, Burlington Road, Dublin 4, Ireland

512

than the time-dynamics on an algebra of observables. One such version based on time-dependent Green's functions was elaborated in [5], [6] for equilibrium states and in [7] in a general set-up. However, to check the assumptions on an initial state $\mathcal{S}$ which were formulated in [7] is not an easy matter.

The present paper deals with an alternative approach to the problem. For an appropriate initial state $\mathcal{S}$ we construct directly the time-evolved state $\mathcal{S}_t$ by setting

$$\mathcal{S}_t(A) = \lim \mathcal{S}(e^{itH_\Lambda} A e^{-itH_\Lambda}) \tag{1.1}$$

for any local observable A. We consider a system of interacting bosons on the one-dimensional lattize $\mathbf{Z}^1$.

The (formal) Hamiltonian H of the infinite system is

$$H = -\frac{1}{2} \sum_{j \in \mathbf{Z}^1} a_j^+ (\Delta a)_j + \sum_{j,j' \in \mathbf{Z}^1} \Phi(|j - j'|) n_j n_{j'}, \tag{1.2}$$

where Δ stands for the second difference operator,

$$(\Delta a)_j = \frac{1}{2}(a_{j+1} + a_{j-1} - 2a_j), j \in \mathbf{Z}^1,$$

a_k^+ and a_k are respectively, the creation and annihilation operators at a point $k \in \mathbf{Z}^1, \Phi$ is a real-valued function on $\mathbf{Z}_+^1$, the set of non-negative integers, with bounded support and $n_k = a_k^+ a_k$ is the particle number operator at a point $k \in \mathbf{Z}^1$. In principle, one can admit infinite values for the potential Φ (hard-core type interaction), but to emphasize the "bosonic" character of the system under consideration, we shall assume that $|\Phi(r)| < \infty$ for any $r \in \mathbf{Z}_+^1$.

Given a finite "box" Λ (an interval, or a segment of the lattice $\mathbf{Z}^1$), one can consider the finite-volume version H_Λ of the Hamiltonian (1.2); more precisely, one takes the self-adjoint extension corresponding to a standard boundary condition on $\partial\Lambda$, and for definiteness we shall deal with the Dirichlet condition. Then the $*$-automorphism group

$$A \longmapsto e^{itH_\Lambda} A e^{-itH_\Lambda}, t \in \mathbf{R}^1, \tag{1.3}$$

determines the time-dynamics of an abservable A. For a class of states $\mathcal{S}$ which we call *diagonal*, we establish the existence of the limit (1.1), as $\Lambda \nearrow \mathbf{Z}^1$, for any local observable A. This yields the family $\{\mathcal{S}_t, t \in \mathbf{R}^1\}$ of locally normal states of the C^*-algebras of the infinite system which describes the time-evolution of the initial state $\mathcal{S}$.

In a separate paper (also in this volume) we shall prove that the family $\{\mathcal{S}_t\}$ provides a unique solution of the infinite-volume Liouville equation as well as a solution to the corresponding BBGKY hierarchy.

The condition of diagonality on the initial state is used to simplify technicalities; we hope to drop it in a later publication.

The method of proof relies heavily on the concrete form (1.2) of the Hamiltonian as well as on the one-dimensional character of the system. The problem of the recovering dynamics of continuous systems of interacting quantum particles by passing to the limit when the lattice spacing tends to zero is reserved for future study.

In conclusion we remark that the main idea of this paper is inspired by Sinai's approach to the construction of a cluster-dynamics for classical systems (see [8], [9]).

2. Preliminaries and results[1]

For any $k \in \mathbf{Z}^1$ consider a copy $\mathcal{H}_k$ of the separable Hilbert space $\ell_2(\mathbf{Z}_+^1)$ with standard orthonormal basis $\{e_s^{(k)}, s \in \mathbf{Z}_+^1\}$. Letting $\mathcal{B}_k$ be the C^*-algebra of bounded operators in $\mathcal{H}_k$, we set $\mathcal{B} = \otimes \mathcal{B}_j$. The creation and annihilation operators a_k^+ and a_k act in $\mathcal{H}_k$ as $a_k^+ e_s^{(k)} = (s+1)^{\frac{1}{2}} e_{s+1}^{(k)}$ and $a_k e_s^{(k)} = s^{1/2} e_{s-1}^{(k)}, s \in \mathbf{Z}_+^1 (a_k e_0^{(k)}$ is set to be zero), $k \in \mathbf{Z}_+^1$. Given a finite lattice interval $\wedge \subset \mathbf{Z}^1$, the Hamiltonian $H_\wedge$ acts in the Hilbert space $\mathcal{H}_\wedge = \bigotimes_{k \in \wedge} \mathcal{H}_k$ and is defined via (1.2) with the Dirichlet boundary condition. Hence, $e^{\pm it H_\wedge}$ is unitary and (1.3) defines the $*$-automorphism group on $\mathcal{B}$, the time-dynamics in the volume $\wedge$.

A locally normal state $\mathcal{S}$ of the C^*-algebra $\mathcal{B}$ is determined by its values $\mathcal{S}(E_{x,x'}^{(\wedge^0)})$; here $\wedge^0$ is an arbitrary bounded interval of the lattice, x and x' stand for occupation number configurations in $\wedge^0$, i.e., for functions $\wedge^0 \to \mathbf{Z}_+^1$ (we shall use notation $x, x' \dot\subset \wedge^0$) and $E_{x,x'}^{(\wedge^0)}$ is the corresponding matrix unit in the standard orthonormal basis $\{e_y^{(\wedge^0)}, y \dot\subset \wedge^0\}$ in $\mathcal{H}_{\wedge^0}$:

$$E_{x,x'}^{(\wedge^0)} e_y^{(\wedge^0)} \begin{aligned} &= e_x^{(\wedge^0)}, &&\text{provided that} \quad \text{y=x'}, \\ &= 0, &&\text{otherwise.} \end{aligned} \qquad (2.1)$$

A locally normal state $\mathcal{S}$ is called gauge-invariant if and only if $\mathcal{S}(E_{x,x'}^{(\wedge^0)}) = 0$ provided $|x| \neq |x'|$ where $|y|$ is the total particle number in $y : |y| = \sum_j y(j)$. Likewise, $\mathcal{S}$ is called diagonal if and only if $\mathcal{S}(E_{x,x'}^{(\wedge^0)}) = 0$ provided $x \neq x'$.

Diagonal states are in one-to-one correspondence with probability measures on the measurable space $(M, \mathcal{M})$ where M is the space of occupation number configurations in $\mathbf{Z}^1$, i.e. of functions $X : \mathbf{Z}^1 \to \mathbf{Z}_+^1$, and $\mathcal{M}$ is the σ-algebra of subsets of M generated by "cylinders" $B_x^{(\wedge^0)}$:

$$B_x^{(\wedge^0)} = \{X \in M : X|\wedge^0 = x\}.$$

The correspondence between a diagonal state $\mathcal{S}$ and the measure on $(M, \mathcal{M})$, which we denote by the same symbol $\mathcal{S}$, is established by the formula

$$\mathcal{S}(E_{x,x}^{(\wedge^0)}) = \mathcal{S}(B_x^{(\wedge^0)}).$$

---

[1] A detailed C^*-algebra background for this section may be found in [10]. For details of a probabilistic character see, e.g., [11].

A simple, but useful, example of a probability measure on $(M, \mathcal{M})$ (and hence, of a diagonal state) is the product (Bernoulli) measure $\prod_j p^{(j)}$ where $p^{(j)}(= p)$ is a fixed probability distribution on $\mathbf{Z}_+^1$ with $p(0) > 0$ (e.g., a geometric distribution:

$$p(s) = (1 - q)q^s, s \in \mathbf{Z}_+^1, \text{ where } q \in (0, 1)).$$

A Bernoulli measure has the following property which will play an important role in the sequel: let $\{S_m, m \in \mathbf{Z}_+^1\}$ be a monotone increasing sequence of positive integers and set

$$\begin{aligned}
I_m &= [-S_{m+1}, S_{m+1}], \\
I_m^- &= [-S_{m+1} - 1, -S_m], \\
I_m^+ &= [S_m, S_{m+1} - 1]
\end{aligned} \tag{2.2}$$

(all the space intervals considered here and below are on the lattice $\mathbf{Z}^1$). Denote $\gamma_{m+1} = S_{m+1} - S_m$ and consider the event (from $\mathcal{M}$) that neither on I_m^- nor on I_m^+ can one find a subinterval of length greater than or equal α_m, $\alpha_m < \gamma_m$, which is free of particles. The probability of that event does not exceed

$$2(1 - p(0)^{\alpha_m})^{\gamma_m/\alpha_m} \leq 2\exp(-p(0)^{\alpha_m}\gamma_m/\alpha_m).$$

If we assume that $p(0)^{\alpha_m}\gamma_m/\alpha_m \nearrow \infty$ as $m \nearrow \infty$ so that

$$\sum_{m \geq 1} \exp(-p(0)^{\alpha_m}\gamma_m/\alpha_m) < \infty,$$

then, by the Borel-Cantelli Lemma, $\mathcal{S}$-a.e. $X \in M$ one can find $m_0 = m_0(X)$ with the property that for any $m \geq m_0$ there exist subintervals $J^\pm \subset I_m^\pm$ of length greater than or equal to α_m such that $X(j) = 0$ for $j \in J^- \cup J^+$. Moreover, the following bound holds:

$$\mathcal{S}(m_0(X) \geq \overline{m}) \leq 2 \sum_{m \geq \overline{m}} \exp(-p(0)^{\alpha_m}\gamma_m/\alpha_m), \overline{m} \in \mathbf{Z}_+^1. \tag{2.3}$$

Now consider an arbitrary measure $\mathcal{S}$ on $(M, \mathcal{M})$. We shall suppose that the foregoing conditions hold for $\mathcal{S}$ with $S_m = 2^{m+1}, \gamma_m = 2^m$ and $\alpha_m = [m^\delta]$, where $\delta \in (0,1)$ is fixed. In that case we denote the property under consideration by (d^*). In probabilistic terms property (d^*) means that, for a measure $\mathcal{S}$,

$$\mathcal{S}\left(\bigcup_{l=1}^{\infty} \bigcap_{m \geq l} M(m)\right) = \lim_{l \to \infty} \mathcal{S}\left(\bigcap_{m \geq l} M(m)\right) = 1 \tag{2.4}$$

Here $M(m)$ stands for the event

$$\begin{aligned}
&\{X \in M : \exists \text{ intervals } J^\pm = J^\pm(X, m) \subset I_m^\pm, \text{ of} \\
&\text{length } \geq [m^\delta], \text{ such that } X(j) = 0 \quad \forall j \in J^- \cup J^+\}.
\end{aligned} \tag{2.5}$$

In what follows we shall think of $J^{\pm}$ as the longest intervals possessing the properties listed (if such intervals are not unique, we take the left-most (resp., the right-most) of them). So far we have checked that the property (d^*) holds for Bernoulli measures but the class of probability measures on $(M, \mathcal{M})$ for which (d^*) holds is in fact much larger; it includes positively recurrent Markov chains and DLR measures corresponding to classical superstable interactions (not necessarily of a finite range).

Another property of an initial state which will be used heavily in the sequel is again inspired by the example of Bernoulli measures. This property is denoted below by (d^{**}) and is a combination of the two conditions, $(d_1 **)$ and $(d_2 **)$.

$(d_1 **)$ There exists constants $c, \sigma, b > 0$ and a value $s_0 \in \mathbf{Z}_+^1$ such that for any bounded interval $\wedge \subset \mathbf{Z}^1$ with $|\wedge| \geq s_0$ we have

$$\mathcal{S}(\prod_{\geq s}^{(\wedge)}) \leq c \int_{\overline{s}}^{\infty} du \exp(-u^2/2), \tag{2.6}$$

where

$$\overline{s} = (\sigma|\wedge|^{1/2})^{-1}(s - b|\wedge|)_+$$
$$r_+ = \max\{0, r\}, r \in \mathbf{R}^1,$$

and $\prod_{\geq s}^{(\wedge)}$ is the orthogonal projection in $\mathcal{H}_\wedge$ onto the subspace generated by occupation number configurations $x \dot{\subset} \wedge$ with $|x| \geq s$.

$(d_2 **)$ There exist constants $c_1 > 0$ and $\rho \in (0, 1)$ such that for any pair of finite occupation number configurations y, y' we have:

$$\mathcal{S}(\overline{B}_{y+y'}) \leq c_1^{R(y)} \rho^{|y|} \mathcal{S}(\overline{B}_{y'}), \tag{2.7}$$

where $R(y) = \sum_j \chi(y(j) \geq 1)$ and we have used the notation

$$\overline{B}_x = \{X \in M : X \geq x\};$$

the sum $y + y'$ means here and below summation of functions.

As before, the conditions $(d_1 **)$ and $(d_2 **)$ hold for a large class of probability measures on $(M, \mathcal{M})$. Property $(d_1 **)$ establishes a kind of limit theorem estimation while $(d_2 **)$ gives a kind of stability bound, both are natural from the point of view of statistical mechanics.

Remark: We have stated properties $(d*)$ and $(d**)$ in a form chosen to emphasize their probabilistic character; as will be seen from what follows, this is not necessary. For instance, the inequality (2.6) may be replaced with the following relation: for any bounded interval $\wedge^0 = [v^0, v^1) \subset \mathbf{Z}^1$

$$\lim_{s \to \infty} \mathcal{S}(\prod_{\geq s}^{(\wedge^0([s^{1/2}]))}) = 0,$$

where $\wedge^0(r) = [v^0 - r, v^1 + r], r \in \mathbf{Z}_+^1$.

We are now able to formulate the results of the paper:

Theorem 1.
Suppose that a diagonal state $\mathcal{S}$ has property (d^*). Then, for any bounded interval $\wedge^0 \subset \mathbf{Z}^1$, any pair of occupation number configurations $x, x' \dot{\subset} \wedge^0$ and any $t \in \mathbf{R}^1$ the following limit exists:

$$\mathcal{S}_t(E_{x,x'}^{(\wedge^0)}) = \lim_{\wedge \nearrow \mathbf{Z}^1} \mathcal{S}_{\wedge,t}(E_{x,x'}^{(\wedge^0)}), \tag{2.8}$$

where

$$\mathcal{S}_{\wedge,t}(A) = \mathcal{S}(e^{itH_\wedge} A e^{-itH_\wedge}), A \in \mathcal{B}.\triangleleft^{1)} \tag{2.9}$$

Theorem 2
Suppose that a diagonal state $\mathcal{S}$ has property $(d**)$. Then, for any bounded interval $\wedge^0 \subset \mathbf{Z}^1$ and any $t \in \mathbf{R}^1$ the following relation holds:

$$\lim_{s \to \infty} \sup_{\wedge \supset \wedge^0} \mathcal{S}_{\wedge,t}(\prod_{\geq s}^{(\wedge^0)}) = 0.\triangleleft \tag{2.10}$$

From these two statements we derive immediately the following result:

Theorem 3
Suppose that a diagonal state $\mathcal{S}$ has both properties (d^*) and (d^{**}). Then, given $t \in \mathbf{R}^1$, for any local $A \in \mathcal{B}$ the following limit exists:

$$\mathcal{S}_t(A) = \lim_{\wedge \nearrow \mathbf{Z}^1} \mathcal{S}_{\wedge,t}(A), \tag{2.11}$$

where $\mathcal{S}_{\wedge,t}(A)$ is defined by (2.9). The limit (2.11) determines a locally normal state $\mathcal{S}_t$ of the C^*-algebra $\mathcal{B}.\triangleleft$

Proof of Theorem 3 (given Theorems 1 and 2): Let A be localized in an interval $\wedge^0 \subset \mathbf{Z}^1$. Writing

$$\mathcal{S}_{\wedge,t}(A) = \mathcal{S}_{\wedge,t}(\prod_{\geq s}^{(\wedge^0)} A \prod_{\geq s}^{(\wedge^0)}) + \mathcal{S}_{\wedge,t}((1 - \prod_{\geq s}^{(\wedge^0)})A \prod_{\geq s}^{(\wedge^0)})$$

$$+ \mathcal{S}_{\wedge,t}(\prod_{\geq s}^{(\wedge^0)} A(1 - \prod_{\geq s}^{(\wedge^0)})) + \mathcal{S}_{\wedge,t}((1 - \prod_{\geq s}^{(\wedge^0)})A(1 - \prod_{\geq s}^{(\wedge^0)})),$$

$^{1)}$ The symbol $\triangleleft$ indicates the end of a statement. The end of a proof is indicated by the symbol $\square$.

we estimate from above the first term in the RHS by $\mathcal{S}_{\Lambda,t}(\prod_{\geq s}^{(\Lambda^0)})$ and the second and third ones by $(\mathcal{S}_{\Lambda,t}(1-\prod_{\geq s}^{(\Lambda^0)})\mathcal{S}_{\Lambda,t}(\prod_{\geq s}^{(\Lambda^0)}))^{1/2}$. Theorem 2 asserts that all these estimates may be made small uniformly in Λ whereas Theorem 1 says that the fourth term has a limit when $\Lambda \nearrow \mathbf{Z}^1$. Hence, the whole RHS of (2.11) tends to a limit.

The problem of verifying that $\mathcal{S}_t$ is a locally normal state is reduced to checking the equality

$$\sum_{x \dot{\subset} \Lambda^0} \mathcal{S}_t(E_{x,x}^{(\Lambda^0)}) = 1 \tag{2.12}$$

for any bounded interval $\Lambda^0 \subset \mathbf{Z}^1$. To do this, we again use Theorem 2 for estimating the difference

$$1 - \mathcal{S}_{\Lambda,t}(1 - \prod_{\geq s}^{(\Lambda^0)})$$

and then Theorem 1 for doing the same for

$$1 - \sum_{x \dot{\subset} \Lambda^0 : |x| \leq s} \mathcal{S}_t(E_{x,x}^{(\Lambda^0)}).$$

These estimates vanish when $s \to \infty$. This yields the assertion of Theorem 3. $\square$

The proofs of Theorems 1 and 2 are based on the representation of the matrix elements $(e^{\pm it H_\Lambda})_{y,y'}$, in terms of integrals over the paths of a Markov jump process on $\mathbf{Z}^1$. Precisely, let $P_j^\tau, j \in \mathbf{Z}^1, \tau \geq 0$, denote the path distribution up to time τ of the process starting at the point j which, after spending the mean one exponential time at a given site, jumps to one of the nearest neighbour sites, each with probability $\frac{1}{2}$.

Suppose a triple of occupation number configurations $z, y, y' \dot{\subset} \Lambda$ with $|y| = |y'|$ is given. By $O_{y,y'}$, we denote the set of all matchings between y and y'. Every matching is identified with a function $\Gamma : \Lambda^0 \times \Lambda^0 \to \mathbf{Z}_+^1$ with the following properties: $(i)\Gamma(j,j') = 0$ provided $y(j)y'(j') = 0, (ii) \sum_{k' \in \Lambda^0} \Gamma(j,k') = y(j)$ and $\sum_{k \in \Lambda^0} \Gamma(k,j') = y(j')$ for any $j,j' \in \Lambda^0$. Given $\tau \geq 0$, let P_y^τ denote the product measure $P_y^\tau = \prod_{j:y(j)\geq 1} (P_j^\tau)^{xy(j)}$. This describes the development up to time τ of $|y|$ copies of the process starting from the points occupied by y. Furthermore, given a matching $\Gamma \in O_{y,y'}$, denote by W_Γ^τ the set of families $\Omega = \{\omega\}$ of paths which start at the points occupied by y and arrive, at the epoch τ, at the corresponding points occupied by y'. Next, given a family $\Omega = \{\omega\}$, we set

$$N(\Omega) = \sum_{\omega \in \Omega} N(\omega), \quad \chi_\Lambda(\Omega) = \prod_{\omega \in \Omega} \chi_\Lambda(\omega), \tag{2.13}$$

$$U(\Omega|z) = \sum_{\substack{\omega,\omega' \in \Omega: \\ \omega \neq \omega'}} U(\omega,\omega') + \sum_{\substack{\omega \in \Omega, \\ k \in \Lambda}} U(\omega,k)z(k), \tag{2.14}$$

where

$$N(\omega) \text{ is the number of jumps of a path } \omega, \tag{2.15}$$

$$\chi_\Lambda(\omega) = 1, \text{ if } \omega(u) \in \Lambda \text{ for any } u \in [0,\tau],$$
$$= 0, \text{ otherwise}, \tag{2.16}$$

$$U(\omega,\omega') = \int_0^\tau du\,\Phi(|\omega(u) - \omega'(u)|), \tag{2.17a}$$

$$U(\omega,k) = \int_0^\tau du\,\Phi(|\omega(u) - k|). \tag{2.17b}$$

Finally, given a pair, $\Omega = \{\omega\}, \Omega' = \{\omega'\}$, of families of paths we set

$$U(\Omega \cup \Omega'|z) = \sum_{\substack{\omega,\omega'\in\Omega:\\ \omega\neq\omega'}} U(\omega,\omega') + \sum_{\substack{\omega,\omega'\in\Omega':\\ \omega\neq\omega'}} U(\omega,\omega')$$
$$+ \sum_{\omega\in\Omega,\omega'\in\Omega'} U(\omega,\omega') + \sum_{\omega\in\Omega\cup\Omega',k\in\Lambda} U(\omega,k)z(k). \tag{2.18}$$

For later use, it is convenient to consider Hamiltonians of a slightly more general form - those which include a term representing an external field generated by a fixed occupation number configuration. Precisely, let $H_{\Lambda,z}$ denote the operator in $\mathcal{H}$ given by

$$H_{\Lambda,z} = -1/2 \sum_{j\in\Lambda} a_j^+(\Delta a)_j + \sum_{j,j'\in\Lambda} \Phi(|j - j'|)n_j n_{j'}$$
$$+ \sum_{j,k\in\Lambda} \Phi(|j - k|)n_j z(k), \tag{2.19}$$

and $\tilde{H}_{\Lambda,z}$ be the operator in $\mathcal{H}_\Lambda \otimes \mathcal{H}_\Lambda$ of the form

$$\tilde{H}_{\Lambda,z} = H_{\Lambda,z} \otimes 1 + 1 \otimes H_{\Lambda,z} + V, \tag{2.20}$$

where V is the cross interaction energy

$$V = \sum_{k,k'\in\Lambda} \Phi(|k - k'|)n_k \otimes n_{k'}. \tag{2.21}$$

Lemma 2.1
Given a bounded interval $\Lambda \subset \mathbf{Z}^1$, a quadruple of occupation number configurations y, y', y_1, y_1' in Λ with $|y| = |y'|, |y_1| = |y_1'|$ and $t \geq 0$, we have:

a) the matrix elements of the operators $e^{\pm itH_{\Lambda,z}}$ are given by

$$(e^{\pm itH_{\Lambda,z}})_{y,y'} = e^{t(1\pm i)|y|} \sum_{\Gamma\in O_{y,y'}} \int_{W_\Gamma^t} P_y^t(d\Omega)\chi_\Lambda(\Omega)$$
$$\times i^{\mp N(\Omega)} \exp(\pm iU(\Omega)|z)), \tag{2.22}$$

b) the matrix elements of the operators $e^{\pm it\tilde{H}_{\Lambda,z}}$ are given by

$$(e^{\pm it\tilde{H}_{\Lambda,z}})_{(y,y_1),(y',y_1')} = e^{t(1\pm i)(|y|+|y_1|)} \sum_{\Gamma\in O_{y,y'},\Gamma_1\in O_{y_1,y_1'}} \times \qquad (2.23)$$

$$\int_{W_\Gamma^t} P_y^t(d\Omega) \int_{W_{\Gamma_1}^t} P_{y_1}^t(d\Omega_1)\chi_\Lambda(\Omega)\chi_\Lambda(\Omega_1)i^{\mp(N(\Omega)+N(\Omega_1))}\exp(\pm iU(\Omega\cup\Omega_1|z)).\lhd$$

Notice that if $|y| \neq |y'|$ or $|y_1| \neq |y_1'|$, the matrix elements of $e^{\pm itH_{\Lambda,z}}$ and $e^{\pm it\tilde{H}_{\Lambda,z}}$ vanish due to the gauge invariance of the Hamiltonians $H_{\Lambda,z}$ and $\tilde{H}_{\Lambda,z}$.

The proof of Lemma 2.1 is standard: Observe that for each choice of sign the RHS's of (2.22) and (2.23) obey the operator semigroup rule and then calculate the infinitesimal generators.

3. Proof of Theorem 1

For definiteness, assume that $t > 0$. Notice that the RHS of (2.8) vanishes if $|x| \neq |x'|$. Hence, we can assume that $|x| = |x'|$. Denote

$$S_{\Lambda,t,m}^0(E_{x,x'}^{(\Lambda^0)}) = S(\prod^{(\Lambda)}(m)e^{itH_\Lambda}E_{x,x'}^{(\Lambda^0)}e^{-itH_\Lambda}\prod^{(\Lambda)}(m)), \qquad (3.1)$$

where $\prod^{(\Lambda)}(m)$ is the orthogonal projector in $\mathcal{H}_\Lambda$ onto the subspace generated by those basis vectors $e_y^{(\Lambda)}$ for which the occupation number configuration y belong to $M(m)$ (see (2.5)).

Lemma 3.1. The following equality holds true:

$$\lim_{m\to\infty}\sup_{\Lambda\supset I_m}|S_{\Lambda,t}(E_{x,x'}^{(\Lambda^0)}) - S_{\Lambda,t,m}^0(E_{x,x'}^{(\Lambda^0)})| = 0, \qquad (3.2)$$

where I_m is defined in (2.2). $\lhd$

Proof. Let $\Lambda \supset I_m$ (this will be always assumed in the proof of intermediate assertions which follow). We have

$$|S_{\Lambda,t}(E_{x,x'}^{(\Lambda^0)}) - S_{\Lambda,t,m}^0(E_{x,x'}^{(\Lambda^0)})| \leq |S((1-\prod^{(\Lambda)}(m))e^{itH_\Lambda}\times$$

$$E_{x,x'}^{(\Lambda^0)}e^{-itH_\Lambda}(1-\prod^{(\Lambda)}(m)))| + |S((1-\prod^{(\Lambda)}(m))e^{itH_\Lambda}\times \qquad (3.3)$$

$$E_{x,x'}^{(\Lambda^0)}e^{-itH_\Lambda}\prod^{(\Lambda)}(m))| + |S(\prod^{(\Lambda)}(m)e^{itH_\Lambda}E_{x,x'}^{(\Lambda^0)}e^{-itH_\Lambda}(1-\prod^{(\Lambda)}(m)))|$$

The first term in the RHS of (3.3) is upper-bounded by $\mathcal{S}(1 - \prod^{(\wedge)}(m))$ whereas both the second and the third ones do not exceed $(\mathcal{S}(1 - \prod^{(\wedge)}(m))\mathcal{S}(\prod^{(\wedge)}(m)))^{1/2}$. Finally, we get that the LHS of (3.3) is less than or equal to

$$\mathcal{S}(M(m)^c) + 2\mathcal{S}(M(m)^c)^{1/2}, \tag{3.4}$$

where C denote the set-theoretical complement. Due to (2.4), this yields the proof. $\square$

The next approximation to $\mathcal{S}_{\wedge,t}(E_{x,x'}^{(\wedge^0)})$ is provided by the following quantity:

$$\mathcal{S}_{\wedge,t,m}^1(E_{x,x'}^{(\wedge^0)}) = \mathcal{S}(\overset{(\wedge)}{\prod}(m) D_{t,\wedge,m}^L E_{x,x'}^{(\wedge^0)} D_{-t,\wedge,m}^R \overset{(\wedge)}{\prod}(m)), \tag{3.5}$$

where (cf. (2.22) for the particular case $z = 0$)

$$(D_{\pm t,\wedge,m}^{L/R})_{y,y'} = e^{t(1\pm i)|y|} \sum_{r \in O_{y,y'}} \int_{W_\Gamma^t} P_y^t(d\Omega)$$

$$\times \chi_\wedge(\Omega) i^{\mp N(\Omega)} \exp(\pm i U(\Omega)) \chi_{(m)}^{L/R}(\Omega).$$

New indicators $\chi_{(m)}^L$ and $\chi_{(m)}^R$ are relating to the free intervals $J^\pm$ (see (2.5)). We recall that, because of presence of the projections $\prod^{(\wedge)}(m)$ in (3.5), we have $y \in M(m)$ when we deal with D^L and $y' \in M(m)$ when we deal with D^R. So,

$$\chi_{(m)}^{L/R}(\Omega) = \prod_{\omega \in \Omega} \chi_{(m)}^{L/R}(\omega). \tag{3.7}$$

Here, for given path ω from a family $\Omega \in W_\Gamma^t$, we have set

$$\chi_{(m)}^L(\omega) = 1, \text{ if } \omega(u) \notin (\hat{J}^- \cup \hat{J}^+) \text{ for any } u \in [0,t], \tag{3.8}$$
$$= 0, \text{ otherwise,}$$

where the intervals $\hat{J}^\pm \subset J^\pm(y,m)$ (see (2.5)) are defined as follows: if $J^\pm(y,m) = [v_1^\pm, v_2^\pm]$, then

$$\hat{J}^\pm = [v_1^\pm + [m^\delta/3], v_2^\pm - [m^\delta/3]] \tag{3.9}$$

(distinguish parenthesis for the integer part of a positive number from that for an interval of the lattice $\mathbf{Z}^1$). Recall that the lengths $v_2^\pm - v_1^\pm$ of $J^\pm$ s are $\geq [m^\delta]$. Likewise,

$$\chi_{(m)}^R(\omega) = 1, \text{ if } \omega(u) \notin (\hat{J}^- \cup \hat{J}^+) \text{ for any } u \in [0,t], \tag{3.10}$$
$$= 0, \text{ otherwise,}$$

where the intervals $\hat{J}^\pm \subset J^\pm(y',m)$ (see (2.5) again) are defined in the similar way.

Physically speaking, in the integral in the RHS of (3.6) we forbid the paths to go "too far" into free intervals $J^{\pm}$. As a result, the paths which are on one hand side of $J^{\pm}$ will not interact with those which are on the other hand side. Such an interaction breakdown means, as we shall see, that $\mathcal{S}^1_{\Lambda,t,m}(E^{(\Lambda^0)}_{x,x'})$ depends "very weakly" on Λ.

We now want to estimate the diffeence between $\mathcal{S}^0_{\Lambda,t,m}$ and $\mathcal{S}^1_{\Lambda,t,m}$:

Lemma 3.2
The following equality takes place:

$$\lim_{m\to\infty}\sup_{\Lambda\supset I_m}|\mathcal{S}^0_{\Lambda,t,m}(E^{(\Lambda^0)}_{x,x'}) - \mathcal{S}^1_{\Lambda,t,m}(E^{(\Lambda^0)}_{x,x'})| = 0. \triangleleft \qquad (3.11)$$

Proof. Comparing (2.22) (where z is taken to be zero) and (3.6), we shall write the difference between $\mathcal{S}_{\Lambda,t,m}$' s in a form where the integration over the trajectories which obey the taboo imposed in (3.6) is separated from that over the trajectories which violate it. Denote by M_Λ the set of occupation number configurations in Λ:

$$M_\Lambda = \{X \in M : X(j) = 0 \; \forall j \notin \Lambda\}$$

and set $M_\Lambda(m) = M_\Lambda \cap M(m)$. To simplify the notations, we set $\mathcal{S}(E^{(\Lambda)}_{y,y}) = \mathcal{S}^{(\Lambda)}(y)$ and omit the indicator χ_Λ in all the integrands. Denoting by $w\lceil\Lambda^\sim$ the restriction of a function w to a set $\Lambda^\sim$, we have

$$\mathcal{S}^0_{\Lambda,t,m}(E^{(\Lambda^0)}_{x,x'}) - \mathcal{S}^1_{\Lambda,t,m}(E^{(\Lambda^0)}_{x,x'}) = \sum_{\substack{y^0,y^1,y^2\in M_\Lambda(m):\\ |y^0|+|y^1|+|y^2|\geq 1}} e^{t(2|y^0|+|y^1|+|y^2|)} \times$$

$$\sum_{\substack{z^1,z^2\in M_\Lambda:\\ |y^0|+|y^1|=|z^1|,|z^2|=|y^2|+|y^0|}} \sum_{\substack{\Gamma^1\in O_{y^0+y^1,z^1}\\ \Gamma^2\in O_{z^2,y^2+y^0}}} \sum_{\substack{y_1,y_2\in M_\Lambda:\\ y^0+y^1+y_1=y_2+y^2+y^0\in M_\Lambda(m)}} e^{t(|y_1|+|y_2|)}$$

$$\sum_{\substack{z_1,z_2\in M_\Lambda:\\ (z^1+z_1)\lceil\Lambda^0=x,x'=(z^2+z_2)\lceil\Lambda^0,\\ (z^1+z_1)\lceil\Lambda\backslash\Lambda^0=(z^2+z_2)\lceil\Lambda\backslash\Lambda^0,\\ |z_1|=|y_1|,|z_2|=|y_2|}} \sum_{\substack{\Gamma_1\in O_{y_1,z_1},\\ \Gamma_2\in O_{z_2,y_2}}} \int_{W^t_{\Gamma^1}} P^t_{y^0+y^1}(d\Omega^1)i^{-N(\Omega^1)}\overline{\chi}^L_{(m)}(\Omega^1)\times \qquad (3.12)$$

$$\int_{W^t_{\Gamma^2}} P^t_{z^2}(d\Omega^2)i^{N(\Omega^2)}\overline{\chi}^R_{(m)}(\Omega^2) \int_{W^t_{\Gamma_1}} P^t_{y_1}(d\Omega_1)i^{-N(\Omega_1)}\chi^L_{(m)}(\Omega_1)\times$$

$$\int_{W^t_{\Gamma_2}} P^t_{z_2}(d\Omega_2)i^{N(\Omega_2)}\chi^R_{(m)}(\Omega_2) \times \exp(iU(\Omega^1\cup\Omega_1) - iU(\Omega^2\cup\Omega_2))\mathcal{S}^{(\Lambda)}(y^0+y^1+y_1).$$

Here Ω^1 is the collection of paths from the (e^{itH_Λ})-integral which ignore the taboo, and Ω^2 is that of those from the (e^{-itH_Λ})-integral. On the other hand, Ω_1 is the collection of taboo-obeying paths from $e^{itH_\wedge}$ and Ω_2 is that of those from $e^{-itH_\wedge}$. Correspondingly, the indicator $\overline{\chi}^\bullet_{(m)}$ is totally disjoint from $\chi^\bullet_m$:

$$\overline{\chi}^\bullet_{(m)}(\Omega^\alpha) = \prod_{\omega \in \Omega^\alpha} (1 - \chi^\bullet_{(m)}(\omega)), \alpha = 1, 2. \tag{3.13}$$

Finally, notice that the taboo under question is established relating to the occupation number configuration $y^0 + y^1 + y_1 = y^0 + y^2 + y_2$ which, by assumption, belongs to $M_\wedge(m)$.

We shall now rewrite the RHS of (3.12) in a slightly more convenient form. For further simplifications, we omit from the notations (but of course, keep in mind) regulations concerning occupation number configurations which are in the RHS of (3.12). We also omit, whenever possible, the references to z^1, z^2, z_1, z_2, because these occupation number configurations may be recovered from matchings $\Gamma^1, \Gamma^2, \Gamma_1$ and Γ_2. So, the RHS of (3.12) equals

$$\sum_{\substack{J_*^+ \subset I_m^+, J_*^- \subset I_m^- : \\ |J_*^\pm| \geq [m^\delta]}} \sum_{\substack{y^0, y^1, y^2 : \\ y^0 + y^1 + y^2 \lceil J_*^+ \cup J_*^- = 0}} e^{t(2|y^0| + |y^1| + |y^2|)} \sum_{\Gamma^1, \Gamma^2}$$

$$\int_{W_{\Gamma^1}^t} P^t_{y^0 + y^1}(d\Omega^1) i^{-N(\Omega^1)} \overline{\chi}_{(\hat{J}_*^\pm)}(\Omega^1) \int_{W_{\Gamma^2}^t} P^t_{z^2}(d\Omega^2) i^{N(\Omega^2)} \overline{\chi}_{(\hat{J}_*^\pm)}(\Omega^2)$$

$$\times \sum_{\substack{y_1, y_2 : \\ J^\pm(y^0 + y^1 + y_1, m) = J_*^\pm}} e^{t(|y_1| + |y_2|)} \sum_{\Gamma_1, \Gamma_2} \int_{W_{\Gamma_1}^t} P^t_{y_1}(d\Omega_1) i^{-N(\Omega_1)} \chi_{(\hat{J}_*^\pm)}(\Omega_1) \tag{3.14}$$

$$\times \int_{W_{\Gamma_2}^t} P^t_{z_2}(d\Omega_2) i^{N(\Omega_2)} \chi_{(\hat{J}_*^\pm)}(\Omega_2)$$

$$\times \exp(iU(\Omega^1 \cup \Omega_1) - iU(\Omega^2 \cup \Omega_2)) S^{(\wedge)}(y^0 + y^1 + y_1).$$

Here $|J_*^\pm|$ stands for the length $(v_{2*}^\pm - v_{1*}^\pm)$ of an interval $J_*^\pm = [v_{1*}^\pm, v_{2*}^\pm]$ and $\hat{J}_*^\pm$ denotes, as before (cf. (3.10)), the corresponding taboo subinterval,

$$\hat{J}_*^\pm = [v_{1*}^\pm + [m^\delta/3], v_{2*}^\pm - [m^\delta/3]]. \tag{3.15}$$

Correspondingly, $\overline{\chi}_{(\hat{J}_*^\pm)}$ is the indicator which vanishes provided at least one trajectory ω from $\Omega^\alpha, \alpha = 1, 2$, obeys the taboo and $\chi_{(\hat{J}_*^\pm)}$ is the totally disjoint indicator which equals 1 provided all the trajectories ω from $\Omega_\alpha, \alpha = 1, 2$, obey the taboo. The sum $\sum_{y_1, y_2}$ is now restricted to those pairs of occupation number configurations y_1, y_2 for which the reference intervals $J_*^\pm$ are just the free intervals for the full configuration $y^0 + y^1 + y_1$.

Now we estimate (3.14) from above by

$$
\sum_{J_*^\pm} \sum_{y^0,y^1,y^2} e^{t(2|y^0|+|y^1|+|y^2|)} \sum_{\Gamma^1,\Gamma^2} \int\limits_{W_{\Gamma^1}^t} P_{y^0+y^1}^t(d\Omega^1)\overline{\chi}_{(J_*^\pm)}(\Omega^1)
$$

$$
\times \int\limits_{W_{\Gamma^2}^t} P_{z^2}^t(d\Omega^2)\overline{\chi}_{(J_*^\pm)}(\Omega^2) | \sum_{y_1,y_2} e^{t(|y_1|+|y_2|)} \sum_{\Gamma_1,\Gamma_2} \int\limits_{W_{\Gamma_1}^t} P_{y_1}^t(d\Omega_1)i^{-N(\Omega_1)} \times
$$

$$
\chi_{(J_*^\pm)}(\Omega_1) \int\limits_{W_{\Gamma^2}^t} P_{z_2}^t(d\Omega_2)i^{N(\Omega_2)}\chi_{(J_*^\pm)}(\Omega_2)\times
$$

$$
\exp(iU(\Omega_1|\Omega^1) - iU(\Omega_2|\Omega^2))\mathcal{S}^{(\wedge)}(y^0+y^1+y_1)|. \tag{3.16}
$$

Here we continue the policy to omit from the notations various sum conditions. Notice where the absolute value appears: this will play the crucial role just below. The potential energy exponent is defined by

$$
U(\Omega_\alpha|\Omega^\alpha)(= U(\Omega_\alpha \cup \Omega^\alpha) - U(\Omega^\alpha)) =
$$

$$
= \sum_{\substack{\omega,\omega'\in\Omega_\alpha: \\ \omega\neq\omega'}} U(\omega,\omega') + \sum_{\substack{\omega\in\Omega_\alpha, \\ \omega'\in\Omega^\alpha}} U(\omega,\omega'); \tag{3.17}
$$

this is related to a "conditional" energy of a "domestic" path family Ω_α in the environment created by a "wild" path family Ω^α.

Proposition 3.3.
The quantity under the absolute value sign in (3.16) is equal to the following trace-like sum

$$
\sum_{\substack{y\in M_\wedge: \\ y^0+y^1+y\in M_\wedge(m), \\ J^\pm(y^0+y^1+y,m)=J_*^\pm}} \mathcal{S}^{(\wedge)}(y+y^1+y^0)(\mathcal{E}_t(\Omega^1;\wedge\setminus(\hat{J}_*^- \cup \hat{J}_*^+)))E_{x-z^1\lceil\wedge^0,x^1-z^2\lceil\wedge^0}^{(\wedge^0)}
$$

$$
\times \prod_{\substack{\cdot+z^1\lceil\wedge\setminus\wedge^0=\cdot+=z^2\lceil\wedge\setminus\wedge^0}}^{\wedge\setminus\wedge^0} \mathcal{E}_{-t}(\Omega^2;\wedge\setminus(\hat{J}_*^- \cup \hat{J}_*^+)) \prod_{\substack{\cdot+y^2+y^0=\cdot+y^0+y^1}}^{(\wedge)})_{y,y}, \tag{3.18}
$$

where:

$$
\text{(a) } \mathcal{E}_t(\Omega^1;\wedge\setminus(\hat{J}_*^- \cup \hat{J}_*^+)) \text{ and } \mathcal{E}_{-t}(\Omega^2;\wedge\setminus(\hat{J}_*^- \cup \hat{J}_*^+))
$$

are the products of unitary operators in $\mathcal{H}_\wedge$:

$$
\mathcal{E}_t(\Omega^1;\wedge\setminus(\hat{J}_*^- \cup \hat{J}_*^+)) = \prod_{j:0\to N(\Omega^1)} e^{iT_j(\Omega^1)H_{\wedge\setminus((\hat{J}_*^-\cup\hat{J}_*^+),j,\Omega^1}}, \tag{3.19}
$$

$$\mathcal{E}_{-t}(\Omega^2; \wedge \setminus (\hat{J}_*^- \cup \hat{J}_*^+)) = \prod_{j:0 \to N(\Omega^2)} e^{-iT_j(\Omega^2)H_{\wedge \setminus (\hat{J}_*^- \cup \hat{J}_*^+),j,\Omega^2}}, \tag{3.20}$$

(b) $\prod_{\cdot + z^1 \lceil \wedge \setminus \wedge^0 = \cdot + z^2 \lceil \wedge \setminus \wedge^0}^{(\wedge \setminus \wedge^0)}$ is the partial isometry operator in $\mathcal{H}_{\wedge \setminus \wedge^0} = \bigotimes_{k \in \wedge \setminus \wedge^0} \mathcal{H}_k$ with the entries

$$\left(\prod_{\cdot + z^1 \lceil \wedge \setminus \wedge^0 = \cdot + z^2 \lceil \wedge \setminus \wedge^0}^{(\wedge \setminus \wedge^0)} \right)_{z_1, z_2} = 1, \text{ if } z_1 + z^1 \lceil \wedge \setminus \wedge^0 = z_2 + z^2 \lceil \wedge \setminus \wedge^0,$$
$$= 0 \text{ otherwise}, \tag{3.21}$$

c) likewise, $\prod_{\cdot + y^2 + y^0 = \cdot + y^0 + y^1}^{(\wedge)}$ is the partial isometry operator in $\mathcal{H}_\wedge$ with

$$\left(\prod_{\cdot + y^2 + y^0 = \cdot + y^0 + y^1}^{(\wedge)} \right)_{y_1, y_2} = 1, \text{ if } y_1 + y^0 + y^1 = y_2 + y^2 + y^0,$$
$$= 0 \text{ otherwise}. \tag{3.22}$$

Furthermore, $T_j(\Omega^\alpha) = \tau_{j+1}(\Omega^\alpha) - \tau_j(\Omega^\alpha)$ is the time between the epochs of the $(j+1)$th and jth jump for the path family Ω^α (we set $\tau_0(\Omega^\alpha) = 0, \tau_{N(\Omega^\alpha)+1}(\Omega^\alpha) = t), \alpha = 1,2$, and the self-adjoint operators $H_{\wedge \setminus (\hat{J}_*^- \cup \hat{J}_*^+),j,\Omega^\alpha}$ are the Hamiltonians of the motion in the external potential field created by the occupation number configuration $\Omega^\alpha(u)$ (i.e., by the time section of this wild path family Ω^α at at moment u), $\tau_j(\Omega^\alpha) < u < \tau_{j+1}(\Omega^\alpha), 0 \leq j \leq N(\Omega^\alpha), \alpha = 1,2$:

$$H_{\wedge \setminus (\hat{J}_*^- \cup \hat{J}_*^+),j,\Omega^\alpha} = -1/2 \sum_{k \in \wedge \setminus (\hat{J}_*^- \cup \hat{J}_*^+)} a_k^+ (\Delta a)_k$$
$$+ \sum_{k,k' \in \wedge \setminus (\hat{J}_*^- \cup \hat{J}_*^+)} \Phi(|k - k'|) n_k n_k,$$
$$+ \sum_{\substack{k \in \wedge \setminus (\hat{J}_*^- \cup \hat{J}_*^+), \\ \omega \in \Omega^\alpha}} \Phi(|k - \omega(u)|) n_k, \tag{3.23}$$

with the Dirichlet boundary condition on $\partial(\wedge \setminus (\hat{J}_*^- \cup \hat{J}_*^+))$. ◁

Proof of proposition 3.3 - by inspection, using Lemma 2.1 a).

Since all the operators in (3.18) are of norm ≤ 1, we conclude from Proposition 3.3 that the absolute value in (3.16) is upper-bounded by

$$\sum_{\substack{\tilde{y} \in M_\wedge(m): \\ \tilde{y} \geq \max[y^1, y^2] + y^0, \\ J^\pm(\tilde{y}, m) = J_*^\pm}} \mathcal{S}^{(\wedge)}(\tilde{y}) \leq \sum_{\substack{y \in M_\wedge(m): \\ J^\pm(y, m) = J_*^\pm}} \mathcal{S}^{(\wedge)}(y). \tag{3.24}$$

Now we pass to estimate the internal sum $\sum_{\Gamma^1,\Gamma^2}$ in (3.16):

$$\sum_{\Gamma^1,\Gamma^2} \int_{W^t_{\Gamma^1}} P^t_{y^1}(d\Omega^1)\overline{\chi}_{(J^{\pm}_*)}(\Omega^1) \int_{W^t_{\Gamma^2}} P^t_{z^2}(d\Omega^2)\overline{\chi}_{(J^{\pm}_*)}(\Omega^2) \tag{3.25}$$

for fixed intervals $J^{\pm}_*$ and occupation number configurations y^0, y^1, y^2. This is obviously less than the product

$$\prod_{j\in\Lambda}(P^t_0(\sup_{u\in[0,t]}|\omega(u)| \geq [m^{\delta}/3] + \mathrm{dist}(j, J^-_*)) \\ + P^t_0(\sup_{u\in[0,t]}|\omega(u)| \geq [m^{\delta}/3] + \mathrm{dist}(j, J^+_*)))^{2y^0(j)+y^1(j)+y^2(j)}. \tag{3.26}$$

Making the summation over y^0, y^1 and y^2 yields the upper bound

$$\sum_{n=1}^{\infty} 3^n e^{2tn}(4 \sum_{r\geq[m^{\delta}/3]} P^t_0(\sup|\omega(u)| \geq r))^n, \tag{3.27}$$

whereas the sum over $J^{\pm}_*$ gives a value ≤ 1 due to (3.24). Assuming that m is chosen so large that

$$24e^t \sum_{r\geq[m^{\delta}/3]} P^t_0(\sup|\omega(u)| \geq r) < 1, \tag{3.28}$$

we get that (3.16) is less than LHS of (3.28). This goes to zero because

$$P^t_0(\sup|\omega(u)| \geq r) \leq \sum_{l\geq r} \frac{t^l}{l!}e^{-t} \leq \frac{t^r}{r!}. \tag{3.29}$$

Lemma 3.2 is proved.

$\square$

Now write

$$S^1_{\Lambda,t,m}(E^{(\Lambda^0)}_{x,x'}) = \sum_{\substack{J^{\pm}_*\subset I^{\pm}_m: \\ |J^{\pm}_*|\geq[m^{\delta}]}} \sum_{\substack{y\in M_\Lambda(m): \\ J^{\pm}(y,m)=J^{\pm}_*}} S^{(\Lambda)}(y)(D^L_{t,\Lambda,m}E^{(\Lambda^0)}_{x,x'}D^R_{-t,\Lambda,m})_{y,y} \tag{3.30}$$

and notice that for $y \in M_\Lambda(m)$ and m large enough

$$(D^L_{t,\Lambda,m}E^{(\Lambda^0)}_{x,x'}D^R_{-t,\Lambda,m})_{y,y} = (D^L_{t,I_m,m}E^{(\Lambda^0)}_{x,x'}D^R_{-t,I_m,m})_{y\lceil I_m,y\lceil I_m}, \tag{3.31}$$

since both LHS and RHS coinside with the diagonal entry of the operator

$$\exp(itH_{(I_m\setminus(J^-_*\cup J^+_*))_{\mathrm{int}}})E^{(\Lambda^0)}_{x,x'}\exp(-itH_{(I_m\setminus(J^-_*\cup J^+_*))_{\mathrm{int}}}), \tag{3.32}$$

where $I_m \setminus (\hat{J}_*^- \cup \hat{J}_*^+))_{\text{int}}$ stands for the "internal connected component" of the set-theoretical difference $I_m \setminus (\hat{J}_*^- \cup \hat{J}_*^+)$.

Having this in mind and combining the arguments from the proof of Lemmas 3.1 and 3.2, one can get

Lemma 3.4. The following relation is valid:

$$\lim_{m \to \infty} \sup_{\Lambda \supset I_m} |\mathcal{S}^1_{\Lambda,t,m}(E^{(\Lambda^0)}_{x,x'}) - \mathcal{S}_{I_m,t}(E^{(\Lambda^0)}_{x,x'})| = 0. \quad \lhd \tag{3.33}$$

This finishes the proof of Theorem 1.

$\square$

4. Proof of Theorem 2.

Assume again that $t > 0$. The proof of Theorem 2 proceeds along a similar way. First of all, we write

$$\mathcal{S}_{\Lambda,t}(\prod_{\geq s}^{(\Lambda^0)}) = \mathcal{S}(\prod_{\geq s}^{(\overline{\Lambda}^0([s^{1/2}]))} e^{itH_\Lambda} \prod_{\geq s}^{(\Lambda^0)} e^{-itH_\Lambda} \prod_{\geq s}^{(\overline{\Lambda}^0([s^{1/2}]))}) \\ + \mathcal{S}(\prod_{<s}^{(\overline{\Lambda}^0([s^{1/2}]))} e^{itH_\Lambda} \prod_{\geq s}^{(\Lambda^0)} e^{-itH_\Lambda} \prod_{<s}^{(\overline{\Lambda}^0([s^{1/2}]))}), \tag{4.1}$$

where $\prod_{<s}^{(\overline{\Lambda}^0([s^{\frac{1}{2}}]))}$ denotes the complementary projector:

$$\prod_{<s}^{(\overline{\Lambda}^0([s^{1/2}]))} = 1 - \prod_{\geq s}^{(\overline{\Lambda}^0([s^{1/2}]))}$$

and $\overline{\Lambda}^0([s^{\frac{1}{2}}])$ is obtained by stretching the interval $\Lambda^0 = [v_1^0, v_2^0]$:

$$\overline{\Lambda}^0([s^{1/2}]) = [v_1^0 - [s^{1/2}], v_2^0 + [s^{1/2}]].$$

Due to condition (d_1^{**}), the first term in the RHS of (4.1) does not exceed, in the absolute value,

$$\mathcal{S}(\prod_{\geq s}^{(\overline{\Lambda}^0([s^{1/2}]))}) \leq c \int_{\overline{s}}^{\infty} du \exp(-u^2/2), \tag{4.2}$$

where

$$\overline{s} = (\sigma(|\Lambda^0| + 2s^{1/2})^{1/2})^{-1}(s - (|\Lambda^0| + 2s^{1/2})b)_+. \tag{4.3}$$

We are now going to estimate the second term in the RHS of (4.1). For the sake of simplicity we write $\overline{\Lambda}^0$ instead of $\overline{\Lambda}^0([s^{1/2}])$.

                                             A. Shuhov et al

Lemma 4.1. The following bound holds true

$$|\mathcal{S}(\prod_{<s}^{(\overline{\wedge}^0)} e^{itH_\wedge} \prod_{\ge s}^{(\wedge^0)} e^{-itH_\wedge} \prod_{<s}^{(\overline{\wedge}^0)}) \le \psi_{\wedge^0,t}(s),$$

where $\lim_{s\to\infty} \psi_{\wedge^0,t}(s) = 0$ for any $\wedge^0$ and t. ◁

Proof. The proof of Lemma 4.1 will resemble that of Lemma 3.2. We have (cf. (3.13)):

$$\mathcal{S}(\prod_{<s}^{(\overline{\wedge}^0)} e^{itH_\wedge} \prod_{\ge s}^{(\wedge^0)} e^{-itH_\wedge} \prod_{<s}^{(\overline{\wedge}^0)}) =$$

$$= \sum_{\substack{z^0,z^1,z^2 \in M_{\wedge^0}: \\ |z^0|+|z^\alpha|\ge 1, \alpha=1,2}} e^{t(2|z^0|+|z^1|+|z^2|)} \sum_{\substack{y^1,y^2 \in M_{\wedge\setminus\overline{\wedge}^0}: \\ |y^1|=|z^0|+|z^1|, \\ |y^2|=|z^0|+|z^2|}}$$

$$\sum_{\substack{\Gamma^1 \in O_{z^0+z^1,y^1} \\ \Gamma^2 \in O_{y^2,z^0+z^2}}} \int_{W^t_{\Gamma^1}} P^t_{z^0+z^1}(d\Omega^1)\chi_\wedge(\Omega^1)i^{N(\Omega^1)} \int_{W^t_{\Gamma^2}} P^t_{y^2}(d\Omega^2)\chi_\wedge(\Omega^2)i^{-N(\Omega^2)} \times$$

$$\sum_{\substack{z_1,z_2 \in M_{\wedge^0}: \\ z^0+z^1+z_1=z_2+z^2+z^0, \\ |z^0|+|z^1|+|z_1|\ge s}} e^{t(|z_1|+|z_2|)} \sum_{\substack{y_1,y_2 \in M_{\overline{\wedge}^0}: \\ |y_\alpha|=|z_\alpha|, \alpha=1,2}} \sum_{\substack{\Gamma_1 \in O_{z_1,y_1}, \\ \Gamma_2 \in O_{y_2,z_2}}}$$

$$\int_{W^t_{\Gamma_1}} P^t_{z_1}(d\Omega_1)\chi_\wedge(\Omega_1)i^{N(\Omega_1)} \int_{W^t_{\Gamma_2}} P^t_{y_2}(d\Omega_2)\chi_\wedge(\Omega_2)i^{-N(\Omega_2)} \sum_{z \in M_{\wedge\setminus\wedge^0}} e^{2t|z|} \times \qquad (4.4)$$

$$\sum_{\substack{y,y' \in M_\wedge: \\ y^1+y_1+y=y'+y_2+y^2, \\ |y|=|z|=|y'|}} \sum_{\substack{\Gamma \in O_{z,y}, \\ \Gamma' \in O_{y',z}}} \int_{W^t_z} P^t_z(d\Omega)\chi_\wedge(\Omega)i^{N(\Omega)} \int_{W^t_{y'}} P^t_{y'}(d\Omega')\chi_\wedge(\Omega')i^{-N(\Omega')} \times$$

$$\exp(-iU(\Omega^1 \cup \Omega_1 \cup \Omega) + iU(\Omega^2 \cup \Omega_2 \cup \Omega'))\mathcal{S}^{(\wedge)}(y^1 + y_1 + y).$$

As before, we omit henceforth the standard indicator $\chi_\wedge$ and do not write regulations for summations. We estimate the RHS of (4.4) from above by

$$\sum_{z^0,z^1,z^2} e^{t(2|z^0|+|z^1|+|z^2|)} \sum_{y^1,y^2} \sum_{\Gamma^1,\Gamma^2} \int_{W^t_{\Gamma^1}} P^t_{z^0+z^1}(d\Omega^1) \int_{W^t_{\Gamma^2}} P^t_{y^2}(d\Omega^2) \times$$

$$|\sum_{z_1,z_2} e^{t(|z_1|+|z_2|)} \sum_{y_1,y_2} \sum_{\Gamma_1,\Gamma_2} \int_{W^t_{\Gamma_1}} P^t_{z_1}(d\Omega_1) i^{N(\Omega_1)} \int_{W^t_{\Gamma_2}} P^t_{y_2}(d\Omega_2) i^{-N(\Omega_2)} \times \tag{4.5}$$

$$\sum_{z,z'} e^{2t|z|} \sum_{y,y'} \sum_{\Gamma,\Gamma'} \int_{W^t_{\Gamma}} P^t_z(d\Omega) i^{N(\Omega)} \int_{W^t_{\Gamma'}} P^t_{y'}(d\Omega') i^{-N(\Omega)} \times$$

$$\exp(-iU(\Omega_1 \cup \Omega|\Omega^1) + iU(\Omega_2 \cup \Omega'|\Omega^2)) S^{(\wedge)}(y^1 + y_1 + y)|.$$

It is again important to notice where the absolute value sign appears in (4.5). The conditional energies $U(\Omega_2 \cup \Omega'|\Omega^2)$ and $U(\Omega_1 \cup \Omega|\Omega^1)$ are defined in analogy with (3.17).

Proposition 4.2. The quantity under the absolute value sign in (4.5) is equal to the following trace-like sum

$$\sum_{\substack{y_1 \in M_{\tilde{\wedge}^0}, \\ \nu \in M_\wedge}} S^{(\wedge)}(y^1 + y_1 + y) \Big(\widetilde{\prod}^{(\wedge)}_{\cdot+y^1+\cdot=\cdot+y^2+\cdot}\Big) \Big(\prod^{(\wedge)}_{\tilde{\wedge}^0} \otimes 1\Big) \tilde{\mathcal{E}}_t(\Omega^2;\wedge) \tilde{\mathcal{E}}_t(\Omega^2;\wedge) \times$$

$$\Big(\prod^{(\wedge)}_{\cdot+z^0+z^2=\cdot+z^0+z^1} \otimes \prod^{(\wedge)}_{\wedge\backslash\wedge^0}\Big) \times \Big(\Big(\prod^{(\wedge)}_{\wedge^0} \prod^{(\wedge)}_{|\cdot|+|z^0|+|z^1|\geq s}\Big) \otimes 1\Big) \times \tag{4.6}$$

$$\tilde{\mathcal{E}}_{-t}(\Omega^1;\wedge) \Big(\prod^{(\wedge)}_{\wedge^0} \otimes 1\Big)_{(y_1,y),(y_1,y)},$$

where

(a) $\widetilde{\prod}^{(\wedge)}_{\cdot+y^1+\cdot=\cdot+y^2+\cdot}$ is the partial isometry operator in the tensor square $\mathcal{H}_\wedge \otimes \mathcal{H}_\wedge$ with the entries

$$\Big(\widetilde{\prod}^{(\wedge)}_{\cdot+y^1+\cdot=\cdot+y^2+\cdot}\Big)_{(y_1,y),(y_2,y')} = 1, \text{ if } y_1 + y^1 + y = y_2 + y^2 + y', \tag{4.7}$$

$$= 0 \text{ otherwise,}$$

(b) $\prod^{(\wedge)}_{\tilde{\wedge}}$ is the orthogonal projector in $\mathcal{H}_\wedge$ confining an occupation number configuration to a subset $\tilde{\wedge} \subset \wedge$,

(c) $\prod^{(\wedge)}_{\cdot+z^0+z^2=\cdot+z^0+z^1}$ is the partial isometry operator in $\mathcal{H}_\wedge$ with the entries

$$\Big(\prod^{(\wedge)}_{\cdot+z^0+z^2=\cdot+z^0+z^1}\Big)_{z_2,z_1} = 1, \text{ if } z_2 + z^0 + z^2 = z_1 + z^0 + z^1, \tag{4.8}$$

$$= 0, \text{ otherwise,}$$

(d) $\prod^{(\Lambda)}_{|\cdot|+|z^0|+|z^1|\geq s}$ is the orthogonal projector in $\mathcal{H}_\Lambda$ onto the space corresponding to occupation number configurations z_1 with $|z_1| + |z^0| + |z^1| \geq s$,

(e) $\tilde{\mathcal{E}}_{-t}(\Omega^1; \Lambda)$ and $\tilde{\mathcal{E}}_t(\Omega^2; \Lambda)$ are the products of unitary operators in $\mathcal{H}_\Lambda \otimes \mathcal{H}_\Lambda$:

$$\tilde{\mathcal{E}}_{-t}(\Omega^1; \Lambda) = \prod_{j:0\to N(\Omega^1)} e^{-iT_j(\Omega^1)\tilde{H}_{\Lambda,j,\Omega^1}}, \tag{4.9}$$

$$\tilde{\mathcal{E}}_t(\Omega^2; \Lambda) = \prod_{j:0\to N(\Omega^2)} e^{iT_j(\Omega^2)\tilde{H}_{\Lambda,j,\Omega^2}}. \tag{4.10}$$

Here, as in Proposition 3.3, $T_j(\Omega^\alpha)$ is the time between the epochs $\tau_j(\Omega^\alpha)$ and $\tau_{j+1}(\Omega^\alpha)$ of subsequent jumps for the path family $\Omega^\alpha, j = 0, \cdots, N(\Omega^\alpha), \alpha = 1, 2$. The self-adjoint operators $\tilde{H}_{\Lambda,j,\Omega^\alpha}$ in $\mathcal{H}_\Lambda \otimes \mathcal{H}_\Lambda$ are of the form

$$\tilde{H}_{\Lambda,j,\Omega^\alpha} = H_{\Lambda,j,\Omega^\alpha} \otimes 1 + 1 \otimes H_{\Lambda,j,\Omega^\alpha} + V, \tag{4.11}$$

where

$$\begin{aligned} H_{\Lambda,j,\Omega^\alpha} = -1/2 \sum_{k\in\Lambda} a_k^+(\Delta a)_k + \sum_{k,k'\in\Lambda} \Phi(|k - k'|)n_k n_{k'}, \\ + \sum_{k\in\Lambda} \Phi(|k - \omega(u)|)n_k, \end{aligned} \tag{4.12}$$

where $\tau_j(\Omega^\alpha) < u < \tau_{j+1}(\Omega^\alpha), \alpha = 1, 2$ (cf. (3.23)), and V is the cross interaction energy (2.21). ◁

The proof of Proposition 4.2, like that of Proposition 3.3, is done by inspection, using Lemma 2.1 b).

We use Proposition 4.2 and condition (d_2^{**}) to conclude that the absolute value in (4.5) is less than

$$\begin{aligned} \sum_{\substack{y_1\in M_{\Lambda\to 0}:\\ |y_1|+|y^1|\geq s}} \sum_{y\in M_\Lambda} \mathcal{S}^{(\Lambda)}(y_1 + y^1 + y) = \sum_{\substack{y_1\in M_{\Lambda\to 0}:\\ |y_1|+|y^1|\geq s}} \overline{\mathcal{S}}(y_1 + y^1) \\ \leq c_1^{2[s^{1/2}]+|\Lambda^0|}\rho^{|y^1|} \sum_{\substack{y_1\in M_{\Lambda\to 0}:\\ |y_1|\geq (s-|y^1|)_+}} \overline{\mathcal{S}}(y_1), \end{aligned} \tag{4.13}$$

where we have used the notation

$$\overline{\mathcal{S}}(y^\sim) = \mathcal{S}(\overline{B}_{y^\sim}). \tag{4.14}$$

In the same way one can check that the absolute value in question is less than

$$c_1^{2[s^{1/2}]+|\Lambda^0|}\rho^{|y^2|} \sum_{\substack{y_2\in M_{\Lambda\to 0}:\\ |y_2|\geq (s-|y^2|)_+}} \overline{\mathcal{S}}(y_2).$$

Hence, this absolute value is less than

$$c_1^{2[s^{1/2}]+|\wedge^0|}\rho^{(|y^1|+|y^2|)}\Big(\sum_{\substack{y_1\in M_{\widehat{\wedge}^0}:\\ |\nu_1|\geq(s-|y^1|)_+}}\overline{S}(y^1)\cdot\sum_{\substack{y_2\in M_{\widehat{\wedge}^0}:\\ |\nu_2|\geq(s-|y^2|)_+}}\overline{S}(y^2)\Big)^{1/2}.$$

Now we estimate from above the sum $\displaystyle\sum_{y_1,y_2}\sum_{\Gamma^1,\Gamma^2}$ in (4.5) by

$$\prod_{j\in\wedge^0}(P_0^t(|\omega(t)|\geq[s^{1/2}]+j-v_1^0)+P_0^t(|\omega(t)|\geq[s^{1/2}]+v_2^0-j))^{(2z^0+z^1+z^2)(j)}\times$$

$$c_1^{2[s^{1/2}]+|\wedge^0|}\rho^{(|z^1|+|z^2|)/2+|z^0|}\Big(\sum_{\substack{y_1\in M_{\widehat{\wedge}^0}:\\ |\nu_1|\geq(s-|z^0|-|z^1|)_+}}\overline{S}(y^1)\cdot\sum_{\substack{y_2\in M_{\widehat{\wedge}^0}:\\ |\nu_2|\geq(s-|z^0|-|z^2|)_+}}\overline{S}(y)\Big)^{1/2}$$

Making the summation over z^0, z^1, z^2 yields, by virtue of (d_2^{**}), the following upper bound for (4.5):

$$\Big(\sum_{n=1}^{\infty}n^2e^{2tn}(4\sum_{r\geq[s^{1/2}]}P_0^t(|\omega(t)|\geq r))^n\rho^{\frac{n}{2}}(c_1^{|\wedge^0|+2s^{1/2}})^2\overline{\rho}^{|\wedge^0|+2s^{1/2}})^2,$$

where

$$\overline{\rho}=(1-\rho)^{-1}=\sum_{m\geq0}\rho^m.$$

The final remark is similar to that in the end of the proof of Lemma 3.2. This finishes the proof of Lemma 4.1 and hence, that of Theorem 2. $\square$

Acknowledgements. One of the authors (Yu. M. Suhov) thanks Professor J.T. Lewis and Dublin Institute for Advanced Studies for the warm hospitality. The authors thank Catherine McAuley for typing the first version of the manuscript.

References

1 Robinson D.W. Statistical Mechanics of Quantum Spin Systems, II. Commun. Math. Phys., 7, 337 (1968).

2 Streater R.F. On certain Non-Relativistic Quantized Fields. Commun. Math. Phys., 7, 93 (1968).

3 Dubin D.A. Solvable Models in Algebraic Statistical Mechanics. Oxford: Clarendon Press. 1974; see also the references therein.

4 Majevski W.A. Time Development of Bose Systems, Journ. Math. Phys., 22, 2921-2925 (1980); see also the references therein.

5 Ruelle D. Definition of Green's functions for dilute Fermi Gas. Helv. Phys. Acta 45, 215-219 (1972).

6 Park Y.M. Quantum Mechanics for Superstable Interactions: Bose-Einstein Statistics. Journ. Stat. Phys., 40, 259 (1985); see also the reference therein.

7 Bratteli O, Robinson D. Green's functions, Hamiltonians and Modular Automorphisms. Commun. Math. Phys., 50, 133-156 (1976).

8 Sinai Ya.G. Construction of the Dynamics for one-Dimensional Systems of Statistical Mechanics. Teoret. Matem. Fiz., 12, 487-497 (1972) (in Russian).

9 Sinai Ya.G. Construction of Cluster Dynamics for Dynamical Systems of Statistical Mechanics. Vestnik Mosc. Gos. Univ, ser 1 (Matem., Mech.), 29, 152-159 (1974) (in Russian).

10 Bratteli O., Robinson D.W. Operator Algebras and Quantum Statistical Mechanics, Vols. 1,2. Berlin et al: Springer-Verlag, 1979-1981.

11 Liggett, T. Interacting Particle Systems. New York et al: Springer-Verlag, 1985.

Towards Time-Dynamics for Bosonic Systems in Quantum Statistical Mechanics, 2

by

A.G. Shuhov[1] and Yu.M. Suhov[1,2]

Abstract

This is a continuation of a study of time-evolution of a state of an infinitely-extended one-dimensional boson lattice system, which was initiated in the paper [1] under the same title. We consider here finite range, pair interaction potentials and prove that the family of the time-evolved states provides a solution to the corresponding Liouville equation and BBGKY hierarchy.

Key words and phrases: one-dimensional lattice boson system, diagonal state, finite-range interaction potential, time-evolution, Liouville's equation, BBGKY hierarchy.

1 Introduction

This paper is a continuation of [1]. We studied in the paper [1] the non-equilibrium time-evolution problem for a "locally-unbounded" phase space quantum model where the "traditional" approach [2] based on a limiting *-automorphism group of an underlying C^*-algebra does not work. The model was on the one-dimensional lattice $\mathbf{Z}^1$ and was given by its formal Hamiltonian

$$H = -\frac{1}{2} \sum_{j \in \mathbf{Z}^1} a_j^+ (\Delta a)_j + \sum_{j,j' \in \mathbf{Z}^1} \Phi(|j - j'|) n_j n_{j'}, \tag{1.1}$$

where a_j^+, a_j and $n_j = a_j^+ a_j$ are, respectively, the bosonic creation, annihilation and particle number operators in the Hilbert space $\mathcal{H}_j$ which is realized as $l_2(\mathbf{Z}_+^1)$ with the standard orthonormal basis $\{e_s^{(j)}, s \in \mathbf{Z}_+^1\}$. Let $\mathcal{B}_j$ denote the C^*-algebra of bounded operators in $\mathcal{H}_j$ and $\mathcal{B} = \otimes_{j \in \mathbf{Z}^1} \mathcal{B}_j$ be the corresponding quasilocal C^*-algebra. We have constructed in [1] a family $\{\rho_t, t \in \mathbf{R}^1\}$ of states of $\mathcal{B}$ which describes the time-evolution of an initial state $\rho = \rho_0$ (in [1] states are denoted by $\mathcal{S}$ instead of ρ). The states ρ_t are given by

$$\rho_t(A) = \lim_{\Lambda \nearrow \mathbf{Z}^1} \rho_{\Lambda,t}(A), \quad A \in \mathcal{B}, \, t \in \mathbf{R}^1, \tag{1.2}$$

---

[1]Institute for Problems of Information Transmission, USSR Academy of Sciences, 19 Yermolova Str., GSP – 4 Moscow, 101447 USSR

[2]Dublin Institute for Advanced Studies, 10 Burlington Road, Dublin 4, Ireland

where

$$\rho_{\Lambda,t}(A) = \rho(e^{itH_\Lambda} A e^{-itH_\Lambda}) \tag{1.3}$$

and H_Λ is the Hamiltonian (1.1) confined to the finite volum $\Lambda \subset \mathbf{Z}^1$ (bounded interval on the lattice).

In the paper [1] we assumed that the (locally normal) initial state ρ is diagonal. This condition means that the density matrix of the state ρ in any bounded interval $\Lambda \subset \mathbf{Z}^1$ is diagonal in the reference basis $\{e_x^{(\Lambda)}\}$ in the space $\mathcal{H}_\Lambda = \otimes_{j \in \Lambda} \mathcal{H}_j$. Here x is an arbitrary occupation number configuration in Λ, i.e., a function $\Lambda \to \mathbf{Z}_+^1$ (as in [1], we shall use the notation $x \dot{\subset} \Lambda$). The diagonality of a state ρ means in physical terms that it describes a classical particle system on the lattice. Of course, this property is destroyed by the time-evolution, but it makes possible to formulate further assumptions on the initial state in natural probabilistic terms. For example, the basic assumption that the state ρ has long space "holes" which are free of particles is related to Borel-Cantelli type assertions (see [1]).

In the situation under consideration we are able to prove that the states $\rho_t, t \in \mathbf{R}^1$, given by (1.2), (1.3) provide a solution to the infinite-volume Liouville equation

$$\frac{d}{dt} p_t^{(\Lambda)} = itr_{\Lambda_+ \backslash \Lambda}[p_t^{(\Lambda+)}, \bar{H}_{\Lambda_+}^0] +$$
$$+ i \sum_{\substack{j,j' \in \mathbf{Z}^1: \\ \Lambda \cap \{j,j'\} \neq \emptyset}} tr_{\{j,j'\} \backslash \Lambda}[p_t^{(\Lambda \cup \{j,j'\})}, \Phi(|j-j'|)n_j n_{j'}]. \tag{1.4}$$

Equation (1.4) is on a family $\{p_t^{(\Lambda)}\}$ of density matrices in finite volumes $\Lambda \subset \mathbf{Z}^1$, which obey the standard consistency condition $tr_{\Lambda' \backslash \Lambda} p_t^{(\Lambda')} = p_t^{(\Lambda)}$, $\Lambda' \supset \Lambda$. The term $[p_t^{(\Lambda+)}, \bar{H}_{\Lambda_+}^0]$ in the RHS of (1.4) corresponds to the kinetic energy: here Λ_+ is $[v_1 - 1, v_2 + 1]$ for $\Lambda = [v_1, v_2]$; for a general case Λ_+ is defined as $\{j \in \mathbf{Z}^1 : \mathrm{dist}(j, \Lambda) \leq 1\}$. Namely,

$$\bar{H}_{\Lambda_+}^0 = -\frac{1}{2} \sum_{\substack{j \in \Lambda_+: \\ j-1 \in \Lambda_+}} a_j^+ a_{j-1} - \sum_{\substack{j \in \Lambda_+: \\ j+1 \in \Lambda_+}} a_j^+ a_{j+1} + \sum_{j \in \Lambda} n_j \tag{1.5}$$

(the notation $\bar{H}_{\Lambda_+}^0$ is used to stress that no boundary condition is involved in the RHS of (1.5)). The partial trace $tr_{\Lambda^\sim}, \Lambda^\sim \subset \mathbf{Z}^1$, is taken in $\mathcal{H}_{\Lambda^\sim}$; in the case $\Lambda^\sim = \emptyset$ (which occurs when $\{j, j'\} \subset \Lambda$ in the second-term sum in the RHS of (1.4)) no trace is taken.

Equivalently, equation (1.4) may be rewritten in the following form

$$\frac{d}{dt} p_t^{(\Lambda)} = itr_{\Lambda_{+r} \backslash \Lambda}[p_t^{(\Lambda+r)}, \bar{H}_{\Lambda_{+r}}]. \tag{1.6}$$

Here and below $\Lambda_{+r} = \{j \in \mathbf{Z}^1 : \mathrm{dist}(j, \Lambda) \leq r\}$, $r = \frac{1}{2}\mathrm{diam}\ \mathrm{supp}\ \Phi$. The precise formulation of the result is given in the next section.

Another object related to the time-evolution in statistical mechanics is the BBGKY hierarchy:

$$\frac{d}{dt}\rho_t^{(n)} = i[\rho_t^{(n)}, H^{(n)}] + \mathbf{A}\rho_t^{(n+1)}, \quad n \geq 0, \, t \in \mathbf{R}^1. \tag{1.7}$$

This is a system of equations for a sequence $\{\rho_t^{(n)}, n \geq 0\}$ of operators $\rho_t^{(n)} : \mathcal{H}^{(n)} \to \mathcal{H}^{(n)}$ where $\mathcal{H}^{(n)} = (\mathcal{H}^{(1)})_{\mathrm{symm}}^{\otimes n}$ is the symmetrized tensor product of n copies of the underlying Hilbert space $\mathcal{H}^{(1)} = l_2(\mathbf{Z}^1)$ ($\mathcal{H}^{(0)}$ is set to be the one-dimensional complex space). Thereby the matrix elements of $\rho_t^{(n)}$ are labeled by pairs (x, x') of occupation number configurations $x, x' \subset \mathbf{Z}^1$ with $|x| = |x'| = n$ (by $|y|$ we denote, as in [1], the number of particles in an occupation number configuration $y : |y| = \sum_j y(j)$). The first term in the RHS of (1.7) is the commutator with the n-particle Hamiltonian $H^{(n)}$ corresponding to (1.1). Like $\rho_t^{(n)}$, the operator $H^{(n)}$ acts in $\mathcal{H}^{(n)}$ and its matrix elements are again labeled by pairs (x, x') where $x, x' \subset \mathbf{Z}^1$, $|x| = |x'|$. They are given by

$$(H^{(n)})_{x',x} = -\frac{1}{4} \sum_{j,j' \in \mathbf{Z}^1} \Phi(|j - j'|)x(j)x(j'),$$

if occupation number configurations x and x' coincide,

$$(H^{(n)})_{x',x} = -\frac{1}{4}x(j)^{1/2}(x(j \mp 1) + 1)^{1/2}, \quad \text{if } x' = x - \delta_j + \delta_{j \mp 1},$$

and

$$(H^{(n)})_{x',x} = 0 \quad \text{otherwise.}$$

Here and below δ_j denotes the one-particle configuration concentrated at a site $j \in \mathbf{Z}^1$. As to the second term in the RHS of (1.7), this is the following operator in $\mathcal{H}^{(n)}$.

$$\mathbf{A}\rho_t^{(n+1)} = i \sum_{j \in \mathbf{Z}^1} [U(j), \rho_t^{(n+1)}(j)]. \tag{1.8}$$

Here $\rho_t^{(n+1)}(j)$ is an operator

$$(\rho_t^{(n+1)}(j))_{x,x'} = (\rho_t^{(n+1)})_{x+\delta_j, x'+\delta_j}, \quad x, x' \subset \mathbf{Z}^1, \, |x| = |x'| = n,$$

(recall that $\rho_t^{(n+1)}$ acts in $\mathcal{H}^{(n+1)}$ and therefore its matrix elements are labeled by pairs (y, y') with $|y| = |y'| = n+1$). Likewise, $U(j)$ is an operator in $\mathcal{H}^{(n)}$ with

$$U(j))_{x,x'} = \sum_{k \in \mathbf{Z}^1} \Phi(|j - k|)x(k),$$

if occupation number configurations x and x' coincide,

$$(U(j))_{x,x'} = 0, \quad \text{otherwise.}$$

In our situation we prove that a solution to (1.7) is given by the family of operators with

$$(\rho_t^{(n)})_{x,x'} = \rho_t\Big(\prod_{j\in \mathbf{Z}^1} a_j^{+^{x'(j)}} \prod_{k\in \mathbf{Z}^1} a_k^{x(k)} \Big), \tag{1.9}$$

$$x, x' \subset \mathbf{Z}^1,\ |x| = |x'| = n,\ n \geq 0,\ t \in \mathbf{R}^1.$$

Of course, the existence of the expectations in the RHS of (1.9) should be guaranteed. For the precise formulation of this result, see Section 3.

Notice that we do not state that our solutions to the Liouville and BBGKY equations are unique: one must be able for this to construct the time evolution for a much larger family of initial states ρ.

2   The Liouville equation

Theorem 1. Suppose that a diagonal state ρ has properties (d^*) and (d^{**}) from [1]. Then, for any bounded interval Λ^0 and any operator $A \in \mathcal{B}$ localized in Λ^0 $(A \in \mathcal{B}_{\Lambda^0})$, the function $t \mapsto \rho_t(A)$ is smooth and the following equation takes place

$$\frac{d}{dt}\rho_t(A) = i\rho_t([\bar{H}_{\Lambda^0_{+r}}, A]), \quad t \in \mathbf{R}^1 \tag{2.1}$$

where the RHS of the equality is defined as the limit

$$\lim_{s\to\infty} \rho_t\Big(\overset{(\Lambda^0_{+r})}{\underset{<s}{\prod}} [\bar{H}_{\Lambda^0_{+r}}, A] \overset{(\Lambda^0_{+r})}{\underset{<s}{\prod}} \Big), \tag{2.2}$$

$\prod_{<s}^{(\Lambda)}$ is the orthogonal projection in $\mathcal{H}_\Lambda$ onto the subspace generated by the occupation number configurations $x \subset \Lambda$ with $|x| < s$. ◁

Equation (2.1) may be considered as a weak form of the infinite-volume Liouville equation. From this point of view, a family of states ρ_t, $t \in \mathbf{R}^1$, is called sometimes a weak solution of the Liouville equation.

Proof of Theorem 1. At the first step we shall check equation (2.1) for $A = E_{x,y}^{(\Lambda^0)}$, $|x| = |y|$.

Lemma 1.1. Given bounded $\Lambda \supset \Lambda^0$, the convergence

$$\rho_{\Lambda,t}([\bar{H}_{\Lambda^0_{+r}}, E_{x,y}^{(\Lambda^0)}]) =$$

$$= \lim_{s\to\infty} \rho_{\Lambda,t}\Big(\overset{(\Lambda^0_{+r})}{\underset{<s}{\prod}} [\bar{H}_{\Lambda^0_{+r}}, E_{x,y}^{(\Lambda^0)}] \overset{(\Lambda^0_{+r})}{\underset{<s}{\prod}} \Big) \tag{2.3}$$

is uniform for $\Lambda \supset \Lambda^0_{+r}$ and t in compacts, and the function

$$t \mapsto \rho_{\Lambda,t}([\bar{H}_\Lambda, E_{x,y}^{(\Lambda^0)}]) = \rho_{\Lambda,t}([\bar{H}_{\Lambda^0_{+r}}, E_{x,y}^{(\Lambda^0)}])$$

is bounded on compacts uniformly for $\Lambda \supset \Lambda^0_{+r}$. Moreover,

$$\rho_{\Lambda,t}([\bar{H}_{\Lambda^0_{+r}}, E^{(\Lambda^0)}_{x,y}]) =$$
$$= \lim_{s\to\infty} \rho_{\Lambda,t}\left(\prod_{<s}^{(\Lambda)} [\bar{H}_{\Lambda^0_{+r}}, E^{(\Lambda^0)}_{x,y}] \prod_{<s}^{(\Lambda)} \right), \tag{2.4}$$

and the convergence (2.4) is also uniform for t in compacts. $\triangleleft$

Proof. Writing

$$\rho_{\Lambda,t}\left(\prod_{<s}^{(\Lambda^0_{+r})} [\bar{H}_{\Lambda^0_{+r}}, E^{(\Lambda^0)}_{x,y}] \prod_{<s}^{(\Lambda^0_{+r})} \right) =$$
$$= \sum_{j\in\Lambda^0_{+r}} \rho_{\Lambda,t}\left(\prod_{<s}^{(\Lambda^0_{+r})} [a^+_j(\Delta a)_j, E^{(\Lambda^0)}_{x,y}] \prod_{<s}^{(\Lambda^0_{+r})} \right) + \tag{2.5}$$
$$+ \sum_{j,j'\in\Lambda^0_{+r}} \Phi(|j-j'|)\rho_{\Lambda,t}\left(\prod_{<s}^{(\Lambda^0_{+r})} [n_j n_{j'}, E^{(\Lambda^0)}_{x,y}] \prod_{<s}^{(\Lambda^0_{+r})} \right),$$

we see that it suffices to check that every term in the RHS of (2.5) converges uniformly to the corresponding quantity as $s\to\infty$.

We restrict ourselves to the term

$$\rho_{\Lambda,t}\left(\prod_{<s}^{(\Lambda^0_{+r})} [n_j n_{j'}, E^{(\Lambda^0)}_{x,y}] \prod_{<s}^{(\Lambda^0_{+r})} \right), \quad j\in\Lambda^0, \ j'\notin\Lambda^0. \tag{2.6}$$

To prove the convergence of (2.6) it suffices to check that

$$\left| \rho_{\Lambda,t}\left(\prod_{s,s'}^{(\Lambda^0_{+r})} n_{j'} \prod_{s,s'}^{(\Lambda^0_{+r})} \right) \right| \overset{s\to\infty}{\longrightarrow} 0, \tag{2.7}$$

where $s' > s$, $\prod_{s,s'}^{(\Lambda^0_{+r})} = \prod_{>s}^{(\Lambda^0_{+r})} \prod_{<s'}^{(\Lambda^0_{+r})}$. The LHS of (2.7) is estimated by

$$\left| \rho_{\Lambda,t}\left(\prod_{s,s'}^{(\Lambda^0_{+r})} n_{j'} \prod_{<\ln s}^{(j')} \prod_{s,s'}^{(\Lambda^0_{+r})} \right) \right| + \left| \rho_{\Lambda,t}\left(\prod_{s,s'}^{(\Lambda^0_{+r})} n_{j'} \prod_{\geq\ln s}^{(j')} \prod_{s,s'}^{(\Lambda^0_{+r})} \right) \right| \leq$$
$$\leq \left\| \prod_{<\ln s}^{(j')} n_{j'} \prod_{<\ln s}^{(j')} \right\| \cdot \left| \rho_{\Lambda,t}\left(\prod_{s,s'}^{(\Lambda^0_{+r})} \right) \right| + \sum_{k\geq\ln s} \left| \rho_{\Lambda,t}\left(\prod_{s,s'}^{(\Lambda^0_{+r})} n_{j'} \prod_{=k}^{(j')} \prod_{s,s'}^{(\Lambda^0_{+r})} \right) \right| \leq$$
$$\leq \ln s \left| \rho_{\Lambda,t}\left(\prod_{>s}^{(\Lambda^0_{+r})} \right) \right| + \sum_{k\geq\ln s} k \left| \rho_{\Lambda,t}\left(\prod_{\geq k}^{(j')} \right) \right| <$$
$$< c_1 \ln s \cdot s^{-(2+\alpha)} + c_2 \sum_{k\geq\ln s} k \cdot k^{-(2+\alpha)} \tag{2.8}$$

where c_1, c_2, α are some positive constants. The last bound in (2.8) is the consequence of bounds obtained in the proof of Theorem 2 from [1].

The proof of equality (2.4) proceeds along the same scheme. This finishes the proof of Lemma 1.1. $\square$

Notice that a simple modification of these arguments gives that for any $j_1, \ldots, j_l \in \mathbf{Z}^1$ and $\alpha_1, \ldots, \alpha_l \in \mathbf{Z}^1_+$

$$\rho_{\Lambda,t}(n_{j_1}^{\alpha_1} \cdots n_{j_l}^{\alpha_l}) < \infty \tag{2.9}$$

uniformly for t in compacts.

From Lemma 1.1 one obtains that, for fixed Λ, $\rho_{\Lambda,t}([\bar{H}_{\Lambda^0_{+r}}, E_{x,y}^{(\Lambda^0)}])$ is a continuous function of the variable t and the following equality is valid

$$\rho_{\Lambda,t}(E_{x,y}^{(\Lambda^0)}) = \rho(E_{x,y}^{(\Lambda^0)}) + i \int_0^t \rho_{\Lambda,t'}([\bar{H}_{\Lambda^0_{+r}}, (E_{x,y}^{(\Lambda^0)}])dt'. \tag{2.10}$$

Lemma 1.2. The following limits exist uniformly for t in compacts

$$\rho_t([\bar{H}_{\Lambda^0_{+r}}, E_{x,y}^{(\Lambda^0)}]) =$$

$$= \lim_{s \to \infty} \rho_t \left(\prod_{<s}^{(\Lambda^0_{+r})} [\bar{H}_{\Lambda^0_{+r}}, E_{x,y}^{(\Lambda^0)}] \prod_{<s}^{(\Lambda^0_{+r})} \right)$$

and

$$\lim_{\Lambda \nearrow \mathbf{Z}^1} \rho_{\Lambda,t}([\bar{H}_{\Lambda^0_{+r}}, E_{x,y}^{(\Lambda^0)}]). \tag{2.11}$$

Moreover, the limits are equal. $\triangleleft$

Proof. From Lemma 1.1 one has that the convergence in (2.3) is uniform for $\Lambda \supset \Lambda^0_{+r}$. From [1] (see Theorem 3) one can obtain the convergence

$$\lim_{\Lambda \nearrow \mathbf{Z}^1} \rho_{\Lambda,t} \left(\prod_{<s}^{(\Lambda^0_{+r})} [\bar{H}_{\Lambda^0_{+r}}, E_{x,y}^{(\Lambda^0)}] \prod_{<s}^{(\Lambda^0_{+r})} \right) = \rho_t \left(\prod_{<s}^{(\Lambda^0_{+r})} [\bar{H}_{\Lambda^0_{+r}}, E_{x,y}^{(\Lambda^0)}] \prod_{<s}^{(\Lambda^0_{+r})} \right).$$

Hence, the limits exist and are equal. Lemma 1.2 is proven. $\triangleleft$

From (2.10) and Lemma 1.2 one obtains that $\rho_t([\bar{H}_{\Lambda^0_{+r}}, E_{x,y}^{(\Lambda^0)}])$ is a continuous function of t and

$$\rho_t(E_{x,y}^{(\Lambda^0)}) = \rho(E_{x,y}^{(\Lambda^0)}) + i \int_0^t \rho_{t'}([\bar{H}_{\Lambda^0_{+r}}, E_{x,y}^{(\Lambda^0)}])dt'.$$

Lemma 1.3. Let $A \in \mathcal{B}$ be an operator localized in Λ^0. Then, for any $u \in \mathbf{Z}_+^1$

$$t \mapsto \rho_{\Lambda,t}\left(\left[\bar{H}_{\Lambda_{+r}^0}, \prod_{>u}^{(\Lambda^0)} A \prod_{>u}^{(\Lambda^0)}\right]\right) \tag{2.12}$$

is a continuous function which is bounded uniformly for t in compacts. The value (2.12) converges to 0 as $u \to \infty$ uniformly in Λ. $\triangleleft$

The proof of Lemma 1.3 is given by a combination of arguments from the proofs of Lemma 1.1 and Lemma 1.2. Here one uses the following bound

$$\left|\rho_{\Lambda,t}\left(n_j n_{j'} \prod_{>u}^{(\Lambda^0)} A \prod_{>u}^{(\Lambda^0)}\right)\right| \le [\rho_{\Lambda,t}(n_j^2 n_{j'}^2)]^{1/2} \cdot \|A\| \left[\rho_{\Lambda,t}\left(\prod_{>u}^{(\Lambda^0)}\right)\right]^{1/2}.$$

The details are omitted. $\square$

From Lemma 1.3 one obtains

$$\rho_t(A) = \rho(A) + i \int_0^t \rho_{t'}([\bar{H}_{\Lambda_{+r}^0}, A]) dt' \tag{2.13}$$

for any $A \in \mathcal{B}$ localized in Λ^0. The last remark is that (2.13) is equivalent to (2.1). Theorem 1 is proven. $\square$

Corollary 1. Assume that a diagonal state ρ satisfies the conditions of Theorem 1. Let $p_t^{(\Lambda^0)}$ be the density matrix of the state ρ_t in a volume Λ^0. The family of matrices $p_t^{(\Lambda)}$, $t \in \mathbf{R}^1$, provides a solution to the infinite-volume Liouville equation written in the operator form

$$\frac{d}{dt}\left(p_t^{(\Lambda^0)}\right) = i \, tr_{\Lambda_{+r}^0 \setminus \Lambda^0}[p_t^{(\Lambda_{+r}^0)}, \bar{H}_{\Lambda_{+r}^0}], \quad t \in \mathbf{R}^1.$$

The derivative in the LHS is understood in the weak operator topology. $\triangleleft$

     Equivalently, this equation may be rewritten in the form (1.4).

Proof. From Lemma 1.2 and bounds (2.8) we have

$$\rho_t([\bar{H}_{\Lambda_{+r}^0}, E_{x,y}^{(\Lambda^0)}]) = \lim_{s \to \infty} \rho_t\left(\prod_{<s}^{(\Lambda_{+r}^0)} [\bar{H}_{\Lambda_{+r}^0}, E_{x,y}^{(\Lambda^0)}] \prod_{<s}^{(\Lambda_{+r}^0)}\right) =$$

$$= \sum_{z,w \in \Lambda_{+r}^0} (p_t^{(\Lambda_{+r}^0)})_{z,w}([\bar{H}_{\Lambda_{+r}^0}, E_{x,y}^{(\Lambda^0)}])_{w,z} \tag{2.14}$$

and the series in the RHS of (2.14) is absolutely convergent. Hence, (2.14) may be rewritten in the following form

$$\sum_{z,w \in \Lambda_{+r}^0} ([p_t^{(\Lambda_{+r}^0)}, \bar{H}_{\Lambda_{+r}^0}])_{z,w}(E_{x,y}^{(\Lambda^0)} \otimes 1^{(\Lambda_{+r}^0 \setminus \Lambda^0)})_{w,z} =$$

$$= \sum_{z \in \Lambda_{+r}^0 \setminus \Lambda^0} ([p_t^{(\Lambda_{+r}^0)}, \bar{H}_{\Lambda_{+r}^0}])_{z \vee x, z \vee y}$$

and from (2.1) one obtains

$$\frac{d}{dt}\Big((p_t^{(\Lambda^0)})_{x,y}\Big) = i(tr_{\Lambda^0_{+r}\setminus\Lambda_0}[p_t^{(\Lambda^0_{+r})},\bar{H}_{\Lambda^0_{+r}}])_{x,y}\,.$$

From the last equation and Lemma 1.3 it follows that for any $f,g\in\mathcal{H}_{\Lambda^0}$

$$\frac{d}{dt}(\langle p_t^{(\Lambda^0)}f,g\rangle_{\Lambda^0}) = i\langle(tr_{\Lambda^0_{+r}\setminus\Lambda_0}[p_t^{(\Lambda^0_{+r})},\bar{H}_{\Lambda^0_{+r}}])f,g\rangle_{\Lambda^0},\tag{2.15}$$

where $<\cdot,\cdot>_{\Lambda^0}$ is the scalar product in $\mathcal{H}_{\Lambda^0}$. $\square$

From (2.15) one can directly deduce the following

Corollary 2. The equation (2.15) may be rewritten in the following vector form: for any $z\in\mathbf{Z}^1$ and $f\in\prod_{<s}^{(\Lambda^0)}\mathcal{H}_{\Lambda^0}$

$$\frac{d}{dt}p_t^{(\Lambda^0)}f = i\,tr_{\Lambda^0_{+r}\setminus\Lambda^0}[p_t^{(\Lambda^0_{+r})},\bar{H}_{\Lambda^0_{+r}}]f.\quad\triangleleft$$

3   The BBGKY hierarchy

Theorem 2. Let a diagonal state ρ have properties (d^*) and (d^{**}) from [1]. Then, for any $n\geq 0$ and any pair x,x' of occupation number configurations with $|x|=|x'|=n$ the function

$$t\mapsto R_t(x,x')$$

is smooth and the following equation takes place:

$$\frac{d}{dt}R_t^{(n)}(x,x') = i(\mathbf{H}^{(n)}R_t^{(n)})(x,x') + (\mathbf{A}R_t^{(n+1)})(x,x').\tag{3.1}$$

Here $R_t(x,x')$ is defined by

$$R_t(x,x') = \lim_{s\to\infty}\rho_t\Big(\prod_{<s}^{(\Lambda^0)}\prod_{j\in\mathbf{Z}^1}a_j^{+x'(j)}\prod_{k\in\mathbf{Z}^1}a_k^{x(k)}\prod_{<s}^{(\Lambda^0)}\Big),\tag{3.2}$$

$\Lambda^0=\Lambda^0(x,x')=\{l\in\mathbf{Z}^1:x(l)+x'(l)\geq 1\}$ is the support of (x,x'), the projectors $\prod_{<s}^{(\Lambda^0)}$ were introduced before. $\mathbf{H}^{(n)}R_t^{(n)}$ is given by

$$\mathbf{H}^{(n)}R_t^{(n)}(x,x') = \sum_y(R_t^{(n)}(x,y)(H^{(n)})_{y,x'} - (H^{(n)})_{x,y}R_t^{(n)}(y,x'))\tag{3.3}$$

and

$$\mathbf{A}R_t^{(n+1)})_{x,x'} =$$
$$= i\sum_{j,k\in\mathbf{Z}^1}\Phi(|j-k|)(x(k)-x'(k))R_t^{(n+1)}(x+\delta_j,x'+\delta_j).\quad\triangleleft\tag{3.4}$$

Equation (3.1) may be considered as a weak form of the infinite-volume BBGKY hierarchy for the time-evolution under consideration.

Proof of Theorem 2. The proof of Theorem 2 follows the same line of arguments as that of Theorem 1. To avoid repetitions, we shall omit techincal details.

Let $x, y \subset \Lambda^0$ be occupation number configuration in Λ^0. We denote

$$a^+(x) = \prod_{j \in \Lambda^0} a_j^{+x(j)}, \quad a(y) = \prod_{j \in \Lambda^0} a_j^{y(j)}.$$

Lemma 3.1. Given bounded $\Lambda \supset \Lambda^0$, the limits

$$\rho_{\Lambda,t}(a^+(x)a(y)) = R_{\Lambda,t}(y, x) =$$
$$= \lim_{s \to \infty} \rho_{\Lambda,t}\left(\prod_{<s}^{(\Lambda^0)} a^+(x)a(y) \prod_{<s}^{(\Lambda^0)} \right), \tag{3.5}$$
$$\rho_{\Lambda,t}([\bar{H}_{\Lambda_{+r}^0}, a^+(x)a(y)]) =$$
$$= \lim_{s \to \infty} \rho_{\Lambda,t}\left(\prod_{<s}^{(\Lambda_{+r}^0)} [\bar{H}_{\Lambda_{+r}^0}, a^+(x)a(y)] \prod_{<s}^{(\Lambda_{+r}^0)} \right)$$

are uniform for $\Lambda \supset \Lambda^0$ and t in compacts. The functions

$$t \mapsto \rho_{\Lambda,t}([\bar{H}_{\Lambda_{+r}^0}, a^+(x)a(y)]), \quad t \mapsto R_{\Lambda,t}(y, x)$$

are continuous and bounded on compacts uniformly for $\Lambda \supset \Lambda^0$. Moreover, the following limits exist

$$R_{\Lambda,t}(y, x) = \lim_{s \to \infty} \rho_{\Lambda,t}\left(\prod_{<s}^{(\Lambda)} a^+(x)a(y) \prod_{<s}^{(\Lambda)} \right),$$
$$\rho_{\Lambda,t}([\bar{H}_{\Lambda_{+r}^0}, a^+(x)a(y)]) = \tag{3.6}$$
$$= \lim_{s \to \infty} \rho_{\Lambda,t}\left(\prod_{<s}^{(\Lambda)}[\bar{H}_{\Lambda_{+r}^0}, a^+(x)a(y)] \prod_{<s}^{(\Lambda)} \right)$$

and the convergence in (3.6) is uniform for t in compacts. $\lhd$

The proof of Lemma 3.1 proceeds in the same way as Lemma 2.1.

From Lemma 3.1 one obtains that for fixed $\Lambda \supset \Lambda_{+r}^0$ the following equality is valid

$$\rho_{\Lambda,t}(a^+(x)a(y)) =$$
$$= \rho(a^+(x)a(y)) + i \int_0^t \rho_{\Lambda,t'}([\bar{H}_{\Lambda_{+r}^0}, a^+(x)a(y)])dt'. \tag{3.7}$$

Lemma 3.2.   Uniformly for t in compacts the following limits exist:

$$R_t(y,x) = \rho_t(a^+(x)a(y)) =$$
$$= \lim_{s \to \infty} \rho_t\left(\prod_{<s}^{(\Lambda^0)} a^+(x)a(y) \prod_{<s}^{(\Lambda^0)} \right) =$$
$$= \lim_{\Lambda \nearrow \mathbf{Z}^1} R_{\Lambda,t}(y,x)$$

and

$$\rho_t([\bar{H}_{\Lambda^0_{+r}}, a^+(x)a(y)]) =$$
$$= \lim_{s \to \infty} \rho_t\left(\prod_{<s}^{(\Lambda^0_{+r})} [\bar{H}_{\Lambda^0_{+r}}, a^+(x)a(y)] \prod_{<s}^{(\Lambda^0_{+r})} \right) =$$
$$= \lim_{\Lambda \nearrow \mathbf{Z}^1} \rho_{\Lambda,t}[\bar{H}_{\Lambda^0_{+r}}, a^+(x)a(y)]). \quad \lhd$$

The proof of Lemma 3.2 repeats that of Lemma 2.2.

As in Section 2, from (3.7) and Lemma 3.2 one obtains

$$R_t(y,x) = \rho(a^+(x)a(y)) + i \int_0^t \rho_{t'}([\bar{H}_{\Lambda^0_{+r}}, a^+(x)a(y)])dt' \tag{3.8}$$

Hence, the function $t \mapsto R_t(y,x)$ is smooth, and direct computation gives that the equality (3.8) may be rewritten in the form (3.1). Theorem 2 is proven. $\square$

References

[1] Shuhov, A. G., Suhov, Yu. M., Teslenko, A. V.: Towards time-dynamics for bosonic systems in quantum statistical mechanics, *this volume.*

[2] Bratteli, O., Robinson, D. W.: Operator algebras and quantum statistical mechanics, *Vols 1, 2. Berlin et al.: Springer-Verlag, 1979, 1981.*